Control of Heavy Metals
in the Environment

Offering a broad coverage of advanced principles and applications, *Control of Heavy Metals in the Environment series* provides chemical and environmental engineers with the most complete resources available on the treatment of heavy metal contaminants with an emphasis on advanced and alternative approaches. It investigates a variety of environmental pollution sources and waste characteristics that require a multitude of remediation methods. It covers metal oxide nanoparticle pollution and nanotechnology applications for remediation. The authors delve into costs and effluent standards and offer several illustrative case histories to illustrate the regional and global effects of key pollution control practices.

Features:

- Provides technical information for industrial and hazardous waste treatment.
- Explores the newest methods of clean production and waste minimization.
- Covers topics related to environmental geochemistry.
- Includes numerous figures, tables, examples, and case histories.

Advances in Industrial and Hazardous Wastes Treatment
Series Editors: Lawrence K. Wang and Mu-Hao Sung Wang

Control of Heavy Metals in the Environment, Volume 1
Edited by Lawrence K. Wang, Mu-Hao Sung Wang, Yung-Tse Hung, and J. Paul Chen

Control of Heavy Metals in the Environment, Volume 2
Edited by Lawrence K. Wang, Mu-Hao Sung Wang, Yung-Tse Hung, and J. Paul Chen

Control of Heavy Metals in the Environment, Volume 3
Edited by Lawrence K. Wang, Mu-Hao Sung Wang, Yung-Tse Hung, and J. Paul Chen

Control of Heavy Metals
in the Environment
Recent Advances in Metal Toxicity, Pollution Control, and Remediation Techniques

Edited by
Lawrence K. Wang, Mu-Hao Sung Wang,
Yung-Tse Hung, and J. Paul Chen

CRC Press is an imprint of the
Taylor & Francis Group, an **informa** business

Designed cover image: Shutterstock

First edition published 2025
by CRC Press
2385 NW Executive Center Drive, Suite 320, Boca Raton, FL 33431

and by CRC Press
4 Park Square, Milton Park, Abingdon, Oxon, OX14 4RN

CRC Press is an imprint of Taylor & Francis Group, LLC

© 2025 selection and editorial matter, Lawrence K. Wang, Mu-Hao Sung Wang, Yung-Tse Hung, and J. Paul Chen individual chapters, the contributors

ISBN: 978-1-032-89184-2 (hbk)
ISBN: 978-1-032-89185-9 (pbk)
ISBN: 978-1-003-54161-5 (ebk)

DOI: 10.1201/9781003541615

Typeset in Times
by codeMantra

Contents

Preface

Environmental managers, engineers, and scientists who have had experience with industrial and hazardous waste management problems have noted the need for a handbook series that is comprehensive in its scope, directly applicable to daily waste management problems of specific industries, and widely acceptable by practicing environmental professionals and educators. Taylor & Francis and CRC Press have developed this timely book series entitled *Advances in Industrial and Hazardous Wastes Treatment*, which emphasizes an in-depth presentation of environmental pollution sources, waste characteristics, control technologies, management strategies, facility innovations, process alternatives, costs, case histories, effluent standards, and future trends for each industrial or commercial operation, and an in-depth presentation of methodologies, technologies, alternatives, regional effects, and global effects of each important industrial pollution control practice that may be applied to all industries.

Control of Heavy Metals in the Environment, Volumes 1 to 3, is an independent mini-series in the *Advances in Industrial and Hazardous Wastes Treatment* series and is the second edition of an old book, *Heavy Metals in the Environment*. The importance of metals, such as lead, chromium, cadmium, zinc, copper, nickel, iron, and mercury, is discussed in detail in the mini-series. They could be important constituents of most living animals, plants, and microorganisms and many nonliving substances in the environment. Some of them are essential for the growth of biological and microbiological lives. Their absence could limit the growth of small microorganisms to large plants or animals. However, the presence of any of these heavy metals in excessive quantities will be harmful to human beings and will interfere with many beneficial uses of the environment due to their toxicity and mobility. Therefore, it is frequently desirable to measure and control the heavy metal concentrations in the environment.

In a deliberate effort to complement other industrial waste treatment and hazardous waste management texts published by Taylor & Francis and CRC Press, this book, *Control of Heavy Metals in the Environment, Volume 3,* covers the following subjects: (a) environmental metal research trends and advances; (b) toxicity and sources of Pb, Cd, Hg, Cr, As, and radionuclides in the environment; (c) environmental behavior and effects of engineered metal, and metal oxide nanoparticles; (d) heavy metal removal with exopolysaccharide-producing cyanobacteria; (e) environmental geochemistry of high-arsenic aquifer systems; (f) nanotechnology application in metal ion adsorption; (g) biosorption of metals onto granular sludge; (h) arsenic pollution: occurrence, distribution, and technologies; (i) treatment of metal-bearing effluents: removal and recovery; (j) advances in the management and treatment of acid pickling wastes containing heavy metals; (k) advances in the treatment and management of metal-finishing industry wastes; (l) use of biosorption of agro-industrial wastes and microorganisms for the removal of potentially toxic metals; (m) advances in the removal of heavy metals from contaminated soil; (n) advances in the remediation of metal-finishing brownfield sites; (o) biological treatment of mercury-polluted wastewater; (p) bioassay technology, terminologies, applications, and examples: heavy metals and phenol investigations; and (q) pollutants, technologies, and terminologies for emission control in the transportation industry.

This book's first sister book, *Control of Heavy Metals in the Environment, Volume 1,* covers the following subjects: (a) innovative process systems, including dissolved air flotation, protein separation, flue gas reuse, sulfide precipitation, and chromium removal for tannery waste treatment; (b) titanium dioxide recovery, filler retention and white water treatment using flotation and membrane filtration; (c) adsorptive amputation of hexavalent chromium from aqueous solution and textile wastewater by *Heinsia crinita* seed coat biomass; (d) protection of biosystem organisms in heavy metals-contaminated environment by using molecular hydrogen; (e) application of ion exchange resin in the removal of pollutant from wastewater; (f) treatment of nickel-chromium plating wastes; (g) heavy metal pollution resulting from industrial wastes used for soil stabilization in groundwater;

(h) sustainable water and wastewater treatment systems consisting of magnesium coagulation–precipitation, dissolved air flotation, recarbonation, and filtration; (i) sludge treatment using pressurized oxygenation-ozonation, dissolved gas flotation, and dewatering for heavy metals removal; (j) hazardous wastewater treatment in the petroleum refining industry using best available flotation and filtration technologies; (k) using electroflotation and filtration for lake restoration, pollution prevention, and environmental conservation; (l) pretreatment or total treatment of scallop processing wastewater by independent physicochemical wastewater treatment systems; (m) treatment of cold, low alkalinity, and super-high turbidity, industrial waters using polymeric metallic coagulants; (n) chemical oxygen demand (COD) determination and cleaning solution preparation using potassium permanganate for hazardous waste minimization; (o) innovative process system consisting of dissolved air flotation, recarbonation, and filtration and using lime and sodium aluminate for the removal of hardness, iron, and manganese from groundwater; and (p) electro-Fenton, heterogeneous electro-Fenton process, and treatment of non-biodegradable wastewater.

This book's second sister book, *Control of Heavy Metals in the Environment, Volume 2*, covers the following subjects: (a) metals and trace elements on earth; (b) strategies for bioremediation of heavy metals-contaminated industrial effluents; (c) surface engineering of colloidal quaternary chalcogenide nanocrystals via ligand exchange for photo-induced charge separation applications and water remediation; (d) removal of arsenic, oil and grease, chemical oxygen demand and color from stormwater runoff by continuous dissolved air flotation and filtration; (e) utilization of fungal bioprocess in treating heavy metal-contaminated water ; (f) depletion of hexavalent chromium to trivalent chromium from tannery industry effluent using natural bacterial biomass; (g) reduction of high color, humic substances, turbidity, coliform bacteria, trihalomethane maximum formation potential (THMMFP), ultraviolet absorbance, cysts, chlorine demand, and metals from unpredictable water source by dissolved air flotation and filtration; (h) treating tannery waste using stack flue gas recycle, sulfide precipitation, chromium removal, ferrous sulfide recycle, ferrous ion recycle, filtration, flotation, membrane, and bioreactor; (i) sludge drying bed technology; (j) ecologically sustainable industrial development, solid and hazardous wastes management, and DAF landfill leachate pretreatment; (k) removal of iron, copper, coliform bacteria and hardness from cooling tower water by dissolved air-ozone flotation; (l) recycling and disposal of hazardous solid wastes containing heavy metals and other toxic substances; (m) environmental management of electroplating and metal finishing operations; (n) reduction of color, turbidity, odor, humic acid, metals, EDB, coliform and TTHM by adsorption, flotation, and filtration; and (o) control and management of air emissions from the transportation industry.

Special efforts were made to invite experts to contribute chapters in their own areas of expertise. Since the area of hazardous industrial waste treatment is very broad, no one can claim to be an expert in all heavy metals and their related industries; collective contributions are better than a single author's presentation for a book of this nature.

The entire mini-series, *Control of Heavy Metals in the Environment, Volumes 1 to 3*, is to be used as a college textbook as well as a reference book for the environmental professional. It features the major hazardous heavy metals in air, water, land, and facilities that have significant effects on public health and the environment. Professors, students, and researchers in environmental, civil, chemical, sanitary, mechanical, and public health engineering and science will find valuable educational materials here. The extensive bibliographies for each heavy metal or metal-related industrial waste treatment or practice should be invaluable to environmental managers or researchers who need to trace, follow, duplicate, or improve on a specific industrial hazardous waste treatment practice.

A successful modern heavy metal control program for a particular industry will include not only traditional water pollution control but also air pollution control, soil conservation, site remediation, groundwater protection, public health management, solid waste disposal, and combined industrial and municipal heavy metal waste management. In fact, it should be a total environmental control program. Another intention of this mini-series is to provide technical and economical information on the development of the most feasible total heavy metal control program that can benefit both industry and local municipalities. Frequently, the most economically feasible methodology is a combined industrial–municipal heavy metal management.

Lawrence K. Wang, New York, USA
Mu-Hao Sung Wang, New York, USA
Yung-Tse Hung, Ohio, USA
J. Paul Chen, Singapore

Editors

Lawrence K. Wang has served the society as a professor, inventor, chief engineer, chief editor, and public servant (UN, USEPA, New York State) for 50+ years, with experience in the entire field of environmental science, technology, engineering, and mathematics (STEM). He is a licensed professional engineer, a certified laboratory director, a licensed water operator, and an Occupational Safety and Health Administration hazardous waste management instructor. He has special passion and expertise in developing various innovative technologies, educational programs, licensing courses, international projects, academic publications, and humanitarian organizations, all for his dream goal of promoting world peace. He is a retired acting president/professor of the Lenox Institute of Water Technology, USA; a senior advisor of the United Nations Industrial Development Organization (UNIDO), Vienna, Austria; and a former professor/visiting professor of Rensselaer Polytechnic Institute, Stevens Institute of Technology, University of Illinois, National Cheng Kung University, Zhejiang University, and Tongji University. He is the author of 750+ papers and 50+ books and is credited with 29 invention patents. He holds a BSCE degree from National Cheng Kung University, Taiwan, ROC; an MSCE degree from the University of Missouri-Rolla; an MS degree from the University of Rhode Island; and a PhD degree from Rutgers University, USA. Currently, he is the book series editor of CRC Press, Springer Nature Switzerland, Lenox Institute Press, World Scientific Singapore, and John Wiley and Sons. He has been a delegate of the People to People International Foundation; a diplomate of the American Academy of Environmental Engineers; a member of American Society of Civil Engineers, American Institute of Chemical Engineers, Water Environment Federation, American Water Works Association, and Overseas Chinese Environmental Engineers and Scientists Association (OCEESA) and a recipient of Water Environment Federation Kenneth Research Award (NY), Five-Star Innovative Engineering Award (first dissolved air flotation drinking water plant in Americas), Korean Pollution Control Association Award (transfer of flotation technology to South Korea), and OCEESA Medal of Honor.

Mu-Hao Sung Wang has been an engineer of the New York State Department of Environmental Conservation; an editor of CRC Press, Springer Nature Switzerland, and Lenox Institute Press; and a university professor at the Stevens Institute of Technology, National Cheng Kung University, Tongji University and the Lenox Institute of Water Technology. Totally, she has been a government official and an educator in the USA and Taiwan for over 50 years. She is a licensed professional engineer and a diplomate of the American Academy of Environmental Engineers (AAEE). Her publications have been in the areas of water quality, modeling, environmental sustainability, solid and hazardous waste management, National Pollutant Discharge Elimination System, flotation technology, industrial waste treatment, and analytical methods. She is the author of over 50 publications and an inventor of 14 US and foreign patents. She holds a BSCE degree from National Cheng Kung University, Taiwan, ROC; an MS degree from the University of Rhode Island, RI, USA; and a PhD degree from Rutgers University, NJ, USA. She is the co-series editor of the *Handbook of Environmental Engineering* series (Springer Nature Switzerland), and the *Advances in Industrial and Hazardous Wastes Treatment* series (CRC Press of Taylor & Francis Group), and the coeditor of the *Environmental Science, Technology, Engineering and Mathematics* series (Lenox Institute Press). She is a member of American Water Works Association, New England Water Works Association, New England Water Environment Association, Water Environment Federation, and Overseas Chinese Environmental Engineers and Scientists Association.

Yung-Tse Hung has been a professor of civil engineering at Cleveland State University, Cleveland, Ohio, USA from 1981 to 2024. He is a fellow of the American Society of Civil Engineers, a licensed professional engineer in Ohio and North Dakota, and a diplomate of the American Academy of Environmental Engineers. He has taught at 16 universities in eight countries. His research interests and publications have been involved with biological treatment processes, solid wastes, hazardous waste management, and industrial waste treatment. He is credited with over 470 publications and presentations, 28 books, and 159 book chapters in water and wastewater treatment. He received his BSCE and MSCE degrees from National Cheng Kung University, Taiwan, ROC, and his PhD degree from the University of Texas at Austin, USA. He is the editor-in-chief of the *International Journal of Environmental Pollution Control and Management,* the *International Journal of Environmental Engineering,* and the *International Journal of Environmental Engineering Science,* and the coeditor of the *Advances in Industrial and Hazardous Wastes Treatment* series (CRC Press of Taylor & Francis Group), and the *Handbook of Environmental Engineering* series (Springer). He is also the chief editor of the *Handbook of Environment and Waste Management* series (World Scientific Singapore), and the permanent executive director and the ex-president of OCEESA.

J. Paul Chen was a professor at the National University of Singapore (NUS) since 1998. He has worked in R&D, education, and consultancy across separation, water treatment, safety, carbon capture, and biomedical engineering areas with more than 100 projects leading to great impacts (>200 papers and books/book chapters with citation >22000, H-index = 75, 12 patents, 2 NUS spin-off). Through working internationally on industrial projects (e.g., UK, US, UAE, Norway, and Malaysia), he has established strong networks with industries, universities, and governments. Since 2023, he has served as Chair of the Management Committee, IWA Nano, and Water Specialist Group. Additionally, he has served as a member of several IWA management committees as well as vice chair of IChemE Singapore. He is a recipient of the IChemE Sustainable Technology Award. He is an elected fellow of IWA, RSC, IChemE, IET, and IMarEST.

Contributors

Changlun Chen
CAS Key Laboratory of Photovoltaic and
  Energy Conservation Materials
Institute of Plasma Physics, Chinese Academy
  of Sciences
Hefei, P. R. China

Jorge Diniz de Oliveira
Center for Exact, Natural and Technological
  Sciences
State University of the Tocantina Region of
  Maranhão
Imperatriz, Brazil

Roberto De Philippis
Department of Agrifood Production and
  Environmental Sciences
University of Florence
Piazzale delle Cascine, Italy

Helayne Santos de Sousa
Graduation in Chemistry Licentiate
Center for Exact, Natural and Technological
  Sciences University of the Tocantina
Imperatriz, Brazil

Yamin Deng
School of Environmental Studies, China
  University of Geosciences
Wuhan, P. R. China

Jéssica Mesquita do Nascimento
Department of Environmental Engineering
Federal University of Rondônia
Ji-Paraná Campus
Ji-Paraná, RO, Brazil

Mohammad I. El-Khaiary
Chemical Engineering Department, Alexandria
  University
Alexandria, Egypt

Veysel Eroglu
Department of Environmental Engineering
Minister of Environment and Forestry
Istanbul Technical University
Istanbul, Turkey

Ferruh Erturk
Department of Environmental Engineering
Yildiz Technical University
Istanbul, Turkey

Wen-Xin Gong
School of Environmental Science and
  Engineering
Shandong University
Jinan, P. R. China

Saurabh Gupta
Department of Microbiology
Mata Gujri College
Fatehgarh Sahib, India

Yuh-Shan Ho
Trend Research Centre
Asia University
Taichung, Taiwan, Republic of China

Kaihou Huang
College of Chemistry and Environmental
  Engineering
Shenzhen University
Shenzhen, P. R. China

Yung-Tse Hung
Department of Civil and Environmental
  Engineering
Cleveland State University
Cleveland, Ohio

TengHui Jin
College of Chemistry and Environmental
  Engineering
Shenzhen University
Shenzhen, P. R. China

Jonas Juliermerson Silva Otaviano
Master Student Department of Geosciences
Aveiro University
Aveiro, Portugal

Amanpreet Kaur
Department of Microbiology
Mata Gujri College
Fatehgarh Sahib, India

Huijuan Liu
School of Environment
Tsinghua University
Beijing, P. R. China

Ruiping Liu
School of Environment
Tsinghua University
Beijing, P. R. China

Xiao Liu
College of Chemistry and Environmental
    Engineering
Shenzhen University
Shenzhen, P. R. China

Zihui Lu
College of Chemistry and Environmental
    Engineering
Shenzhen University
Shenzhen, P. R. China

Yue Ma
School of Environmental Science and
    Engineering
Shandong University
Jinan, P. R. China

Ernesto Micheletti
Department of Agrifood Production and
    Environmental Sciences
University of Florence
Firenze, Italy

Ghinwa M. Naja
Department of Chemical Engineering
McGill University
Montreal, Quebec, Canada

Bernd Nowack
Empa-Swiss Federal Laboratories for Materials
    Science and Technology
Technology and Society Laboratory
Duebendorf, Switzerland

Jiuhui Qu
School of Environment
Tsinghua University
Beijing, P. R. China

Nazih K. Shammas
Lenox Institute of Water Technology
Latham, New York

Satnam Singh
Department of Political Science
Mata Gujri College
Fatehgarh Sahib, India

Vijay Singh
Department of Botany
Mata Gujri College
Fatehgarh Sahib, India

Xue Fei Sun
School of Environmental Science and
    Engineering
Shandong University
Jinan, P. R. China

Bohumil Volesky
Department of Chemical Engineering
McGill University
Montreal, Quebec, Canada

Lawrence K. Wang
Lenox Institute of Water Technology
Latham, New York

Mu-Hao Sung Wang
Lenox Institute of Water Technology
Latham, New York

Shu-Guang Wang
School of Environmental Science and
    Engineering
Shandong University
Jinan, P. R. China

Xiangke Wang
College of Environmental Science and
    Engineering
North China Electric Power University
Beijing, P. R. China

Yanxin Wang
School of Environmental Studies
China University of Geosciences
Wuhan, P. R. China

Hui Wen
College of Chemistry and Environmental
 Engineering
Shenzhen University
Shenzhen, P. R. China

Yi Yang
Beijing Normal University
Zhuhai, P. R. China

Zi-Yi Yang
School of Environmental Science and
 Engineering
Shandong University
Jinan, P. R. China

Gaosheng Zhang
School of Environmental Science and
 Engineering
Guangzhou University
Guangzhou, P. R. China

1 Environmental Metal Research Trends and Advances

Mu-Hao Sung Wang, Lawrence K. Wang,
Yuh-Shan Ho, and Mohammad I. El-Khaiary

ACRONYMS

ACH	aluminum chlorohydrate
AMMTO	Advanced Materials & Manufacturing Technologies Office
BAT's	best available technologies
BOD	biochemical oxygen demand
CAS	Chinese Academy of Sciences
COD	chemical oxygen demand
CP	international collaborative publication
CSIC	Consejo Superior de Investigaciones Científicas (Spanish National Research Council)
DAF	dissolved air flotation
DT	detention time
FP	first-author publication
HK	Hong Kong
IE	ion exchange
IF	impact factor
IIN	international identity number
IIT	Indian Institute of Technology
ISI	Institute for Scientific Information
JCR	Journal Citation Reports
LIWT	Lenox Institute of Water Technology
MF	membrane filtration
PAC	polyaluminum chloride
PC-SBR	physicochemical sequencing batch reactor
PIC	polyiron chloride
POTW	publicly owned treatment work
R	ranking
RCRA	Resource Conservation and Recovery Act
RP	corresponding author publication
SCI	Science Citation Index
SCM	stream current monitor
SP	independent publication
TP	total publications
UK	United Kingdom
USEPA	US Environmental Protection Agency

DOI: 10.1201/9781003541615-1

1.1 INTRODUCTION

1.1.1 METALS IN THE ENVIRONMENT

Metals and metalloids are electropositive elements found in all ecosystems, with varying natural concentrations based on local geology. Human and animal activities can redistribute and concentrate metals in areas not naturally enriched. Unlike sediment and nutrient impairments, metal contamination often lacks visible evidence.

All metals, even those abundant and relatively nontoxic metal salts, can be toxic at some level, even those common metals or essential metals as nutrients. Impairments of heavy metals occur when these metals are biologically available at toxic concentrations, affecting aquatic organisms' survival, reproduction, and behavior.

The RCRA 8 metals, also known as the Resource Conservation and Recovery Act 8 metals, are a group of highly toxic heavy metals that require special management due to their potential harm. Here they are in alphabetical order: (a) Arsenic (As): Used in electrical circuits, wood preservatives, and pesticides. While not harmful in small amounts, larger quantities can be fatal; (b) Barium (Ba): A heavy metal that can leach into soil or groundwater when landfilled; (c) Cadmium (Cd): Implicated in gastrointestinal and kidney dysfunction, nervous system disorders, and cancer; (d) Chromium (Cr): Also associated with serious health issues, including skin lesions and immune system dysfunction; (e) Lead (Pb): Known for its harmful effects on the nervous system, especially in children; (f) Mercury (Hg): Extremely toxic and can cause severe health problems if ingested or absorbed; (g) Selenium (Se): Although not a true metal, it's monitored due to its potential toxicity; (h) Silver (Ag): Another heavy metal that requires careful management to prevent harm. If industrial facility operations involve any of these metals in its waste stream, the facility manager must handle them as hazardous waste from "cradle-to-grave," meaning from initial generation to ultimate disposal. The US Environmental Protection Agency (USEPA) has established the maximum concentration of contaminants (toxic metals and organics) for the toxicity characteristics (Table A1).

Some heavy metals (such as iron, zinc, copper, and manganese) are essential metals to human health at low concentrations but become toxic heavy metals at higher concentration levels. Some heavy metals (such as arsenic, cadmium, lead, thallium, and mercury) serve no known biological role.

1.1.2 METAL RESEARCH TRENDS

It has long been known that, in the right concentrations, many metals are essential to life and ecosystems [1–4]; chronic low exposures to metals can lead to severe environmental and health effects. Similarly, in excess, these same metals can be poisonous [5–9]. The main metal threats are associated with heavy metals such as lead, arsenic, cadmium, and mercury. Unlike many organic pollutants, which eventually degrade to carbon dioxide and water, heavy metals tend to accumulate in the environment, especially in lakes, estuarine, or marine sediments [10]. Metals can also be transported from one environment compartment to another [11], which complicates the containment and treatment problem.

Heavy metals are closely connected with environmental deterioration and the quality of human life and thus have aroused concern all over the world. More and more countries have signed treaties to monitor and reduce heavy metal pollution [12]. Moreover, this field of research has been receiving increasing scientific attention due to its negative effects on life [9,10,13,14]; it was found that metals accumulate in animal and plant cells, leading to severe negative effects. The transport and accumulation of heavy metals by air [15], water [14,16], and soil [17,18] have also been a hot topic for research. It was found that in some cases, contamination was circulated on a global range. Another related research topic was the monitoring of metal pollution and predicting critical levels and loads, which distilled into national and international regulations such as the European Union's Dangerous

Substances Directive [19], the USEPA [20] for water, the EU Air Quality Framework Directive [21], and the World Health Organization [22] for air.

A large body of research that deals with the treatment of metal pollution by different methods such as adsorption [23–26], activated sludge [27,28], phytoextraction [29–32], electrokinetic methods [33], electroosmosis [34], and ion exchange [35] has also been published.

Today, researchers are carrying out more comprehensive studies on metal pollution, leading to the unusual breadth of topics. Despite increasing interest, there have been few attempts at gathering systematic data on the nature and extent of metal pollution research. Garfield indicated that recent research focus could be detected by publication output [36]. A common research tool is the bibliometric method, which has already been widely applied to many fields of science and engineering. Furthermore, the Science Citation Index (SCI), from the Institute for Scientific Information (ISI), Web of Science databases, is the most important and frequently used source database of choice for a broad review of scientific accomplishment [37–44].

In the second metal research trend review period of 2000–2019, the overall growth trend of publications increased by nearly 1,700%, while the growth rate for the last decade (2010–2019) increased by nearly 350% due to environmental awareness. In the second period 2000–2019, China, the USA, Japan, and Germany were the most productive countries in the overall metal research area, but China, India, the USA, Turkey, and South Korea were the first, second, third, fourth, and fifth most productive countries, respectively, in the area of aqueous heavy metals research [45–49].

Specific metal research trends reviews covered in this publication are (a) heavy metal pollution in Antarctica and the southern hemisphere; (b) urban soil heavy metal pollution; (c) heavy metal health risk assessment; (d) geoaccumulation index of heavy metal pollution affected by industrialization and urbanization; (e) heavy metals detection and monitoring; (f) heavy metals removal technologies; (g) thallium toxicity, detoxification, remediation, removal and call for research; (h) critical metals, precious metals, transition metals, and rare metals; (i) innovative and poly-metal coagulants; and (j) toxic heavy metals substitution [45–134].

1.2   DATA SOURCES AND METHODOLOGY

Data were based on the online version of the SCI and Web of Science. SCI is a multidisciplinary database of the ISI, Philadelphia, USA. According to *Journal Citation Reports* (JCR), it indexes 6,166 major journals with citation references across 172 scientific disciplines in 2006. One hundred and ninety-five journals listed in the three ISI subject categories of environmental engineering ($n=35$), environmental sciences ($n=144$), and water resources ($n=57$) were considered in this study. The online version of SCI was searched under the keyword "metal or metals" to compile a bibliography of all papers related to metal research from 1991 to 2006. Articles originating from England, Scotland, Northern Ireland, and Wales were reclassified as being from the UK. Besides, the reported impact factor (IF) of each journal was obtained from the 2006 JCR. Collaboration type was determined by the addresses of the authors, where the term "single country" was assigned if the researchers' addresses were from the same country. The term "international collaboration" was assigned to those articles that were coauthored by researchers from multiple countries.

1.3   RESULTS AND DISCUSSION

The total number of publications that met the selection criteria was 25,449. These publications were divided into 13 document types. The most frequently used document type (96%) was articles (24,409), followed distantly by reviews (609; 2.39%). Other document types of less significance were notes (108; 0.42%), letters (108; 0.42%), and editorial material (104; 0.41%). Since peer-reviewed journal articles represent the majority of documents within this field, 24,409 articles were further analyzed in this study. The emphasis of the following discussion is to determine the pattern of scientific production; research activity trends that consist of authorship, institutes, and countries; and also the trends in the research subjects addressed.

1.3.1 Language of Publication

Written languages of all metal-related articles in the environmental field were grouped. The results showed that English had a clear monopoly, making up 99% of all article publications. Other languages were French (0.26%), German (0.25%), and Spanish (0.025%). French articles were published in *Environmental Technology* (n=32), *Houille Blanche-Revue Internationale de l Eau* (n=14), *Water Quality Research Journal* of Canada (n=8), *Journal of Environmental Engineering and Science* (n=6), *Water Research* (n=3), and *Science of the Total Environment* (n=1). German articles were published in *Gefahrstoffe Reinhaltung der Luft* (n=34), *Acta Hydrochimica et Hydrobiologica* (n=27), and *Isotopes in Environmental and Health Studies* (n=1). Only one journal published Spanish articles: *Ingenieria Hidraulica en Mexico* (n=6). For all practical purposes, English was the international language of choice in metal research, at least according to the SCI database. We leave the debate of whether or not English was the lingua franca of international scientific communication to other commentators [40].

1.3.2 Article Output and Distribution in Journals

Figure 1.1 shows the article output results from 1991 to 2006. The number of articles per year increased from 550 in 1991 to 2871 in 2006, reflecting the increasing interest in this field of research. More than 55% of the records were published during the period 2001–2006. In total, 24,409 articles were published in 173 journals [44].

The overall growth trend of publications from 2000 to 2019 in the field of heavy metals removal from water and wastewater increased by nearly 1,700%, while this growth rate for the last decade (2010–2019) was nearly 350%. The increase in the number of technical papers in the past two decades can be due to various factors such as the overall increase in scientific output in recent years. But more specifically, this can be due to increased environmental awareness [47].

Six core journals contained 34% of the total articles. Figure 1.2 shows the trend of article publication in these six journals from 1991 to 2006. It is noticed that *Applied Catalysis A—General* rose from the sixth rank in 1991 to the first rank in 2006; also *Chemosphere* rose from the fifth rank in 1991 to the second rank in 2006.

The following is an update of the research trend in the field of environmental heavy metals research. In the past two decades (2000–2019), and the narrow area of metal research related to incineration, *Waste Management* was the most productive journal, with 1,262 publications. Other productive journals included *Chemosphere, Environmental Science and Technology*, and *Journal of Hazardous Materials*, each with more than 540 publications in the same period [45].

1.3.3 Publication Performance: Countries, Institutes, and Authorship

Among the 24,409 environmental heavy metals articles produced by 145 countries in 1991–2006, the top 20 most active countries produced 23,062 articles (95%), whereas the remaining 125 countries produced 1,347 articles. Table 1.1 shows that the most active country was the USA (6,081; 25%). In the same 1991–2006 period, the USA also produced the most independent publications (4,859; 24%). Moreover, the seven most industrialized countries (G7: Canada, France, Germany, Italy, Japan, the UK, and the USA) collectively held the major portion (59%) of the world's publications in the field of environmental heavy metals research. Figures 1.3 and 1.4 show the trend of article production in the top 10 countries from 1991 to 2006. The numbers of articles produced per

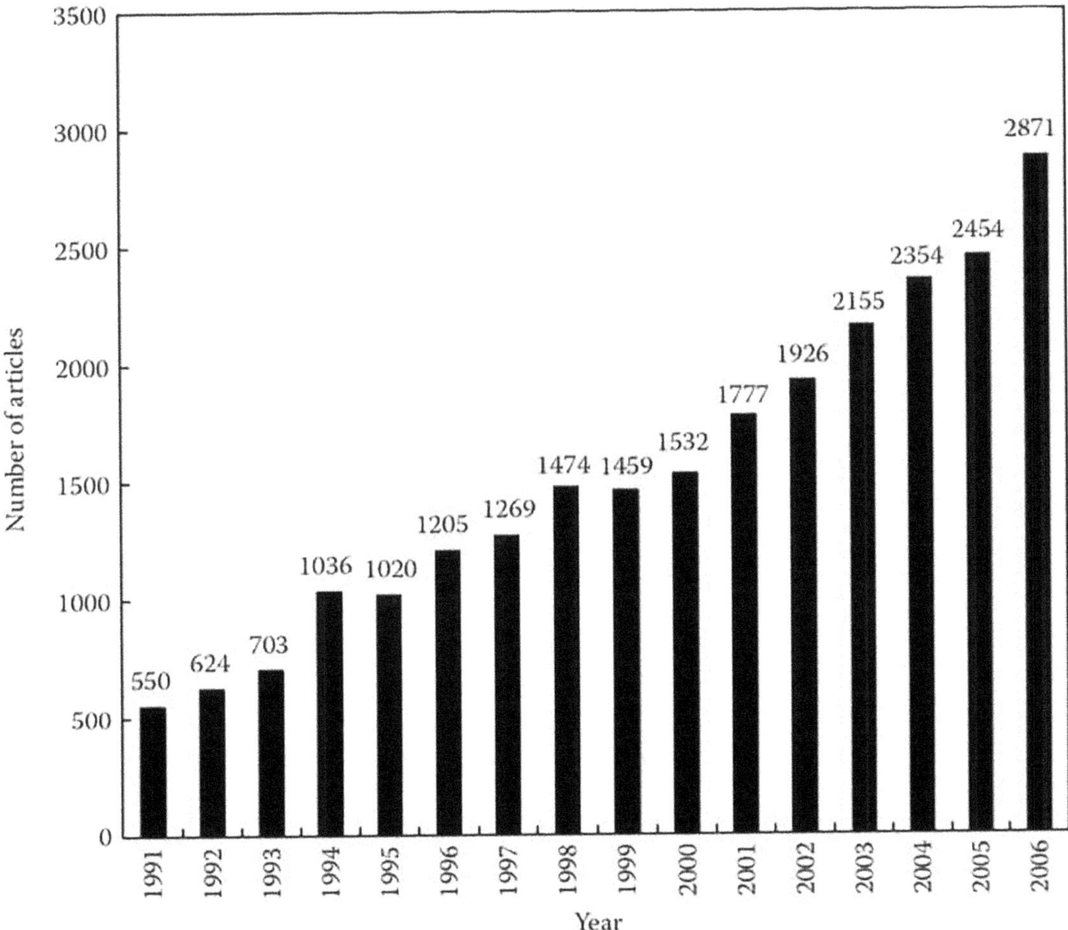

FIGURE 1.1 Publication outputs per year for the period 1991–2006 [44]. (Update: The overall growth trend of publications from 2000 to 2019 in the field of heavy metals removal from water and wastewater increased by nearly 1,700%, while this growth rate for the last decade (2010–2019) was nearly 350% [47].)

year seem to increase at similar rates for most countries, except for China (whose rank changed from tenth in 1991 to second in 2006) and Spain (whose rank changed from ninth in 2001 to third in 2006) [44].

In the past two decades (2000–2019), the country performance ranking changed significantly [45–47]. China played a significant role in contributing to the total number of publications related to environmental metals. It ranks first in terms of research output during the past two decades since 2000. In the research area of environmental metal related to incineration, the USA, Japan, and Germany are the second, third, and fourth most productive countries, respectively, after the first-ranked China [45].

Nazaripour [47] reviewed the status of research in the field of heavy metal removal from water and wastewater in the 21st century (2000–2019). According to the results of the analysis, a considerable increase in research on the heavy metals topic is evident, not limited to a specific region or country. The countries from which the top cited papers originate are China, India, the USA, Turkey, and South Korea. China stands in first place by a large margin.

The top 20 most productive institutes in 1991–2006 are listed in Table 1.2. There are seven institutes from the USA, three from Canada, two from China, and one each from Spain, Italy,

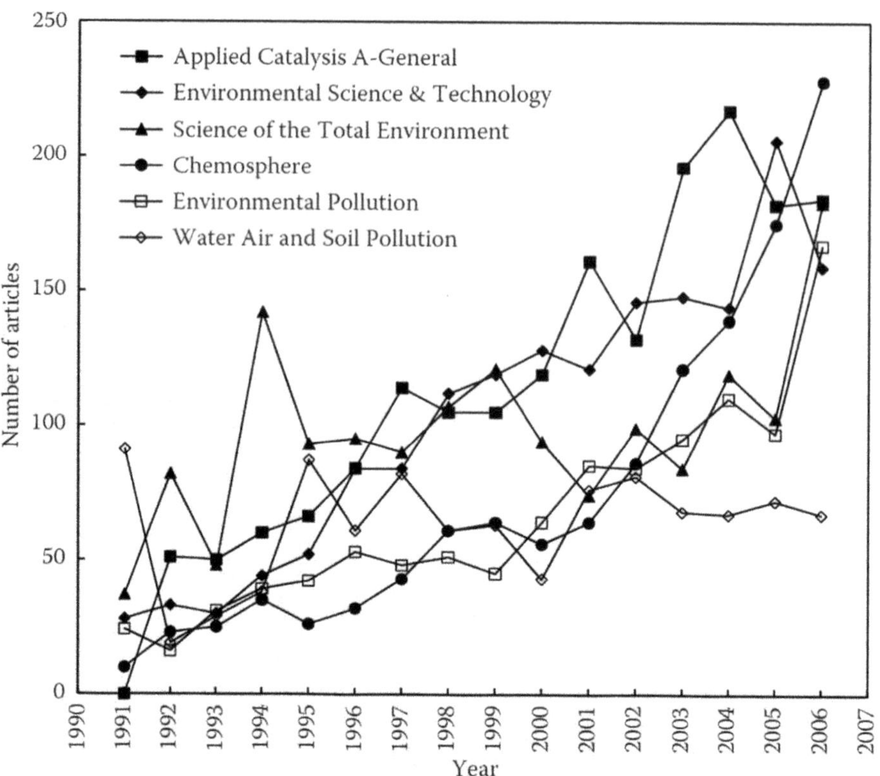

FIGURE 1.2 Comparison of the growth trends of articles in the top six active journals during the period 1991–2006 [44]. (Update: In the past two decades (2000–2019), and the narrow area of metal research related to incineration, *Waste Management* was the most productive journal, and *Chemosphere, Environmental Science and Technology*, and *Journal of Hazardous Materials* were the next top active journals after *Waste Management* in the same period [45].)

France, Taiwan [44], Hong Kong, the UK, Belgium, and India. The highest production came from the USEPA (375; 1.5%), followed closely by the Chinese Academy of Sciences (CAS; 330; 1.4%) and CSIC from Spain (316, 1.3%). The high activity of the latter two institutes is partially responsible for the recent increase in the rank of China and Spain [44].

Collaboration plays an important role in contemporary scientific research, which is manifested in internationally coauthored papers traced by bibliometric tools [41]. Among the 24,365 articles with author address published from 1991 to 2006, 13,080 (54%) were produced by single institutions, whereas 7,304 (30%) were produced by intranational collaboration and 4,025 (17%) by international collaboration (CP). The USA produced the largest number of CP (1,227), which amounts to 20% of the total articles from the USA and 30% of CP produced by all countries. However, Hong Kong ranked number one in the percentage of its publications produced by international collaboration (42%), followed by Germany (41%) and China (40%). Other countries that produce a good part of their articles by CP are France (36%), the Netherlands (35%), Sweden (32%), and Finland (30%) [44].

It is usually assumed that the corresponding author is the senior most among the research group. Hence, articles without corresponding author address information on the ISI Web of Science were excluded from the analysis. The analysis comprised a total of 22,698 articles with 13,310 corresponding authors. Among these corresponding authors, 9,222 (41%) published only one article, and 4,348 (19%) published two articles as corresponding authors. The most active corresponding author was Burger J. from Rutgers State University, NJ, USA, who published 68 articles as corresponding author, 68 articles as first author, and 78 articles in total. This was followed by Wang W.X. from

TABLE 1.1

Top 20 Most Productive Countries of Articles during 1991–2006 [44]

Country	TP	%TP	SP	R (%)	CP	R (%)	FP	R (%)	RP	R (%)
United States	6,081	25	4,854	1 (24)	1,227	1 (30)	5,448	1 (22.4)	4,925	1 (22)
United Kingdom	1,809	7.4	1,140	2 (5.6)	669	2 (17)	1,432	2 (5.9)	1,285	2 (5.7)
Canada	1,569	6.4	1,103	3 (5.4)	466	5 (12)	1,353	3 (5.6)	1,212	3 (5.3)
France	1,298	5.3	822	7 (4.0)	476	3 (12)	1,037	6 (4.3)	941	7 (4.1)
Italy	1,297	5.3	1,042	4 (5.1)	255	9 (6.3)	1,159	4 (4.8)	1,126	4 (5.0)
Spain	1,263	5.2	939	6 (4.6)	324	7 (8.0)	1,101	5 (4.5)	1,038	5 (4.6)
Germany	1,154	4.7	682	9 (3.4)	472	4 (12)	881	8 (3.6)	828	8 (3.6)
India	1,109	4.6	950	5 (4.7)	159	19 (4.0)	1,035	7 (4.2)	945	6 (4.2)
China	1,076	4.4	642	10 (3.2)	434	6 (11)	861	9 (3.5)	826	9 (3.6)
Japan	1,010	4.1	730	8 (3.6)	280	8 (7.0)	853	10 (3.5)	795	10 (3.5)
Australia	647	2.7	432	13 (2.1)	215	11 (5.3)	528	12 (2.2)	501	12 (2.2)
Netherlands	638	2.6	417	15 (2.1)	221	10 (5.5)	505	15 (2.1)	445	16 (2.0)
Sweden	634	2.6	429	14 (2.1)	205	13 (5.1)	527	13 (2.2)	496	13 (2.2)
Turkey	574	2.4	516	11 (2.5)	58	31 (1.4)	549	11 (2.3)	536	11 (2.4)
Poland	568	2.3	399	16 (2.0)	169	15 (4.2)	499	16 (2.0)	492	14 (2.2)
Taiwan	546	2.2	480	12 (2.4)	66	28 (1.6)	510	14 (2.1)	491	15 (2.2)
Belgium	498	2.0	288	19 (1.4)	210	12 (5.2)	380	17 (1.6)	369	17 (1.6)
South Korea	454	1.9	293	18 (1.4)	161	17 (4.0)	367	18 (1.5)	359	18 (1.6)
Hong Kong	434	1.8	252	22 (1.2)	182	14 (4.5)	367	18 (1.5)	350	19 (1.5)
Finland	403	1.7	281	20 (1.4)	122	22 (3.0)	340	20 (1.4)	328	20 (1.4)

CP, international collaborative publication; FP, first-author publication; R, ranking; RP, corresponding author publication; SP, independent publication; TP, total publications; %TP, share in publication [44].

Hong Kong University of Science & Technology, who published 52 articles as corresponding author, 12 articles as first author, and 63 articles as author. Regarding corresponding author countries, the USA ranked at the top (4,925; 22%) followed by the UK (1,285; 5.7%), Canada (1,212; 5.3%), and Italy (1,126; 5.0%). Table 1.2 shows that the ranking of institutes according to the number of corresponding author articles is not the same as the ranking according to the total number of articles. The CAS (233; 1.0%) and the USEPA (227; 1.0%) are almost equally at the top, followed closely by CSIC of Spain (205; 0.90%) (Figure 1.5) [44].

Based on the assumption that the first author of an article performs most of the research, a distribution of first authors was undertaken. The USA produced the largest number of first-author articles (5,448; 22%), followed by the UK (1,432; 5.9%) and Canada (1,353; 5.6%). However, when it comes to institutes, the top-ranking institute in regard to the first author was not from the USA. The institute with the highest number of first-author papers was the CAS (241; 0.99%), followed by CSIC (221; 0.91%) and then by the USEPA (207; 0.85%) [44].

A bias would appear in authorship analysis if any two or more authors have the same name or if authors use different names in their publications (e.g., name changes due to marriage). In addition, authors could work for different institutions or countries over time or within the same period of time, which increases the difficulties of analyzing authorship. Therefore, it is strongly recommended that an "international identity number (IIN)" is created, which is offered to an individual person for all authors when they publish their first paper in an ISI-listed journal. We believe that assigning and tracing the IIN will be the only way of accurately recording authorship. Similarly, a bias would also appear because both the Chinese Academy of Sciences (CAS) and the Indian Institute of Technology (IIT) have branches in many cities.

In this study, the publications of these two institutes were pooled under one heading; dividing the publications among the branches would have given different rankings.

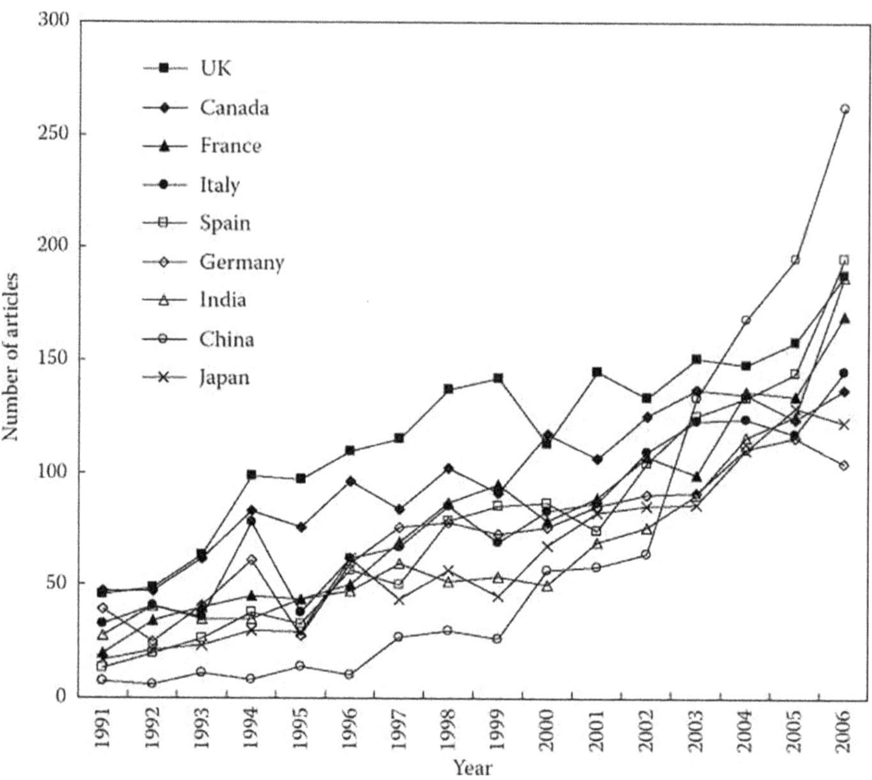

FIGURE 1.3 Comparison of the growth trends of articles in active countries (ranking as second to ninth) during the period 1991–2006. Note that the USA was ranked as the first country and its growth trend is given in Figure 1.4, in the period 1991–2006 [44]. (Update: In the field of total environmental metals research, China, the USA, Japan, and Germany were the first, second, third, and fourth most productive countries, respectively, in the period 2000–2019 [45]. In the field of aqueous heavy metals research, China, India, the USA, Turkey, and South Korea were the first, second, third, fourth, and fifth most productive countries, respectively, in the same period 2000–2019 [47].)

1.3.4 Research Emphasis: Author Keywords and Keywords Plus

The statistical analysis of keywords aimed at discovering the directions of science [42], and proved to be important for monitoring the development of science and programs. The bibliometric method concerning author keyword analysis has been found in recent years [38], whereas using author keywords to analyze the trend of research has been much more infrequent [43]. The examination of author keywords in the period of this study revealed that 32,167 author keywords were used. Of these, 23,741 (74%) appeared only once, 3,706 (12%) appeared twice, and 1,422 (4.4%) appeared thrice. The large number of once-only author keywords probably indicated a lack of continuity in research and a wide disparity in research focus [39]. Table 1.3 shows the distributions of the top 30 [44] most active author keywords used during 1991–2006 and also in four-year periods. Besides "heavy metals" (16%) and "metals" (5.9%), the most frequently used keywords for all periods were the names of certain heavy metals: cadmium (5.9%), copper (5%), lead (4.9%), and zinc (3.4%). It is interesting to note that the sum of using keywords "adsorption," "sorption," and "biosorption" is 5.2%, which reflects the dominance of sorption-related techniques for the treatment of metal pollution. The closest treatment method is "leaching" at 1.1%, whereas other methods of metal removal do not appear in the top 30 list of keywords. Generally, the ranking of most author keywords did not fluctuate distinctly, showing that metal research in the environmental field was basically steady in the past 16 years. However, several keywords such as "bioavailability," "arsenic," "adsorption," "sorption," and "biosorption" have increased in ranking of

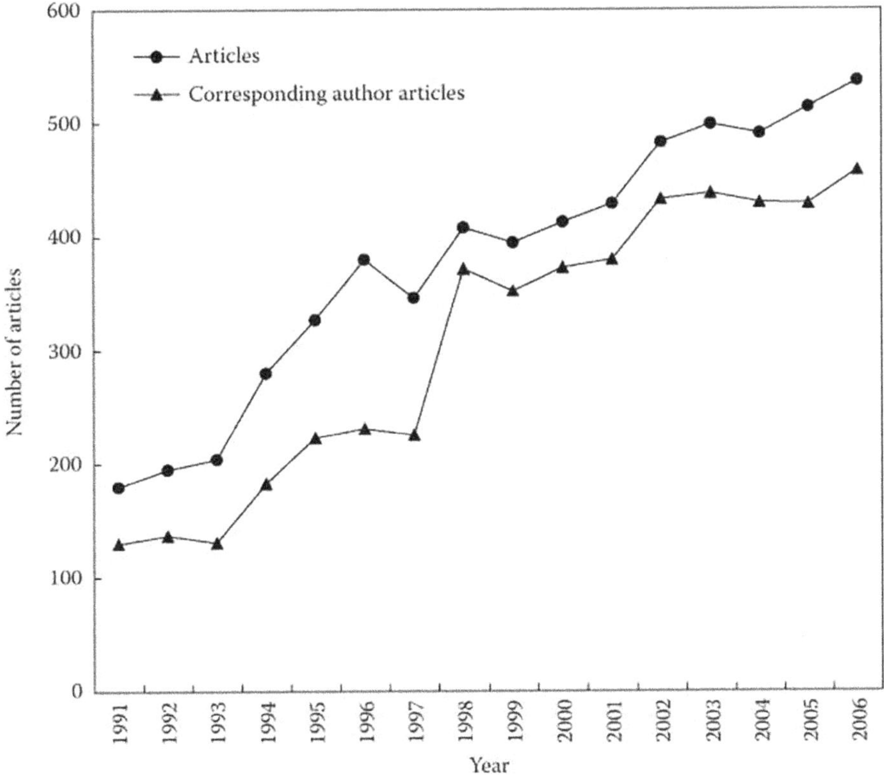

FIGURE 1.4 The growth trends of articles in the USA during the period 1991–2006 [44]. (Update: In the period 2000–2019, China and the USA were the first and second, respectively, most productive countries in the total environmental heavy metals research area related to incineration [45] and were the first and third, respectively, most productive countries in the aqueous heavy metals research area [47].)

frequency, which might be identified as current environmental research hotspots. On the other hand, the use of keywords such as "sediments," "speciation," and "nickel" has steadily declined from 1991 to 2006, which might indicate well-established disciplines or that the research trend has moved away from these topics [44].

Furthermore, keywords plus, which supplied additional search terms, was extracted from the titles of papers cited by authors in their bibliographies and footnotes in the ISI database [42]. Keywords plus analysis is an independent supplement that reveals article contents with more details. In source title analysis, as we segment the title into single words, the result is not repeated and can be statistically analyzed by rule and line; however, it breaks the integrality of phrases in the title. In author keywords analysis, we preserve the intact words that the authors want to convey. Although it makes the same single word or phrase appear in different author keywords, we can compare the discrimination between author keywords or sum up dissimilar keywords with a common phrase or single word for further study. Keywords plus substantially augmented title word and author keyword indexing. In all, 21,783 articles were found to include keywords plus information. Table 1.4 shows the 30 most frequently used keywords plus their rankings and percentages [44].

Keywords plus analysis as an independent supplement reveals article contents with more details. There are some similar and dissimilar trends between their statistical results in the study period. The keywords plus "heavy metals," "metals," "cadmium," and "copper" were at the top of the list, in accordance with the frequency of author keywords and title words. It is interesting to note that in the author's keywords analysis, "cadmium," "copper," and "lead" received more or less the same focus in research (5.9%, 5%, and 4.9%, respectively). However, keywords plus tell another story:

TABLE 1.2

Top 20 Most Productive Institutes of Articles during 1991–2006 [44]

Institute	TP	TP R (%)	SP R (%)	CP R (%)	FP R (%)	RP R (%)
US Environmental Protection Agency, USA	375	1 (1.5)	4 (0.7)	1 (2.5)	3 (0.85)	2 (1)
Chinese Academy of Sciences, P.R. China	330	2 (1.4)	1 (0.99)	2 (1.8)	1 (0.99)	1 (1.0)
Consejo Superior de Investigaciones Cientfficas (CSIC), Spain	316	3 (1.3)	2 (0.96)	3 (1.7)	2 (0.91)	3 (0.9)
United States Geological Survey (USGS), USA	237	4 (0.97)	3 (0.76)	5 (1.2)	4 (0.64)	4 (0.72)
Consiglio Nazionale delle Ricerche (CNR), Italy	211	5 (0.87)	15 (0.4)	4 (1.4)	6 (0.53)	5 (0.57)
Environment Canada, Canada	177	6 (0.73)	26 (0.31)	6 (1.2)	9 (0.46)	7 (0.45)
University of Quebec, Canada	174	7 (0.71)	5 (0.62)	11 (0.82)	5 (0.55)	6 (0.45)
University of Florida, USA	165	8 (0.68)	8 (0.46)	10 (0.93)	7 (0.47)	7 (0.45)
University of Delaware, USA	158	9 (0.65)	19 (0.37)	7 (0.97)	8 (0.46)	11 (0.41)
Rutgers, The State University of New Jersey, USA	149	10 (0.61)	29 (0.30)	7 (0.97)	14 (0.38)	12 (0.41)
University of Georgia, USA	136	11 (0.56)	15 (0.40)	13 (0.74)	15 (0.36)	18 (0.32)
Le Centre National de la Recherche Scientifique (CNRS), France	131	12 (0.54)	98 (0.16)	7 (0.97)	25 (0.29)	24 (0.28)
McGill University, Canada	128	13 (0.53)	13 (0.41)	16 (0.65)	13 (0.39)	13 (0.38)
University of Maryland, USA	123	14 (0.50)	19 (0.37)	14 (0.66)	20 (0.32)	19 (0.31)
National Taiwan University, Taiwan, Republic of China	122	15 (0.50)	18 (0.39)	17 (0.63)	19 (0.34)	16 (0.33)
Hong Kong University of Science and Technology, Hong Kong, P.R. China	120	16 (0.49)	6 (0.60)	67 (0.37)	11 (0.41)	9 (0.43)
Zhejiang University, P.R. China	118	17 (0.48)	12 (0.42)	25 (0.56)	10 (0.41)	9 (0.43)
Imperial College of Science, Technology and Medicine, University of London, UK	117	18 (0.48)	15 (0.40)	22 (0.57)	16 (0.35)	20 (0.30)
Universiteit Gent, Belgium	116	19 (0.48)	9 (0.44)	30 (0.52)	17 (0.34)	15 (0.37)
Indian Institute of Technology, India	115	20 (0.47)	7 (0.56)	67 (0.37)	12 (0.40)	14 (0.37)

CP, international collaborative publication; FP, first-author publication; R, ranking; RP, corresponding author publication; SP, independent publication; TP, total publications.

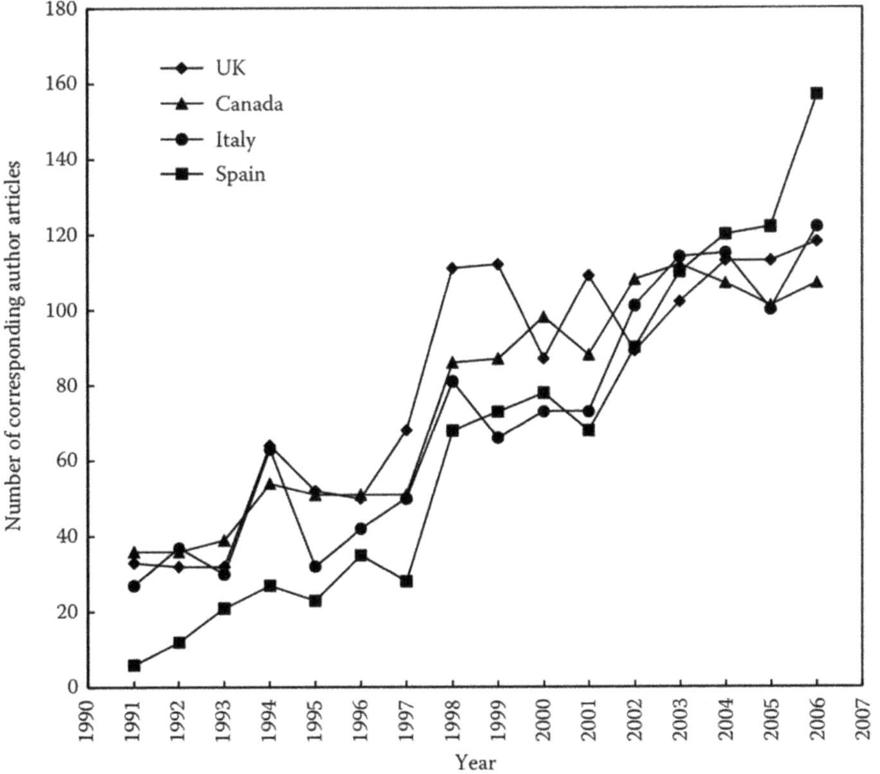

FIGURE 1.5 Comparison of the growth trends of corresponding author articles during the period 1991–2006 in the UK, Canada, Italy, and Spain [44]. (Update: In the period 2000–2019, the growth trend of corresponding author articles in China was ranked first place [45,47]).

"cadmium" ranked second at 12% after "heavy metals" (16%), whereas "copper" and "lead" were relatively distant at 8.6% and 7.2%, respectively. On the other hand, "water," which ranked fifth in keywords plus (7.2%), did not appear at all in the top 30 author keywords. This indicates that the research was more oriented toward metal pollution in aqueous systems, and is corroborated by noticing that the keyword plus "air" ranked 80th (1%). Moreover, the ranking of "water" improved from eighth during 1991–1994 to fifth during 2003–2006, whereas that of air dropped from 55th to 99th in the same time period. Another disagreement was found in the appearance of "mercury": in author keywords its ranking improved from 15th (1991–1994) to 8th (1995–1998) and then steadily declined until it held the 11th rank (2003–2006); on the other hand, its ranking in keywords plus dropped from 16th (1991–1994) to 40th (2003–2006). This may be attributed to the global legislation that reduced/eliminated mercury compounds from many products, thus reducing the environmental problems of mercury and consequently diverting the research focus to other pollutants. Other terms that are frequent in keywords plus but not in the top 30 author keywords are "accumulation" (5.8%), "removal" (4.4%), "growth" (2.7%), and "exposure" (2.5%). The frequency of these words has significance because keywords plus is usually more concerned with the novel research direction than with the mature direction in the field [42]. "Adsorption" and "sorption" increased in keywords plus frequency from 1991 to 2006, corroborating the observed importance of this treatment method. Moreover, the word "removal" increased in ranking and frequency from 24 (2.1%) in 1991–1994 to 10 (6%) in 2003–2006. This increase, in view of the declining frequency of "toxicity," might indicate that research is moving away from assessing the impact of metal pollution and focusing instead on the treatment of polluted bodies. The treatment of metal pollution by "reduction" increased in ranking from 58th (1991–1994) to 28th (2003–2006), suggesting increased interest in this technique [44].

TABLE 1.3

Top 30 Most Frequency of Author Keywords Used during 1991–2006 and 4 Four-Year Periods [44]

Author Keywords	91-06 TP (%)	91-94 R (%)	95-98 R (%)	99-02 R (%)	03-06 R (%)
Heavy metals	2,625 (16)	1 (15)	1 (16)	1 (17)	1 (15)
Metals	996 (5.9)	2 (6.9)	3 (6.0)	2 (6.4)	3 (5.4)
Cadmium	994 (5.9)	3 (6.6)	2 (6.3)	3 (5.9)	2 (5.6)
Copper	844 (5.0)	5 (4.9)	4 (5.2)	4 (5.3)	4 (4.7)
Lead	829 (4.9)	4 (5.6)	4 (5.2)	5 (5.0)	5 (4.6)
Trace metals	585 (3.5)	6 (4.7)	6 (4.0)	6 (4.0)	9 (2.7)
Zinc	582 (3.4)	7 (4.1)	7 (3.4)	7 (3.9)	8 (3.1)
Heavy metal	498 (2.9)	23 (1.7)	16 (2.1)	8 (2.9)	6 (3.5)
Adsorption	471 (2.8)	11 (3.3)	13 (2.3)	15 (2.1)	7 (3.4)
Mercury	448 (2.7)	15 (2.3)	8 (3.1)	9 (2.9)	11 (2.4)
Sediment	431 (2.6)	10 (3.4)	10 (2.6)	10 (2.8)	12 (2.3)
Soil	417 (2.5)	16 (2.2)	12 (2.4)	13 (2.3)	10 (2.6)
Toxicity	406 (2.4)	9 (3.6)	9 (2.7)	11 (2.6)	16 (2.0)
Pollution	392 (2.3)	18 (2.0)	11 (2.5)	12 (2.6)	14 (2.2)
Sediments	334 (2.0)	8 (3.8)	14 (2.2)	14 (2.1)	21 (1.5)
Bioavailability	318 (1.9)	24 (1.4)	24 (1.4)	19 (1.8)	13 (2.2)
Bioaccumulation	307 (1.8)	16 (2.2)	18 (1.8)	16 (2.0)	18 (1.7)
Arsenic	297 (1.8)	30 (1.1)	23 (1.5)	20 (1.7)	15 (2.0)
Nickel	291 (1.7)	12 (3.1)	15 (2.2)	17 (1.8)	27 (1.3)
Trace elements	288 (1.7)	19 (1.9)	19 (1.7)	23 (1.4)	17 (1.9)
Speciation	286 (1.7)	14 (2.7)	17 (2.0)	18 (1.8)	23 (1.4)
Chromium	253 (1.5)	22 (1.8)	19 (1.7)	21 (1.6)	27 (1.3)
Biomonitoring	252 (1.5)	36 (0.92)	28 (1.1)	22 (1.6)	19 (1.6)
Sewage sludge	230 (1.4)	19 (1.9)	21 (1.7)	30 (1.1)	25 (1.3)
Kinetics	205 (1.2)	26 (1.3)	26 (1.2)	32 (1.1)	25 (1.3)
Sequential extraction	200 (1.2)	24 (1.4)	33 (1.0)	24 (1.3)	30 (1.1)
Sorption	198 (1.2)	30 (1.1)	37 (0.93)	33 (1.0)	22 (1.4)
Biosorption	196 (1.2)	97 (0.42)	43 (0.86)	27 (1.2)	24 (1.4)
Leaching	194 (1.1)	53 (0.67)	29 (1.1)	26 (1.2)	29 (1.2)
Fish	187 (1.1)	30 (1.1)	25 (1.2)	25 (1.3)	36 (0.96)

R (%), the rank and percentage of the articles; TP, publications in the study period.

1.4　RESEARCH TRENDS OBSERVATIONS

1.4.1　Metal Research Publications in 1991–2006

In the first part of this study, dealing with metal research in the 195 journals listed in the environmental field SCI papers, the authors obtained some significant points on worldwide research trends throughout the period from 1991 to 2006. The effort provided a systematic structural picture as well as clues on the impacts of metal research. English was by far the dominant language (99%); three other languages were also used, indicating that metal research is globally communicated in English. Apparently, more and more authors, institutes, and countries have been engaged in metal research over the years. The G7, with a longer research tradition in this field, has the absolute superiority of production. Besides, China and Spain (both non-G7 countries) have boosted their research in the last few years and held the second and third ranks, respectively, in 2006.

TABLE 1.4
Top 30 Frequency of Keywords Plus Used and 4 Four-Year Periods [44]

Keywords Plus	91-06 TP (%)	91-94 R (%)	95-98 R (%)	99-02 R (%)	03-06 R (%)
Heavy metals	3,532 (16)	2 (13)	1 (14)	1 (15)	1 (19)
Cadmium	2,664 (12)	1 (13)	2 (14)	2 (12)	2 (12)
Metals	1,904 (8.7)	3 (11)	4 (9.1)	3 (9.0)	4 (8.0)
Copper	1,871 (8.6)	4 (10)	3 (9.3)	4 (8.3)	3 (8.1)
Water	1,574 (7.2)	8 (6.7)	5 (8.4)	6 (6.9)	5 (7.1)
Zinc	1,568 (7.2)	5 (8.7)	6 (7.6)	8 (6.7)	7 (6.9)
Lead	1,522 (7.0)	6 (7.6)	7 (7.0)	5 (7.2)	8 (6.7)
Trace metals	1,462 (6.7)	7 (7.0)	8 (6.8)	7 (6.8)	9 (6.5)
Adsorption	1,375 (6.3)	10 (5.7)	9 (5.9)	11 (5.7)	5 (7.1)
Accumulation	1,264 (5.8)	11 (5.6)	11 (5.6)	10 (5.9)	11 (5.9)
Toxicity	1,205 (5.5)	9 (5.8)	10 (5.7)	9 (6.0)	12 (5.1)
Sediments	1,103 (5.1)	12 (5.2)	12 (5.2)	12 (5.1)	13 (4.9)
Removal	949 (4.4)	24 (2.1)	24 (2.2)	14 (4.3)	10 (6.0)
Pollution	931 (4.3)	13 (3.8)	13 (4.1)	14 (4.3)	14 (4.5)
Speciation	926 (4.3)	14 (3.4)	14 (4.0)	13 (4.5)	15 (4.4)
Soils	816 (3.7)	15 (3.0)	15 (3.4)	16 (3.5)	16 (4.2)
Sorption	670 (3.1)	44 (1.4)	24 (2.2)	17 (3.1)	17 (3.9)
Contamination	618 (2.8)	35 (1.6)	19 (2.3)	19 (2.7)	18 (3.4)
Plants	599 (2.7)	33 (1.8)	27 (2.1)	23 (2.6)	19 (3.4)
Soil	591 (2.7)	18 (2.6)	23 (2.2)	18 (2.7)	21 (2.9)
Growth	590 (2.7)	17 (2.8)	17 (2.8)	21 (2.6)	25 (2.7)
Kinetics	546 (2.5)	30 (1.9)	31 (2.0)	26 (2.2)	20 (3.1)
Exposure	546 (2.5)	35 (1.6)	36 (1.8)	20 (2.7)	22 (2.9)
Oxidation	537 (2.5)	24 (2.1)	18 (2.4)	30 (2.1)	23 (2.8)
Mercury	523 (2.4)	16 (2.8)	16 (3.0)	22 (2.6)	40 (1.9)
Trace elements	508 (2.3)	39 (1.5)	35 (1.8)	26 (2.2)	24 (2.8)
Iron	477 (2.2)	27 (2.0)	21 (2.3)	25 (2.3)	31 (2.1)
Sewage sludge	454 (2.1)	24 (2.1)	22 (2.3)	32 (2.0)	33 (2.0)
Bioavailability	452 (2.1)	46 (1.3)	44 (1.5)	28 (2.2)	27 (2.5)
Reduction	451 (2.1)	58 (1.1)	31 (2.0)	34 (2.0)	28 (2.4)

R(%), the rank and percentage of the articles; TP, publications in the study period.

The USA produced the largest number of internationally collaborative articles, but five other countries had more than a third of their production by international collaboration (Hong Kong, Germany, China, France, and the Netherlands). The most frequently used keywords were "cadmium," "copper," and "lead," which reflects stability in this research field. Among the metal removal methods investigated, adsorption was the most frequent. Other significant methods of removal in metal research are "oxidation" and "reduction" [1–44].

1.4.2 METAL RESEARCH PUBLICATIONS IN 2000–2019

The second stage of this metal research trend investigation covers the period of 2000–2019. The overall growth trend of publications from 2000 to 2019 in the field of heavy metals removal from water and wastewater increased by nearly 1,700%, while this growth rate for the last decade (2010–2019) was nearly 350%. The increase in the number of heavy metal papers in the past two decades can be due to the overall increase in scientific output in recent years. But more specifically, this can be due to increased environmental awareness [47].

In the past two decades (2000–2019), and the narrow area of metal research related to incineration, *Waste Management* was the most productive journal, and *Chemosphere, Environmental Science and Technology*, and *Journal of Hazardous Materials* were the next top active journals after *Waste Management* in the same period [45].

In the field of total environmental metals research, China, the USA, Japan, and Germany were the first, second, third, and fourth most productive countries, respectively, in the period 2000–2019 [45]. In the field of aqueous heavy metals research, China, India, the USA, Turkey, and South Korea were the first, second, third, fourth, and fifth most productive countries, respectively, in the same period 2000–2019 [47].

1.4.3 RESEARCH TRENDS OF HEAVY METAL POLLUTION SOURCES

The following are some recent research trends related to environmental heavy metals pollution sources. Human activities, from burning fossil fuels and fireplaces to the contaminated dust produced by mining and industrial developments, have altered Earth's atmosphere in countless ways. Records of these environmental impacts over a long period of time have been preserved in everlasting polar ice that serves as a sort of time capsule, allowing researchers, such as J. M. McConnell's international team at the Desert Research Institute (DRI) to link Earth's history with that of human activities [62]. Toxic heavy metal pollution has been discovered in the Southern Hemisphere according to the DRI report [62]. Ice cores from Antarctica reveal that lead and other toxic heavy metals linked to mining activities have been polluting the Southern Hemisphere since the 13th century. Early Andean cultures and later Spanish Colonial mining contributed to detectable lead pollution in Antarctica, even 9,000 km away. Thallium, bismuth, and cadmium pollution were also assessed for the first time.

The toxicity and pollution sources of lead, cadmium, mercury, chromium, arsenic, and radionuclides in the environment have been investigated by Naja et al. [112], while the environmental geochemistry and toxic arsenic source in groundwater have been investigated by Wang et al. [113].

Sources of soil pollution caused by the toxic heavy metals are reported by Zorluter and Hung [114] and Semalti et al. [115].

Sources of air pollution caused by the toxic heavy metals, organics, and particulates are reported by Wang et al. [87,116].

1.4.4 RESEARCH TRENDS OF URBAN SOIL HEAVY METAL POLLUTION

Urban soil heavy metal pollution has been studied extensively by Tang et al. [63]. Developmental trajectories, research hotspots, and emerging trends in this field are being explored by Tang's research team using a comprehensive bibliometric analysis of 1,247 articles sourced from the Web of Science Core Collection Database (WoSCC) spanning the period from 2000 to 2022. Their analysis revealed a significant upward trend in the number of publications during the period 2000–2022, a trend expected to persist. The journal, *Science of the Total Environment*, held the most influence, while China led in the number of publications, with the CAS as the foremost contributor. The research predominantly focused on source apportionment of urban soil heavy metal pollution, pollution risk assessment, and the application of environmental magnetism.

1.4.5 RESEARCH TRENDS OF HEAVY METAL HEALTH RISK ASSESSMENT

Due to the widespread presence and harmfulness of heavy metals in the environment, Zhang team's research [64] evaluated the exposure characteristics and health risks of heavy metals using bibliometric analysis tools to conduct a scientometric analysis of the literature related to the health risk assessment of heavy metals in the Web of Science database from 2000 to 2022. Their analysis results indicate that research related to heavy metal health risk assessment is rapidly developing in both developed and developing countries. China's significant international influence in this field is worth noting, as there are many publications and highly cited documents related to China. France and other developed countries also play an important role in this field due to their high centrality and strong bursts. The results of co-citation cluster analysis and keyword co-occurrence analysis indicate that in the past two decades, the primary research domains and hotspots of heavy metal health risk assessment have been the study of heavy metals in soil, dust, drinking water, vegetables, fish, and sediment. There is a specific focus on bioaccumulation, bioavailability, source apportionment, and spatial distribution of heavy metals. The main types of heavy metals studied are lead, cadmium, mercury, and zinc. Future research trends may focus more on the health risks of heavy metals in different functional areas of cities.

1.4.6 RESEARCH TRENDS OF GEOACCUMULATION INDEX OF HEAVY METAL POLLUTION AFFECTED BY INDUSTRIALIZATION AND URBANIZATION

Industrial wastes contain toxic heavy metals and volatile organic compounds (VOCs), and in turn, pollute the world's land, water, and air. Heavy metal contamination potential has been assessed by Swan [65] using geoaccumulation index, enrichment factor, and other criteria. According to Swan's assessment [65], heavy metal concentrations have increased due to industrial wastes, geochemical shifts, agriculture, and mining. Heavy metals can harm human health and cause cancer by modifying the cell structure. Comprehensive monitoring is needed to control the heavy metals in the environment. This review reports the nature, origin, and extent of heavy metal contamination.

1.4.7 RESEARCH TRENDS OF HEAVY METALS DETECTION

The detection and removal of heavy metal ions have become critical due to their impact on environmental protection and human health. Some current trends in this field explore the following: (a) Investigation of advanced functional materials for detection and removal: these advanced functional materials include nanomaterials, polymers, porous materials, and biomaterials; (b) Investigation the advanced functional materials that offer several advantages in detection and separation of heavy metals: (b1) real-time heavy metals detection: they enable real-time monitoring of heavy metal ions; (b2) excellent heavy metals separation efficiency: functional materials efficiently capture and separate heavy metals; (b3) anti-interference heavy metals detection: they are designed to minimize interference from other substances; (b4) quick response in heavy metals detection: functional systems respond rapidly to changes in metal ion concentrations; (b5) high heavy metals selectivity: they selectively bind to specific metal ions for detection or separation; and (b6) low limit of heavy metals detection: functional materials can detect even trace amounts of heavy metals [66–70]. The water treatment process for removing heavy metals from polluted sources is explained in detail. Punia et al. [66] have reviewed various metal detection and removal techniques and their modeling for heavy metal removal. Improved detection of heavy metals is possible by several optical, spectroscopic, and electrochemical methods. Due to their high efficiencies and accurate evaluation capabilities, several spectroscopic as well as other techniques like neutron activation analysis and X-ray fluorescence are researched and found highly significant.

1.4.8 RESEARCH TRENDS OF COMMON AND HEAVY METALS REMOVAL

Based on the review of Punia et al. [66], adsorption, photocatalysis, ion exchange (IE), electrochemical methods, MF, chemical precipitation, and forward osmosis have been the most frequently researched process technologies for heavy metal removal in recent years. The design, operation, and key features of all techniques are explained by Punia et al. [66]. The regeneration of spent adsorbents, resins, and other materials is the key step to heavy metals removal from the wastewater.

The technologies reported by Liu et al. [78] for arsenic removal from groundwater include oxidation, coagulation-flocculation, filtration, manganese oxide adsorption, FMBO adsorption, lime softening, and IE.

The technologies reported by Naja et al. [79] for the treatment of metal-bearing effluents include coagulation, flocculation, clarification, sludge thickening, sludge dewatering, sludge disposal, IE, reverse osmosis (RO), adsorption, oxidation-reduction, liquid-liquid extraction, and biosorption. The application of IE resin in removing metal pollutants from wastewater is also reported by Madu et al. [123].

Some noticeable innovative technologies for aqueous common and heavy metals control are nanotechnologies [80,81], activated sludge and granular sludge biosorption [82–84], sulfide precipitation [92], physicochemical sequencing batch reactor (PC-SBR) [100–103], exopolysaccharide-producing cyanobacteria biotechnology [105], fungal biotechnology [118], natural bacterial biomass technology [119], magnesium carbonate coagulation-precipitation [120,124], combined flotation and filtration [121], *Heinsia crinita* seed coat biomass biotechnology [122], and electro-Fenton technology [125].

The pump-and-treat technology, solidification/stabilization, vitrification, soil washing, soil flushing, pyrometallurgy, electrokinetics, phytoremediation, etc. are the technologies used for the removal of heavy metals from contaminated soil or groundwater [85,86].

Elimination or reduction of heavy metals in air can be accomplished by control, management, and treatment of metal emissions from vehicles, airplanes, ships, and industrial plants [87,88].

Minimization and reduction of toxic heavy metals and precious metals in solid wastes can be accomplished by proper e-waste recycle and management (e-waste) [89].

Lenox Institute of Water Technology (LIWT) has developed several dissolved air flotation (DAF)-related process systems for the removal of chromium [90,91], titanium [76], nickel and chromium [92], calcium, magnesium, iron and manganese [93,94], copper [94], heavy metals mixtures [86,95–97,106–111], arsenic [98,99], etc. from surface water, groundwater, wastewater, chemical spills, and/or sludge. Recent and ongoing research projects have been aimed to develop green, efficient, and economical processes, such as DAF and membrane filtration (MF), for heavy metal removal. The ongoing and future research trends in heavy metal removal technologies will be the combination of DAF and MF that are the two best available technologies (BAT). LIWT has also developed the physicochemical sequencing batch reactor (PC-SBR) [92,100–104]. Benefits of these innovative DAF, MF, PC-SBR process approaches include efficient heavy metals removal, small footprint, cost-effectiveness, and environmental friendliness. By understanding the design principles and mechanisms behind these processes and chemicals, researchers can expedite the development of high-performance processes and chemicals and improve environmental pollution control.

1.4.9 RESEARCH TRENDS OF THALLIUM TOXICITY, DETOXIFICATION, REMEDIATION, REMOVAL

Thallium is a bluish-white heavy metal that exists in two oxidation states: thallous (I) and thallic (III). Despite its low natural abundance, it can be found in up to 89 different minerals and is widely distributed in water environments [71]. Thallium is a potent toxin, and its removal from the environment is crucial for safeguarding human health.

Thallium toxicity can cause dermatological problems (such as hair loss), gastrointestinal diseases, and neurological disorders. Thallium exposure may lead to disorders of the nervous system, including tremors, seizures, and peripheral neuropathies. In severe cases, thallium poisoning can result in death.

Sources of thallium exposure include (a) mining and industrial activities: thallium enters the environment through mining, ore smelting, and waste discharge from various industries; and (b) food and beverages: thallium has been detected in beverages, tobacco, and vegetables, highlighting its presence in food sources.

Several technologies are attempted to detoxify, remediate, or remove from the environment: (a) phytoremediation: plants like *Solanum nigrum* and *Callitriche cophocarpa* show promise for removing thallium from contaminated soil; (b) macrocyclic compounds: crown ethers (e.g., 18-crown-6) can absorb thallium from wastewater; (c) chemical precipitation: adding compounds like chloride ions, sulfide ions, potassium iron vanadium, and Prussian blue to thallium-containing wastewater can facilitate removal [72]; (d) adsorption: various adsorbents, including metal oxide materials, have been explored for thallium removal [72]; (e) titanium dioxide (TiO_2): amorphous titanium dioxide has shown high efficiency in removing thallium from water and wastewater [73]. The titanium dioxide prepared at room temperature is amorphous and effective for Tl(I) adsorption, exhibiting high maximal adsorption capacities of 230.3–302.6 mg/g under neutral pH conditions. These values outperform the majority of reported adsorbents.

Recent research trends on understanding the impact of low-dose thallium on public health and its diverse effects on different populations and organs [74]. Studies explore cytotoxic effects, genotoxic potential, and molecular mechanisms related to thallium exposure. Additionally, detoxification therapies are being investigated to mitigate thallium toxicity [75].

Since thallium is a potent toxin, and its removal from the environment is crucial for safeguarding human health, Wang and Wang of the LIWT [76] have conducted research for recovering titanium dioxide from paper-mill and pulp-mill wastewater as an industrial product for thallium spill control. Wang and Wang are calling for three further research studies in the areas of thallium spill control, removal, and analysis: (a) development of simple analytical method for quick determination of thallium content in the field in case of thallium spills. As little as four parts per million of thallium may be detected in clear water, and semi-quantitative determinations should be possible up to 200 parts per million on the basis of opacity and color of the resultant suspension; (b) development of commercial product of the "recovered titanium dioxide" from white water treatment for thallium spill control or removal; and (c) demonstration of soluble thallium (such as Tl_2CO_3) removal from aqueous phase by sulfide precipitation process (using sodium sulfide) and DAF process (instead of sedimentation or centrifugation) because the resulting insoluble flocs of Tl_2S cannot be easily separated by sedimentation or centrifugation clarification. Further removal by membrane filtration (MF) is needed. The combination of DAF-MF can be very effective.

1.4.10 RESEARCH TRENDS OF CRITICAL METALS, PRECIOUS METALS, TRANSITION METALS, AND RARE METALS

Researchers have delved into the fascinating world of critical metals and their research trends in recent years because they are essential metals for various applications, including renewable energy, electronics, and transportation. Critical metals include precious metals (such as gold, silver, platinum, and palladium), transition metals (like iron, cobalt, nickel, and lithium), and rare earth metals (such as lanthanum, yttrium, and neodymium) [46,48,49,50–56].

Liu et al. [46] have conducted a bibliometric analysis of global critical metals' research trends and dynamical development during 1991–2020. They conclude that critical metals research has been steadily increasing significantly since 2010. Universities lead critical metal research, emphasizing interdisciplinary approaches. Popular keywords include "critical metals," "rare metals," "precious metals," and "transition metals." Research covers the entire life cycle, mining, extraction, treatment, and application in the emerging areas of (a) bioleaching: a crucial process for critical metal recovery; (b) urban mining: circular economy concepts focus on recovering critical metals from e-waste and waste landfills; (c) targeted studies: investigating specific critical elements like copper, zinc, gallium, silver, and lithium; (d) rare elements: considered "industry vitamins"; (e) coal deposits and ashes: alternative sources for critical metals [77]; and battery metals: zinc, a transition metal, is both critical and a battery metal, and its annual demand for zinc in batteries is projected to rise significantly by 2030 [78].

1.4.11 RESEARCH TRENDS OF INNOVATIVE AND POLY-METAL COAGULANTS

Effective water treatment involves a combination of coagulation, flocculation, and other processes to ensure safe and clean water for consumption and environmental protection. Inorganic polymeric coagulants are recent innovative coagulants. These are substances used in water and wastewater treatment to aid in the removal of impurities. Polymeric coagulants, both synthetic and natural, show promise as alternatives to traditional inorganic coagulants like alum.

Polyaluminum chloride (PAC) is a widely used inorganic coagulant. It has better decolorizing and COD-reducing properties than other conventional coagulants [126]. Researchers [127] have explored using natural and biocompatible materials like sodium alginate as a coagulant aid to reduce the reliance on PAC. Magnetite significantly reduces settling time under an external magnetic field, achieving equilibrium in just 20 minutes compared to 60 minutes without magnetite, the combination of PAC, sodium alginate, and magnetite (iron oxide nanoparticles) enhances wastewater treatment efficiency. Sodium alginate, when used as a coagulant aid, improves Congo red removal compared to PAC alone [127].

Polyiron chloride (PFC) is another inorganic polymer coagulant that has been synthesized and tested for treating water and wastewater [126,128]. Among various coagulants tested, PAC exhibited the best performance for organics removal from the wastewater. Since PFC contains iron ions, it is also effective for the removal of arsenic (V) from contaminated water. PFC's inherent structural features enhance the coagulation process and improve cost-effectiveness.

Wang and Wang [126] introduced the methods for the production of polymeric metal coagulants, and the case histories of treating cold, low-alkalinity, super-high turbidity waters using polymeric metallic coagulants. Yang et al. [129] treated paper-mill wastewater using a composite inorganic coagulant prepared from steel mill waste pickling liquor.

Wang et al. [130] have investigated a newly developed "green," inorganic, polymeric coagulant, known as aluminum chlorohydrate (ACH). ACH is effective for water treatment especially when the influent water quality is unpredictable, and normally, the poor raw water source is unusable. Stream current monitor (SCM) is a green automatic chemical dosage monitor and controller that is highly useful for treating troubled water that constantly changes its quality. DAF is a green water treatment technology that achieves high effluent quality but cuts detention time (DT) and O&M costs by about 25%. Wang et al. [130] introduced the use of three green technologies of ACH, SCM, and DAF for treating unpredictable water, in turn, saving raw water resources and O&M costs, and improving water quality

1.4.12 RESEARCH TRENDS OF TOXIC HEAVY METALS SUBSTITUTION

Contamination by toxic heavy metals, such as soluble chromium-containing substances, is a serious issue due to their toxicity, persistence, and bioaccumulative nature. When these heavy metals are released into the environment, they can reach the human body causing serious biological and physiological complications.

Wang, Wang, and De Michele [131,132] propose heavy metals substitution or management in the determination of biochemical oxygen demand (BOD) [131] and chemical oxygen demand (COD) [132] in research laboratories. Although laboratory wastewater containing toxic heavy metals and toxic organic substances can be properly managed, collected, treated, and disposed of [133], the best available technology (BAT) should not be an expensive end-of-the-pipe wastewater treatment. The waste containing high concentrations of heavy metals and toxic organics should not be discharged into a sewer system anticipating that these toxic substances would be properly treated by a Publicly Owned Treatment Work (POTW).

Correct hazardous waste management procedures must be implemented for handling and disposal of all wastes containing highly concentrated heavy metals and toxic organics. Heavy metal waste minimization is usually the key to success in environmental protection. The method better than the heavy metal minimization is the heavy metal substitution, so the heavy metal wastes will not be generated at all.

Wang and Wang [134] have proposed the use of potassium permanganate (instead of potassium dichromate) for COD determination and cleaning solution preparation in the laboratories. The heavy metal substitution method is much better because the toxic chromium-containing waste will not be generated at all if potassium dichromate is substituted by potassium permanganate in COD analysis and glassware cleaning solution preparation.

GLOSSARY

Base metals (a) Base metals are metals that are not considered precious. Unlike precious metals (such as gold or silver), base metals are readily available, relatively inexpensive, and tend to tarnish, oxidize, or corrode more easily when exposed to air or moisture. Common examples of base metals include copper, lead, tin, aluminum, nickel, zinc; alloys of these elemental metals, such as brass and bronze, are also considered base metals; (b) There are at least three definitions of base metals: (b1) A base metal can be a common metal (element or alloy) with low value, not used as a basis for currency. Examples include bronze and lead; (b2) It can be the primary metal in an alloy. For instance, iron is the base metal in steel; (b3) It can also refer to a metal or alloy to which a plating or coating is applied. For example, steel or iron serves as the base for galvanized steel. Again in the context of plated metal products, the base metal underlies the plating metal. For instance, copper underlies silver in Sheffield plate; (c) When industrial and economic significance in mining and economics is considered, base metals refer to industrial non-ferrous metals (excluding precious metals), such as copper, lead, nickel, and zinc; (d) The US Customs and Border Protection Agency has an inclusive definition, which includes metals like iron, aluminum, tin, tungsten, and several other transition metals as base metals [57–61].

Bibliometric analysis It is a quantitative method that is used to analyze academic literature, typically focusing on publications such as academic books, journal articles, conference proceedings, and invention patents. Bibliometric analysis involves examining various aspects of these publications, including citation statistical data, citation patterns, authorship, collaboration institutes collaboration countries, and research trends. Researchers use the analysis to explore the impact of an academic field; evaluate the influence of research articles, authors, books, and journals; and identify particularly impactful technical papers within specific academic areas

Chinese Academy of Sciences CAS is the national academy for natural sciences in the People's Republic of China. It serves as the highest consultancy for science and technology in the country. CAS is the world's largest research organization, comprising 100 research institutes and 2 universities. It brings together scientists and engineers from China and around the world to address both theoretical and applied problems. CAS plays a crucial role in advancing scientific knowledge and technological innovation in China and beyond.

Critical metals Researchers have delved into the fascinating world of critical metals and their research trends in recent years because they are essential metals for various applications, including renewable energy, electronics, and transportation. Critical metals include precious metals (such as gold, silver, platinum, and palladium), transition metals (like iron, cobalt, nickel, and lithium), and rare earth metals (such as lanthanum, yttrium, and neodymium).

Essential metals Some heavy metals (such as iron, zinc, copper, and manganese) are essential metals to human health at low concentrations, but they become toxic heavy metals at higher concentration levels.

Impact factor (IF) IF, also known as the journal impact factor (JIF), is a scientometric index calculated by Clarivate that reflects the yearly mean number of citations of articles published in the last two years in a given journal, as indexed by Clarivate's Web of Science. IF can be explained by the following example: If a journal has an impact factor of 51.9 in 2017, it means that, on average, its papers published in 2015 and 2016 received roughly 52 citations each in 2017. Impact factors are often used to assess a journal's relative importance within its field.

Metals and metalloids They are electropositive elements found in all ecosystems, with varying natural concentrations based on local geology. Human and animal activities can redistribute and concentrate metals in areas not naturally enriched. Unlike sediment and nutrient impairments, metal contamination often lacks visible evidence. All metals, even those abundant and relatively nontoxic metal salts, can be toxic at some level, even those common metals or essential metals as nutrients. Impairments of heavy metals occur when these metals are biologically available at toxic concentrations, affecting aquatic organisms' survival, reproduction, and behavior.

RCRA 8 metals They, also known as the Resource Conservation and Recovery Act 8 metals, are a group of highly toxic heavy metals that require special management due to their potential harm. Here they are in alphabetical order: (a) Arsenic (As): Used in electrical circuits, wood preservatives, and pesticides. While not harmful in small amounts, larger quantities can be fatal; (b) Barium (Ba): A heavy metal that can leach into soil or groundwater when landfilled; (c) Cadmium (Cd): Implicated in gastrointestinal and kidney dysfunction, nervous system disorders, and cancer; (d) Chromium (Cr): Also associated with serious health issues, including skin lesions and immune system dysfunction; (e) Lead (Pb): Known for its harmful effects on the nervous system, especially in children; (f) Mercury (Hg): Extremely toxic and can cause severe health problems if ingested or absorbed; (g) Selenium (Se): Although not a true metal, it's monitored due to its potential toxicity; (h) Silver (Ag): Another heavy metal that requires careful management to prevent harm. If industrial facility operations involve any of these metals in its waste stream, the facility manager must handle them as hazardous waste from "cradle-to-grave," meaning from initial generation to ultimate disposal.

Science Citation Index Expanded (SCIE) SCIE was previously known as the Science Citation Index (SCI). It is a comprehensive citation index that covers notable and significant journals across 178 disciplines from 1,900 to the present. It was originally produced by the ISI and created by Eugene Garfield. SCIE provides complete indexing, including cited references, author information, and accurate subject classification, making it a valuable resource for bibliometric analysis. Researchers can access SCIE online within the Web of Science platform, which contains over 53 million records from thousands of academic journals worldwide.

Spanish National Research Council (Spanish: *Consejo Superior de Investigaciones Científicas*, CSIC) CSIC was established on November 24, 1939, and is the largest public institution dedicated to research in Spain and the third largest in Europe. Its main objective is to develop and promote research that will help bring about scientific and technological progress, and it is prepared to collaborate with Spanish and foreign entities to achieve this aim. CSIC has 6% of all the staff dedicated to research and development in Spain, and they generate approximately 20% of all scientific production in the country. So, CSIC plays an important role in scientific and technological policy since it encompasses an area that takes in everything from basic research to the transfer of knowledge to the productive sector.

US Environmental Protection Agency (USEPA) The USEPA is an independent agency of the US government responsible for environmental protection matters. It was proposed by President Richard Nixon on July 9, 1970, and it began its operation on December 2, 1970, after Nixon signed an executive order. USEPA's mission is to protect human health and the environment. They address climate change, invest in America, and provide resources on environmental issues.

REFERENCES

1. Morgan, J.J. and Stumm, W. The role of multivalent metal oxides in limnological transformations as exemplified by iron and manganese. *J. Water Pollut. Control Feder*, 36, 276–277, 1964.
2. Butt, E.M., Nusbaum, R.E., Gilmour, T.C., Mariano, S.X., and Didio, S.L. Trace metal levels in human serum and blood. *Arch. Environ. Health*, 8, 52–57, 1964.
3. Yunice, A.A., Perry, E.F., and Perry, H.M. Effect of desferrioxamine on trace metals in rat organs. *Arch. Environ. Health*, 16, 163–170, 1968.
4. Salanki, J., Licsko, I., Laszlo, F., Balogh, K.V., Varanka, I., and Mastala, Z. Changes in the concentration of heavy-metals in the Zala Minor Balaton-Zala system (water, sediment, aquatic life). *Water Sci. Technol.*, 25, 173–180, 1992.
5. Handovsky, H. The acute and chronic heavy metal poisoning II announcement—Impact of bivalent tin. *Naunyn-Schmiedebergs Arch. Exp. Pathol. Pharmakol.*, 114, 39–46, 1926.
6. Lamb, R. A suggested measure of toxicity due to metals in industrial effluents, sewage and river water. *Air Water Pollut.*, 8, 243–249, 1964.
7. Wahlberg, J.E. Percutaneous toxicity of metal-compounds—A comparative investigation in guinea-pigs. *Arch. Environ. Health*, 11, 201–204, 1965.
8. Hecker, L.H., Allen, H.E., Dinman, B.D., and Neel, J.V. Heavy-metal levels in acculturated and unacculturated populations. *Arch. Environ. Health*, 29, 181–185, 1974.
9. Nriagu, J.O. A silent epidemic of environmental metal poisoning. *Environ. Pollut.*, 50, 139–161, 1988.
10. Long, E.R., Macdonald, D.D., Smith, S.L., and Calder, F.D. Incidence of adverse biological effects within ranges of chemical concentrations in marine and estuarine sediments. *Environ. Manage.*, 19, 81–97, 1995.
11. Christensen, T.H., Kjeldsen, P., Albrechtsen, H.J., Heron, G., Nielsen, P.H., Bjerg, P.L., and Holm, P.E. Attenuation of landfill leachate pollutants in aquifers. *Crit. Rev. Environ. Sci. Technol.*, 24, 119–202, 1994.
12. OECD. Declaration on risk reduction for lead. *Adopted at the Meeting of Environment Ministers*, February 20, 1996.
13. Abernathy, C.O., Liu, Y.P., Longfellow, D., Aposhian, H.V., Beck, B., Fowler, B., Goyer, R., Menzer, R., Rossman, T., Thompson, C., and Waalkes, M. Arsenic: Health effects, mechanisms of actions, and research issues. *Environ. Health Perspect.*, 107, 593–597, 1999.
14. Boening, D.W. Ecological effects, transport, and fate of mercury: A general review. *Chemosphere*, 40, 1335–1351, 2000.
15. Nriagu, J.O. A global assessment of natural sources of atmospheric trace metals. *Nature*, 338, 47–49, 1989.
16. Barrie, L.A., Gregor, D., Hargrave, B., Lake, R., Muir, D., Shearer, R., Tracey, B., and Bidleman, T. Arctic contaminants—Sources, occurrence and pathways. *Sci. Total Environ.*, 122, 1–74, 1992.
17. Evans, L.J. Chemistry of metal retention by soils—Several processes are explained. *Environ. Sci. Technol.*, 23, 1046–1056, 1989.
18. Holmgren, G.G.S., Meyer, M.W., Chaney, R.L., and Daniels, R.B. Cadmium, lead, zinc, copper, and nickel in agricultural soils of the United States of America. *J. Environ. Qual.*, 22, 335–348, 1993.
19. Council of the European Committees. The Directive 76/464/EEC of May 4, 1976, on pollution caused by certain dangerous substances discharged into the aquatic environment of the community, 1976.
20. U.S. EPA. National recommended water quality criteria. *Fed. Reg.*, 63, 68354–68364, 1998.
21. The Council of the European Union. Council Directive 96/62/EC of September 27, 1996, on ambient air quality assessment and management, Official Journal L 296, 21/11/1996 P., 0055–0063, 1996.
22. World Health Organization. *Air Quality Guidelines for Europe*. World Health Organization, Copenhagen, 1987.

23. Pollard, S.J.T., Fowler, G.D., Sollars, C.J., and Perry, R. Low-cost adsorbents for waste and waste-water treatment—A review. *Sci. Total Environ.*, 116, 31–52, 1992.

24. Orhan, Y. and Buyukgungor, H. The removal of heavy-metals by using agricultural wastes. *Water Sci. Technol.*, 28, 247–255, 1993.

25. Namasivayam, C. and Ranganathan, K. Removal of Cd(II) from waste-water by adsorption on waste Fe(III)/Cr(III) hydroxide. *Water Res.*, 29, 1737–1744, 1995.

26. Babel, S. and Kurniawan, T.A. Low-cost adsorbents for heavy metals uptake from contaminated water: A review. *J. Hazard. Mater.*, 97, 219–243, 2003.

27. Oliver, B.G. and Cosgrove, E.G. Efficiency of heavy-metal removal by a conventional activated-sludge treatment plant. *Water Res.*, 8, 869–874, 1974.

28. Brown, M.J. and Lester, J.N. Metal removal in activated-sludge—Role of bacterial extracellular polymers. *Water Res.*, 13, 817–837, 1979.

29. Ebbs, S.D., Lasat, M.M., Brady, D.J., Cornish, J., Gordon, R., and Kochian, L.V. Phytoextraction of cadmium and zinc from a contaminated soil. *J. Environ. Qual.*, 26, 1424–1430, 1997.

30. Baker, A.J.M., McGrath, S.P., Sidoli, C.M.D., and Reeves, R.D. The possibility of *in-situ* heavy-metal decontamination of polluted soils using crops of metal-accumulating plants. *Resour. Conserv. Recycl.*, 11, 41–49, 1994.

31. Dushenkov, V., Kumar, P., Motto, H., and Raskin, I. Rhizofiltration—The use of plants to remove heavy metals from aqueous streams. *Environ. Sci. Technol.*, 29, 1239–1245, 1995.

32. Lasat, M.M. Phytoextraction of toxic metals: A review of biological mechanisms. *J. Environ. Qual.*, 31, 109–120, 2002.

33. Pamukcu, S. and Wittle, J.K. Electrokinetic removal of selected heavy-metals from soil. *Environ. Progr.*, 11, 241–250, 1992.

34. Acar, Y.B., Gale, R.J., Alshawabkeh, A.N., Marks, R.E., Puppala, S., Bricka, M., and Parker, R. Electrokinetic remediation—Basics and technology status. *J. Hazard. Mater.*, 40, 117–137, 1995.

35. Ma, Q.Y., Traina, S.J., Logan, T.J., and Ryan, J.A. Effects of aqueous Al, Cd, Cu, Fe(II), Ni, and Zn on Pb immobilization by hydroxyapatite. *Environ. Sci. Technol.*, 28, 1219–1228, 1994.

36. Garfield, E. Citation indexing for studying science. *Essays Inform. Scientist*, 1, 133–138, 1970.

37. Kostoff, R.N. The underpublishing of science and technology results. *Scientist*, 14, 6, 2000.

38. Chiu, W.T. and Ho, Y.S. Bibliometric analysis of tsunami research. *Scientometrics*, 73, 3–17, 2007.

39. Chuang, K.Y., Huang, Y.L., and Ho, Y.S. A bibliometric and citation analysis of stroke-related research in Taiwan. *Scientometrics*, 72, 201–212, 2007.

40. Garfield, E. The English language: The lingua franca of international science. *Scientist*, 3, 12, 1989.

41. Moed, H.F. and Hesselink, F.T. The publication output and impact of academic chemistry in the Netherlands during the1980s. *Res. Pol.*, 25, 819–836, 1996.

42. Garfield, E. Keywords plus-ISIS breakthrough retrieval method. 1. Expanding your searching power on current-contents on diskette. *Curr. Contents*, 32, 5–9, 1990.

43. Ho, Y.S. Bibliometric analysis of adsorption technology in environmental science. *J. Environ. Protect. Sci.*, 1, 1–11, 2007.

44. Ho, Y.S. and El-Khaiary, M.I. Metal research trends in the environmental field. In *Heavy Metals in the Environment*. Wang, L. K., Chen, J. P., Hung, Y. T., and Shammas, N. K. (editors), CRC Press, Boca Raton, FL, pp. 1–12, 2009.

45. Mostafa Hatami, A., Sabour, M.R., and Nikravan, M.A. Systematic analysis of research trends on incineration during 2000–2019. *Int. J. Environ. Sci. Technol.*, 18, 353–364, 2021. https://doi.org/10.1007/s13762-020-02794-x

46. Liu, W., Li, X., Wang, M., and Liu, L. Research trend and dynamical development of focusing on the global critical metals: A bibliometric analysis during 1991–2020. *Environ. Sci. Pollut. Res.*, 29, 26688–26705, 2022. https://doi.org/10.1007/s11356-021-17816-5

47. Nazaripour, M., Reshadi, M.A.M., Mirbagheri, S.A., Nazaripour, M., and Bazargan, A. Research trends of heavy metal removal from aqueous environments. *J. Environ. Manage.*, 287, 2021. https://doi.org/10.1016/j.jenvman.2021.112322

48. Burton, J. *List of Critical Minerals.* US Geological Survey, US Department of the Interior, Washington DC, USA, February 2022. https://www.usgs.gov/news/national-news-release/us-geological-survey-releases-2022-list-critical-minerals

49. DOE. *Critical Minerals and Materials*, US Department of Energy, The Advanced Materials & Manufacturing Technologies Office's (AMMTO), Washington, DC, 2024. https://www.energy.gov/eere/ammto/critical-minerals-and-materials

50. Bedder, J.C.M. Classifying critical materials: A review of European approaches. *Appl. Earth Sci.*, 124, 207–212, 2015.
51. Das, N. Recovery of precious metals through biosorption — A review. *Hydrometallurgy*, 103, 180–189, 2010.
52. Elshkaki, A. An analysis of future platinum resources, emissions and waste streams using a system dynamic model of its intentional and non-intentional flows and stocks. *Resour. Policy*, 38, 241–251, 2013.
53. Ge, J., Lei, Y., and Zhao L. China's rare Earths supply forecast in 2025: A dynamic computable general equilibrium analysis. *Minerals*, 6, 95, 2016.
54. Jowitt, S.M., Werner, T.T., Weng, Z., and Mudd, G.M. Recycling of the rare earth elements. *Curr. Opin. Green Sustain. Chem.*, 13, 1–7, 2018.
55. Su, J., Zhou, J., Wang, L., Liu, C., and Chen, Y. Synthesis and application of transition metal phosphides as electrocatalyst for water splitting. *Sci. Bull.*, 62, 633–644, 2017.
56. Sun, Z., Cao, H., Xiao, Y., Sietsma, J., Jin, W., Agterhuis, H., and Yang, Y. Toward sustainability for recovery of critical metals from electronic waste: The hydrochemistry processes. *ACS Sustain. Chem. Eng.*, 5, 21–40, 2017.
57. Wang, L.K. and Wang, M.H.S. Metals and trace elements on earth. In *Control of Heavy Metals in the Environment, Volume 2*. Wang, L. K., Wang, M. H. S., Hung, Y. T. and Chen, J. P. (editors), CRC Press, Boca Raton, FL, 1–41, 2025.
58. USEPA. *Methods/Indicators for Determining when Metals are the Cause of Biological Impairments of Rivers and Streams: Species Sensitivity Distributions and Chronic Exposure-Response Relationships from Laboratory Data*, EPA/600/X-05/027. U.S. Environmental Protection Agency, Office of Research and Development, Washington, DC, 2005.
59. USEPA. Metals: *Causal Analysis/Diagnosis Decision Information System (CADDIS)*, 2024. https://www.epa.gov/caddis/metals
60. Swistock, B. and Sharpe, W. *Iron and Manganese in Private Water Systems*. Pennsylvania State University, PA, 2023. https://extension.psu.edu/iron-and-manganese-in-private-water-systems
61. Water Science School. *Hardness of Water*. US Geological Survey, Washington, DC, 2018. https://www.usgs.gov/special-topics/water-science-school/science/hardness-water
62. Desert Research Institute. The first assessment of toxic heavy metal pollution in the Southern Hemisphere over the last 2,000 years. *ScienceDaily*, 2024. www.sciencedaily.com/releases/2024/01/240111162632.htm
63. Tang, S., Wang, C., Song, J., Ihenetu, S.C., and Li, G. Advances in studies on heavy metals in urban soil: A bibliometric analysis. *Sustainability*, 16(2), 860, 2024. https://doi.org/10.3390/su16020860
64. Zhang, Y., Lu, X., Deng, S., Zhu, T., and Yu, B. Bibliometric and visual analysis of heavy metal health risk assessment: Development, hotspots and trends. *Arch. Environ. Prot.*, 50(1), 56–71, 2024. https://doi.org/10.24425/aep.2024.149432
65. Swain, C.K. Environmental pollution indices: A review on concentration of heavy metals in air, water, and soil near industrialization and urbanisation. *Discov. Environ.*, 2, 5, 2024. https://doi.org/10.1007/s44274-024-00030-8
66. Punia, P., Bharti, M.K., Dhar, R., Thakur, P., and Thakur, A. Recent advances in detection and removal of heavy metals from contaminated water. *Chem. Bio Eng. Rev.*, 9(4), 351–369, 2022.
67. Xiao, X.Y., Zhao, Y.H., Li, Y.Y., Song, Z.Y., Chen, S.H., Huang, H.Q., Yang, M., Li, P.H., and Huang, X.J. General strategies to construct highly efficient sensing interfaces for metal ions detection from the perspective of catalysis. *Anal. Chem.*, 94(40), 13631–13641, 2022. https://doi.org/10.1021/acs.analchem.2c01797
68. Onyancha, R.B., Aigbe, U.O., Ukhurebor, K.E., Osibote, O.A., Balogun, V.A., and Kusuma, H.S., Current existing techniques for environmental monitoring. In *Nanobiosensors for Environmental Monitoring: Fundamentals and Application*. Springer Nature, Switzerland, pp. 239–262, 2022.
69. Gupta, P., Rahm, C.E., Griesmer, B., and Alvarez, N.T. Carbon nanotube microelectrode set: Detection of biomolecules to heavy metals. *Anal. Chem.*, 93(20), 7439–7448, 2021. https://doi.org/10.1021/acs.analchem.1c00360
70. Gupta, P., Rahm, C.E., Jiang, D., Gupta, V.K., Heineman, W.R., Justin, G., and Alvarez, N.T. Parts per trillion detection of heavy metals in as-is tap water using carbon nanotube microelectrodes. *Anal. Chim. Acta*, 1155, 338353, 2021. https://doi.org/10.1016/j.aca.2021.338353
71. John Peter, A.L. and Viraraghavan, T. Thallium: A review of public health and environmental concerns. *Environ. Int.*, 31(4), 493–501, 2005. https://doi.org/10.1016/j.envint.2004.09.003

72. Ren, X., Feng, H., Zhao, M., Zhou, X., Zhu, X., Ouyang, X., Tang, J., Li, C., Wang, J., Tang, W., and Tang, L. (2023). Recent advances in thallium removal from water environment by metal oxide material. *Int. J. Environ. Res. Public Health*, 20(5), 3829, 2023. https://doi.org/10.3390/ijerph20053829

73. Zhang, G., Luo, J., Cao, H., Hu, S., Li, H., Wu, Z., Xie, Y. and Li, X. Highly efficient removal of thallium(I) by facilely fabricated amorphous titanium dioxide from water and wastewater. *Sci. Rep.*, 12, 72, 2022. https://www.nature.com/articles/s41598-021-03985-3.pdf

74. Chang, Y. and Chiang, C.K. The Impact of thallium exposure in public health and molecular toxicology: A comprehensive review. *Int. J. Mol. Sci.*, 25(9), 4750, 2024. https://doi.org/10.3390/ijms25094750

75. Genchi, G., Carocci, A., Lauria, G., Sinicropi, M.S., and Catalano, A. Thallium use, toxicity, and detoxification therapy: An overview. *Appl. Sci.*, 11(18), 8322, 2021. https://doi.org/10.3390/app11188322

76. Wang, L.K. and Wang, M.H.S. Titanium dioxide recovery, filler retention and white water treatment using flotation and membrane filtration. In *Control of Heavy Metals in the Environment*. Volume 1, Wang, L. K., Wang, M. H. S., Hung, Y. T. and Chen, J. P. (editors), CRC Press, Boca Raton, FL, 48–71, 2025.

77. Butu, A., Rodino, S., and Butu, M. Global research progress and trends on critical metals: A bibliometric analysis. *Sustainability*, 15(6), 4834, 2023. https://doi.org/10.3390/su15064834

78. Liu, H., Liu, R., Qu, J., Zhang, G, and Wang, L.K., Arsenic pollution: Occurrence, distribution, and technologies. In: *Control of Heavy Metals in the Environment*, Volume 3, Wang, L. K., Wang, M. H. S., Hung, Y. T. and Chen, J. P. (editors),. CRC Press, Boca Raton, FL, pp. 249–272, 2025.

79. Naja, G.M., Volesky, B., and Wang, L.K. Treatment of metal-bearing effluents: Removal and recovery. In: *Control of Heavy Metals in the Environment*, Volume 3, Wang, L. K., Wang, M. H. S., Hung, Y. T. and Chen, J. P. (editors),. CRC Press, Boca Raton, FL, pp. 273–318, 2025.

80. Nowack, B. and Wang, L.K. Environmental behavior and effects of engineered metal, and metal oxide nanoparticles. In: *Control of Heavy Metals in the Environment*, Volume 3, Wang, L. K., Wang, M. H. S., Hung, Y. T. and Chen, J. P. (editors),.. CRC Press, Boca Raton, FL, pp. 80–104, 2025.

81. Xiangke, W., Changlun, C., and Wang, L.K. Nanotechnology application in metal ion adsorption. In: *Control of Heavy Metals in the Environment* Volume 3, Wang, L. K., Wang, M. H. S., Hung, Y. T. and Chen, J. P. (editors), CRC Press, Boca Raton, FL, pp. 174–223, 2025.

82. Wang, S.G., Sun, X.F., Gong, W.X., Ma, Y., and Wang, L.K. Biosorption of metals onto granular sludge. In: *Control of Heavy Metals in the Environment*, Volume 3, Wang, L. K., Wang, M. H. S., Hung, Y. T. and Chen, J. P. (editors), CRC Press, Boca Raton, FL, pp. 224–228, 2025.

83. Nascimento, J.M.D., Otaviano, J.J.S., Sousa, H.S.D., Oliveira, J.D.D., and Hung, Y.T. Use of biosorption of agro-industrial wastes and microorganisms for removal of potentially toxic metals. In: *Control of Heavy Metals in the Environment*, Volume 3, Wang, L. K., Wang, M. H. S., Hung, Y. T. and Chen, J. P. (editors), CRC Press, Boca Raton, FL, pp. 399–456, 2025.

84. Kaur, A., Singh, V., Singh, S., Hung, Y.T., and Gupta, S. Biological treatment of mercury polluted wastewater. In: *Control of Heavy Metals in the Environment* Volume 3, Wang, L. K., Wang, M. H. S., Hung, Y. T. and Chen, J. P. (editors), CRC Press, Boca Raton, FL, pp. 576–592, 2025.

85. Wang, L.K. and Wang, M.H.S. Ecologically sustainable industrial development, solid and hazardous wastes management, and DAF landfill leachate pretreatment. In: *Control of Heavy Metals in the Environment* Volume 2, Wang, L. K., Wang, M. H. S., Hung, Y. T. and Chen, J. P. (editors),. CRC Press, Boca Raton, FL, pp. 269–282, 2025.

86. Wang, M.H.S., Wang, L.K., and Shammas, N.K. Advances in removal of heavy metals from contaminated soil. In: *Control of Heavy Metals in the Environment* Volume 3, Wang, L. K., Wang, M. H. S., Hung, Y. T. and Chen, J. P. (editors),. CRC Press, Boca Raton, FL, pp. 457–515, 2025.

87. Wang, L.K., Balasubramanian, R., He, J., and Wang, M.H.S. Control and management of air emissions from transportation industry. In: *Control of Heavy Metals in the Environment* Volume 2, Wang, L. K., Wang, M. H. S., Hung, Y. T. and Chen, J. P. (editors), CRC Press, Boca Raton, FL, pp. 392–416, 2025.

88. Wang, L.K., Pereira, N.C., and Hung, Y.T. (editors), *Air Pollution Control Engineering*. Humana Press, NJ, p. 504, 2004.

89. Wang, M.H.S. and Wang, L.K. Recycling and disposal of hazardous solid wastes containing heavy metals and other toxic substances. In: *Control of Heavy Metals in the Environment* Volume 2, Wang, L. K., Wang, M. H. S., Hung, Y. T. and Chen, J. P. (editors),. CRC Press, Boca Raton, FL, pp. 299–320, 2025.

90. Wang, L.K. and Wang, M.H.S. Innovative process systems including dissolved air flotation, protein separation, flue gas reuse, sulfide precipitation, and chromium removal for tannery waste treatment. In: *Control of Heavy Metals in the Environment* Volume 1, Wang, L. K., Wang, M. H. S., Hung, Y. T. and Chen, J. P. (editors),. CRC Press, Boca Raton, FL, pp. 1–47, 2025.

91. Wang, L.K., Nagghappan, L.N.S.P., Wang, M.H.S., and Krofta, M. , Treating tannery waste using stack flue gas recycle, sulfide precipitation, chromium removal, ferrous sulfide recycle, ferrous ion recycle, filtration, flotation, membrane and bioreactor. In: *Heavy Metals Remediation, Control of Heavy Metals in the Environment* Volume 2, Wang, L. K., Wang, M. H. S., Hung, Y. T. and Chen, J. P. (editors), CRC Press, Boca Raton, FL, 2025.

92. Wang, L.K., Wang, M.H.S., Shammas, N.K., Aulenbach, D.B., and Selke, W.A. Treatment of nickel-chromium plating wastes. In: *Heavy Metals Management, Control of Heavy Metals in the Environment* Volume 1, Wang, L. K., Wang, M. H. S., Hung, Y. T. and Chen, J. P. (editors),. CRC Press, Boca Raton, FL, 2025.

93. Wang, L.K. and Wang, M.H.S. Innovative process system consisting of dissolved air flotation, recarbonation and filtration and using lime and sodium aluminate for removal of hardness, iron and manganese from groundwater. In: *Control of Heavy Metals in the Environment* Volume 1, Wang, L. K., Wang, M. H. S., Hung, Y. T. and Chen, J. P. (editors), CRC Press, Boca Raton, FL, 2025, pp. 1–47.

94. Wang, L.K. and Wang, M.H.S. Removal of iron, copper, coliform bacteria and hardness from cooling tower water by dissolved air-ozone flotation. In: *Control of Heavy Metals in the Environment* Volume 2, Wang, L. K., Wang, M. H. S., Hung, Y. T. and Chen, J. P. (editors), CRC Press, Boca Raton, FL, 2025, pp. 283–298.

95. Wang, L.K. and Wang, M.H.S. Sludge treatment using pressurized oxygenation-ozonation, dissolved gas flotation and dewatering. In: *Control of Heavy Metals in the Environment* Volume 1, Wang, L. K., Wang, M. H. S., Hung, Y. T. and Chen, J. P. (editors), CRC Press, Boca Raton, FL, 2024.

96. Wang, L.K. and Wang, M.H.S. Hazardous wastewater treatment in petroleum refining industry using best available flotation and filtration technologies. In: *Control of Heavy Metals in the Environment* Volume 1, Wang, L. K., Wang, M. H. S., Hung, Y. T. and Chen, J. P. (editors), CRC Press, Boca Raton, FL, 2024.

97. Wang, L.K. and Wang, M.H.S. Pretreatment or total treatment of scallop processing wastewater by independent physicochemical wastewater treatment systems. In: *Control of Heavy Metals in the Environment* Volume 1, Wang, L. K., Wang, M. H. S., Hung, Y. T. and Chen, J. P. (editors) CRC Press, Boca Raton, FL, 2024.

98. Wang, L.K., Wang, M.H.S., and Shammas, N.K. Treatment of wastewater, storm runoff, and combined sewer overflow by dissolved air flotation and filtration. In: *Handbook of Advanced Industrial and Hazardous Wastes Management* Volume 2, Wang, L. K., Wang, M. H. S., Hung, Y. T., Shammas, N. K. and Chen, J. P. (editors), CRC Press, Boca Raton, FL, pp. 577–610, 2018.

99. Wang, L.K. and Wang, M.H.S. Removal of arsenic, oil and grease, chemical oxygen demand and color from stormwater runoff by continuous dissolved air flotation and filtration. In: *Control of Heavy Metals in the Environment* Volume 2, Wang, L. K., Wang, M. H. S., Hung, Y. T., and Chen, J. P. (editors), CRC Press, Boca Raton, FL, pp. 83–95, 2025.

100. Wang, L.K., Kurylko, L., and Wang, M.H.S. Sequencing batch liquid treatment. US Patent No. 5354458, US Patent and Trademark Office, Washington, DC, 1994.

101. Wang, L.K., Wang, M.H.S., Suozzo, T., Dixon, R.A., and Wright, T.L. Chemical and biochemical technologies for environmental infrastructure sustainability. In: *Evolutionary Progress in Science, Technology, Engineering, Arts, and Mathematics.* Wang, L. K. and Tsao, H. P. (editors), Lenox Institute Press, Lenox, MA, Vol. 5, No. 10A, p. 58, 2023. https://doi.org/10.17613/z30s-gj22

102. Wang, L.K. and Wang, M.H.S. New technologies for water and wastewater treatment. In: *Evolutionary Progress in Science, Technology, Engineering, Arts, and Mathematics.* Wang, L. K. and Tsao, H. P. (editors), Lenox Institute Press, Lenox, MA, Vo. 5, No. 10C, p. 34, 2023. https://doi.org/10.17613/wena-kx42

103. Wang, L.K., Shammas, N.K., Selke, W.A., and Aulenbach, D.B. *Flotation Technology.* Humana Press, Totowa, NJ, p. 680, 2010.

104. Wang, L.K., Wang, M.H.S., Shammas, N.K., and Aulenbach, D.B. *Environmental Flotation Engineering.* Springer Nature, Switzerland, p. 433, 2024.

105. Philippis, R.D., Micheletti, E., and Wang, L.K. Heavy metal removal with exopolysaccharide-producing cyanobacteria. In: *Control of Heavy Metals in the Environment* Volume 3, Wang, L. K., Wang, M. H. S., Hung, Y. T. and Chen, J. P. (editors), CRC Press, Boca Raton, FL, pp. 105–139, 2025.

106. Wang, M.H.S., Wang, L.K., Eroglu, V., and Erturk, F. Advances in the management and treatment of acid pickling wastes containing heavy metals. In: *Control of Heavy Metals in the Environment* Volume 3, Wang, L. K., Wang, M. H. S., Hung, Y. T. and Chen, J. P. (editors), CRC Press, Boca Raton, FL, pp. 319–335, 2025.

107. Wang, M.H.S., Wang, L.K., and Shammas, N.K. Advances in the treatment and management of metal finishing industry wastes. In: *Control of Heavy Metals in the Environment* Volume 3, Wang, L. K., Wang, M. H. S., Hung, Y. T. and Chen, J. P. (editors), CRC Press, Boca Raton, FL, pp. 346–398, 2025.

108. Wang, L.K. and Wang, M.H.S. Bioassay technology, terminologies, applications and examples: Heavy metals and phenol investigation. In: *Control of Heavy Metals in the Environment* Volume 3, Wang, L. K., Wang, M. H. S., Hung, Y. T. and Chen, J. P. (editors), CRC Press, Boca Raton, FL, pp. 593–611, 2025.

109. Wang, M.H.S. and Wang, L.K. Environmental management of electroplating and metal finishing operations. In: *Control of Heavy Metals in the Environment* Volume 2, Wang, L. K., Wang, M. H. S., Hung, Y. T. and Chen, J. P. (editors), CRC Press, Boca Raton, FL, pp. 221–374, 2025.

110. Wang, L.K. and Wang, M.H.S. Reduction of color, turbidity, odor, humic acid, metals, EDB, coliform and TTHM by adsorption, flotation and filtration. In: *Control of Heavy Metals in the Environment* Volume 2, Wang, L. K., Wang, M. H. S., Hung, Y. T. and Chen, J. P. (editors), CRC Press, Boca Raton, FL, pp. 375–392, 2025.

111. Wang, L.K. and Wang, M.H.S. Using electroflotation and filtration for lake restoration, pollution prevention and environmental conservation. In: *Control of Heavy Metals in the Environment* Volume 1, Wang, L. K., Wang, M. H. S., Hung, Y. T. and Chen, J. P. (editors), CRC Press, Boca Raton, FL, pp. 275–300, 2025.

112. Ghinwa, M., Naja, G.M., Volesky, B., and Wang, L.K. Toxicity and sources of Pb, Cd, Hg, Cr, As, and radionuclides in the environment. In: *Control of Heavy Metals in the Environment* Volume 3, Wang, L. K., Wang, M. H. S., Hung, Y. T. and Chen, J. P. (editors), CRC Press, Boca Raton, FL, pp. 29–79, 2025.

113. Wang, Y., Deng, Y., and Wang, L.K. Environmental geochemistry of high-arsenic aquifer systems. In: *Control of Heavy Metals in the Environment* Volume 3, Wang, L. K., Wang, M. H. S., Hung, Y. T. and Chen, J. P. (editors), CRC Press, Boca Raton, FL, pp. 140–173, 2025.

114. Zorluer, I. and Hung, Y.T. Heavy metal pollution resulting from industrial wastes used for soil stabilization in groundwater. In: *Control of Heavy Metals in the Environment* Volume 1, CRC Press, Boca Raton, FL, pp. 193–206, 2025.

115. Katyal, P., Kaur, H., Natt, S.K., and Hung, Y.T. Strategies for bioremediation of heavy metals contaminated industrial effluents. In: *Control of Heavy Metals in the Environment* Volume 2, Wang, L. K., Wang, M. H. S., Hung, Y. T. and Chen, J. P. (editors), CRC Press, Boca Raton, FL, pp. 42–68, 2025.

116. Wang, M.H.S. and Wang, L.K. Pollutants, control technologies and terminologies for emission control of transportation industry. In: *Control of Heavy Metals in the Environment* Volume 2, CRC Press, Boca Raton, FL, pp. 612–632, 2025.

117. Semalti, P., Sharma, S.N., and Hung, Y.T. Surface engineering of colloidal quaternary chalcogenide Cu_2ZnSnS_4 nanocrystals via ligand exchange for photo-induced charge separation applications for water remediation. In: *Control of Heavy Metals in the Environment* Volume 2, Wang, L. K., Wang, M. H. S., Hung, Y. T. and Chen, J. P. (editors), CRC Press, Boca Raton, FL, pp. 69–82, 2025.

118. Gül, U.D. and Hung, Y.T. Utilization of fungal bioprocess in heavy metal contaminated water treatment technologies. In: *Control of Heavy Metals in the Environment* Volume 2, Wang, L. K., Wang, M. H. S., Hung, Y. T. and Chen, J. P. (editors), CRC Press, Boca Raton, FL, pp. 96–117, 2025.

119. Govindaraj, V. and Hung, Y.T. Depletion of hexavalent chromium to trivalent chromium from tannery industry effluent using natural bacterial biomass. In: *Control of Heavy Metals in the Environment* Volume 2, Wang, L. K., Wang, M. H. S., Hung, Y. T. and Chen, J. P. (editors), CRC Press, Boca Raton, FL, pp. 115–141, 2025.

120. Wang, L.K. Humanitarian engineering education of the Lenox Institute of Water Technology and Its new potable water flotation processes. In: *Environmental Flotation Engineering*, Wang, LK, Wang, MHS, Shammas, NK and Aulenbach, DB (eds.). Springer Nature, Switzerland, pp. 1–72, 2021.

121. Wang, L.K. and Wang, M.H.S. Reduction of high color, humic substances, turbidity, coliform bacteria, THMMFP, UV absorbance, cysts, chlorine demand, and metals from unpredictable water source by dissolved air flotation and filtration. In: *Control of Heavy Metals in the Environment* Volume 2, Wang, L. K., Wang, M. H. S., Hung, Y. T. and Chen, J. P. (editors), CRC Press, Boca Raton, FL, pp. 142–170, 2025.

122. Akpan, B.M., Ajibola, A.S., Michael, A.A., and Hung, Y.T. Adsorptive amputation of hexavalent chromium from aqueous solution and textile wastewater by *Heinsia crinita* seed coat biomass. In: *Control of Heavy Metals in the Environment,* Volume 1, Wang, L. K., Wang, M. H. S., Hung, Y. T. and Chen, J. P. (editors), CRC Press, Boca Raton, FL, pp. 72–118, 2025.

123. Madu, IE, Amoo, Oladeji, A., Embrandiri, A., Kamaruddin, M.A.B., and Hung, Y.T. Application of ion exchange resin in the removal of pollutant from wastewater. In: *Control of Heavy Metals in the Environment* Volume 1, Wang, L. K., Wang, M. H. S., Hung, Y. T. and Chen, J. P. (editors), CRC Press, Boca Raton, FL, pp. 137–154, 2025.

124. Wang, L.K., and Wang, M.H.S. Sustainable water and wastewater treatment systems consisting of magnesium coagulation-precipitation, dissolved air flotation, recarbonation and filtration. In: *Control of Heavy Metals in the Environment* Volume 1, Wang, L. K., Wang, M. H. S., Hung, Y. T. and Chen, J. P. (editors), CRC Press, Boca Raton, FL, pp. 207–232, 2025.

125. Nguyen, P.T., Nguyen, D.D.D., Vo, Q.T.K., and Hung, Y.T., Electro-fenton, heterogeneous electro-fenton process for treatment of non-biodegradable wastewater. In: *Control of Heavy Metals in the Environment* Volume 1, Wang, L. K., Wang, M. H. S., Hung, Y. T. and Chen, J. P. (editors), CRC Press, Boca Raton, FL, pp. 358–422, 2024.

126. Wang, L.K. and Wang, M.H.S. Treatment of cold, low alkalinity, and super-high turbidity industrial waters using polymeric metallic coagulants. In: *Control of Heavy Metals in the Environment* Volume 1, Wang, L. K., Wang, M. H. S., Hung, Y. T. and Chen, J. P. (editors), CRC Press, Boca Raton, FL, pp. 317–334, 2024.

127. Suhardi, E.D., Hermawan, F.V., Kristianto, H., Prasetyo, S., and Sugih, A.K. Enhancing water–wastewater treatment efficiency: Synergistic approach using polyaluminum chloride, sodium alginate, and magnetite for Congo red removal. *Chem. Paper*, 78, 3971–3981, 2024. https://doi.org/10.1007/s11696-024-03367-9

128. Kanhaiya, L. and Anurag, G. Effectiveness of synthesized aluminum and iron based inorganic polymer coagulants for pulping wastewater treatment. *J. Environ. Chem. Eng.*, 7(5), 103–204, 2019. https://doi.org/10.1016/j.jece.2019.103204

129. Yang, S., Li, W., Zhang, H., Wen, Y., and Ni, Y. Treatment of paper mill wastewater using a composite inorganic coagulant prepared from steel mill waste pickling liquor. *Sep. Purif. Technol.*, 209, 238–245, 2019.

130. Wang, L.K., Wang, M.H.S., Bien, C.C.J., Wu, S., and Lim, E.W. Treating unpredictable potable water source with green technologies of aluminum chlorohydrate (ACH), streaming current monitor (SCM), and dissolved air flotation (DAF). *Environmental Science, Technology, Engineering, and Mathematics.* Lenox Institute Press, MA, p. 38, 2023. https://doi.org/10.17613/a6x9-qs09.

131. Wang, M.H.S., Wang, L.K., and DeMichele, E. BOD determination, cleaning solution preparation, and waste disposal in laboratories. In: *Handbook of Advanced Industrial and Hazardous Wastes Management.* Wang, L. K., Wang, M. H. S., Hung, Y. T. Shammas N. K. and Chen, J. P. (editors), CRC Press, Boca Raton, FL, pp. 797–808, 2018.

132. Wang, M.H.S., Wang, L.K., and DeMichele, E. Principles, procedures, and heavy metal management of dichromate reflux method for COD determination in laboratories. In: *Handbook of Advanced Industrial and Hazardous Wastes Management.* Wang, L. K., Wang, M. H. S., Hung, Y. T. Shammas N. K. and Chen, J. P. (editors), CRC Press, Boca Raton, FL, pp. 809–818, 2018.

133. Zahari, M.S.M., Jaafar, I., Ismail, S., Harun, M.H.C., Hashim, N.H., Wang, M.H.S., and Wang, L.K. Laboratory waste management and treatment. In: *Industrial Waste Engineering.* Wang, L. K., Wang, M. H. S., Hung, Y. T. Shammas N. K. and Chen, J. P. (editors), Springer Nature, Switzerland, pp. 397–440, 2024.

134. Wang, L.K. and Wang, M.H.S. COD determination and cleaning solution preparation using potassium permanganate for hazardous waste minimization. In: *Control of Heavy Metals in the Environment* Volume 1, Wang, L. K., Wang, M. H. S., Hung, Y. T. and Chen, J. P. (editors), CRC Press, Boca Raton, FL, pp. 335–344, 2024.

TABLE A1

Maximum Concentration of Contaminants for the Toxicity Characteristics

EPA HW No.[a]	Contaminant	CAS No.[b]	Regulatory Level (mg/L)[c]
D004	Arsenic	7440–38–2	5.0
D005	Barium	7440–39–3	100.0
D018	Benzene	71–43–2	0.5
D006	Cadmium	7440–43–9	1.0
D019	Carbon tetrachloride	56–23–5	0.5
D020	Chlordane	57–74–9	0.03
D021	Chlorobenzene	108–90–7	100.0
D022	Chloroform	67–66–3	6.0
D007	Chromium	7440–47–3	5.0
D023	o-Cresol	95–48–7 4	200.0
D024	m-Cresol	108–39–4 4	200.0
D025	p-Cresol	106–44–5 4	200.0
D026	Cresol	420	0.0
D016	2,4-D	94–75–7	10.0
D027	1,4-Dichlorobenzene	106–46–7	7.5
D028	1,2-Dichloroethane	107–06–2	0.5
D029	1,1-Dichloroethylene	75–35–4	0.7
D030	2,4-Dinitrotoluene	121–14–2 3	0.13
D012	Endrin	72–20–8	0.02
D031	Heptachlor (epoxide)	76–44–8	0.008
D032	Hexachlorobenzene	118–74–1 3	0.13
D033	Hexachlorobutadiene	87–68–3	0.5
D034	Hexachloroethane	67–72–1	3.0
D008	Lead	7439–92–1	5.0
D013	Lindane	58–89–9	0.4
D009	Mercury	7439–97–6	0.2
D014	Methoxychlor	72–43–5	10.0
D035	Methyl ethyl ketone	78–93–3	200.0
D036	Nitrobenzene	98–95–3	2.0
D037	Pentrachlorophenol	87–86–5	100.0
D038	Pyridine	110–86–1	35.0
D010	Selenium	7782–49–2	1.0
D011	Silver	7440–22–4	5.0
D039	Tetrachloroethylene	127–18–4	0.7
D015	Toxaphene	8001–35–2	0.5
D040	Trichloroethylene	79–01–6	0.5
D041	2,4,5-Trichlorophenol 95–95–4	400.0	
D042	2,4,6-Trichlorophenol 88–06–2	2.0	
D017	2,4,5-TP (Silvex)	93–72–1	1.0
D043	Vinyl chloride	75–01–4	0.2

Source: USEPA.

[a] Hazardous waste number.

[b] Chemical abstracts service number.

[c] Quantitation limit is greater than the calculated regulatory level. The quantitation limit therefore becomes the regulatory level.

2 Toxicity and Sources of Pb, Cd, Hg, Cr, As, and Radionuclides in the Environment

Ghinwa M. Naja, Bohumil Volesky, Teng Hui Jin,
Xiao Liu, and Lawrence K. Wang

ACRONYMS

ADI	acceptable daily intake
AMD	acid mine drainage
DM	demineralization
FGD	flue gas desulfurization
MAC	maximum acceptable concentration
USNIOSH	U.S. National Institute of Occupational Safety and Health
USEPA	U.S. Environmental Protection Agency
WHO	World Health Organization

2.1 METAL TOXICITY

Out of 118 identified elements, about 80 of them are called metals. These metallic elements can be divided into two groups: those that are essential for survival, such as iron and calcium, and those that are nonessential or toxic, such as cadmium and lead. These toxic metals, unlike some organic substances, are not metabolically degradable, and their accumulation in living tissues can cause death or serious health threats. Furthermore, these metals, dissolved in wastewaters and discharged into surface waters, will be concentrated as they travel up the food chain. Eventually, highly poisonous levels of toxin can migrate to the immediate environment of the public. Metals that seep into groundwaters will contaminate drinking water wells and harm the consumers of that water.

Pollution from man-made sources can easily create local conditions of elevated metal presence, which could lead to disastrous effects on animals and humans. Actually, man's exploitation of the world's mineral resources and his technological activities tend to unearth, dislodge, and disperse chemicals and particularly metallic elements, which have been brought into the environment in unprecedented quantities and concentrations and at extreme rates. Man's new technologies involving nuclear fission opened up a whole new area of hope and concern simultaneously. Radioactive isotopes of elements and, indeed, new elements have been discovered and handled in historically unprecedented quantities and concentrations. The sneaking and deadly danger of radioactivity associated particularly with long-lived and high-radiation isotopes has cast a shadow over our lives. Actually, the disposal problems concerning radioactive isotopes originating directly or indirectly from the operation of nuclear-generating facilities have produced a considerable slowdown in the deployment of this technology, which, after all, may only be a transient phase made obsolete by the dangers it generates.

DOI: 10.1201/9781003541615-2

2.1.1 Selected Heavy Metals

Heavy metals can be defined in several ways. One possible definition is the following: Heavy metals form positive ions in solution, and they have a density five times greater than that of water. They are of particular toxicological importance. Many metallic elements play an essential role in the function of living organisms; they constitute a nutritional requirement and fulfill a physiological role. However, an overabundance of the essential trace elements and particularly their substitution by nonessential ones, such as the case may be for cadmium, nickel, or silver, can cause toxicity symptoms or death. Humans receive their allocation of trace elements from food and water, an indispensable link in the food chain being plant life, which also supports animal life. It is a well-established fact that the assimilation of metals takes place in the microbial world as well as in plants, and these elements tend to get concentrated as they progress through the food chain. It has been shown that spectacular metal enrichment coefficients of the order of 10^5–10^7 can occur in cells [1]. Imbalances or excessive amounts of a metal species along this route lead to toxicity symptoms, disorders in cellular functions, long-term debilitating disabilities in humans, and eventually death.

2.1.1.1 Lead

Lead is the most common of the heavy elements. Several stable isotopes exist in nature, 208Pb being the most abundant. The average molecular weight of lead is 207.2. Lead is a soft metal that resists corrosion and has a low melting point (327°C). From a drinking water perspective, the almost universal use of lead compounds in plumbing fittings and as a solder in water distribution systems is essential. Distribution systems and plumbing installed before 1945 were made from lead pipes [2]. Solid and liquid (sludge) wastes account for more than 50% of the lead discharged into the environment, usually into landfills, but lead has been dispersed more widely in the general environment through atmospheric emissions—particularly from car exhausts. With the introduction of unleaded fuel, lead emissions from this source declined. The annual consumption of lead is in the order of 3 million tons, of which 40% is used in the production of electrical accumulators and batteries, 20% is used in gasoline as alkyl additives, 12% in building construction, 6% in cable coatings, 5% in ammunition, and 17% in other usages. It is estimated that approximately 2 million tons are mined yearly. Probably 10% of this total is lost in the treatment of the ore to produce the concentrate, and a further 10% is lost in making pig lead. The amount of lead discharged into the environment is equal to the amount weathered from igneous rocks. In global lead-level terms, the power storage battery industry may have a relatively low impact on the environment because about 80% of all batteries are recycled.

2.1.1.1.1 Exposure

Lead is present in tap water as a result of dissolution from natural sources or household plumbing systems containing lead in pipes. The amount of lead from the plumbing system that may be dissolved depends on several factors, including acidity (pH), water softness, and standing time of the water [3]. Food can be contaminated by naturally occurring lead in the soil as well as by lead from sources such as atmospheric fallout or water used for cooking.

The total intakes and uptakes of lead from all sources are 29.5 and 12.5 mg/d, respectively, for children and 63.7 and 6.7 mg/d, respectively, for adults in urban areas [4]. The relative contribution of water to average intake is estimated to be 9.8% and 11.3% for children and adults, respectively. The total intake of lead from three of the four major sources—air, food, and dust—appears to have dropped significantly since the mid-1980s as a result of regulatory and voluntary actions to control lead from air (via gasoline) and food (via cans).

Other routes of exposure include ceramic ware, activities involving arts and crafts, ingestion of lead paint chips or flakes (from layers that predate the phase-out of lead-based paints and which may have been painted over), contaminated consumer goods such as imported toys, jewelry, foodstuffs, utensils, cookware, traditional cosmetics (e.g., the eye cosmetic tiro) and medicines (e.g., some ayurvedic preparations), illicitly manufactured drugs containing lead, or hobby exposures (e.g., ingestion of fishing weights or lead shot in game meat), contact with occupational "take home" exposures via family members (*para*-occupational exposure) and renovations resulting in dust or fumes from paint [5]. No allowance was made for the contribution of lead from these sources because they occur on a highly sporadic basis and because no quantitative data are available. It has been pointed out [5] that old paint has been an important source of excess lead intake for inner-city children living in older housing stock in the United States. Although lead pollution from mining activities presents a relatively localized problem, its magnitude is significant, and particularly on the water pollution side, it is compounded by the occurrence of other heavy metals. The obvious danger of pollution from smelting operations has long been recognized. Pollution control practices, however, leave a great deal to be desired. Primary smelters process the ore material and are usually large but few in number, whereas secondary smelters process scrap from old batteries, cable sheathing, etc. and represent more dispersed point sources of heavy metal pollution.

2.1.1.1.2 Health Effects

Lead can be absorbed by the body through inhalation, ingestion, dermal contact (mainly as a result of occupational exposure), or transfer via the placenta. In adults, approximately 10% of ingested lead is absorbed into the body. Young children absorb from 40% to 53% of lead ingested from food. Once lead is absorbed, it enters either a "rapid turnover" biological pool with distribution to the soft tissues (blood, liver, lung, spleen, kidney, and bone marrow) or a "slow turnover" pool with distribution mainly to the skeleton [6]. Of the total body lead, approximately 80%–95% in adults and about 73% in children accumulate in the skeleton. The biological half-life of lead is approximately 16–40 days in blood [6] and about 17–27 years in bones [6].

2.1.1.1.3 Acute and Chronic Exposure

Perhaps no other metal, not even arsenic, has had its toxicology so extensively studied as has lead. Lead poisoning has been actually linked to the fall of the Roman Empire. The high lead content in bones from the Roman period supports the hypothesis that the use of lead containers for wine and other liquids, the use of lead water pipes, and lead-containing ceramic glazing of earthenware containers contributed to the decimation of the ruling class, who were more able to afford the lead containers [7]. The lead poisoning of children has been linked to contemporary earthenware glazed surfaces and pigments of older paints. The toxicology of lead has been extensively studied. Inorganic lead is a general metabolic poison and enzyme inhibitor (like most heavy metals). Organic lead is even more poisonous than inorganic lead. The earliest symptoms of lead poisoning seem to be physical (e.g., excitement, depression, and irritability). Young children are particularly affected and can suffer mental retardation and semipermanent brain damage. One of the most insidious effects of inorganic lead is its ability to replace calcium in bones and remain there to form a semipermanent reservoir for long-term release well after the initial absorption. The usual indicator of the degree of inorganic lead poisoning in humans is the content of this element in whole blood. Different authorities suggest safety levels in the range of 0.2–0.8 ppm. The value 0.2 ppm seems to reflect a worldwide minimum. The disturbing fact is that the natural levels in human blood are already very close to what is considered a reasonable toxicological limit, not leaving us with any margin for exposure to lead.

Lead is a cumulative general poison, with fetuses, infants, children up to 6 years of age, and pregnant women (because of their fetuses) being most susceptible to adverse health effects. Lead can severely affect the central nervous system. Overt signs of acute intoxication include dullness, restlessness, irritability, poor attention span, headaches, muscle tremors, hallucinations, and loss of memory [8], with encephalopathy occurring at blood lead levels of 100–120 pg/dL in adults and 80–100 pg/dL in children. Signs of chronic lead toxicity, including tiredness, sleeplessness, irritability, headaches, joint pain, and gastrointestinal symptoms, may appear in adults with blood lead levels of 50–80 pg/dL [8]. After 1 or 2 years of exposure, muscle weakness, gastrointestinal symptoms, lower scores on psychometric tests, disturbances in mood, and symptoms of peripheral neuropathy were observed in occupationally exposed populations at blood lead levels of 40–60 pg/dL [9]. At levels of 30–50 pg/dL, there were significant reductions in nerve conduction velocity. Renal disease has long been associated with lead poisoning; however, chronic nephropathy in adults and children has not been detected below blood lead levels of 40 pg/dL. Finally, it has been demonstrated that interactions between calcium and lead were responsible for a significant portion of the variance in the scores on general intelligence ratings and that calcium had a significant effect on the deleterious effect of lead [10]. Several lines of evidence demonstrate that both the central and peripheral nervous systems are principal targets for lead toxicity. These include subencephalo-pathic neurological and behavioral effects in adults and electrophysiological evidence of both central and peripheral effects on the nervous system in children with blood lead levels well below 30 pg/dL. The carcinogenicity of lead in humans has been investigated in several epidemiological studies of occupationally exposed workers [11]. The International Agency for Research on Cancer considered the overall evidence for the carcinogenicity of lead in humans to be inadequate [11].

2.1.1.2 Cadmium

Cadmium is a silvery-white, lustrous, but tarnishable metal; it is soft, ductile, and has a relatively high vapor pressure. Cadmium is nearly always divalent; chemically, it closely resembles zinc and occurs in almost all zinc ores by isomorphous replacement [12]. Cadmium is found in natural deposits as ores containing other elements. The most significant use of cadmium is primarily for electroplating, paint pigments, plastics, silver-cadmium batteries, and coating operations, including transportation equipment, machinery and baking enamels, photography, and television phosphors. It is also used in nickel-cadmium batteries, solar batteries, and pigments [13]. In one review, it was noted that the use of cadmium products has expanded in recent years at a rate of 5%–10% annually, and the potential for further growth is very high [14]. The whole world's annual production of cadmium is around 20,000 tons. Discharge of cadmium into natural waters is derived partly from the electroplating industry, which accounts for about 50% of the annual cadmium consumption in the United States. Other sources of water pollution are the nickel-cadmium battery industry and smelter operations, which are more likely to be fewer in number but of a greater point source significance, often affecting the environment at distances of a 100 km order of magnitude [15].

2.1.1.2.1 Occurrence

Cadmium is a relatively rare element. It is uniformly distributed in the Earth's crust, where it is generally estimated to be present at an average concentration of between 0.15 and 0.2 mg/kg [16]. Cadmium occurs in nature in the form of various inorganic compounds and as complexes with naturally occurring chelating agents; organocadmium compounds are extremely unstable and have not been detected in the natural environment. Industrial and municipal wastes are the main sources of cadmium pollution. The solubility of cadmium in water is influenced to a large degree by the acidity of the medium. Dissolution of suspended or sediment-bound cadmium may result when there is an increase in acidity [17]. The need to determine cadmium levels in suspended matter and sediments in order to assess the degree of contamination of a water body has been pointed out. The concentration of cadmium in unpolluted fresh waters is generally less than 0.001 mg/L [16]; the concentration of cadmium in seawater averages about 0.00015 mg/L [17]. Surface waters containing in excess of a few micrograms of cadmium per liter have probably been contaminated by industrial wastes from metallurgical plants, plating works, plants manufacturing cadmium pigments, textile operations, cadmium-stabilized plastics, or nickel-cadmium batteries, or by effluents from sewage treatment plants [17].

High concentrations of cadmium in the air are associated with heavily industrialized cities, notably those with refinery and smelting activities [16], where levels may be several hundred times those found in noncontaminated areas [18]. According to the earlier (1969) data from the U.S. National Air Sampling Network, the annual average cadmium concentrations at 29 nonurban stations were all less than 0.000003 mg/m³; those for the 20 largest cities ranged from 0.000006 to 0.000036 mg/m³ [18].

The presence of cadmium in vegetation may arise from the deposition of cadmium-containing aerosols directly on plant surfaces and by the absorption of cadmium through roots. Plants vary in their tolerance to cadmium in soil and in the amounts they can accumulate. Certain shellfish, such as crabs and oysters, may concentrate cadmium to extremely high levels in certain tissues, even if they inhabit waters containing low levels of cadmium. Reported concentrations of cadmium in foodstuffs vary widely; concentrations in most foods average about 0.05 mg/kg on a wet-weight basis. Other fresh meats generally contain less than 0.05 mg/kg; cadmium concentrations in fish are usually less than 0.02 mg/kg [18]. The greatest exposures are from staple crops that are consumed in large quantities, such as wheat, potatoes, and rice. In cadmium-polluted areas, cadmium levels may be significantly elevated; rice and wheat from contaminated areas of Japan have been found to contain concentrations near 1 mg/kg, at least a factor of 10 higher than those for most parts of the world [18]. In a survey in a metal-mining county in Hunan province in China, 88% of the rice grain exceeded the permissible limit of 0.2 mg Cd/kg dry weight; in that area, the calculated median dietary intake per month was 2.7 times higher than the U.S. Food and Drug Administration and World Health Organization limit of 25 μg Cd/kg body weight for the general population, and 4.6 times higher than the limit for children. In other continents where data exist, for example, Europe, the mean total population Cd intake, mainly from food, excluding those living in highly contaminated areas, is close to or exceeds the tolerable weekly intake [20].

2.1.1.2.2 Health Considerations

Cadmium is not at present believed to be an essential nutrient for animals or humans. Several studies on human subjects indicate that 4%–7% of a single dose of ingested cadmium is absorbed from the intestine. The absorption of cadmium nitrate or cadmium chloride in animal studies ranged from 0.5% to 3% [18]. The total amount absorbed by humans has been estimated as 0.0002–0.005 mg/day [19].

Absorbed cadmium accumulates mainly in the renal cortex and liver. The pancreas, thyroid, gallbladder, and testes can also contain relatively high concentrations. Several studies suggest that the accumulation of cadmium in the human body is a function of age [20]; one author claims that there is a 200-fold increase in the cadmium content of the body in the first 3 years of life and that in this early period, humans accumulate almost one-third of their total body burden. Cadmium accumulates with age until a maximum level is reached at about age 50; the total body burden of a person of 50 years of age ranges from 5 to 40 mg. About half the body burden is found in the kidneys and liver; the cadmium concentration of the cortex of the kidneys ranges from 0.005 to 0.1 mg/g. Concentrations of cadmium in the renal cortex are normally 5–20 times those in the liver [17].

2.1.1.2.3 Toxic Effects

Due to its acute toxicity studied only recently, cadmium has joined lead and mercury in the most toxic "Big Three" category of heavy metals with the greatest potential hazard to humans and the environment. Cadmium is one of the metals most strongly absorbed by living cells accumulated by vegetation. It is also among the most toxic to living organisms and more likely to leach from industrial wastes. The acute oral lethal dose of cadmium for humans has not been established; it has been estimated to be several hundred milligrams [21]. Doses as low as 15–30 mg [21] from acidic foodstuffs stored in cadmium-lined containers have resulted in acute gastroenteritis. The consumption of fluids containing 13–15 mg of cadmium per liter by humans has caused vomiting and gastrointestinal cramps.

Acute cadmium poisoning has occurred following exposure to fumes during the melting or pouring of cadmium metal [22]. Fatalities have resulted from a 5-hour exposure to 8 mg/m³, although some individuals have recovered after exposure to 11 mg/m^3 for 2 hours. Acute pneumonitis resulted from inhalation of concentrations between 0.5 and 2.5 mg/m^3 for 3 days. Symptoms of acute poisoning include pulmonary edema, headaches, nausea, vomiting, chills, weakness, and diarrhea. Cadmium has been established as a very toxic heavy metal. A disease known as "Itai-Itai" in Japan is specifically associated with cadmium poisoning, resulting in multiple fractures arising from osteomalacia [23]. Symptoms of the disease, which occurred most often among elderly women who had many children, are the same as those of osteomalacia (softening of the bone); the syndrome is characterized by lumbar pain, myalgia, and spontaneous fractures with skeletal deformation. It is accompanied by the classical renal effects of industrial cadmium poisoning: proteinuria, glucosuria, and aminoaciduria [18].

Cadmium tends to accumulate in the human body (30 mg in an average American male), with 33% in the kidneys and 14% in the liver. Chronic cadmium poisoning produces proteinuria and causes the formation of kidney stones. There is evidence of a link between cadmium and hypertension. The main problem with cadmium in humans appears to be that the body seldom excretes as much cadmium as is absorbed. There is little general agreement about acceptable safety limits for cadmium intake. In the United States, the safety level of cadmium in drinking water has been set at 10 ppb. Sampling of surface waters revealed some dangerously high cadmium levels. Chronic exposure to airborne cadmium results in a number of toxic effects; the two main symptoms are lung emphysema and proteinuria [22]. Emphysema appears after approximately 20 years of exposure; levels of exposure that result in disability have not been systematically determined. It has been proposed that the minimum critical level of cadmium in the kidney required to produce renal tubular damage is approximately 0.2 mg/g [24]. The World Health Organization (WHO) has recommended that the provisional permissible intake of cadmium not exceed 0.4–0.5 mg per week or 0.057–0.071 mg/d [24].

2.1.1.3 Mercury

Mercury is a dense, silvery-white metal that melts at −38.9°C. Mercury is present in the Earth's crust at an average concentration of 0.08 mg/kg; cinnabar (mercury[II] sulfide, HgS) is the most common mercury ore [25]. Igneous, metamorphic, and sedimentary rocks contain mercury at concentrations up to 0.25, 0.40, and 3.25 mg/kg, respectively [25]. Mercury and its compounds are used in dental preparations, thermometers, fluorescent and ultraviolet lamps, and pharmaceuticals, and as fungicides in paints, industrial process waters, and seed dressings. The pulp and paper industry also consumes mercury in significant amounts in the form of phenyl mercuric acetate, a fungicide, and caustic soda, which may contain up to 5 mg/kg as an impurity.

2.1.1.3.1 Occurrence

Many mercury compounds are volatile, and most decompose to form mercury vapor. Elemental mercury has a substantial vapor pressure even at ambient temperatures but, except at elevated temperatures, does not react readily with oxygen in the air. Mercury can exist as univalent and divalent ions. Mercury(I) is always in 2+, and all of its compounds are in the dimeric form. Mercury(II), Hg^{2+}, forms both covalent and ionic bonds; $HgCl_2$, for example, is covalent. This causes a relatively low solubility of $HgCl_2$ in water and higher solubility in organic solvents. Mercury(II) can also form complexes by accepting pairs of electrons from ligands. The covalent property of mercury(II) allows a stable mercury-carbon bond and the formation of organometallic compounds. The organomercury salts are soluble in organic solvents, and compounds such as dimethyl mercury, $(CH_3)_2Hg$, can easily be separated from inorganic salts and even from $HgCl_2$, as $HgCl_2$ can first be complexed to form 2+ with excess chloride. The distribution of mercury between the three oxidation states is determined by redox potential, pH, and the anions present.

Mercury can enter the atmosphere by simple transport as metallic mercury vapor or as volatilized organic mercury compounds. The formation of volatile organomercurials may occur through microbial, animal, or plant metabolic activity. These natural processes result in the constant circulation of significant quantities of mercury in the atmospheric environment. The U.S. Environmental Protection Agency (USEPA) has estimated rural concentrations of mercury in the air to be 0.000005 mg/m^3, urban concentrations 0.00003 mg/m^3, and indoor concentrations $0.0001–0.0002$ mg/m^3; the average atmospheric concentration was 0.00002 mg/m^3, and it was stated that atmospheric concentrations are unlikely to exceed an average value of 0.00005 mg/m^3 [26]. Mercury in air can be washed out by rain. In industrial areas, mercury concentrations as high as 0.0002 mg/L have been reported in rain. In most surface waters, $Hg(OH)_2$ and $HgCl_2$ are the predominant mercury species. In reducing sediments, however, most of the mercury is immobilized as the sulfide. Concentrations of mercury in surface and drinking waters are generally below 0.001 mg/L [26]. The presence of higher levels of mercury in water is due to effluents from the chlor-alkali industry, the pulp and paper industry, mining, gold, and other ore-recovering processes, and irrigation or drainage of areas in which agricultural pesticides are used.

Inorganic mercury in sediments, under anaerobic conditions, can be transformed by microorganisms into organic mercury compounds, the most common of which is methyl mercury [27]. These compounds can readily associate with suspended and organic matter and be taken up by aquatic organisms. Methyl mercury has a high affinity for lipids and is distributed to the fatty tissues of living organisms [28]. Although methyl mercury is estimated to constitute only 1% of the total mercury content of water, more than 90% of the mercury in biota is in the form of methyl mercury [28]. It has been estimated that about 5,000 tons of mercury are annually released into the environment by man's activities. Mercury is readily scavenged by organic matter. Mercury salts from industrial effluents deposit in river or lake sediments and are then acted upon by anaerobic bacteria, which convert them into toxic methyl mercury and dimethyl mercury. It is in the hydrosphere that the effects of mercury pollution are most significant. Soluble mercury is readily incorporated into organisms in the aquatic environment and ultimately finds its way into higher members of the food chain such as man. The progress of mercury through the food chain successively increases its concentration to such an extent that natural levels in some commercial fish are close to, or exceed, the lowest level now set by the health authorities in many countries. It is therefore obvious that a small additional "pollution" component can be sufficient to cause a public health hazard under certain circumstances. This situation has already been reached for mercury and lead and may soon apply to cadmium. Analyses of the Greenland ice cap revealed that while mercury levels worldwide had been constant since 800 B.C., since 1950 the amounts present seem to have doubled.

2.1.1.3.2 Health Effects

Absorption of metallic mercury following ingestion is negligible; less than 0.01% of an administered dose of metallic mercury was absorbed in animals, for example. In humans, accidental ingestion of several grams of metallic mercury increased blood mercury levels [29], but only rarely did doses of 100–500 g cause clinical illness (stomatitis and diarrhea). Soluble inorganic mercury(II) salts are absorbed to a limited extent, 7%–15% in humans, and sparingly water-soluble mercury(I) salts are absorbed to an even lesser degree. The mercury(I) ion can be biotransformed to the mercury(II) ion in vivo, however. Ingested organic mercury, on the other hand, is readily absorbed [30]; 95% or more is absorbed in humans. Absorption depends on particle size, solubility, and rate of decomposition of the salts in biological fluids. A fraction of inhaled mercury salts will be cleared to the alimentary tract and absorbed by ingestion. Generally, aerosols of inorganic mercury compounds are absorbed to a lesser degree than mercury vapor.

2.1.1.3.3 Distribution and Metabolism

The world's annual consumption of mercury averages about 10,000 tons, and about half of this total is used in the production of chlorine for bleaching paper pulp. As a result of the established mercury threat, the largest man-handling of mercury in chlorine manufacture is being limited by introducing alternative technologies that do not employ mercury electrodes. The next largest mercury consumption (approximately 35%) is in the production of switch gears and batteries.

Inorganic mercury compounds are rapidly accumulated by the kidney, the main target organ for these compounds. Mercury in the kidneys is in the form of a metallothionein-like complex. The binding of the mercury by the protein, metallothionein, is enhanced in the presence of cadmium. Phenyl and methoxyethyl mercuric salts rapidly degrade to mercuric salts and distribute as such in the bodies of humans and animals. The toxicity of these organomercurials is dependent on the rate of their conversion (biotransformation) to inorganic mercury; because this conversion is rapid, the toxicity of these compounds in cases of chronic exposure is similar to that seen after inorganic mercury exposure. Elemental mercury vapor that is inhaled rapidly diffuses through the alveolar membrane; in the body, it is oxidized to mercuric ions, which produce toxic effects.

Absorption of methyl mercury from food (bound to protein) or water (as chloride salt) is almost complete both in animals and in humans. Methyl mercury has considerable stability in the body and circulates for a time unchanged in the blood. It is distributed in high concentrations to the kidney and somewhat less to the liver. In the kidney, 40% is present in the inorganic form. In humans, methyl mercury has a ratio of 20:1 between red blood cells and plasma, in contrast to the 1:1 ratio after exposure to inorganic or phenyl mercury [26]. The most reliable index of exposure to methyl mercury and retention in the central nervous system is the finding of methyl mercury in red blood cells. Hair mercury levels reflect past exposure and are dependent on the rate of hair growth. There is an almost linear relationship between the amount of methyl mercury in blood and that in the hair that was formed during exposure; the ratio of hair to blood levels has been consistently found in the range 230–300:1 [26]. At a steady state, the level of mercury in blood is proportional to the daily intake of methyl mercury; the constant of proportionality for a 70 kg adult has been estimated to be between 0.3 and 1.0 (units of days per liter).

2.1.1.3.4 Toxicity

With the possible exception of lead, mercury as a pollutant has been studied more extensively than any other trace element during the past three decades. Although it had been known for many centuries that mercury is poisonous to animals and humans, it was not until the late 1950s that its extreme toxicity to humans was appreciated as it made headlines worldwide. In 1953, the mysterious death of 52 persons living in fishing villages along Minamata Bay in Japan was unmistakably linked with mercury poisoning. High levels of mercury originating from the nearby plastics factory were found in the shellfish eaten by the villagers. The "minamata disease," mercury poisoning, has been linked to many more deaths around the globe ever since and symptoms of mercury poisoning crippled countless more. Advanced analytical methods made it possible only relatively recently to monitor low levels of mercury in the environment, which, however, are sufficient to cause these serious problems on a large scale.

A particularly disturbing feature of mercury poisoning is that the effects are not immediately obvious. Methyl mercury is particularly toxic because it readily passes from the bloodstream into the cerebellum and cortex, causing damage that is symptomized by numbness, awkwardness of gait, and blurring vision. Clinical tests to determine mercury poisoning are based mainly on the levels of this element in whole blood. Identifiable symptoms of mercury poisoning occur with levels of 0.2–0.6 ppm. Such levels would be reached by a daily intake of 0.3–1.0 mg of mercury by a healthy man. The WHO proposed an acceptable daily intake (ADI) of 0.3 mg Hg, of which not more than 0.2 mg should be in the form of methyl mercury. Since most of the environmental mercury is derived from natural sources, man's addition to this load is of critical importance. The appearance, character, and extent of the toxic effects of mercury depend on a number of factors: the chemical form of the mercury; the mercury compound and its ionization potential; the dose, duration of exposure, and the route of administration; and the dietary levels of interacting elements, especially selenium [30]. When given in acute massive doses, mercury, in whatever chemical form, will denature proteins, inactivate enzymes, and cause severe disruption of any tissue with which it comes into contact in sufficient concentration.

The two major responses to mercury poisoning involve neurological and renal disturbances. The former is characteristic of poisoning by methyl and ethyl mercuric salts, in which liver and renal damages are of relatively little significance. The latter is characteristic of inorganic mercurial poisoning. In general, however, acute lethal toxic doses by ingestion of any form of mercury will result in the same terminal signs and symptoms, which consist of shock, cardiovascular collapse, acute renal failure, and severe gastrointestinal damage. After acute administration of ionizable inorganic salts of mercury to animals or humans, the highest levels of mercury are found in the kidney; although acute oral poisoning results primarily in hemorrhagic gastritis and colitis, the ultimate damage is to the kidney. Clinical symptoms of acute intoxication include pharyngitis, dysphagia, abdominal pain, nausea and vomiting, bloody diarrhea, and shock. Later, swelling of the salivary glands, stomatitis, loosening of the teeth, nephritis, anuria, and hepatitis occur. Ingestion of 500 mg mercuric chloride causes severe poisoning and sometimes death in humans. Acute exposure results from inhalation of air containing mercury vapor in the range of 0.05–0.35 mg/m^3. Exposure for a few hours to a concentration of between 1 and 3 mg/m^3 may give rise to pulmonary irritation and destruction of lung tissue and occasionally central nervous system disorders [30]. Chronic exposure occurs in persons occupationally exposed to large amounts of mercury on occasion and as a result of prolonged therapeutic use. Alkyl compounds of mercury are the most toxic to humans, producing illness, irreversible neurological damage, or death from the ingestion of milligram quantities [31]. Outbreaks of poisonings by these organic derivatives have been the result of accidents or environmental contamination in a number of countries: Iraq, Guatemala, Pakistan, Japan (Minamata and Niigata), and the United States [31].

Symptoms may occur weeks or months after exposure to toxic concentrations of either methyl mercury or ethyl mercury. Therefore, no clear distinction between acute and chronic symptomatology can be made. In cases of severe poisoning, pronounced weight loss can occur with or without intestinal symptoms. Neurological symptoms include mental deterioration, rigidity and hyperkinesia, and salivation and sweating. Alkyl mercury readily crosses the placenta unchanged and concentrates in fetal tissues. As a result, infants born to exposed mothers may suffer from mental retardation, cerebral palsy, and convulsions. The fetus is far more sensitive to methyl mercury poisoning than is the child, and children below 10 years of age are more susceptible than adults. Although methyl mercury acts on basic genetic systems such as the spindle fiber mechanism and DNA, its mutagenic potential appears to be small. No evidence for genetic, teratogenic, or carcinogenic effects has yet been described for inorganic mercury.

2.1.1.4 Chromium

As with any other transition metal, chromium can be found at a degree of oxidation ranging from (−II) to (+VI). However, the most common oxidation states of chromium are (0), (III), and (VI). In natural deposits, chromium is present in complex cubic isomorphic minerals called spinel. Most of the chromium found in nature is in its trivalent state (the most stable one), but small amounts of the hexavalent form have been found along with the divalent oxidation state. Chromium is used in many industrial applications. It can be either used melted with other metals to produce alloys or plated. Chromium steel alloys provide high corrosion resistance and good hardenability. Other applications of chromium range from tanning agents, paint pigments, and catalysts to impregnation solutions for wood or photography. The world production of chromite ore is several millions of tons per year. Ferrochromite is obtained by direct reduction of the ore while chromium metal is produced either by chemical reduction (aluminothermic process) or by electrolysis of either CrO_3 or chrome alum solutions.

2.1.1.4.1 Chemical Pathways

Chromium concentrations in both air and soil are subject to large variations. In air the concentrations range from 0.3 ng/m^3 in remote sites to 50 ng/m^3 in urban areas, and in soil they vary from traces to 250 mg/kg, where it may be from phosphate fertilizers. Chromium concentrations in natural waters are very limited by the low solubility of Cr(III) oxides. Thus a major part of chromium in waters is in the hexavalent state. Main contaminations are generated by industrial wastewaters. Since the trivalent state is predominant in soils, it is unlikely that even heavily polluted farmland would result in chromium accumulation in the food chain via plants. No common plant used as animal feed or food has been reported with a tendency to concentrate chromium.

2.1.1.4.2 *Clinical Effects*

Chromium was recognized to be a hazardous element in the early years after it was discovered. No reports indicate that chromium salts (+III) have severe toxic effects. Hexavalent chromium is considered to be lethal for a dose higher than 3 g for adult humans. The first symptoms are vomiting and persisting diarrhea. After a week, hemorrhagic diathesis and epitasis are commonly observed. Convulsions occur during the final stages of the illness. Repeated occupational inhalation of hexavalent chromium compounds causes perforations of the nasal septum and skin ulceration "chrome holes." The sense of smell and acute irritative dermatitis or allergic eczematous dermatitis have frequently been reported in case of chronic exposure to chromic acid vapors as well as an increased incidence of cancer in the respiratory organs. Bronchial asthma due to chromate dust or chromic acid fumes has been experienced by a number of workers. Environmental contamination with chromium seems trivial compared to mercury or cadmium. Nevertheless, severe toxic effects on plants have been reported at Cr(VI) concentrations of approximately 0.5 mg/L.

2.1.1.4.3 *Monitoring and Legislations*

It is accepted that monitoring both the atmosphere and biological material from exposed workers is essential. Chromium concentrations can be determined using colorimetry, atomic absorption, or emission spectroscopy. The U.S. Standards Institute listed a maximum acceptable concentration (MAC) of 0.1 mg/m^3 for chromic acid. The U.S. National Institute of Occupational Safety and Health (USNIOSH) makes a difference between noncarcinogenic Cr(VI) and carcinogenic Cr(VI). The time-weighted average values at a workplace are 25 mg/m^3 for airborne carcinogenic chromium and 50 mg/m^3 for noncarcinogenic chromium.

2.1.2 RADIONUCLIDES

2.1.2.1 Uranium

Uranium (U) is a hard, silvery-white amphoteric metal and a radioactive element. In the natural state, it consists of isotopes ^{238}U (99.28%), ^{234}U (0.006%), and ^{235}U (0.714%). Uranium occurs naturally in the +2, +3, +4, +5, or +6 valence states, but most commonly in the hexavalent form. In nature, hexavalent uranium is commonly associated with oxygen as the uranyl ion, UO_2^{2+}. There are over 100 uranium minerals: those of commercial importance are the oxides and the oxygeneous salts. Although uranium has a family of 15 radioisotopes, only three occur naturally. The radiation levels from all three are very low due to two factors: (1) the radiation they emit is not very penetrating and is emitted at a low rate and (2) the high density of uranium (1.7 times the density of lead and 2.5 that of steel) acts as a shield against its own radiations. Large quantities of natural uranium can therefore be handled without any special precautions such as shielding or remote handling. Uranium is present in water supplies as a result of leaching from natural deposits, its release in mill tailings, emissions from the nuclear industry, and the combustion of coal and other fuels. Phosphate fertilizers, which may contain uranium at concentrations as high as 150 mg/kg, may also contribute to the uranium content of groundwater.

2.1.2.1.1 *Biological Pathways*

Uranium occurs in the mammalian body in soluble form only as tetravalent uranium or hexavalent uranium in uranyl complexes. Both hexavalent and tetravalent uranium form complexes with carbonate ions and proteins in the body. Oxidation of tetravalent uranium to hexavalent uranium is likely to occur in the organism. Absorption of uranium salts may occur by inhalation or by ingestion; 95% of uranium retained in the body is deposited in bone. Excretion is mainly via the kidney. As all uranium isotopes are radioactive, the hazards of a high intake of uranium are twofold: chemical toxicity and radiological damage. There is no evidence that uranium has any metabolic function in the mammalian organism.

2.1.2.1.2 Chemical Toxicity

The critical organ for chemical toxicity is the proximal tubule of the kidney. Chemical injury reveals itself, in humans, by increased catalase excretion in urine and proteinurea. Such changes are likely to occur when the uranium concentration in the kidney exceeds 1 mg/kg. The concentration of uranium in the kidney is mainly dependent on the solubility of the uranium compound to which the individual is exposed. The limiting daily intake of uranium is in the order of 1.5 mg/d, and it is mainly derived from food items such as vegetables, cereals, and table salt. Occupational exposure involves the inhalation of dust particles of varying size and density containing uranium compounds with different solubility.

The most important effect of uranium is the damage to the kidneys. High doses of uranium cause tissue damage in the kidneys, leading to functional loss as indicated by failure to resorb urinary protein, glucose, catalase, phosphate, citrate, and creatinine, causing slow death by suppression of respiration. A high dosage of uranium also affects blood vasculature throughout the body. Capillary permeability, blood pressure, and edema may increase and clotting ability may decrease. Uranium may damage the capillary membrane, and it is also known to induce some damage to liver and muscle tissue. Its effects on the nervous system may be similar to those from poisoning by other heavy metals. A study of the chemical toxicity of uranium revealed that a body burden of 0.1 mg/kg of body weight produced a definite nephrotoxic effect. The toxic effects of uranium were reviewed by a panel of prominent uranium toxicologists in 1984 and are summarized in Table 2.1.

2.1.2.2 Radon

Radon-222 is a chemically inert gas formed through the radioactive decay of ^{226}Ra. Both are members of the ^{238}U decay series. Although a gas, radon deserves a mention here due to its toxic effects and stealthy presence in habitable structures. Radon is soluble in water, and its solubility decreases rapidly with an increase in temperature (510, 230, and 169 cm^3/kg at 0°C, 20°C, and 30°C, respectively) [32]. Radon is extremely volatile and is readily released from water.

Uranium and radium are present in varying amounts in all rocks and soils. Although most of the radon produced in soil from radium is retained in the Earth, where it decays, a small portion diffuses into the pore spaces and hence into the atmosphere. Other sources of radon include groundwater that passes through radium-bearing rocks and soils, traditional building materials such as wallboard and concrete blocks, uranium tailings, coal residues, and fossil fuel combustion.

TABLE 2.1
Health Effects of Uranium

Health Effect	Uranium/kg Body Weight (mgU/kg)	Uranium (mg) in a 70 kg Person	Uranium Intake (mg) by a 70 kg Person
50% lethality	1.63	114	230
Threshold for permanent renal damage	0.3	21	40
Threshold for transient renal injury or effect	0.058	4.06	8.3
No effect	0.03	2.1	4.3

2.1.2.2.1 Exposure

Radon is the major source of naturally occurring radiation exposure for humans. Exposure occurs via the ingestion of radon dissolved in water and the inhalation of airborne radon. A U.S. survey estimated geometric mean radon levels in public water supplies, public groundwater supplies, and private wells of 2.5, 4.8, and 34 Bq/L. Public wells analyzed by King et al. [33] contained radon at an average concentration of ~40 Bq/L. Nazaroff et al. [34] reported a geometric mean radon concentration of 5.2 Bq/L in public well water supplies in the United States, based on population-weighted statistics. In Finland and Sweden, the population-weighted average for drinking water from private wells has been estimated at 60 and 38 Bq/L, respectively [35].

Outdoor radon concentrations vary seasonally and diurnally and are influenced by height above ground level and meteorological conditions such as wind speed and temperature [35]. Enhanced levels will be found in the vicinity of uranium mines and mill and tailings operations. Indoor radon levels are typically much higher and much more variable than outdoor levels. Radon entry into houses and other buildings is primarily from the soil or rock under the structures. Radon in water, building materials, and natural gas can also contribute to indoor levels [35], particularly in confined spaces with low air change rates (e.g., homes that have been tightly sealed for energy conservation).

The relationship between the concentration of radon in the water supply and the concentration of radon in indoor air depends on several factors, including the rate and type of usage of the water (e.g., drinking water, showers, and laundry), the loss or transfer of radon from the water to the air, and the characteristic ventilation of the house. Nazaroff et al. [34], based on measurements in U.S. homes and water supplies, estimated that public supplies derived from groundwater serving 1,000 or more persons contribute about 2% to the mean indoor radon concentration for houses using these sources. In general, under normal conditions, the intake of radon from indoor and ambient air far surpasses the intake of radon from drinking water via both the ingestion and inhalation routes.

2.1.2.2.2 Health Effects

Radon consumed in water appears to rapidly enter the bloodstream from the stomach, perfusing all the cells of the body [36]. As it is lipid soluble, it does not distribute evenly throughout the body [37]. Clearance of radon from the bloodstream is relatively rapid, with a half-time on the order of minutes.

Hursh et al. [37] demonstrated that radon is removed from the body primarily through exhalation via the lung. Several studies have found that radon is removed from the body with a primary half-time of between 30 and 70 minutes, with a smaller component (possibly that associated with fatty tissue) having a half-time on the order of several hours. The rate of radon elimination from a resting person appears to be slower than that for a physically active person [36]. Most radon inhaled with indoor air is exhaled and remains in the lungs for only a short time. The radon daughter ^{218}Po is very reactive and electrostatically attracted to tiny particulates in the air. These particulates are inhaled and deposited in the lungs. Radon's daughters then decay sequentially, releasing damaging alpha and beta particles. Therefore, it is radon's progeny, not radon, that actually causes damage to the bronchial epithelium, because only the progeny remains in the lungs long enough to decay significantly.

Epidemiological data derived from underground miners of various metal ores have shown a relatively consistent relationship between lung cancer incidence and exposure to radon progeny [38]. Limitations of the miner studies include crude exposure estimates, inadequate follow-up periods [38], and the inability to account for the confounding factor of cigarette smoking. Comparatively few epidemiological studies have investigated the exposure to natural background radon levels, and those that are available show no significant increase in lung cancer death rate from inhalation exposure to normally occurring levels of radon and radon progeny [38]. Also, there are no experimental or epidemiological data available that link ingested radon with any known health impacts in humans [38]. It has been concluded that there is no need to establish a MAC for radon in drinking water. However, anyone whose indoor air radon concentrations exceed acceptable levels (800 Bq/m as an annual average concentration in the normal living area) should investigate the possibility that their groundwater also contains high levels of radon. Individuals who attempt to remove radon from their water supply using point-of-use devices containing activated carbon should be cautioned regarding the difficulties of disposing of the used radioactive carbon.

2.1.3 ARSENIC POLLUTION

Highly poisonous arsenic is widely distributed in nature and occurs in the form of inorganic or organic compounds. The most toxic form of arsenic is its trivalent cation As^{3+}. Arsenic contamination has been reported from many parts of the world, including the United States, United Kingdom, Canada, and Australia; however, in terms of the severity of the problem, Bangladesh tops the list, followed by India and China. In what has been dubbed "the largest poisoning in the history of mankind," an estimated 40–60 million people suffer from different degrees of acute arsenic poisoning in Bangladesh and Eastern India alone. Well-known health problems caused by acute arsenic poisoning exist on a large scale in those parts of the world where high levels of arsenic are naturally present in a widespread aquifer tapped for drinking and irrigation.

2.1.3.1 Arsenic Speciation and Toxicity

Arsenic is a poisonous chemical that is widely distributed in nature and occurs in the form of inorganic or organic compounds. It is ranked as twentieth in abundance among the elements in the Earth's crust. Arsenic can exist in four valence states: –3, 0, +3, and +5. Under reducing conditions, arsenite, As(III), is the dominant form; arsenate, As(V), is generally the stable form in oxygenated environments. Elemental arsenic is not soluble in water. Arsenic salts exhibit a wide range of solubilities, depending on pH and ionic environment.

Inorganic compounds consist of water-soluble arsenite (As III), the most toxic form, and arsenate (As V), the less toxic form, and such pollutants have been associated with many health problems such as skin lesions, keratosis (skin hardening), lung cancer, and bladder cancer [39–41] (Figures 2.1 and 2.2). Organic arsenic species, abundant in seafood, are much less harmful to health and are readily eliminated by the body. The release of arsenic into the environment occurs in a variety of ways through industrial effluents, pesticides, wood preservative agents, combustion of fossil fuels, and mining activity [39–41]. Indeed, arsenical insecticides have been used in agriculture for centuries, and particularly lead arsenate was quite extensively used in Australia, New Zealand, Canada, and the United States [41,42].

2.1.3.2 Arsenic-Contaminated Countries

Arsenic contamination has been reported from many parts of the world, including the United States, United Kingdom, Canada, and Australia; however, in terms of the severity of the problem, Bangladesh tops the list, followed by India and China (Figure 2.3) [40]. In these countries, arsenic has been released in the groundwater by oxidation of the arsenopyrites/pyrites (arsenic is present in more than 200 mineral species, the most common of which is arsenopyrite) from the subsoil or oxyhydroxide reduction. It has been estimated that about one-third of the atmospheric flux of arsenic is of natural origin. Volcanic action is the most important natural source of arsenic, followed by low-temperature volatilization. Inorganic arsenic of geological origin is found in groundwater used as drinking water in several parts of the world, for example, Bangladesh [41].

Besides the drinking of contaminated groundwater, people in such countries use this water for crop irrigation. Therefore, arsenic compounds find their way into soils used for rice (*Oryza sativa*) cultivation through polluted irrigation water, and through historic contamination with arsenic-based pesticides [43]. Arsenic contamination poses a particular challenge, as this pollutant can enter plants through their phosphate transporter [44] and its contamination is invisible and has no taste or smell.

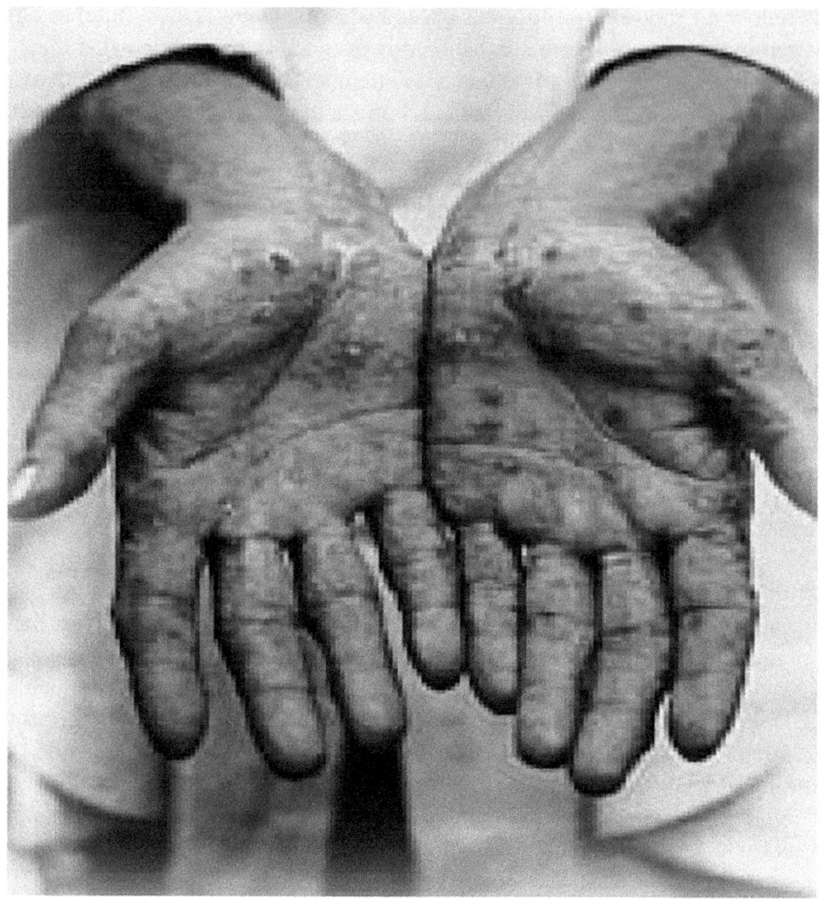

FIGURE 2.1   Health effects of arsenic. Skin lesions and keratosis due to arsenic poisoning (hands).

2.1.3.3   Clinical Effects

Skin disorders, including hyper/hypopigmentation changes and keratosis, are the most common external manifestations, although skin cancer has also been identified. Around 5,000 patients have been identified with As-related health problems in West Bengal (including skin pigmentation changes), although some estimates put the number of patients with arsenicosis at more than 200,000 [45]. In some areas in Bangladesh, groundwater arsenic concentrations can reach 2 mg/L (2 ppm) [41,46], whereas the WHO provisional guideline value for drinkable water is only 0.01 mg/L (10 ppb) [43]. An estimated Bangladesh population of 65 million is exposed to the threat of arsenic poisoning through drinking water [40], and surprisingly at least 32 million Americans consume water containing more than 2 ppb of arsenic. The U.S. EPA is now considering a new standard in the range of 2–20 ppb.

2.2   METALS IN GROUNDWATERS

While the interest in groundwater has mainly focused on the supply of water, a shift of concern to groundwater quality has occurred. At present, the reservoir of fresh water found beneath the surface is gradually degrading due to man's activities. This is placing a strain on the drinking water supply, especially in rural areas where tapping into the ground for water is a common practice. Regulations

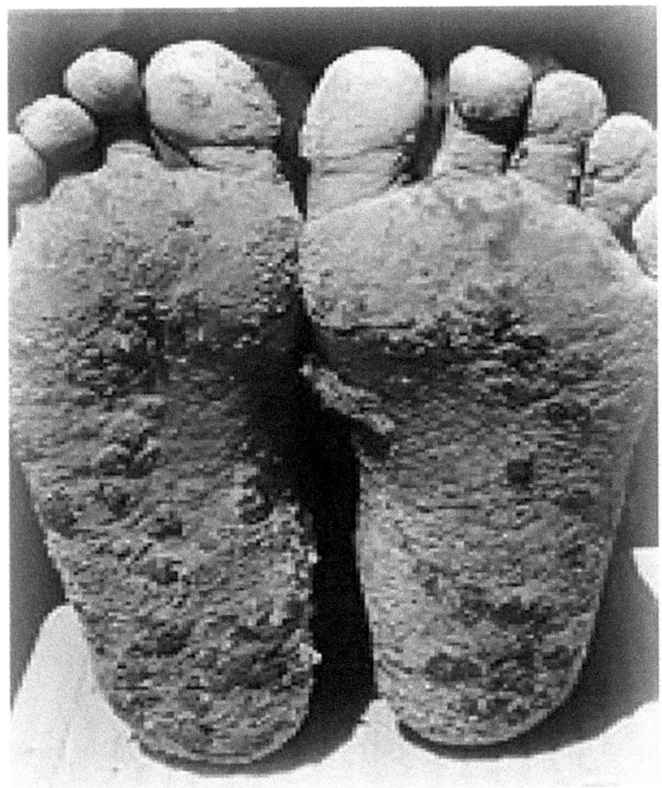

FIGURE 2.2   Health effects of arsenic. Skin lesions and keratosis due to arsenic poisoning (feet).

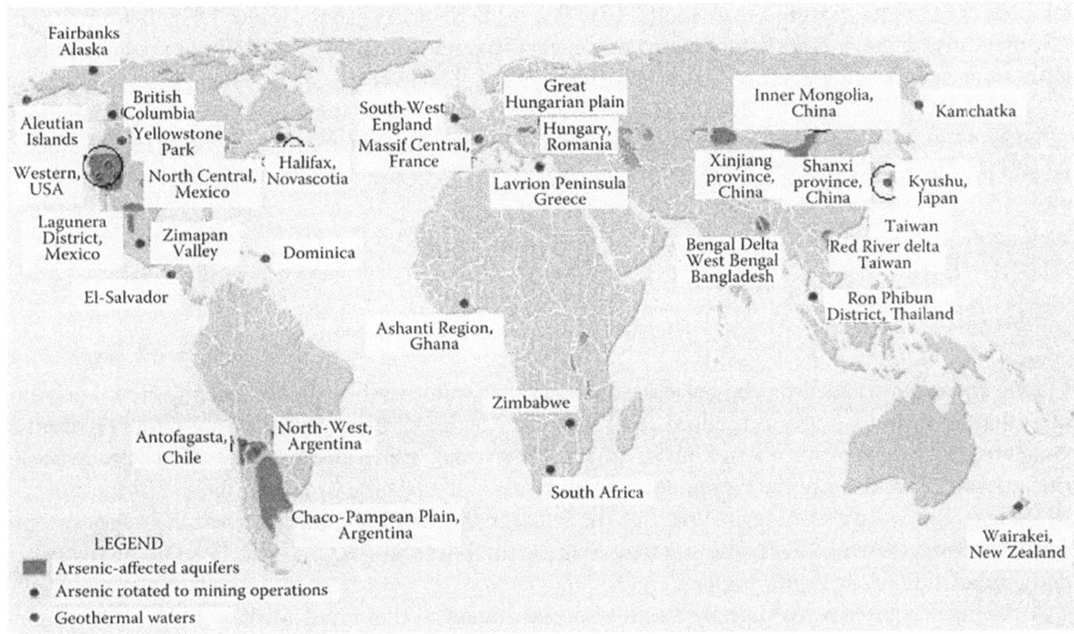

FIGURE 2.3   Arsenic geographic occurrence.

and the introduction of standards concerning water quality have been counteracting this deterioration. However, groundwater contamination goes more frequently undetected or is largely undetectable due to heterogeneities underground, until the damage is widespread. Hope for the future lies in understanding the movement of water and contaminants in aquifers, a porous and permeable type of geological formation that holds and conducts the flow of groundwater.

Focus on inorganic contaminants has been relatively recent since organics are more used in industry and seem to have posed a more eminent problem. Many inorganic contaminants, specifically heavy metals, are toxic and pose a great health and environmental concern in quite low concentrations. Due to the progressive mobilization of heavy metals above the water table caused by man's increasing technological activities, the metals reach underground aquifers in increasing quantities. While some metals may be partially removed by ion exchange with the soil components before they reach the aquifer, the danger of contaminating the latter has been well established. The movement of heavy metals in aquifers depends on how the heavy metals act in aqueous environment. These considerations involve pH, hydrolysis, redox potential, and formation of complexes. Metal mobility generally tends to decline with pH where a solid, typically a metal hydroxide, a metal carbonate, a metal sulfide, or other complexes, becomes a more dominant phase.

2.2.1 Heavy Metals in Aquifers

The transport of solutes in porous materials, such as aquifers, can be considered as flow through a fixed volume element that can be reflected in the following mass balance:

$$(\text{net rate of change of mass within element}) = (\text{flux of solute out of the element})$$

$$-(\text{flux of solute into the element})$$

$$\pm(\text{loss or gain of solute mass due to reaction}).$$

Processes that dictate the flux in or out are advection and hydrodynamic dispersion. Advection is a component of solute movement attributed to transport by flowing groundwater. Dispersion refers to the spreading of the contaminant caused by the fact that not all of the contaminant actually moves at the same speed as the average linear velocity [47–49]. The average linear velocity ($\bar{v}$) is given by (v/n), where v is the specific discharge and n is the porosity. The dispersion–advection equation, Equation (2.1), describes the transport of dissolved constituents that are reactive in saturated isotropic and porous media:

$$\frac{d}{dx}\left(T_x \frac{dh}{dx}\right) + \frac{d}{dy}\left(T_y \frac{dh}{dy}\right) = S\frac{dh}{dt} - R + L, \tag{2.1}$$

where $L = (-K'h_{\text{source}} - h)/b'$ [48].

This equation is based on the basic assumption that water in the aquifer tends to flow horizontally, that is, in the x and y directions, and vertically as leaks through confining beds. Hydraulic conductivity is the ability of an aquifer to transmit water and transmissivity (T) is the average transmission. T_x and T_y are the components of transmissivity, h is the hydraulic head, S is the storage coefficient, R is a sink/store term, and L is the leakage through the confining bed. K is the vertical hydraulic conductivity of the confining bed, b' is the thickness, and h_{source} is the head in the reservoir on the other side of the confining bed.

A confined aquifer is overlain by a unit of porous material that tends to slow down water movement and its transmissivity will remain constant if the aquifer is both uniform in thickness and homogeneous in nature, which rarely is the case. To simplify the advection–dispersion further, the case of flow for nonreactive dissolved constituents in saturated, homogeneous isotropic material at a steady state is taken. The advection–dispersion equation then becomes

$$D_1 \frac{d^2 C}{dl^2} - \bar{v}_t \frac{dC}{dl} = \frac{dC}{dl}, \tag{2.2}$$

where l is a curvy linear coordinate taken to be in the directions of the flow line, D_1 is the coefficient of hydrodynamic dispersion in the longitudinal direction, which depends on the dispersivity and properties of the porous media, and C is the solute concentration. Even with this simplified equation, the flow of contaminants with water through the aquifer is dependent on many variables.

A good visualization of the one-dimensional advection–dispersion equation is the passing of a nonreactive tracer (C_0) through a homogeneous granular medium and looking at its relative concentration in the outflow as seen in Figures 2.4 and 2.5. The assumption of plug flow would have the solute exiting as a step function, indicated by the line marked "position of advection front." In reality, mechanical dispersion and molecular diffusion cause flow to deviate from this; some molecules will move faster and some slower than the average linear velocity ($\bar{v} = Q/nA$), which is the flowrate divided by the product of porosity and cross-sectional area. The greater the distance of flow, the greater the spread of the contaminant. At low velocities diffusion is dominant, and at high velocities dispersion is dominant.

As seen in Figure 2.6, diffusion of contaminants can translate to movement of great distances even through low-permeable rock and can be considered an important factor in the movement of inorganic toxic contaminants. The overall spread of the species will depend on whether the source of contamination is continuous or instantaneous, as depicted in Figure 2.7. Despite the fact that the medium is isotropic, dispersion is anisotropic. It is stronger in the direction of flow than normal to it. The continuous contaminant spreads more from the source as time passes. The point source

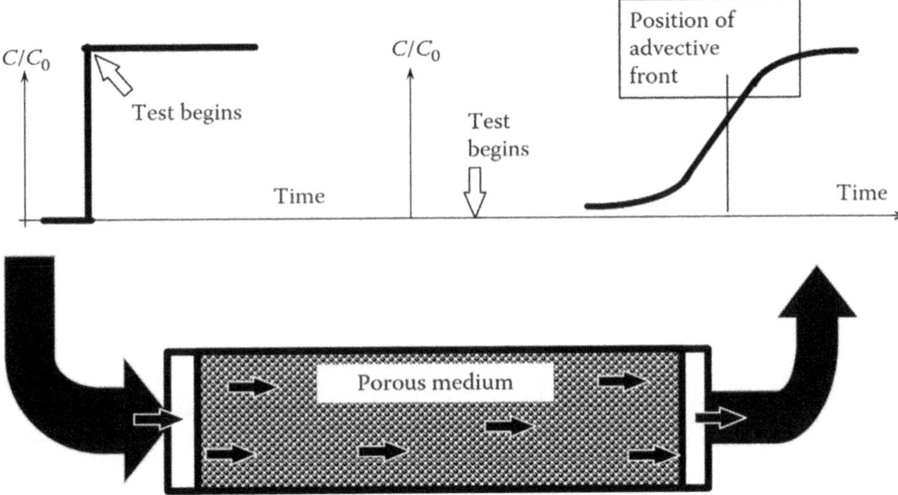

FIGURE 2.4 Experimental apparatus to illustrate dispersion in a column. The test begins with a concentration input of tracer $C/C_0 = 1$ at the inflow end. The relative concentration versus time function at the outflow characterizes dispersion in the column.

Porous medium being penetrated.

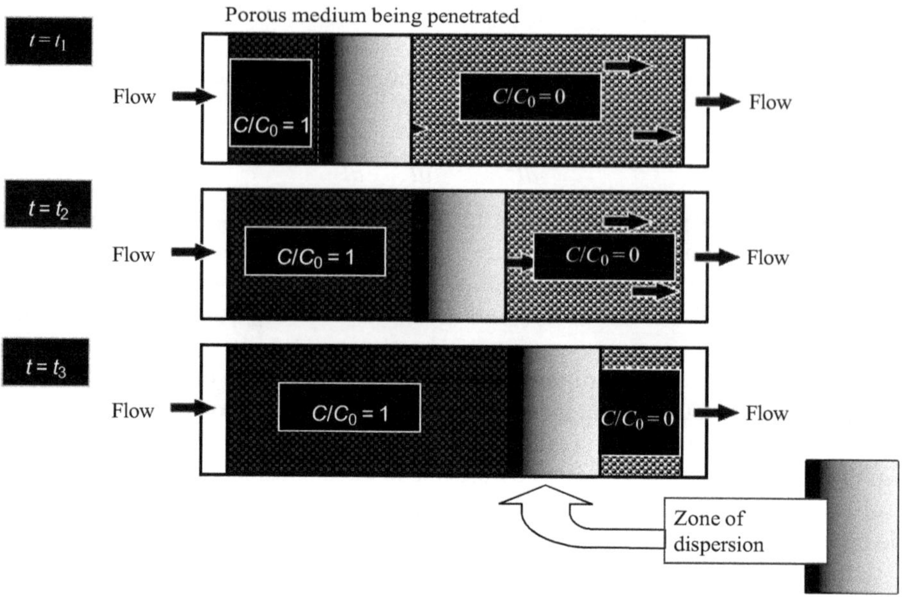

FIGURE 2.5 Schematic representation of dispersion within the porous medium at three different times. A progressively larger zone of mixing forms between the two fluids ($C/C_0 = 1$ and $C/C_0 = 0$) displacing one another.

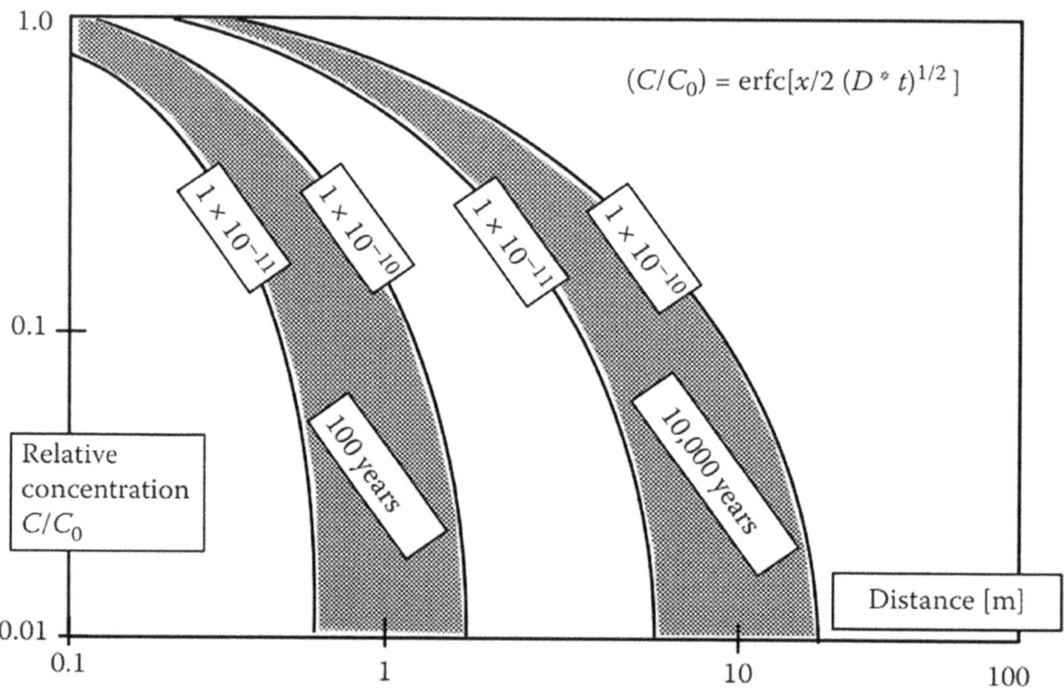

FIGURE 2.6 Positions of the contaminant front migrating by molecular diffusion away from a source where $C = C_0$ at $t > 0$. Migration times are 100 and 10,000 years.

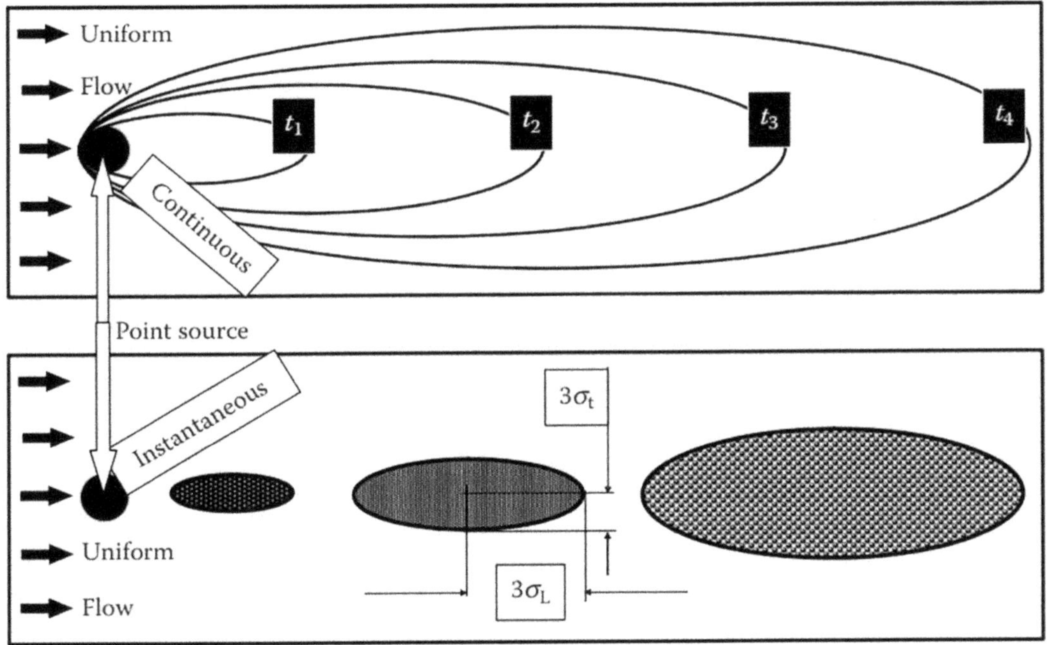

FIGURE 2.7   Spreading a tracer in a two-dimensional uniform flow field in an isotropic sand. (a) Continuous tracer feed with step-function initial condition; (b) instantaneous point source. (From Domenico, P.A., Physical and Chemical Hydrogeology, Wiley, Canada, 1990.)

contaminant contains a fixed mass, yet it is spread over a larger volume with time. Equations dictating this movement are used as a preliminary estimate. The contaminant concentration at a given point at a given time is

$$C(x,y,z,t) = \frac{M}{8(\pi t)^{3/2}D_xD_yD_z}\,\frac{exp-X^2-Y^2-Z^2}{4D_xt4D_yt4D_zt},\tag{2.3}$$

where M is the mass of the contaminant, D is the coefficient of dispersion in the x, y, or z direction, and $X = (x - vt)$, $Y = y$, $Z = z$ [49].

Density plays a key role in the downward movement of the contaminant. As density increases, relative to water, the contaminant plume which was shallow and close to the water table will sink into the groundwater. The assumption of homogeneous media is not true to reality; most geological media contain much heterogeneity. Figure 2.8 shows variations in the heterogeneous media. K_1 represents coarse gray sand and K_2 smaller grained sand in (c). A thin horizontal layer of higher conductivity extends through the original domain, most of the flow occurs in this layer, and overall travel time is one-fifth that in (d). Diagram (d) represents a discontinuous layer of flow conductivity, contamination moves the first and under the second lens, and (e) indicates the discontinuity of a thin and high conductivity layer. The stratospheric differences are important to understand the movement of water in the aquifer.

Heterogeneities are usually determined by careful drilling and mapping. This is more applicable to large-scale heterogeneities. Changes in hydraulic conductivity can be attributed to small-scale heterogeneities such as changes in granular aquifers (changes in silt or clay content). Such heterogeneities cannot be determined from drillings from borehole to borehole. This fact leads to errors in dispersity by up to an order of magnitude from those determined experimentally; hence

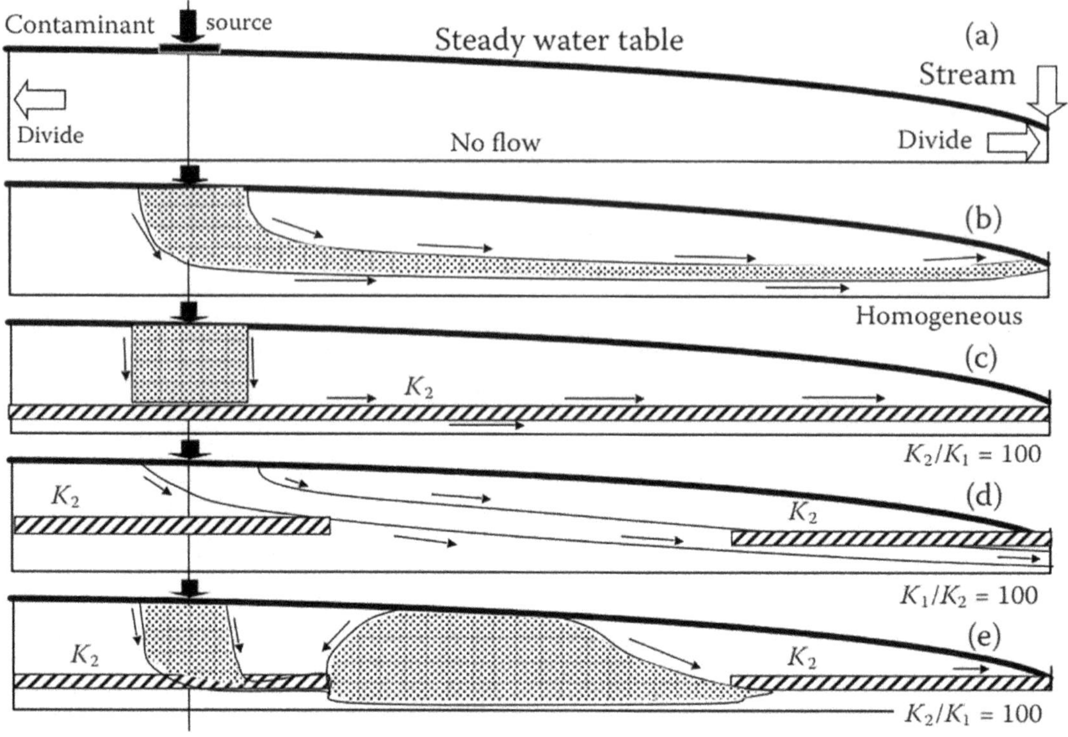

FIGURE 2.8 Effect of layers and lenses on flow paths in shallow steady-state groundwater flow systems. (a) Boundary conditions; (b) homogeneous case; (c) single higher-conductivity layer; (d) two lower-conductivity lenses; (e) two higher-conductivity lenses. (Adapted from Freeze, A.R. and Cherry, J.A., *Groundwater*, Prentice-Hall, New Jersey, 1979.)

dispersion plays a great role in contamination transport. Added complexity occurs when reactions within groundwater during the transport of the contaminants occur, varying the concentrations substantially. These include adsorption/desorption reactions, solution precipitation reactions, oxidation/reduction reactions, ion-pairing or complexation reactions, and microbial cell synthesis. With the influence of reactions, specifically adsorption/desorption, the one-dimensional advection–dispersion equation becomes

$$D_1 \frac{d^2C}{dl^2} - \bar{v}_1 \frac{dC}{dl} + \frac{p^b}{n} \frac{dS}{dt} = \frac{dC}{dt},$$ (2.4)

where p is the bulk mass density of the porous medium, n is the porosity, S is the mass of the chemical adsorbed on the porous medium, and $p/n \, (dS/dT)$ is the change in concentration due to adsorption/desorption following:

$$-\frac{dS}{dt} = -\frac{dS * dC}{dCdt} \quad \text{and} \quad -\frac{p_b}{n} \frac{dS * dC}{dCdt}$$ (2.5)

The term dS/dC represents a partitioning of the contaminant between solution and solids. In the lab, the mass adsorbed per unit mass of dry solids is plotted against the concentration of the constituent in solution in a log/log graph.

The resulting expression follows Equation (2.6):

$$\log S = b \log C + \log K \quad \text{or} \quad S = K_b C^b \, (\text{Freundlich equation}),$$ (2.6)

which represents the relation between the mass of the solute species adsorbed (S) and the solute concentration (C) where K and b are constants that depend on solute species, type of porous media, and other conditions.

Reactions generally slow the rate with which the front of contaminants moves. The retardation equation for reactions is given by

$$\frac{\bar{v}}{\bar{v}_0} = 1 + p_b \frac{k_b}{n},$$ (2.7)

where $\bar{v}$ is the average linear velocity and $\bar{v}_0$ is the velocity of the $C/C_0 = 0.5$ point.

For example, for unconsolidated granular deposits, porosity typically is between 0.2 and 0.4 and average mass density is 2.65 for unconsolidated deposits. For the porosity, given the bulk mass densities (p) are 1.6–2.1 g/cm^3, p_b/n values from 4 to 10 g/cm^3 are to be used. If $K_d = 1$ mL/g then the flow of groundwater would be slowed by a factor of 5–11 times. If $K_d = 10$ mL/g then the contaminating species would not move far from its point of input. A mixture of contaminants will separate into zones after time, given by $\bar{v}/\bar{v}_0$ for each species. These zones will travel at their own velocities. Later, when the contaminant discharge is discontinued, lower-concentration water will pass through and the adsorbed contaminant will be transferred into the liquid phase. If reactions are reversible, then in time, all evidence of contamination will be negligible; although this may be irreversible in a realistic time period when contaminants react slower than the actual movement of water, the retardation time will be even slower than that for the case of a fast reaction. Information on the movement of contaminants in the porous media is hard to come by and the use of the general retardation equation will yield errors in the prediction of the rates of migration of contaminants.

Heavy metals in contaminated water rarely occur at concentrations above 1 mg/L. Concentrations are low due to solubility, taking into account that other minerals are dissolved in this same water as well as the possibility of adsorption onto clay minerals or on hydrous oxides of iron and manganese or organic material. Isomorphous substitution of coprecipitation with minerals or amorphous solids can also be of some interest. Most heavy metals form hydrolyzed species and form complex species by combining with inorganic anions, such as HCO_3^-, CO_3^{2-}, SO_4^{2-}, Cl^-, F^-, and NO_3^-. Complexing with organic compounds may be important where present.

An equation that sums all amounts of the particular heavy metal complexed with various anions will give the total amount of metals present in all forms. If the total is known, then the amount of each species can be calculated using mass action equations. The hydrolyzed and inorganic species with mercury, for example, include $HgOH+$, $Hg(OH)_2$, $HgCl_2$, HgS, HgO, and $Hg(HS)_2$. In assessing the mobility of mercury and any other heavy metal, a knowledge of these and other species must be looked at. As mentioned before, the increasing pH of groundwater leads to increased hydrolysis of the heavy metals. As their concentration in aquifers increases, so does their probability of complexing with anions present such as CN^- and HS^-. Heavy metal concentration can be calculated using concentration data obtained from laboratory analysis. Being able to predict the mobility of heavy metals in groundwater depends on the ease with which the concentration of the most dominant complexes formed can be predicted.

A generalized table of the mobilities of heavy metals in soil is shown in Table 2.2. This directly applies to aquifers since it is comparative and no numbers have been given.

Almost all the trace metals found in groundwater are influenced by redox conditions, due to changes in the oxidation state of the metal complex. Redox conditions, in a way, may influence the concentration of trace metals in the solid phase of the porous medium that causes adsorption of the trace metal. The diagram of Eh versus pH for heavy metals in water shows the main stability regions of the particular heavy metal complexes.

In anaerobic groundwaters, the insolubility of sulfide minerals can limit trace metals to very low concentrations. In groundwater, which is nonacidic and has a high concentration of dissolved inorganic carbon, the solubility of certain carbonate materials will, if equilibrium is achieved, limit

TABLE 2.2
Relative Mobilities of Heavy Metals

	Conditions			
Relative Mobility	Oxidizing	Acid	Neutral (Alkaline)	Reducing
Very high			Se	
High	Se	Se, Hg		
Medium	Hg, As, Cd	As, Cd	As, Cd	Tl
Low	Pb, As, Sb, Tl	Pb, Bi, Sb, Tl	Pb, Bi, Sb, Tl, In	
Very low to immobile	Te	Te	Te, Hg	Te, Se, Hg, As, Cd, Pb, Bi, Tl

Source: Ferguson, J.E., *The Heavy Elements: Chemistry, Environmental Impact and Health Effects.* Pergamon Press, New York, 1990.

the levels of trace metals at low concentrations, that is, Cd and Pb. This is true if excessive amounts of inorganic or organic substances, which tend to be complex with heavy metals in water, are not present. Adsorption is the key mechanism that tends to keep concentrations far below those dictated merely by solubility. It occurs due to the presence of clay minerals, organic matter, crystalline solids, and other amorphous solids in the porous material. Certain oxides, for instance, Fe and Mn, not only control but also enhance adsorption onto the medium for which they form a coat.

2.2.2 Cases and Remediation

In 1947, the city of Babylon in New York saw the beginning of landfilling. Disposal included urban refuse, incinerated garbage, cesspool waste, and industrial refuse. The refuse was placed below the water table and the cesspool was treated and placed in lagoons. The surface sand aquifer was about 27.5 m thick and had a hydraulic conductivity of 1.7×10^{-3} m/s. Contaminants in the aquifer included the major ions Ca^{2+}, Mg^{2+}, Na^+, K^+, HCO_3^-, SO_4^{2-}, Cl^-, NH_4^+, and NO_3^-, heavy metals (particularly iron and manganese), and organic compounds. The Cl^- plume began 9.1 m below the water table and continued 12.1 m below. It did not react, so the mass transfer continued. The advection velocity was determined at 2.9×10^{-4} cm/s, which is the range expected for a sand aquifer. Dispersivities in the x, y, and z directions were 18.6, 3.1, and 0.6 m, respectively. Tracer tests estimated that the value in the x-direction would be six times those in the y-direction and the distance in the z-direction would be considerably smaller. The amount of Cl^- away from the source declined constantly, suggesting that the point source was continuous as indicated in Figure 2.7. Most of the nitrogen was present as NH_4^+, indicating reducing conditions near the source. Mixing brought oxygen to the plume, producing NO_3^- - N as the distance from the source increased. Tracking the nitrogen species allowed one to assess the redox conditions.

The area that was reduced is an explanation for the mobility of the heavy metals, iron, and manganese. The Eh-pH diagram shows that Fe^{2+} was the stable form of iron at moderate reducing conditions and a pH of 6; this was true for Mn^{2+} as well. The gradual increase in oxidizing conditions down the plume decreased mobility with the formation of solids [i.e., $Fe(OH)_3$ and $MnO(OH)$] [47]. It is a common practice to apply similar cases to areas that have minimum information: for instance, in the area around the Saint-Laurent Basin (Quebec, Canada), where four types of aquifers exist and one is similar to that mentioned in the Babylon case study. This is one of unconsolidated sand and gravel deposits. Methods of remediation are available for removing inorganics from groundwater; they include chemical addition, removal of suspended solids, ion exchange, and polymeric binding with microfiltration. Changing the pH of water with chemical addition will cause the precipitation of heavy metals. To adjust acidic water, pass through a limestone bed mixed with lime slurries, add caustic soda (NaOH), or add soda ash (Na_2CO_3). To adjust alkaline water, bubble carbon dioxide in the water or add a strong acid. It is rare to have water that is too alkaline. Acids are added to adjust the pH back to normal after a high pH results in precipitation. Some metals do not precipitate out of solution at high pH and may need to be precipitated as a sulfide rather than a hydroxide since they are soluble [50].

Heavy metals, such as hexavalent chromium, are soluble in water at high pH. This heavy metal is used in industrial operations and is not naturally found in groundwater. Its anthropogenic spread through natural aquifers could have serious health consequences in the affected population (as depicted in the feature film, *Erin Brockovich* [51]). Hexavalent chromium is best reduced to its less soluble trivalent form for removal. The pH is reduced to pH 2, a chemical reducing agent such as sulfur dioxide is added, the pH is raised, and trivalent chromium is precipitated. Mercury is precipitated with sulfide addition. The lower treatability limit for mercury is 10–20 mg/L by sulfide precipitation, 1–5 mg/L with ion exchange, 1–10 mg/L with alum coagulation, 0.5–5 mg/L with iron coagulation, and 0.25 mg/L with activated carbon. Arsenic in groundwater may be present in arsenite (AsO^{2-}) or arsenate (AsO^{4-}). Oxygen will oxidize the arsenite to arsenate and most surface water contamination will be in the arsenate form. If contamination occurs in the deep and likely anaerobic aquifer, the arsenite or arsenic form should be somehow oxidized since arsenate is easier to remove. To remove arsenic, a floc must be formed. A polyvalent metallic coagulant must be added to produce a hydroxide floc. A relatively new method of precipitation is the addition of iron to water by electrochemical methods to enhance the precipitation of other inorganics. The system uses sacrificial electrodes to produce an insoluble ferrous ion, which absorbs and precipitates heavy metals.

Suspended solids can be removed by flocculation, chemical addition, and pH adjustment, which convert inorganic contaminants to non-soluble forms. This together with one of several types of settlers can be employed in the removal of suspended solids. Ion exchange is the exchange of an ion with a high affinity to the sorbent for an ion with a lower affinity. All of the heavy metals, in an aqueous environment, are in the divalent or trivalent state, except hexavalent chromium. A home sodium-ion-exchange unit will remove all these compounds. Ion exchange is not cheap and the brine with heavy metals will have to be disposed of. The ionex process is put to the best use in low concentrations as the final treatment before potable use. Polymeric binding and microfiltration is a two-step method that selectively removes metals from groundwater. First, the addition of a water-soluble polymer binds the metal sand, which is then followed by microfiltration. The favored polymer for heavy metals is polyethylene-imine at low pH.

2.3 HEAVY METAL POLLUTION SOURCES

Due to man's industrial activity, the concentrations of some heavy metals have reached high levels, posing a danger to public health. Whereas the nature of domestic wastewater is relatively constant, the extreme diversity of industrial effluents calls for an individual approach to handling each type of industry and often entails the use of specific treatment processes. Therefore, a thorough understanding of the metal handling upstream production processes and of the overall production system organization is fundamental.

2.3.1 ACID MINE DRAINAGE

The major source of liquid waste in the mining industry is acid mine drainage (AMD). AMD is by far the most extensive and most severe environmental problem associated with mining activities, both current as well as past.

Precious metal and uranium mines contain sulfide minerals, either in the ore or in the surrounding waste rock. AMD is common in areas where mining openings intersect the water table and where rocks contain pyrite and/or other sulfides. When these sulfide minerals, particularly pyrite and pyrrhotite, are exposed to oxygen and water, a process of conversion of sulfide to sulfate takes place. Water in contact with these oxidizing minerals is made acidic and water carries with it toxic metals and elevated levels of dissolved salts. As the reactions proceed, temperature and acidity increase, resulting in an increased rate of reaction. Between pH levels of 2 and 4, bacteria and ferric iron catalyze the reaction rate. Rainfall and snowmelt flush the toxic solutions from the waste sites into the downstream environment. If acidic drainage is left uncollected and untreated, it could contaminate groundwater and local water courses, damaging the health of plants, wildlife, and fish and eventually posing a threat to human health, particularly through the toxicity of the heavy metals that it carries.

At active mine sites (and some inactive mine sites), mining companies operate comprehensive systems to collect and treat effluents and seepage from all sources. These facilities, when well operated and maintained, are sufficient to prevent downstream environmental impact. However, acid generation may persist for hundreds of years following mine closure. The problems are compounded by the demise and disappearance of the original mine operators.

The operation of treatment plants for very long periods of time is clearly not desirable and counter to the principles and goals of sustainable development. In addition, the conventional lime treatment process produces sludges that contain a very low percentage by weight of solids. In some severe cases, in a few decades, the volume of lime sludge will exceed the volume of tailings or waste rock producing acidic drainage.

2.3.1.1 Chemistry of Acid Mine Water

Nowadays, the principles of AMD generation are fairly well understood. Pyrite and other sulfide minerals on exposure to oxygen and water, and in the presence of oxidizing bacteria such as *Thiobacillus ferrooxidans*, oxidize to produce dissolved metals and acidity (sulfuric acid) according to the following steps [52].

The first of these reactions is the oxidation of pyrite:

$$2FeS_2 + 2H_2O + 7O_2 \Leftrightarrow 2FeSO_4 + 2H_2SO_4 \qquad (2.8a)$$

The next step is the oxidation of ferrous ions to ferric ions:

$$4FeSO_4 + 2H_2SO_4 + O_2 \Leftrightarrow 2Fe_2(SO_4)_3 + 2H_2O \qquad (2.8b)$$

This process occurs very slowly at the low pH values found in acidic mine water. Below pH 3.5, the iron oxidation is catalyzed by the iron-oxidizing bacterium *T. ferrooxidans*, and in the pH range of 3.5–4.5, it may be catalyzed by a variety of *Metallogenium*, a filamentous iron-oxidizing bacteria. Other bacteria that may be involved in acid mine water formation are *T. thiooxidans* and *Ferrobacillus ferrooxidans*.

The ferric ion further dissolves pyrite:

$$7Fe_2(SO_4)_3 + FeS_2 + 8H_2O \Leftrightarrow 15FeSO_4 + 8H_2SO_4 \qquad (2.8c)$$

which in conjunction with reaction 2.8b constitutes a cycle for the dissolution of pyrite. At pH values much above 3, iron(III) precipitates as the hydrated iron(III) oxide:

$$Fe^{3+} + 3H_2O \Leftrightarrow Fe(OH)_3 + 3H^+ \qquad (2.8d)$$

The beds of streams afflicted with AMD are often covered with "yellowboy," an unsightly deposit of amorphous, semigelatinous $Fe(OH)_3$. The most damaging component of acid mine water, however, is sulfuric acid. It is directly toxic and has other undesirable effects. While the devastation potential of AMD is very high, its dispersed nature and large areas involved in its generation represent a formidable problem in devising suitably effective control of it. This is the reason why different techniques devised to treat AMD met with only little success. The techniques are based on the following methodologies:

Chemical: limestone/lime application to enhance alkalinity; sulfide precipitation or removal; application of bactericides.
Physicochemical: ion exchange barrier and application of a vegetative and/or geological membrane (cover) to prevent oxygen diffusion.
Biological: other than wetlands, not many biological treatments are known so far.

One approach to eliminating excess acidity involves the use of carbonate rocks. When acid mine water is treated with limestone, the following reaction occurs:

$$CaCO_3 + H_2SO_4 \Leftrightarrow CaSO_4 + H_2O + CO_2 \tag{2.8e}$$

Unfortunately, because iron(III) is generally present, reaction 2.8d occurs as the pH is raised. The hydrated iron(III) oxide that forms as a result of elevated pH soon covers the particles of carbonate rock with a relatively impermeable layer. This armoring effect prevents further neutralization of the acid.

2.3.1.2 Extent of the Damage

The acidic water eventually interacts with other minerals and solubilizes heavy metals such as lead, copper, zinc, cadmium, and nickel that may even be present in minute quantities. Table 2.3 lists the amounts of major cations and anions contained in the mining water and solids of a typical AMD.

AMD is generated not only in and around abandoned mine sites but also at currently active operations. The principal sources of AMD at active mine sites are piles of waste rock containing

TABLE 2.3

Quantitative Analysis by ICP-AES and Ionic Chromatography of the Major Cations and Anions in Mining Water and Solids

Element	Inflow (µg/L)	Outflow (µg/L)	P-deposit (Inflow) (mg/kg)	P'-deposit (Outflow) (mg/kg)
Fe (mg/L)	98.57	34.10	600×10^3	610×10^3
Fe^{2+}/Fe^{3+}	88.00	22.00	—	—
Cr	2.79	1.98	25.80	82.00
As	47.42	10.61	438.00	460.00
Ba	5.55	12.99	39.00	456.00
Nd	9.14	9.29	18.70	41.23
Ce	32.07	33.85	32.43	95.63
La	13.83	14.33	12.25	38.39
U	561.80	634.00	91.55	213.70
Cu	4.38	7.40	8.80	27.60
Pb	1.23	4.39	12.00	21.40
Zn	107.63	142.00	15.20	88.90
Ni	62.87	75.22	13.70	31.90
Co	44.87	56.51	4.27	17.50
SO_4^{2-} (mg/L)	1,938.00	1,888.0	nd	nd
Cl^- (mg/L)	177.00	157.00	nd	nd
F^- (mg/L)	25.00	24.20	nd	nd

pyrite, which is exposed to the atmosphere, to precipitation, and to springwater runoff. According to Kalin [53], 113 million cubic meters of contaminated water are produced annually from mining waste management areas in Canada. Filion et al. [55] have estimated the cost of remedial action at operating and abandoned mine sites across Canada in the order of 4 billion Can\$ over the next 20 years. The rate of generation of heavy metal pollution is likely to further increase since the large-scale mining of low-grade ores uses the method of open pits, which often cut into the groundwater streams. During mining, the water is pumped out of the mine; however, once a mine is closed, water flows into the pit (filling it) and forms a lake. The seriousness of this problem is demonstrated by the example of Berkeley Pit in Montana (http://formontana.net/pit.html), which extends over an area of 77 ha, is 542 m deep, and is filling at a rate of 23 million liters per day [56]. Currently, the water contains approximately 180 mg/L of Cu and 500 mg/L of Zn in combination with 1,000 mg/L of Fe and has an acidic pH of 2.8. This site has also been designated as the experimental site of the U.S. "environmental superfund" for testing metal removal/recovery techniques.

The movement of tailings-derived water away from mining sites and into adjoining surface and groundwater flow systems constitutes an ever-increasing environmental problem of truly monumental dimensions. The major problem is that it is difficult, in many instances almost impossible, to contain or curtail AMD. It is the most persistent and, unfortunately in many instances, NONPOINT source pollution problem of mining regions. There is no typical acid mine water, the ferric to ferrous ion ratio may vary, and several other ions such as silica, aluminum, calcium, or magnesium may be present in significant quantities in addition to the dissolved other toxic heavy metals. Untreated AMD pollutes receiving streams and aquifers. The impact on the environment can be severe, leading to a virtual disappearance of aquatic life, a low pH of the water, and a coating of river bottoms with a layer of rust-like particles. Milling operations include the comminution and concentration of the ore. Waste rock and the process water from these operations are usually discharged into large basins called "tailing ponds." Although most of the heavy metals are present in the form of suspended solids that settle to the bottom of the basins, the overflow from these ponds still contains low but significant concentrations of toxic metals [52]. Furthermore, the dams of these ponds are often constructed from waste rock; thus seepage from the tailings contributes to AMD [53].

2.3.1.3 Radioactive AMD

The uranium mining industry produces a large volume of low-level radioactive waste material, which, following milling, extraction, and neutralization processes, is deposited in extensive tailings impoundments. As mentioned, in many cases the ore bodies are associated with metal sulfides such as pyrite, marcasite, and pyrrhotite, which are not desired products and are released to the tailings as part of the mill wastes. Upon weathering, these metal sulfides are readily oxidized, producing AMD conditions from these tailings piles, with subsequent leaching of the tailings material resulting in highly acidic pore water containing significant concentrations of iron, sulfate, heavy metals, and trace radionuclides. The migration of such poor-quality tailings water by either surface runoff or subsurface groundwater can lead to serious deterioration in the quality of adjacent natural water systems.

The mining industry is thus faced with the difficult task of devising long-term abandonment schemes that minimize pyrite oxidation and prevent the release of contaminants to the environment. These schemes should be cost-effective and should require very little future maintenance or monitoring. Some radionuclides found in water, particularly uranium, thorium, and radium, originate from natural sources, particularly leaching from minerals. The levels of radionuclides found in water typically are measured in units of picoCuries per liter (one picoCurie is equal to 2.2 disintegrations per minute). The U.S. Public Health Service specifications stipulate that water supplies should not contain more than 3 picoCuries per liter of naturally occurring radium-226. The uranium tailings will require proper management technology, without much human involvement in the form

TABLE 2.4

Persistent Radionuclides in Water

Radionuclide half-life reaction, source, comment from reactor, and weapons fission:

Strontium-90: 28 years Fission products radioisotopes of highest significance

Cesium-131: 30 years because of their high yields

Iodine-131: 8 days and biological activity

Cobalt-60: 5.25 years from nonfission neutron reactions in reactors

Iron-55: 2.7 years $^{56}Fe(n,2n)^{55}Fe$, from high-energy neutrons acting on iron in hardware

Manganese-54: 310 days from nonfission neutron reactions in reactors

Plutonium-239: 24,300 years $^{238}U(n,g)^{239}Pu$, neutron capture by uranium

Barium-140: 13 days these fission products

Zirconium-90: 65 days are listed here

Cerium-141: 33 days in generally decreasing

Strontium-89: 51 days order of

Ruthenium-103: 40 days fission yield

Naturally occurring from ^{238}U series:

Radium-40: 1,620 years diffusion from sediments, atmosphere

Lead-210: 21 years ^{226}Ra—6 steps—^{210}Pb

Thorium-230: 75,200 years ^{238}U—3 steps—^{230}Th produced *in situ*

Thorium-234: 24 days ^{238}U—^{234}Th produced *in situ*

Naturally occurring and from cosmic radiation:

Carbon-14: 5,730 years $^{1}4N(n,p)^{14}C$,* thermal neutrons from cosmic or nuclear weapon sources reacting with N_2

Silicon-32: 300 years $^{40}Ar(p,s)^{32}Si$, nuclear splitting of the nucleus of atmospheric argon by cosmic-ray protons

Potassium-40: 1.4×10^9 years 0.0119% of natural potassium

of continuous treatment and monitoring, in limiting the release of contaminants to the environment. The mining and milling of uranium-bearing ores results in four types of waste: waste rock, mine water, process effluents, and solid wastes (tailings). Liquid effluents appear to be the most serious waste disposal problem from operating mines and mills, and the chemicals added to the milling process are of particular concern. The solid wastes represent a problem on account of both their magnitude and their radioactivity. On abandonment, the containment of the radioactivity is of utmost concern because of the hazards of long-lived radioisotopes in liquid effluents. A brief list of radioactive elements and their respective half-lives is presented in Table 2.4.

2.3.1.4 Treatment of AMD

The control and prevention of radionuclides and heavy metals contamination of surface and groundwater is best made during the ore treatment procedure and prior to waste and effluent disposal to tailings impoundments. Ideally, due to their persistent nature in the environment and toxicity, the ultimate treatment of heavy metals in effluents would be either their very stable mineralization and deposition or a complete recovery (and recycle) [57]. However, as these alternatives could hardly be cost-feasible and reasonably expected, carefully crafted regulatory limits on the discharge of metals into the environment need to be crafted and adhered to with strict enforcement.

2.3.1.4.1 Lime Addition

Upon disposal of effluent waste to tailings impoundments, lime, and barium chloride are added to reduce the acidity and promote the chemical precipitation of heavy metals and radionuclides. However, this treatment mode is a nonpermanent solution because tailings are subject to continuous infiltration by rainfall and an oxygen supply that increases the acidity of water due to sulfide oxidation and promotes the formation of heavy metal and radionuclide leachates. Lime therefore has to be added continuously; thus increasing sludge volumes with time, sludge volumes produced by lime addition will be greater than the tailing waste volumes! Meanwhile, the conventional method of treatment of AMD with lime or limestone demands more landfill space for waste disposal and the leaching of toxic metals into the soil and groundwater may threaten the environment [54]. This is clearly not the best long-term treatment method to abate AMD.

2.3.1.4.2 Vegetation Cover

The most promising long-term compromise solution has been establishing a vegetation cover directly on the tailings material [58]. The vegetative cover on tailings has provided greater surface stability by controlling erosion and has improved the general aesthetics of the area. Its overall effect on acid generation and tailings area water quality, however, cannot be established to date. There has been no evidence of improvement in the quality of water that leaves a tailings area even though a site may have been vegetated for the last 10 years.

2.3.1.4.3 Wet Barriers and Wetlands

Wet barriers, such as water and/or wetland cover on tailings, are believed to be effective in preventing acid generation by cutting off the oxygen supply to the tailings. The anoxic conditions so produced further support the growth of anaerobic heterotrophs such as sulfate reducers, which, with the breakdown of sulfates, produce hydrogen sulfide, thereby precipitating dissolved metals as sulfides [59]. Wet barriers are therefore artificial whereas wetlands are natural. Analysis of algae from a wetland that was removing manganese from mine water demonstrated phenomenal plant uptake, and Mn concentrations as high as 56,000 ppm (dry weight) were recorded. Iron and manganese-oxidizing bacteria are also very active in these acidic wetlands; so much so that some researchers believe they are the most critical aspect of metal removal in cattail marshes [60]. In the organic-rich substrate, other bacteria are active. *Desulfovibrio desulfuricans*, for example, converts the sulfate component of mine water into hydrogen sulfide. This, in turn, reacts with dissolved metals, adsorbed metals, and precipitated metals to form insoluble metal sulfides. This is the eventual fate of all the metals removed from acid water, as dead vegetation sinks to the bottom and is replaced by new vegetation. Absorption by the organic substrate (especially peat) can also be very high. Finally, there are geochemical removal mechanisms occurring in the wetland that may be significant. For example, the cattail marshes that are most successful in removing manganese all have an incorporated layer of limestone beneath the organic substrate. This produces an environment of near-neutral pH and high carbon dioxide concentrations (due to neutralization reactions) and may result in the precipitation of manganese carbonate. Artificial wetlands constructed for the treatment of wastewater or AMD have different design considerations than those for the control of flood, storm, or wildlife habitat management.

2.3.2 Metal Finishing and Surface Treatment Operations

Surface finishing consists of various chemical and physical processes, including electroplating, that change the surface of a product or enhance its appearance, increase its corrosion resistance, or produce surface characteristics essential for subsequent operations. The major surface finishing operations consist of pretreatment, electroplating, electroless plating, anodizing, chromating (conversion coating), cyanide hardening, and quenching. The most prevalent surface finishing operations are electroplating, anodizing, and hot dip galvanizing. Table 2.5 gives a breakdown of metal finishing by specific industry.

Surface treatment is applied mainly to metal parts, but also to certain synthetic materials. It involves the following:

- a preliminary preparation of the surface (degreasing and pickling),
- a coating using electroplating,
- a coating by chemical means.

TABLE 2.5
Surface Finishing Market Breakdown

Industry	% of Market (Canada, 1983)
Automobile parts	26.0
Steel strip mills	14.0
Hardware	12.0
Electrical appliances	10.0
Wire goods	10.0
Plumbing fixtures electrical	6.0
Electrical equipment	5.0
Furniture	5.0
Pole hardware and heavy steel	5.0
Electronics	4.0
Engine and worn parts	2.0
Hollowware and flatware	0.5
Jewelry	0.5

Automobile parts: In recent years the use of electroplated metals for decoration in the automobile industry has fallen drastically. The reasons for this change are a trend toward smaller cars and a designer's preference for materials other than nickel and chrome. Bumper manufacturers are the major clients for decorative surface finishing services. Although the trend to soft plastic coatings and paint finishing on bumpers has reduced this market, the demand in the United States for this plating service is larger than the present capacity. Consequently, increased nickel and chrome plating in Canada is expected for the next few years. Demand for functional finishes has increased with pressure to reduce corrosion and extend the life of vehicles. As a result, zinc plating will continue to grow at about 10%–20% per year as usage extends to more automobile parts.

Steel strip mills: Cold-rolled steel in strip form is plated with either zinc or tin. The zinc-plated strip is used in the fabrication of parts that require added corrosion protection on interior surfaces. Tin-plated strips are used primarily in can manufacturing; however, aluminum is replacing tin-plated cans in more and more areas.

Hardware: The demand for hardware products comes directly from the housing and construction industry and the quantity of surface finishing required fluctuates accordingly. The types of surface finishes required are nickel-chrome, zinc, brass, and bronze.

Electrical appliances: The major appliance industry, including manufacturers of stoves and refrigerators, has a constant requirement for surface finishing, primarily for decorative nickel-chrome plating on handles and trim. No increase in the surface finishing demand is expected from this industry.

Wire goods: These products include display racks, shelving, and shopping carts. The typical surface finish is electroplated nickel, chrome, and brass. Applications for wire goods are numerous and growth potential is in the order of 5%–8% per year.

Plumbing fixtures: These products include taps and bathroom and kitchen fixtures. Base metals are either zinc- or chrome-plated. Demand is related to the housing and construction industries.

Electrical equipment: Products included in this group are service boxes, conduit pipes, and transformer parts. Zinc and tin electroplating is used. Surface finishing production rates for these goods fluctuate with the housing and construction industries.

Furniture: Steel furniture is usually plated with nickel and chrome, nickel and brass, or brass only. Growth in this area is highly variable and difficult to predict as it depends on regional trends and designer's preferences.

Pole hardware and heavy steel: Products such as highway guard rails, transmission towers, and some heavy steel structures used in construction are included here. Hot dip galvanizing is used since the parts are too big to electroplate.

Electronics: The production of printed circuit boards requires primarily copper plating but nickel, gold, tin, and tin-lead processes are also used. While production rates for printed circuit boards are expected to increase over the next few years, the reduction in the surface areas actually plated and the trend to reduced water usage and metal recovery suggest that pollutant generation rates from this industry will not increase.

Engine and worn parts: Parts requiring surface finishing come from pumps, diesel engines, gasoline engines, paper mill rolls, etc. The parts vary in size and require heavy deposits of chromium to restore the original dimensions and provide a durable surface coating.

Hollowware and flatware: Hollowware products include coffee pots, teapots, ice bowls, cream and sugar pots, gravy boats, and flower holders. These products are often plated with silver using a silver cyanide bath. There is little growth expected in this market. Flatware includes tableware such as knives, forks, and spoons. These products are made either from stainless steel or from cold-rolled steel, which is plated in a silver cyanide solution. Demand for the product is primarily dependent on population growth and the formation of new households; consequently, little growth is expected in the future.

Jewelry: All costume jewelry falls under this heading. The typical surface finish is a flash-coating of precious metals such as gold or rhodium on a nickel base. Annual growth is expected to be about 1%–3%.

Effluents from a finishing industry must be separated into three categories:

1. concentrated spent baths,
2. wash waters containing an average concentration of substances likely to precipitate (soaps, greases, and metallic salts),
3. dilute rinse water that may be recyclable after treatment.

To secure and facilitate treatment, acidic and chromate-laden effluents must be separated from alkaline and cyanide effluents. The average composition of wastewaters from a surface finishing plant is presented in Table 2.6.

The pollutants may be divided into several families:

- toxic pollutants such as CN^-, Cr(VI), and F^-,
- pollutants that change the pH, that is, acidic or basic substances,
- pollutants that raise the level hydroxides, carbonates, and phosphates,
- pollutants covered by a particular regulation, S^{2-} and Fe^{2+},
- organic pollutants (EDTA, etc.), especially from degreasing.

All the constituents of baths are found in the rinse water, which may also contain metallic ions (Ni, Cr, Cu, Zn, Sn, Cd, Au, Ag, Pb, Fe, and some others) dissolved from the parts treated. The discharge conditions standards vary greatly, depending on the country, and are rapidly becoming increasingly strict as to pollution concentration as well as the flow of rinse water. A large portion of these effluents is received by municipal sanitary sewers (78%). The heavy metals are generally removed

TABLE 2.6

Principal Constituents of Untreated Effluents from Major Metal-Finishing Processes

Composition of Wastewater from Surface Finishing Industries

Species	Plating on Steel	Plating on Zinc	Plating on Brass	Plating on Plastic	Anodizing	Concentration (mg/L)
Fe^{2+}	x					1–10
Cu^{2+}	x	x	x	x		5–50
Ni^{2+}	x	x	x	x	x	2–15
Cr^{6+}	x	x	x	x	x	10–120
Cr^{3+}	x	x	x	x	x	0.1–1
Zn^{2+}	x	x				10–50
Cd^{2+}	x					10–50
Sn^{2+}	x			x		0.1–20
CN^-	x	x	x			1–50
SO_4^{2-}	x	x	x	x	x	15–25
Cl^-	x	x	x			1–250
CO_3^{2-}	x	x	x	x		10–50
Si^{2-}	x	x	x	x		30–50
PO_4^{3-}	x	x	x		x	20–50
Organics	x	x	x	x		0.1–1

Source: Environment Canada, Environmental Protection Service. *Overview of the Surface Finishing Industry: Status of the Industry and Measures for Pollution Control.* EPS 2/SF/1, Ottawa, ON, Canada, p. 43, 1987.

Note: The cross x is stated whenever the species was detected in the water effluent.

TABLE 2.7

France Guidelines for Metal-Finishing Liquid Effluents (mg/L)

Cr (VI)	0.1	Zn	5.0
Cr (III)	3.0	Fe	5.0
Cd	0.2	Al	5.0
Ni	5.0	Pb	1.0
Cu	2.0	Sn	2.0

by the activated sludge process and become a part of the resulting biological sludge. The beneficial properties of the sludge, which is often used as a fertilizer in agriculture, are then limited because of the heavy metal content [61].

The general guidelines to limit pollutant discharges from surface finishers depend on the laws of each country. In France, for example (decree of November 8, 1985), the metal concentration of $(Zn+Cu+Ni+Al+Fe+Cr+Cd+Pb+Sn)$ combined should be less than 15 mg/L. In particular, the thresholds summarized in Table 2.7 should not be exceeded (mg/L)

The Canadian federal guidelines [62] for metal-finishing liquid effluents presented in Table 2.8 provide a baseline standard for water discharges.

A quick comparison of Tables 2.6–2.8 shows that the metal concentrations in untreated plating shop effluents may be as much as 120 times higher than those permissible by law. Clearly, untreated metal-plating effluent streams may contribute a large amount of metal pollutants to the environment. For example, chromium is most likely to be found at high concentrations in the effluent streams.

TABLE 2.8
Federal Guidelines for Metal-Finishing Liquid Effluents

Metal	Cu	Zn	Cd	Cr	Ni	TSS
Maximum total concentration (mg/L)	1.0	2.0	1.5	1.0	2.0	30.0

Source: Environment Canada, Environmental Protection Service. *Overview of the Surface Finishing Industry: Status of the Industry and Measures for Pollution Control.* EPS 2/SF/1, Ottawa, ON, Canada, p. 43, 1987.

2.3.2.1 A Typical Electroplating Process

Electroplating is the electro-deposition of an adherent metallic coating upon an electrode, which is the workpiece, for the purpose of obtaining a surface with properties or dimensions different from those of the basic metal. These properties may include improvement of appearance, corrosion protection, wear resistance, and so on. The operation takes place in aqueous solutions containing the metal ion to be plated. The workpiece is cathodic, and in most instances, the metal ion is constantly replenished from an anode containing the metal. A notable exception is chromium where the anode is insoluble and metal ions are replenished by the addition of chromic acid. Electroplating must be preceded by cleaning and activating operations, and a typical sequence of operations involved would be the following:

1. Vapor degrease or soak clean in an emulsion or detergent cleaner.
2. Spray clean in a detergent cleaner.
3. Electroclean in an alkaline cleaner. (The function of electrocleaning is to remove remaining soil and to make the surface chemically active. The operation takes place in an alkaline solution and the work may be anodic or cathodic.)
4. Sulfuric acid dip.
5. Electroplate.
6. Electroclean.
7. Sulfuric acid dip.
8. Second electroplate. Rinsing would follow each process step except 1.

Surface finishers produce and discharge a variety of waste streams, including process wastewaters, spent process solutions, sludges, and air emissions. For proper plating to occur, the parts must be clean and free of contamination from previous processes. Considerable quantities of raw water are thus used to rinse the parts. Depending on the process for which the rinsing takes place, the wastewater produced may be acidic or alkaline and may contain particular metals or combinations, solvents or cleaning solutions, and/or particulates. Another source of wastewater contamination comes from floor drains. Often, through poor housekeeping, the plating solution is allowed to drip as the rack or barrel is passed from tank to tank, and this solution subsequently finds its way into the plant sewer system. Several waste streams, including spent process solutions and sludges, are considered hazardous. Any substance or mixture being discarded is considered hazardous if it is flammable, carcinogenic, toxic, corrosive, and explosive or meets other criteria developed by a (Canadian) federal-provincial working group. Spent process solutions include the following:

a. Acidic waste from pickling, etching, bright dipping, and electropolishing.
b. Alkaline cleaning baths and electrocleaning baths.
c. Solvent degreasing waste.
d. Salt bath descaling solution.
e. Spent baths themselves when they can no longer be rehabilitated.

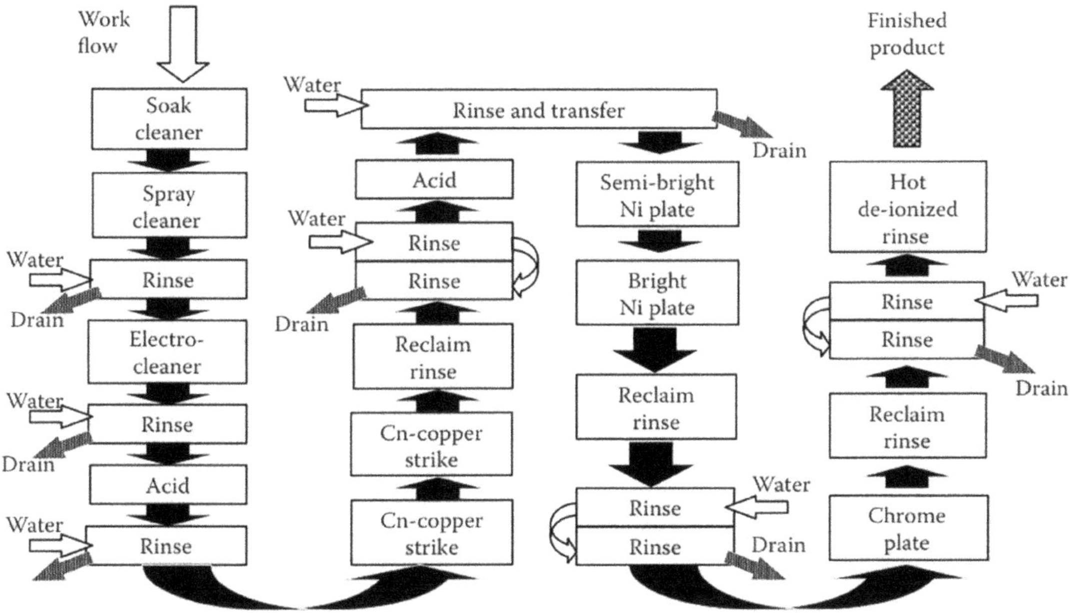

FIGURE 2.9   A schematic flow sheet of a copper-nickel-chrome plating system without waste treatment.

The acidic waste contains a high level of dissolved metals, oils, and suspended particles. The large number of different compounds and additives in cleaning solutions often make recovery of metals or chemicals from the spent solutions impractical. Solvent waste contains soil and oily buildup. Spent plating and coating solutions are generated during electroplating, electroless plating, hot dip coating, anodizing, and chemical conversion coating operations. These wastes, and the related rinse waters, may be acidic or alkaline and may contain hexavalent and trivalent chrome, cyanide, and other toxic compounds.

A number of metal-finishing operations leave sludges on the bottoms of plating bath tanks. Large amounts of sludge are also formed during cleaning, painting, and effluent treatment. Sludge from effluent treatment is only 1%–5% solids and can be dewatered to reduce its volume. Sludges usually contain hazardous materials, which could upset the municipal treatment plant if discharged into the sewer. Figure 2.9 shows a schematic flow sheet of a copper-nickel-chrome plating system without waste treatment. One should note the numerous opportunities that the wastewater has to pick up pollutants.

2.3.2.2   Future Trends in Electroplating

Since pressures to reduce impact and liability will continue, if not increase, for the foreseeable future, prudent firms that expect to be using conventional processes are working toward optimizing them. Two popular trends include:

1. **Approaching zero exposure:** isolating employees from contact with materials or effluents in process operations, thus approaching zero risk conditions.
2. **Approaching zero discharge:** maximizing material utilization and recovery, thus minimizing the impact on the environment from wastewater, air emissions, and concentrated waste streams (spent process solutions and treatment sludges and solids).

The techniques that firms employ to achieve near-zero exposure and discharge from wet processes depend on the specific process and production situation but can involve:

- Enclosing process lines (a common practice in the printed wiring board and semiconductor industries).
- Reducing and recovering drag out.
- Using process solution and rinse purification and recycling technologies.
- Using racking and fixturing off-line to reduce operator exposure and using configurations that optimize process efficiency and yield and minimize waste.
- Using process automation and control systems to optimize material usage and yield.
- Modeling processes for optimization.

Examples of processes using metals with environmental health and safety (EH&S) concerns that can approach near-zero discharge include:

- Chromium plating.
- Chromic anodizing.
- Nickel plating.
- Electroless nickel plating.
- Cadmium plating.
- Lead plating.
- Tin-lead plating.

Hundreds of surface finishing facilities have already implemented process optimization projects that have resulted in near-zero discharge. The improvements have typically yielded cost savings, since the optimized processes exhibit better performance, along with lower material usage and reduced waste generation. In addition, a small fraction of existing surface finishing facilities have enclosed automated process lines with ventilation and air emissions control systems that provide near-zero exposure risk. Industries such as printed wiring board manufacture provide examples where such systems have been successfully implemented.

2.3.3 LEATHER TANNING PROCESS

Leather tanning is the process of converting raw hides or skins into leather. Hides and skins have the ability to absorb tannic acid and other chemical substances that prevent them from decaying, make them resistant to wetting, and keep them supple and durable. The surface of hides and skins contains the hair and oil glands and is known as the grain side. The flesh side of the hide or skin is much thicker and softer. The three types of hides and skins most often used in leather manufacture are cattle, sheep, and pigs. Tanning is essentially the reaction of collagen fibers in the hide with tannins, chromium, alum, or other chemical agents. The most common tanning agents used are trivalent chromium and vegetable tannins extracted from specific tree bark. Alum, syntans (man-made chemicals), formaldehyde, glutaraldehyde, and heavy oils are other tanning agents. The process of leather tanning generates wastewater effluents containing chromium, which, when emitted at high concentrations, can be toxic to the environment. Public awareness of the dangers of harmful effluents has grown during the past decade and the need for stricter environmental regulations has forced many process-related industries to dramatically refine their dangerous polluting effluents. The industry's growth in the Western world has stagnated since the 1980s due to the influence of more affordable synthetic leather substitutes. Western regulations regarding wastewater controls are more stringent than those of developing countries and this translates into higher relative production costs for the tanners in developed countries.

2.3.3.1 Description of the Chromium Tanning Process

The modern process of chrome tanning dates back to its discovery by Federick Knapp in 1958. The purpose of chrome tanning is to transform a hide or skin into a finished leather product that is insusceptible to putrefaction. By varying the specifics of a process, it is possible to obtain leather with the required grain, temper, break, and strength. Specifically, tanning is the reaction of the collagen protein fibers of the hide with chromium. The most widely believed mechanism is that of a coordination of the protein carboxyl groups with a chromium complex [63]. The tanning process can be divided into two divisions:

1. **Beamhouse:** Hides are first trimmed and soaked to remove salt and other solids and to restore moisture lost during curing. After soaking, the hides are fleshed to remove excess tissue, impart uniform thickness, and remove muscles or fat adhering to the hide. Hides are then dehaired to ensure that the grain is clean and the hair follicles are free of hair roots. Liming is the most common method of hair removal, but thermal, oxidative, and chemical methods also exist. The normal procedure for liming is to use a series of pits or drums containing lime liquors (calcium hydroxide) and sharpening agents. Following liming, the hides are dehaired by scraping or by machine. Deliming is then performed to make the skins receptive to vegetable tanning, which is a long-drawn process (~3 weeks) starting with a low chemical concentration that is gradually increased as tannage proceeds.

2. **Tanyard:** Chrome-tanned leather tends to be softer and more pliable, has higher thermal stability, is very stable in water, and takes less time to produce than vegetable-tanned leather. Almost all leather that is made from lighter-weight cattle hides and from the skin of sheep, lambs, goats, and pigs is chrome-tanned. The first steps of the process (soaking, fleshing, liming/dehairing, deliming, bating, and pickling) and the drying/finishing steps are essentially the same as vegetable tanning. However, in chrome tanning, the additional processes of retanning, dyeing, and fat liquoring are usually performed to produce usable leathers, and a preliminary degreasing step may be necessary when using animal skins, such as sheepskin.

Chrome tanning in the United States is performed using a one-bath process that is based on the reaction between the hide and a trivalent chromium salt, usually a basic chromium sulfate. In the typical one-bath process, the hides are in a pickled state at a pH of 3 or lower, the chrome tanning materials are introduced, and the pH is raised. Following tanning, the chrome-tanned leather is piled down, wrung, and graded for thickness and quality, split into flesh and grain layers, and shaved to the desired thickness. Grain leathers from the shaving machine are then separated for retanning, dyeing, and fatliquoring. Leather that is not subject to scuffs and scratches can be dyed on the surface only. For other types of leather (i.e., shoe leather), the dye must penetrate further into the leather. Typical dyestuffs are aniline-based compounds that combine with the skin to form an insoluble compound.

Fat liquoring is the process of introducing oil into the skin before the leather is dried to replace the natural oils lost in beam house and tanyard processes. Fat liquoring is usually performed in a drum using an oil emulsion at temperatures of about 60°C–66°C (140–150°F) for 30–40 minutes. After fat liquoring, the leather is wrung, set out, dried, and finished. The finishing process refers to all the steps that are carried out after drying. Chromium (trivalent) tanning agents are added in the tanning step. Any unfixed tanning agents are removed from the leather in the wringer. The products at this point are referred to as blue hides. Tanneries often perform only the beam house and tanyard processes and sell their "blue" hides to retanners.

2.3.3.1.1 Beamhouse Operations

Cured hides received from the market must undergo pretreatment before they can be processed into leather. The objective of beamhouse operations is to prepare the hides. Figure 2.10 shows a schematic of the process. In the side-and-trim step, the hides are cut into two sides and any unwanted sections of the hide are trimmed off. The wash-and-soak step involves soaking of the hide in water for 8–20 hours. The hides absorb water to make up for the moisture lost in the curing process. Washing removes nonfibrous proteins, dirt, salt, blood, and manure from the hide. Fleshing is the mechanical removal of excess flesh, fat, and muscle from the hide. This is done in cold water to ensure that the fat remains congealed. Alkaline chemicals (Na_2S and $NaOH$) are then added to dissolve any hair and destroy hair roots.

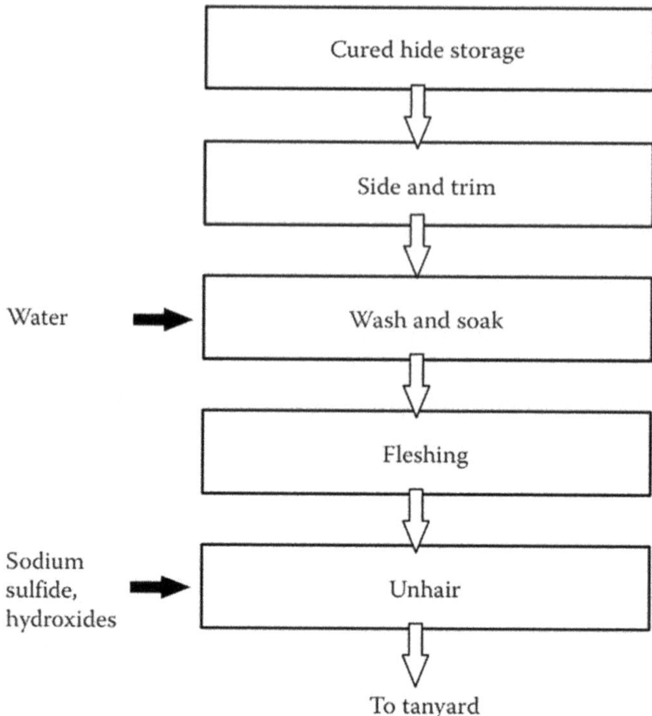

FIGURE 2.10 Leather tanning—a box schematic diagram for the beamhouse process.

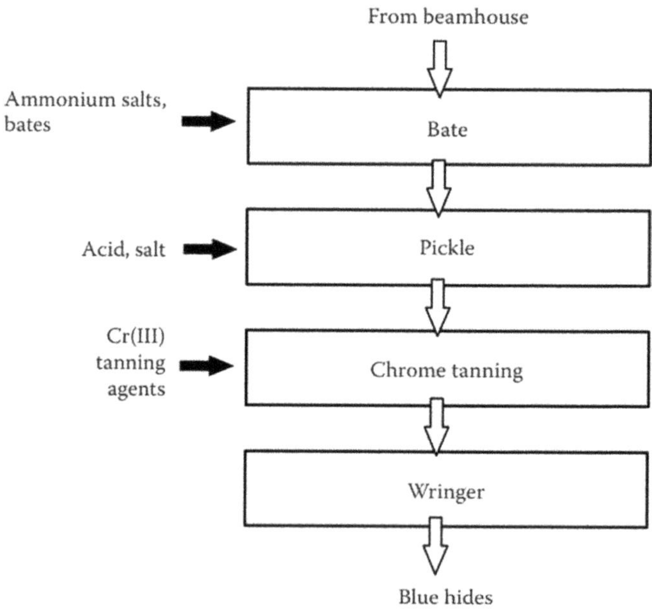

FIGURE 2.11 Leather tanning—a box schematic diagram for the tanyard process.

2.3.3.1.2 *Tanyard Process*

A schematic for the tanyard process is shown in Figure 2.11. Sulfated/chlorinated ammonium salts are added in the bating step to solubilize any alkaline material present from the beamhouse process. Bates (enzymes) are added to further destroy hair roots and pigments and to prepare the collagen

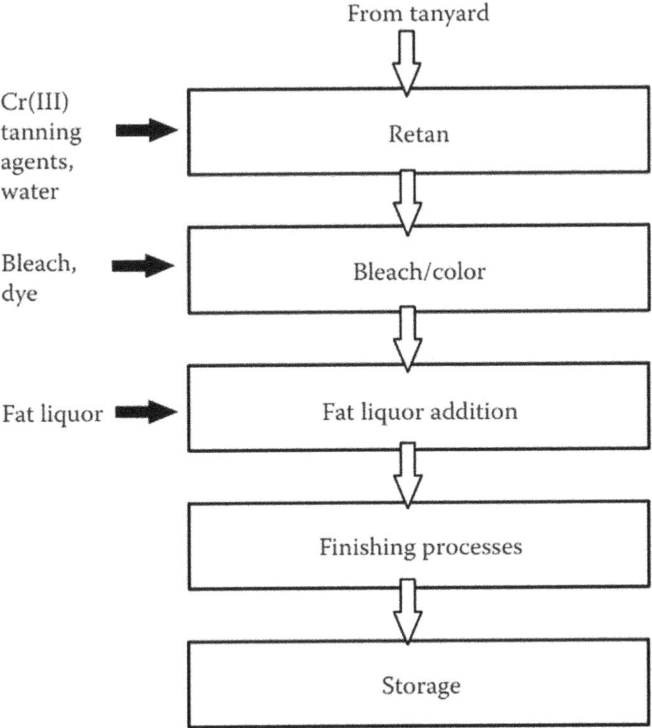

FIGURE 2.12 Leather tanning—a box schematic diagram for the retanning process.

fibers for their reaction with the tanning agents. An acidic medium is required for chrome tanning; hence the addition of H_2SO_4 in the pickling step ensures that all alkaline material has been washed away. Salt is also added in the pickling step as it prevents "acid swelling" by reducing excess moisture. Chromium (trivalent) tanning agents are added in the tanning step. Any unfixed tanning agents are removed from the leather in the wringer. The products at this point are referred to as blue hides.

2.3.3.1.3 Retanning

Retanning is necessary to convert the blue hides to leathers suited for specific use. A schematic for the process is shown in Figure 2.12. Chromium is usually used as the retanning agent. The leather quality is often upgraded by the introduction of phenols and other complementary tanning agents. The leather is then bleached and dyed as desired. The addition of fat liquor replaces any natural oils lost in the beam house and tanyard processes. Final finishing includes drying, conditioning, buffing, and plating.

2.3.3.2 Wastes Generated in the Chromium Tanning Process

The beamhouse process normally accounts for about 40% of wastewater volume in a tannery [64]. Beamhouse wastes have a high pH of 10–12 and contain high amounts of proteins and sulfides. Nitrogen, BOD, and TSS are also very prevalent in beamhouse wastes. Wastes from the bating step have high ammonia concentrations and also contain proteins and dissolved hair. The pickling step generates a highly acidic waste (pH 2.5–3.5) containing salts. Toxic levels of trivalent chromium at elevated temperatures characterize the acidic waste from the tanning and retaining steps. The retanning process also yields wastes containing dyes and sulfonated oils at elevated temperatures. Tanyards and retanning effluents have considerably high levels of COD. These elevated COD levels are caused by high ammonia concentrations from the bating step. In addition to nitrogenous COD demand, if a phenol retanning step is used in the process it may account for up to 30% of the total

COD [65]. The retanning finishing step also contributes to higher levels of COD. This is due to the introduction of organic dyes, sulfonated oils, pigments, and coatings.

2.3.3.2.1 A Leather Tannery Wastewater Case

Production levels, seasonal variations, process variations, and the batch nature of tannery operations result in large variations in the wastewater sampled parameters. A medium-sized tannery may typically process 500–1,000 kg of hides in one batch. About 50,000–150,000 L of water are used for each batch, thus requiring approximately 200,000–500,000 L of water per day. Table 2.9 shows a typical case of raw effluent composition.

It is evident from the above table that extremely high chromium concentrations are present. This is due to the washing of the chromium from the original tanning step. It can be seen that high concentrations of undesirables are present. Table 2.10 shows the parameters for river outfall. High levels of chromium, phenols, COD, BOD, and oil and greases are present following a simple settling pretreatment. The highly acidic nature of the effluent (pH 3.3) should be raised to a more moderate level. High levels of COD exert a heavy load on the receiving water body and require special treatment to reduce them. The case is similar for chromium content.

TABLE 2.9

Original Wash Sampling Results[a] (pH 5.0)

Parameters	#1	#2	#3
Total solids	5,444	—	2,520
TFS	1,480	140	127
Oil and grease	0.42	88.5	94.7
BOD	536	120	410
COD	3,047	1,445	1,032
Pb	1.1	7.5	7.5
Cr	122	140	120
Phenols	0.14	2.5	1.95

[a] All results are in mg/L (ppm).

TABLE 2.10

River Outfall Sampling Results[a] (pH 3.3)

Parameters	#1	#2	#3
Total solids	1,540	—	1,300
TFS	—	88	165
Oil and grease	0.01	44	98.7
BOD	454	238	904
COD	2,500	1,875	3,016
Phosphates	0.25	3.7	—
Pb	0.25	7.75	7.5
Cr	91	40	130
Phenols	0.52	1.55	0.40
NH_4	—	0.91	1.61

[a] All results are in mg/L (ppm).

2.3.3.3   Effluent Treatment

There are various options for reducing chromium effluent concentration in a leather tanning process. Chrome recycling within the process can account for large economic savings with respect to the amount of chemicals required and will allow for a less concentrated chromium effluent. Ideally, direct recycling could account for reductions of 40% in "suspended" solids, 50% in BOD, and 80% in toxicity [65].

2.3.3.3.1   Changing Process Chemicals

The following are three forms of chemical changes capable of improving effluent quality: organic acid pickling, "low-use" chrome compounds, and synthetic tanning agents. As mentioned in the above description of the chromium tanning process, the pickling step usually employs sulfuric acid to prepare the hides for chrome tanning. Substitution of H_2SO_4 for organic acids can substantially reduce (approximately 20%) the chrome required in the tanning process. Organic acids show a reduced affinity for collagen fibers, thus enabling acid penetration and chrome fixation to occur more rapidly [65]. Employing low-chrome compounds is the most widely used method for chromium effluent reduction. These low-chrome compounds are commercial tanning agents (e.g., Chromotan, Blancoral, and Baychrome 2403) capable of producing quality tanned leather at lower chrome concentrations. Low-chrome compounds combined with low float (liquor-to-hide ratio) techniques are capable of lowering chrome and water consumption [65]. Substitution of chromium tanning complexes for synthetic tannins would totally eliminate chromium from the process. However, this is not possible because product quality would suffer. The most common of the new synthetic tannins are those of complex aluminum salts. Research is presently being done to find new chrome-free tanning agents consistent with a fine quality of leather [65].

For example, a leather tanning process has been developed in which animal skins are treated with a tanning agent comprising a mixed complex of aluminum(III) ions and titanium(IV) ions, and as a masking compound a salt of a polyhydroxymonocarboxylic acid. Titanyl sulfate solution, prepared by the dissolution of hydrous titanium oxide in sulfuric acid, can be mixed with aluminum sulfate in the desired proportions, treated with a masking agent and basified to the appropriate acidity, and then used in the "cleaner" tanning process.

2.3.3.3.2   Treatment Improvements

Process effluent streams containing alkaline materials are segregated from acid streams (containing chromium). High sulfide concentrations in the alkaline streams are first targeted by screening and then by sulfide oxidation. The oxidation takes place in an aerated tank and uses manganese sulfate as a catalyst. Oxidation may also be forced using peroxide addition. The effluent from the oxidation step is mixed with the segregated acid effluent in an equalization tank. Chromium precipitation and subsequent recycling is another means of reducing chromium in the effluent. This is done by raising the pH to an optimum level (8.5) using a hydroxide. The chromium precipitates as chromium hydroxide $(Cr(OH)_3)$. Coagulants and flocculants are also added to aid in substrate removal. The hydroxide precipitate is then filtered and redissolved in sulfuric acid to form chrome sulfate, which is subsequently recycled [65]. The application of ion exchange resins to remove undesirable ions from wastewater (i.e., chromium) has to be carefully considered in combination with appropriate pretreatment since the organic components in wastewater can seriously affect resin performance. Metal biosorption metal removal/recovery effluent treatment alternative deserves special attention because of the low cost of biosorbent materials. The introduction of newer technologies in both leather tanning processes and the treatment of wastewater can greatly lower the pollution load normally posed by conventional tannery effluent disposal.

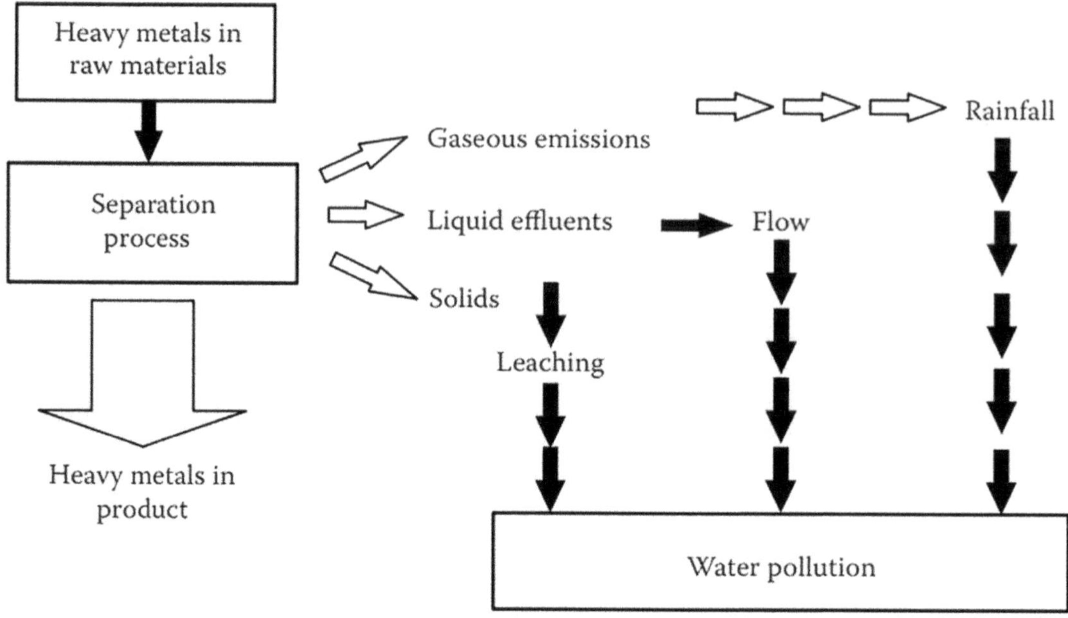

FIGURE 2.13   Heavy metal sources from metal production.

2.3.4 FERROUS METAL INDUSTRIES

The production of metals via extraction from metal ores is one of the oldest metallurgical processes. In the Roman Empire, metal production was used to make dishes, tools, and weapons. During the following centuries, more metal was required as a result of industrialization, and the metallurgical industry was born. The modem metallurgical industry can be separated into two main categories: ferrous and nonferrous. Ferrous metallurgy deals with the production of iron and its alloys, whereas nonferrous metallurgy deals with the production of other metals such as copper, nickel, lead, and zinc. Previously, there has been little concern about heavy metal emissions from the industry, owing to ignorance of the impact of heavy metals on the environment. Today, the toxicity of heavy metals is relatively well established and key pollution regulations have been legislated for the mining and metal production industries. Studies have demonstrated that the heavy metal pollution arising from metal production usually becomes water pollution. For example, heavy metals in air emissions are brought down into water during rainfall, they are leached from solids into surface streams and rivers, and effluents from industries are often discharged directly into rivers or other receiving surface water bodies (Figure 2.13).

2.3.4.1 Ferrous Metal Processing

Ferrous metal processing can be divided into three categories:

1. Primary iron and steel production.
2. Ferroalloy production.
3. Ferrous foundry production.

The process used in the production of ferrous metals is similar for all three types of production categories and can be summarized in general as illustrated by the flow sheet in Figure 2.14.

Depending on the category of industry, the furnace and the reactants involve change. Air emissions arise from the recuperation of heavy metals that are in exhaust gas. The recovery is not 100% effective; hence heavy metals are discharged into the atmosphere. The air pollution control

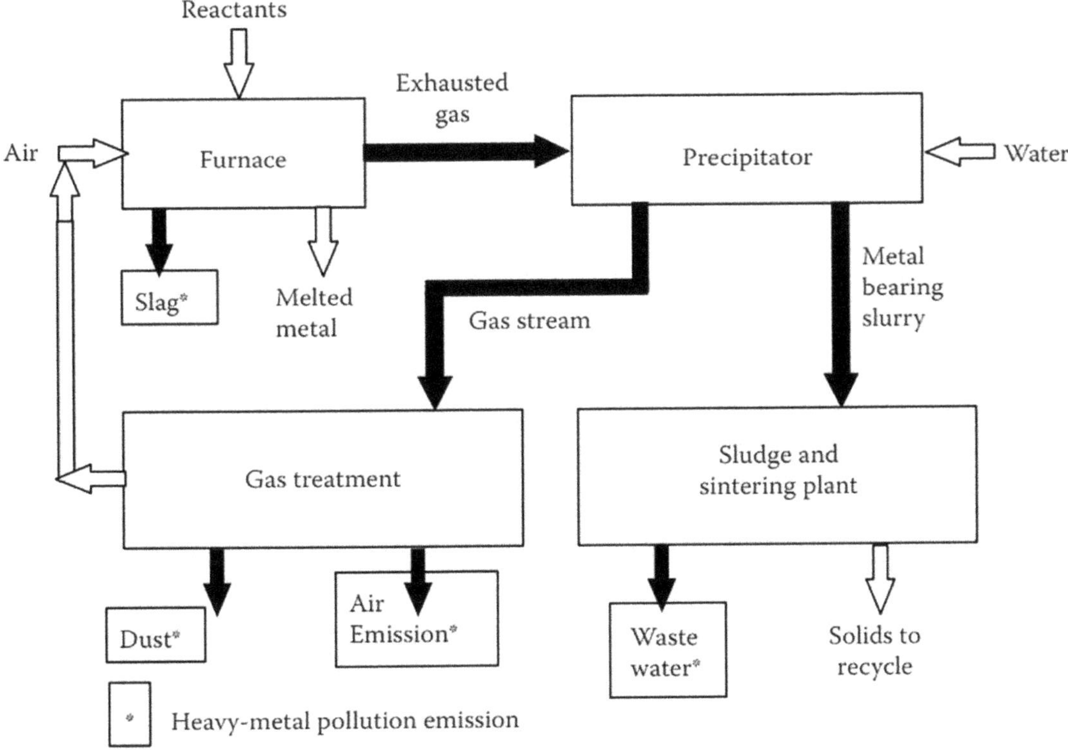

FIGURE 2.14   Flow sheet of a typical ferrous metal production plant.

TABLE 2.11

Heavy Metals in Wastewater from Ferrous Industries

	Sb	Bi	Cd	Cr	Co	Cu	Fe	Pb	Mn	Hg	Mo	Ni	Se	Zn
Primary iron	—	—	x	x	x	x	x	x	—	x	—	x	—	x
Ferroalloy	—	—	x	x	—	x	x	x	—	x	—	x	—	x
Foundries	x	x	x	x	x	x	x	x	x	x	x	x	x	—

equipment used will be different, depending on the type of furnace used. Solid wastes from ferrous processing industries are present in the form of slags and dusts. Dusts that cannot be recycled back into the process are disposed of continuously. Slag comprises waste oxides and is usually dumped in a slag dump. Generally, water pollution comes from process water, contact cooling water, and wash-down water. A sludge plant is commonly used to treat process wastewater. Table 2.11 shows the most common heavy metals that can be found in wastewaters from the three different ferrous industries.

2.3.4.1.1   Primary Iron and Steel Production

Iron is produced by a blast furnace process. Steel is produced from pig iron originating from the blast furnace [66]. In 1976, the Canadian production of steel amounted to 10,916,929 tons from basic oxygen and open hearth furnaces and 1,665,880 tons from electric arc furnaces [66]. Usually, iron and steel production is integrated into the same plant. Total emissions from primary iron and steel production are estimated at 83 tons per year of copper and 51 tons per year of nickel [66]. Heavy

TABLE 2.12

Amount of Wastewater, Air Emissions, and Solid Disposals for an Example of an Iron-Making Plant

Pollutant	Wastewater Effluents (kg/h)	Air Emissions (kg/h)	Solid Disposals (kg/h)
Cd	<0.023	—	—
Cr	<0.023	—	3
Cu	<0.023	1	7
Fe	0.517	—	422
Pb	<0.025	7.23	62
Mn	—	2.71	36
Ni	<0.023	0.41	—
Zn	<0.025	—	160

TABLE 2.13

Concentrations of Metals in Effluents from a Ferrochrome Production Plant

Pollutant	Stream #1 (mg/L)	Stream #2 (mg/L)	Stream #3 (mg/L)
Fe	<0.1	<0.1	0.3
Cr	<0.1	<0.1	<0.1
Cu	<0.01	0.02	<0.01
Pb	<0.11	<0.1	<0.1
Mg	24	59	1.5
Hg	<0.0005	<0.0005	<0.0005
Cd	0.019	<0.005	<0.005
Zn	61	0.3	0.02
Ni	0.06	0.01	<0.05

metals pollution results from three kinds of emission: water, air, and solids. Table 2.12 gives a comparison between water pollution, air emissions, and solid disposals from an iron and steel-making plant. From this table, it can be seen that solid disposals are more polluting than water discharges.

2.3.4.1.2 Ferroalloy and Ferrous Foundry Production

A ferroalloy is an alloy consisting of iron and one or more other metals. Various kinds are produced in the world, mainly ferrosilicon, ferromanganese, silicomanganese, and ferrochromium. Table 2.13 shows the composition of the effluent of a ferrochrome production plant.

Foundries range in size from small operations to large production types, which turn out tonnage casting. Some plants produce special wear- and heat-resistant castings, which are usually high alloy. Other plants produce the normal gray iron and ductile iron castings, which may weigh from a few ounces to several tons. Because of the various operation conditions (capacity and charge), no typical data have been found that could be considered representative of the industry branch.

In conclusion, in ferrous metal production, solid wastes contain the highest amounts of heavy metals followed by air emissions and liquid effluents. Solid wastes are considered relatively inert. The air emissions and liquid effluents containing heavy metals are the most damaging to the environment since these heavy metal concentrations sand discharge rates are very high and can be easily brought into the water table or accumulate directly in the aquatic environment and biota.

2.3.5 COAL-FIRED POWER GENERATION

The whole thermal-power industrial sector, including both conventional and nuclear power-generating plants, withdrew 64% of the total water intake in 1996. Next to fuels, water is the most important resource used in large-scale thermal-power production. The production of 1 kWh of electricity requires 140L of water for fossil fuel plants and 205L for nuclear power plants. Some of the water is converted to steam, which drives the generator producing the electricity. Most of the water, however, is used for condenser cooling because today's processes can only convert 40% of the fuel's energy into usable electricity. The rest is wasted. This shows the double cost of inefficient energy use: first, in the wasted energy, and then in the water required to cool the wasted heat to the temperature where it can be released safely into the environment. This requires a continuous flow of cooling water circulating through the condenser. All the cooling water is therefore returned to the environment much warmer. However, the temperature can be reduced using cooling towers and other such devices.

Nonnuclear thermal electric power-generating stations are a somewhat less known source of a large amount of metal emissions yearly. This is due to the large flowrates of water involved in the operation of the plants. On the other hand, their consumption of coal that brings in trace metal impurities is truly gigantic. A large coal train called a "unit train" maybe 2km (over a mile) long, containing 100 cars with 100 tons of coal in each one, for a total load of 10,000 tons. A large plant under full load requires at least one coal delivery this size every day. Plants may receive as many as three to five trains a day, especially in "peak season," during the summer months when power consumption is high. A large thermal-power plant (e.g., Nanticoke, Ontario, Canada) could store several million tons of coal for winter use when the supply faces interruptions.

Metals from coal, as they do not combust, end up in their gaseous, aqueous, and solid waste streams. Coal also contains low levels of uranium, thorium, and other naturally occurring radioactive isotopes whose release into the environment leads to radioactive contamination. While these substances are present as very small trace impurities, enough coal is burned that significant amounts of these substances are released. A 1,000 MW coal-burning power plant could release as much as 5.2 tons per year of uranium (containing 74 pounds of uranium-235) and 12.8 tons per year of thorium. The radioactive emission from this coal power plant is 100 times greater than a comparable nuclear power plant with the same electrical output; including processing output, the coal power plant's radiation output is over three times greater [67].

Trace amounts of mercury exist in coal and other fossil fuels [68]. When these fuels burn, toxic mercury is released, which accumulates in the food chain and is especially harmful to aquatic ecosystems. Worldwide emissions of mercury from both natural and human sources were estimated at 5,500 tons in 1995 [68]. U.S. coal-fired plants emit an estimated 48 tons annually, which is approximately one-third of all mercury emitted into the air by human activity in the United States [68]. In contrast, China's coal-fired power plants emitted an estimated 68 tons of mercury in 1999, which was about one-eighth of Chinese human-generated mercury emissions [69].

Of the three types of plants (coal-fueled, oil-fueled, and mixed), coal-fueled plants are the worst polluters due to the high-sulfur coal burned and their high water consumption rates. It is possible to significantly reduce metal emissions if appropriate and currently available technologies are incorporated into the power generation operation schemes. Unfortunately, too few plants have made any efforts to address this particular problem, which remains as overwhelming as ever.

2.3.5.1 Coal-Fired Station Types

There are two types of generic stations. The waste stream compositions of each model depend on coal composition, scrubber design, and system operation. For instance, the "Eastern" type of power plant uses high-sulfur bituminous coal fuel, while the "Western" type uses low-sulfur lignite coal fuel. Unfortunately, available sources list only the emission of iron from streams, specifically, grouping all other metals present into one category. The other metals that may be present are aluminum, arsenic, barium, beryllium, boron, cadmium, chromium, cobalt, copper, lead, lithium, manganese, mercury, molybdenum, nickel, selenium, silver, strontium, thorium, titanium, vanadium, zinc, and zirconium. However, the metals usually present in the largest concentrations are aluminum, manganese, iron, nickel, copper, zinc, and vanadium [70].

First Model (Eastern) – Typical Characteristics:

- 400 MW generating capacity,
- high-sulfur bituminous coal fuel,
- dry fly ash handling system,
- recirculating bottom ash handling system,
- once-through seawater cooling,
- limestone-based flue gas desulfurization,
- combined fly and bottom ash disposal area,
- separate flue gas desulfurization (FGD) sludge disposal area.

Periodic wastewater streams contain metals from the substances removed during the annual boiler and equipment cleaning. These streams are high in metal concentrations (Table 2.14). Some continuous wastewater streams contain high metal concentrations. These are produced in the daily operation of the station from the condenser cooling blowdown, the ash transport systems, and FGD systems. While metal concentrations in other continuous streams may be relatively low, from 1 to 10 mg/L, some of these streams have very large flowrates [70].

Periodic wastewater streams annually emit 8,980 kg of iron and 2,195 kg of other metals. Continuous streams emit 715,957 kg/y of iron and 4,091,957 kg/y of other metals. This is an annual total of almost 5 million kilograms per year of metals emitted into the environment.

TABLE 2.14

First Model (Eastern) of a Power-Generating Station—Stream Data

Source	Flow (L/y)	Iron (mg/L)	Other Metals (mg/L)
Periodic:			
Air preheater wash	500,000	7,560	1,320
Boiler fireside wash	1,000,000	1,060	935
Boiler streamside wash	600,000	6,900	1,000
Continuous:	**L/s**		
Coal pile runoff	0.6	5,320	610
Filter backwash	0.9	0.7	5.2
Ionex regeneration	1.3		
Cation		0.3	7.1
Anion		6.2	0.5
Fly/bottom ash disposal	1.5	0.7	17
Bottom ash blowdown	23	23	20
Boiler room sump	20	13	60
Condensate polisher	0.4	0.2	<0.1
Cooling water	17,000	1.1	7.5
FGD sludge disposal	1.5	0.4	20
FGD system blowdown	11	1.0	14

Source: Environment Canada, Environmental Protection Service. *Significance and Treatment of Dissolved Solids in Wastewaters from Canadian Steam Electric Stations.* Ottawa, ON, Canada, 1985.

Second Model (Western) - Typical characteristics include:

- 400 MW generating capacity,
- low-sulfur lignite coal fuel,
- combined fly and bottom ash transport with lagoon disposal,
- recirculating cooling tower,
- no FGD.

The Western stations tend to be more modern and employ more advanced technology, particularly in the cleaning of streams. Unfortunately, no general composition data are readily available for periodic wastewater streams. Continuous wastewater discharge streams show a significant reduction in metal concentrations due to the use of lower-sulfur coal and the presence of cooling towers that reduce the water flows (Tables 2.15 and 2.16). With the cooling towers, less fresh water is used and greater concentrations of metals can be removed before emission. Unfortunately, there is still the blowdown water, which is taken from the recirculating water to control the buildup of suspended and dissolved solids in the system.

The annual amount of iron released into the environment from continuous streams is 22,889 kg; the amount of other metals is 94,027 kg. This is a reduction of 96.8% and 97.7%, respectively. The majority of this reduction is due to the use of cooling towers and more efficient cleaning prior to emission through the use of reverse osmosis [70].

2.3.5.2 Generating Station Water Use

The steam electric power-generating industry is one of the major point source users of surface waters. Earlier, water was used on a once-through basis for each subsystem of the plant. However, with increased competition for available clean surface water, new designs had to be implemented for plants to reduce water intake. Plants were to maximize the internal reuse of service water streams

TABLE 2.15

Second Model of a Power-Generating Station—Stream Data

Source	Flow (L/y)	Iron (mg/L)	Other Metals (mg/L)
Periodic:			
Air preheater wash	730,000		
Boiler fireside wash	600,000		
Boiler streamside wash	500,000		
Continuous:	**L/s**		
Coal pile runoff	0.3	530	1,670
Filter backwash	0.9	0.7	5.2
Ionex regeneration	3.4		
Cation		0.3	7.1
Anion		6.2	0.5
Combined ash lagoon overflow	30	0.4	15
Boiler room sump	20	1.6	
Condensate polisher	0.4	0.2	0.1
Reverse osmosis reject brine	1.5		
Cooling tower blowdown	100	5	40

Source: Environment Canada, Environmental Protection Service. *Significance and Treatment of Dissolved Solids in Wastewaters from Canadian Steam Electric Stations.* Ottawa, ON, Canada, 1985.

TABLE 2.16

Environment Canada Criteria for Discharge of Wastewaters from Thermal-Power-Generating Stations

Parameter	Acceptable Limits
pH	6.0–9.5
Total suspended solids	25.0 mg/L
Chromium	0.5 mg/L
Copper	0.5 mg/L
Iron	1.0 mg/L
Nickel	0.5 mg/L
Zinc (system without recycle)	0.5 mg/L
Zinc (system with recycle)	0.2 mg/L

TABLE 2.17

Composition of Specific Metal Concentration Present in Station Waste Stream versus in Nature

Parameter	Natural Concentration (mg/L)	Waste Concentration(mg/L)
Aluminum	0.08	7.4
Manganese	0.05	15.0
Iron	0.047	1,300
Nickel	0.005	55
Copper	0.003	8.6
Zinc	0.001	4.3
Barium	0.10	7.9
Vanadium		20.0

Source: Environment Canada, Environmental Protection Service. *Significance and Treatment of Dissolved Solids in Wastewaters from Canadian Steam Electric Stations.* Ottawa, ON, Canada, 1985.

through recirculation. With lower volumes of wastewater to treat, removal of contaminants could be maximized. Metals in the water streams cause corrosion and scale problems. Most water intake requirements are for major condenser cooling. Water is also used for cooling auxiliary equipment, as boiler makeup water, and for cleaning boilers and other equipment. Particularly in coal-fired plants, water is necessary for coal spray and for the removal and transportation of combustion wastes. The metals present in station effluents, many exceeding environmental guidelines, come from demineralization of water, corrosion of equipment and scaling as well as coal pile runoff—a large source of dissolved metals—and from the cleaning of FGD or the removal of fly ash solids [70]. The various major wastewater streams in the power-generating plant are given below (Table 2.17).

Boiler blowdown: Impurities entering the steam-generating system with boiler makeup water and corrosion prevention chemicals are concentrated as steam in the boiler. To avoid excessive buildup of the impurities, the boiler side is bled as a boiler blowdown stream. The boiler blowdown water is discharged over ash dykes.

Demineralization (DM) plant neutralization effluent: Raw water from the clarifier is treated in the DM plant with the help of ion exchange resin beds. During regeneration and backwashing, effluents are generated, which are collected in a neutralization pit. Effluents during the regeneration of anionic and cationic beds are mixed together along with other effluents and discharged after pH adjustment into ash ponds. The DM plant neutralization effluent may also be eventually discharged into ash ponds, where it is treated along with other effluents before its discharge into the environment (ash pond overflow).

Cooling tower blowdown: In this system, hot water from the condenser is cooled in the cooling tower and the cooled water is recirculated to the condenser by CW pumps through the water conductor system. The blowdown from the cooling water system is taken to an ash water sump to meet the requirements of the ash handling plant.

Ash pond overflow waters: The ash slurry from the generating units is dumped in the ash dyke area. This area consists of two compartments; the first one consists of 80% of the area where most of the ash is settled and water with fine ash flows into the second compartment (stilling pond). In the stilling pond, the final ash is settled. The water from the stilling pond is partially recycled and partially discharged in an open-lined channel for irrigation after proper treatment.

Oxidation pond outlet: The wastewater from residential colonies along with sewage generated at thermal plants is led to the oxidation plants. The treated water from the oxidation ponds may be used for irrigation and green belt development.

Collected coal-heap runoff: The mixed runoff generated in the coal-heap area originates from atmospheric deposition (rain and snowmelt), washing, and mainly from coal-heap spraying (to minimize dust and also self-ignition). Due to the action of acid-generating microorganisms, the runoff tends to be acidic (like AMD), effectively leaching trace metal content of the huge quantities of coal throughput. The combined collected runoff of low pH is usually neutralized (lime) and the resulting precipitates settle out as toxic sludge. The water effluent may be discharged or recycled, if feasible. Its residual overall amount of toxic heavy metals in this stream, however, can be quite significant.

CONCLUSIONS

Thermal stations using fossil fuels have always been linked in the public mind with heavy air pollution. The liquid effluents from these plants have never been given much attention; hence the public has been left with a false sense of security about the purity of the water around thermal power-generating plants. However, it is clear that the waste streams being emitted could use much cleaning. Many plants in North America exceed the regulatory norms for the concentration of heavy metals in the environment. Unfortunately, those norms are deceptive since the environmental impact of heavy metals is measured more in terms of quantities of metals in the water bodies rather than effluent concentrations. With the immense amounts of water discharged by those plants, even if they met the norms they would still have a considerable impact on nature. A better solution would be to make plants comply with a set amount of metal emissions per day rather than with a specific stream concentration.

GLOSSARY

Acid mine drainage (AMD) It refers to the acidic water that forms when iron sulfides from mining activities interact with water and air. Here's how it happens: (a) AMD Cause: Iron sulfides, typically pyrite, a solid waste byproduct of coal mining, react with water and air; (b) AMD Process: This interaction creates sulfuric acid, which is highly corrosive and breaks down surrounding rocks. Toxic metals then enter the water and eventually dissolve; (c) AMD Prevalence: Acid mine drainage is believed to be the most common issue related to abandoned mines. Over 12,000 miles of streams in the United States have been affected by it; and (d) AMD Environmental Impact: The heavy metals produced by acid mine drainage don't biodegrade, posing risks to humans, wildlife, and ecosystems. It contaminates drinking water, corrodes infrastructure, and harms aquatic life.

Metal toxicity or metal poisoning It occurs when certain metals, in specific forms and doses, have a toxic effect on living organisms, including humans, animals, and plants. Here are some key points about metal toxicity: (a) Types of Metals Involved: (a1) Toxic Metals: Some metals are toxic when they form poisonous soluble compounds. Examples include arsenic, lead, mercury, cadmium, chromium, etc; (a2) Essential Metals: Not all heavy metals are toxic. Some, like iron, are essential metals for biological processes, but are toxic only when their concentrations are very high; (a3) Trace Elements: Abnormally high doses of trace elements may also be toxic; (b) Symptoms and Impact: (b1) Toxic metals can interfere with metabolic processes, leading to illness; (b2) Chronic toxicity is common, especially with radioactive heavy metals like radium, which can mimic calcium and become incorporated into human bones; (b3) Lead and mercury poisoning also have serious health implications; (c) Treatment Options: Only soluble metal-containing compounds are toxic, and metal complexes can be detoxified by converting them to insoluble derivatives or using chelating agents; (c1) Chelation therapy involves administering chelating agents to remove metals from the body of humans or animals; (c2) Phytoremediation uses plants to extract and lower toxic heavy metal concentrations in soil for environmental conservation.

Radionuclide A radionuclide is also known as a radioactive nuclide, radioisotope, or radioactive isotope that is a type of nuclide having an excess of either neutrons or protons, resulting in excess nuclear energy and making it unstable. Radionuclides play essential roles in various fields, including medicine, industry, and scientific research. Here are some key points about radionuclides: (a) Decay Modes: Radionuclides undergo radioactive decay, during which they emit radiation. This decay can occur in three ways: (a1) Gamma Radiation: The emission of gamma rays from the nucleus; (a2) Conversion Electron: Transfer of excess energy to an electron, causing its release; (a3) Particle Emission: Creation and emission of new particles (alpha or beta particles) from the nucleus; (b) Random Process: Radioactive decay is a random process at the level of individual atoms. It's impossible to predict when a specific atom will decay; (c) Natural and Artificial: Radionuclides occur naturally or are artificially produced in nuclear reactors, cyclotrons, particle accelerators, or radionuclide generators; (d) Variety: There are over 730 radionuclides with half-lives longer than 60 minutes. Some are primordial (created before Earth's formation), while others are detectable in nature or produced artificially; (e) Health Impact: Unplanned exposure to radionuclides can harm living organisms, including humans. The degree of harm depends on factors like radiation type, exposure, and biochemical properties.

Toxicologists They are the professionals who investigate poisons, their effects, and related problems, such as clinical, industrial, or legal issues. They play a crucial role in ensuring public safety by assessing environmental toxins in water, soil, air, and food, as well as evaluating the risks associated with various substances.

Toxicology It is a scientific field that investigates or researches the adverse effects of chemical substances on living organisms. It involves understanding the harmful impact of chemicals, substances, or situations on people, animals, and the environment.

REFERENCES

1. Volesky, B., Sears, M., Neufeld, R.J., and Tsezos, M. Recovery of strategic elements by biosorption. In: Venkatasubramanian, K., Constantinides, A., and Vieth, W.R. (Eds.), *Biochemical Engineering 3, Annals NY Acad. Sci.* New York Academy of Science, New York, p. 310, 1983.
2. Quinn, M.J. and Sherlock, J.C. The correspondence between U.K. "action levels" for lead in blood and in water. *FoodAddit. Contam.*, 7(3), 387–424, 1990.
3. Moore, M.R. Plumbosolvency of waters. *Nature*, 243, 222–223, 1973.
4. World Health Organization (WHO). *Report of the 30th Meeting of the Joint FAO/WHO.* Expert Committee on Food Additives. World Health Organization, Geneva and Rome, 1987.

5. Roberts, D.J., Bradberry, S.M., Butcher, F., and Busby, A. Lead Exposure in Children. *BMJ*, 063950, 2022.

6. Rabinowitz, M.B., Wetherill, G.W., and Kopple, J.D. Kinetic analysis of lead metabolism in healthy humans. *J. Clin. Invest.*, 58, 260–270, 1976.

7. Gilfillan, S.C. Lead poisoning and the fall of Rome. *J. Occup. Med.*, 7, 53, 1965.

8. Syracuse Research Corporation Agency for Toxic Substances and Disease Registry (ATSDR). *Toxicological Profile for Lead.* Public Health Service/U.S. Environmental Protection Agency, Washington DC, USA, 1990.

9. Baker, E.L., Feldman, R.G., White, R.A., Harley, J.P., Niles, C.A., Dinse, G.E., and Berkey, C.S. Occupational lead neurotoxicity: A behavioral and electrophysiological evaluation. Study design and year one result. *Br. J. Ind. Med.*, 41, 352–361, 1984.

10. Lester, M.L., Horst, R.L., and Thatcher, R.W. Protective effects of zinc and calcium against heavy metal impairment of children's cognitive function. *Nutr. Behav.*, 3, 145–161, 1986.

11. International Agency for Research on Cancer (IARC). Chemicals, industrial processes and industries associated with cancer in humans. In: *IARC Monographs on the Evaluation of the Carcinogenic Risk of Chemicals to Humans*, vol. 1–29, p. 149, 1982.

12. Cotton, F.A. and Wilkinson, G. Zinc, cadmium and mercury. In: Cottohn, F.A. (Ed.), *Advanced Inorganic Chemistry,* Interscience Publishers, London, p. 503, 1972.

13. Riihimaki, V. Cadmium. In: *The Hazards to Health of Persistent Substances in Water. Long Term Program in Environmental Pollution Control in Europe* (Annex to a report). World Health Organization, Geneva, Switzerland, 1972.

14. Lymburner, D.B. The production, use and distribution of cadmium in Canada. In: *Environmental Contaminants Inventory Study No. 2.* Report series no. 39, Centre for Inland Waters (Directorate), Ottawa, ON, Canada, 1974.

15. Butterworth, J., Lester, P., and Nickless, G. Distribution of heavy metals in the Severn Estuary. *Mar. Poll. Bull.*, 3, 72, 1972.

16. Hiatt, V. and Huff, J.E. The environmental impact of cadmium: An overview. *Int. J. Environ. Stud.*, 7, 277–285, 1975.

17. Fleischer, M., Sarofim, A.F., Fassett, D.W., Hammond, P., Shacklette, H.T., Nisbet, I.C., and Epstein, S. Environmental impact of cadmium: A review by the panel on hazardous trace substances. *Environ. Health Perspect.*, 7, 253–323, 1974.

18. Friberg, L., Piscator, M., Nordberg, G.F., and Kjellstrom, T. *Cadmium in the Environment*, 2nd Edition. CRC Press, Cleveland, OH, 1974.

19. Bernard, A. and Lauwerys, R. Cadmium in human populations. *Experientia*, 40, 143–152, 1984.

20. Schroeder, H.A. and Balassa, J.J. Abnormal trace metals in man—Cadmium. *J. Chronic Dis.*, 14, 236–258, 1961.

21. McGrath, S.P. Keeping toxic cadmium out of the food chain. *Nat Food*, 3(8), 569–570, 2022.

22. World Health Organization. *Environmental Health Criteria for Cadmium*, WHO, Geneva, Switzerland, 1974.

23. Murata, I., Hirano, T., Saeki, Y., and Nakagawa, S. Cadmium enteropathy, renal osteomalacia ("Itai-Itai" disease) in Japan. *Bull. Soc. Int. Chir.*, 1, 34, 1970.

24. World Health Organization. Technical documents on arsenic, cadmium, lead, manganese and mercury. In: *The Hazards to Health of Persistent Substances in Water. Long Term Program in Environmental Pollution Control in Europe* (Annex to a report), World Health Organization, Copenhagen, 1972.

25. Jonasson, I.R. and Boyle, R.W. *Geochemistry of Mercury. Mercury in Man's Environment.* The Royal Society of Canada, Ottawa, ON, Canada, p. 22, 1971.

26. U.S. Environmental Protection Agency (U.S. EPA). *Drinking Water Criteria Document for Mercury.* Final draft, Environmental Criteria and Assessment Office. Cincinnati, OH, p. 22, 1985.

27. Jensen, S. and Jernelov, A. Biological methylation of mercury in aquatic organisms. *Nature*, 223, 753–754, 1969.

28. Jernelov, A., Lander, R.L., and Larrson, T. Swedish perspectives on mercury pollution. *J. Water Pollut. Control Fed.*, 47, 810–822, 1975.

29. Suzuki, T. and Tanaka, A. Absorption of metallic mercury from the intestine after rupture of Miller-Abbot Balloon. *Ind. Med.*, 13, 52–58, 1971.

30. Skerfving, S. and Vostal, J. Symptoms and signs of intoxication. In: Friberg, L. and Vostal, J. (Eds.), *Mercury in the Environment.* CRC Press, Cleveland, OH, p. 93, 1972.

31. Bakir, F. Methylmercury poisoning in Iraq. *Science*, 181, 230–241, 1973.

32. National Council on Radiation Protection and Measurements. *Measurement of Radon and Radon Daughters in Air.* NCRP report no. 97, Bethesda, MD, 1988.

33. King, P.T., Michel, J., and Moore, W.S. Ground water geochemistry of 228Ra, 226Ra and 222Rn. *Geochim. Cosmochim. Acta*, 46, 1173–1182, 1982.

34. Nazaroff, W.W., Doyle, S.M., Nero, A.V., and Sexton, R.G. Potable water as a source of airborne 222Rn in U.S. dwellings: A review and assessment. *Health Phys.*, 52, 281–295, 1987.

35. United Nations Scientific Committee on the Effects of Atomic Radiation (UNSCEAR). *Sources, Effects and Risks of Ionizing Radiation.* Report to the General Assembly, New York, 1988.

36. Gosink, T.A., Baskaran, M., and Holleman, D.F. Radon in the human body from drinking water. *Health Phys.*, 59(6), 919–924, 1990.

37. Hursh, J.B., Morken, D.A., Davis, R.P., and Lovass, A. The fate of radon ingested by man. *Health Phys.*, 11, 465–468, 1965.

38. Cross, F.T., Harley, N.H., and Hofmann, W. Health effects and risks from 222Rn in drinking water. *Health Phys.*, 48(5), 649–670, 1985.

39. Abernathy, C., Calderson, R.L., and Chappel, W.R. *Arsenic Exposure and Health Effects*, Elsevier, London, UK, 1999.

40. Podgorski, J. and Berg, M. Global threat of arsenic in groundwater. *Science*, 368(6493), 845–850, 2020.

41. World Health Organization (WHO). Arsenic and arsenic compounds. In: *Environmental Health Criteria 224*, 13, pp. 5940–5948, 2005. Available at: https://www.inchem.org/documents/ehc/ehc/ehc224.htm.

42. Peryea, F.J. and Kammereck, R. Phosphate-enhanced movement of arsenic out of lead arsenate-contaminated topsoil and through uncontaminated subsoil. *Water, Air, Soil Pollut.*, 93(1–4), 243–254, 1997.

43. Abedin, M.J., Feldmann, J., and Meharg, A.A. Uptake kinetics of arsenic species in rice plants. *Plant Physiol.*, 128(3), 1120–1128, 2002.

44. Lee, R.B. Selectivity and kinetics of ion uptake by barley plants following nutrient deficiency. *Ann. Bot.*, 50, 429–449, 1982.

45. Smith, A.H., Lingas, E.O., and Rahman, M. Contamination of drinking-water by arsenic in Bangladesh: A public health emergency. *Bull. World Health Organ.*, 78, 1093–1103, 2000.

46. Tondel, M., Rahman, M., Magnuson, A., Chowhury, I.A., Faruquee, M.H., and Ahmad, S.A. The relationship of arsenic levels in drinking water and the prevalence rate of skin lesions in Bangladesh. *Environ. Health Perspect.*, 107, 727–729, 1999.

47. Domenico, P.A. *Physical and Chemical Hydrogeology.* Wiley, Canada, 1990.

48. Anderson, M.P. and Woessener, W.W. *Applied Groundwater Modeling.* Academic Press, California, 1992.

49. Freeze, A.R. and Cherry, J.A. *Groundwater.* Prentice-Hall, New Jersey, 1979.

50. Ferguson, J.E. *The Heavy Elements: Chemistry, Environmental Impact and Health Effects.* Pergamon Press, New York, 1990.

51. *Erin Brockovich*, directed by Steven Soderbergh, Universal Pictures, 2000.

52. Williams, R. *Waste Production and Disposal in Mining, Milling and Metallurgical Industries.* Miller Freeman Publications, California, 1975.

53. Kalin, M. The role of applied biotechnology in decommissioning mining operations. In: *Proceedings of the 13th Annual General Meeting of BIOMINET,* Ottawa, ON, Canada, pp. 103–120, 1997.

54. Kefeni, K.K., Msagati, T.A.M., and Mamba, B.B. Acid mine drainage: Prevention, treatment options, and resource recovery: A review. *J. Clean. Prod.*, 151, 475–493, 2017.

55. Filion, M.P., Sirois, L.L., and Ferguson, K. Acid mine drainage research in Canada. *CIM Bull.*, 83, 33–44, 1990.

56. Hammack, R.W., Edenborn, H.M., and Dvorak, D.H. Treatment of water from open-pit copper mine using biogenic sulfide and limestone: A feasibility study. *Water Res.*, 28, 2321–2329, 1994.

57. Naja, G., Mustin, C., Volesky, B., and Berthelin, J. Biosorption study in a mining wastewater reservoir. *Int. J. Environ. Pollut.*, 34(1/2/3/4), 14–27, 2008.

58. Murray, D.R. *Soil Profi le Development in Vegetated Uranium Tailings.* Division report, MRP-MRL, Natural Resources Canada, Energy Technology Center CANMET, pp. 81–126, 1981.

59. Hedin, R.S. Treatment of acid coal mine drainage with constructed wetlands. In: Sopper, W. E. (Ed.), *Wetlands Ecology, Productivity and Values; Emphasis on Pennsylvania.* Pennsylvania Academy of Sciences, PA, 1989.

60. Singer, P.C. and Stumm, W. Acidic mine drainage: The rate determining step. *Science*, 167, 1121–1123, 1970.

61. Wang, Y., Zhang, Z., Li, Y., Liang, C., Huang, H., and Wang, S. Available heavy metals concentrations in agricultural soils: Relationship with soil properties and total heavy metals concentrations in different industries. *J. Hazard. Mater.*, 471, 134410, 2024.
62. Environment Canada, Environmental Protection Service. *Overview of the Surface Finishing Industry: Status of the Industry and Measures for Pollution Control.* EPS 2/SF/1, Ottawa, ON, Canada, p. 43, 1987.
63. O'Flaherty, F., Roddy, W.T., and Lollar, R.M. *The Chemistry and Technology of Leather*, Vol. II. Reinhold, New York, p. 293, 1958.
64. U.S. Environmental Protection Agency (U.S. EPA). *Guidance Manual for Leather Tanning and Finishing Pretreatment Standards.* US Environmental Protection Agency, Washington DC, USA, pp. 2–3, 1986.
65. Di Perno, N. *Physico-Chemical and Resource Management Options for a Canadian Leather Retanner.* Mechanical Engineering Thesis. McGill University, Montreal, Canada, 1991.
66. Environment Canada, Environmental Protection Service. *National Inventory of Sources and Emissions of Copper and Nickel.* Air Pollution Control Directorate (91976), Report EPS-3-AP, pp. 81–84, 1981.
67. Gabbard, A. Coal combustion: Nuclear resource or danger. *ORNL Rev.*, 26, 3–4, 1993.
68. Department of Energy. U.S. Mercury Emissions Control R&D. U.S. Department of Energy (2006-01-18), Washington, DC, 2006.
69. Streets, D.G., Hao, J., Wu, Y., Jiang, J., Chan, M., Tian, H., and Feng, X. Anthropogenic mercury emissions in China. *Atmos. Environ.*, 39(40), 7789–7806, 2005.
70. Environment Canada, Environmental Protection Service. *Significance and Treatment of Dissolved Solids in Wastewaters from Canadian Steam Electric Stations.* Environment Canada, Industrial Programs Branch, Ottawa, ON, Canada, 1985.

3 Heavy Metal Removal with Exopolysaccharide-Producing Cyanobacteria

Roberto De Philippis, Ernesto Micheletti,
Xiao Liu, and Lawrence K. Wang

ACRONYMS

BET Brunauer-Emmett-Teller
ED electro-dialysis
EPS Exopolysaccharides
IX ion exchange
LPS lipopolysaccharides
NF nanofiltration
RO reverse osmosis
RPS released polysaccharides
UF ultrafiltration

3.1 INTRODUCTION

3.1.1 Heavy Metal Pollution in Water Bodies: An Increasing Concern for Human Health

Heavy metals are one of the most widespread causes of pollution, both in water and soil, and the presence of increasing levels of these metals in the environment is causing serious concern in public opinion owing to the toxicity shown by most of them [1,2].

Heavy metals are usually defined as metals with a density greater than 4–5 gcm^{-3}, but in the literature, it is possible to find so many different definitions that recently, IUPAC defined the term "heavy metal" as a confusing and misleading one [3]. Generally speaking, metals are natural components of the Earth's crust and some of them (e.g., copper, selenium, and zinc) are essential as trace elements to maintain the metabolism of the human body even if, at higher concentrations, they may have toxic effects. Many other metals (e.g., mercury, cadmium, and lead) have direct toxic effects on human health [4]. Owing to their chemical characteristics, metals remain in the environment, in many cases only changing from one chemical state to another one and eventually accumulating in the food chain. These pollutants enter the environment through a variety of human activities, such as mining, refining, and electroplating industries [5]. The effluents produced by these industries contain a variety of heavy metals, such as cadmium, copper, chromium, nickel, lead, and zinc, and their release in water bodies may significantly contribute to the increased presence of toxic heavy metals in aquatic environments. Owing to their high water solubility, heavy metals can be easily absorbed by living organisms, and due to their mobility in natural water ecosystems and their toxicity to living forms, they have been ranked as major inorganic contaminants in surface and ground waters. Even if they may be present in dilute, almost undetectable quantities, their recalcitrance to degradation and consequent persistence in water bodies implies that, through natural processes such as biomagnification, their concentration may

DOI: 10.1201/9781003541615-3

become elevated to such an extent that they begin exhibiting toxic effects. Of the 35 metals considered dangerous for human health, 23 have been defined as heavy metals: antimony, arsenic, bismuth, cadmium, cerium, chromium, cobalt, copper, gallium, gold, iron, lead, manganese, mercury, nickel, platinum, silver, tellurium, thallium, tin, uranium, vanadium, and zinc [6]. However, the main threats to human health from heavy metals are associated with exposure to lead, cadmium, mercury, and arsenic (this element is a metalloid, but it is usually defined as a heavy metal). Large amounts of any of these metals may cause acute or chronic toxicity (poisoning), resulting in damaged or reduced mental and central nervous functions, modifying blood composition, and damaging the lungs, kidneys, liver, and other vital organs. Long-term exposure to the above-mentioned heavy metals may result in slowly progressing physical, muscular, and neurological degenerative processes that mimic Alzheimer's disease, Parkinson's disease, muscular dystrophy, and multiple sclerosis. Allergies are not uncommon, and repeated long-term contact with some metals or their compounds may even cause cancer [7]. Heavy metals may enter the human body through food, water, and air or may be absorbed through the skin when they enter into contact with humans in agriculture and manufacturing, pharmaceutical, industrial, or residential settings [8].

Although several adverse health effects of heavy metals have been known for a long time, exposure to these metals is continuing and even increasing in some parts of the world. Thus, the control of heavy metal dumplings and the removal of toxic heavy metals from waters has become a challenge for the 21st century.

3.1.2 Heavy Metal Removal: Conventional Technologies

A number of techniques for the treatment of heavy metal-containing wastewaters have been developed in recent years, in order to both decrease the amount of metal-containing wastewaters produced by industrial activities and improve the quality of treated effluents. Various treatments, such as chemical precipitation, coagulation-flocculation, flotation, membrane filtration, ion exchange, and electrochemical treatment techniques, can be utilized to remove heavy metals from contaminated wastewaters, each with their own inherent advantages and limitations in application.

3.1.2.1 Chemical Precipitation

Chemical precipitation is the most common technology used to remove dissolved (ionic) metals from water solutions, such as process wastewaters containing toxic metals. The ionic metals are converted to an insoluble form (particle) by the chemical reaction between the soluble metal compounds and the precipitating reagent. Typically, the metal precipitated from the solution is in the form of hydroxide. The conceptual mechanism of heavy metal removal by chemical precipitation is represented as

$$\mathrm{M}^{n+} + n\left(\mathrm{OH}^-\right) \leftrightarrow \mathrm{M(OH)}_n \downarrow, \tag{3.1}$$

where $\mathrm{M(OH)}_n$ is insoluble metal hydroxide and M^{n+} and OH^- represent dissolved metal ions and precipitants, respectively. Therefore, the optimum pH for precipitation of one metal may cause another metal to solubilize or start to go back into solution. Most process wastewaters contain mixed metals, and hence, precipitating these different metals as hydroxides can be a tricky process. Chemical precipitation requires large amounts of chemicals to reduce the concentration of metals to an acceptable level for discharging wastewaters into the environment. Other drawbacks of this method are related to the excessive sludge production that requires further treatments, the cost of sludge disposal, the slow kinetics of metal precipitation, the poor settling of metal hydroxides, the aggregation of metal precipitates, and the long-term environmental impact of sludge disposal [9].

3.1.2.2 Coagulation–Flocculation

Coagulation and flocculation occur in successive steps intended to overcome the forces stabilizing suspended particles, allowing particle collision and growth as flocks. If the first step is incomplete, the following one will be unsuccessful. The coagulation process destabilizes colloidal particles, constituting metal-containing compounds, by adding a coagulant, resulting in the sedimentation of these compounds. To increase the particle size, coagulation is followed by the flocculation of unstable particles into bulky floccules. The general approach for this technique includes a preliminary pH adjustment and involves the addition of ferric/aluminum salts as coagulants to overcome the repulsive forces between colloidal particles. Like chemical precipitation, a pH ranging from 11.0 to 11.5 has been found to be the most effective in maximizing the heavy metal removal obtained using this process. Improved sludge settling, dewatering characteristics, capability of inactivating bacterial activities, and sludge stability are reported to be the major advantages of lime-based coagulation. In spite of its advantages, coagulation-flocculation has limitations, such as high operational costs due to the use of large amounts of chemicals. In addition to this drawback, the large volume of sludge generated from the coagulation-flocculation process may hinder its adoption as a global strategy for wastewater treatment [10].

3.1.2.3 Flotation

This technology separates solids or dispersed liquids from a liquid phase using bubble attachment. Flotation can be divided into five different kinds of processes: (a) dispersed-air flotation, (b) dissolved air flotation (DAF), (c) vacuum air flotation, (d) electro-flotation, and (e) biological flotation. Among the various types of flotation, DAF is the one most frequently used for treating metal-containing wastewaters. Adsorptive bubble separation utilizes foaming to separate the metal impurities.

3.1.2.4 Membrane Filtration

Membrane technology has become a dignified separation technology over the past decennia. The main force of membrane technology is the fact that it works without the addition of chemicals, with relatively low-energy use and easy and well-arranged process conductions. This technology can remove not only suspended solids and organic compounds but also inorganic contaminants such as heavy metals. Depending on the size of the particles that have to be retained, various types of filtration techniques [i.e., ultrafiltration (UF), nanofiltration (NF), and reverse osmosis (RO)] can be utilized for the removal of heavy metals from wastewaters. UF utilizes permeable membranes to separate heavy metals, macromolecules, and suspended solids from inorganic solutions based on the size of the pores (5–20 nm) and the molecular weight (1,000–100,000 Da) of the compounds that have to be separated. NF has unique properties between UF and RO membranes. In NF, the separation mechanism involves steric (sieving) and electrical (Donnan) effects. A Donnan potential is created between the charged anions in the NF membrane and the metal co-ions in the effluent, thus creating conditions for the rejection of the latter. NF membranes, generally, can treat inorganic effluents with a metal concentration of 2,000 mg/L. In RO, the metal-containing solution is forced, by applying pressure on the fluid, to pass through a membrane: heavy metals are retained and accumulated on one side while purified water is recovered on the other side of the membrane. In general, compared with UF and NF, RO is more effective for heavy metal removal from inorganic solutions, as indicated by the rejection percentage of over 97% with a metal concentration ranging between 20 and 200 mg/L, but its major drawback is related to the high operational costs [11].

3.1.2.5 Ion Exchange

Ion exchange (IX) is a reversible chemical reaction wherein an ion present in solution is exchanged with a similarly charged ion bound to a stationary solid phase (resin). Ion exchange can also be used for recovering valuable heavy metals from inorganic effluents. After separating the loaded resin, metals can be recovered in a more concentrated solution by eluting with suitable reagents.

Since the acidic functional groups of the resins consist of sulfonic acid, it is assumed that the physicochemical interactions occurring during metal removal can be expressed as follows:

$$\left(n\text{RSO}_3^- \text{-H}^+\right) + \text{M}^{n+} \leftrightarrow \left(n\text{RSO}_3^- \text{-M}^{n+}\right) + n\text{H}^+,$$

$$\text{(resin)} \qquad \text{(solution)} \quad \text{(resin)} \qquad \text{(solution)} \tag{3.2}$$

where (nRSO$_3^-$) and M^{n+} represent the anionic group attached to the ion-exchange resin and the metal cation, respectively, while m is the coefficient of the reaction component, depending on the oxidation state of metal ions. Selecting the optimum dosage level depends mainly on the quality of finished water required, considering economic and operating factors. Depending on the characteristics of the ion exchanger, heavy metal removal by ion exchange is effective in acidic conditions with pH ranging from 2 to 6. However, ion exchange also has some limitations in treating wastewaters containing heavy metals. Prior to ion exchange, appropriate pretreatment systems for secondary effluent such as removing suspended solids from wastewater are required. In addition, suitable ion-exchange resins are unavailable for all heavy metals, and the capital and operational costs are high [12].

3.1.2.6 Electrochemical Treatment Techniques

Electro-dialysis (ED) is a membrane separation technique in which ionized species in the solution are passed through an ion-exchange membrane by applying an electric potential. The membranes are thin sheets of plastic materials with either anionic or cationic charge. When a solution containing ionic species passes through the cell compartments, the anions migrate toward the anode, and the cations migrate toward the cathode, crossing the anion-exchange and cation-exchange membranes. Since ED is a membrane process, it requires clean feed, careful operation, and periodic maintenance to prevent any damage to the membranes.

In conclusion, it is possible to say that physicochemical treatments offer various advantages but their benefits are counterbalanced by a number of drawbacks such as their high operational costs due to the chemicals used, high energy consumption, and handling costs for sludge disposal.

3.1.3 HEAVY METAL REMOVAL: USE OF MICROORGANISMS

The property of nonliving microbial biomass to accumulate heavy metal ions, a not metabolically driven process called *biosorption*, has been known for several decades [13]. This process occurs on the cell surface, including physical adsorption [14], ion exchange [15], and surface complexation [16]. It may be driven by the chemical osmotic gradient of the biological cytoplasmic membrane.

In contrast, the term *bioaccumulation* describes an active process where the removal of metals requires the metabolic activity of a living organism, including the internal accumulation of cells and surface microprecipitation [17,18]. Research on the mechanisms of biosorption has intensified because biomass can be profitably utilized to sequester heavy metals from industrial effluents (e.g., produced by mining or electroplating industries) or to recover precious metals from processing solutions [13].

Microbial cells may represent excellent biosorbents owing to their high surface-to-volume ratios and a high content of potentially active chemo-sorption sites [19].

The use of biomass as an adsorbent for the control of pollution due to heavy metals can generate revenue for those microbiological industries that are currently wasting the biomass derived from the industrial process and can also ease the burden of the costs associated with the disposal of the waste biomass produced. Alternatively, the biomass can also be grown using unsophisticated fermentation techniques and inexpensive growth media [20]. The use of dead biomass in industrial applications offers certain advantages over living cells. Systems using living cells are likely to be more sensitive to metal ion concentration (toxicity effects) and adverse operating conditions (pH and temperature). Furthermore, a constant nutrient supply is required for systems using living cells, resulting in high operating costs for waste streams devoid of nutrients, and the recovery of metals and the regeneration of biosorbents are more complex for living cells. Dead biomass can be obtained from industrial sources as a waste product derived from fermentation processes. Microbial cells can be killed by physical treatment methods using heat treatment [21,22], autoclaving, and vacuum drying [23,24], by using chemicals such as acids, alkalis, detergents [23,25–31], and other organic compounds (e.g., formaldehyde) [32], or by mechanical disruption [30,33].

TABLE 3.1

Most Important Heavy Metal-Binding Groups Present in the External Layers of Microbial Cells

Binding Group	pK_a	Occurrence in Selected Biomolecules
Hydroxyl	9.5–13	PS, UA, SPS, AA
Carbonyl (ketone)	—	Peptide bond
Carboxyl	1.7–4.7	UA, AA
Sulfhydryl (thiol)	8.3–10.8	AA
Sulfonate	1.3	SPS
Thioether	—	AA
Amine	8–11	Cto, AA
Secondary amine	13	Cti, PG, peptide bond
Amide	—	AA
Imine	11.6–12.6	AA
Imidazole	6.0	AA
Phosphonate	0.9–2.1/6.1–6.8	PL
Phosphodiester	1.5	TA, LPS

Source: Volesky, B. *Water Res.,* 41, 4017–4029, 2007.

AA, amino acids; Cti, chitin; Cto, chitosan; LPS, lipopolysaccharides; PG, peptidoglycan; PL, phospholipids; PS, polysaccharides; SPS, sulfated PS; TA, teichoic acid; UA, uronic acids.

The biosorption of metals is based on several mechanisms that quantitatively and qualitatively differ according to the species used, the origin of the biomass, and their method of processing. Metal sequestration follows complex mechanisms, mainly based on ion exchange, chelation, adsorption by physical forces, and ion entrapment in inter- and intrafibrillar capillaries and spaces of the structural polysaccharidic network of external cell layers. There are several chemical groups that could attract and sequester the metals in biomass, depending on a number of external environmental factors as well as on the type of metal, its toxic form in solution, and the type of active binding site responsible for the sequestration of the metal [34]. The most important binding compounds are acetamido groups of chitin, structural polysaccharides of fungi, amino and phosphate groups in nucleic acids, amino, amido, sulfhydryl, and carboxyl groups in proteins, and carboxyls, hydroxyls, and sulfates in polysaccharides (Table 3.1). However, it should be stressed that the presence of a specific functional group does not guarantee its accessibility for sorption, owing to possible steric and conformational barriers. Research on biosorption of heavy metals has led to the identification of a number of microbial biomass types that are extremely effective in concentrating metals. Some types of biomass are the waste of large-scale industrial fermentations (e.g., the bacterium *Bacillus subtilis*). Other metal-binding microbial types, such as certain abundant cyanobacterial strains, can be readily harvested from specific environments, such as lakes or the sea, where they form superficial blooms. These biomass types can accumulate considerable amounts of heavy metals, such as Pb, Cd, U, Cu, Zn, Cr, and others.

Large quantities of waste microbial biomass are produced in many industries, such as those producing citric acid or penicillin. Berkeley [35] estimated that fermentation industries produce some 790,000 tons of microbial waste each year, with 41,000 tons resulting from citric acid production by *Aspergillus miger.* The use of dead rather than live biomass eliminates the problems of waste toxicity and nutrient requirements.

Some types of biosorbents could be broad range, binding and collecting the majority of heavy metals with no specific priority, whereas others can even be specific for certain types of metals [36,37].

An important consideration in studying biosorption is the solution chemistry of metals, particularly as it relates to their hydration and hydrolysis reactions. Biosorption of heavy metals usually leads to the acidification of solutions because the biomass usually acts as a weakly acidic ion exchanger. Since deprotonation of each functional group depends on pH, the process of biosorption is strongly pH-dependent. This was confirmed in the biosorption experiments carried out at different pH values. The contribution of functional groups in the biosorption process was confirmed by chemical modification of the groups. Chemically blocked groups did not show either biosorption or ion exchange capabilities.

3.1.4 MODELING MICROBIAL BIOSORPTION

Sorption can be defined, in general terms, as a process in which a solid, named sorbent, interacts with a chemical species (sorbate) dissolved in a liquid phase (usually water); as a consequence of this interaction, the sorbate is removed from the liquid phase with efficiency depending on its affinity for the sorbent. When the sorbent is a biological material (i.e., microbial cells, vegetable residues, and wood residues), this process is defined as *biosorption* [38].

To evaluate the quality of a sorbing material and its capability of removing from water solutions, the sorbate of interest is generally compared with the performances obtained with other sorbents previously reported in the literature. The parameter most frequently used for describing the performances of sorbing materials is the metal uptake q, defined as the amount of sorbate bound per unit of sorbent. In studies aimed at defining a practical process for removing heavy metals, q is generally expressed in terms of milligrams of metal removed per gram of dry weight of the sorbent. When the stoichiometry of the process has to be defined, the amount of metal removed can be better expressed as millimoles (i.e., milligrams of sorbate removed divided by its atomic or molecular weight) or milliequivalents (i.e., millimoles removed divided by the ion valence of the sorbate) per gram of dry weight of sorbent.

In many papers, it is possible to find other parameters, such as the percentage of sorbate removed on the amount initially present in the solution; however, these parameters are of limited significance for carrying out a sound comparison with sorption processes performed under different conditions or with different sorbents. In addition to these considerations, it has to be stressed that many factors affect the sorption processes (see Section 3.3.3), thus suggesting that one carefully check the operating conditions utilized before comparing the results obtained in different laboratories. Indeed, the sorption process is affected by the temperature; thus all the experiments must be carried out under controlled temperature and it is only possible to compare experiments carried out at the same temperature. Moreover, the q values are comparable only if they were calculated when the equilibrium between the amount of sorbate on the sorbent and in the liquid phase was attained. Finally, the sorption process is strongly affected by the pH and the ionic strength of the metal solution; thus, any of these parameters should be maintained under strict control during the experiments aimed at characterizing the sorption performances of a new biosorbent.

The data obtained in the processes of metal biosorption with microorganisms are usually expressed utilizing the Freundlich or the Langmuir isotherms, the two most utilized adsorption models for single-solute systems [39].

3.1.4.1 Freundlich Adsorption Isotherm

The Freundlich isotherm relationship is an empirical exponential equation with the following form:

$$q = K\, C_e^{1/n} \tag{3.3}$$

which can be linearized as

$$\ln q = \ln K + n^{-1} \ln C_e \tag{3.4}$$

where q is the solute (metal) adsorbed per mass of adsorbent, typically expressed as mg/g, K is a constant related to the adsorbent capacity, n is a constant related to the energy of sorption, and C_e is the equilibrium (final) concentration of solute (metal) in the solution, typically expressed as mg/L.

The Freundlich isotherm, not indicating a finite uptake capacity, should only be utilized at low or intermediate values of the final metal concentration C_f [40].

3.1.4.2 Langmuir Adsorption Isotherm

The Langmuir adsorption isotherm has the following form:

$$q = bC_e \, q_{max} (1 + bCe)^{-1} \tag{3.5}$$

which can be rearranged as

$$C_e q^{-1} = \left(bq_{max}\right)^{-1} + C_e \left(q_{max}\right)^{-1} \tag{3.6}$$

where q is the solute (metal) adsorbed per mass of adsorbent, q_{max} is the metal adsorbed, at saturation, per mass of adsorbent, b is a constant, related to the energy of adsorption, and C_e is the equilibrium (final) concentration of solute (metal) in the solution.

The Langmuir isotherm is based on a number of assumptions: (a) the sorption process involves only a monolayer on the surface of the sorbent; (b) there is only a fixed number of active sites on the sorbent; (c) all the active sites are chemically equivalent; (d) only one sorbate is present; (e) one active site reacts with only one molecule; and (f) there are no interactions among the sorbed molecules.

It has to be stressed that in many, if not all, practical cases, these assumptions are far from reality; however, the experimental data of most biosorption processes so far described in the literature fit rather well in the Langmuir isotherm and thus, the use of this model allows the calculation of three very useful parameters such as q, q_{max}, and b. The value of q_{max} can be considered as the maximum amount of sorbate that may theoretically be sorbed by the sorbent (i.e., the total number of binding sites available on the sorbent), whereas q is the number of binding sites really occupied by the molecules of sorbate; b is a parameter inversely related to the affinity of the sorbent for the sorbate.

Other, more complex isotherm relationships have been utilized in order to obtain more detailed models of the sorption process, even if none of them is actually related to the real physicochemical mechanism of the sorption process. In particular, the Brunauer-Emmett-Teller (BET) model, which is based on the assumption that sorption is a multilayer process where the binding sites are independent one from the other, has been frequently utilized.

3.1.4.3 BET Adsorption Isotherm

The BET adsorption isotherm has the following form:

$$q = BQC_f \left\{ \left(C_s - C_f\right) \left[1 + (B-1)\frac{C_f}{C_s} \right] \right\}^{-1}, \tag{3.7}$$

where q is the solute (metal) adsorbed per mass of adsorbent (mg/g), B is a constant related to the energy of interaction of the sorbate with the surface of the sorbent, Q is the number of moles of sorbate adsorbed per gram of sorbent in the formation of a complete monolayer, and C_s is the saturation constant of the sorbate.

Other more sophisticated models have been proposed, but they are empirical models. Indeed, the sorption process is very complex and not ascribable to a unique phenomenon; in addition, in the case of biosorption of heavy metals on biological materials, it has been suggested that the occurrence of a process of formation of gels and the presence of a mechanism of exchange between the newly sorbed ions and those coming out from the gel. For the above-reported reasons, the isotherms thus far proposed can only be considered useful models for calculating the most important sorption parameters when the experimental data nicely fit one of them [40].

3.2 CYANOBACTERIA

3.2.1 GENERAL CHARACTERISTICS OF CYANOBACTERIA

Cyanobacteria are a wide group of Gram-negative, phototrophic prokaryotes characterized both by the capability of performing oxygenic, plant-like photosynthesis and by autotrophy as their main mode of nutrition, even if the additional capability of growing under heterotrophic and/or mixotrophic growth conditions has been reported for a number of species [41]. In addition, some cyanobacterial strains were found to be capable of growing in the dark on organic substrates [42,43] or, under anaerobic conditions, of carrying out anoxygenic photosynthesis utilizing sulfide as the electron donor [44]. A large number of species possess the capability of fixing atmospheric dinitrogen, which is reduced to ammonia and utilized for the biosynthesis of nitrogen-containing cellular compounds. As a consequence of these metabolic features, cyanobacteria can be considered as the microbial group possessing the simplest nutritional requirements of all microorganisms [45,46]. Such trophic independence, along with their competitive ability and the relative ease of cultivation, makes cyanobacteria well suited for use in wastewater treatment.

Cyanobacteria are characterized by great morphological diversity. They are included in the group of both unicellular and filamentous species, with large differences in their cell volume that range over more than five orders of magnitude [47]. They have been found in most of the natural illuminated environments, both aquatic and terrestrial, as a consequence of a broad spectrum of physiological properties and tolerance to environmental stress [48].

3.2.2 CHARACTERISTICS OF THE CYANOBACTERIAL CELL WALL

Cyanobacteria are classified as Gram-negative prokaryotes with regard to the organization of their cell walls. These microorganisms possess a four-layered cell wall that consists of a peptidoglycan layer (murein) surrounded by a lipopolysaccharidic layer (outer membrane), as is typical for Gram-negative bacteria [49]. However, there are some peculiarities in the cell walls of cyanobacteria compared to those of other Gram-negative bacteria. Indeed, the peptidoglycan layer is considerably thicker and with a higher degree of cross-linking between the polysaccharidic chains, reaching values close to those reported for Gram-positive bacteria [50]. On the other hand, most cyanobacteria showed the presence of the typical Gram-negative bacterial *meso*-diaminopimelic acid in the peptides linking the polysaccharidic chains.

Some other peculiarities of the cell walls of cyanobacteria, related to the composition of lipopolysaccharides (LPSs) and the presence of unusual constituents (e.g., carotenoids), distinguish cyanobacterial cell walls from the typical cell wall of Gram-negative cyanobacteria.

3.2.3 CHARACTERISTICS OF THE POLYSACCHARIDIC EXTERNAL STRUCTURES IN CYANOBACTERIA

Several cyanobacteria possess, outside their outer membrane, additional surface structures mainly of polysaccharidic nature, which are usually defined as sheaths, capsules, and slimes. The sheath is defined as a thin layer that surrounds single cells, single filaments, or cell groups, reflecting their shape and microscopically visible without any staining (Figure 3.1); the sheath shows, in most cases, a microfibrillar substructure and is characterized by considerable mechanical and physicochemical stability [51–54]. The capsule is a polysaccharidic, gelatinous layer, associated with the cell surface and characterized by sharp outlines (Figure 3.2); it is structurally coherent to exclude particles, and this characteristic is used for its negative staining with Indian ink [54]. The slime is an amorphic, mucilaginous shroud characterized by a non-well-defined structure; it may enclose a certain number

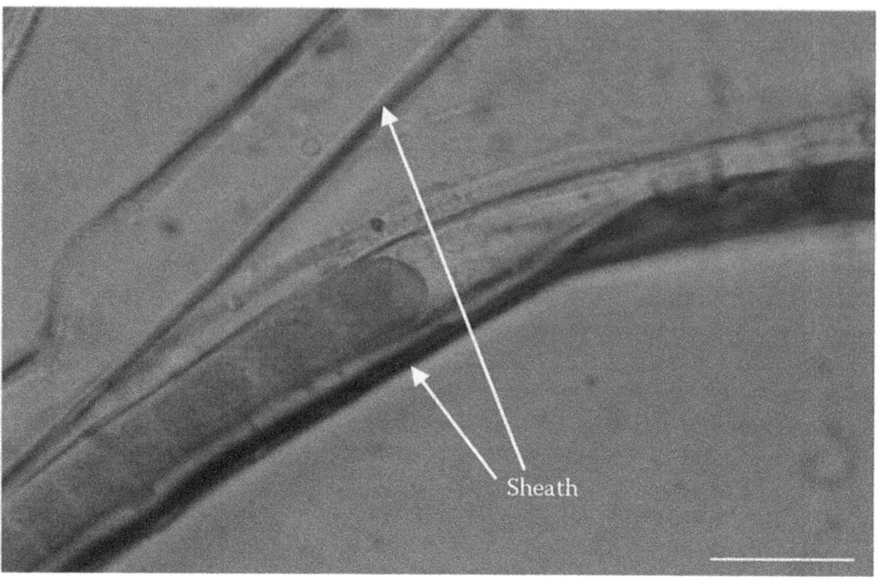

FIGURE 3.1   Example of the sheath in cyanobacteria: Nomarski differential interference contrast photomicrograph of *Lyngbya* sp. (scale bar = 10 mm).

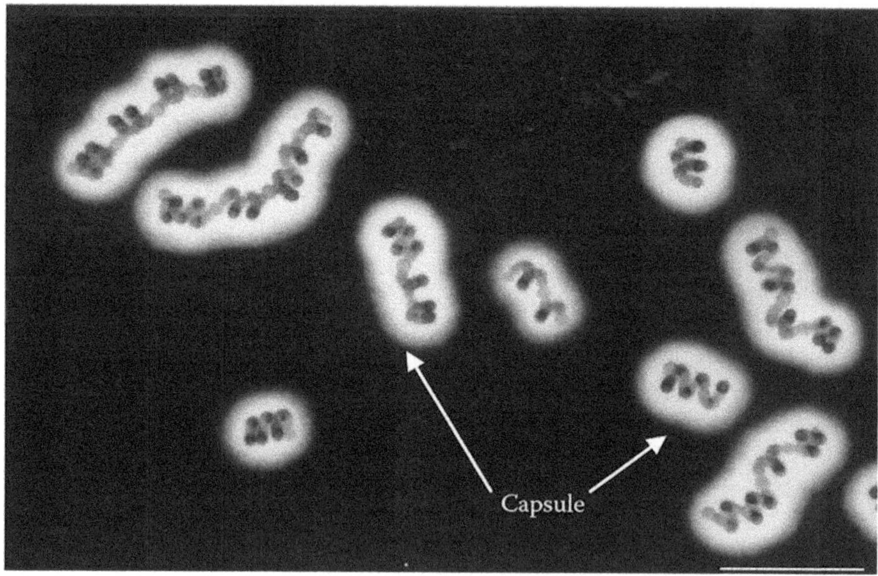

FIGURE 3.2   Example of capsule in cyanobacteria: Nomarski differential interference contrast photomicrograph of *Cyanospira capsulata* filaments stained with India ink (scale bar = 100 mm).

of cells or filaments, showing the additional presence of capsules (Figure 3.3) [54–56]. During cell growth in batch cultures, aliquots of the polysaccharidic material of sheaths, capsules, and slimes may be solubilized into the surrounding medium, constituting the so-called released polysaccharides (RPSs), which, depending on their molecular weight and concentration, may cause a significant increase in the viscosity of the culture medium.

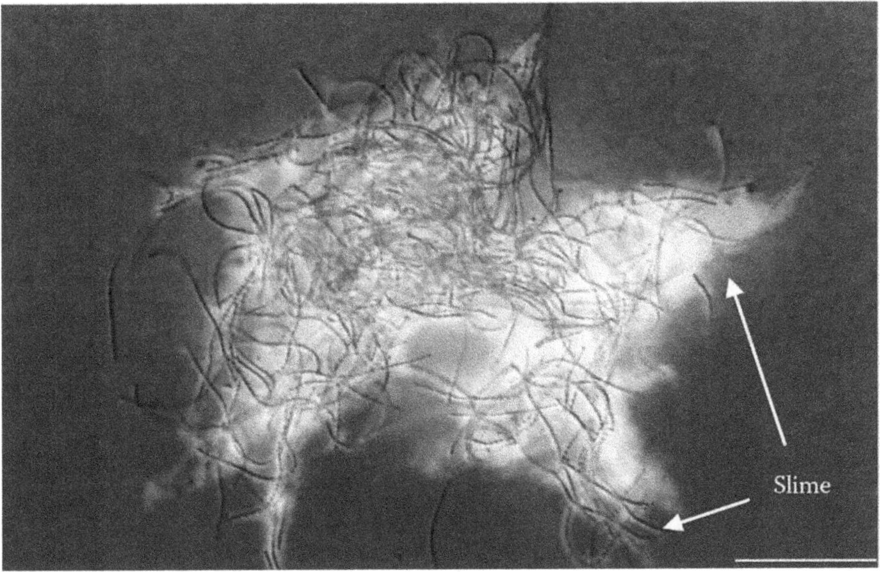

FIGURE 3.3 Example of slime in cyanobacteria: Nomarski differential interference contrast photomicro graph of *Nostoc* PCC7936 filaments stained with India ink (scale bar = 150 mm).

From the data so far available in the literature, based on the characterization of more than 150 cyanobacterial RPSs, it appears that they can be defined as complex heteropolymers containing three to fourteen different kinds of monosaccharides. Indeed, about 80% of the polymers described so far showed the presence of six or more different monomers; a feature peculiar to this group of microorganisms is that the polysaccharides produced by other microbial groups are generally constituted by a much lower number of monomers [57]. Another important feature of cyanobacterial polysaccharides represents their anionic character, with more than 90% of them being characterized by the presence of one or two uronic acids, namely, galacturonic and/or glucuronic acid. In about half of the RPSs studied, the amount of uronic acids is higher than 20% of the RPS dry weight. A certain number of charged, nonsaccharidic constituents, such as sulfate and ketal-linked pyruvyl groups, have also been found in many cyanobacterial RPSs, contributing to the global negative charge of these macromolecules [58]. A rather large number of cyanobacterial RPSs also showed the presence of peptidic moieties, which, depending on their amino acidic composition, may contribute either to the electrostatic charge of the macromolecules or to their hydrophobicity [54,59].

Owing to the presence of a large number of negative charges on the external layers of the exopolysaccharide-producing cyanobacteria, these microorganisms and the polymers that they produce have been considered very promising chelating agents for the removal of positively charged heavy metal ions from water solutions [58,59].

3.3 HEAVY METAL REMOVAL WITH CYANOBACTERIA

3.3.1 Putative Mechanisms of Metal Biosorption with Cyanobacteria

The ability of microorganisms to accumulate and remove heavy metals from water has been widely reported in the literature [5,19,34]. The complexity of the structure of microbial cells implies that there are many possible ways for the metal to be captured by the cell. Sorption mechanisms are therefore various and in some cases still not completely understood; however, generally speaking, it is possible to classify these processes as (a) active, metabolically driven or (b) passive, not metabolically driven processes.

In active processes, the sorption, named *bioaccumulation*, is due to the transport of the metal across the cell membrane, with a consequent intracellular accumulation of the metal depending on cell metabolism. Consequently, bioaccumulation occurs only with viable cells and is often associated with an active defense system of microorganisms, which reacts in the presence of a toxic metal. In this case, metal sorption is not immediate since it requires time for the reaction of the microorganism to the presence of the metal; thus, the occurrence of bioaccumulation is evidenced by the kinetics of the metal sorption, which is slower in this metabolically driven process in comparison with the mere physicochemical adsorption on the cell surface. When both processes occur, the slope of the line representing metal uptake versus time shows an inflexion point, which is due to the temporal sequence of passive adsorption, characterized by a steep slope, followed by active bioaccumulation, characterized by a less steep slope.

In the case of physicochemical interactions (based on physical adsorption, ion exchange, and complexation) between the metal and functional groups present on the cell surface, the sorption process is not dependent on the metabolism. Cell walls, mainly composed of polysaccharides, proteins, and lipids, may offer particularly abundant sites for the binding of metals. These not metabolically driven, physicochemical processes are relatively fast and usually reversible [37]. In the presence of such mechanisms, the microbial biomass acts as an ion-exchange resin. Ion exchange is an important concept in biosorption because it explains many observations made during heavy metal uptake experiments. However, it should be pointed out that the term ion exchange does not explicitly identify the binding mechanism. The precise mechanism(s) may range from physical (i.e., electrostatic or London-van der Waals forces) to chemical binding (i.e., ionic and covalent).

Like other microbial biomasses tested, cyanobacteria have a cell wall capable of passively adsorbing high levels of dissolved metals, usually via a charge-mediated attraction. Cell walls present, in fact, functional groups, such as carboxylate, hydroxyl, sulfate, phosphate, and amino groups (see Sections 3.2.2 and 3.2.3). In addition, many strains are characterized by the presence of outermost polysaccharidic envelopes, often coupled with the capability of releasing exocellular polysaccharides (EPSs) into the culture medium during cell growth [54]. Most of these polymers are characterized by anionic nature, owing to the presence of uronic acids and/or other charged groups that contribute to give these macromolecules a rather high anion density [54]. For this reason, the biomass of EPS-producing cyanobacteria can be considered very promising as a chelating agent for removing heavy metals from water [58], with a larger number of binding sites for metal ions compared with the biomass of other microorganisms.

In EPS-producing cyanobacteria, it is possible to hypothesize a complex mechanism of interaction with positively charged metal ions due to the presence of more than one negatively charged external cell structure. Indeed, the metal cations can be taken up by the negatively charged groups present on the cell wall (Figure 3.4), those present on the polysaccharidic layers surrounding the cell wall (sheath, capsule, and/or slime) (Figure 3.5), and those present in the RPSs released by the cells into the surrounding environment (Figure 3.6).

3.3.2 Heavy Metal Removal with Cyanobacteria

The ability of cyanobacterial biomass to accumulate and remove heavy metals from water solutions has been widely reported in the literature. Up to now, a very large number of cyanobacterial strains have been screened with regard to their capability of removing heavy metals from metal solutions, mostly operating under laboratory conditions with pure metal solutions. As is evident from Table 3.2, a large number of papers describe experiments performed using cyanobacteria for the removal of metals with a high level of toxicity, such as Cd, Hg, Ni, Zn, and Pb. *Tolypothrix tenuis* and *Calothrix parietina* showed a very high capability of removing Hg [60], *Scytonema schmidlei*, *Anabaena cylindrica*, and *A. torulosa* removed 96%–98% of Cd from 1 mg/L solutions of this metal [61], and *Gloeocapsa* sp., *Nostoc paludosum*, *N. piscinale*, *N. punctiforme*, *N. commune*, *Oscillatoria agardhii*, *Phormidium molle*, and *Tolypothrix* removed 90%–96% of Pb from 1 mg/L

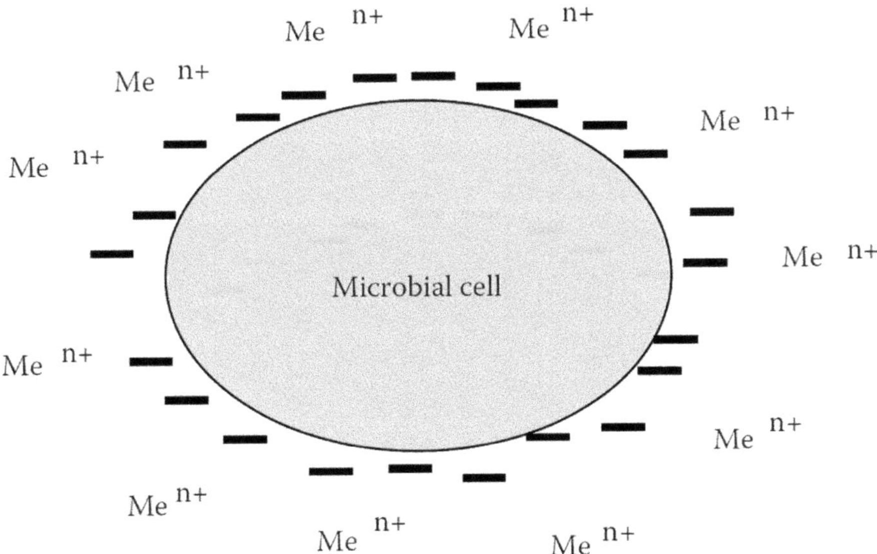

FIGURE 3.4   Interactions of metal cations with negative charges on the cell surface.

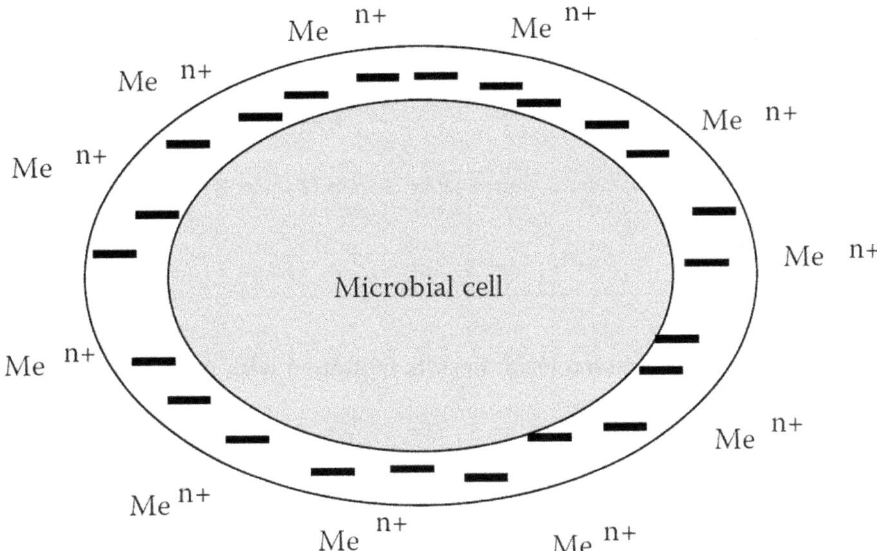

FIGURE 3.5   Interactions of metal cations with negative charges on the polysaccharidic exocellular layers (sheath or capsule).

solutions [62]. In addition to studies on the removal of toxic metals, the biosorption of nontoxic, but economically valuable, metals was also studied, often using the same cyanobacterial strains utilized with toxic heavy metals (Table 3.2). However, as already discussed in the previous sections, one of the problems arising when one intends to make a sound comparison among the performances of cyanobacteria reported in the literature comes from the different experimental conditions frequently utilized by various authors and from the uncertainty, in many cases, that the equilibrium in the sorption process was actually attained, a condition indispensable for giving sense to the comparison of the results obtained by different authors [63].

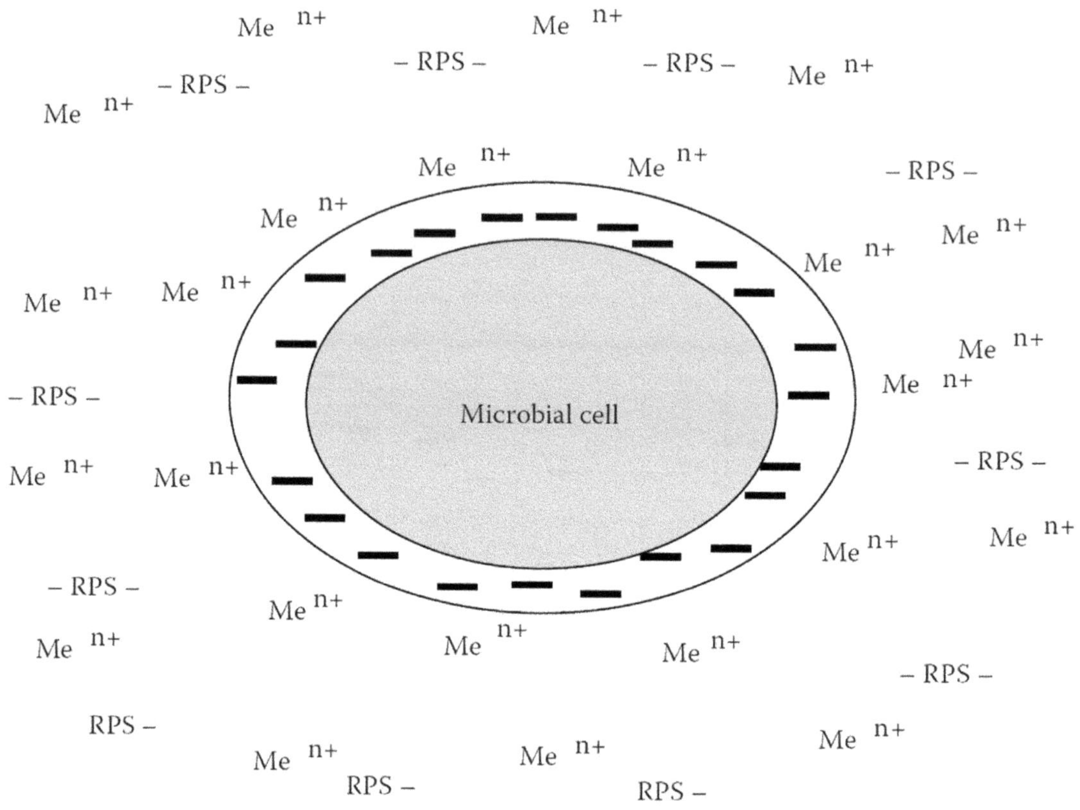

FIGURE 3.6 Interactions of metal cations with negative charges of the polysaccharide released (RPS) into the water medium.

TABLE 3.2

Heavy Metal Removal by Cyanobacteria: Results Obtained with the Cyanobacteria and the Metals So Far Tested

Metal	Cyanobacterium	Sorption (mmol/g)[a] or Metal Removal Efficiency (%)[b]	References
Al	*Spirulina platensis*	0.0045[a,c]	[63]
	Spirulina maxima	0.0145[a,c]	[63]
	Mastigocladus laminosus #113	0.0284[a,c]	[63]
Co	*Oscillatoria angustissima*	0.26[a]	[75]
	Spirulina sp.	0.0002[a]	[76]
	Phormidium valderianum BDU 30501	n.r.	[77]
Cd	*Anabaena nodosum*	0.087[a]	[78]
	Anaabena nodosum	0.81[a]	[79]
	Anabaena sp. BCC 2	85[b]	[61]
	Anabaena variabilis NIES 23	57[b]	[72]
	Anacystis nidulans	n.r.	[80]
	Aphanocapsa halophytia	n.r.	[81]

(Continued)

TABLE 3.2 (*Continued*)

Heavy Metal Removal by Cyanobacteria: Results Obtained with the Cyanobacteria and the Metals So Far Tested

Metal	Cyanobacterium	Sorption (mmol/g)[a] or Metal Removal Efficiency (%)[b]	References
	Calothrix sp.	n.r.	[82]
	Calothrix marchica BCC 4	57[b]	[61]
	Calothrix marchica BCC 6	87[b]	[61]
	Calothrix parietina TISTR 8093	0.70[a]	[61]
	Calothrix sp. BCC 8	88[b]	[61]
	Calothrix sp. BCC 10	89[b]	[61]
	Calothrix sp. TISTR 8130	82[b]	[61]
	Chlorococcus paris	n.r.	[83]
	Cylindrospermum sp. BCC 20	65[b]	[61]
	Gloeocapsa sp. BCC 25	96[b]	[61]
	Gloeothece magna	3.78[a,c]	[84]
	Hapalosiphon hibernicus BCC 27	90[b]	[61]
	Hapalosiphon sp. BCC 30	62[b]	[61]
	Hapalosiphon welwitschii BCC 34	75[b]	[61]
	Lyngbya hieronymusii BCC 41	97[b]	[61]
	Lyngbya spiralis BCC 42	80b	[61]
	Lyngbya taylorii	0.37[a]	[85]
	Mastigocladus laminosus #113	0.0048[a,c]	[63]
	Mastigocladus sp. BCC 36	78[b]	[61]
	Nostoc rivularis	n.r.	[86]
	Microcystis sp.	1.28[a]	[87]
	Microcystis aeruginosa f. aeruginosa NIES 44	95.3[b]	[72]
	Microcystis aeruginosa f. fl os-aquae C3-40	1.23[a,c]	[88]
	Nostoc linckia	n.r.	[86]
	Nostoc sp. BCC 50	94[b]	[61]
	Nostoc commune. BCC 76	69[b]	[61]
	Nostoc micropicum BCC 77	72[b]	[61]
	Nostoc piscinale. BCC 47	82[b]	[61]
	Nostoc punctiforme BCC 48	73[b]	[61]
	Nostoc punctiforme BCC 49	84[b]	[61]
	Oscillatoria amoena BCC 53	83[b]	[61]
	Oscillatoria jasorvensis BCC 56	94[b]	[61]
	Oscillatoria agardhii BCC 52	90[b]	[61]
	Phormidium angustissimus BCC 68	87[b]	[61]
	Phormidium molle BCC 7193	95[b]	[61]
	Phormidium valderianum BDU 30501	83[b]	[77]
	Rivularia sp. BCC 80	88[b]	[61]
	Spirulina platensis	0.0035[a,c]	[63]
	Spirulina maxima	0.0017[a,c]	[63]

(Continued)

TABLE 3.2 (*Continued*)

Heavy Metal Removal by Cyanobacteria: Results Obtained with the Cyanobacteria and the Metals So Far Tested

Metal	Cyanobacterium	Sorption (mmol/g)[a] or Metal Removal Efficiency (%)[b]	References
	Spirulina platensis	0.33[a]	[89]
	Spirulina platensis	1.07[a]	[90]
	Spirulina sp.	0.004[a]	[76]
	Spirulina vulgaris	1.00[a]	[91]
	Stigonema sp. BCC 90	89[b]	[61]
	Stigonema sp. BCC 92	80[b]	[61]
	Synechococcus sp. PCC 7942	0.06[a]	[92]
	Tolypotrix tenuis TISRT 8063	0.80[a]	[61]
	Tolypotrix tenuis	0.18[a]	[72]
	Tolypotrix tenuis BCC 100	53[b]	[61]
Cr	*Aphanocapsa halophytia*	0.29[a]	[81]
	Cyanospyra capsulata	0.36[a]	[93]
	Cyanothece CE 4	1.83[a]	[93]
	Cyanothece ET 5	0.59[a]	[93]
	Cyanothece PE 14	0.26[a]	[93]
	Cyanothece TI 4	1.29[a]	[93]
	Cyanothece VI 13	0.93[a]	[93]
	Cyanothece VI 22	1.08[a]	[93]
	Cyanothece 16Som 2	3.77[a]	[93]
	Nostoc PCC 7936	0.07[a]	[93]
	Spirulina sp.	0.19[a]	[76]
	Synechococcus sp. PCC 7942	0.10[a]	[92]
Cu	*Anabaena variabilis* NIES 23	35.4[b]	[72]
	Anacystis nidulans	n.r.	[80]
	Aphanocapsa halophytia	n.r.	[81]
	Calothrix sp.	n.r.	[82]
	Chroococcus paris	n.r.	[83]
	Cyanospyra capsulata PCC 9502	3.65[a]	[94]
	Cyanospyra capsulata	1.97[a]	[59]
	Cyanospyra capsulata	2.25[a]	[93]
	Cyanothece CE 4	0.52[a]	[93]
	Cyanothece ET 5	1.78[a]	[93]
	Cyanothece PE 14	0.16[a]	[93]
	Cyanothece TI 4	0.50[a]	[93]
	Cyanothece VI 13	0.95[a]	[93]
	Cyanothece VI 22	0.98[a]	[93]
	Cyanothece 16Som 2	3.17[a]	[93]
	Gloeothece sp. PCC 6909	0.41[a]	[95]
	Microcystis sp.	2.91[a]	[87]

(Continued)

TABLE 3.2 (*Continued*)
Heavy Metal Removal by Cyanobacteria: Results Obtained with the Cyanobacteria and the Metals So Far Tested

Metal	Cyanobacterium	Sorption (mmol/g)[a] or Metal Removal Efficiency (%)[b]	References
	Microcystis aeruginosa f. fl os-aquae C3-40	4.1a,c	[88]
	Nostoc PCC 7936	1.465[a]	[59]
	Oscillatoria angustissima	4.22[a]	[96]
	Phormidium laminosum	97.5[c]	[69]
	Spirulina, sp.	0.19[a]	[76]
	Synechococcus sp. PCC 7942	0.18[a]	[92]
	Tolypotrix tenuis TISTR 8063	0.18[a]	[72]
Fe	*Phormidium laminosum*	94.8[b]	[69]
Hg	*Anabaena sp.* BCC 2	68[b]	[61]
	Calothrix marchica BCC 4	84[b]	[61]
	Calothrix sp. BCC 8	86[b]	[61]
	Calothrix sp. BCC 10	92[b]	[61]
	Calothrix parietina TISTR 8093	50[b]	[61]
	Calothrix sp. TISTR 8130	40[b]	[61]
	Cylindrospermum sp. BCC 20	83[b]	[61]
	Gloeocapsa sp.BCC 25	50[b]	[61]
	Hapalosiphon hibernicus BCC 27	84[b]	[61]
	Hapalosiphon welwitschii BCC 34	85[b]	[61]
	Lyngbya hieronymusii BCC 41	92[b]	[61]
	Lyngbya spiralis BCC 42	96[b]	[61]
	Mastigocladus laminosus #113	0.005a,c	[63]
	Mastigocladus sp. BCC 36	89[b]	[61]
	Nostoc sp. BCC 50	86[b]	[61]
	Nostoc commune sp. BCC 76	43[b]	[61]
	Nostoc micropicum BCC 77	26[b]	[61]
	Nostoc piscinale sp. BCC 47	22[b]	[61]
	Nostoc punctiforme sp. BCC 48	66[b]	[61]
	Nostoc punctiforme sp. BCC 49	49[b]	[61]
	Oscillatoria amoena BCC 53	12[b]	[61]
	Oscillatoria jasorvensis BCC 5689	89[b]	[61]
	Oscillatoria agardhii BCC 52	73[b]	[61]
	Phormidium angustissimus BCC 68	74[b]	[61]
	Phormidium molle BCC 71	93[b]	[61]
	Rivularia sp. BCC 80	86[b]	[61]
	Spirulina maxima	0.00125a,c	[63]
	Spirulina platensis	0.00055a,c	[63]
	Stigonema sp. BCC 90	92[b]	[61]
	Stigonema sp. BCC 92	94[b]	[61]
	Tolypotrix tenuis BCC 100	94[b]	[61]

(Continued)

TABLE 3.2 (*Continued*)
Heavy Metal Removal by Cyanobacteria: Results Obtained with the Cyanobacteria and the Metals So Far Tested

Metal	Cyanobacterium	Sorption (mmol/g)[a] or Metal Removal Efficiency (%)[b]	References
Mn	*Microcystis aeruginosa f. fl os-aquae* C3-40	2.84[a,c]	[88]
Ni	*Aphanocapsa halophytia*	n.r.	[81]
	Cyanospyra capsulata	1.41[a]	[93]
	Cyanothece CE 4	1.24[a]	[93]
	Cyanothece ET 5	0.24[a]	[93]
	Cyanothece PE 14	0.19[a]	[93]
	Cyanothece TI 4	0.62[a]	[93]
	Cyanothece VI 13	0.39[a]	[93]
	Cyanothece VI 22	0.59[a]	[93]
	Cyanothece 16Som 2	0.96[a]	[93]
	Lyngbya taylorii	0.65[a]	[85]
	Microcystis aeruginosa	4.26[a]	[97]
	Nostoc PCC 7936	0.032[a]	[93]
	Phormidium laminosum	85[b]	[69]
	Spirulina sp.	0.003[a]	[76]
	Synechococcus sp. PCC 7942	0.05[a]	[92]
	Synechocystis sp.	3.23[a]	[98]
Pb	*Anabaena* sp. BCC 2	29[b]	[61]
	Anabaena variabilis NIES 23	48.92[b]	[72]
	Aphanothece halophytica	22[b]	[99]
	Calothrix sp. BCC 8	59[b]	[61]
	Calothrix sp. BCC 10	13[b]	[61]
	Calothrix sp. TISTR 8130	86[b]	[61]
	Calothrix sp.	n.r.	[82]
	Cylindrospermum sp. BCC 20	52[b]	[61]
	Gloeocapsa sp. BCC 25	96[b]	[61]
	Hapalosiphon hibernicus BCC 27	13[b]	[61]
	Hapalosiphon welwitschii BCC 34	47[b]	[61]
	Lyngbya heironymusii BCC 41	80[b]	[61]
	Lyngbya spiralis BCC 42	73[b]	[61]
	Lyngbya taylorii	1.47[a]	[85]
	Mastogocladus sp. BCC 36	29[b]	[61]
	Microcystis aeruginosa f. fl os-aquae C3-40	1.50[a,c]	[88]
	Nostoc sp. BCC 50	58[b]	[61]
	Nostoc commune BCC 76	94[b]	[61]
	Nostoc piscinale BCC 47	94[b]	[61]
	Nostoc punctiforme BCC 48	98[b]	[61]
	Nostoc punctiforme BCC 49	51[b]	[61]
	Oscillatoria agardhii BCC 52 73	96[b]	[61]

(*Continued*)

TABLE 3.2 (*Continued*)
Heavy Metal Removal by Cyanobacteria: Results Obtained with the Cyanobacteria and the Metals So Far Tested

Metal	Cyanobacterium	Sorption (mmol/g)[a] or Metal Removal Efficiency (%)[b]	References
	Oscillatoria amoena BCC 53	89[b]	[61]
	Oscillatoria jasorvensis BCC 56 89	85[b]	[61]
	Phormidium angustissimum BC 68	77[b]	[61]
	Phormidium molle BCC 71	90[b]	[61]
	Rivularia sp. BCC 80	76[b]	[61]
	Spirulina maxima	84[b]	[100]
	Spirulina platensis	35[b]	[99]
	Spirulina platensis	0.08[a]	[78]
	Spirulina, sp.	0.00005[a]	[76]
	Stigonema sp. BCC 90	52[b]	[61]
	Stigonema sp. BCC 92	59[b]	[61]
	Synechococcus sp. PCC 7942	0.15[a]	[92]
	Tolypothrix tenuis TISTR 8063	88[b]	[61]
	Tolypothrix tenuis BCC 100	90[b]	[61]
	Tolypothrix tenuis TISTR 8063	0.15[a]	[72]
Sn	*Aphanocapsa halophytia*	82[b]	[81]
Zn	*Anabaena variabilis* NIES 23	67.7[b]	[72]
	Anacystis nidulans	n.r.	[80]
	Aphanocapsa halophytia	n.r.	[81]
	Aphanothece halophytia	2.03[a]	[101]
	Chroococcus paris	n.r.	[83]
	Lyngbya taylorii	0.49[a]	[85]
	Mastigocladus laminosus #113	0.00856[a,c]	[63]
	Microcystis aeruginosa f. fl os-aquae C3-40	1.23[a,c]	[88]
	Microcystis sp.	15.29[a]	[97]
	Nostoc linckia	n.r.	[86]
	Nostoc rivularis	n.r.	[86]
	Oscillatoria angustissima	0.33[a]	[102]
	Oscillatoria angustissima	9.81[a]	[103]
	Phormidium laminosum	78.2[b]	[69]
	Spirulina maxima	0.0023[a,c]	[63]
	Spirulina platensis	0.0046[a,c]	[63]
	Spirulina platensis	0.11[a]	[78]
	Spirulina sp.	0.003[a]	[76]
	Tolypotrix tenuis TISTR 8063	0.14[a]	[72]

[a] Sorption, expressed as millimoles of metal removed per gram of dry biomass.
[b] Metal removal efficiency, expressed as percent of metal removed on its initial concentration.
[c] Bio-removal carried out with solutions of the pure polysaccharide produced by the cyanobacterium. n.r. = quantitative data not reported.

In spite of the large number of data available on the metal removal capability of cyanobacteria, the presence and the possible role of their extracellular polysaccharidic layers in the biosorption process were almost neglected until a few years ago. Indeed, even if since the late 1980s, it was hypothesized that the metal-binding process occurs by the complexation of metal ions with the polysaccharidic, mucilaginous material covering the cell wall or released by the cell surface [64,65], most of the studies on the role of polysaccharides in metal uptake date from the beginning of this century.

The comparison between metal biosorption in capsulated and decapsulated cells of field samples and laboratory-grown cultures of the unicellular cyanobacterium *Microcystis* showed a higher biosorption capability of the cells surrounded by the polysaccharidic capsular layer, pointing out a significant role of the polysaccharides in heavy metal removal [66]. Similar results were obtained in experiments of Cu biosorption with laboratory monocultures of microalgae and cyanobacteria, demonstrating the importance of surface mucilage in the adsorption process [67]. Mucilaginous algal species (e.g., *Anabaena* spp. and *Eudorina elegans*) showed a higher copper sorption in comparison with nonmucilaginous ones (*Chlorella vulgaris*), suggesting that this gelatinous layer increased the number of binding sites [67].

Inthorn et al. [61] reported that the cyanobacterium *Calothrix marchica* was able to remove Pb from wastewater, and Ruangsomboon et al. [68] demonstrated that cells of this cyanobacterium, covered with a mucilaginous sheath, were able to remove Pb ions also owing to the contribution of polysaccharidic material. The accumulation of heavy metals in capsular polysaccharide was also reported for *Phormidium laminosum*, which showed the capability of adsorbing Pb, Fe, Cd, Cu, Zn, and Ni [69]. Capsular polysaccharides of *Microcystis aeruginosa* and *M. flos-aquae* showed the capability of adsorbing iron ions [70,71]. Nagase et al. [72] found that the microorganisms Cyanobacterium treated with sodium hydroxide had better adsorption capacity for metal cations in the presence of other interfering cations.

A wide difference in the metal-binding capacity of capsulated and uncapsulated cyanobacterial strains has been demonstrated [73]. A cyanobacterial strain, characterized by the presence of a thick capsule surrounding the cells, showed higher metal-binding capacity than strains devoid of or with only a thin capsule [59,74]. However, the presence of gelatinous capsules surrounding cyanobacterial cells may also cause the slowing down of the diffusion rate of metal ions into the chelating matrix of the cell wall [66].

The role of RPSs was demonstrated by the work carried out by De Philippis et al. [104], who pointed out the significant contribution of the solubilized polysaccharidic fraction in the bio-removal of copper, by two cyanobacteria, *Cyanospira capsulata* and *Nostoc* PCC 7936, comparing the results previously obtained with the sole biomass with those obtained with the whole cultures confined to dialysis tubings [59]. It is an effective method for heavy metal ions removal by the combination of two or more biomass in the same contaminated water system [105].

Tien et al. [67] confirmed, by scanning electron microscopy x-ray microanalysis, that the higher binding capacity of mucilaginous cyanobacterial species is consistent with the hypothesis of the role of the polysaccharidic layer in metal uptake, having shown that about 40% of the adsorbed copper ions are bound to the mucilage. The longer adsorption time and the higher copper adsorption per unit of surface area of mucilaginous algae compared to *Chlorella* also supported this conclusion [67]. Parker et al. [88] and Li et al. [106] have found that the mucilaginous sheaths isolated from *M. aeruginosa* and *Aphanothece halophytica* exhibited a strong affinity for metal ions. High copper uptake by two mucilaginous cyanobacteria, *C. capsulata* and *Nostoc* PCC7936, was also shown by De Philippis et al. [59] in comparison with other microbial biomass.

A number of papers have been dedicated to the identification of the functional groups involved in the uptake of metal cations. The functional groups involved in metal sorption cyanobacteria have been identified by FTIR spectroscopy, pH and potentiometric titrations, and specific chemical modification of the biomass, with the block of functional groups with specific chemical reagents. There is a general agreement on the very important role carried out by carboxyl groups in the binding of positively charged metal ions by cyanobacterial cells [72,95,107], and this role is of particular importance in the polysaccharide-producing cyanobacteria, which are in most cases characterized by the presence of significant amounts of galacturonic and/or glucuronic acids (see Section 3.2.3). In the capsulated *Microcystis*, it was shown that the slime interacts with Cd, Ni, and Cu cations due to the presence of galacturonic acid [108]. The predominance of galacturonic acid in the capsule of *M. flos-acquae* suggested that the binding of Fe cation was due to carboxyl groups present in this polysaccharidic layer [71]. However, other groups have also been reported to be involved in the heavy metal-binding process by cyanobacterial cells. For instance, infrared (IR) data [100] showed that amino and hydroxyl groups play a predominant role (at high pH) in the binding of Pb by *Spirulina maxima*. Chojnacka [76] found that the process of Cr removal by *Spirulina* sp. was hindered when carboxyl and phosphate groups were esterified, showing the important role of these groups in the biosorption process. An insignificant decrease in the biosorption properties of the biomass after methylation of hydroxyl groups showed that these groups do not significantly contribute to the biosorption process.

The above-reported results strongly support the conclusion, reported in Section 3.3.3, that at low pH values, carboxyl groups of EPSs play a predominant role in heavy metal sorption by algae and cyanobacteria. Other functional groups, such as sulfonate and amino groups, play a relatively minor role in metal sorption, even if at high pH values, amino groups start to significantly contribute to metal uptake.

In any case, it is worth stressing that the mere determination of the quantity of charged groups is not sufficient for anticipating the actual metal-binding capability of a polymer because, depending on the conformation of the macromolecules, some of the charged groups might hardly be accessible by metal ions. Hence, it is possible to say that the uptake of metal ions depends on charge density, their distribution on the polymers, and their accessibility by metal ions. Indeed, Tien [109] reported that *Oscillatoria limnetica*, even if it did not produce large amounts of surface mucilage in comparison with *Anabaena spiroides* and *Eudorina elegans*, which produced large amounts of dense and thick mucilages, showed the highest Cu and Pb uptake capacity per volume unit of mucilage. As a consequence of these results, the author infers that the metal-binding activity of surface mucilage depends on the characteristics and not on the amount of mucilage produced. These findings were confirmed by a recent study carried out by Micheletti et al. [93], who investigated the role of outermost polysaccharidic investments in the process of copper removal by using the sheathed unicellular cyanobacterium *Gloeothece* PCC 6909 and its sheathless mutant PCC 6909/1. In this study, both the wild type and the sheathless mutant could remove copper ions from metal solutions, but the specific metal uptake of the sheathed cells was slightly higher than the unsheathed mutant strain. On the other hand, the polysaccharide released by the two strains into the culture medium showed a different metal uptake, with the specific metal removal of the polysaccharide released by the mutant strain being higher than that of the polysaccharide released by the wild type. These results are discussed by the authors in terms of the type and number of binding sites present in the polysaccharide released by the mutant strain, underlining that the metal removal process depends not only on the amount of EPS produced but also on its quality and structure.

Thus, it is possible to conclude that, although various potential functional groups have been demonstrated to be present on the external cell layers of microalgae and cyanobacteria, their mere presence is insufficient for demonstrating their participation in the biosorption of metals. Indeed, steric hindrance, conformational changes, or cross-linking may prevent some of the surface functional groups from being involved in the binding of metal ions [110]. Consequently, the only way to evaluate the potential of a cyanobacterial strain in metal removal is the experimental evaluation of its performances.

3.3.3 FACTORS AFFECTING METAL UPTAKE BY CYANOBACTERIA

Metal biosorption has been described as being dependent on the cyanobacterial species used and on the differences in the composition of their cell walls. However, many other factors may affect the biosorption of metals in addition to cell wall composition and structure, beginning from differences in cell size and shape between species [40,63]. Also, the concentration of biomass in solution seems to influence specific uptake, that is specific metal uptake, expressed as mg of metal removed per gram of biomass, the highest in correspondence with the lowest biomass concentration [75]. The decrease in the specific metal uptake observed with the increase in biomass concentration was explained by hypothesizing that an increase in biomass concentration leads to a negative interference between binding sites [59].

The pH of the metal solution is one of the most important factors affecting the biosorption process: it influences the chemistry of the metal in solution, the activity of the functional groups present on the biomass, and the competition among metallic ions for the binding sites [30]. The amount of ions bound at a given pH is, in fact, determined by the affinity constants of the metal-binding functional groups. The value of these constants depends on the kind and valence of the metal ion and on the pK_a value of the functional group (see Table 3.1). Depending on pH, different functional groups participate in the metal-binding process: at pH ranging from 2 to 5, the active functional groups are carboxyl groups, at pH ranging from 5 to 9, the active functional groups are carboxyl and phosphate groups, at pH ranging from 9 to 12, the active functional groups are carboxyl, phosphate, and hydroxyl (or amine) groups [76,95].

The simultaneous presence of more than one metal in solution usually depresses the biosorption of the metal of interest or the biosorption of other cations. Such an effect can be explained in terms of competition between metal ions for the same binding sites on the biomass. Studies on the effect of this competition and the selectivity of biosorption in mixtures of metal cations will be discussed in Section 3.4.3.

Temperature does not seem to significantly affect biosorption performances in the range 20°C–35°C [77]; for this reason, most biosorption studies with microalgae and cyanobacteria were performed at room temperature, without a strict control of this parameter [1]. However, if the aim of the experiments is to build up the adsorption isotherms of the metal removal process and also to allow a sound comparison among experimental data obtained in different laboratories, the experiments should be carried out in a thermostatic chamber in order to maintain a constant temperature.

It also has to be stressed that the physiological state of the biomass significantly affects its metal removal capability; a large difference between the amounts of metal uptake by dead or live microbial biomass has been reported by many authors. In most studies, biosorption of metals is greater with dead biomass than with equal amounts of living biomass. When biomass is in the dead state, the cells are permeable, and this permeability allows metals to enter the cells and bind on intracellular constituents and surfaces as well as on the external surface, thus increasing the amount of metal uptake [72]. In addition, metal uptake is not affected by the toxicity of the metal, which could affect the active transport of metal ions into living cells.

Other important factors that significantly affect metal uptake by microbial biomass are the electric charge and the chemical characteristics of the metal ions that interact with the cells [1]. Indeed, it has been demonstrated that metal ions with small ionic radius are sorbed onto a fixed surface area of a sorbent more quickly than metal ions characterized by larger ionic radii [24]. In accordance with these results, it was found that Cu (ionic radius=0.72 A) was taken up in larger amounts than Zn (ionic radius=0.74 A) and Cd (ionic radius=0.96 A) by biomass of *Microcystis* sp. [74]. These results are in good accordance with previous results obtained by Cho et al. [111] and the hypothesis formulated by Lee [112].

Another very important factor affecting the performances of microbial biomass in the removal of heavy metals is the physical or chemical pretreatment of cells before their contact with metal ions. Not much research has been dedicated to this aspect with cyanobacterial biomass, but it was clearly demonstrated that the pretreatment may significantly improve the metal removal capability of biomass. In the case of *Oscillatoria* sp. biomass, an acid pretreatment with 0.01 N HCl induced a 10% increase in Cu adsorption in comparison with the metal uptake obtained with native biomass. Sampedro et al. have reported increased biosorption in the filamentous cyanobacterium *P. laminosum* after NaOH treatment, whereas HCl treatment was ineffective. An extensive study carried out with several different pretreatment agents (i.e., with acid, alkaline, or chelating compounds) showed that, in the case of the biomass of two capsulated filamentous cyanobacteria, *C. capsulata* and *Nostoc* PCC 7936, the best performances were obtained after a pretreatment with a 0.1 N NaOH solution, which induced a 10% increase in the Cu removal of the treated biomass [94]. Moreover, the same authors demonstrated that the biomass of *C. capsulata* can withstand eight sorbing-desorbing cycles, carried out with 0.1 N HCl or 0.1 N NaOH as desorbing agents for removing adsorbed Cu ions during the previous sorbing cycle, without any degradation and while continuing to maintain a high sorbing capacity, averaging about 90% with both desorbing agents, as was previously found with other cyanobacteria [69,113,114].

3.4 METAL REMOVAL IN MULTIMETAL SOLUTIONS

3.4.1 MODELING MICROBIAL BIOSORPTION IN MULTIMETAL SOLUTIONS

Biosorption has often been studied in simple sorption systems, usually containing only one heavy metal. This is an inappropriate simplification for the understanding of the real process, because industrial wastewaters usually simultaneously contain more than one metal ion. For this reason, in studies aimed at realistically representing the situation of metal-containing wastewaters, it is necessary to develop models on the metal-binding process in multimetal solutions.

Metal sorption in binary metal systems may be described by the three-parameter extended Langmuir equation given below [115]:

$$q_1 = \frac{q_{\max 1} b_1 C_{f1}}{1 + b_1 C_{f1} + b_2 C_{f2}} \tag{3.8}$$

where q_1 is the equilibrium sorption (mmol/g) of the first metal; $q_{\max 1}$ is the maximum sorption of the first metal at equilibrium (final) concentration of metal in solution (C_{f1} and C_{f2}, expressed as mM); b_1 and b_2 are constants for the first and second metal, respectively; and subscripts 1 and 2 represent the metal of primary interest and the second metal, respectively.

In multimetal solutions, q (mmol/g) can be calculated by using the multicomponent extended Langmuir model with four parameters, as reported by Chong and Volesky [75]:

$$q(\mathrm{M}_i) = \frac{\left(q_{\max}/K_i\right)C_f[\mathrm{M}_i]}{1 + (1/K_1)C_f[\mathrm{M}_1] + (1/K_2)C_f[\mathrm{M}_2] + (1/K_3)C_f[\mathrm{M}_3]} \tag{3.9}$$

where i (= 1, 2, 3) represents the metal of interest and $C_f[\mathrm{M}_1]$, $C_f[\mathrm{M}_2]$, and $C_f[\mathrm{M}_3]$ are the residual concentrations of metals 1, 2, and 3, respectively. The four parameters ($q_{\max}$, K_1, K_2, and K_3) of the above model could be evaluated using an appropriate software package. Although the above model is empirical, it is very often used to describe the equilibrium isotherm data of multicomponent systems.

A model represented in Equation (3.10) has also been proposed; it is a combination of the conventional Langmuir and Freundlich models [115]:

$$q_1 = \frac{q_{\max 1} b_1 C_{f1}^{n1}}{1 + b_1 C_{f1}^{n1} + b_{12} C_{f2}^{n2}}. \tag{3.10}$$

The parameters of the above model ($q_{\max}$, b_1, b_2, n_1, and n_2) for the sorption of metal ions from a multicomponent system may be obtained by fitting the data using suitable software. In the same way, the sorption of the second metal (q_2) may also be represented.

Models focused on the ion-exchange mechanism are a closer representation of metal sorption, confirming the experimental evidence that cyanobacterial biomass is protonated or contains other cations that are released when heavy metal ions are taken up by the biomass. The ion exchange isotherm model represents an ion-exchange reaction between the metal species, M_1 and M_2 (metal ions or cations bound on biosorbent (X) [116]. The ion-exchange reaction is given by

$$X - mM_2^{+n} + nM_1^{+m} \rightleftarrows X - nM_1^{+m} + mM_2^{+n}$$

For this reaction, the equilibrium constant is represented as

$$K_{M_zM_2} = \frac{q_{M_1}^n C_{fM_2}^m}{q_{M_2}^m C_{fM_2}^n} \tag{3.11}$$

By assuming the above ion-exchange reaction, the ion-exchange isotherm equation can be derived [117] as

$$q_{M_1} = \frac{Q}{1 + C_{fM_2}/K_{M_1M_2}C_{fM_1}},$$

$$q_{M_2} = \frac{Q}{1 + C_{fM_2}/K_{M_1M_2}C_{fM_2}} \tag{3.12}$$

where M_1 and M_2 are heavy metal ions to be sorbed and already bound on the biosorbent, respectively; m and n are the valences of M_1 and M_2, respectively; Q is the total number of binding sites in algal biomass; C_{fM1} and C_{fM2} are the equilibrium (final) concentrations of metals M_1 and M_2 in solution, respectively; and q_M and q_M are the contents of M_1 and M_2 in the biosorbent. In this model, M_2 can be replaced by proton [H] if biomass is protonated.

However, it has to be stressed that important variables (such as metal and biomass concentration, pH, and ionic strength) are not taken into account in such models.

There are many equilibrium models that incorporate metal ion concentration, pH, ionic strength, and biomass swelling [118,119]. One of them is the two-site model for predicting ion-exchange and pH effect [120]. This model is based on the fact that in the ion-exchange process, 2H$^+$ ions are released from previously protonated biomass for each divalent metal ion bound. The two-site ion-exchange model assumes two types of binding sites, carboxyl (C) and sulfate (S) groups, to be present in the biomass. Considering metal and proton binding on these two sites, Schiewer and Volesky [117] derived an isotherm equation that is able to predict the binding of protons and metal ions as a function of metal concentration and pH:

$$qH = C_t\left(K_{CH}[H]\right)\Big/\left(1 + K_{CH}[H] + \left(K_{CM}[M]\right)^{0.5}\right)$$

$$+ S_t\left(K_{SH}[H]\right)\Big/\left(1 + K_{SH}[H] + \left(K_{SM}[M]\right)^{0.5}\right) \tag{3.13}$$

$$qM = C_t\left(K_{CM}[M]^{0.5}\right)\Big/\left(1 + K_{CH}[H] + \left(K_{CM}[M]\right)^{0.5}\right)$$

$$+ S_t\left(K_{SM}[M]^{0.5}\right)\Big/\left(1 + K_{SH}[H] + \left(K_{SM}[M]\right)^{0.5}\right) \tag{3.14}$$

where C_t and S_t are the total number of carboxylic and sulfate sites, respectively. Constants K_{CH} and K_{SH} are equal to $10^{4.8}$ (pK_a for carboxyl group, 4.8) and $10^{2.5}$ (pK_a for sulfate group, 2.5) [118]. K_{CM} and K_{SM} represent metal-binding constants for the binding of metal (M) on carboxyl and sulfate groups, respectively. qH and qM represent proton and metal uptake, respectively, by biosorbent. Values of C_t and S_t need to be determined for different algal biosorbents. Constants K_{CH} and K_{SH} can be used for binding any metal on a biosorbent. Parameters K_{CM} and K_{SM} have to be determined for binding individual metals. Although the applicability of this model has not been tested other than by its developer, Schiewer [118] pointed out that being pH sensitive, the two-site model could successfully predict the binding of both protons and metal at different pH values and metal concentrations. Because the two-site model considers the effect of pH and proton binding, it has a better predictive power than the Langmuir model.

3.4.2 Metal Biosorption by Cyanobacteria in Multimetal Solutions

Many industrial wastewaters contain high concentrations of several different metal ions. For instance, the wastewater arising from a fur cleaning and dyeing industry has been shown to contain 7.04, 20.14, 0.17, 1.73, and 0.12 mg/L of Cu, Cr, Ni, Zn, and Cd, respectively [121]. Similarly, Cr and Cu are frequently encountered together in industrial wastewaters, for example, from mining, metal cleaning, plating, electroplating, metal processing, dyeing, and oil industries. In metal cleaning, plating, and metal processing industries, Cu and Cr concentrations may approach 100–120 mg/L and 10–270 g/L, respectively [122]. Effluents from mining operations have Cu, Zn, Pb, Cr, As, and Se together [123]. Cr, Ni, Cd, and Zn are reported to occur together in effluents generated during electroplating operations [123]. Most of the industrial effluents have high concentrations of Al along with other metal ions. Although Al is not a major environmental problem, its ubiquitous presence in solution interferes with the sorption of many other metals [124]. While the accumulation of single species of heavy metal ions by cyanobacterial biomass has been extensively studied, less attention was given to the study of multimetal solutions. The presence of a multiplicity of metals leads to interactive effects on physiological and biochemical processes and the growth and metal uptake of various organisms [125]. Studies on multimetal systems clearly revealed that sorption may be species and metal-specific as well as variable in single- and multimetal solutions, showing, in most cases, competitive interactions among metals for the binding on sorption sites.

Singh et al. [74] examined the competitive biosorption of Pb^{2+}, Cu^{2+}, Zn^{2+}, and Cd^{2+} in single, bi-, tri-, and multimetal mixtures by using as biosorbents biomasses derived from field-grown and laboratory-grown *Microcystis* cultures. In single-metal solutions, the cyanobacterial strains exhibited the following order of metal biosorption: Pb > Cu > Zn > Cd. In sharp contrast, the biosorption of Cd was greater than Zn in tri- and multimetal solutions. All bimetal and trimetal solutions tested with any of the biomass showed antagonistic behavior for Cu, Zn, and Cd biosorption, resulting in a 12%–49% reduction in the amount of metal uptake in comparison with removal in single-metal solutions. In multimetal biosorption tests, the decrease ranged from 33.8% to 59% compared to the single-metal systems. It has to be stressed that the biosorption of Pb was less affected, or not affected at all, by the presence of other metals.

Pradhan and Rai [87] provided information on the biosorption of Cu, Zn, and Cd by *Microcystis* sp. in single-, bi-, and tri-metal solutions. The highest biosorption was observed for Cu, followed by Zn and Cd, both in single- and multimetal solutions. This observation was confirmed by the Freundlich adsorption isotherm: Cu showed the highest K_f value ($K_f = 45.18$), followed by Zn ($K_f = 16.71$) and Cd ($K_f = 15.63$) in single-metal solutions. The decrease in Cu, Zn, and Cd biosorption observed in bi- and tri-metal solutions suggested the presence of competition among metal ions for binding sites on the cell surface.

The next study on *Microcystis* sp. demonstrated that this cyanobacterium possesses the highest affinity for Fe, followed by Ni and Cr, in single-, bi-, and tri-metal solutions [107]. Fe was not only preferentially adsorbed from tri-metal solutions, but the presence of the other two metals did not cause any decrease in its biosorption. The experimental data of Fe biosorption fitted well in the Langmuir model, suggesting monolayer sorption and the existence of constant sorption energy under the experimental conditions utilized. Maximum Fe, Ni, and Cr uptake, at the equilibrium concentration, was 240, 100, and 65 mg/(g dry wt), respectively. Fe biosorption was not much affected by Ni and Cr in bi- and tri-metal solutions. In contrast, Ni biosorption decreased in multimetal in comparison with single-metal solutions. The presence of Ni and Cr did not affect Fe sorption, whereas the presence of Fe reduced Ni and Cr sorption. Interestingly, in contrast with Fe and Ni biosorption, Cr biosorption increased in tri-metal solutions of Cr-Ni-Fe in comparison with Cr-Ni solutions [107].

Parker et al. [88] demonstrated that the order in which metals are added to solutions may play an important role in the biosorption capacity exhibited by the biosorbent. In this study, the sorption of cadmium, copper, lead, manganese, and zinc by purified capsular polysaccharide from the cyano-bacterium *M. aeruginosa f. flos-aquae* strain C3–40 was evaluated. Competition between bimetal solutions was tested with simultaneous and sequential additions of metal. The results showed that cadmium and lead, as well as lead and zinc, competed relatively equally and reciprocally for polymer-binding sites. In contrast, manganese strongly inhibited the binding of cadmium and lead but was itself not substantially inhibited by either the prior or simultaneous adsorption of cadmium or lead. This result suggests that manganese can indirectly affect the access of lead and cadmium to their sites, perhaps causing an altered polymer conformation or a cross-linking [88].

Another way to study biosorbent behavior in multimetal solutions is to consider the total amount of metal ions sorbed in these conditions and to compare this result with the highest q value obtained in single-metal solutions. Micheletti et al. [93] tested nine EPS-producing cyanobacteria for their ability to remove Cr, Cu, and Ni in both single- and multimetal solutions. The comparison between the total number of millimoles of the three metals adsorbed per gram of dry biomass in multimetal solutions and the number of millimoles of the metal removed with the highest q value in single-metal solutions (usually Cu) pointed out different behaviors among the cyanobacterial strains. Two strains removed the same millimoles of metals in single- and tri-metal solutions, showing noninteractive action between the metal ions, which seem to progressively saturate the available binding sites on the biomass without any preferential order and without any reciprocal hindrance. Four rains showed antagonistic action between metal ions, which was explained by hypothesizing the setting up of competitive interaction among metals for the binding sites on the biomass. In the other three strains, a synergistic action between metals was observed, which caused an increase in the global amount of metal ions adsorbed. This behavior was explained by hypothesizing that the sorption on the biomass of ions with the highest affinity has induced a modification of the specific binding sites for ions with the lowest affinity, increasing their sorption capability.

The above-reported studies clearly point out that the sorption process by cyanobacterial biomass can be species and metal-specific and variable in the single- and multimetal solutions, the combined effects of two or more metals on the metal uptake capability of the microorganisms depending on the number of metals competing for binding sites, metal combination, levels of metal concentration, order of metal addition, and also on the surface-specific properties of the biosorbent utilized.

3.4.3 SELECTIVITY IN METAL REMOVAL IN MULTIMETAL SOLUTIONS

One of the difficulties most frequently faced in the removal of heavy metals from industrial wastewaters is connected with the possible contemporaneous presence of different metal ions in the waste stream that has to be treated. This situation may have significant consequences on the treatment of heavy metal-containing wastewaters because the presence of several different metals can reduce the capability of the biosorbent to take up the desired element. Recent research in the area of heavy metal removal from wastewaters and sediments has focused on the development of biomaterials with increased affinity, capacity, and selectivity for target metals.

As stated earlier, many functional groups, such as hydroxyl, phosphoryl, amino, carboxyl, sulfhydryl, and so on, present on the cell surface or on exocellular structures, may confer negative charge on the cells, giving them the capability of binding metal cations in solution (see Section 3.1.3).

The affinity of various cyanobacterial species for binding to metal ions shows different hierarchies. In general, metal ions with greater electronegativity and smaller ionic radii are preferentially sorbed by algal biomass. In most cases, the lowest affinity was found for Ni, with the exception of *M. aeruginosa,* which possesses a very high Ni sorption capability [97].

In a screening carried out with nine exopolysaccharide-producing cyanobacteria tested with the aim of assessing their capability of selectively removing Cu, Cr, or Ni, an interesting behavior was observed for the strain *Nostoc* PCC 7936 [93]. In the tri-metal solutions, this strain removed Cu at the same q value observed in the absence of Cr and Ni, showing a very high selectivity of its binding sites for copper. The metal affinity of the nine cyanobacteria tested generally decreased in the order Cu > Cr > Ni. The specific metal uptake was, for some of the strains, very high, in particular toward copper and chromium. Another strain, *Cyanothece* 16Som 2, showed a very high affinity for both Cu and Cr, either in single or in tri-metal solutions, while the q value for Ni dramatically decreased in the presence of the other two metals, pointing out a selective affinity of the cultures of this strain for Cu and Cr. 16Som 2 can be considered a very promising biosorbent for the selective removal of Cu and Cr from aqueous solutions, having shown q values toward these two metals that are among the highest so far reported for other cyanobacteria and microorganisms [74,76,109].

Also, *Microcystis* showed high affinity for Cu in bi- and tri-metal solutions, probably due to the existence of a greater number of active sites with high specificity for this metal [73]. Biosorption of Cu and Cd from Cu-Cd solutions indicated a greater affinity for Cu and a strong inhibition of Cd biosorption caused by the presence of Cu. High-affinity sites only include carboxyl groups, which are rapidly occupied by Cu [73]. In a subsequent study, *Microcystis* has been tested with Fe, Ni, and Cr in single-, bi-, and tri-metal solutions [107]. This study demonstrated that the specific metal uptake of *Microcystis* was greatest with Fe, followed by Ni and Cr. IR spectroscopy, utilized to determine the functional groups present on the surface of *Microcystis*, pointed out the role of a large number of -COO⁻ groups and of a limited number of amino groups in the metal removal process. Extra peaks present in the spectra obtained with the cells analyzed after the contact with Ni and Cr suggested that amino groups were responsible for Ni and Cr biosorption, whereas carboxyl groups were responsible for Fe biosorption. This study suggests the excellent biosorption potential of naturally abundant and cheap microbial biomass, which can be successfully and economically used in the selective removal of Fe from wastewaters.

Singh et al. [74], using the technique of differential pulsed anodic stripping voltammetry, showed a significant difference in Pb, Cu, Zn, and Cd binding efficiencies of the macrophyte *Lemna*, of the colonial cyanobacterium *Microcystis* and of the filamentous green alga *Spirogyra*. The order of metal biosorption was Pb > Cu > Zn > Cd. Differences in metal biosorption efficiencies were due to the presence of various functional groups on the surface of the biosorbents, which were identified by IR spectra. In this work, it was also demonstrated that the binding process of different metal ions on biomaterials having different functional groups depends on the properties of the metals (electronegativity, ionization potential, ionic radius, and redox potential). It was suggested that the higher the electronegativity of a metal, the higher its affinity toward negatively charged sorbents, as indicated by the smaller uptake of Cu (electronegativity 1.75, Pauling scale) in comparison with Pb (electronegativity 1.8, Pauling scale). Tobin et al. [24] suggested that ions with smaller ionic radii can be more quickly sorbed onto a fixed surface area of sorbent.

Parker et al. [88], using bi- and tri-metal solutions of Cd, Cu, Mn, Pb, and Zn, showed that the order in which metals are added to water solution (simultaneously or with subsequent additions) may influence the affinity and selectivity of biosorbent toward metals. These authors showed that lead and cadmium, at saturation conditions, compete fairly equally for similar or overlapping sites on the capsular polysaccharide. On the other hand, roughly 80% of the manganese-binding sites on the capsule are not the same as or overlap substantially with, those of lead or cadmium. This effect was explained by suggesting that manganese indirectly affects the access of lead and cadmium to their sites, perhaps by means of an alteration of the polymer conformation or by a cross-linking of the binding sites. Similar phenomena, involving conformational changes, cation bridging, and modified solvation, have been invoked to explain the effects of these same metals on the viscosity of the capsular polysaccharide produced by strain C3–40 [108]. De Philippis et al. [104] tested the metal removal capacity of the cultures of two exopolisaccharide-producing cyanobacteria, *Cyanospira capsulata* and *Nostoc* PCC7936. Part of this study focused on the interference of Zn(II) and Ni(II) ions with copper removal. Both cyanobacteria removed metals from the multimetal solutions containing Cu, Zn, and Ni, but copper was removed in larger amounts than Zn or Ni, showing a great degree of affinity toward this ion.

3.5   FIELD APPLICATIONS: CASE STUDIES

In the literature, a large number of studies on the metal sorption carried out by microbial biomasses operating under laboratory conditions in single-metal systems are available. On the other hand, only a limited number of studies have been dedicated to the competitive interactions occurring among metals when the microbial biomass operates in multimetal solutions (see Section 3.4), even if these systems are very similar to what occurs in many kinds of industrial wastes. These studies were mostly dedicated to testing metal bio-removal operating in pure multimetal solutions under well-controlled, laboratory conditions, whereas very few studies have been dedicated to metal bio-removal from real industrial wastewaters by using microbial biomass. In particular, the use of cyanobacteria for metal uptake from industrial wastewaters was almost completely neglected, and only in the last few years, a limited number of patents for this application have been reported (for a review of the U.S. patents, see [126]), together with some preliminary case studies.

3.5.1   CASE STUDY 1: *CYANOSPIRA CAPSULATA* FOR THE REMOVAL OF METALS FROM INDUSTRIAL WASTEWATERS

The EPS-producing filamentous, heterocystous cyanobacterium *C. capsulata* was tested with regard to its capability of removing heavy metals from two kinds of industrial wastewaters [127]. The experiments were carried out with the percolating of two different industrial solid wastes stored outdoors for several months. The two percolates, named A and B by the authors, contained respectively six and ten different metals at various concentrations, ranging from less than 1 to about 60 mg/L (Table 3.3), and were characterized by COD values of about 12–15,000 mg/L. For carrying out the tests, various amounts of fresh biomass of the cyanobacterium, obtained from the centrifugation of cultures carried

TABLE 3.3

Metal Composition of the Percolates of Two Wastes Derived from Industrial Fermentations Utilized for Experiments of Biosorption with the EPS-Producing Cyanobacterium *Cyanospira capsulata*

Metal	Metal Concentration (mg/L)	
	Wastewater A	Wastewater B
Ba	1.87	1.39
Cr	1.60	3.84
Mn	6.54	6.32
Fe	44.00	57.10
Ni	0.61	0.80
Al	8.99	27.40
Cu	Not present	9.39
Pb	Not present	6.49
Zn	Not present	6.54
Hg	Not present	5.20

Source: De Philippis et al., In *Proceedings of the European Symposium on Environmental Biotechnology*, Verstraete, W., Ed., A.A. Balkema Publishers, Leiden, pp. 459–462, 2004.

out under laboratory conditions, were directly suspended in the wastewater at pH 5.5. When the biomass of *C. capsulata* was suspended, at a concentration of 140 mg (dry weight)/L, into wastewater A, a reduction of about 60% of the initial concentration of each metal was observed, with the exception of Ba, which showed a reduction of about 40%. An increase in the biomass concentration, up to 190 and 280 mg (dry weight)/L, increased the global amount of metals removed. However, the correlation between the increase in the amount of biosorbent and the increase in metal removal was not linear, a phenomenon that was explained by hypothesizing an increase in the competitive interactions among the binding sites due to their increased proximity. For increasing the effectiveness of the treatment, the authors treated the wastewater using three consecutive cycles carried out with small amounts of fresh cyanobacterial biomass operating on the same water sample. After the second cycle, the amount of metals removed reached values of about 80%–90% of the amount present before the biological treatment, and after the third cycle, the removal reached values corresponding to about 95%–97% of the initial metal concentrations, thus demonstrating the good efficiency of *C. capsulata* in the biosorption of metals contained in wastewater A. Among the metals present, the highest value of specific metal uptake was observed for iron (20.4 mg of Fe removed per gram of biomass dry weight).

Better results were obtained when *C. capsulata* was utilized for the removal of heavy metals from wastewater B. Under this condition, the amount of metals removed at the end of the first cycle already reached values higher than 80% of the initial metal concentration. After the second cycle, the removal attained values higher than 97% for each metal, thus demonstrating that the system was very efficient in spite of the presence of a large number of different metals. In such a complex system, the highest value of specific metal uptake was again found for iron, with a value of 24.8 mg/Fe (g biomass dry wt).

3.5.2 Case Study 2: EPS-Producing Cyanobacteria for the Removal of Cr(VI) from Wastewaters of Plating Industries

The aim of this experimentation was to assess the capability of some EPS-producing cyanobacteria to remove chromate from the wastewaters of a plating industry in laboratory and in semi-pilot devices [128]. A preliminary screening, carried out under laboratory conditions with seven EPS-producing cyanobacteria, pointed out the good efficiency of one *Nostoc* strain for the removal of chromate from wastewaters of a plating industry. In laboratory experiments, a complex role of the cyanobacterial biomass of this strain in the removal of Cr(VI) was observed: the cells, when subjected to treatments removing the external polysaccharidic structures, carried out the reduction of Cr(VI) to Cr(III), while the polysaccharidic fraction, previously released and solubilized by cells in the culture medium, was capable of removing the Cr(III) cations formed in the reduction process. This behavior was confirmed in field experiments carried out with the cyanobacterial biomass confined to three different experimental devices (a filter press, a filter column, and a dialysis membrane system) and operating with 50 L of Cr(VI)-containing wastewaters of a plating industry. The results obtained, presented at the last Congress of the International Society for Applied Phycology held in Galway, Ireland, in June 2008, showed the potential of using EPS-producing cyanobacteria in the removal of heavy metals present in anionic form in water solutions.

3.6 CONCLUSIONS

The above-reported results, obtained by a large number of research groups utilizing cyanobacterial biomass for the removal of heavy metals from water solutions, pointed out very promising performances for some of the species tested. In particular, some of the EPS-producing cyanobacteria showed very high uptake of positively charged metal ions in comparison with other microbial sorbents previously utilized. For instance, the results obtained in the removal of copper ions with the cultures of *Cyanothece* 16Som 2, a unicellular cyanobacterium characterized by a thick polysaccharidic capsule surrounding the cells (Figure 3.7), were the highest so far reported for bacterial, algal, and fungal biosorbents (Table 3.4).

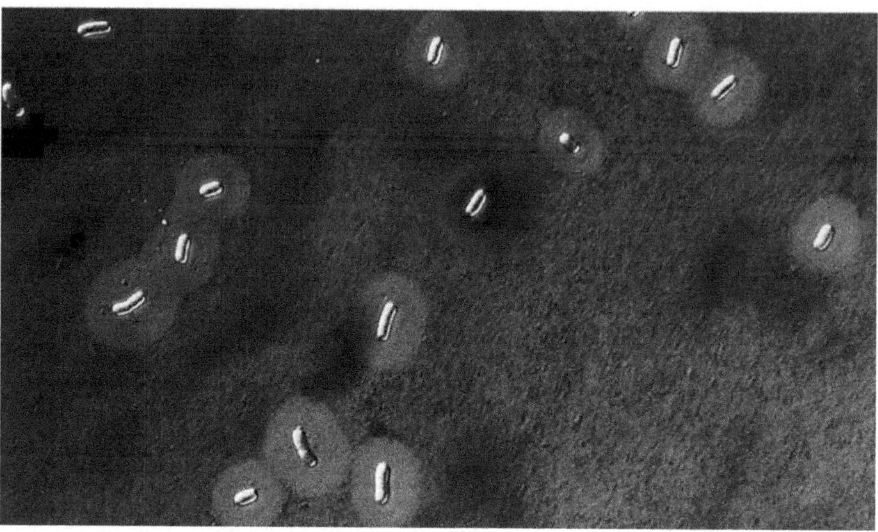

FIGURE 3.7 Nomarski differential interference contrast photomicrographs of the capsulated strain *Cyanothece* 16Som 2 stained with India ink (scale bar = 10 μm).

TABLE 3.4

Comparison of the Best Results Obtained in the Bioremoval of Cu with Microorganisms and Macroalgae

Species	q_{max} (mg/g)	References
Cyanothece 16Som 2	201	[93]
Spirulina sp.	196	[76]
Oscillatoria sp.	165	[96]
Arthrobacter sp.	148	[129]
Cyanospira capsulata	143	[93]
Sargassum sp.	140	[38]
Candida tropicalis	107	[66]

However, as is evident from Section 3.5, very few case studies have been published, showing that this kind of biomass is still far from being utilized at the industrial level in the treatment of wastewaters containing heavy metals. Two main reasons can be suggested to explain this situation: (a) the production costs of cyanobacterial biomass, which are still too high to be attractive for industrial applications of this kind of biosorbent, and (b) a lack of research-oriented to the setting up of a suitable device that could be profitably used with exopolysaccharide-producing cyanobacteria for metal bio-removal. A possible solution to the former problem may reside in the use of bloom-derived biomass, which is freely available at no production cost. Indeed, in many water bodies, there is a periodic unrestrained growth of microalgal and cyanobacterial blooms due to the introduction of inorganic nutrients into the environment. Recently, preliminary experiments carried out by our research group together with the group headed by Professor Liu Yong Ding (Institute of Hydrobiology, Chinese Academy of Sciences, Wuhan, China) on the use of bloom-derived biomass withdrawn from Lake Dian Chi, Yunnan Province, China, showed a promising capability of this biomaterial to act as sorbent toward metals in aqueous solutions. Studies are now in progress for the optimization of operational conditions. Another interesting development that could attract interest in the use of EPS-producing cyanobacterial biomass is the use of this biosorbent for the recovery of metals with high economical value such as gold, ruthenium, palladium, and copper. In this connection, our research group is carrying out a study with a large number of EPS-producing cyanobacterial strains for the removal of these valuable metals from wastewaters and for finding the best conditions for recovering the metals from the biomass.

As a general conclusion, it is possible to say that interactions between cyanobacteria and metals are very complex, depending on a large number of factors that are related to the chemical and morphological features of microbial cells, to the chemical and physical properties of metals to be removed, and to the operational conditions utilized in the treatment. For these reasons, the selectivity, a characteristic that may have a great interest for many industrial applications, as well as the good metal uptake capability shown by some of the cyanobacteria tested in the laboratory, must be confirmed under conditions reproducing the specific industrial application that is of interest. In any case, the promising results so far obtained, the increase in the prices of chemical products, the possibility of using free cyanobacterial biomass derived from microalgal blooms, and the increasing public perception of the interest in using green biotechnologies for the resolution of problems derived from the pollution of water bodies seem to open up new perspectives in the use of cyanobacteria for the removal of heavy metals from wastewaters.

GLOSSARY

Cyanobacteria These are photosynthetic bacteria commonly found in aquatic environments. They play a crucial role in oxygen production and nutrient cycling.

Exopolysaccharide-producing cyanobacteria They are photosynthetic bacteria that are capable of producing exopolysaccharides (EPS) and can effectively adsorb heavy metals from polluted waters.

Exopolysaccharides (EPS) These are complex, long-chain carbohydrates produced by cyanobacteria. EPS can be found either intra-cellularly (within the cell) or extracellularly (outside the cell).

Extracellular exopolysaccharides (EEPS) These polysaccharides stick to the outer side of the cell wall and can also exist in the supernatant (not connected to the cell wall). They serve various purposes, including protection, adhesion, and resource storage.

ACKNOWLEDGMENTS

The authors gratefully acknowledge all the members of the research group operating at the Department of Agricultural Biotechnology of the University of Florence (Dr. Giovanni Colica and the graduate/undergraduate students Giacomo Bertini, Pier Cesare Mecarozzi, Simone Petreni, and Federico Rossi) for their enthusiastic activity in the laboratory and in the field, which made it possible to obtain the unpublished results mentioned in this review. The authors also thank Dr Claudio Sili (ISE, CNR, Firenze, Italy) for some of the photomicrographs shown in this chapter, Federico Rossi for his contribution in editing the manuscript, and Ente Cassa di Risparmio di Firenze, which supported, in the framework of the *Firenze Hydrolab* project, the purchase of the culture devices and analytical instruments utilized for all the research that was carried out at the Department of Agricultural Biotechnology.

REFERENCES

1. Forster, C.F. and Wase, D.A.J. Biosorption of heavy metals: An introduction. In: *Biosorbents for Metal Ions*, Wase, D.A.J. and Forster, C., Eds., Taylor & Francis, London, 1997.
2. Ledin, M. Accumulation of metals by microorganisms-processes and importance for soil systems. *Earth Sci. Rev.*, 51, 1–31, 2000.
3. Duffus, J.H. "Heavy metals"—A meaningless term? *Pure Appl. Chem.*, 74(5), 793–807, 2002.

4. Zhou, Q., Yang, N., Li, Y., Ren, B., Ding, X., Bian, H., and Yao, X. Total concentrations and sources of heavy metal pollution in global river and lake water bodies from 1972 to 2017. *Glob. Ecol. Conserv.*, 22, e00925, 2020.

5. Tiemann, K.J., Gardea-Torresdey, J.L., Gamez, G., Dokken, K., Cano-Aguilera, I., Renner, M.W., and Furenlid, L.R. Effects of oxidation state on metal ion binding by *Medicago sativa* (alfalfa): Atomic and X-ray absorption spectroscopic studies with Fe(II) and Fe(III). *Environ. Sci. Technol.*, 34(4), 693–698, 2000.

6. Glanze, W. *The Signet Mosby Medical Encyclopedia*, Revised edition, Penguin Books, New York, 1996.

7. International Occupational Safety and Health Information Centre. Metals. In: *Basics of Chemical Safety*. International Labour Organization, Geneva, Chapter 7, 1999.

8. Bharti, R. and Sharma, R. Effect of heavy metals: An overview. *Mater. Today: Proc.*, 51, 880–885, 2022.

9. Hunsom, M., Pruksathorn, K., and Damronglerd, S., Vergnes, H. and Duverneuil, P., Electrochemical treatment of heavy metals (Cu2+, Cr6+, Ni2+) from industrial effluent and modeling of copper reduction. *Water Res.*, 39(4), 610–616, 2005.

10. Aziz, A., Agamuthu, P., and Fauziah, S.H. Removal of bisphenol A and 2, 4-Di-tert-butylphenol from landfill leachate using plant-based coagulant. *Waste Manag. Res.*, 36(10), 975–984, 2018.

11. Zhang, L., Yanjun, W.U., Xiaoyan, Q.U., Zhenshan, L., and Jinren, N. Mechanism of combination membrane and electro-winning process on treatment and remediation of Cu2+ polluted water body. *J. Environ. Sci.*, 21(6), 764–769, 2009.

12. O'Connell, D.W., Birkinshaw, C., O'Dwyer, T.F. Heavy metal adsorbents prepared from the modification of cellulose: A review. *Bioresour. Technol.*, 99(15), 6709–6724, 2008.

13. Volesky, B. and Holan, Z.R. Biosorption of heavy metals. *Biotechnol. Prog.*, 11, 235–250, 1995.

14. Won, S.W., Yun, H.J., and Yun, Y.S. Effect of pH on the binding mechanisms in biosorption of Reactive Orange 16 by Corynebacterium glutamicum. *J. Colloid Interface Sci.*, 331(1), 83–89, 2009.

15. Sajjad, M., Aziz, A., and Kim, K.S. Biosorption and binding mechanisms of Ni2+ and Cd2+ with aerobic granules cultivated in different synthetic media. *Chem. Eng. Technol.*, 40(12), 2179–2187, 2017.

16. Qu, J., Meng, X., Jiang, X., You, H., Wang, P., and Ye, X., Enhanced removal of Cd (II) from water using sulfur-functionalized rice husk: Characterization, adsorptive performance and mechanism exploration. *J. Clean. Prod.*, 183, 880–886, 2018.

17. Janyasuthiwong, S., Rene, E.R., Esposito, G., and Lens, P.N. Effect of pH on Cu, Ni and Zn removal by biogenic sulfide precipitation in an inversed fluidized bed bioreactor. *Hydrometallurgy*, 158, 94–100, 2015.

18. Liu, M., Dong, F., Zhang, W., Nie, X., Wei, H., Sun, S., Zhong, X., Liu, Y., and Wang, D. Contribution of surface functional groups and interface interaction to biosorption of strontium ions by Saccharomyces cerevisiae under culture conditions. *RSC Adv.*, 7(80), 50880–50888, 2017.

19. Beveridge, T.J. Role of cellular design in bacterial metal accumulation and mineralization. *Ann. Rev. Microbiol.*, 43, 147–171, 1989.

20. Kuyucak, N. Feasibility of biosorbents application. In: *Biosorption of Heavy Metals*, Volesky, B., Ed., CRC Press, Boca Raton, FL, pp. 371–378, 1990.

21. Yin, P., Yu, Q., Jin, B., and Ling, Z. Biosorption removal of cadmium from aqueous solution by using pretreated fungal biomass cultured from starch wastewater. *Water Res.*, 33(8), 1960–1963, 1999.

22. Galun, M., Keller, P., and Malki, D. Removal of uranium (VI) from solution by fungal biomass and fungal wall related biopolymers. *Science*, 219, 285–286, 1983.

23. Huang, C.P., Westman, D., Quirk, K., and Huang, J.P. The removal of cadmium (II) from dilute aqueous solutions by fungal adsorbent. *Water Sci. Technol.*, 20, 369–376, 1988.

24. Tobin, J.M., Cooper, D.G., and Neufeld, R.J. Uptake of metal ions by *Rhizopus arrhizus* biomass. *Appl. Microbiol.*, 47, 821–824, 1984.

25. Azab, M.S. and Peterson, P.J. The removal of cadmium from water by the use of biological sorbents. *Water Sci. Technol.*, 21, 1705–1706, 1989.

26. Brady, D., Stoll, A., and Duncan, J.R. Biosorption of heavy metal cations by non-viable yeast biomass. *Environ. Technol.*, 15, 429–438, 1994.

27. Muzzarelli, R.A.A., Tanfani, F., and Scarpini, G. Chelating, film-forming and coagulating ability of the chitosan-glucan complex from *Aspergillus niger* industrial wastes. *Biotechnol. Bioeng.*, 22, 885–896, 1980.

28. Muzzarelli, R.A.A., Tanfani, F., Scarpini, G., and Tucci, E. Removal and recovery of cupric and mercuric ions from solutions using chitosan-glucan from *Aspergillus niger*. *J. Appl. Biochem.*, 2, 54–59, 1980.

29. Wales, D.S. and Sagar, B.F. Recovery of metal ions by microfungal filters. *J Chem. Technol. Biotechnol.*, 49(4), 345–355, 1990.

30. Tsezos, M. and Volesky, B. Biosorption of uranium and thorium. *Biotechnol. Bioeng.*, 23, 583–604, 1981.

31. Gadd, G.M. Accumulation of metal by micro-organisms and algae. In *Biotechnology: A Complete Treatise*, Vol. 6b, Rehm, H., Ed., Special microbial processes, Vol. 4, VCH, Verlagsgesellschaft, Weinheim, pp. 401–430, 1988.

32. Huang, C.P., Westman, D., Quirk, K., Huang, J.P., and Morehart, A.L. Removal of cadmium (II) from dilute solutions by fungal biomass. *Particulate Sci. Technol.*, 6, 405–419, 1988.

33. Yakubu, N.A. and Dudeney, A.W.L. Biosorption of uranium with *Aspergillus niger*. In *Immobilization of Ions by Biosorption*, Eccles, H.H. and Hunt, S., Eds., Ellis Horwood, Chirchester, UK, pp. 183–200, 1986.

34. Volesky, B., Biosorption and me. *Water Res.*, 41, 4017–4029, 2007.

35. Berkeley, R.C.W. Chitin, chitosan and their degradative enzymes. In *Microbial Polysaccharides.* Berkeley, R.C.W., Gooday, C.W., and Elwood, D.C., Eds., Academic Press, New York, 205–236, 1979.

36. Hosea, M., Greene, B., McPherson, R., Henzl, M., Alexander, M.D., and Darnall, D.V. Accumulation of elemental gold on the alga *Chlorella vulgaris. Inorg. Chim. Acta*, 123(3), 161–165, 1986.

37. Kuyucak, N. and Volesky, B. Biosorbents for recovery of metals from industrial solutions. *Biotechnol. Lett.*, 10(2), 137–142, 1988.

38. Davis, T.A., Volesky, B., and Mucci, A. A review of the biochemistry of heavy metal biosorption by brown algae. *Water Res.*, 37, 4311–4330, 2003.

39. Volesky, B. Advances in biosorption of metals: Selection of biomass types. *FEMS Microbiol. Rev.*, 14(4), 291–302, 1994.

40. Volesky, B. *Sorption and* Biosorption, BV Sorbex Inc., Montreal-St. Lambert, Canada, 2004.

41. Whitton, B.A. and Potts, M. *The Ecology of Cyanobacteria, Their Diversity in Time and Space*, 1st Edition, Kluwer Academic Publishers, Dordrecht, the Netherlands, 2000.

42. Smith, A.J. Modes of cyanobacterial carbon metabolism. *Ann. Microbiol. (Inst. Pasteur)*, 134b, 93–113, 1983.

43. Stal, L.J. and Moezelaar, R. Fermentation in cyanobacteria. *FEMS Microbiol. Rev.*, 21, 179–211, 1997.

44. Cohen, Y., Jørgensen, B.B., Revsbech, N.P., and Paplawski, R. Adaptation to hydrogen sulfide of oxi- genic and anoxigenic photosynthesis among cyanobacteria. *Appl. Environ. Microbiol.*, 51, 398–407, 1986.

45. Fay, P. Oxygen relations of nitrogen fixation in cyanobacteria. *Microbiol. Rev.*, 56, 340–373, 1992.

46. Bergman, B., Gallon, J.R., Rai, A.N., and Stal, L.J. N2 fixation by non-heterocystous cyanobacteria. *FEMS Microbiol. Rev.*, 19, 139–185, 1997.

47. Whitton, B.A. Diversity, ecology and taxonomy of the cyanobacteria. In *Photosynthetic Prokariotes*, Mann, N.H. and Carr, N.G., Eds., Plenum Press, New York, pp. 1–51, 1992.

48. Tandeau de Marsac, N. and Houmard, J. Adaptation of cyanobacteria to environmental stimuli: New steps towards molecular mechanisms. *FEMS Microbiol. Rev.*, 104, 119–190, 1993.

49. Tomaselli, L. The microalgal cell. In *Microalgal Culture: Biotechnology and Applied Phycology*, Richmond, A., Ed., Blackwell Science, Oxford, UK, pp. 3–19, 2004.

50. Hoiczyk, E. and Hansel, A. Cyanobacteria cell walls: News from an unusual prokariotic envelope. *J. Bacteriol.*, 182(5), 1191–1199, 2000.

51. Castenholz, R.W. and Phylum, B.X. Cyanobacteria. Oxygenic photosynthetic bacteria. In *Berge's Manual of Systematic Bacteriology. Vol. 1: The Archaea and the Deeply Branching and Phototropic Bacteria*. 2nd Edition. Garrity, G., Boone, D.R., and Castenholz, R.W., Eds., Springer-Verlag, New York, pp. 473–599, 2001.

52. Hoiczyc, E. Structural and biochemical analysis of the sheath of *Phormidium uncinatum. J. Bacteriol.*, 180, 3923–3932, 1998.

53. Hoiczyc, E. and Baumeister, W. Envelope structure of four gliding filamentous cyanobacteria. *J. Bacteriol.*, 117(9), 2387–2395, 1995.

54. De Philippis, R. and Vincenzini, M. Exocellular polysaccharides from cyanobacteria and their possible applications. *FEMS Microbiol. Rev.*, 22, 151–175, 1998.

55. Drews, J. and Weckesser, J. Function, structure and composition of cell walls and external layers. In *The Biology of Cyanobacteria*, Carr, N.G. and Whitton, B.A., Eds., Blackwell Scientific, Oxford, pp. 333–357, 1982.

56. Martin, T.J. and Wyatt, J.T. Extracellular investments in blue-green algae with particular emphasis on genus *Nostoc. J. Phycol.*, 10, 204–210, 1974.

57. Sutherland, I.W. Biotechnology of microbial exopolysaccharides. In *Biotechnology of Microbial Exopolysaccharides*, Cambridge University Press, Cambridge, pp. 12–117, 1990.

58. De Philippis, R., Sili, C., Paperi, R., and Vincenzini, M. Exopolysaccharide-producing cyanobacteria and their possible exploitation: A review. *J. Appl. Phycol.*, 13, 293–299, 2001.

59. De Philippis, R., Paperi, R., Sili, C., and Vincenzini, M. Assessment of the metal removal capacity of two capsulated cyanobacteria, *Cyanospira capsulata* and *Nostoc* PCC 7936. *J. Appl. Phycol.*, 15, 155–161, 2003.

60. Inthorn, D. Removal of heavy metal by using microalgae. In *Photosynthetic Microorganisms in Environmental Biotechnology*, Kojima, H. and Lee, Y.K., Eds., Springer-Verlag, Hong Kong, pp. 111–135, 2001.

61. Inthorn, D., Siditoon, N., Silapanuntakul, S., and Incharoensakdi, A. Sorption of mercury, cadmium and lead by microalgae. *ScienceAsia*, 28, 253–261, 2002.

62. Drungkokkruad, N. Removal of Cadmium and Lead in Aqueous Solution by *Nostoc paludosum* and *Phormidium anguistissimum*. Masters Degree Thesis, Mahidol University, 2002.

63. Kratochvil, D. and Volesky, B. Advances in the biosorption of heavy metals. *Trends Biotechnol.*, 16, 291–300, 1998.

64. Tease, B.E. and Walker, R.W. Comparative composition of the sheath of the cyanobacterium *Gloeothece* ATTC 27152 cultured with and without combined nitrogen. *J. Gen. Microbiol.*, 133, 3331–3339, 1987.

65. Xue, H.B. and Sigg, L. Binding of Cu (II) to algae in a metal buffer. *Water Res.*, 24, 1129–1136, 1990.

66. Singh, S.P., Pradhan, S., and Rai, L.C. Comparative assessment of Fe3+ and Cu2+ biosorption by field and laboratory-grown *Microcystis*. *Process Biochem.*, 33, 495–504, 1998.

67. Tien, C.J., Sigee, D.C., and White, K.N. Copper adsorption kinetics of cultured algal cells and freshwater phytoplankton with emphasis on cell surface characteristics. *J. Appl. Phycol.*, 17, 379–389, 2005.

68. Ruangsomboon, S., Chidthaisong, A., Bunnag, B., Inthorn, D., and Harvey, N.W. Lead adsorption characteristics and sugar composition of capsular polysaccharides of cyanobacterium *Calothrix marchica*. *Songklanakarin. J. Sci. Technol.*, 29(2), 529–541, 2007.

69. Blanco, A., Sanz, B., Lama, M.J., and Serra, L.J. Biosorption of heavy metals to immobilized *Phormidium laminosum* biomass. *J. Biotechnol*, 69, 227–240, 1999.

70. Nakagawa, M., Takamura, Y., and Yagi, O. Isolation and characterization of the slime from a cyanobacterium, *Miicrocystis aeruginosa* K-3A. *Agric. Biol. Chem.*, 51, 329–337, 1987.

71. Plude, J.L., Parker, D.L., Schommer, O.J., Timmerman, R.J., Hagstrom, S.A., Joers, J.M., and Hnasko, R. Chemical characterization of polysaccharide from the slime layer of the cyanobacterium *Microcystis fl os-aquae* C3–40. *Appl. Environ. Microbiol.*, 57, 1696–1700, 1991.

72. Nagase, H., Inthorn, D., Oda, A., Nishimura, J., Kajiwara, Y., Park, M.O., Hirata, K. and Miyamoto, K. Improvement of selective removal of heavy metals in cyanobacteria by NaOH treatment. *J. Biosci. Bioeng.*, 99(4), 372–377, 2005.

73. Pradhan, S. and Rai, L.C. Optimization of flow rate, initial metal ion concentration and biomass density for maximum removal of Cu (II) by immobilized *Microcystis*. *World J. Microbiol. Biotechnol.*, 16, 579–584, 2000.

74. Singh, S., Pradhan, S., and Rai, L.C. Metal removal from single and multimetallic systems by different biosorbent materials as evaluated by differential pulse anodic stripping voltammetry. *Process Biochem.*, 36, 175–182, 2000.

75. Fourest, E. and Roux, J.C. Heavy metal biosorption by fungal mycelial by-products: Mechanisms and influence of pH. *Appl. Microbiol. Biotechnol.*, 3, 399–403, 1992.

76. Chojnacka, K., Chojnacki, A., and Gorecka, H. Biosorption of Cr3+, Cd2+ and Cu2+ by blue-green algae *Spirulina* sp.: Kinetics, equilibrium and ions the mechanism of the process. *Chemosphere*, 59, 75–84, 2005.

77. Karna, R.R., Uma, L., Subramanian, G., and Mohan, P.M. Biosorption of toxic metal ions by alkali-extracted biomass of a marine cyanobacterium, *Phormidium valderianum* BDU 30501. *World J. Microbiol. Biotechnol.*, 15, 729–732, 1999.

78. Sandau, E., Sandau, P., Pultz, L.O., and Zimmermann, M. Heavy metal sorption by marine algae and algal by-products. *Acta Biotechnol.*, 16, 103–119, 1996.

79. Chong, K.H. and Volesky, B. Metal biosorption equilibria in a ternary system. *Biotechnol. Bioeng.*, 49, 629–638, 1996.

80. Takatera, K. and Watanabe, T. Individual and synergistic effects of heavy-metal ions on the induction of cyanobacterial metallothionein examined by high-performance liquid-chromatography inductively coupled plasma mass-spectrometry. *Anal. Sci.*, 9, 19–23, 1993.

81. Takashi, N., Noriyuki, T.H.N., and Tadashi, M. Heavy metal removal by biosorption using halophilic cyanobacterium *Aphanocapsa halophytia. Nippon Kagakkai Baiotekunoroji Bukai Shinpojiumu Koen Yoshishu*, 3, 59, 1998.

82. Yee, N., Phoenix, V.R., Konhauser, K.O., Benning, L.G., and Ferris, F.G. The effect of cyanobacteria on silica precipitation at neutral pH: Implications for bacterial silicification in geothermal hot springs. *Chem. Geol.*, 199, 83–90, 2003.

83. Les, A. and Walker, W.R. Toxicity and binding of copper, zinc, and cadmium by blue-green alga, *Chroococcus paris. Water Air Soil Pollut.*, 23, 129–139, 1998.

84. Mohamed, Z.A. Removal of cadmium and manganese by a non-toxic strain of the freshwater cyanobacterium *Gloeothece magna. Water Res.*, 35, 4405–4409, 2001.

85. Klimmek, S., Stan, H.J., Wilke, A., Bunke, G., and Buchholz, R. Comparative analysis of the biosorption of cadmium, lead, nickel and zinc by algae. *Environ. Sci. Technol.*, 35, 4283–4288, 2001.

86. El-Enany, A.E. and Issa, A.A. Cyanobacteria as a biosorbent of heavy metals in sewage water. *Environ. Toxicol. Pharmacol.*, 8, 95–101, 2000.

87. Pradhan, S. and Rai, L.C. Biotechnological potential of *Microcystis* sp. in Cu, Zn and Cd biosorption from single and multimetallic systems. *Biometals*, 14, 67–74, 2001.

88. Parker, D.L., Mihalick, J.E., Plude, J.L., Plude, M.J., Clark, T.P., Egan, L., Flom, J.J., Rai, L.C., and Kumar, H.D. Sorption of metals by extracellular polymers from the cyanobacterium *Microcystis aeruginosa* fo. *flos-aquae* strain C3–40. *J. Appl. Phycol.*, 12, 219–224, 2000.

89. Rangsayatorn, N., Pokethitiyook, P., Upatham, E.S., and Lanza, G.R. Cadmium biosorption by cells of *Spirulina platensis* TISTR8217 immobilized in alginate and silica gel. *Environ. Int.*, 30, 57–63, 2004.

90. Zhou, J.L., Huang, P.L., and Lin, R.G. Sorption and desorption of Cu and Cd by macroalgae and microalgae. *Environ. Pollut.*, 10, 67–75, 1998.

91. Ofer, R., Yerachmiel, A., and Shmuel, Y. Marine macroalgae as biosorbent for cadmium and nickel in water. *Water Environ. Res.*, 75, 246–253, 2003.

92. Gardea-Torresdey, J.L., Arenas, J.L., Francisco, N.M.C., Tiemann, K.J., and Webb, R. Ability of immobilized cyanobacteria to remove metal ions from solution and demonstration of the presence of metallothionein genes in various strains. *J. Hazard. Sub. Res.*, 1, 1–18, 1998.

93. Micheletti, E., Colica, G., Viti, C., Tamagnini, P., and De Philippis, R. Selectivity in the heavy metal removal by exopolysaccharide-producing cyanobacteria. *J. Appl. Microbiol.*, 105, 88–94, 2008.

94. Paperi, R., Micheletti, E., and De Philippis, R. Optimization of copper sorbing-desorbing cycles with confined cultures of the exopolysaccharide-producing cyanobacterium *Cyanospyra capsulata. J. Appl. Microbiol.*, 101, 1351–1356, 2006.

95. Micheletti, E., Pereira, S., Mannelli, F., Moradas-Ferreira, P., Tamagnini, P., and De Philippis, R. Sheathless mutant of the cyanobacterium *Gloethece* sp. PCC 6909 with increased capacity to remove copper ions from aqueous solutions. *J. Appl. Environ. Microbiol.*, 74(9), 2797–2804, 2008.

96. Ahuja, P., Gupta, R., and Saxena, R.K. *Oscillatoria angustissima*: A promising Cu2+ biosorbent. *Curr. Microbiol.*, 35, 151–154, 1997.

97. Pradhan, S., Singh, S., Rai, L.C., and Parker, D.L. Evaluation of metal biosorption efficiency of laboratory- grown *Microcystis* under various environmental conditions. *J. Microbiol. Biotechnol.*, 8, 53–60, 1998.

98. Donmez, G., Aksu, Z., Oztourk, A., and Kutsal, T.A. Comparative study on heavy metal biosorption characteristics of some algae. *Process Biochem.*, 34, 885–892, 1999.

99. Kitjaharn, P. and Incharoensakdi, A. Factors affecting the accumulation of lead by *Aphanothece halophytica. J. Sci. Chula*, 17, 141–147, 1992.

100. Gong, R., Ding, Y.D., Liu, H., Chen, Q., and Liu, Z. Lead biosorption by intact and pretreated *Spirulina maxima* biomass. *Chemosphere*, 58, 125–130, 2005.

101. Incharoensakdi, A. and Kitjaharn, P. Zinc biosorption from aqueous solution by a halotolerant cyanobacterium *Aphanothece halophytica, Curr. Microbiol.*, 45, 261–264, 2002.

102. Mohapatra, H. and Gupta, R. Concurrent sorption of Zn (II), Cu (II) and Co (II) by *Oscillatoria angustis- sima* as a function of pH in binary and ternary metal solutions. *Bioresour. Technol.*, 96, 1387–1398, 2005.

103. Ahuja, P., Gupta, R., and Saxena, R.K. Zn biosorption by *Oscillatoria anguistissima. Process Biochem.*, 34, 77–85, 1999.

104. De Philippis, R., Paperi, R., and Sili, C. Heavy metal sorption by released polysaccharides and whole cultures of two exopolysaccharide-producing cyanobacteria. *Biodegradation*, 18, 181–187, 2007.

105. Qin, H., Hu, T., Zhai, Y., Lu, N., and Aliyeva, J. The improved methods of heavy metals removal by biosorbents: A review. *Environ. Poll.*, 258, 113777, 2020.

106. Li, P., Liu, Z., and Xu, R. Chemical characterization of the released polysaccharide from the cyanobacterium *Aphanothece halophytica* GR02. *J. Appl. Phycol.*, 13, 71–77, 2001.

107. Pradhan, S., Singh, S., and Rai, L.C. Characterization of various functional groups present in the capsule of *Microcystis* and study of their role in biosorption of Fe, Ni and Cr. *Bioresour. Technol.*, 98, 595–601, 2007.

108. Parker, D.L., Schram, B.R., Plude, J.L., and Moore, R.E. Effect of metal cations on the viscosity of a pectin-like capsular polysaccharide from the cyanobacterium *Microcystis fl os-aquae* C3–40. *Appl. Environ. Microbiol.*, 62, 1208–1213, 1996.

109. Tien, C.J. Biosorption of metal ions by freshwater algae with different surface characteristics. *Process Biochem.*, 38, 605–613, 2002.

110. Adhiya, J., Cai, X., Sayre, R.T., and Traina, .J. Binding of aqueous cadmium by the lyophilized biomass of *Chlamydomonas Reinhardtii*. *Physicochem. Eng. Asp.*, 210, 1–11, 2002.

111. Cho, D.Y., Lee, S.T., Park, S.W., and Chung, A.S. Studies on the biosorption of heavy metals onto *Chlorella vulgaris*. *J. Environ. Sci. Health, Part A*, 29, 389–409, 1994.

112. Lee, J.D. *Concise Inorganic Chemistry*, ELBS with Chapman and Hall, London, 1998.

113. Lo, W., Ng, L.M., Chua, H., Yu, P.H.F., Sin, S.N., and Wong, P.K. Biosorption and desorption of copper (II) ions by *Bacillus* sp. *Appl. Biochem. Biotechnol.*, 105, 581–591, 2003.

114. Ferraz, A.I., Tavares, T., and Teixera, J.A. Cr (III) removal and recovery from *Saccharomyces cerevisiae*. *Chem. Eng. J.*, 105, 11–20, 2004.

115. Al-Asheh, S. and Duvnjak, Z. Binary metal sorption by pine bark: Study of equilibria and mechanisms. *Sep. Sci. Technol.*, 33, 1303–1329, 1998.

116. Kratochvil, D., Volesky, B., and Demopoulos, G. Optimizing Cu removal/recovery in a biosorption column. *Water Res.*, 31, 2327–2339, 1998.

117. Figuera, M., Volesky, B., and Ciminelli, V.S.T. Assessment of interference in biosorption of heavy metals. *Biotechnol. Bioeng.*, 54, 334–350, 1997.

118. Schiewer, S. Modeling complexation and electrostatic attraction in heavy metal biosorption by *Sargassum* biomass. *J. Appl. Phycol.*, 11, 79–87, 1999.

119. Schiewer, S. and Wong, M.H. Ionic strenght effects in biosorption of metals by marine algae. *Chemosphere*, 41, 271–282, 2000.

120. Schiewer, S. and Volesky, B. Modeling of the proton-metal ion exchange in biosorption. *Environ. Sci. Technol.*, 29, 3049–3058, 1995.

121. Klein, L.A., Lang, M., Nash, N., and Kirschner, S.L. Sources of metals in New York city watewaters. *J. Water Pollut. Control. Fed.*, 46, 2653–2662, 1974.

122. Sag, Y. and Kutsal, T., Fully competitive biosorption of Cr (VI) and Fe (III) ions from binary metal mixtures by *R. arrhizus*: Use of competitive Langmuir model. *Process Biochem.*, 31, 573–585, 1996.

123. Volesky, B. Detoxification of metal-bearing effluents: Biosorption for the next century. *Hydrometallurgy*, 59, 203–216, 2001.

124. Lee, H.S., Suh, J.H., Kim, B.I., and Yoon, T. Effect of aluminium in two-metal biosorption by an algal biosorbent. *Miner. Eng.*, 17, 487–493, 2004.

125. Ting, Y.P., Lawson, F., and Prince, I.G. Uptake of cadmium and zinc by the alga *Chlorella vulgaris*. II. Multi-ion situation. *Biotechnol. Bioeng.*, 37, 445–455, 1991.

126. Sekar, S. and Paulraj, P. Strategic mining of cyanobacterial patents from the USPTO patent database and analysis of their scope and implication. *J. Appl. Phycol.*, 19, 277–292, 2007.

127. De Philippis, R., Paperi, R., and Vincenzini, M. Metal bioremoval from industrial waste waters with an EPS-producing cyanobacterium. In Proceedings of the European Symposium on Environmental Biotechnology, Verstraete, W., Ed., A.A. Balkema Publishers, Leiden, pp. 459–462, 2004.

128. Colica, G. Personal Communication, 2009.

129. Veglid, F., Beolchini, F., Gasbarro, A., Lora, S., Corain, B., and Toro, L. Polyhydroxoethylmethacrylate (polyHEMA)-trimethylolpropanetrimethacrylate (TMPTM) as a support for metal biosorption with *Arthrobacter* sp. *Hydrometallurgy*, 44, 317–320, 1997.

Yanxin Wang, Yamin Deng, Kaihou Huang,
Xiao Liu, and Lawrence K. Wang

4 Environmental Geochemistry of High-Arsenic Aquifer Systems

ACRONYMS AND NOMENCLATURES

(Co,Fe)AsS	glaucodot
As	arsenic
As$_2$O$_3$	arsenolite
As$_2$S$_2$	realgar
As$_2$S$_3$	orpiment
As$_4$S$_4$	realgar
AVS	acid volatile sulfides
CA	citric acid
CAO	chemolithoautotrophic arsenite oxidizers
CoAs	safflorite
CoAsS	cobaltite
Cu$_3$AsS$_4$	enargite
DARPs	dissimilatory arsenate-respiring prokaryotes
DMA(III)	dimethylarsinous acid
DMA(V)	dimethylarsinic
DMAA	dimethylarsinic acid
FA	fulvic acids
Fe(As,S)$_2$	arsenian pyrite
Fe$_2$O$_3$	hematite
Fe$_3$O$_4$	magnetite
FeAs$_2$	loellingite
FeAsS	arsenopyrite
FeCO$_3$	siderite
FeRB	Iron-reducing bacteria
FTIR	Fourier transform infrared spectroscopy analyses
HA	humic acid
HAO	heterotrophic arsenite oxidizers
HFO	hydrous ferric oxides
MCL	maximum concentration limit
MMA(III)	monomethylarsonous acid
MMA(V)	monomethylarsonic acid
NiAs	niccolite
NiAs$_2$	rammelsbergite
NiAsS	gersdorffite
NOM	natural organic matter

DOI: 10.1201/9781003541615-4

TDS	total dissolved solids
TMA(III)	trimethylarsine
TMAO	trimethylarsenic oxide
USEPA	U.S. Environmental Protection Agency
WHO	World Health Organization
α-FeOOH	goethite
β-FeOOH	akaganeite
γ-FeOOH	lepidocrocite

4.1 INTRODUCTION

Arsenic (As) is a carcinogenic, mutagenic, and teratogenic element [1]. Most As compounds are odorless and tasteless and readily dissolve in water. As poisoning can cause skin diseases, cardiovascular, neurological, hematological, renal, and respiratory diseases, as well as lung, bladder, liver, kidney, and prostate cancers [2]. The World Health Organization (WHO) has set a provisional guideline limit of 10 mg/L for As in drinking water [3], which was subsequently adopted by the European Union [4]. The United States Environmental Protection Agency (USEPA) lowered the maximum contaminant level (MCL) for As in drinking water from 50 to 10 mg/L, effective in January 2006 [5]. The provisional guideline value for As concentration in drinking water of China has been lowered from 50 to the WHO's recommendation of 10 mg/L since 2007. In Bangladesh and India, the guideline value for As in drinking water is still 50 mg/L.

As is mobilized in the environment through a combination of natural processes such as weathering reactions, biological activities, and volcanic emissions as well as through a range of anthropogenic activities such as mining and pesticide production. Most waterborne As problems, however, are the result of As mobilization in shallow aquifers under natural conditions in different parts of the world [6]. In Asia, for example, the presence of elevated concentrations of As in groundwaters has become a major threat to the health of people in Bangladesh, India [7–10], China [11–19], Vietnam [20–23], Cambodia [24,25], Nepal [26], Pakistan [27], and Indonesia [28].

To assess the environmental and toxicological effects of As it is important to ascertain the source of As in groundwaters and to understand its geochemical behavior in aquifers. In this chapter, we will first review the fundamentals of As geochemistry that control the abundance, aqueous speciation, mineralogy, and geochemical behavior of As in groundwater. The complex controls that govern As mobilization and transport in groundwaters in South and Southeast Asia and Northern China are then discussed. More detailed information about the hydrogeochemistry of As can be found in a series of previous publications as well [6,29–35,37].

4.2 OCCURRENCE, DISTRIBUTION, AND SOURCE OF AS IN THE NATURAL ENVIRONMENT

4.2.1 Occurrence and Distribution of As in Rocks, Sediments, and Soils

As is a natural constituent of the Earth's crust and ranks 20th in abundance in relation to the other elements, with an average As content in continental crust varying between 2 and 3 mg/kg [29]. As is found in more than 200 minerals, which include arsenates (60%), sulfides, and sulfosalts (20%), and minor amounts of arsenides, arsenates, oxides, silicates, and As in its native form [34].

As occurs at crustal concentrations in many rock-forming minerals because it can substitute for Si^{4+}, Al^{3+}, Fe^{3+}, and Ti^{4+} in their structures [6]. Major As-containing primary minerals are arsenopyrite (FeAsS), realgar (As_4S_4), and orpiment (As_2S_3). Realgar (As_4S_4) and orpiment (As_2S_3) are the two common reduced forms of As. As occurs in oxidized form in the mineral arsenolite (As_2O_3). Other naturally occurring As-bearing minerals include loellingite ($FeAs_2$), safforlite (CoAs), niccolite (NiAs), rammelsbergite ($NiAs_2$), arsenopyrite (FeAsS), cobaltite (CoAsS), enargite (Cu_3AsS_4), gersdorffite (NiAsS), glaucodot ((Co,Fe)AsS), and elemental As [35].

TABLE 4.1

Natural Abundance of As (mg/kg) in Crustal Materials

Rock Type	As Concentration Average (and/or Range)
Igneous rocks	
Ultrabasics	1.5 (0.03–15.8)
Basalts	2.3 (0.18–113)
Andesites	2.7 (0.5–5.8)
Granites/silicic rocks	1.3 (0.2–13.8)
Sedimentary rocks	
Shales and clays	3–15 (up to 490)
Phosphorites	21 (0.4–188)
Sandstones	4.1 (0.6–120)
Limestones/dolomite	2.6 (0.1–20.1)
Iron formations and Fe-rich sediment	(1–2,900)
Evaporites (gypsum and anhydrite)	3.5 (0.1–10)
Metamorphic rocks	
Quartzite	5.5 (2.2–7.6)
Hornfels	5.5 (0.7–11)
Phyllite/slate	18 (0.5–143)
Schist/gneiss	1.1 (<0.1–18.5)
Amphibolite and greenstone	6.3 (0.4–45)
Coal	
Bituminous[a]	9.0±0.8
Lignites[a]	7.4±1.4
Peat	(16–340)

Source: Modified from Smedley, P.L. and Kinniburgh, D.G. *Appl. Geochem.*, 17, 517–568, 2002 and Nriagu, J.O. et al., In: Bhattacharya, P. et al., (Eds), *Trace Metals and Other Contaminants in the Environment*, Vol. 9, pp. 3–60. Elsevier, 2007.

Note: Coal Clarke value of As.

[a] Yudovich, Y.E. and Ketris, M.P., *Int. J. Coal Geol.*, 61, 141–196, 2005 [38].

The interaction of groundwater with host rocks drives the multiphase cycling of As in aquifer systems. Typical concentrations of As in crustal rocks are presented in Table 4.1. Relative to igneous rocks, As concentrations are significantly higher in fine-grained and organic-rich sedimentary rocks and their metamorphic equivalents. As concentrations in sedimentary rocks can be more variable. The highest As concentrations (20–200 mg/kg) are typically found in organic- and sulfide-rich shales, sedimentary ironstones, phosphate rocks, and some coals [6]. Although As concentrations in coals can reach up to 3.5×10^4 mg/kg, concentrations in the range from <1 to 17 mg/kg are more typical [36].

In sedimentary rocks, As is concentrated in clays and other fine-grained sediments, especially those rich in sulfide minerals, organic matter (OM), secondary iron oxides, and phosphates. The average concentration of As in shale is an order of magnitude greater than that in sandstone, limestone, and carbonate rocks. As is strongly sorbed by oxides of iron, aluminum, and manganese as well as by some clay minerals, leading to its enrichment in ferromanganese nodules and mangani-ferous deposits.

The assessment of arsenic content in sediments from the Ice Seu'um geothermal display area in Aceh, Indonesia, showed that the presence of arsenic was detected in all sediment samples, ranging from 2.56 to 6.86 mg/kg [39]. The median As concentration in stream sediments from 20 study areas across the United States collected as part of the National Water Quality Assessment

(NAWQA) program was 6.3 mg/kg [40]. The As concentration in soils shows a range similar to that in sediments, except in places contaminated by industrial or agricultural activities.

4.2.2 Occurrence and Distribution of As in Groundwaters

The concentration of As in most groundwaters is <10 mg/L [41,42] and often below the detection limit of routine analytical methods. A survey of groundwaters used for public supply in the United States showed that only 7.6% exceeded 10 mg/L, and 64% contained As <1 mg/L [43]. Nonetheless, naturally occurring high-As groundwaters have been found in aquifers in many parts of the world and their As concentrations occasionally reach the mg/L level [6], in a range of four orders of magnitude, from <0.5 to 5,000 mg/L [44]. Industrial activities can also give rise to very high dissolved As concentrations in groundwater, but the affected areas are usually localized. For example, Kuhlmeier [45] found concentrations of As up to 4.08×10^5 mg/L in groundwater close to a herbicide plant in Texas [45].

Although most high-As groundwater provinces result from natural occurrences of As, high concentrations of As have been found in groundwater in different environments. These include both oxidizing and reducing aquifers and areas affected by geothermal, mining, and industrial activities (Table 4.2). Cases of mining-induced As pollution are numerous in the literature but are mostly localized. As mobilization and enrichment in groundwater are discussed in more detail in Section 5.4.

Reducing conditions favorable for As mobilization have been reported most frequently from Quaternary basins and/or deltaic regions where strong neotectonic processes have resulted incomplex patterns of sedimentation and rapid burial of large amounts of sediment containing fresh OM during basin/delta development. Thick sequences of young sediments are quite often the sites of high As concentrations in groundwater. The most notable example of these conditions is the Bengal Basin, which includes Bangladesh and West Bengal [46]. Other examples include Nepal, Myanmar, Cambodia [33], parts of northern China [11–19,72], the Great Hungarian Plain of Hungary and Romania [49,73], the Red River floodplain of Vietnam [20,23], and parts of the western USA [42]. Recent groundwater extraction in many of these areas, either for public supply or for irrigation, has induced an increase in both groundwater flow and As transport in aquifer systems [74].

High concentrations of naturally occurring As are also found in oxidizing conditions where groundwater pH values are high (usually >8) [6]. In such environments, inorganic As(V) predominates, and As concentrations are positively correlated with those of other anion-forming species such as HCO_3^-, F^-, H_3BO_3, and $H_2VO_4^-$. Examples include San Joaquin Valley, California [75], the Lagunera region of Mexico [76], the Antofagasta area of Chile [50], and the Chaco-Pampean Plain of Argentina [44] (Table 4.2). These high-As groundwater provinces are usually located in arid or semiarid regions where groundwater salinity is high. Evaporation has been suggested to be an important additional cause for As accumulation in groundwater from arid areas [75].

TABLE 4.2

Concentration Ranges of As in Various Aquifers of the World

Country/Region Location	As Concentration: Average or Range (µg/L)	References
Bangladesh	<0.5–2,500	[46]
West Bengal, India	<10–3,700	[9]
China		
Taiwan, Chianan Plain	<10–1,300	[47]
Xinjiang, Tianshan Plain	40–750	[48]

(Continued)

TABLE 4.2 (*Continued*)
Concentration Ranges of As in Various Aquifers of the World

Country/Region Location	As Concentration: Average or Range (µg/L)	References
Inner Mongolia, Huhhot Basin	<1–1,480	[17]
Inner Mongolia, western Hetao Plain	76–1,093	[18]
Shanxi, Datong Basin	9–1,530	[12]
Red River Delta, Vietnam	1–3,050	[20]
Mekong Plain, Cambodia	1–1,700	[25]
Terai Basin, Nepal	<10–740	[26]
Kalalanwala area, Punjab, Pakistan	32–1,900	[27]
Lowlands of Sumatra, Indonesia	<0.1–65	[28]
Pannonian Basin, Hungary	0–300	[49]
La Pampa Province, central Argentina	<4–5,300	[44]
Rio Loa (Second Region), northern	100–1,000	[50]
Zimapan Valley, Mexico	<50–1,100	[51]
United States		
California, Owens Dry Lake	100–96,000	[52]
Southeast Michigan	0.5–278	[53]
Central New Hampshire	1.95–397.5	[54]
Eastern Wisconsin, Fox River Valley	2–12,000	[55]
Central Arizona, Verde Valley	10–210	[56]
Central Illinois	<1–266	[57]
Western Nevada	Nondetectable-6,000	[58]
Maine, Northport	0.75–1,900	[59]
New England	<5–1,100	[60]
Okavango Delta, NW Botswana	1.8–112	[61]
Various areas in Canada	0.5–11,000	[62]
Fukuoka Prefecture, Japan	1–293	[63]
Mineralized area, Bavaria, Germany	<10–150	[64]
Western Anatolia, Turkey	0–10,500	[65]
South-western Uruguay	0.1–58	[66]
Duero Cenozoic Basin, Central Spain	0.42–613.45	[67,68]
Volcanic areas in southern Italy	0.1–6,940	[69]
Porphyry Cu deposits, Chile	<10–278	[70]
Coastal sand aquifer, Australia	52–337	[71]

The arsenic (As) concentration and geochemical analysis of groundwater and core sediments from 20 tube wells in the Chapai Nawabganj area of Bangladesh show that the concentration range of As in the upper layer of groundwater (depth 20–40 m) is 2.34 to 586.96 µg/L, the geochemical conditions of groundwater oxidize into gradually decreasing oxides with increasing arsenic concentration and depth. The highest sediment arsenic content (25 mg/kg) was found at a shallow depth of 15 m [77]. Similar cases have been reported in the United States, Canada, Poland, and Austria [6,62,71]. Groundwater As problems in areas of unexploited ore deposits are less common. But Boyle et al. [78] found concentrations of up to 580 mg/L in groundwater from the sulfide mineralized areas of Bowen Island, British Columbia [78]. Heinrichs and Udluft [64] also found As concentrations up to 150 mg/L in groundwater from a mineralized sandstone aquifer in Bavaria.

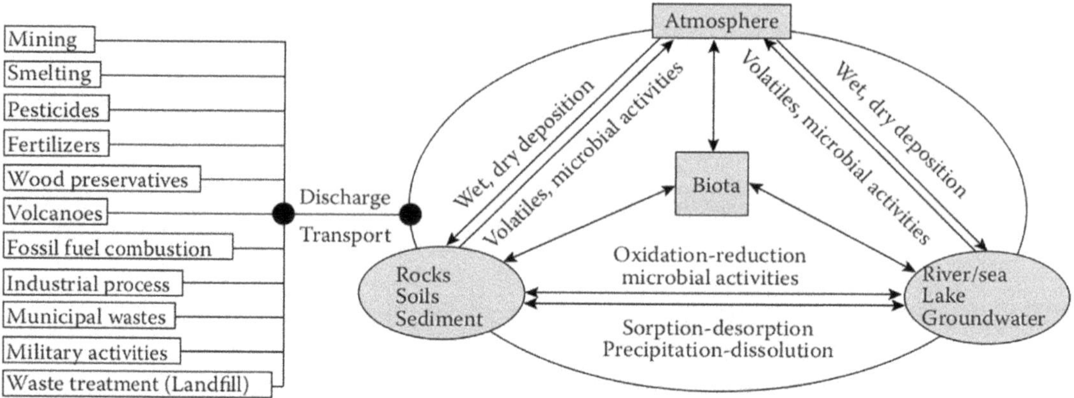

FIGURE 4.1 Sources and distribution pathways for As in the environment. (Modified from Wang, S.L. and Mulligan, C.N., *Sci. Total Environ.*, 366, 701–721, 2006.)

4.2.3 SOURCE AND BIOGEOCHEMICAL CYCLE OF AS IN THE ENVIRONMENT

As is ubiquitous in different compartments of the ecosystem, as is shown in Figure 4.1. The most common primary sources of As in the natural environment are volcanic rocks, marine sedimentary rocks, hydrothermal ore deposits and associated geothermal waters, and fossil fuels [6]. As is released from these sources into the environment through various natural processes such as weathering, volcanic eruption, geothermal activity, and forest fires. Wind-blown dust and sea salt spray may transport As over long distances as suspended particulates or gases through the atmosphere.

The conceptual model for biogeochemical cycling of As species in the environment is shown in Figure 4.2. The occurrence of As in natural water is dependent on the geology, hydrogeology, and geochemistry of the aquifer, climate conditions, and human activities. Natural sources of As in water have been attributed to several natural geochemical processes, including oxidation of As-bearing sulfides, desorption of As from (hydro)oxides (e.g., iron, aluminum, and manganese oxides), reductive dissolution of As-bearing iron (hydro)oxides, release of As from geothermal water, and evaporative concentration, as well as leaching of As from sulfides by carbonates [6,79].

4.3 AS SPECIES IN NATURAL WATERS

There are four valence states (−3, 0, +3, and +5) of As. Metallic As rarely occurs and the (-III) oxidation state is found only in extremely reduced environment. Arsenate ions (As(V)) occur under aerobic conditions, whereas arsenite ions (As(III)) occur in anaerobic environments. Methylated As species, such as monomethylarsonous acid (MMA(III)), monomethylarsonic acid (MMA(V)), dim-ethylarsinous acid (DMA(III)), dimethylarsinic (DMA(V)), trimethylarsine (TMA(III)), and trimethylarsenic oxide (TMAO), can be formed through biomethylation by microorganisms under favorable conditions [29,80]. Arsenobetaine and arsenosugar occur commonly in marine animals and show no potential toxicity at any dose [81]. Generally, inorganic forms are more toxic than organoarsenic species, while arsenite is considered to be more toxic than arsenate.

Inorganic speciation is important since the varying protonation and charge of the As species present at different oxidation states have a strong effect on their behavior, for example, their adsorption [6]. By contrast, the concentrations of organic As species are generally low or negligible in most groundwaters [82]. For instance, Bednar et al. [83] tested various As species in more than 100 surface water, groundwater, and acid mine drainage samples and found that methylated As species concentrations were less than 100 mg/L and were detected only in some surface water samples [83].

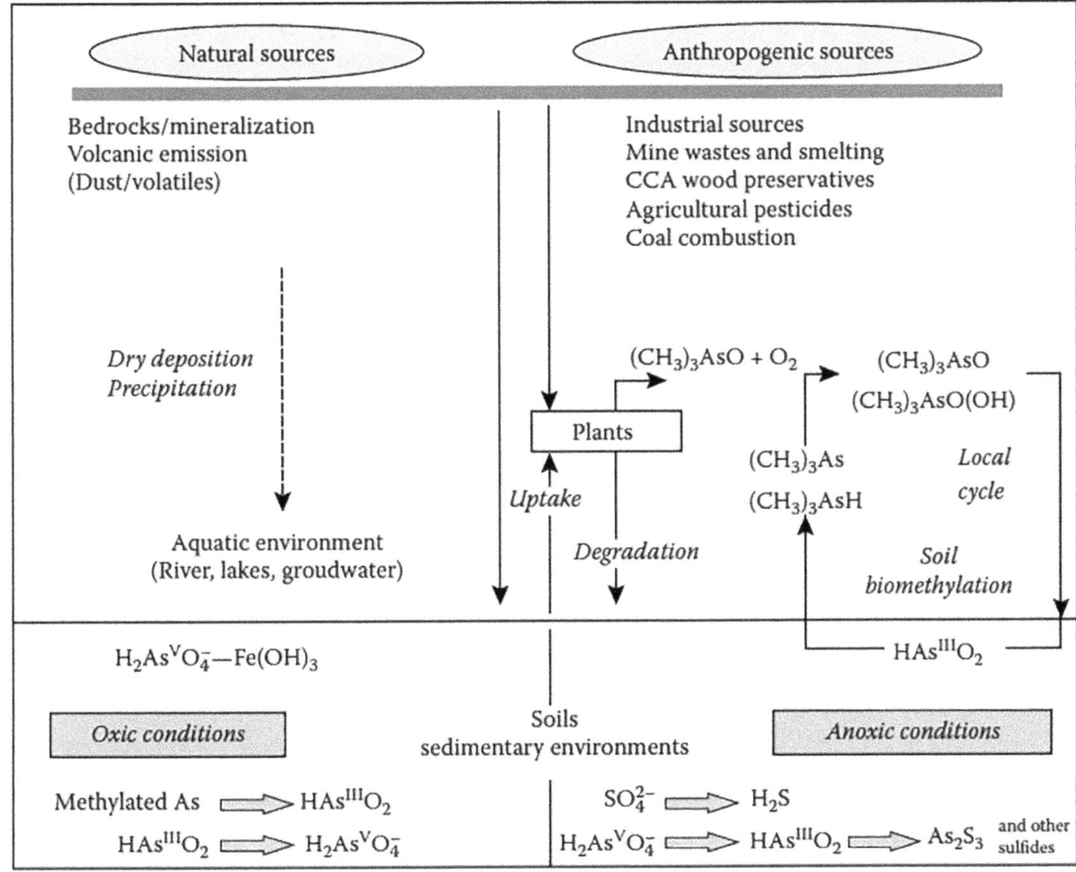

FIGURE 4.2 Conceptual model of As biogeochemical cycling in the environment. (Modified from Nriagu, J.O. et al., In: Bhattacharya, P. et al. (Eds), *Trace Metals and Other Contaminants in the Environment*, Vol. 9, pp. 3–60. Elsevier, 2007.)

The mobility of As in groundwater environment is usually controlled by redox conditions, pH, biological activity, and adsorption-desorption reactions, rather than solubility control of As. Factors such as pH, Eh (thermodynamic redox potential), solution composition, competing and complexing ions, mineralogy, reaction kinetics, and hydraulics of aquifer systems can all potentially affect As speciation and concentrations. The major As species in groundwater are inorganic arsenite (As(III)) and arsenate (As(V)). Under oxidizing conditions, $H_2AsO_4^-$ is dominant at low pH (less than about pH 6.9), whereas at higher pH, $HAsO_4^{2-}$ becomes dominant ($H_3AsO_4^0$ and AsO_4^{3-} may be present in extremely acidic and alkaline conditions, respectively). Under reducing conditions at pH less than 9.2, the uncharged arsenite species $H_3AsO_3^0$ will predominate [84,85,86] (Figure 4.3). The distribution of the species as a function of pH is shown in Figure 4.4.

The Eh-pH diagrams for the As-O-S system are useful since they necessarily simplify highly complex natural systems. In practice, most studies in the literature report speciation data without consideration of the degree of protonation. In the presence of extremely high concentrations of reduced S, dissolved As-sulfide species can be significant. In sulfide-concentrated solutions, dissolved thioarsenic species such as $H_2AsOS_2^-$ and $H_2AsS_3^-$ can be formed under near-neutral to alkaline conditions [87]. Reducing and acidic conditions favor the precipitation of orpiment (As_2S_3),

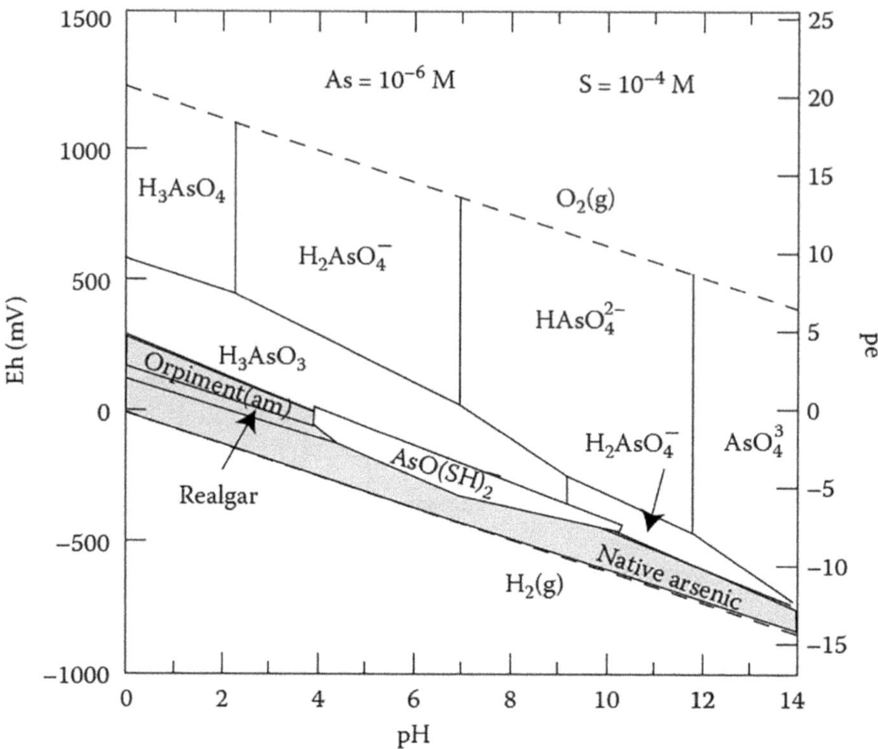

FIGURE 4.3 Eh-pH stability diagram for the As-O-S system at 25°C and 1 bar total pressure. The dashed lines indicate stability field for water. The gray area represents a solid phase. (Modified from Plant, J.A. et al., In: Holland, H.D. and Turekian, K.K. (Eds), *Treatise on Geochemistry*, Chapter 9.02, pp. 17–66. Elsevier, 2004.)

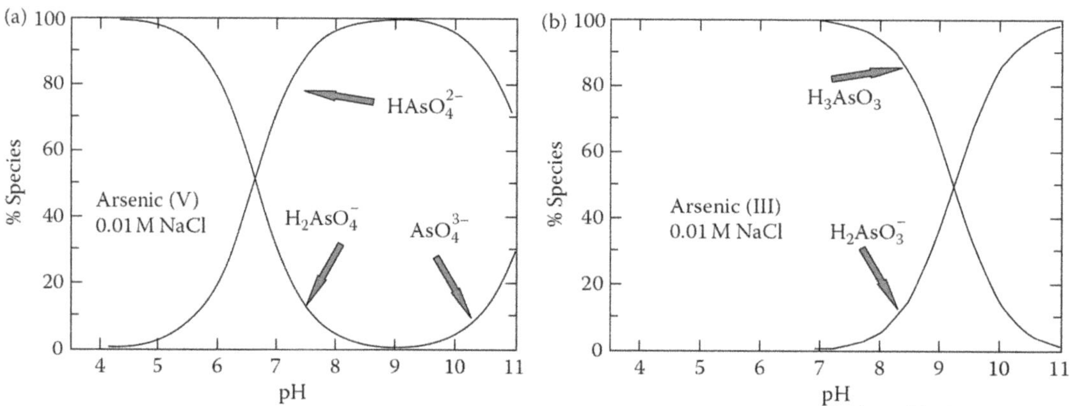

FIGURE 4.4 Speciation of: (a) As(V) and (b) As(III) in a 0.01 M NaCl medium as a function of pH at 25°C. (Modified from Smedley, P.L. and Kinniburgh, D.G., *Appl. Geochem.*, 17, 517–568, 2002.)

realgar (As_4S_4), or other sulfide minerals containing coprecipitated As [29]. Therefore, high-As waters are not expected where there is a high concentration of free sulfide [88]. In reducing waters containing free sulfide, a large fraction of the dissolved As can be present in the form of soluble As-S compounds [89]. Four currently unidentified As-S species encountered are probably monomeric thioarsenites or thioarsenates, which have been found to play a key role in the dissolution of As-sulfide minerals under alkaline conditions [89,90].

Sulfur and iron play important roles in influencing As speciation and its transport. As has a high adsorption affinity on iron oxides. Sulfides can be a source or sink for As under different conditions [42]. Dissolution of sulfide minerals or iron oxides may contribute As to groundwater. The precipitation of sulfide minerals, which can occur in sulfide-rich water or sorption by iron oxides, can in turn remove As from water.

The Eh-pH diagram for the As-Fe-S-O-H_2O system is shown in Figure 4.5. It can be seen that an overwhelmingly large field exists for the ionic arsenate and arsenite species, but mostly combined

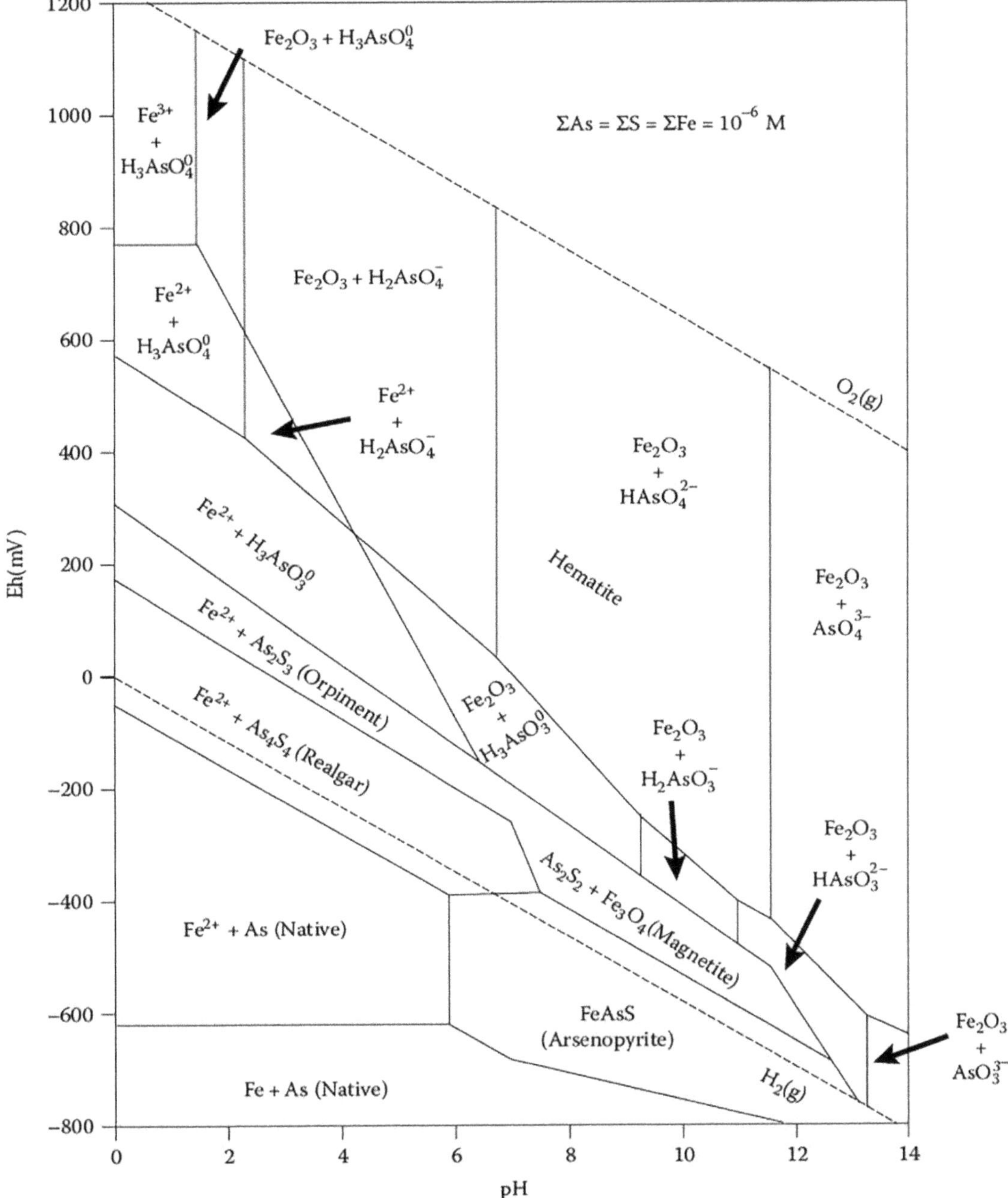

FIGURE 4.5 Eh-pH stability diagram of the As-Fe-S-O-H_2O system at 25°C and 1 bar total pressure with the total concentration of As of 10^{-6} mol/L, as well as S and Fe. (Modified from Vink, B.W., *Chem. Geol.*, 130, 21–30, 1996.)

with hematite (Fe_2O_3). Under reducing alkaline conditions, realgar (As_2S_2) coexists with magnetite (Fe_3O_4), whereas arsenopyrite (FeAsS), the most common mineral of As, is stable only under extremely reducing and alkaline conditions.

4.4 GEOCHEMICAL PROCESSES AND BEHAVIOR OF AS IN AS-AFFECTED AQUIFERS

As with most trace metals, the concentration of As in aquifer systems is controlled by mineral–water interactions. Knowing the types of processes involved is important not only for understanding the response of As to changes in groundwater chemistry but also for determining the modeling approach to predict the trend of change in As concentration. Mineral–water interactions can be divided into four major types from a geochemical point of view: precipitation-dissolution, adsorption-desorption, oxidation and reduction, and microbial activities. Besides, natural organic matter (NOM) and microbes have a strong impact on the behavior of As in various biogeochemical processes.

4.4.1 PRECIPITATION AND DISSOLUTION

Precipitation-dissolution reactions involve the growth or erosion of a mineral structure and therefore only involve structural ions, that is, the elements included in the chemical formula of the mineral. The solubility of minerals can in principle be described quite well by a solubility product, although this may vary with the particle size and crystallinity of the mineral. Moreover, the rate of dissolution or precipitation can be very slow, and thermodynamic equilibrium is therefore often not attained in practical time scales.

Coprecipitation is also a common natural process whereby minor constituents are incorporated or scavenged into a mineral structure as it forms. For instance, As can be coprecipitated during the formation of pyrite. Furthermore, both phosphate and As are coprecipitated during the formation of iron oxides.

Arsenian pyrite, $Fe(As,S)_2$, is probably the most significant reservoir of mineral As in nature. Pyrite can incorporate up to a maximum of 8 wt% As into its structure, and it was speculated that pyrite may form a solid solution with arsenopyrite, FeAsS [91,92]. The so-called authigenic pyrite is typically formed when OM is introduced into an iron- and sulfur-rich system, which causes redox potentials to drop. Highly insoluble ferric iron (iron oxide) is first reduced by the OM to soluble ferrous iron, and then sulfate is reduced to sulfide. The dissolved As concentrations then decline as pyrite forms through a complex process that coprecipitates As [93].

4.4.2 ADSORPTION AND DESORPTION

Under oxidizing conditions, As concentrations in water are mostly controlled by adsorption rather than mineral solubility. Sorption of As to solid phases has been proposed as a principal control on its mobility (Table 4.3), which can transfer soluble or mobile As to particulate phases, thus immobilizing it. Especially, (hydro)oxides of Fe, Al, and Mn are ubiquitous in soils and sediments, either as discrete particles or as coatings on other mineral solids with corresponding high surface areas, and are potentially the most important As adsorbents [94].

Spectroscopic studies have confirmed that both As(III) and As(V) may form inner-sphere complexes on the surfaces of the (hydro)oxides and clay minerals through ligand exchange with OH and OH_2^+ surface functional groups [98]. As(III) may also form outer-sphere complexes by simple coulombic (electrostatic) interactions on the surface of amorphous Al hydroxides and sulfide minerals. Inner-sphere complex bonds are much stronger than outer-sphere complex bonds, resulting in stronger adsorption, which makes the immobilization more persistent [95–97]. Although surface complexation models developed to predict arsenate adsorption behavior have included the formation

TABLE 4.3

Effects of the Main Aquifer Materials on As Immobilization and Transformation in Adsorption-Desorption Processes

Aquifer Materials	Main Affecting Mechanism
Fe hydroxides	As(V), $CH_3AsO_2(OH)^-$, and $(CH_3)_2AsOOH$ adsorption at pH 4–7, maximized around pH 4
	As(III) adsorption at pH 7–10, maximized around pH 7
	Desorption when pH increases
	Amorphous phases of a higher adsorption capacity than crystalline phases
	Releasing sorbed As during chemical and microbial reductive dissolution
	As(III) oxidization, catalyzed by light or H_2O_2 in alkaline pH
Al hydroxides	As(V), $CH_3AsO(OH)_2$, and $(CH_3)_2AsOOH$ adsorption up to pH 7 and decreases significantly at higher pH
	As(III) adsorption at pH 6–9.5 and decreases at higher pH
	Amorphous phases of a higher adsorption capacity than crystalline phases
Mn hydroxides	Negligible As(V) adsorption, but increased in the presence of other divalent cations
	Slightly greater As(III) oxidation at low pH (pH 4)
	Poorly crystalline phases with high surface areas are more efficient
Clay	As(V) adsorption up to pH 7 and decreases with pH increases
	Low As(III) adsorption at low pH and increases with pH
	Clays with high surface areas showing a higher adsorption capacity
	As(III) oxidation in the presence of trace amounts of impurities such as Fe or Mn oxides, iodide, or TiO_2
Sulfides	As substitution for S in sulfides, forming of As sulfide precipitates in the reduced environment Releasing As during chemical and microbial oxidation of As-bearing sulfides
Calcite[a]	Equilibrium data of As(III) sorption onto calcite depicts an S-shaped sorption isotherm
	At low concentrations, $As(OH)_3$ is adsorbed by complexation to surface Ca surface sites
	Sorbed As increases linearly with solution concentration, up to the saturation of As with respect to the precipitation of $CaHAsO_3(s)$
Muscovite and mica[b]	Function of solution pH (pH 3-8 for muscovite, and 3-11 for biotite), amount of As adsorbed increases with increasing pH, exhibiting a maximum value, before decreasing at higher pH values
	Maxima correspond to 3.22 ± 0.06 mmol/kg As(V) at pH 4.6-5.6 and 2.86 ± 0.05 mmol/kg As(III) at pH 4.1–6.2 for biotite, and 3.08 ± 0.06 mmol/kg As(III) and 3.13 ± 0.05 mmol/kg As(V) at pH 4.2–5.5 for muscovite
NOM	Enhancing As release mainly through competition for active adsorption sites, forming aqueous complexes, and changing the redox chemistry of site surfaces and As species
	Inhibiting As mobility by serving as a binding agent and/or by forming insoluble complexes, especially when saturated with metal cations
Anions	Competition for active adsorption sites, influenced by pH and concentration ratios between anion and As
Cations	Enhancing As sorption by increasing the amount of positive charge on the oxide surfaces and/or forming a positively charged surface

Source: Modified from Wang, S.L. and Mulligan, C.N., *J. Hazard. Mater. B*, 138, 459-470, 2006.

[a] Roman-Ross, G., Cuello, G.J., Turrillas, X., Fernandez-Martfnez, A., and Charlet, L., *Chem. Geol.*, 233, 328–336, 2006.

[b] Chakraborty, S., Wolthers, M., Chatterjee, D., and Charlet, L., J. *Colloid Interface Sci.*, 309, 392–401, 2007.

of only inner-sphere species (e.g., [99–102]), Catalano et al. [103] observed arsenate adsorption on corundum and hematite using resonant surface X-ray scattering measurement, and the results demonstrated that arsenate surface complexation is unexpectedly bimodal, adsorbing simultaneously as inner- and outer-sphere species [103].

4.4.2.1   As Sorption to Fe, Al, and Mn (Hydro)oxides

Iron hydroxides, such as goethite (α-FeOOH), akagenéite (β-FeOOH), and lepidocrocite (γ-FeOOH), have high isoelectric points of about 8.6 and possess net positive charges in most geological environments, showing high affinities for As species. Their adsorption affinity is higher for As(V) and As(III) at lower and higher pH values, respectively [6].

Al (hydro)oxides are ubiquitous in acidic soils and aquatic environments. Amorphous $Al(OH)_3$ has an isoelectric point of 8.5 and thus is an extremely efficient adsorbent to immobilize As. Previous studies indicated that As(III) and As(V) adsorption on an Al oxide (g-Al_2O_3) and gibbsite (Al_2O_3 $3H_2O$) formed inner-sphere complexes [95].

Mn hydroxides have an isoelectric point of about 2.3 and therefore carry a net negative charge at pH 3–9 in ordinary natural waters. In other words, it is difficult for them to adsorb As anions. It was observed that the negatively charged As(V) species, $H_2AsO_4^-$, was adsorbed negligibly onto the negatively charged birnessite surface at a pH range of 4–7 [104]. However, Mn hydroxides have the capacity to oxidize As(III) to As(V) [105,106], which may create fresh adsorption sites for As(V) on the oxide surface.

The crystallinity and surface area of the (hydro)oxides have demonstrated significant effects on their sorption capacity. Generally, poorly crystalline hydroxides with a higher surface area show a higher As sorption capacity by providing more active sorption sites [95]. For example, the adsorption of As(III) and As(V) was two or three times greater (on a surface area basis) on ferrihydrite than on goethite [107].

However, desorption and remobilization of the sorbed As from the (hydro)oxides may occur when biogeochemical conditions change with time. Significant As(V) remobilization from Fe hydroxides can occur at pH above 8 due to the increase of electrostatic repulsion on the negatively charged oxide surface, and the rate of As(V) desorption can be quite high [108]. In reducing sediments, As sorbed on Fe(III) hydroxides could be remobilized and released into groundwater as a result of the microbial reduction of Fe(III) to Fe(II) and the reduction of As(V) to As(III) [74,109,110]. But there is no clear correlation between the concentrations of dissolved As and Fe as can be expected from this hypothesis. Horneman et al. [111] suggested that this poor As-Fe correlation could be due to the formation of secondary Fe(II) or mixed Fe(II)/Fe(III) phases such as siderite ($FeCO_3$), green rust, and magnetite (Fe_3O_4) [111]. Jonsson and Sherman [112] found that As(V) can be sorbed to fougerite, magnetite, and siderite by forming inner-sphere surface complexes, and no evidence for As(V) reduction was found [112]. Their work also demonstrated that As(III) can form inner-sphere surface complexes on magnetite and fougerite but only a (presumably) weak outer-sphere complex on siderite, and no evidence for As(III) oxidation was found. Besides, As(V) desorbs from magnetite, fougerite, and siderite at pH > 8; however, As(III) sorption to these minerals is enhanced with increasing pH.

4.4.2.2   As Sorption to Clay Minerals

Ubiquitous in the terrestrial environment, clay minerals largely consist of aluminosilicates with alternating layers of silica oxide and aluminum oxide. Fourier transform infrared spectroscopy analyses (FTIR) indicated that the retention of As(V) by halloysite was likely due to the formation of fulvic acids (FAs) [117,120–122]. As(V) and phosphorus sorption on hydrous ferric oxides (HFOs) n of hydroxy-As(V) interlayers in crystals. It was also reported that nearly all As(III) and As(V) were physiosorbed to smectite, and only a portion of them was chemisorbed on kaolinite forming inner-sphere complexes [95].

As(V) adsorption to kaolinite, montmorillonite, illite, halloysite, and chlorite occurs up to pH 7 and then decreases with a pH increase [95]. As(III) adsorption by the same clay minerals is minimal at low pH and increases with increasing pH. As(V) is adsorbed to a greater extent than As(III) on all clay minerals at a pH below 7 [95]. At higher pH values, the adsorption behaviors of As(V) and As(III) are more comparable. Moreover, the poorly crystallized clay minerals of a larger surface area have higher As sorption capacity, and the presence of mineral impurities such as Fe species may further enhance the capacity [113].

4.4.2.3 Enhanced Sorption by Cations

Cations, such as Ca^{2+} and Fe^{2+}, may increase As adsorption by increasing the amount of positive charge on the oxide surface and/or forming a positively charged surface [114]. Ghosh and Teoh [115] observed that the adsorption of As(V) onto Al oxides was enhanced in the presence of Ca^{2+} at pH above 8 [115]. The addition of Ca^{2+} also increased As(V) adsorption onto ferrihydrite at pH 9 [116]. Meng et al. [117] reported that the addition of Ca^{2+} and Mg^{2+} to the suspension of ferrihydrite negated part of the competitive effect of silicate on As adsorption [117]. The formation of $CaCO_3$ minerals can restrict the development of high pH, thus inhibiting As(V) release from oxides and clays.

Once the oxides have an adsorbed load, any change in their surface chemistry or the solution chemistry can lead to release of adsorbed As, and thereby increase of As concentration in groundwater. The extremely high solid/solution ratio of soils and aquifers makes them highly sensitive to such changes [118,119].

4.4.2.4 Competing Sorption by Anions

A complicating factor in As adsorption is competition by other oxyanions. In reducing groundwaters, these include phosphate, silicate, bicarbonate, and FAs [117,120–122]. As(V) and phosphorus sorption on HFOs are broadly similar, although there is usually a slight preference for phosphorus. Not surprisingly, As(V) is much more strongly affected by phosphate competition than As(III) [120]. Sulfate has essentially no effect on As(V) adsorption but may compete with As(III) adsorption when the pH is below 7 [117,120]. Generally, carbonate exhibits little effect on As(III) and As(V) adsorption [117], but the presence of bicarbonate (HCO_3^-) can facilitate As mobilization from As-containing sulfides such as orpiment in both oxic and anoxic environments [79]. As mobilization increases with increasing HCO_3^- concentrations and pH. Silicate reduces the adsorption of As(III) and As(V) on ferrihydrite at pH 6.8 [117]. Competitive adsorption between arsenate and molybdate was also observed [123]. The competing sorption between As species and other anions can inhibit As sorption and thus increase As mobility.

Anion displacement as a mechanism for As release has received less attention compared with the reductive dissolution of iron oxides, although several laboratory and field studies have demonstrated the ability of naturally occurring (in)organic ions to displace absorbed As or inhibit its absorption. Most studies focused on phosphate-promoted desorption of As due to the increasing use of fertilizers. Fewer researchers evaluated the potential roles of silicate, carbonate, and sulfate, the three ubiquitous anions in groundwaters. For sulfate, it is not an effective desorbent [6,124]. The results of studying the influence of bicarbonate on As desorption have shown that in a neutral to basic pH environment of groundwaters, bicarbonate concentrations exceed the reactive surface site densities of iron (hydro)oxides and may promote As desorption [114,125]. Latest research on the importance of silicate on displacing As from goethite demonstrated that silicate can reduce arsenite adsorption rates, block potential adsorption sites, and irreversibly displace 0.3%–1.5% absorbed arsenite, resulting in As concentration increase from 9 to 266 mg/L [126].

4.4.3 Oxidation and Reduction

Redox reactions are important for controlling the behavior of many major and minor species in natural waters, including that of As [6]. In practice, redox equilibrium is hardly achieved, and the redox potential tends to be controlled by major redox-sensitive elements (O, C, N, S, and Fe), and redox-sensitive minor and trace elements such as As respond to these changes rather than control them.

A well-known sequence of reduction reactions occurs when lakes, sediments, and aquifers become anaerobic [127,128]. The processes causing changes in Fe redox chemistry are particularly important since they can directly affect the mobility of As. One of the principal causes of high As concentrations in subsurface waters is the reductive dissolution of hydrous Fe oxides and/or the release of adsorbed or combined As [6]. This sequence begins with the consumption of O_2 and an increase in dissolved CO_2 from the decomposition of OM. Next, NO_3^- decreases by reduction to NO_2^- and the gases N_2O and N_2. Insoluble manganese oxides dissolve by reduction to Mn^{2+} and HFOs by reduction to Fe^{2+}. These processes are followed by SO_4^{2-} reduction to S^{2-}, then CH_4 production from fermentation and methanogenesis, and finally reduction of N_2 to NH_4^+. During SO_4^{2-} reduction, the consequent S^{2-} reacts with any available Fe^{2+} to produce FeS and ultimately pyrite, FeS_2. As(V) reduction would normally be expected to occur after Fe(III) reduction but before SO_4^{2-} reduction [6].

Bhattacharya et al. [129] postulated that the principal redox reactions controlling As release in the groundwater of the Bengal Delta are given by the following scheme [129]:

(OM oxidation by O_2)

$$CH_2O + O_2 \Rightarrow CO_2 + H_2O$$

(Dissolution and hydrolysis)

$$CO_2 + H_2O \Rightarrow H_2CO_3 \quad \left(H^+ + HCO_3^-\right).$$

(Denitrification)

$$5CH_2O + 4NO_3^- \Rightarrow 2N_2 + 4HCO_3^- + CO_2 + 3H_2O.$$

(Sulfate reduction)

$$2CH_2O + SO_4^{2-} \Rightarrow 2HCO_3^- + H_2S.$$

(Reductive dissolution Fe oxides)

$$4Fe^{III}OOH + CH_2O + 7H_2CO_3 \Rightarrow 4Fe^{II} + 8HCO_3^- + 6H_2O.$$

The As sorbed onto iron oxides is vulnerable to release by changing geochemical conditions, especially a decrease in redox potential. When the redox potential drops, the ferric iron is reduced to the highly soluble ferrous iron, and As is released from the surface of oxides into the solution. The reduction is generally driven by the introduction of OM into the aquifer as well as by microbial activities [74,110].

Saunders et al. [130,131] and J.A. Saunders et al. [132] proposed that when optimizing anaerobic bacterial metabolism by providing electron donors and acceptors, As moves under Fe reduction conditions and remains stationary under SO_4 reduction conditions, while arsenopyrite may be a stable mineral phase formed under SO_4 reduction conditions, rather than a pure As-S phase, such as realgar or orpiment. Kirk et al. [133] reached a similar conclusion in studying As-rich groundwater in Illinois [133]. Further, Saunders et al. [131] proposed that As-bearing pyrite should be the most important solid As phase formed under SO_4-reducing conditions in natural systems [130,131]. In contrast, O'Day et al. [134] and O'Day [135] reported pure As-S phases including realgar (As_4S_4) and orpiment (As_2S_3) from an industrial As-contaminated site, and proposed that these As solid phases would have formed under reducing conditions. Recently, As-bearing pyrite has been found in alluvial sediments in Bangladesh [136] and West Bengal, India [137], as previously in the United States [130,138,139].

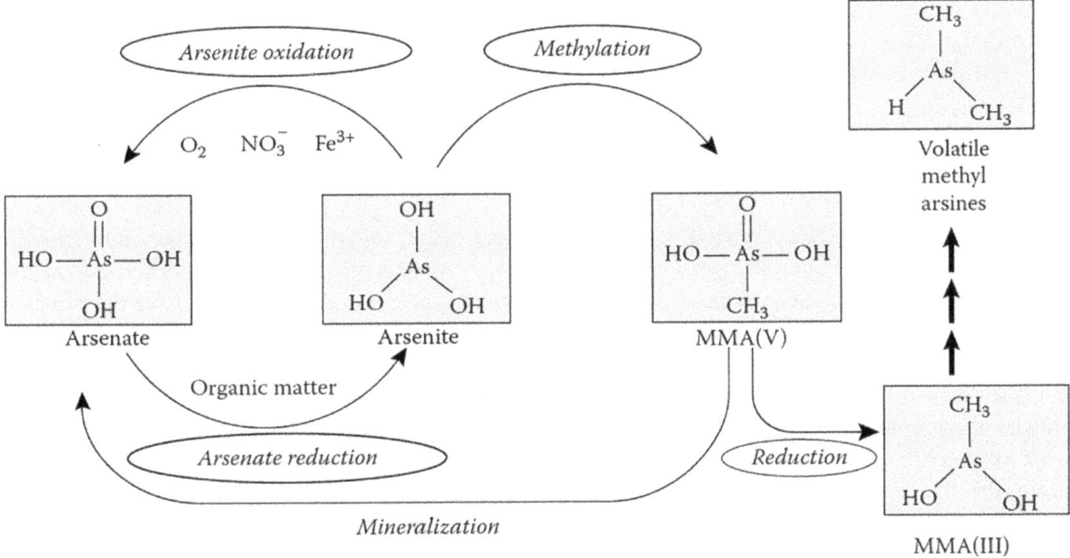

FIGURE 4.6 Microbial transformation in As cycling in the environment. (Modified from Lopez, I.C. Microbial transformation of arsenic and organoarsenic compounds in anaerobic environments. PhD Dissertation Thesis, University of Arizona, 236 pp., 2007.)

4.4.4 Microbial Transformation

As with most redox reactions in the natural environment, both the oxidation of arsenite and the reduction of arsenate can be bacterially catalyzed. Microbial activity plays an important role in dictating the fate of As in the environment, as shown in Figure 4.6.

4.4.4.1 Microbial Arsenite Oxidation

The microbial oxidation of As(III) is of considerable environmental importance because it can impact the mobility and speciation of As in the environment. Oxidation of As(III) can be performed by heterotrophic arsenite oxidizers (HAOs) and chemolithoautotrophic arsenite oxidizers (CAOs). As(III) oxidation by HAOs microorganisms is utilized as a detoxification mechanism, and occurs in the cell's outer membrane [141]. CAOs utilize arsenite as an electron donor to fix CO_2 for growth. Oxidation of As(III) coupled to the reduction of oxygen [142], nitrate to nitrite [143], or nitrate to nitrogen gas [144] has been reported.

4.4.4.2 Microbial Arsenate Reduction

There are two main mechanisms responsible for the reduction of As(V) to As(III) [140]: (a) As(V) reduction as a detoxification pathway and (b) by the dissimilatory reduction of As(V).

4.4.4.2.1 Arsenate Reduction as a Detoxification Mechanism

Similarly to As(III) oxidation, the reduction of As(V) is implicated in the mechanisms of detoxification developed by a wide variety of microorganisms. A well-studied mechanism of detoxification is the *ArsC* system developed in *Escherichia coli* and *Staphylococcus aureus* [145]. *ArsC* is an enzyme that reduces As(V) to the more toxic As(III) in the cytoplasm. Reduced glutathione serves as an electron donor for the As(V) reduction. Although As(III) is more toxic, it serves as a substrate for the *ArsB* transport protein. As(V) conversion seems counterproductive, but it is a mechanism to differentiate As(V) from PO_4^{3-} and avoid extrusion of PO_4^{3-} from the cell. In the absence of significant levels of microbial $SO_4^{(2-)}$ reduction, water flooding resulted in an increase in the concentrations of Fe (II) and As (III) within 10 weeks, which was consistent with a decrease in microbial Fe (III)$^-$ and As (V)$^-$. Due to the formation of FeS, the reduction of microbial $SO_4^{(2-)}$ leads to a decrease in the concentration of Fe (II) in pore water. The newly formed FeS chelates a large amount of As, and As (V) is reduced to As (III), indicating that in the presence of FeS, a portion of the solid phase As can be retained as As_2S_3 like complexes related to tetragonal FeS [146].

4.4.4.2.2 Dissimilatory Arsenate Reduction

Although As is normally associated with poisoning and death, this metalloid also serves to support bioenergetic reactions in certain types of microorganisms. As described above, many microorganisms have evolved capabilities of resistance involving redox reactions to protect themselves from the negative effect of As [147]. Some anaerobic bacteria and archaea can actually conserve the energy gained via the oxidation of organic compounds or H_2 coupled to the reduction of As(V) to As(III). This process is termed dissimilatory arsenate reduction, and it is a means of anaerobic respiration that supports the growth of a number of microorganisms [148]. Dissimilatory arsenate-respiring prokaryotes (DARPs) are considered a key driving force for solid-phase arsenic reduction and dissolution. A new DARP strain, Priestia sp. F01 was mentioned in this study. Through a series of studies, it has been proven that excessive phosphate significantly increases the abundance of the Arr gene, enhances the transcription level of the ArrA gene, and enhances the As (V)$^-$ respiratory activity of bacteria in F01 cells [147].

4.4.4.3   As Biomethylation

As can be biomethylated by algae, fungi, and a wide variety of bacteria mainly as a detoxification mechanism. Volatile (methylarsine, dimethylarsine, and trimethylarsine) and nonvolatile As compounds (mainly MMA^V and DMA^V) are formed by biomethylation. Microbial methylation is a process that has been reviewed recently [141,149,150].

4.4.4.4   Microbial Mobilization of Sorbed As

As sorbed onto iron hydroxides can be mobilized under anaerobic conditions by the reductive dissolution of iron hydroxides, allowing the release of As(V) into the aqueous phase, followed by a rapid reduction of aqueous As(V) to As(III) via either biotic or abiotic pathways. Alternatively, As(V) may be reduced to As(III) on the surface and then released upon reductive dissolution of the iron hydroxide phase [151,152].

Liqin Duan studied the migration of arsenic in the Yangtze River Estuary SWI under three different seasons and benthic redox conditions. Arsenate (As (V)) is preferentially reduced to arsenite (As (III)) and subsequently re adsorbed onto newly formed crystalline Fe oxides, limiting the release of As in the As (V) reduction layer. The enhancement of Fe (III) reduction in the Fe (III) reduction layer facilitates the release of As, while the presence of low As high Fe high SO levels leads to the removal of As by adsorption onto pyrite in the sulfate reduction layer [154].Korte [153] first proposed that the As enrichment of alluvial aquifer groundwater was caused by the codeposition of HFOs containing sorbed As and NOM in river floodplain alluvium, and that the OM caused reductive dissolution of HFOs, releasing both Fe(II) and As to groundwater [153]. The scale of As enrichment of groundwater was recognized in Southeast Asia in the mid-to-late 1990s. Saunders et al. [130], Penny et al. [154], and Lee et al. [155] extended the geochemical model of Korte [153] to include the metabolic effects of Fe-reducing bacteria (FeRB) and Mn-reducing bacteria for releasing Fe and As (and other trace elements such as Mn, Co, Ni, Ba, V, and rare earth elements) in alluvial aquifers in the United States. Further, Saunders et al. [130] showed that SO_4-reducing bacteria removed As, Fe, Co, and Ni by coprecipitating them in biogenic pyrite [130]. Later, other researchers showed that FeRB were apparently responsible for causing As enrichment of Holocene alluvial aquifers in Bangladesh and India (e.g., [8,156,157]). Laboratory investigations by Islam et al. [158] on sediment cores from Southeast Asia showed that FeRB such as *Geobacter* could liberate As from minerals [158]. Field investigations by Saunders et al. [131] at Korte's [153] discovery location (Kansas City, MO, USA) showed that FeRB (of the genus *Geobacter*) were abundant in groundwaters containing elevated As and absent in groundwaters without As. Sulfate reducing bacteria (principally *Desulfovibrio desulfuricans*) were also present in As-enriched groundwater. Saunders et al. [131] proposed that such bacteria were also important in As geochemical cycling in Southeast Asia [131].

As-respiring bacteria are capable of reducing not only soluble arsenate but also adsorbed arsenate [159] and arsenate within solids such as scorodite [160], with recent evidence suggesting that reduction proceeds first through a dissolution step [161]. Dissimilatory arsenate reduction may enhance the solubility of As, particularly in environments with low iron (hydro)oxide content. Zobrist et al. [119], for example, showed that reduction of As(V) adsorbed or coprecipitated on amorphous aluminum hydroxides by *Sulfospirillium barnesii* greatly increased dissolved As(III) concentrations.

4.4.5 ROLE OF NOM

NOM such as humic acid (HA) and FA is an inherently complex mixture of polyfunctional organic acids derived from the decomposition of terrestrial and aquatic animals and plants. Prevalent in the subsurface, NOM is highly reactive toward both metals and surfaces, and therefore may play an important role in governing the mobility and bioavailability of As [162]. NOM has great potential in influencing As sorption behavior and its speciation by interacting with mineral surfaces and/or with As itself, and thus may play a major role in the release of As from soils and sediments into the groundwater [162,163].

Ubiquitous in aquatic environments, NOM may interact strongly with As. A number of studies have been performed to elucidate the effects of NOM on the sorption behavior of As species and the consequential influence on their mobility [107,110,163–170]. The interactions between NOM and As can be influenced by various factors such as pH, redox potential, As speciation and concentration, other competing ions and complexing ligands, aquifer mineralogical properties, and reaction kinetics [171].

The presence of NOM may enhance As release mainly through three main pathways [162]: (a) competition for available adsorption sites; (b) change of the redox chemistry of site surface and As species; and (c) formation of aqueous complexes.

4.4.5.1 Competitive Adsorption of As and NOM

Organic acids such as HA and FA may compete strongly with As(III) and As(V) for active adsorption sites on mineral surfaces. Competition between organic acids and As species can be an important factor in influencing the mobilization of As. The competition for active-binding sites on mineral surfaces between organic acids and As species may lower the levels of As retention, especially under acidic conditions.

Grafe et al. [107] studied the adsorption of As(III) and As(V) on goethite in the presence of a peat HA, a Suwannee River FA, and citric acid (CA) [107]. HA inhibited As(V) adsorption starting at pH 9, reaching a maximum at pH 6.5. The inhibition effects of FA were slighter, starting at pH 5 and increasing as the pH decreased. CA showed no effect on As(V) adsorption. As(III) adsorption is inhibited by HA starting at pH 7 and increased when pH decreased, whereas FA and CA reduced As(III) adsorption starting at pH 8, with a continuous reduction as pH decreased. It was also observed that the adsorption of all three organic acids on the goethite surface was reduced in the presence of both As(V) and As(III). Grafe et al. [166] subsequently examined the effects of the organic acids on the adsorption of As(V) and As(III) on ferrihydrite [166]. However, some researchers reported that the presence of OM might increase the adsorption of As, thus reducing As mobility [164,165,168].

4.4.5.2 Effect of Site Surface and As Species on Redox Chemistry

It has been shown that NOM may catalyze both the oxidation and reduction reactions among chemical species, in part by the quinone-mediated formation of free radicals [167,172]. They may serve as an electron shuttle between kinetically inert redox species or between microorganisms and As species [172]. Hydroxides may act as a surface catalyst or as an electron-transfer intermediate. As release may be enhanced by the redox reactions among NOM, As, and substrates, resulting in As(V) reduction to the more labile and mobile form, As(III).

Redman et al. [167] observed the reduction of As(V) to free As(III) by the Inangahua River NOM, and the reversed process of As(III) oxidation to free As(V) by all other experimental NOM samples [167]. The authors postulated that hematite might have acted as a surface catalyst or as an electron-transfer intermediate in this process.

Palmer et al. [173] found that inorganic arsenates were reduced to arsenite by homogeneous aqueous solutions of several HAs and FAs [173]. The fraction of arsenate that was reduced initially increased with humic concentration, but leveled off as the reduction potential decreased at higher concentrations. Reoxidation of As(III) in humic solutions could be achieved by extended bubbling with air.

Tongesayi and Smart [174] investigated the reduction of inorganic As(V) with Suwannee River FA in aqueous solutions where the pH and concentrations of FA, As(V), As(III), and Fe(III) were independently varied. The results demonstrated that FA can significantly reduce As(V) to As(III), and both dark and light conditions promote reduction. Besides, adding Fe(III) speeded up the reduction reaction.

4.4.5.3 NOM-As Complexation

NOM may form both aqueous and insoluble surface inner-sphere complexes with metal cations due to their strong affinity to metal cations and metal oxides. Aqueous NOM-metal complexes may, in turn, associate strongly with other dissolved anions, presumably by metal-bridging mechanisms, diminishing the tendencies of such anions to form surface complexes [167,169]. Factors influencing the quantity of ions bound by NOM include pH, ionic strength, molecular weight, and functional group content as well as the presence of competing ions [167,169]. The formation of soluble As complexes with NOM may enhance As solubility and therefore As release into groundwater. Moreover, complexation may also affect the partitioning of As to suspended solids in the water column and the sequestration of As to sediments [162].

Infrared spectroscopy studies have confirmed that COOH groups play a predominant role in the complexation of metal ions by HA and FA [175]. Some evidence indicates that OH, C=O, and NH groups may also be involved [165,168].

Mukhopadhyay and Sanyal [176] studied As-humic/fulvic complexation equilibrium and proposed that the equilibrium process depends on the nature and properties of the humic substance, which would affect the retention/release of As from the sediments matrix [176].

Warwick et al. [170] characterized the interaction of As with HA in a system consisting of HA with As(III) and As(V) and dimethylarsinic acid (DMAA) [170]. The interaction is postulated to involve bridging metals and deprotonated functional groups within the HA.

4.5 MECHANISM OF AS MOBILIZATION AND RELEASE IN SHALLOW AQUIFERS: CASE STUDIES IN SOUTHERN ASIA AND NORTHERN CHINA

4.5.1 As in Shallow Reducing Aquifers in Southern Asia

In southern Asia, major alluvial and deltaic plains composed of Quaternary sediments are prone to developing groundwater As problems, particularly around the perimeter of the Himalayan mountain range. High concentrations have been found in the groundwater from such aquifers in the Bengal Basin of Bangladesh and eastern India, the lowland Terai region of Nepal, the Mekong Valley of Cambodia, the Red River Delta of Vietnam, and the Irrawaddy Delta of Myanmar [21,33,177,178]. Similar problems may occur in similar alluvial and deltaic environments elsewhere in the world. Unfortunately, such flat-lying fertile plains are often densely populated, and poor groundwater quality can have a major impact on large numbers of people. At present, the Ganges- Brahmaputra-Meghna Plain and Delta in India and Bangladesh represent the most acutely As-contaminated sites in the world, with concentrations sometimes >4,000 mg/L [179,180]. It has been estimated that more than 50 million people ingested As-contaminated water in this area alone [6]. Consequently, most of the groundwater As studies in recent years have been concentrated in this area (Table 4.4).

TABLE 4.4

Summary of Documented Cases of Naturally Occurring As Problems in Southern Asian Aquifers

Country/Region	Area (km^2)	PopulationExposed	As Concentration Ranges (µg/L)	Aquifer Type	Groundwater Conditions	Sediments/As Contents	References
Bangladesh	150,000	35,000,000	<1–2,300	Holocene alluvial/deltaic sediments. Abundance of solid OM	Strongly reducing, neutral pH, high alkalinity, occurrence of CH_4, slow groundwater flow rates	Fine- to medium-sized sand shallow: brown to gray; deep: orange; As: 2–10 mg/g	[46]
West Bengal, India	23,000	5,000,000	<10–3,200	As Bangladesh	As Bangladesh	Similar to those in Bangladesh	[181]
Vietnam Red River Delta	1,200	10,000,000	1–3,050	Holocene alluvial/deltaic sediments	Reducing, high Fe, Mn, NH_4, high alkalinity, Ca-HCO_3 or Mg-HCO_3 type, rich in CH_4	Brown to black-brown clay 6–33 mg/g As; gray clay 2–12 mg/g; brown to gray sand 0.6–5 mg/g; average 7.5 mg/g As	[20,21]
Cambodia	3,700	120,000	1–1,340	Unconsolidated alluvial deposits of the Quaternary	Neutral pH, low Eh, high Fe and bicarbonate contents, moderate salinities, and low dissolved sulfate	As-rich area: gray clays, silts, and sands. As and Fe decrease with depth; near surface, As(V) dominated; 35%–60% As reduced to As(III) by 17 m depth	[182,183]
Terai Region, Nepal	30,000	550,000	1.7–404	Thick sand and gravel deposits of Holocene age. Interlocked with alluvium flood plains	High HCO_3^- and low SO_4^{2-}, high Fe and Mn	Average As: 9 mg/g; iron oxide, titanium oxide, and calcium oxide concentration were 5%, 0.7%, and 3.9%, respectively	[26,184]
Punjab, Pakistan	?	3040	32–1,900	Quaternary alluvial and deltaic sediments	Alkaline pH (7.3–8.7); high F, As, SO_4^{2-}; water type: Na-HCO_3 or Na-HCO_3-SO_4 dominant	Mostly coarse sand, containing high percentage of fine to very fine sand and silt	[27]

4.5.1.1 Common Hydrological and Hydrochemical Features of These Aquifers

The Ganges-Meghna-Brahmaputra Delta [46], the Mekong Basin near Phnom Penh [24,25], and the Red River Basin near Hanoi [20] have three features in common [33]: (a) River drainage from the rapidly weathering Himalayas. Although groundwater contamination has also been reported upstream, for example, in Nepal and Bihar, India, the major concerns are the delta areas of these river systems because of the high populations they support and will continue to support in the near future. (b) In these delta areas, the three systems are characterized by rapidly buried OM-bearing, relatively young (ca. Holocene) sediments. (c) Very low, basin-wide hydraulic gradients. In all three systems, the spatial heterogeneity of As content distribution occurs on very small scales, with high concentrations (>50 ppb) commonly found within tens of meters of low concentrations. This may reflect the complex lithological structures of the aquifers, which include highly permeable channel fills as well as very low-permeability scoured channels filled with organic-rich overbank deposits. The interconnectivity of highly permeable channel sand units may be critical to determining groundwater As distribution because of its control on groundwater flow patterns [32] and on the rate of flushing of As from the system.

A common feature of all these As-contaminated waters is their anoxia. This anoxia may be inherited from the conditions prevailing during sediment deposition, as discussed below, or may result from the surface input of organic material. As Eh decreases, the classical cascade of electron acceptors involved in the oxidation of this organic material leads to the successive appearance of Mn^{2+}, Fe^{2+}, NH_4^+, acid volatile sulfides (AVSs), and methane.

The most common feature of all these groundwaters is their high concentration of Fe^{2+}, often close to the level required for siderite saturation, in many of the Bengali, Cambodian, and Vietnamese shallow aquifers [8,20,24].

4.5.1.2 As Mobilization in Aquifers

Groundwater As enrichment in these deltas has been largely attributed to coupled redox cycling of As-bearing iron (oxy)hydroxides [6]. Reductive dissolution of Fe(III) phases may release sorbed As. Aqueous Fe concentrations and redox potential are important indicators of aqueous As enrichment in many environments [74,157].

There are a variety of proposed mechanisms for As release within these reducing aquifers. For example, the oxidation and breakdown of As-bearing pyrites by the drawdown of oxygenated water [185–187] was one possible mechanism, although this process has been contested, as the products of the breakdown of pyrite under the neutral conditions of the aquifer would act as a sink for As as opposed to releasing it to the groundwater [7,8]. The association of As with Fe(III) oxides within As-rich aquifers of Bangladesh and West Bengal [7,8,129,157,185–189], coupled with the reducing conditions of the aquifers, could also lead to As release. Microbial degradation of the naturally occurring OM drives the aquifer to anoxia where, under such conditions, Fe(III) oxides become unstable and dissolve, releasing As as well as Fe and HCO_3^- into the groundwaters. Such a process would be expected to produce correlations between these three components [6–8,46]. In addition, microbes can play a more direct role in mediating As release rather than just altering the redox conditions. Microcosm-based geomicrobiological studies on sediments from West Bengal and Bangladesh show that stimulation of the indigenous microbial community leads to increased As release and Fe(III) reduction [74,158,189–191].

Although the mechanism of release is generally agreed upon, other natural processes obscure a simple relationship between dissolved Fe and As levels [111,190]. Among these additional natural processes, sulfur redox cycling may impact Fe and As geochemistry [133,135]. A mechanistic understanding of these processes is necessary to explain the spatial heterogeneity and extent of As enrichment.

Iron mineralogical transformations are intrinsically linked to the cycling of other components, including OM, which drives reduction [183]. Sulfur cycling, which can also undergo redox cycling to form sulfide ions as a product of dissimilatory reduction of SO_4^{2-}, may also affect the fate of As by leading to the formation of As sulfides [135,192]. Although Fe reduction is more thermodynamically favorable than SO_4^{2-} reduction, the latter often precedes Fe reduction for kinetic reasons [193] and, as a result, sulfide and Fe(III) oxides often occur together [194] and can thus react chemically in these environments. OM-driven reduction of Fe (oxy)hydroxides via sulfide is, therefore, common in anoxic settings [183].

Fluvial geomorphological processes influence sediment OM content, composition, and mineralogy, all of which may influence the distribution of As in the environment [25,131,182,195]. Widespread and heterogeneous As enrichment is commonly associated with the large deltas of South Asian rivers. These high flux, sediment-laden, tropical to subtropical river systems have the potential for rapid and recent major sediment deposition. Deltaic, localized depositional environments such as scroll sequences, avulsions, oxbows, abandoned chutes, and channel islands can capture suspended sediment [183]. Suspended OM as well as surficial plants and woody debris will be incorporated into these rapidly buried sediment strata. Sedimentary OM from overbank deposits, channel deposits, and suspended or dissolved OM may differ considerably in composition and reactivity. These differences may impact the fate of As indirectly by influencing Fe and S reduction and the sequestration of As in the solid phase [183].

Ultimately, a complex array of geochemical processes may impact the fate of As. For example, the mechanism of Fe mineral dissolution, whether direct or indirect or the extent of competing sequestration reactions through secondary (diagenetic) mineral precipitation, may conceal the root cause of As enrichment [183].

4.5.2 FORMATION OF SODA WATERS AND GROUNDWATER AS PROBLEMS IN NORTHERN CHINA

The presence of endemic As poisoning has been recognized in China since the 1980s, and today, the scale of the problem is known to be large. As poisoning due to long-term intake of high-As groundwater from Quaternary aquifers was first identified in Taiwan and Xinjiang, and more recently in parts of Inner Mongolia and Shanxi Province. Up to now, cases of waterborne endemic As poisoning have been reported in Taiwan, Xinjiang, Inner Mongolia, Shanxi, Ningxia, Jilin, Qinghai, and Anhui provinces and some villages close to Beijing [196]. Most of these areas are located in large Quaternary basins under arid and semiarid conditions. Groundwater in the As-affected regions appears to be strongly reducing. The population exposed to drinking water with concentrations in excess of 50 mg/L (the old Chinese standard) has been estimated to be around 3 million [197]; if accounted for according to the latest data and national standard (10 mg/L), the population may reach 14.66 million [198].

Datong Basin in Shanxi and Hetao Basin in Inner Mongolia are two representative As poisoning-affected areas in China. They are both located in an arid/semiarid region of northwestern China, with a mean annual rainfall of less than 400 mm and a mean annual evaporation rate exceeding 2,000 mm.

Endemic disease due to long-term intake of high-As groundwater was identified in these two basins in the early 1990s. Since the late 1990s, great effort has been made to characterize the main hydrogeochemical features and understand the mechanisms responsible for the natural As enrichment in the groundwater of these two basins through detailed hydrogeological and hydrochemical investigation [11–19,72,199,200].

The hydrochemical features of Datong Basin and the western Hetao Basin are summarized in Table 4.5. The common hydrochemical features of the high-As groundwater from these two areas can be summarized as follows: the groundwaters are near-neutral to weakly alkaline with high alkalinity; most water samples contain high total dissolved solids (TDS), with HCO_3^- (and Cl) as the dominant anion(s) and Na as the dominant cation; and As and fluoride are enriched simultaneously in groundwater.

TABLE 4.5

Comparison of the Hydrochemical Features of High-As Aquifers between Datong Basin and Western Hetao Basin

Hydrochemical Parameter	Datong Basin (2005)	Western Hetao Basin (2007)
pH	7.2–9.1	7.3–8.3
TDS (mg/L)	262–8,870	444–7,460
HCO_3^- (mg/L)	72–1,080	284–1,290
Cl^- (mg/L)	11–3,170	37–4,531
SO_4^{2-} (mg/L)	0.7–2,240	<0.1–1,130
Na^+ (mg/L)	59–2,395	74–1,834
As (mg/L)	8–1,550	1–1,093
F (mg/L)	0.2–9.2	0.3–6.1
HPO_4^{2-} (mg/L)	<0.1–3.8	<0.1–3.21
Fe (mg/L)	<0.01–5.44	<0.01–5.26
DOC (mg/L)	nd	0.73–35.7
NH_4^+ (mg/L)	nd	<0.2–10.5
Dissolved sulfide (mg/L)	nd	<0.01–0.37
Water chemical type	Na-HCO_3	Na(-Mg)-Cl or Na(-Mg)-Cl-HCO_3 or Na(-Mg)-HCO_3-Cl
Note	Strong odor of H_2S, occasionally high contents of methane	Strong odor of H_2S, occasionally high contents of methane (up to 5110 g/L)

It is interesting to observe that most of the high As and F groundwater are soda waters, which refer to waters with $Na/(Cl^-+SO_4^{2-})$(meq) ratios greater than 1 [14] (Figure 4.7). The occurrence of soda water is regarded as one stage of interaction of water with aluminosilicate minerals, when the groundwater is saturated with calcite. Generally speaking, saturation usually occurs when the salinity of water is more than 0.6 g/L and pH is more than 7.4 [201]. After saturation, Ca and Mg in solution encounter geochemical barriers of calcite and montmorillonite that inhibit further enrichment of Ca and Mg in the aqueous phase and facilitate Na concentration under conditions of retarded subsurface flow and enhanced salinity and alkalinity of groundwater. Hydrolysis of aluminosilicate minerals and evapotranspiration have been proposed to be responsible for the genesis of soda waters in regions with similar conditions [14].

As shown in Table 4.5, soda waters at Datong and Hetao basins are weakly alkaline to alkaline. As concentration in groundwaters is affected by many factors, with pH and redox conditions being the most important. Under the aerobic and acidic to near-neutral conditions typical of many natural environments, As is strongly adsorbed by (hydro)oxide minerals as the arsenate ion. Adsorption protects many natural environments from widespread As toxicity problems [6]. As pH increases, As desorbs from the oxide surfaces, thereby increasing its concentration in solution. It can be seen from Figure 4.8 that As concentration is positively correlated with pH at Datong Basin, with a substantial increase above pH 8.

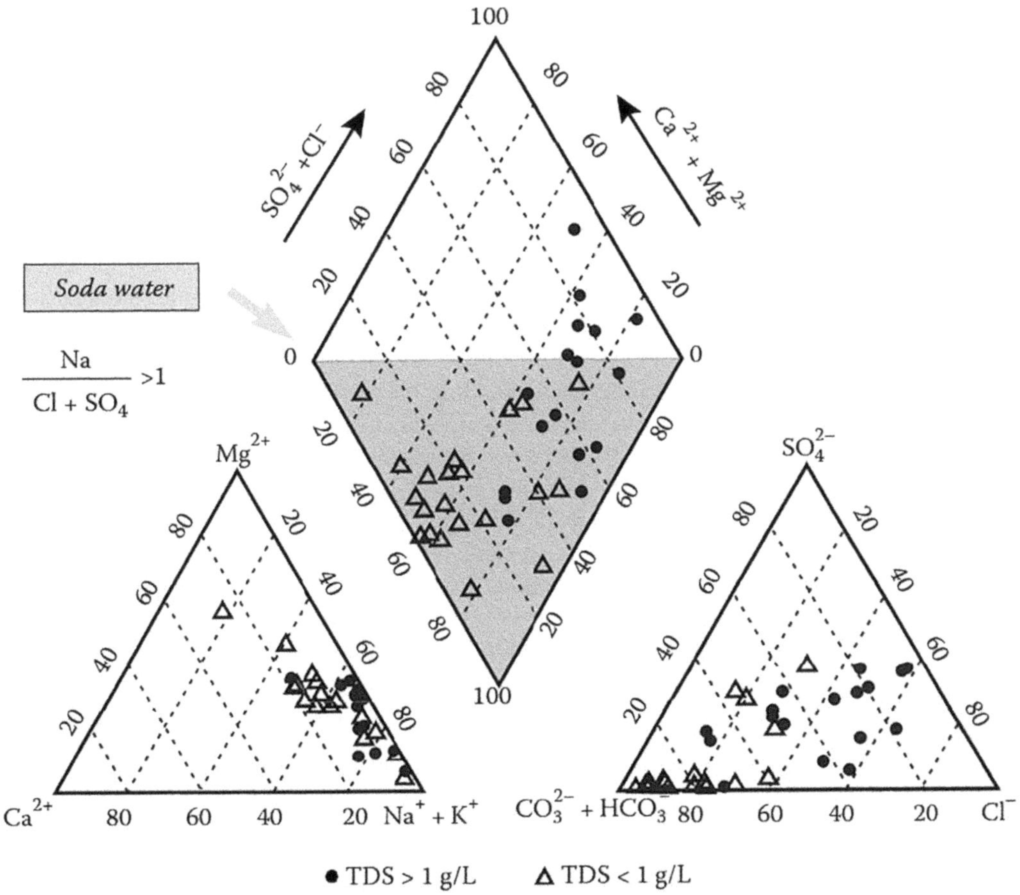

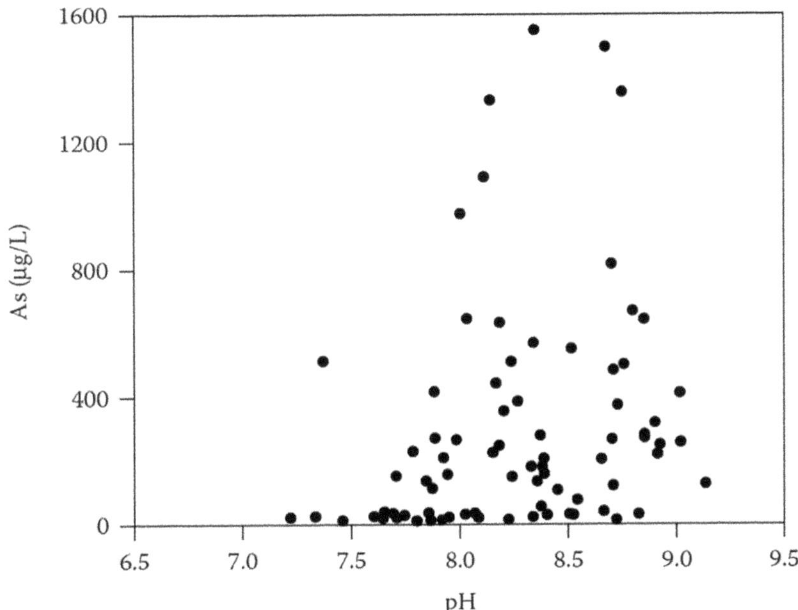

FIGURE 4.7 Piper diagram of high As and fluoride groundwaters in Datong Basin, Northern China.

FIGURE 4.8 Relationship between As concentration and pH of groundwaters in Datong Basin, Northern China.

GLOSSARY

Aluminosilicate It refers to materials containing anionic Si-O-Al linkages. Commonly, the associate cations are sodium (Na^+), potassium (K^+), and protons (H^+). These materials occur as minerals and synthetic compounds, often in the form of zeolites.

Arsenic contamination of groundwater (a) Serious water pollution Arsenic contamination in groundwater is a significant issue, often caused by naturally high concentrations of arsenic in deeper levels of groundwater and/or by infiltration of industrial storm run-off water. The use of deep tube wells for water supply in regions like the Ganges Delta has led to serious arsenic poisoning in large populations. Dangerously high arsenic levels have been found in drinking-water wells in over 25 U.S. states, potentially affecting 2.1 million people. The worst case occurred in Bangladesh, where over 100 million people were poisoned by arsenic in groundwater supplies; Understanding the sources and distribution of arsenic in aquifers is crucial for managing water resources and safeguarding public health; (b) How arsenic gets into groundwater: (b1) Arsenic occurs naturally as a trace element in rocks and sediments; (b2) Whether it's released into groundwater depends on: the chemical form of arsenic, the geochemical conditions in the aquifer, and the biogeochemical processes that occur; (b3) Human activities (such as mining, industrial use, wood preservation, and pesticides) can also release arsenic into groundwater; and (b4) In drinking-water supplies, arsenic is toxic even at low levels and is a known carcinogen; and (c) Water quality standards on arsenic: The U.S. Environmental Protection Agency (USEPA) lowered the MCL for arsenic in public-water supplies to 10 micrograms per liter (μg/L) from 50 μg/L in 2001.

Environmental chemistry (a) It is an academic field that focuses on the chemical and biochemical phenomena occurring in natural environments; (b) Researchers study how chemicals interact in soil, water, air, plants, animals, and even humans; and (c) It plays a crucial role in understanding pollution, nutrient cycling, and the impact of contaminants on ecosystems.

Environmental geochemistry It is an academic field that explores the sources, distribution, and interactions of chemical elements within various Earth systems. Environmental geochemistry informs our understanding of soil quality, water pollution, and human health risks. Scientists investigate the distribution of elements (both major and trace) across different landscapes, helping us address environmental challenges. The science field can be broken down into (a) Geochemistry: (a1) Geochemistry uses the tools and principles of chemistry to explain the mechanisms behind major geological systems. These systems include the Earth's crust and its oceans; (a2) It encompasses the study of elements, isotopes, and compounds in rocks, minerals, soils, water, and the atmosphere; and (b) Environmental Chemistry: (b1) Environmental chemistry focuses on the chemical and biochemical phenomena occurring in natural environments; (b2) Researchers study how chemicals interact in soil, water, air, plants, animals, and even humans; and (b3) It plays a crucial role in understanding pollution, nutrient cycling, and the impact of contaminants on ecosystems; and (c) arsenic in groundwater can be removed by chemical coagulation, dissolved air flotation, filtration, granular activated carbon [202], oxidation, coagulation/sedimentation, lime softening, ion exchange, and membrane filtration [203].

Geochemistry It is an academic field that studies or uses the tools and principles of chemistry to explain the mechanisms behind major geological systems. These systems include the Earth's crust and its oceans; It encompasses the study of elements, isotopes, and compounds in rocks, minerals, soils, water, and the atmosphere.

Humic substances (a) These are colored, recalcitrant organic compounds that naturally form during the long-term decomposition and transformation of biomass residues. Humic substances are vital components of ecosystems, and their solubility characteristics vary based on their specific type. They represent a major part of OM in soil, peat, coal, and sediments. Additionally, they are essential components of dissolved NOM in lakes, rivers, and seawater. Humic substances include: (a1) HA: Soluble in water at neutral and alkaline pH but insoluble at acidic pH (<2); (a2) Fulvic Acid (FA): Soluble in water at any pH; (a3) Humin: Not soluble in water at any pH; (a4) Hymatomelanic Acid: Part of HA that is soluble in ethanol; and (b) Extraction: HAs and FAs may be extracted from soil and other solid-phase sources using a strong base (such as NaOH or KOH).

Microbial transformation (a) Microorganisms play a crucial role in the biogeochemical cycle by actively participating in natural processes like biomineralization, bioweathering, and bioleaching, in turn, they directly or indirectly influence mineral behavior, affecting the fate of pollutants in the environment. These tiny microorganisms, ranging from bacteria to fungi and algae, interact with minerals, affecting their dissolution, aggregation, and secondary mineral formation. The first step for microbes is to contact and adsorb onto mineral surfaces by bio-adsorption. Additionally, microorganisms secrete HAs and other bioactive molecules that enhance adsorption and pollutant retention. Bio-adsorption is a natural process where a substance (such as a mycotoxin) interacts with a surface (the binder) through non-covalent bonds (e.g., Van der Waals interactions). In the context of mycotoxins, adsorption reduces the bioavailability of these harmful compounds in food commodities.

Oxidation-reduction (redox) reactions Redox reactions involve the transfer of electrons between chemical species. Redox reactions are fundamental for energy production and chemical transformations. Three types of redox reactions are introduced: (a) Burning of fuels: Combustion reactions involve oxidation (loss of electrons) of the fuel and reduction (gain of electrons) of oxygen; (b) Corrosion of metals: Metals oxidize (lose electrons) when they react with oxygen or other substances; and (c).Photosynthesis and cellular respiration: These biological processes rely on redox reactions·

Soda waters They refer to waters with $Na/(Cl^- + SO_4^{2-})(meq)$ ratios greater than 1. Most of the high As and F groundwater are soda waters. The occurrence of soda water is regarded as one stage of interaction of water with aluminosilicate minerals, when the groundwater is saturated with calcite. Saturation usually occurs when the salinity of water is more than $0.6\,g/L$ and pH is more than 7.4. After saturation, Ca and Mg in solution encounter geochemical barriers of calcite and montmorillonite that inhibit further enrichment of Ca and Mg in the aqueous phase and facilitate Na concentration, under conditions of retarded subsurface flow and enhanced salinity and alkalinity of groundwater. Hydrolysis of aluminosilicate minerals and evapotranspiration have been proposed to be responsible for the genesis of soda waters in regions with similar conditions.

Sorption Sorption refers to the physicochemical process by which a compound is removed from a solution and accumulates on a solid phase (the sorbent). There are three types of sorptives: (a) Anionic sorptives: These are negatively charged species (e.g., phosphate ions, PO_4^{3-}); (b) Cationic sorptives: These are positively charged species (e.g., calcium ions, Ca^{2+}, and lead ions, Pb^{2+}); and (c) Organic sorptives: These can be uncharged and exhibit varying polarities (e.g., benzene, C_6H_6, and sucrose, $C_{12}H_{22}O_{11}$).

Zeolites They are both naturally occurring and synthesized aluminosilicates with porous structures, making them valuable for various industrial applications, such as ion exchange for heavy metals removal, water purification or wastewater treatment.

ACKNOWLEDGMENTS

Our research work on As hydrogeochemistry in the past 10 years has been financially supported by National Natural Science Foundation of China (Grant Nos 49832005, 40425001, and 40830748), Ministry of Science and Technology of China (2008DFA20950), and Ministry of Education of China (111 project).

REFERENCES

1. National Research Council. *Arsenic in Drinking Water.* National Academy Press, Washington, DC, 1999.
2. Smith, A.H., Lopipero, P.A., Bates, M.N., and Steinmaus, C.M. Arsenic epidemiology and drinking water standard. *Science*, 296, 2145–2146, 2002.
3. World Health Organization. *Guidelines for Drinking-Water Quality. Health Criteria and Other Supporting Information*, 2nd Edition, WHO, Geneva, Switzerland, pp. 940–949, 1996.
4. European Commission. Directive related with drinking water quality intended for human consumption, 98/83/EC, Brussels, Belgium, 1998.
5. EPA Office of Groundwater and Drinking Water. Implementation guidance for the arsenic rule. EPA report-816-D-02-005, U.S. Environmental Protection Agency, Washington, DC, 2002.
6. Smedley, P.L. and Kinniburgh, D.G. A review of the source, behaviour and distribution of arsenic in natural waters. *Appl. Geochem.*, 17, 517–568, 2002.
7. Nickson, R.T., McArthur, J.M., Burgess, W.G., Ahmed, K.M., Ravenscroft, P., and Rahman, M. Arsenic poisoning of Bangladesh groundwater. *Nature*, 338, 395, 1998.
8. Nickson, R.T., McArthur, J.M., Ravenscroft, P., Burgess, W.G., and Ahmed, K.M. Mechanism of arsenic release to groundwater, Bangladesh and West Bengal. *Appl. Geochem.*, 15, 403–413, 2000.
9. Stuben, D., Berner, Z., Chandrasekharam, D., and Karmakar, J. Arsenic enrichment in groundwater of West Bengal, India: Geochemical evidence for mobilization of As under reducing conditions. *Appl. Geochem.*, 18, 1417–1434, 2003.
10. van Geen, A., Zheng, Y., Versteeg, R., Stute, M., Horneman, A., Dhar, R., Steckler, M., Gelman, A., Small, C., Ahsan, H., Graziano, J., Hussein, I., and Ahmed, K.M. Spatial variability of arsenic in 6000 tube wells in a 25 km² area of Bangladesh. *Water Resour. Res.*, 39, 1140, 2003. doi: 10.1029/2002WR001617.
11. Guo, H.M., Wang, Y.X., Shpeizer, G.M., and Yan, S.L. Natural occurrence of arsenic in shallow groundwater, Shanyin, Datong Basin, China. J. Environ. Sci. *Health Part A - Toxic/Hazardous Subst. Environ. Eng.*, 38, 2565–2580, 2003.
12. Guo, H.M. and Wang, Y.X. Geochemical characteristics of shallow groundwater in Datong Basin, northwestern China. *J. Geochem. Explor.*, 87, 109–120, 2005.
13. Wang, Y.X. and Shvartsev, S.L. Major hydrogeochemical processes controlling arsenic enrichment in groundwater of the Datong Basin, Northern China. In: Bullen, T.D. and Wang Y.X. (Eds), *Proceedings of 12th International Symposium on Water-Rock Interaction (WRI-12)*, Taylor & Francis Group, London, pp. 1123–1126, 2007.
14. Wang, Y., Shvartsev, S.L., and Su, C. Genesis of arsenic/fluoride-enriched soda water: A case study at Datong, northern China. *Appl. Geochem.*, 24(4), 641–649, 2009.
15. Xie, X.J., Wang, Y.X., Su, C.L., Liu, H.Q., Duan, M.Y., and Xie, Z.M. Arsenic mobilization in shallow aquifers of Datong Basin: Hydrochemical and mineralogical evidences. *J. Geochem. Explor.*, 98, 107–115, 2008.
16. Lin, N.F., Tang, J., and Bian, J.M. Characteristics of environmental geochemistry in the arseniasis area of the Inner Mongolia of China. *Environ. Geochem. Health*, 24, 249–259, 2002.
17. Smedley, P.L., Zhang, M., Zhang, G., and Luo, Z. Mobilisation of arsenic and other trace elements in fluviolacustrine aquifers of the Huhhot Basin, Inner Mongolia. *Appl. Geochem.*, 18, 1453–1477, 2003.
18. Deng, Y.M., Wang, Y.X., Ma, T., and Gan, Y.Q. Speciation and enrichment of arsenic in strongly reducing shallow aquifers at western Hetao Plain, northern China. *Environ. Geol.*, 56, 1467–1477, 2009.
19. Guo, H.M., Yang, S.Z., Tang, X.H., and Shen, Z.L. Groundwater geochemistry and its implications for arsenic mobilization in shallow aquifers of Hetao Basin, Inner Mongolia. *Sci. Total Environ.*, 393, 131–144, 2008.
20. Berg, M., Tran, H.C., Nguyen, T.C., Pham, T.C., Schertenleib, R., and Giger, W. Arsenic contamination of groundwater and drinking water in Vietnam: A human health threat. *Environ. Sci. Technol.*, 35, 2621–2626, 2001.
21. Berg, M., Stengel, C., Trang, P.T.K., Viet, P.H., Sampson, M.L., Leng, M., Samreth, S., and Fredericks, D. Magnitude of arsenic pollution in the Mekong and Red River deltas—Cambodia and Vietnam. *Sci. Total Environ.*, 372, 413–425, 2007.
22. Berg, M., Trang, P.T.K., Stengel, C., Buschmann, J., Viet, P.H., Dan, N.V., Giger, W., and Stuben, D. Hydrological and sedimentary controls leading to arsenic contamination of groundwater in the Hanoi area, Vietnam: The impact of iron-arsenic ratios, peat, river bank deposits, and excessive groundwater abstraction. *Chem. Geol.*, 249, 91–112, 2008.

23. Postma, D., Larsen, F., Hue, N.T.M., Duc, M.T., Viet, P.H., Nhan, P.Q., and Jessen, S. Arsenic in groundwater of the Red River floodplain, Vietnam: Controlling geochemical processes and reactive transport modeling. *Geochim. Cosmochim. Acta*, 71, 5054–5071, 2007.
24. Polya, D.A., Gault, A.G., Bourne, N.J., Lythgoe, P.R., and Cooke, D.A. Coupled HPLC-ICP-MS analysis indicates highly hazardous concentrations of dissolved arsenic species are present in Cambodian well waters. *R. Soc. Chem. Special Publ.*, 288, 127–140, 2003.
25. Polya, D.A., Gault, A.G., Diebe, N., Feldmann, P., Rosenboom, J.W., Gilligan, E., Fredericks, D., Milton, A.H., Sampson, M., Rowland, H.A.L., Lythgoe, P.R., Jones, J.C., Middleton, C., and Cooke, D.A. Arsenic hazard in shallow Cambodian groundwaters. *Mineral. Mag.*, 69, 807–823, 2005.
26. Gurung, J.K., Ishiga, H., and Khadka, M.S. Geological and geochemical examination of arsenic contamination in groundwater in the Holocene Terai Basin, Nepal. *Environ. Geol.*, 49, 98–113, 2005.
27. Farooqi, A., Masuda, H., and Firdous, N. Toxic fluoride and arsenic contaminated groundwater in the Lahore and Kasur districts, Punjab, Pakistan and possible contaminant sources. *Environ. Pollut.*, 147, 839–849, 2007.
28. Winkel, L., Berg, M., Stengel, C., and Rosenberg, T. Hydrogeological survey assessing arsenic and other groundwater contaminants in the lowlands of Sumatra, Indonesia. *Appl. Geochem.*, 23, 3019–3028, 2008.
29. Cullen, W.R. and Reimer, K.J. Arsenic speciation in the environment. *Chem. Rev.*, 89, 713–764, 1989.
30. Mandal, B.K. and Suzuki, K.T. Arsenic round the world: A review. *Talanta*, 58, 201–235, 2002.
31. Bissen, M. and Frimmel, F.H. Arsenic—a review. Part I: Occurrence, toxicity, speciation, mobility. *Acta Hydrochim. Hydrobiol.*, 31, 9–18, 2003.
32. Harvey, C.F. and Beckie, R.D. Arsenic: Its biogeochemistry and transport in groundwater. *Metal Ions Biol. Syst.*, 44, 145–169, 2005.
33. Charlet, L. and Polya, D.A. Arsenic in shallow reducing groundwaters in southern Asia: An environmental health disaster. *Element*, 2, 91–96, 2006.
34. Baur, J.W. and Onishi, B.M.H. Arsenic. In: Wedepohl, K.H. (Ed.), *Handbook of Geochemistry*, Chapter 33. Springer, Berlin, 1969.
35. Greenwood, N.N. and Earnshaw, A. *Chemistry of the Elements*. Pergamon Press, New York, 1989.
36. Palmer, C.A. and Klizas, S.A. The chemical analysis of Argonne premium coal samples. U.S. Geological Survey Bulletin 2144, U.S. Geological Survey, 1997.
37. Nriagu, J.O., Bhattacharya, P., Mukherjee, A.B., Bundschuh, J., Zevenhoven, R., and Loeppert, R.H. Arsenic in soil and groundwater environment. In: Bhattacharya, P., Mukherjee, A. B., Bundschuh, J., Zevenhoven, R., and Loeppert, R. H. (Eds), *Trace Metals and Other Contaminants in the Environment*, Vol. 9, Elsevier, Amsterdam, pp. 3–60, 2007.
38. Yudovich, Y.E. and Ketris, M.P. Arsenic in coal: A review. *Int.J. Coal Geol.*, 61, 141–196, 2005.
39. Irnawati, I., Idroes, R., Zulfiani, U., Akmal, M., Suhartono, E., Idroes, G.M., and Jalil, Z. Assessment of arsenic levels in water, sediment, and human hair around Ie Seu'um geothermal manifestation area, Aceh, Indonesia. *Water*, 13(17), 2343, 2021.
40. Rice, K.C. Trace-element concentrations in streambed sediment across the conterminous United States. *Environ. Sci. Technol.*, 33, 2499–2504, 1999.
41. Edmunds, W.M., Cook, J.M., Kinniburgh, D.G., Miles, D.L., and Trafford, J.M. Trace-element occurrence in British groundwaters. British Geological Survey, Research Report SD/89/3, 1989.
42. Welch, A.H., Westjohn, D.B., Helsel, D.R., and Wanty, R.B. Arsenic in ground water of the United States: Occurrence and geochemistry. *Ground Water*, 38, 589–604, 2000.
43. Focazio, M.J., Welch, A.H., Watkins, S.A., Helsel, D.R., and Horng, M.A. A retrospective analysis of the occurrence of arsenic in groundwater resources of the United States and limitations in drinking water supply characterizations. U.S. Geological Survey, Water Resources Investigation Report 99–4279, 1999.
44. Smedley, P.L., Nicolli, H.B., Macdonald, D.M.J., Barros, A.J., and Tullio, J.O. Hydrogeochemistry of arsenic and other inorganic constituents in groundwaters from La Pampa, Argentina. *Appl. Geochem.*, 17, 259–284, 2002.
45. Kuhlmeier, P.D. Partitioning of arsenic species in fine grained soils. *J. Air Waste Manage.*, 47, 481–490, 1997.
46. BGS and DPHE. Arsenic contamination of groundwater in Bangladesh (four volumes). In: Kinniburgh, D.G. and Smedley, P.L. (Eds), *BGS Technical Report WC/00/19*. British Geological Survey, Keyworth, 2001.
47. Wang, S.W., Liu, C.W., and Jang, C.S. Factors responsible for high arsenic concentrations in two groundwater catchments in Taiwan. *Appl. Geochem.*, 22, 460–476, 2007.
48. Wang, L. and Huang, J. Chronic arsenism from drinking water in some areas of Xinjiang, China. In: Nriagu, J.O. (Ed), *Arsenic in the Environment, Part II: Human Health and Ecosystem Effects*. Wiley, New York, pp. 159–172, 1994.

49. Varsanyi, I. and Kovacs, L. Arsenic, iron and organic matter in sediments and groundwater in the Pannonian Basin, Hungary. Appl. *Geochem.*, 21, 949–963, 2006.

50. Sancha, A.M. and Castro, M. Arsenic in Latin America: Occurrence, exposure, health effects and remediation. In: Chappell, W.R., Abernathy, C.O., and Calderon, R.L. (Eds), *Arsenic Exposure and Health Effects IV*. Elsevier, Amsterdam, pp. 87–96, 2001.

51. Rodrfguez, R., Ramos, J.A., and Armienta, A. Groundwater arsenic variations: The role of local geology and rainfall. *Appl. Geochem.*, 19, 245–250, 2004.

52. Ryu, J.H., Gao, S.D., Dahlgren, R.A., and Zierngberg, R.A. Arsenic distribution, speciation and solubility in shallow groundwater of Owens Dry Lake, California. Geoch. *Cosmochim. Acta*, 66, 2981–2994, 2002.

53. Kim, M.J., Nriagu, J., and Haack, S. Arsenic species and chemistry in groundwater of southeast Michigan. *Environ. Pollut.*, 120, 379–390, 2002.

54. Peters, S.C. and Blum, J.D. The source and transport of arsenic in a bedrock aquifer, New Hampshire, USA. *Appl. Geochem.*, 18, 1773–1787, 2003.

55. Schreiber, M.E., Simo, J.A., and Freiberg, P.G. Stratigraphic and geochemical controls on naturally occurring arsenic in groundwater, eastern Wisconsin, USA. *Hydrogeol. J.*, 8, 161–176, 2000.

56. Foust, Jr., R.D., Mohapatra, P., Compton-O'Brien, A.M., and Reifel, J. Groundwater arsenic in the Verde Valley in central Arizona, USA. *Appl. Geochem.*, 19, 251–255, 2004.

57. Kelly, W.R., Holm, T.R., Wilson, S.D., and Roadcap, G.S. Arsenic in glacial aquifers: Source and geochemical controls. *Ground Water*, 43, 500–510, 2005.

58. Steinmaus, C.M., Yuan, Y., and Smith, A.H. The temporal stability of arsenic concentration in well water in western Nevada. *Environ. Res.*, 99, 164–168, 2005.

59. Lipfert, G., Reeve, A.S., Sidle, W.C., and Marvinney, R. Geochemical patterns of arsenic-enriched ground water in fractured, crystalline bedrock, Northport, Maine, USA. *Appl. Geochem.*, 21, 528–545, 2006.

60. Robinson, Jr., G.R. and Ayotte, J.D. The influence of geology and land use on arsenic in stream sediments and ground waters in New England, USA. *Appl. Geochem.*, 21, 1482–1497, 2006.

61. Huntsman-Mapila, P., Mapila, T., Letshwenyo, M., Wolski, P., and Henmond, C. Characterization of arsenic occurrence in the water and sediments of the Okavango Delta, NW Botswana. *Appl. Geochem.*, 21, 1376–1391, 2006.

62. Wang, S.L. and Mulligan, C.N. Occurrence of arsenic contamination in Canada: Sources, behavior and distribution. *Sci. Total Environ.*, 366, 701–721, 2006.

63. Kondo, H., Ishiguro, Y., Ohno, K., Nagase, M., Toba, M., and Takagi, M. Naturally occurring arsenic in the groundwaters in the southern region of Fukuoka Prefecture, Japan. *Water Res.*, 33, 1967–1972, 1999.

64. Heinrichs, G. and Udluft, P. Natural arsenic in Triassic rocks: A source of drinking-water contamination in Bavaria, Germany. *Hydrogeol. J.*, 7, 468–476, 1999.

65. Dogan, M. and Dogan, A.U. Arsenic mineralization, source, distribution, and abundance in the Kutahya region of the western Anatolia, Turkey. *Environ. Geochem. Health*, 29, 119–129, 2007.

66. Manganelli, A., Goso, C., Guerequiz, R., Fernandez Turiel, J.L., Valles, M.G., Gimeno, D., and Perez, C. Groundwater arsenic distribution in South-western Uruguay. *Environ. Geol.*, 53, 827–834, 2007.

67. Garrfa-Sanchez, A., Moyano, A., and Mayorga, P. High arsenic contents in groundwater of central Spain. *Environ. Geol.*, 47, 847–854, 2005.

68. Gomez, J.J., Lillo, J., and Sahun, B. Naturally occurring arsenic in groundwater and identification of the geochemical sources in the Duero Cenozoic Basin, Spain. *Environ. Geol.*, 50, 1151–1170, 2006.

69. Aiuppa, A., Alessandro, W.D., Federico, C., Palumbo, B., and Valenza, M. The aquatic geochemistry of arsenic in volcanic groundwaters from southern Italy. *Appl. Geochem.*, 18, 1283–1296, 2003.

70. Leybourne, M.I. and Cameron, E.M. Source, transport, and fate of rhenium, selenium, molybdenum, arsenic, and copper in groundwater associated with porphyry-Cu deposits, Atacama Desert, Chile. *Chem. Geol.*, 247, 208–228, 2008.

71. Smith, J.V.S., Jankowski, J., and Sammut, J. Vertical distribution of As(III) and As(V) in a coastal sandy aquifer: Factors controlling the concentration and speciation of arsenic in the Stuarts Point groundwater system, Northern New South Wales, Australia. *Appl. Geochem.*, 18, 1479–1496, 2003.

72. Luo, Z.D., Zhang, Y.M., Ma, L., Zhang, G.Y., He, X., Wilson, R., Byrd, D.M., Griffiths, J.G., Lai, S., He, L., Grumski, K., and Lamm, S.H. Chronic arsenicism and cancer in Inner Mongolia—consequences of well-water arsenic levels greater than 50 mg/l. In: Abernathy, C.O., Calderon, R.L., and Chappell, W.R. (Eds), *Arsenic Exposure and Health Effects*. Chapman and Hall, London, pp. 55–68, 1997.

73. Gurzau, E.S. and Gurzau, A.E. Arsenic exposure from drinking groundwater in Transylvania, Romania: An overview. In: Chappell, W.R., Abernathy, C.O., and Calderon, R.L. (Eds), *Arsenic Exposure and Health Effects IV*. Elsevier, Amsterdam, pp. 181–184, 2001.

74. Harvey, C.F., Swartz, C.H., Badruzzaman, A.B.M., Keon-Blute, N., Yu, W., Ali, M.A., Jay, J., Beckie, R., Niedan, V., Brabander, D., Oates, P.M., Ashfaque, K.N., Islam, S., Hemond, H.F., and Ahmed, M.F. Arsenic mobility and groundwater extraction in Bangladesh. *Science*, 298, 1602–1606, 2002.

75. Gao, S., Ryu, J., Tanji, K.K., and Herbel, M.J. Arsenic speciation and accumulation in evapoconcentrating water of agricultural evaporation basins. *Chemosphere*, 67, 862–871, 2007.

76. Del Razo, L.M., Arellano, M.A., and Cebrian, M.E. The oxidation states of arsenic in well-water from a chronic arsenicism area of northern Mexico. *Environ. Pollut.*, 64, 143–153, 1990.

77. Islam, M.E., Reza, A.S., Sattar, G.S., Ahsan, M.A., Akbor, M.A., and Siddique, M.A.B. Distribution of arsenic in core sediments and groundwater in the Chapai Nawabganj district, Bangladesh. *Arab. J. Geosci.*, 12, 1–19, 2019.

78. Boyle, D.R., Turner, R.J.W., and Hall, G.E.M. Anomalous arsenic concentrations in groundwaters of an island community, Bowen Island, British Columbia. *Environ. Geochem. Health*, 20, 199–212, 1998.

79. Kim, M.J., Nriagu, J.O., and Haack, S.K. Carbonate ions and arsenic dissolution by groundwater. *Environ. Sci. Technol.*, 34, 3094–3100, 2000.

80. Le, X.C., Lu, X.F., and Li, X.F. Arsenic speciation. *Anal. Chem.*, 1, 27–33, 2004.

81. Francesconi, K.A. and Kuehnelt, D. Arsenic compounds in the environment. In: Frankenberger, W. (Ed), *Environmental Chemistry of Arsenic*, Chapter 3. Marcel Dekker, New York, pp. 51–94, 2002.

82. Chen, S.L., Yeh, S.J., Yang, M.H., and Lin, T.H. Trace element concentration and arsenic speciation in the well water of a Taiwan area with endemic Blackfoot disease. *Biol. Trace Element Res.*, 48, 263–274, 1995.

83. Bednar, A.J., Garbarino, J.R., Burkhardt, M.R., Ranville, J.F., and Wildeman, T.R. Field and laboratory arsenic speciation methods and their application to natural-water analysis. *Water Res.*, 38, 355–364, 2004.

84. Brookins, D.G. *Eh-pH Diagrams for Geochemistry*. Springer, Berlin, 1988.

85. Yan, X.P., Kerrich, R., and Hendry, M.J. Distribution of arsenic(III), arsenic(V) and total inorganic arsenic in porewaters from a thick till and clay-rich aquitard sequence, Saskatchewan, Canada. *Geochim. Cosmochim. Acta*, 64, 2637–2648, 2000.

86. Plant, J.A., Kinniburgh, D.G., Smedley, P.L., Fordyce, F.M., and Klinck, B.A. Arsenic and selenium. In: Holland, H.D. and Turekian, K.K. (Eds), *Treatise on Geochemistry*, Chapter 9.02. Elsevier, Amsterdam, pp. 17–66, 2004.

87. Belzile, N. and Tessier, A. Interactions between arsenic and iron oxyhydroxides in lacustrine sediments. *Geochim. Cosmochim. Acta*, 54, 103–109, 1990.

88. Moore, J.N., Ficklin, W.H., and Johns, C. Partitioning of arsenic and metals in reducing sulfidic sediments. *Environ. Sci. Technol.*, 22, 432–437, 1988.

89. Wallschlager, D. and Stadey, C.J. Arsenic geochemistry in reducing environments—influence of arsenic-sulfide interaction on mobility. Fate of arsenic, antimony and similar elements in the environment (Abstract). *Geochim. Cosmochim. Acta*, 68(Suppl.), A513, 2004.

90. Vink, B.W. Stability relations of antimony and arsenic compounds in the light of revised and extended Eh-pH diagrams. *Chem. Geol.*, 130, 21–30, 1996.

91. Morse, J.W. and Luther, G.W. Chemical influences on metal-sulfide interactions in anoxic sediments. *Geochim. Coschim. Acta*, 63, 3373–3378, 1999.

92. Savage, K.S., Tingle, T.N., O'Day, P.A., Waychunas, G.A., and Bird, D.K. Arsenic speciation in pyrite and secondary weathering phases, Mother Lode Gold District, Tuolumne County, California. *Appl. Geochem.*, 15, 1219–1244, 2000.

93. Richard, D., Schoonen, M.A., and Luther, G.W. Chemistry of iron sulfides in sedimentary environments. In: Vairavamurthy, M.A. and Schoonen, M.A.A. (Eds), *Geochemical Transformation of Sedimentary Sulfur*, Chapter 9, Vol. 612. pp. 168–193. American Chemical Society Symposium Series, Washington, DC, 1995.

94. Stollenwerk, K.G. Geochemical processes controlling transport of arsenic in groundwater: A review of adsorption. In: Welch, A.H. and Stollenwerk, K.G. (Eds), *Arsenic in Ground Water: Geochemistry and Occurrence*. Kluwer Academic Publishers, Boston, MA, pp. 67–100, 2003.

95. Wang, S.L. and Mulligan, C.N. Natural attenuation processes for remediation of arsenic contaminated soils and groundwater. *J. Hazardous Mater. B*, 138, 459–470, 2006.

96. Roman-Ross, G., Cuello, G.J., Turrillas, X., Fernandez-Martinez, A., and Charlet, L. Arsenic sorption and co-precipitaion with calcite. *Chem. Geol.*, 233, 328–336, 2006.

97. Chakraborty, S., Wolthers, M., Chatterjee, D., and Charlet, L. Adsorption of arsenite and arsenate onto muscovite and biotite mica. *J. Colloid Interface Sci.*, 309, 392–401, 2007.

98. Manning, B.A., Fendorf, S.E., and Goldberg, S. Surface structures and stability of arsenic(III) on goethite: Spectroscopic evidence for innersphere complexes. *Environ. Sci. Technol.*, 32, 2383–2388, 1998.

99. Goldberg, S. Chemical modeling of arsenate adsorption on aluminum and iron oxide minerals. *Soil Sci. Soc. Am. J.*, 50, 1154–1157, 1986.

100. Dzombak, D.A. and Morel, F.M.M. *Surface Complexation Modeling: Hydrous Ferric Oxide*. Wiley, New York, 1990.

101. Hiemstra, T. and van Riemsdijk, W.H. Surface structural ion adsorption modeling of competitive binding of oxyanions by metal (hydr)oxides. *J. Colloid Interface Sci.*, 210, 182–193, 1999.

102. Dixit, S. and Hering, J.G. Comparison of arsenic(V) and arsenic(III) sorption onto iron oxide minerals: Implications for arsenic mobility. *Environ. Sci. Technol.*, 37, 4182–4189, 2003.

103. Catalano, J.G., Park, C., Fenter, P., and Zhang, Z. Simultaneous inner- and outer-sphere arsenate adsorption on corundum and hematite. Geochim. *Coschim. Acta*, 72, 1986–2004, 2008.

104. Scott, M.J. and Morgan, J.J. Reactions at oxides surfaces. 1. Oxidation of As(III) by synthetic birnessite. *Environ. Sci. Technol.*, 29, 1898–1905, 1995.

105. Manning, B.A., Fendorf, S.E., Bostick, B., and Suarez, D.L. Arsenic(III) oxidation and arsenic(V) adsorption reactions on synthetic birnessite. *Environ. Sci. Technol.*, 36, 976–981, 2002.

106. Foster, A.L., Brown, Jr., G.E., and Parks, G.A. X-ray absorption fine structure study of As(V) and Se(IV) sorption complexes on hydrous Mn oxides. *Geochim. Cosmochim. Acta*, 67, 1937–1953, 2003.

107. Grafe, M., Eick, M.J., and Grossl, P.R. Adsorption of arsenate (V) and arsenite (III) on goethite in the presence and absence of dissolved organic carbon. *Soil Sci. Soc. Am. J.*, 65, 1680–1687, 2001.

108. Fuller, C.C., Davis, J.A., and Waychunas, G.A. Surface-chemistry of ferrihydrite. 2. Kinetics of arsenate adsorption and coprecipitation. *Geochim. Cosmochim. Acta*, 57, 2271–2282, 1993.

109. Langner, H.W. and Inskeep, W.P. Microbial reduction of arsenate in the presence of ferrihydrite. *Environ. Sci. Technol.*, 34, 3131–3136, 2000.

110. McArthur, J.M., Banerjee, D.M., Hudson-Edwards, K.A., Mishra, R., Purohit, R., Ravenscroft, P., Cronin, A., Howarth, R.J., Chatterjee, A., Talukder, T., Lowry, D., Houghton, S., and Chada, D.K. Natural organic matter in sedimentary basins and its relation to arsenic in anoxic ground water: The example of West Bengal and its worldwide implications. *Appl. Geochem.*, 19, 1255–1293, 2004.

111. Horneman, A., van Geen, A., Kent, D.V., Mathe, P.E., Zheng, Y., Dhar, R.K., O'Connell, S., Hoque, M.A., Aziz, Z., Shamsudduha, M., Seddique, A., and Ahmed, K.M. Decoupling of As and Fe release to Bangladesh groundwater under reducing conditions. Part I: Evidence from sediment profiles. *Geochim. Cosmochim. Acta*, 68, 3459–3473, 2004.

112. Jonsson, J. and Sherman, D.M. Sorption of As(III) and As(V) to siderite, green rust (fougerite) and magnetite: Implications for arsenic release in anoxic groundwaters. *Chem. Geol.*, 255, 173–181, 2008.

113. Lin, Z. and Puls, R.W. Adsorption, desorption, oxidation of arsenic affected by clay minerals and aging process. *Environ. Geol.*, 39, 753–759, 2000.

114. Appelo, C.A.J., Van der Weiden, M.J.J., Tournassat, C., and Charlet, L. Surface complexation of ferrous iron and carbonate on ferrihydrite and the mobilization of arsenic. *Environ. Sci. and Technol.*, 36, 3096–3103, 2002.

115. Ghosh, M.M. and Teoh, R.S. Adsorption of arsenic on hydrous aluminum oxide. In: *Proceedings of Seventh Mid-Atlantic Industrial Waste Conference*, Lancaster, PA, pp. 139–155, 1985.

116. Wilkie, J.A. and Hering, J.G. Adsorption of arsenic onto hydrous ferric oxide: Effects of adsorbate/adsorbent ratios and co-occurring solutes. *Colloids Surfaces A*, 107, 97–110, 1996.

117. Meng, X.G., Bang, S., and Korfiatis, G.P. Effects of silicate, sulfate, and carbonate on arsenic removal by ferric chlorite. *Water Res.*, 34, 1255–1261, 2000.

118. Meng, X.G., Korfiatis, G.P., Jing, C.Y., and Christodoulatos, C. Redox transformations of arsenic and iron in water treatment sludge during aging and TCLP extraction. *Environ. Sci. Technol.*, 35, 3476–3481, 2001.

119. Zobrist, J., Dowdle, P.R., Davis, J.A., and Oremland, R.S. Mobilization of arsenite by dissimilatory reduction of adsorbed arsenate. *Environ. Sci. Technol.*, 34, 4747–4753, 2000.

120. Jain, A. and Loeppert, R.H. Effect of competing anions on the adsorption of arsenate and arsenite by ferrihydrite. *J. Environ. Qual.*, 29, 1422–1430, 2000.

121. Meng, X.G., Korfiatis, G.P., Bang, S.B., and Bang, K.W. Combined effects of anions on arsenic removal by iron hydroxides. *Toxicol. Lett.*, 133, 103–111, 2002.

122. Goldberg, S. Competitive adsorption of arsenate and arsenite on oxides and clay minerals. *Soil Sci. Soc. Am. J.*, 66, 413–421, 2002.

123. Manning, B.A. and Goldberg, S. Modeling competitive adsorption of arsenate with phosphate and molybdate on oxide minerals. *Soil Sci. Soc. Am. J.*, 60, 121–131, 1996.

124. Mahoney, J., Langmuir, D., Gosselin, N., and Rowson, J. Arsenic readily released to pore waters from buried mill tailings. *Appl. Geochem.*, 20, 947–959, 2005.

125. Anawar, H.M., Akai, J., and Sakugawa, H. Mobilization of arsenic from subsurface sediments by effect of bicarbonate ions in groundwater. *Chemosphere*, 54, 753–762, 2004.

126. Luxton, T.P., Eick, M.J., and Rimstidt, D.J. The role of silicate in the adsorption/desorption of arsenite on goethite. *Appl. Geochem.*, 252, 125–135, 2008.

127. Stumm, W. and Morgan, J.J. *Aquatic Chemical: Chemical Equilibria and Rates in Natural Waters.* Wiley- Interscience, New York, 1995.

128. Langmuir, D. *Aqueous Environment Geochemistry.* Prentice-Hall, Upper Saddle River, NJ, 1997.

129. Bhattacharya, P., Jacks, G., Ahmed, K.M., Routh, J., and Khan, A.A. Arsenic in groundwater of the Bengal Delta Plain aquifers in Bangladesh. *Bull. Environ. Contamin. Toxicol.*, 69, 538–545, 2002.

130. Saunders, J.A., Pritchett, M.A., and Cook, R.B. Geochemistry of biogenic pyrite and ferromanganese stream coatings: A bacterial connection? *Geomicrobiol. J.*, 14, 203–217, 1997.

131. Saunders, J.A., Mohammad, S., Korte, N.E., Lee, M.K., Fayek, M., Castle, D., and Barnett, M.O. Groundwater geochemistry, microbiology, and mineralogy of two arsenic-bearing Holocene alluvial aquifers from the USA. In: O'Day, P.A., Vlassopoulos, D., Meng, X., and Benning, L.G. (Eds), *Advances in Arsenic Research: Integration of Experimental and Observational Studies and Implications for Mitigation*, Vol. 915. American Chemical Society Symposium Series, Washington, DC, pp. 191–205, 2005.

132. Saunders, J.A., Lee, M.K., Shamsudduha, M., Dhakal, P., Uddin, A., Chowdury, M.T., Ahmed, K.M. Geochemistry and mineralogy of arsenic in (natural) anaerobic groundwaters. *Appl. Geochem.*, 23(11), 3205–3214, 2008.

133. Kirk, M., Holm, T., Park, J., Jin, Q., Sanford, R., Fouke, B., and Bethke, C. Bacterial sulfate reduction limits natural arsenic contamination in groundwater. *Geology*, 32, 952–956, 2004.

134. O'Day, P.A., Vlassopoulos, D., Root, R., and Rivera, N. The influence of sulfur and iron on dissolved arsenic concentrations in the shallow subsurface under changing redox conditions. *Proc. Nat. Acad. Sci. U.S.A.*, 101, 13703–13708, 2004.

135. O'Day, P.A. Chemistry and mineralogy of arsenic. *Elements*, 2, 77–83, 2006.

136. Lowers, H.A., Breit, G.N., Foster, A.L., Whitney, J., Yount, J., Uddin, M.N., and Muneem, A.A. Arsenic incorporation into authigenic pyrite, Bengal Basin sediment, Bangladesh. *Geochim. Cosmochim. Acta*, 71, 2699–2717, 2007.

137. Acharyya, S.K., Lahiri, S., Raymahashay, B.C., and Bhowmik, A. Arsenic toxicity of groundwater in parts of the Bengal Basin in India and Bangladesh: The role of Quaternary stratigraphy and Holocene sea level fluctuation. *Environ. Geol.*, 47, 1127–1137, 2000.

138. Huerta-Diaz, M.A. and Morse, J.W. Pyritization of trace metals in anoxic marine sediments. *Geochim. Cosmochim. Acta*, 56, 2681–2702, 1992.

139. Southam, G. and Saunders, J.A. Geomicrobiology of ore deposits. *Econ. Geol.*, 100, 1067–1084, 2005.

140. Lopez, I.C. Microbial transformation of arsenic and organoarsenic compounds in anaerobic environments. PhD Dissertation Thesis, University of Arizona, 236 pp., 2007.

141. Mukhopadhyay, R., Rosen, B.P., Phung, L.T., and Silver, S. Microbial arsenic: From geocycles to genes and enzymes. *FEMS Microbiol. Rev.*, 26, 311–325, 2002.

142. Santini, J.M., Stolz, J.F., and Macy, J.M. Isolation of a new arsenate-respiring bacterium—Physiological and phylogenetic studies. *Geomicrobiol. J.*, 19, 41–52, 2002.

143. Oremland, R.S. and Stolz, J.F. The ecology of arsenic. *Science*, 300, 939–944, 2003.

144. Rhine, E.D., Garcia-Dominguez, E., Phelps, C.D., and Young, L.Y. Environmental microbes can speciate and cycle arsenic. *Environ. Sci. Technol.*, 39, 9569–9573, 2005.

145. Jones, E.J.P., Nadeau, T.L., Voytek, M.A., and Landa, E.R. Role of microbial iron reduction in the dissolution of iron hydroxysulfate minerals. *J. Geophys. Res.-Biogeosci.*, 111, G01012, 2006.

146. Edward, D.B., Scott, G.J., and Benjamin, D.K. Arsenic mobility during flooding of contaminated soil: The effect of microbial sulfate reduction. *Environ. Sci. Technol.*, 48(23), 13660–13667, 2014.

147. Shi, W.X., Xu, Y.F., Wu, W.W., and Zeng, X.C. Biological effect of phosphate on the dissimilatory arsenate-respiring bacteria-catalyzed reductive mobilization of arsenic from contaminated soils. *Environ. Pollut.*, 308, 119698, 2022.

148. Oremland, R.S. and Stolz, J.F. Arsenic, microbes and contaminated aquifers. *Trends Microbiol.*, 13, 45–49, 2005.

149. Thomas, D.J., Waters, S.B., and Styblo, M. Elucidating the pathway for arsenic methylation. *Toxicol. Appl. Pharmacol.*, 198, 319–326, 2004.

150. Sakurai, T. Biomethylation of arsenic is essentially detoxicating event. *J. Health Sci.*, 49, 171–178, 2003.

151. Oremland, R.S., Kulp, T.R., Blum, J.S., Hoeft, S.E., Baesman, S., Miller, L.G., and Stolz, J.F. A microbial arsenic cycle in a salt-saturated, extreme environment. *Science*, 308, 1305–1308, 2005.

152. Francesconi, K.A. and Kuehnelt, D. Determination of arsenic species: A critical review of methods and applications, 2000–2003. *Analyst*, 129, 373–395, 2004.

153. Korte, N.E. Naturally occurring arsenic in groundwaters of the midwestern United States. *Environ. Geol.*, 18, 137–141, 1991.

154. Duan, L.Q., Song, J.M., Zhang, Y.T., Yin, M.L., Yuan, H.M., and Li, X.G. Unraveling seasonal shifts in microbial and geochemical mediated arsenic mobilization at the estuarine sediment-water interface under redox changes. *Sci. Total Environ.*, 912, 168939, 2024.

155. Lee, M.-K., Griffin, J., Saunders, J.A., Wang, Y., and Jean, J. Reactive transport of trace elements and isotopes in Alabama coastal plain aquifers. *J. Geophys. Res.*, 112, G02026, 2007.

156. McArthur, J.M., Ravenscroft, R., Safiullah, S., and Thirlwall, M.F. Arsenic in groundwater: Testing pollution mechanisms for sedimentary aquifers in Bangladesh. *Water Res.*, 37, 109–117, 2001.

157. Dowling, C.B., Poreda, R.J., Basu, A.R., and Peters, S.L. Geochemical study of arsenic release mechanisms in the Bengal Basin groundwaters. *Water Resource Res.*, 38, 1173–1190, 2002.

158. Islam, F., Gault, A., Boothman, C., Polya, D., Charnock, J., Chatterjee, D., and Lloyd, J. Role of metal reducing bacteria in arsenic release from Bengal Delta sediments. *Nature*, 430, 68–71, 2004.

159. Herbel, M.J. and Fendorf, S. Transformation and transport of arsenic within ferric hydroxide coated sands upon dissimilatory reducing bacterial activity. In: O'Day, P.A., Vlassopoulos, D., Meng, X., and Benning, L. (Eds), *Advances in Arsenic Research: Integration of Experimental and Observational Studies and Implications for Mitigation*, Vol. 915. American Chemical Society Symposium Series, Washington, DC, pp. 77–90, 2005.

160. Newman, D.K., Kennedy, E.K., Coates, J.D., Ahmann, D., Ellis, D.J., Lovley, D.R., and Morel, F.M. Dissimilatory arsenate and sulfate reduction in Desulfotomaculum auripigmentum sp. nov. *Arch. Microbiol.*, 168, 380–388, 1997.

161. Saltikov, C.W. and Newman, D.K. Genetic identification of a respiratory arsenate reductase. *Proc. Nat. Acad. Sci. U.S.A.*, 100, 10983–10988, 2003.

162. Wang, S.L. and Mulligan, C.N. Effect of natural organic matter on arsenic release from soil and sediments into groundwater. *Environ. Geochem. Health*, 28, 197–214, 2006.

163. Anawar, H.M., Akai, J., Komaki, K., Terao, H., Yoshioka, T., Ishizuka, T., Safiullah, S., and Kato, K. Geochemical occurrence of arsenic in groundwater of Bangladesh: Source and mobilization processes. *J. Geochem. Explor.*, 77, 109–131, 2003.

164. Xu, H., Allard, B., and Grimvall, A. Effects of acidification and natural organic materials on the mobility of arsenic in the environment. *Water, Air, Soil Pollut.*, 57/58, 269–278, 1991.

165. Cornu, S., Saada, A., Breeze, D., Gauthier, S., and Baranger, P. The influence of organic complexes on arsenic adsorption onto kaolinites. *Comptes Rendus Acad Sci.—Series IIA—Earth Planet Sci.*, 28, 877–881, 1999.

166. Grafe, M., Eick, M.J., Grossl, P.R., and Saunders, A.M. Adsorption of arsenate and arsenite on ferrihydrite in the presence and absence of dissolved organic carbon. *J. Environ. Qual.*, 31, 1115–1123, 2002.

167. Redman, A.D., Macalady, D., and Ahmann, D. Natural organic matter affects arsenic speciation and sorption onto hematite. *Environ. Sci. Technol.*, 36, 2889–2896, 2002.

168. Saada, A., Breeze, D., Crouzet, C., Cornu, S., and Baranger, P. Adsorption of arsenic (V) on kaolinite-humic acid complexes: Role of humic acid nitrogen groups. *Chemosphere*, 51, 757–763, 2003.

169. Lin, H.T., Wang, M.C., and Li, G.C. Complexation of arsenate with humic substance in water extract of compost. *Chemosphere*, 56, 1105–1112, 2004.

170. Warwick, P., Inam, E., and Evans, N. Arsenic interaction with humic acid. *Environ. Chem.*, 2, 119–124, 2005.

171. Yong, R.N. and Mulligan, C.N. *Natural Attenuation of Contaminants in Soils*. CRC Press, Boca Raton, 2004.

172. Scott, D.T., McKnight, D.M., Harris, B.E.L., Kolesar, S.E., and Loveley, D.R. Quinone moieties act as electron acceptors in the reduction of humic substances by humic-reducing microorganisms. *Environ. Sci. Technol.*, 32, 2984–2989, 1998.

173. Palmer, N.E., Freudenthal, J.H. and von Wandruszka, R. Reduction of arsenate by humic materials. *Environ. Chem.*, 3, 131–136, 2006.

174. Tongesayi, T. and Smart, R.B. Arsenic speciation: Reduction of As(V) to As(III) by fulvic acid. *Environ. Chem.*, 3, 137–141, 2006.

175. Gu, B., Schmitt, J., Chen, Z., Liang, L., and McCarthy, J.F. Adsorption and desorption of natural organic matter on iron oxide: Mechanisms and models. *Environ. Sci. Technol.*, 28, 38–46, 1994.

176. Mukhopadhyay, D. and Sanyal, S.K. Complexation and release isotherm of arsenic in arsenic-humic/fulvic equilibrium study. *Australian J. Soil Res.*, 42, 815–824, 2004.

177. Feldman, P.R. *Drinking Water Quality Assessment in Cambodia*. WHO, Phenom Penh, 2001.

178. Rowland, H.A.L., Polya, D.A., Gault, A.G., Charnock, J.M., Pederick, R.L., and Lloyd, J.R. Microcosm studies of microbially mediated arsenic release from contrasting Cambodian sediments. *Geochim. Cosmochim. Acta*, 68, A390, 2004.

179. Ghosh, A.R. and Mukherjee, A. Arsenic contamination of groundwater and human health impacts in Burdwan District, West Bengal, India. Abstract with Program—Geological Society of America, Vol. 34, 2002.

180. Rahman, M.M., Sengupta, M.K., Chowdhury, U.K., Lodh, D., Das, B., Ahamed, S., Mandal, D., Hossain, M.A., Mukherjee, S.C., Pati, S., Saha, K.C., and Chakraborti, D. Arsenic contamination incidents around the world. In: Naidu, R., Smith, E., Owens, G., Bhattacharya, P., and Nadebaum, P. (Eds), *Managing Arsenic in the Environment: From Soils to Human Health.* CSIRO Publishing, Melbourne, pp. 3–30, 2006.

181. PHED/UNICEF Joint Plan of Action to Address Arsenic Contamination of Drinking Water. Government of West Bengal and UNICEF, Public Health Engineering Department, Government of West Bengal, 1999.

182. Buschmann, J., Berg, M., Stengel, C., and Sampon, M. Arsenic and manganese contamination of drinking water resources in Cambodia: Coincidences of risk areas with low relief topography. *Environ. Sci. Technol.*, 41, 2146–2152, 2007.

183. Quicksall, A.N., Bostick, B.C., and Sampson, M.L. Linking organic matter deposition and iron mineral transformations to groundwater arsenic levels in the Mekong Delta, Cambodia. *Appl. Geochem.*, 23, 3088–3098, 2008.

184. Pokhrel, D., Bhandari, B.S., and Viraraghavan, T. Arsenic contamination of groundwater in the Terai region of Nepal: An overview of health concerns and treatment options. *Environ. Int.*, 35, 157–161, 2009.

185. Mandal, B.K., Chowdhury, T.R., Samanta, G., Basu, G.K., Chowdhury, P.P., Chanda, C.R., Lodh, D., Karan, N.K., Dhar, R.K., Tamili, D.K., Das, D., Saha, K.C., and Chakraborti, D. Arsenic in groundwater in seven districts of West Bengal, India—the biggest arsenic calamity in the world. *Curr. Sci.*, 70, 976–986, 1996.

186. Das, D., Samanta, G., Mandal, B.K., Chowdhury, T.R., Chanda, C.R., Chowdhury, P.P., Basu, G.K., and Chakraborti, D. Arsenic in groundwater in six districts of West Bengal, India. *Environ. Geochem. Health*, 18, 5–15, 1996.

187. Gault, A.G., Polya, D.A., Charnock, J.M., Islam, F.S., Lloyd, J.R., and Chatterjee, D. Preliminary EXAFS studies of solid phase speciation of As in a West Bengali sediment. *Mineral. Mag.*, 67, 1183–1191, 2003.

188. Swartz, C.H., Blute, N.K., Badruzzman, B., Ali, A., Brabander, D., Jay, J., Besancon, J., Islam, S., Hemond, H.F., and Harvey, C.F. Mobility of arsenic in a Bangladesh aquifer: Inferences from geochemical profiles, leaching data, and mineralogical characterization. *Geochim. Cosmochim. Acta*, 68, 4539–4557, 2004.

189. Akai, J., Izumi, K., Fukuhara, H., Masuda, H., Nakano, S., Yoshimura, T., Ohfuji, H., Anawar, H.M., and Akai, K. Mineralogical and geomicrobiological investigations on groundwater arsenic enrichment in Bangladesh. *Appl. Geochem.*, 19, 215–230, 2004.

190. van Geen, A., Rose, J., Thoral, S., Garnier, J.M., Zheng, Y., and Bottero, J.Y. Decoupling of As and Fe release to Bangladesh groundwater under reducing conditions Part II: Evidence from sediment incubations. *Geochim. Cosmochim. Acta*, 68, 3475–3486, 2004.

191. Gault, A.G., Islam, F.S., Polya, D.A., Charnock, J.M., Boothman, C., and Lloyd, J.R. Microcosm depth profiles of arsenic release in an arsenic contaminated aquifer, West Bengal. *Mineral. Mag.*, 69, 855–863, 2005.

192. Bostick, B.C., Chen, C., and Fendorf, S. Arsenite retention mechanisms within estuarine sediments of Pescadero CA. *Environ. Sci. Technol.*, 38, 3299–3304, 2004.

193. Roychowdhury, T., Tokunaga, H., and Ando, M. Survey of arsenic and other heavy metals in food composites and drinking water and estimation of dietary intake by the villagers from an arsenic-affected area of West Bengal, India. *Sci. Total Environ.*, 308, 15–35, 2003.

194. Motelica-Heino, M., Naylor, C., Zhang, H., and Davison, W. Simultaneous release of metals and sulfide in lacustrine sediment. *Environ. Sci. Technol.*, 37, 4374–4381, 2003.

195. van Geen, A., Zheng, Y., Cheng, Z., Aziz, Z., Horneman, A., Dhar, R.K., Mailloux, B., Stute, M., Weinman, B., Goodbred, S., Seddique, A.A., Hope, M.A., and Ahmed, K.M. A transect of groundwater and sediment properties in Araihazar, Bangladesh: Further evidence of decoupling between As and Fe mobilization. *Chem. Geol.*, 228, 85–96, 2006.

196. Xia, Y. and Liu, J. An overview on chronic arsenicism via drinking water in PR China. *Toxicology*, 198, 25–29, 2004.

197. Sun, G. Arsenic contamination and arsenicosis in China. *Toxicol. Appl. Pharmacol.*, 198, 268–271, 2004.

198. Sun, G., Li, X., Pi, J., Sun, Y., Li, B., Jin, Y., and Xu, Y. Current research problems of chronic arsenicosis in China. *J. Health Popul. Nut.*, 24, 176–181, 2006.

199. Ma, H.Z., Xia, Y.J., Wu, K.G., Sun, T.Z., and Mumford, J.L. Arsenic exposure and health effects in Bayingnormen, Inner Mongolia. In: Chappell, W.R., Abernathy, C.O., and Calderon, R.L. (Eds), *Arsenic Exposure and Health Effects*. Elsevier, Amsterdam, pp. 127–131, 1999.

200. Zhang, H., Ma, D., and Hu, X. Arsenic pollution in groundwater from Hetao area, China. *Environ. Geol.*, 41, 638–643, 2002.

201. Shvartsev, S.L. *Hydrogeochemistry of Supergene Zones*, 2nd Edition, Nedra, Moscow (in Russian), p. 367, 1998.

202. Wang, L.K., Wang, M.H.S., and Shammas, N.K. Treatment of wastewater, storm runoff, and combined sewer overflow by dissolved air flotation and filtration. In: Wang, L. K., Wang, M. H. S., Hung, Y. T., Shammas, N. K., Chen, J. P. (Eds), *Handbook of Advanced Industrial and Hazardous Wastes Management*. CRC Press, FL, pp. 577–609, 2018.

203. Liu, H., Liu, R., Qu, J., Zhang, G. Yang, Y., *Arsenic Pollution: Occurrence, Distribution and Technologies, Heavy Metals in the Environment*. CRC Press, FL, pp. 225–246, 2009.

5 Nanotechnology Application in Metal Ion Adsorption

Xiangke Wang, Changlun Chen, Hui Wen,
Xiao Liu, and Lawrence K. Wang

ACRONYMS

CNT	carbon nanotubes
FTIR	Fourier transform infrared
MWCNT	multiwalled carbon nanotubes
NP	nanoparticles
SDBS	sodium dodecylbenzene sulfonate
SEM	scan electron microscopy
SWCNT	single-walled carbon nanotubes
TEM	transmission electron microscopy
USEPA	U.S. Environmental Protection Agency
XPS	X-ray photoelectron spectroscopy

5.1 INTRODUCTION

With the development of science and technology, the application of nanomaterials and nanotechnology in environmental pollution management has been intensively interesting in the last decades. Materials in the nanosized range are considered the best candidates for the removal of organic and inorganic pollutants from the environment because of their unique physicochemical properties, such as large surface area, high reactivity, and high adsorption capacity. In all kinds of nanomaterials, carbon nanomaterials, including active carbon, carbon nanotubes (CNTs), and fullerenes, are studied extensively in terms of potential applications in catalyst supports, environmental remediation, etc. Their adsorption may also affect the fate, transformation, and transfer of toxic substances in the environment. Therefore, an understanding of the adsorption and desorption behavior of toxic substances is critical in evaluating environmental and health impacts.

CNTs [1] include single-walled carbon nanotubes (SWCNTs) and multiwalled carbon nanotubes (MWCNTs), depending on the number of layers. The monomer structure of fullerene is a closed graphite ball, whereas CNTs are rolled-up graphite sheets forming a coaxial tube (Figure 5.1) [2].

Water pollution due to the indiscriminate disposal of metal ions has been causing worldwide concern. Wastewater from many industries, such as metallurgical, tannery, chemical manufacturing, mining, and battery manufacturing industries, contains one or more toxic metal ions. It is necessary to remove these metal ions from the wastewaters before releasing into the environment, because there is a possibility of toxic metal ions entering into the food chain through waste discharges into water bodies. Adsorption is one of the conventional methods being used to remove metal ions. CNTs are considered promising materials for use as adsorbents in adsorption technology. There are a number of recent publications on the sorption of various metal ions (i.e., Cd^{2+}, Cu^{2+}, Ni^{2+}, Pb^{2+}, Zn^{2+}, $Cr(VI)$, etc.) from aqueous solution by raw and surface-oxidized CNTs, which discuss their sorption capacities, mechanisms, process parameters, desorption, and further research works [3–14]. They showed that metal ion sorption equilibrium data are commonly correlated with the Langmuir and/or the Freundlich equations, as well as an increase of maximum sorption capacity with the application

DOI: 10.1201/9781003541615-5

"

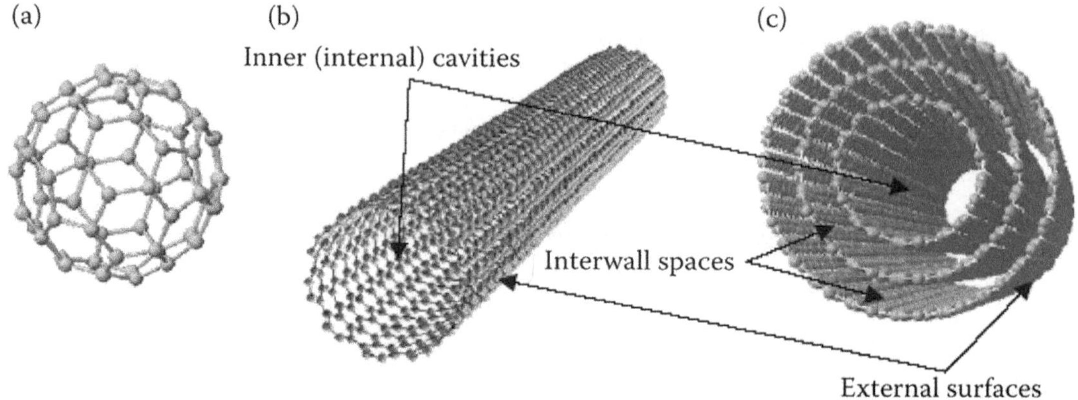

FIGURE 5.1 Schematic structures of fullerene (a), SWCNTs (b), and MWCNTs (c), showing inner cavities, interwall spaces, and external surfaces. Fullerene (C_{60}) has only an external surface. (From Yang, K. and Xing, B.S., *Environ. Pollut.*, 145, 529–537, 2007 [2]. With permission.)

of the oxidized CNTs. An increase in the removal of metal ions with increasing solution pH was also demonstrated and reached a maximum at pH > 8 for most metal ions. Generally, the adsorption of metal ions on oxidized CNTs is a little higher than that of metal ions on raw CNTs. During the oxidization process, many functional groups (such as –COOH, –CH, and –COH) are introduced to the surfaces of oxidized CNTs. In the removal of Cr(VI) from aqueous solutions, the adsorption of Cr(VI) decreased with increasing pH values. At low pH values, part of the Cr(VI) removal was attributed to the reduction of Cr(VI) to Cr(III) on the surfaces of oxidized CNTs [15,16].

Assessing the risks imposed by the use of nanomaterials in commercial products and environmental applications requires a better understanding of their mobility, bioavailability, and toxicity. For nanomaterials to comprise a risk, there must be both a potential for exposure and a hazard that results after exposure. The important processes and pathways of nanoparticles (NPs) in the environment are depicted in Figure 5.2. Release of NPs may come from point sources such as production facilities, landfills, or wastewater treatment plants or from nonpoint sources such as wear from materials containing NPs. Accidental release of NPs during production or transport is also possible. In addition to the unintentional release, there are also NPs released intentionally into the environment. Whether the particles are released directly into water/soil or the atmosphere, they all end up in soil or water, either directly or indirectly, for instance, via sewage treatment plants, waste handling, or aerial deposition. In the environment, the formation of aggregates and therefore of larger particles that are trapped or eliminated through sedimentation affects the concentrations of free NPs (Figure 5.3). Humans can be influenced either directly by NPs through exposure to air, soil, or water or indirectly by consuming plants or animals that have accumulated NPs. Aggregated or adsorbed NPs will be less mobile, but uptake by sediment-dwelling animals or filter feeders is still possible [17].

In the removal of heavy metal ions from the natural environment, many conventional methods have been used, including oxidation, reduction, precipitation, membrane filtration, ion exchange, sorption, etc. Among the above methods, the promising process for the removal of metal ions from water and wastewater is sorption. In this chapter, we will discuss the application of nanomaterials, especially carbon nanomaterials, in the removal of heavy metal ions from wastewater.

5.2 CHARACTERIZATION OF NANOMATERIALS

The raw CNT is a long cylinder made of a hexagonal honeycomb lattice of carbon, bound by two pieces of fullerenes at the ends. The fullerene structure at the end of the raw tube can be removed during treatment with nitric acid or other acids [18]. Generally, 3 mol/L HNO_3 is used to remove

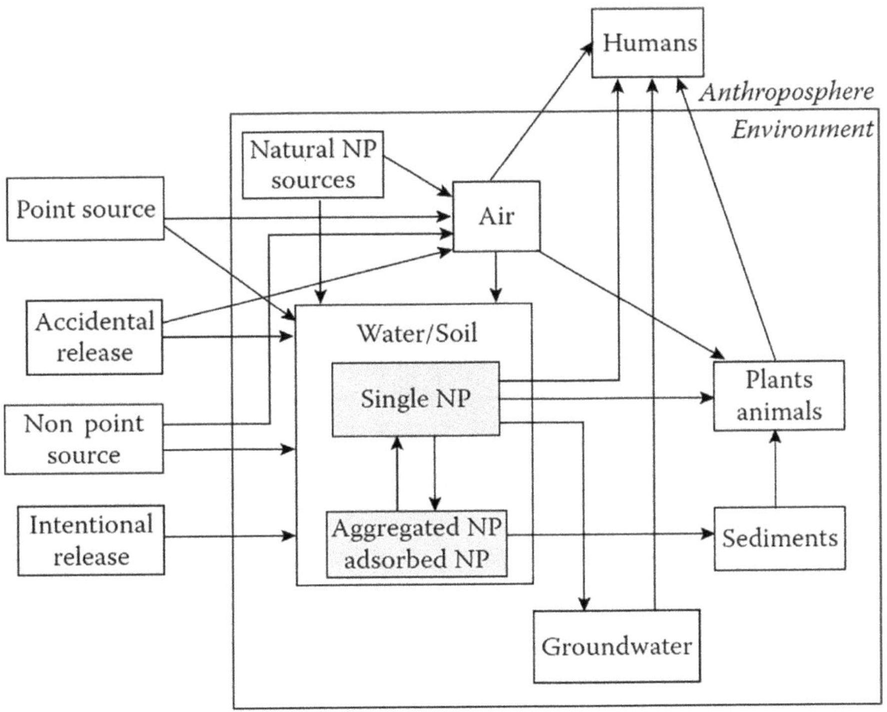

FIGURE 5.2 NP pathways from the anthroposphere into the environment, reactions in the environment, and exposure of humans. (From Nowack, B., and Bucheli, T.D., *Environ. Pollut.*, 150, 5, 2007 [17]. With permission.)

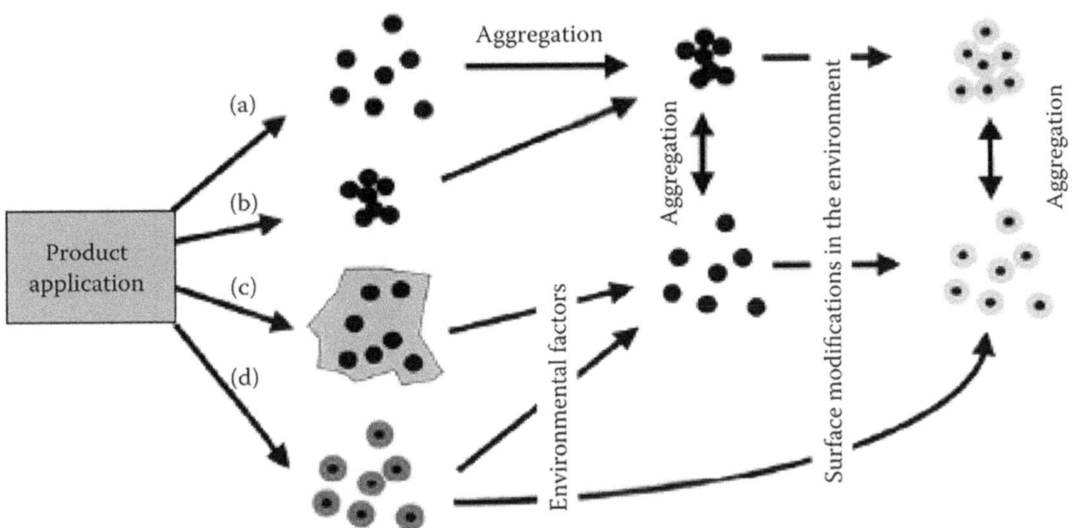

FIGURE 5.3 Release of NPs from products and (intended or unintended) applications: (a) release of free NPs, (b) release of aggregates of NPs, (c) release of NPs embedded in a matrix, and (d) release of functionalized NPs. Environmental factors (e.g., light, microorganisms) result in the formation of free NPs that can undergo aggregation reactions. Moreover, surface modifications (e.g., coating with natural compounds) can affect the aggregation behavior of NPs. (From Nowack, B., and Bucheli, T.D., *Environ. Pollut.*, 150, 5, 2007 [17]. With permission.)

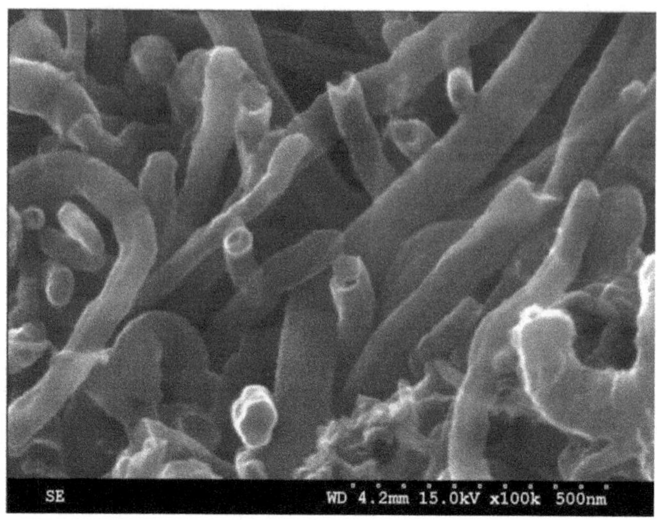

FIGURE 5.4   SEM image of oxidized MWCNTs.

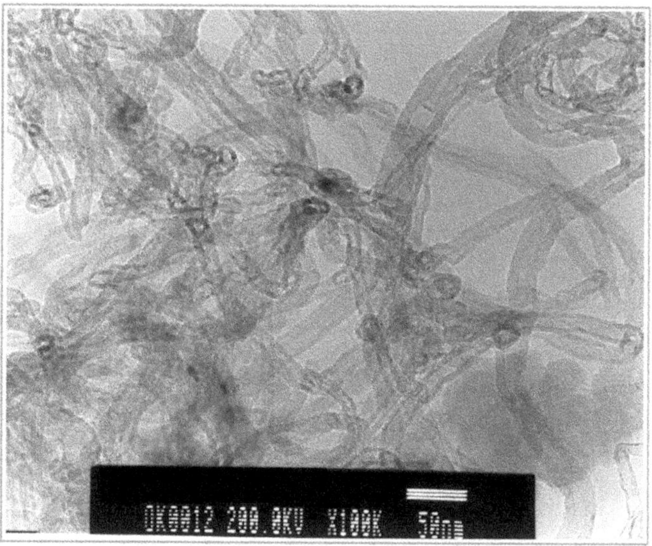

FIGURE 5.5   TEM image of oxidized MWCNTs.

the cap of the tubes and to generate some functional groups at the ends and on the defective sites of CNTs. CNTs and HNO_3 are mixed and ultrasonically stirred for a period of time and then rinsed with distilled water to remove the acid in the system until the pH of the solution reaches about neutral. Removal of the cap from the end of the nanotube makes it possible for water molecules or metal ions to diffuse into the inner channel of the tube [19,20]. The morphological structures of the nanomaterial can be determined by scanning electron microscopy (SEM). SEM analysis (Figure 5.4) shows that the cap of CNTs is removed after oxidation, whereas the hollow structure of CNTs can be observed by transmission electron microscopy (TEM) (Figure 5.5).

The surface properties and the functional groups on the surfaces of oxidized MWCNTs can be characterized by Fourier transform infrared (FTIR) spectra analysis. Generally, the FTIR measurement is mounted on a KBr pellet at room temperature. Figure 5.6 shows an FTIR analysis of

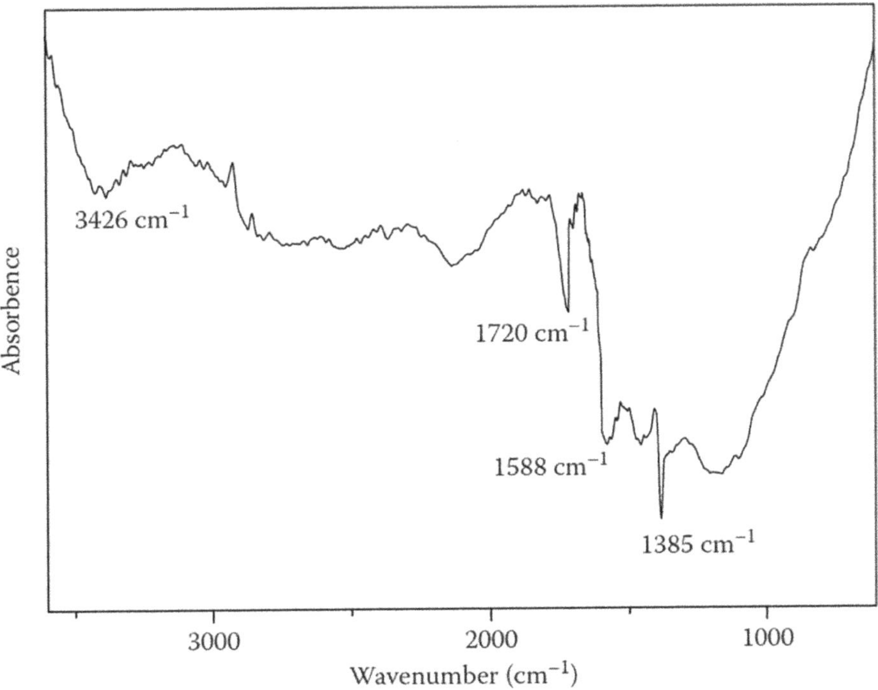

FIGURE 5.6 FTIR of oxidized MWCNTs. (From Chen, C.L. et al., *J. Colloid Interface Sci.*, 323, 33–41, 2008 [22]. With permission.)

oxidized MWCNTs and indicates that this acid treatment generates functional groups on the surfaces of oxidized MWCNTs as a hydroxyl group (−COOH, 3,426 cm⁻¹) and a carboxyl group (−OH, 1,720 cm⁻¹), and the bands of ~1,600 and ~1,380 cm⁻¹ correspond to asymmetric and symmetric −COO⁻ stretching [21,22]. FTIR analysis indicates that many functional groups are generated after the oxidation treatment, and these functional groups are helpful in bonding metal ions from solution to oxidized MWCNTs.

X-ray photoelectron spectroscopy (XPS) can be used to analyze functional groups and an existing form of an element. Figure 5.7a through c shows the XPS survey spectrum, C1s and O1s high-resolution XPS spectra (using an X-ray photoelectron spectrometer, Kratos Axis Ultra DLD) of acidified MWCNTs by concentrated nitric acid [23]. The binding energies obtained in the XPS analysis were corrected by referencing the C1s line to 284.6 eV. Several peaks attributable to carbon and oxygen are present in the survey spectra of acid-oxidized MWCNTs (Figure 5.7a). There is no N peak in the spectra, which indicates that no N was introduced into the acidified MWCNTs after MWCNTs were treated with concentrated nitric acid. The high-resolution C1s spectrum of acidified MWCNTs has been resolved into four individual component peaks at 284.7, 285.3, 286.8, and 289.1 eV, respectively. These peaks represent graphic carbon (284.7±0.1 eV), carbon linked to hydroxyl groups (285.3±0.2 eV), carbon in carbonyl groups (286.9±0.1 eV), and carboxyl groups (289.1±0.1 eV) according to the curve-fitting procedures (Figure 5.7b) [24–26].

The high-resolution O1s spectrum of acidified MWCNTs can be fitted to two peaks (Figure 5.7c). The binding energies at 531.8 and 533.4 eV can be assigned to the O in the C=O and alcoholic C–O groups, respectively [27]. So, incorporating the high-resolution C1s, it can be deduced that the functional groups introduced onto the surface of acidified MWCNTs are mainly hydroxyl, carbonyl, and carboxyl groups. The amounts of functional groups formed on acidified MWCNTs, including hydroxyl, carbonyl, and carboxyl, were determined by Boehm's titration method, and the results are shown in Table 5.1. In Table 5.1, the amounts of hydroxyl and carboxyl first increase and then

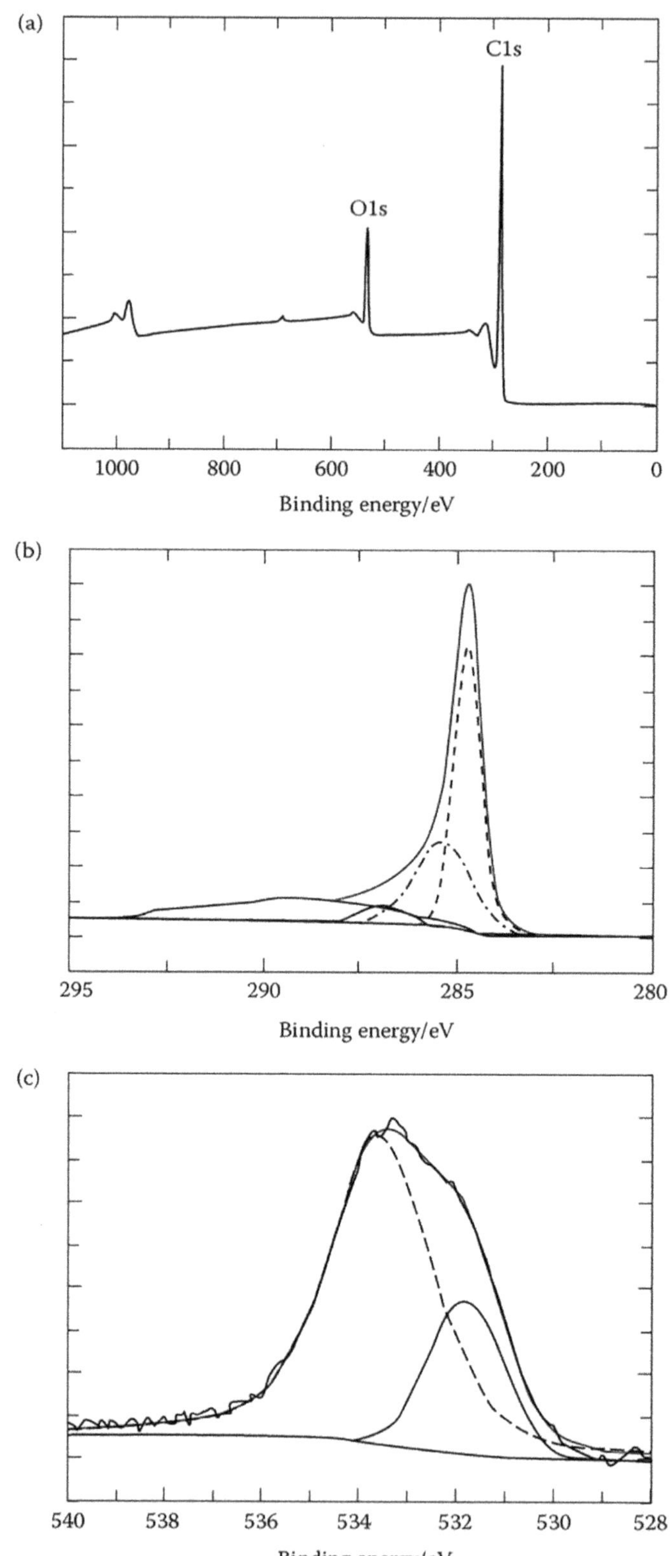

FIGURE 5.7 XPS survey (a), C1s (b), and O1s (c) high-resolution spectra of acidified MWCNTs. (From Wang, H. et al., *J. Colloid Interface Sci.*, 316, 277–283, 2007 [23]. With permission.)

TABLE 5.1

Amount of Functional Groups on MWCNTs after Different Acidification Treatment Time

Acidification Time (h)	Amount of Hydroxyl (mmol/g)	Amount of Carbonyl (mmol/g)	Amount of Carbonyl (mmol/g)	Amount of All the Groups (mmol/g/ wt%)
0	-[a]	-	-	-/-
1	0.44	0.59	0.58	1.61/5.01
2	0.61	0.19	1.04	1.84/6.25
6	0.99	0.46	1.37	2.82/9.13
10	0.77	1.34	0.95	3.06/9.34
Annealed MWCNTs[b]	0.01	0.31	-	0.32/0.89

Source: Wang, H. et al., *J. Colloid Interface Sci.,* 316, 277–283, 2007 [23]. With permission.

[a] Represents nondetectable.

[b] The 6 hour-acidified MWCNTs with 800°C annealed for 2 hours in an argon flow.

decrease, but the amounts of carbonyl change irregularly during the treatment time. During the initial stage of the acidified treatment, the defective sites and the ends of MWCNTs are easily oxidized to alcohol and aldehyde, which causes the increase of hydroxyl and carbonyl groups. The alcohol and aldehyde can be further oxidized into acid, which causes the increase of carboxyl. With increasing time, the oxidation rate of MWCNTs to alcohol and aldehyde slows down, and the alcohol and aldehyde continue to be oxidized to acid, which causes the decrease of alcohol and aldehyde. Although alcohol and aldehyde are still oxidized to acid because the formed acid was finally oxidized to carbon dioxide and the production rate is less than the consumption rate, the amount of carboxyl decreases. Due to the different abilities of MWCNTs being oxidized to alcohol, aldehyde, and acid, the amount of functional groups changes differently and even the carbonyl changes irregularly.

5.3 APPLICATION OF NANOMATERIALS IN THE REMOVAL OF METAL IONS

5.3.1 REMOVAL OF HEAVY METAL IONS

5.3.1.1 Removal of Cr(VI)

Chromium exists in the environment mainly in two states: trivalent Cr(III) and hexavalent Cr(VI). Cr(III) is an essential element in humans and is much less toxic than Cr(VI). Cr(VI) is primarily present in the form of chromate (cro_4^{2-}) and dichromate $\left(Cr_2O_7^{2-}\right)$ ions. Cr(VI) is one of the extremely toxic heavy metals found in various industrial waters and can cause health problems such as liver damage, pulmonary congestion, vomiting, and severe diarrhea [28,29]. Due to the severe toxicity of Cr(VI), the EU Directive, WHO, and the U.S. Environmental Protection Agency (USEPA) have set the maximum contaminant concentration of Cr(VI) in domestic water supplies at 0.05 mg/L [15,30].

In the study of Cr(VI) removal from wastewater, pH is one of the most important parameters controlling the sorption process. The effect of pH on the removal of Cr(VI) is investigated by testing three values of ionic strengths (i.e., 0.1, 0.01, and 0.001 M $NaClO_4$) and two contact times (i.e., 65 and 165 hours). Figure 5.8 shows that the maximum adsorption of Cr(VI) occurs at the pH range below 2, and then the removal of Cr(VI) decreases with increasing pH values, and Cr(VI) is not adsorbed on oxidized MWCNTs at pH>6 [31]. The adsorption of Cr(VI) from an aqueous solution to oxidized MWCNTs is independent of ionic strength and dependent on pH values. The ionic strength can influence the double-layer thickness and interface potential, thereby affecting the

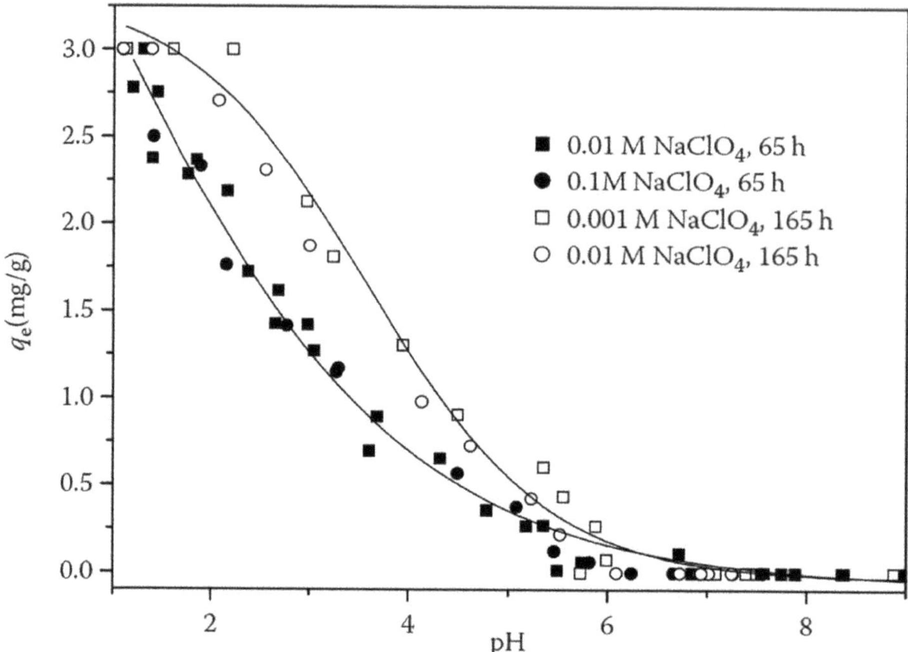

FIGURE 5.8 Removal of Cr(VI) from solution to oxidized MWCNTs as a function of pH values at different ionic strengths and contact times. m/V=0.1 g/L, C[Cr(VI)]$_{initial}$=3.0 mg/L, T=20±2°C. (From Hu, J. et al., *J. Hazard. Mater.*, 162, 1542, 2009 [31]. With permission.)

binding of adsorbed species. Outer-sphere complexes are expected to be more susceptible to ionic strength variations than inner-sphere complexes since the background electrolyte ions are placed in the same plane for outer-sphere complexes. In general, the sorption mechanism of surface complexation is significantly affected by pH, whereas the sorption mechanism of ion exchange is influenced by ionic strength. The strong pH-dependent and ionic strength-independent adsorption of Cr(VI) to oxidized MWCNTs suggests that the adsorption of Cr(VI) is mainly dominated by surface complexation rather than ion exchange. It is well known that Cr(VI) exists mainly as the species of $HCrO_4^{2-}$ and in solution at pH<6, and as CrO_4^{2-} species at pH>6 [28,29]. The functional groups, such as $-OH$ and $-COCr_2O_7^{2-}OH$, on the surface of the oxidized MWCNTs, are supposed that oxidized MWCNTs are the carbonaceous material as C_xOH, where C_x=carbon. It is necessary to note that the hydroxylated surface groups vary at different pH values because of the protonation/deprotonation processes:

$$\text{At low} \quad \text{pH: } C_xOH + H^+ \leftrightarrow C_xOH_2^+. \tag{5.1}$$

$$\text{At high} \quad \text{pH: } C_xOH \rightarrow C_xO^- + H^+. \tag{5.2}$$

The negatively charged Cr(VI) is easily adsorbed on oxidized MWCNTs at low pH values and difficult to adsorb at high pH values. At high pH values, the higher the valence of adsorbed anions, the more negative the surface becomes, thus inhibiting the further adsorption of anions. Two types of chemical treatments are currently used for Cr(VI) removal: the first type removes Cr(VI) anions directly, whereas the second type relies on the reduction of Cr(VI) to Cr(III). It is known [15,16] that Cr(VI) is reduced to Cr(III) in the presence of a reducing substrate (C_xOH) on the occurrence of redox reactions between the surface groups and the Cr(VI) at low pH values [14]:

$$3C_xOH + Cr_2O_7^{2-} + 4H^+ \Leftrightarrow 3C_xO^- + Cr^{3+} + HCrO_4^- + 3H_2O, \tag{5.3}$$

$$3C_xOH + HCrO_4^- + 4H^+ \Leftrightarrow 3C_xO + Cr^{3+} + 4H_2O. \tag{5.4}$$

The total concentrations of Cr(VI) and Cr(III) were analyzed in solution, and it was found that Cr was mainly in the trivalent form at pH < 3 (Figure 5.9) [15]. The presence of functional groups (such as C–H, –OH, and –COOH) on the surface of oxidized MWCNTs ensures the capture of metallic cations by surface complexation and cation exchange mechanisms. The adsorption of Cr(VI) anions is complicated, but it can be related to surface complexation reactions with protonated sites or by electrostatic attraction with electrophilic surface sites. Cr(VI) anions are adsorbed by the protonated groups and the electrophilic surface groups of MWCNTs, and then part of Cr(VI) is reduced to Cr(III) cations, which are partly released back into the solution. Cr(III) cations are captured by sorption and ion exchange on the weak acid surface groups or the basal plane sites on MWCNTs [29,32].

The characteristics and the concentration of ionic species in solution obviously represent the driving force for adsorption phenomena. Hence, speciation analysis of the equilibrium solution is the first step in understanding experimental data. Trivalent and hexavalent chromium speciations are calculated using chromium mass balance and electric neutrality equations coupled with equations' representative of chemical equilibria [33]. Trivalent and hexavalent chromium speciations are shown in Figure 5.10 for the case of a total chromium concentration of 0.5 mM (~25 mg/L). Speciation analyses show that a pH greater than 7 Cr(VI) is mainly present as , whereas at pH lower than 3 trivalent chromium is mainly represented by Cr^{3+} [15].

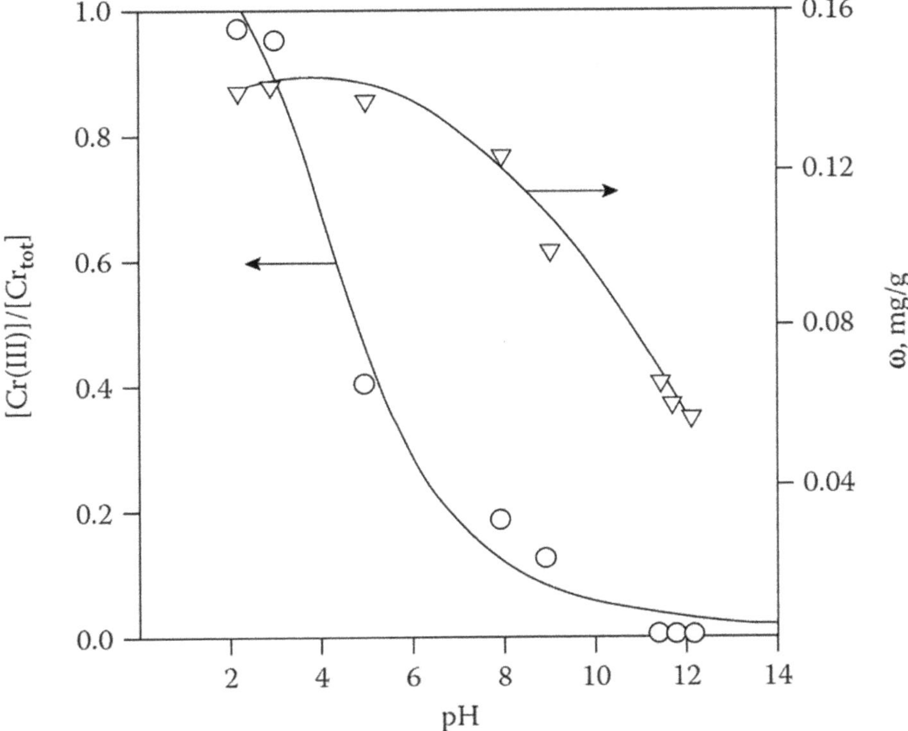

FIGURE 5.9 Total chromium adsorption capacity on activated carbon and trivalent chromium percentage for aqueous solution with initial concentration of Cr(VI) 25 mg/L at different equilibrium pH levels. (From Di Natale, F. et al., *J. Hazard. Mater.*, 145, 381–390, 2007 [15]. With permission.)

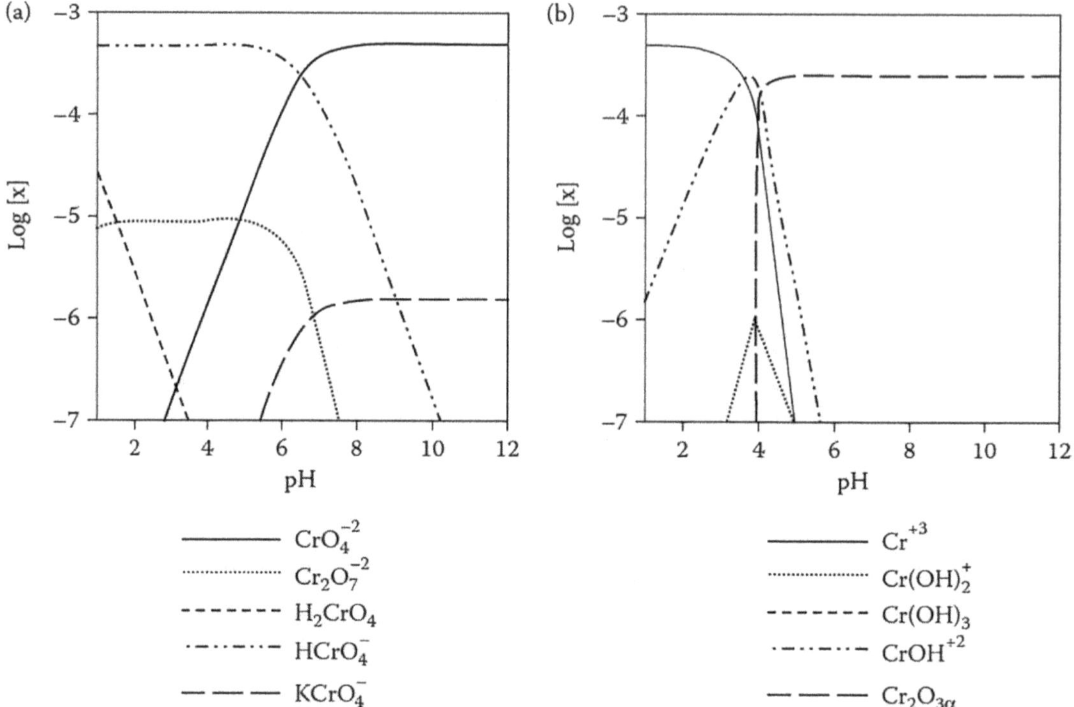

FIGURE 5.10 Speciation of hexavalent (a) and trivalent (b) chromium as a function of pH for a total chromium concentration of 0.5 mM obtained from chemical equilibrium analysis. (From Di Natale, F., et al., *J. Hazard. Mater.*, 145, 381–390, 2007 [15]. With permission.) (Values of the thermodynamics constants for ion complexation are taken from Stumm, W. and Morgan, J.J., *Aquatic Chemistry*, 3rd Edition, John Wiley and Sons, New York, 1996 [33].)

The adsorptions of Cr(VI) on raw and oxidized MWCNTs are shown in Figure 5.11 as a comparison. After the oxidization treatment, the functional groups at oxidized MWCNT surfaces obviously increase and thereby enhance the adsorption of Cr(VI). Figure 5.11 shows that the removal percentage of Cr(VI) on raw MWCNTs is lower than that of Cr(VI) on oxidized MWCNTs under the same experimental conditions, which is also evidence of the presence of oxygen-containing functional groups on the external and internal surfaces of oxidized MWCNTs. Carboxylate groups have also been introduced onto the surfaces of CNTs by reaction with strong oxidizing agents, and hemispherical caps on the nanotubes have been removed. These functional groups are active in forming complexes with Cr(VI) on the surfaces of oxidized MWCNTs.

To further understand the removal mechanism of Cr(VI) on MWCNTs, the sample after reaction with Cr(VI) at pH 2.85 was analyzed with XPS. The XPS spectrum (Figure 5.12) shows two Cr2p3/2 and Cr2p1/2 peaks that are, respectively, centered at 577.5 and 587.2 eV, which are consistent with Cr(III) and Cr(VI), respectively [29,34]. XPS analysis indicates that Cr is adsorbed on MWCNTs as Cr(III) and Cr(VI), which also suggests that some adsorbed Cr(VI) anions are reduced to Cr(III) on the surfaces of MWCNTs.

5.3.1.2 Removal of Bivalent Cations

5.3.1.2.1 Removal of Ni(II)

Nickel is a toxic metal ion that is present in wastewater. More than 40% of the nickel produced is used in steel factories and nickel batteries and in the production of some alloys, which causes an increased Ni(II) burden on the ecosystem and deterioration of water quality. Wastewater dissolved

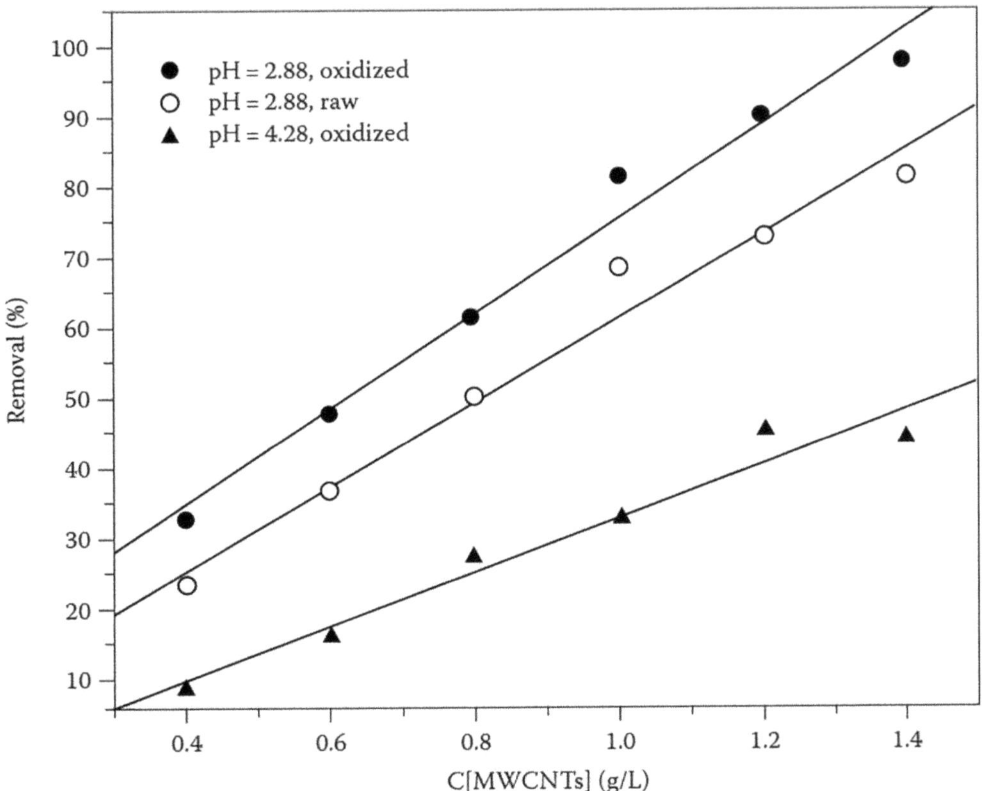

FIGURE 5.11   Removal of Cr(VI) as a function of MWCNT content. $C[Cr(VI)]_{initial}=3.0$ mg/L, $I=0.01$ MNaClO$_4$, $T=20\pm2°C$, contact time $=165$ hours. (From Hu, J. et al., *J. Hazard. Mater.*, 162, 1542, 2009 [31]. With permission.)

or dispersed with Ni(II) is harmful and causes vomiting, chest pain, and shortness of breath. Various technologies exist for removing Ni(II), which include filtration, surface complexation, chemical precipitation, ion exchange, adsorption using activated carbon, electrode position, and membrane process. Adsorption is considered one of the most attractive processes for Ni(II) removal from solution since adsorbents are generally easy to handle and can be used for various situations without large apparatuses.

Because of the highly porous and hollow structure, large specific surface area, light mass density, and strong adsorption ability, CNTs are considered to be a promising material in the removal of organic or inorganic pollutants. The application of CNTs in the removal of Ni(II) was investigated, and the results indicated that CNTs are efficient in the removal of Ni(II) from an aqueous solution [3,12]. In the application of CNTs to remove Ni(II), the adsorption velocity is quite important because it is necessary for the use of CNTs on a large scale. Figure 5.13 shows the effect of contact time and initial Ni(II) concentration on Ni(II) adsorption on oxidized MWCNTs. Equilibrium was reached within 40 minutes for all concentrations of Ni(II) used in the study. This result is very interesting because equilibrium time is one of the parameters for economical wastewater treatment plant applications. The implication is that oxidized MWCNTs could be suitable for a continuous-flow system.

In Figure 5.13, the pseudo-second-order rate model was expressed as

$$\frac{t}{q_t} = \frac{1}{2K' \cdot q_e^2} + \frac{t}{q_e} \tag{5.5}$$

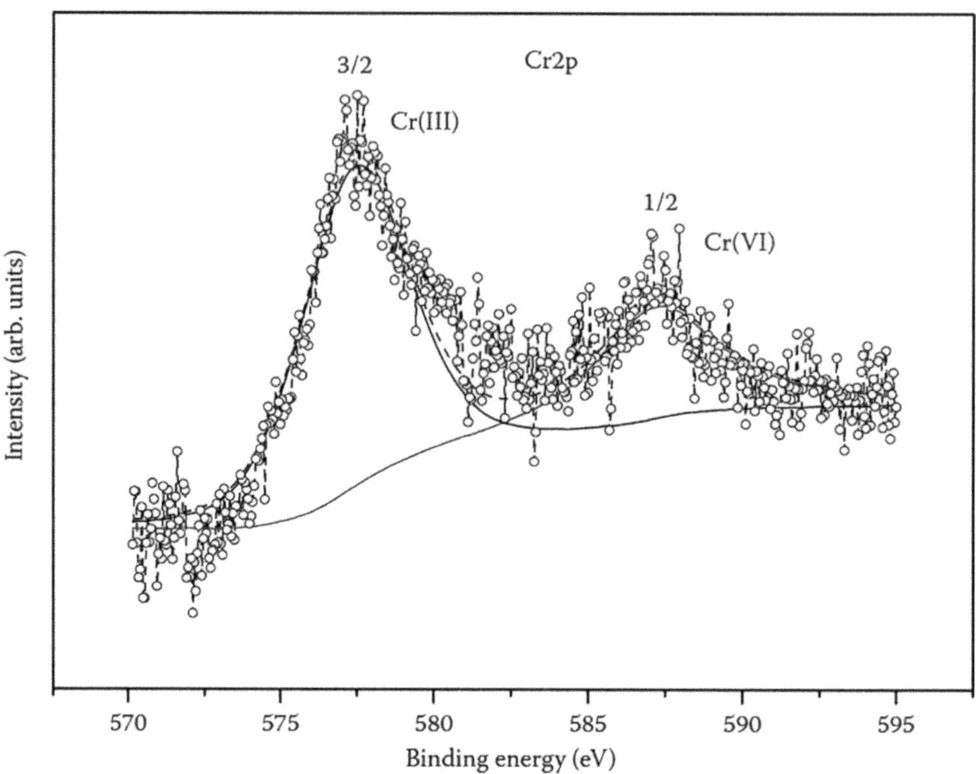

FIGURE 5.12 XPS spectrum of oxidized MWCNT sample after Cr(VI) sorption at pH 2.85. The solid lines represent the Cr(III) and Cr(VI) components. The dashed line is the fit envelope, and the solid circles show the data. (From Hu, J. et al., *J. Hazard. Mater.*, 162, 1542, 2009 [31]. With permission.)

where K' (g/mg/min) is the pseudo-second-order rate constant of adsorption, q_t (mg/g of dry weight) is the amount of Ni(II) adsorbed on the surface of the adsorbent at time t (min), and q_e (mg/g of dry weight) is the equilibrium adsorption capacity. Equation (5.5) was used to simulate the experimental data, and the results indicated that the experimental data were modeled by the pseudo-second-order rate equation very well.

Information about the effect of temperature on the removal of metal ions is important to understand the mechanism. Figure 5.14 shows that the distribution coefficients, K_d (mL/g), increase with an increase in temperature. Bikerman [35] attributed this phenomenon to a negative temperature coefficient of solubility of solutes or a steep simultaneous decrease of real adsorption of solvent. The values of enthalpy, $\Delta H°$, and entropy, $\Delta S°$, were calculated from the slope and intercept of the plot of $\ln K_d$ versus $1/T$ (Figure 5.15) by using the equation

$$\ln K_d = \frac{\Delta S°}{R} \frac{\Delta H°}{RT}.$$ (5.6)

The Gibbs free energy, $\Delta G°$, of specific adsorption was calculated from the equation

$$\Delta G° = H° - T\Delta S°,$$ (5.7)

where R (8.3145 J/mol/K) is the ideal gas constant and T (K) is the temperature. Relevant data calculated from Equations (5.6 and 5.7) are tabulated in Table 5.2. It is evident that the values of $\Delta H°$ are positive, that is, endothermic. One possible interpretation of the endothermicity of the enthalpy

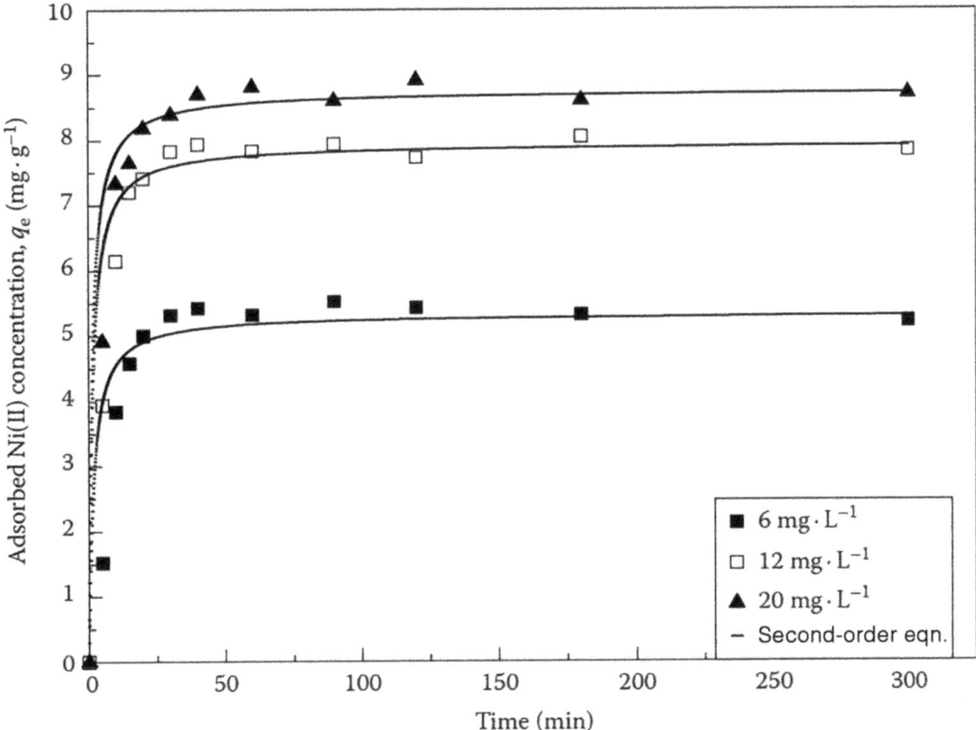

FIGURE 5.13 Effect of contact time on Ni(II) adsorption rate for various initial Ni(II) concentrations onto oxidized MWCNTs: experimental data and pseudo-second-order rate model fit (pH = 6.55 ± 0.02, I = 0.01 M KNO$_3$, m/V = 0.75 g/L, T = 291 ± 1 K). (From Chen, C.L. and Wang, X.K., *Ind. Eng. Chem. Res.*, 45, 9144–9149, 2006 [12]. With permission.)

of adsorption is that ions such as Ni(II) are well-solvated in water. For these ions to adsorb, they are to some extent denuded of their hydration sheath, and this dehydration process of ions needs energy. It is assumed that this energy of dehydration exceeds the exothermicity of ions attaching to the surface. The removal of water molecules from ions is essentially an endothermic process, and it appears that the endothermicity of the desolation process exceeds that of the enthalpy of adsorption to a considerable extent. The Gibbs free energy change ($\Delta G°$) is negative (as expected) for a spontaneous process under the conditions applied. The decrease in $\Delta G°$ with an increase in temperature indicates more efficient adsorption at higher temperatures. At higher temperatures, ions are readily desolvated, and therefore, their adsorption becomes more favorable. The positive values of entropy change ($\Delta S°$) reflect the affinity of oxidized MWCNTs toward Ni(II) ions in aqueous solutions and may suggest some structure changes in the adsorbents [36,37].

However, in the natural environment, both organic and inorganic pollutants are present in wastewater. A knowledge of the influence of organic pollutants on the adsorption of metal ions is helpful in understanding the application of nanomaterials in environmental pollution cleaning. Generally, surfactants are used extensively in household products, detergents, oil recovery, paint technology, flotation, and water treatment. Considerable amounts of surfactants are released into the environment, causing serious pollution problems in rivers, lakes, and oceans. The accumulation of surfactants can form large foam masses, affect aquatic organisms, and interfere with the removal processes of insoluble and soluble substances [38,39]. For all these reasons, many environmental and public health regulatory authorities have fixed stringent limits for anionic detergent as standard 0.5 mg/L for drinking water and relaxable up to 1.0 mg/L for other purposes [40]. At low concentrations, surfactants exist solely as monomers; above a critical aqueous concentration, which is specific for each

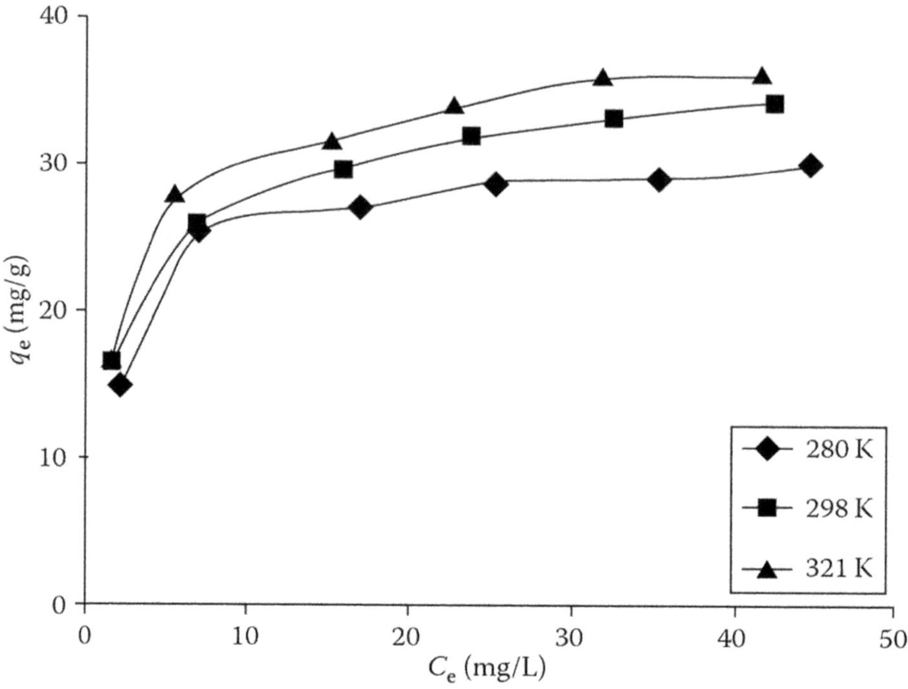

FIGURE 5.14 Plot of distribution coefficient K_d versus temperature for Ni(II) adsorption onto oxidized MWCNTs (pH=6.55±0.02, I=0.01 M KNO3, m/V=0.75 g/L). (From Chen, C.L. and Wang, X.K., *Ind. Eng. Chem. Res.*, 45, 9144–9149, 2006 [12]. With permission.)

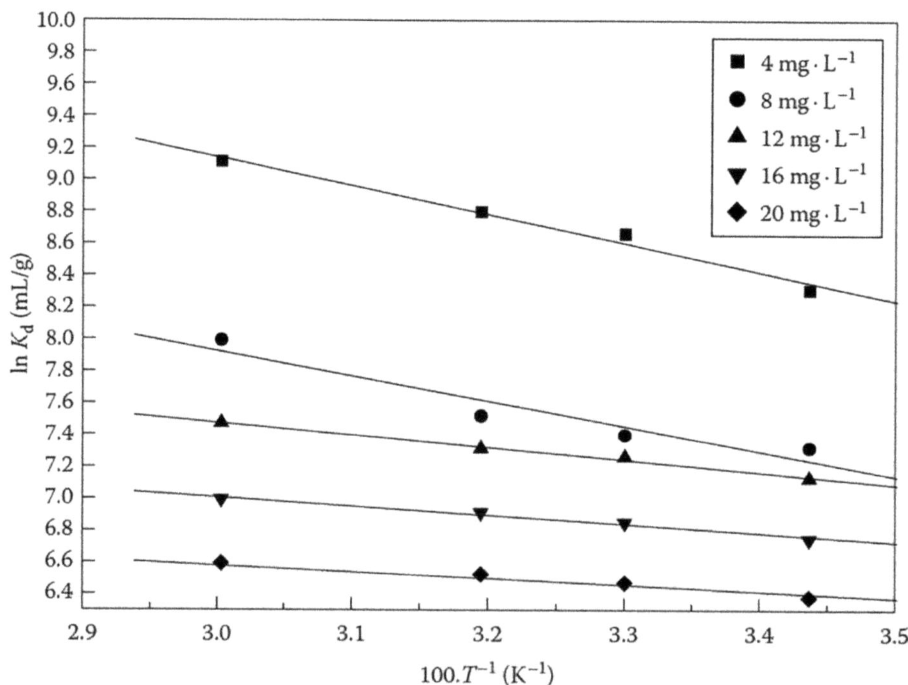

FIGURE 5.15 Semilogarithmic plot of distribution coefficient K_d versus reciprocal temperature for various concentrations of Ni(II) adsorption onto MWCNTs (pH=6.55±0.02, I=0.01 M KNO$_3$, m/V=0.75 g/L). (From Chen, C.L. and Wang, X.K., *Ind. Eng. Chem. Res.*, 45, 9144–9149, 2006 [12]. With permission.)

TABLE 5.2

Thermodynamic Parameters for Ni(II) Adsorption onto Oxidized MWCNT

C_0 (Ni(II))			$-\Delta G°$ (kJ/mol)			
(mg/L)	$\Delta H°$ (J/mol)	$\Delta S°$ (J/mol/K)	$T=292$ K	$T=303$ K	$T=313$ K	$T=333$ K
4	14.93	120.75	35.12	36.57	37.78	40.19
8	13.04	104.99	30.54	31.80	32.85	34.95
12	6.45	81.47	23.70	24.68	25.50	27.12
16	4.62	72.09	20.97	21.84	22.56	24.00
20	4.07	67.09	19.52	20.32	20.99	22.34

Source: Chen, C.L. and Wang, X.K., *Ind. Eng. Chem. Res.*, 45, 9144–9149 [12], 2006. With permission.

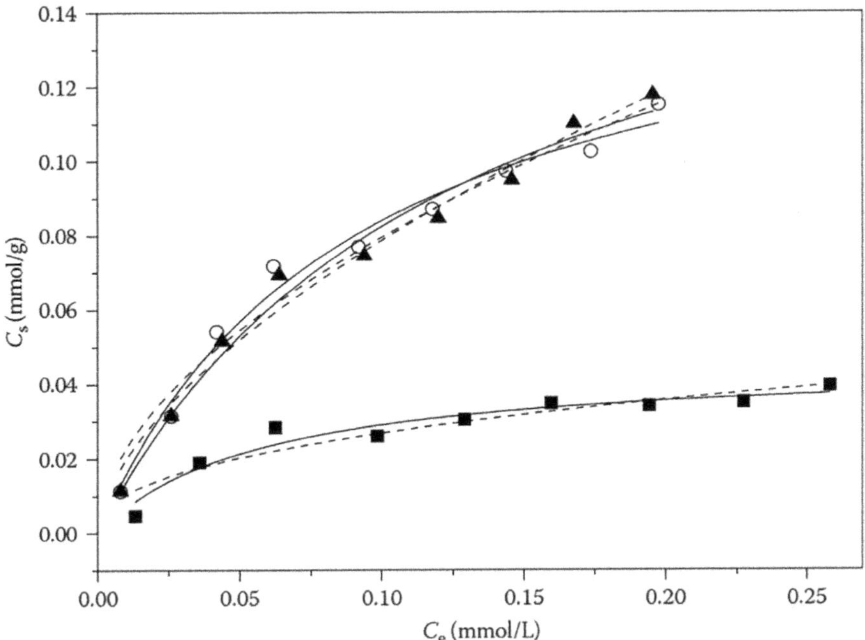

FIGURE 5.16 Adsorption isotherms of nickel to MWCNTs in the presence of SDBS. pH=5.5±0.1, $T=20±2°C$, $I=0.01$ mol/L NaClO4, $m/V=0.8$ g (MWCNTs)/L, C(SDBS)$_{initial}=0.98$ mmol/L. Solid line: Langmuir model; dashed line: Freundlich model; (■) MWCNTs+nickel; (O) MWCNTs and SDBS are pre-equilibrated for 2 days before the addition of nickel (batch 1); (▲) nickel and SDBS are pre-equilibrated for 2 days before the addition of MWCNTs (batch 2). (From Tan, X.L. et al., *Carbon*, 46, 1741, 2008 [43]. With permission.)

surfactant, called the critical micelle concentration (CMC), aggregation of surfactant monomers occurs, and hydrophobic interactions between the hydrocarbon chains of surfactant molecules are balanced by hydration and the electrostatic repulsive effects of the hydrophilic groups. It is recognized that surfactant properties strongly depend on counterion species. Anionic surfactant molecules, which are negatively charged, can bind to positively charged metal ions. Once the adsorption of surfactant molecules on particle surfaces is established, self-organization of the surfactant into micelles (aggregative structures of surfactants) is expected to occur above the CMC [41,42].

The adsorption isotherms of nickel on oxidized MWCNTs are shown in Figure 5.16. To gain a better understanding of the mechanism and to quantify the adsorption data, the Langmuir [$C_s=abC_e/$

TABLE 5.3

Parameters of Adsorption Models for Nickel on Oxidized MWCNTs

	Langmuir Model			Freundlich Model		
System	a (mmol/g)	b (L/mmol)	R^2	K_F (mmol/g)	$1/n$	R^2
MWCNTs+nickel	0.0456	17.56	0.9373	0.0692	2.43	0.8849
Batch 1	0.1608	10.94	0.9850	0.2784	1.84	0.9700
Batch 2	0.1852	8.012	0.9823	0.3132	1.67	0.9821

Source: Chen, C.L. and Wang, X.K., *Ind. Eng. Chem. Res.*, 45, 9144–9149, 2006 [12]. With permission.

$(1+bC_e)]$ and Freundlich $(C_s=)$ isotherm models are used to simulate the adsorption of nickel on MWCNTs. Herein, a is the maximum adsorpt$K_f C_e^{1/n}$ion capacity, b is the Langmuir adsorption constant, and K_f and $1/n$ are the Freundlich constants. The Langmuir and Freundlich constants obtained by fitting the adsorption equilibrium data are listed in Table 5.3. The values of the correlation coefficient (R^2) derived from the Langmuir model are closer to 1 than those derived from the Freundlich model, which indicates that the adsorption of nickel on MWCNTs is better simulated by the Langmuir model than by the Freundlich model. It can also be seen that the adsorption isotherms of nickel in the presence of sodium dodecylbenzene sulfonate (SDBS) are much higher than those of nickel in the absence of SDBS. The maximum adsorption capacity of nickel adsorption on SDBS-bound MWCNTs is higher than that of nickel adsorption on bare MWCNTs, which also indicates that SDBS-MWCNT hybrids have a stronger affinity for nickel adsorption than bare MWCNTs.

The adsorption isotherms of SDBS on oxidized MWCNTs are shown in Figure 5.17. The adsorption of SDBS is not affected by the addition of nickel in ternary systems. The linear adsorption isotherms of SDBS for different additional sequences suggest that the adsorption of SDBS on MWCNTs is far from saturation. The adsorption isotherms of SDBS to river sediment and soils have also been shown to be near-linear [44,45]. The adsorption of SDBS to MWCNTs is far from saturation. MWCNTs are very suitable in the removal of SDBS from aqueous solutions.

To give a clear illustration of SDBS and nickel adsorption on MWCNTs, a diagram of the adsorption mechanism is shown in Figure 5.18. It is commonly believed that the chemical interaction between the metal ions and the surface functional groups of MWCNTs is the major adsorption mechanism in the absence of organic materials. In the absence of a surfactant, protons in the carboxylic and phenolic groups of MWCNTs can exchange with the metal ions in the aqueous phase. The adsorption of metal ions on oxidized MWCNTs is mainly attributed to surface complexation, ion exchange, and electrostatic attraction. In the presence of a surfactant, MWCNT surfaces become hydrophilic due to the hydrophilic heads of the surfactant oriented toward the bulk solution. Surfactant monomers accumulate at the MWCNT-water interfaces and increase the contact angle between MWCNTs and nickel. SDBS molecules adsorbed on the surfaces of MWCNTs cause an attraction between the anionic head groups and the positively charged cations and thereby promote the adsorption of nickel to MWCNTs.

5.3.1.2.2 Removal of Pb(II)

Lead in the natural environment arises from both natural and anthropogenic sources and is detrimental to humans and living things. Long-term drinking water containing a high level of lead will cause serious disorders, such as anemia, kidney disease, and mental retardation [13]. The enrichment and bioavailability of Pb(II) by plants and crops can transfer Pb(II) from the natural environment to humans. Metal ions are nonbiodegradable, and therefore must be removed from water to eliminate the potential danger to humans and the environment. Many conventional methods have

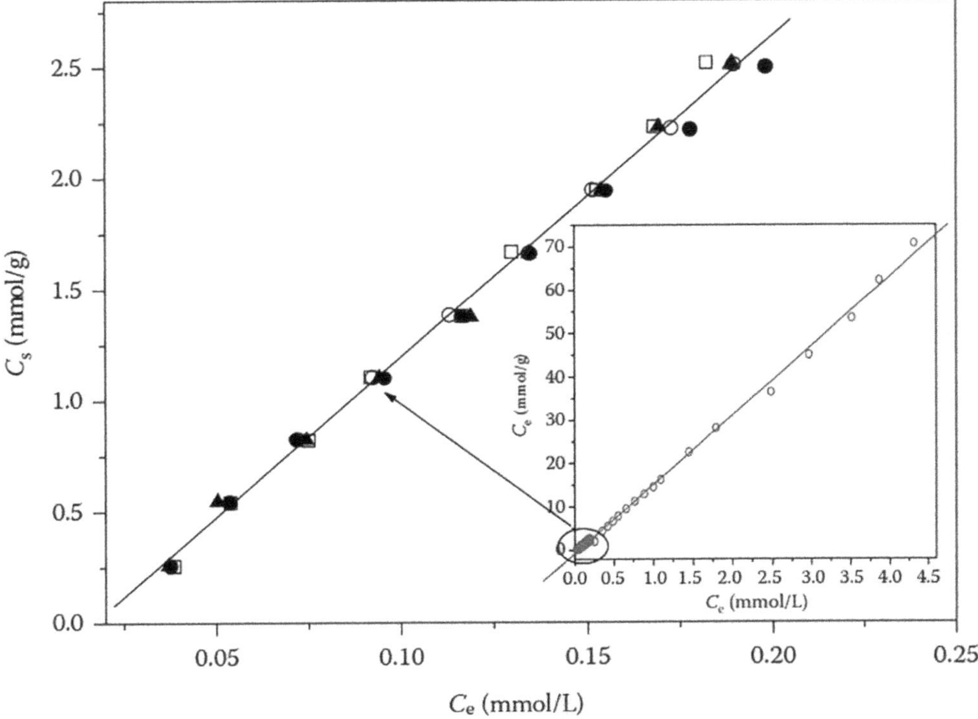

FIGURE 5.17 Adsorption isotherms of SDBS to MWCNTs in the presence of nickel. pH$=5.5\pm0.1$; $T=20\pm2°C$; $I=0.01$ mol/LNaClO$_4$; $m/V=0.8$ g (MWCNTs)/L; C(nickel)$_{initial}=0.15$ mmol/L; ($\square$) MWCNTs+SDBS; (O) MWCNTs and SDBS are pre-equilibrated for 2 days before the addition of nickel; ($\blacktriangle$) nickel and SDBS are pre-equilibrated for 2 days before the addition of MWCNTs; () nickel, SDBS, and MWCNTs are added in solution simultaneously. (From Tan, X.L. et al., *Carbon*, 46, 1741, 2008 [43]. With permission.)

been used to remove metal ions from aqueous solutions, including oxidation, reduction, precipitation, membrane filtration, ion exchange, and sorption. Of these methods, a promising one for the removal of metal ions from water and wastewater is sorption.

Lead in the natural environment arises from both natural and anthropogenic sources and is detrimental to humans and living things. Long-term drinking water containing a high level of lead will cause serious disorders, such as anemia, kidney disease, and mental retardation [13]. The enrichment and bioavailability of Pb(II) by plants and crops can transfer Pb(II) from the natural environment to humans. Metal ions are nonbiodegradable, and therefore must be removed from water to eliminate the potential danger to humans and the environment. Many conventional methods have been used to remove metal ions from aqueous solutions, including oxidation, reduction, precipitation, membrane filtration, ion exchange, and sorption. Of these methods, a promising one for the removal of metal ions from water and wastewater is sorption.

Figure 5.19 shows the adsorption of Pb(II) on oxidized MWCNTs in the presence of 0.01M NaClO$_4$, KClO$_4$, NaCl, and NaNO$_3$, respectively, as a function of pH values. The pH of the solution plays an important role in the sorption characteristics of Pb(II) to oxidized MWCNTs. The removal of Pb(II) increases very quickly from about 10% to 90% at pH 6–7, remains constant with increasing pH values at pH 7–10, and then decreases steeply at pH > 10. It is known that lead species are present as Pb^{2+}, Pb(OH)$^+$, Pb(OH)$_2^0$, and Pb(OH)$_3^-$ at different pH values (Table 5.4 and Figure 5.20). At pH < 6, the predominant lead species is Pb^{2+}, and its removal is mainly accomplished by sorption reaction.

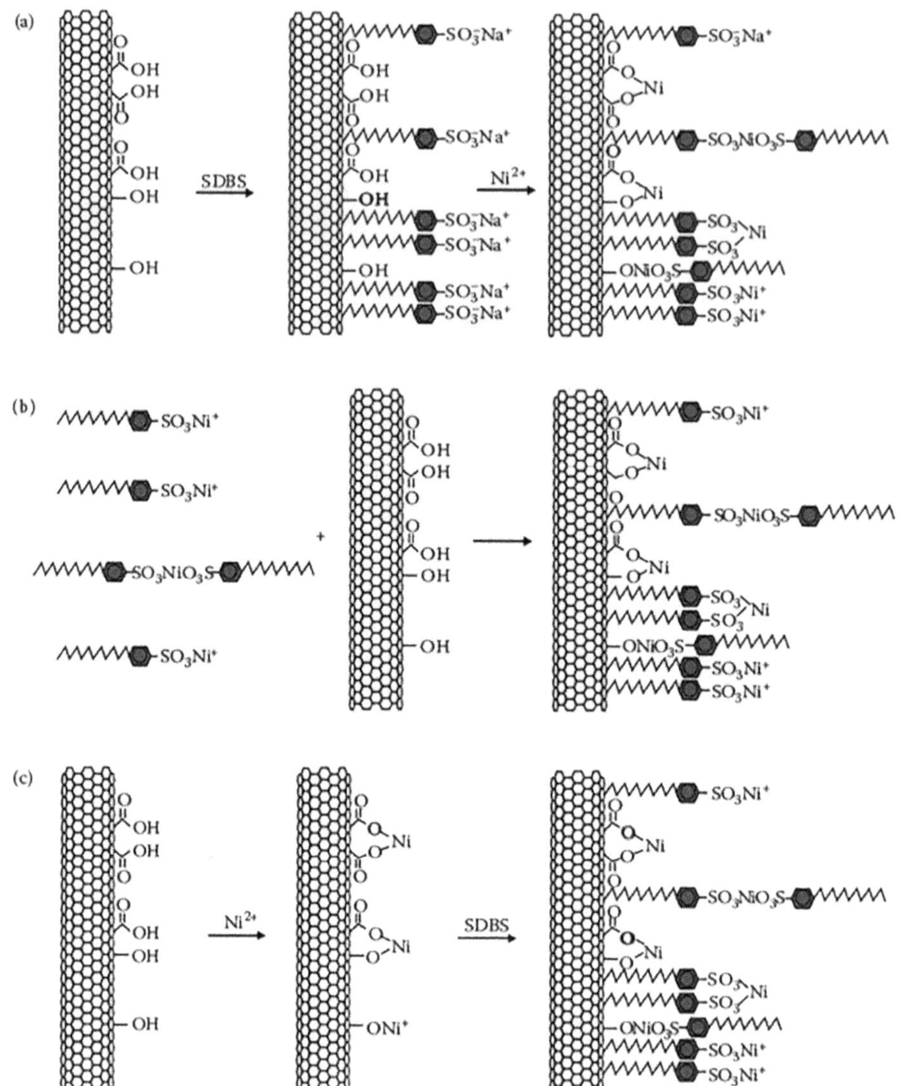

FIGURE 5.18 Diagram of the major mechanism for adsorption of nickel and SDBS on MWCNTs in the presence of SDBS/nickel. (From Tan, X.L. et al., *Carbon*, 46, 1741, 2008 [43]. With permission.)

The low Pb^{2+} sorption that takes place at low pH can be attributed partly to the competition between the H^+ and Pb^{2+} ions on the surface sites. At the pH range 7–10, the removal of Pb remains constant and reaches a maximum. The main species at pH 7–10 are $Pb(OH)^+$ and $Pb(OH)2$, and thus the removal of Pb is possibly accomplished by simultaneous precipitation of $Pb(OH)_2^0$ and sorption of $Pb(OH)^+$. At the pH range 10–12, the predominant lead species are $Pb(OH)_2^0$ and $Pb(OH)_3^-$. The decrease in Pb(II) sorption to oxidized MWCNTs at pH 10–12 can be attributed in part to competition between OH^- and $Pb(OH)_3^-$; the negative $Pb(OH)_3^-$ is difficult to be adsorbed on the negative surface charged oxidized MWCNTs at high pH values [5]. The precipitation constant of $Pb(OH)_{2(s)}$ is 1.2×10^{-15}, and the precipitation curve of lead at a concentration of 4.83×10^{-5} mol/L is also shown in Figure 5.19. Lead begins to form a precipitate at pH 8.7 if no lead is adsorbed on the oxidized MWCNTs. However, ~90% of the lead is adsorbed on oxidized MWCNTs at pH 7, and thereby,

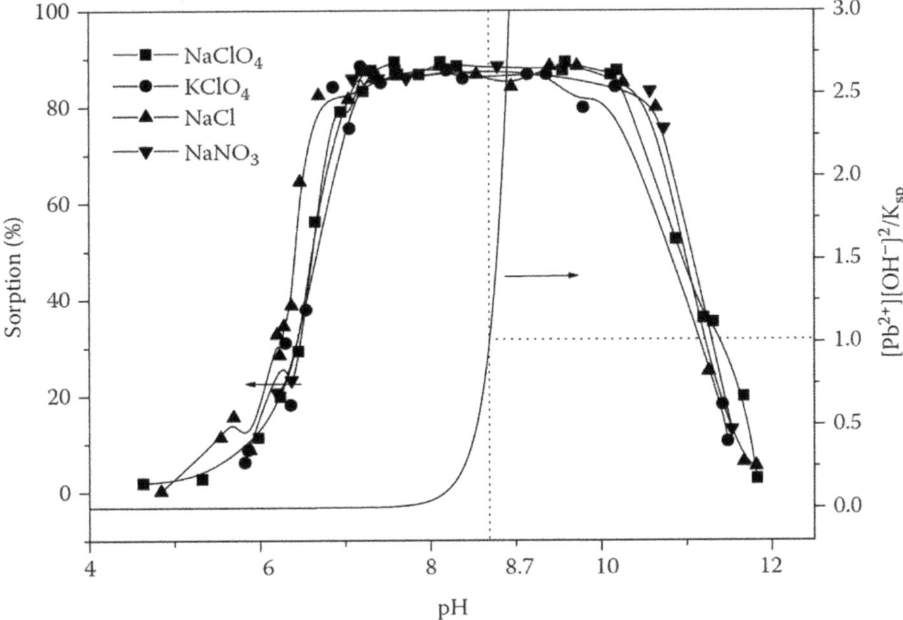

FIGURE 5.19 Variations in sorption of Pb(II) to oxidized MWCNTs as a function of equilibrium pH and foreign ions. $C[Pb(H)]_{initial}=4.83\times10^{-5}$ mol/L, $m/V=1.0\,g/L$, $I=0.01$ M, $T=293$ K, $K.=1.2\times10^{-15}$. (From Xu, D. et al., Removal of Pb(II) from aqueous solution by oxidized multiwalled carbon nanotubes. *J. Hazard. Mater.*, 154, 407–416, 2008. With permission.)

TABLE 5.4

Equilibrium Constants (log K) for Pb(II) Hydrolysis Reactions

	log *K*
Equilibrium	*I*=0.01 M, *T*=298 K
$Pb^{2+}+H_2O=Pb(OH)^{+}+H^{+}$	6.48
$Pb^{2+}+2\,H_2O=Pb(OH)_2+2\,H^{+}$	11.16
$Pb^{2+}+3\,H_2O=Pb(OH)_3^{-}+3H^{+}$	14.16

Source: Weng, C.H., *J. Colloid Interface Sci.*, 272, 262–270, 2004 [46].
With permission.

---

it is impossible to form a precipitate because of the very low concentration of lead remaining in the solution. Therefore, the abrupt sorption of Pb(II) on oxidized MWCNTs at pH 6–7 is not attributed to the precipitation of $Pb(OH)_2$. The precipitation of $Pb(OH)_2$ does not play any role in the removal of Pb(II) from solution to oxidized MWCNTs at the whole pH range. The species of Pb(II) in solution at different pH values is most important for the removal of Pb(II) from an aqueous solution to oxidized MWCNTs.

Figure 5.19 also shows that the removal of Pb(II) from an aqueous solution to oxidized MWCNTs is not influenced by the background electrolyte foreign cations and anions. Heavy metal ions in solution will compete for interactions with the surface functional groups of oxidized MWCNTs, and Pb(II) ions have a higher affinity to the surfaces of oxidized MWCNTs than alkali metal ions; therefore, the competition of alkali ions on Pb(II) uptake to oxidized MWCNTs is almost negligible.

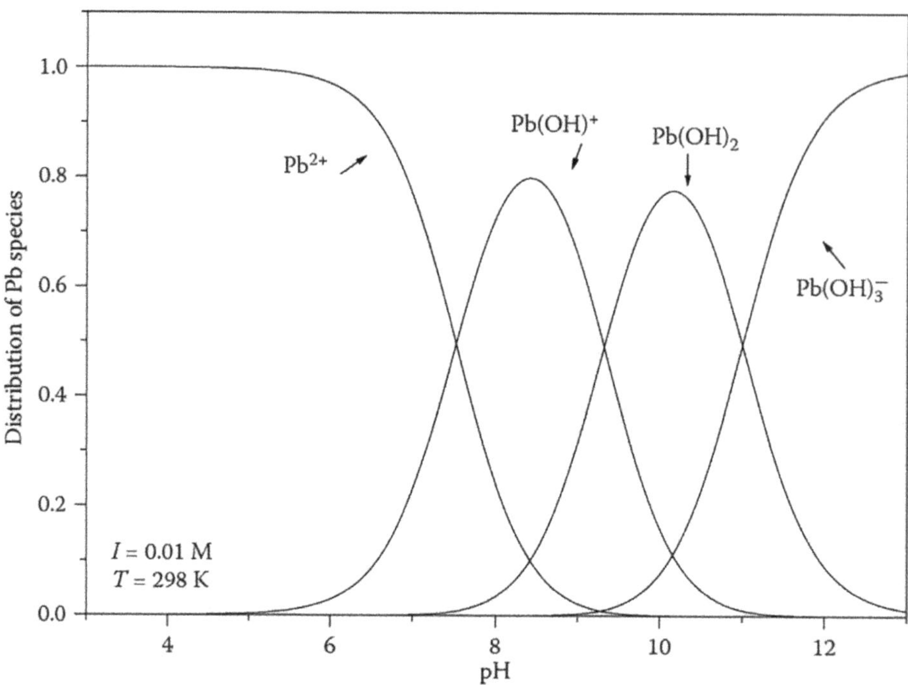

FIGURE 5.20 Distribution of Pb(II) species as a function of pH based on the equilibrium constants. (From Xu, D. et al., Removal of Pb(II) from aqueous solution by oxidized multiwalled carbon nanotubes. *J. Hazard. Mater.*, 154, 407–416, 2008. With permission.)

Although the radii of hydration of $K^+ = 2.32$ Å and $Na^+ = 2.76$ Å [47] are different, the difference in the radii of hydration on bivalent Pb(II) sorption is still very weak. The inorganic acid radicals radium order is $Cl^- < NO_3^- < ClO_4^-$; the negatively charged anions may form complexes with the oxygen-containing functional groups on the surfaces of oxidized MWCNTs. However, the effect of Cl^-, NO_3^-, and ClO_4^- on Pb(II) sorption to oxidized MWCNTs is still very weak, which suggests that surface complexes are formed on oxidized MWCNT surfaces. The effect of foreign ions on Pb(II) removal from solution to oxidized MWCNTs can be negligible.

The initial concentration of Pb(II) and solid oxidized MWCNT content (i.e., m/V) are the same for all data in Figure 5.19. The removal of Pb(II) is different at different pH values. The amounts of Pb(II) adsorbed on a solid (q) and remaining in solution (C_{eq}) will change with changing pH. To illustrate the variation and relationship of pH, C_{eq}, and q, experimental data of Pb(II) sorption at pH 5–10 in 0.01 M $NaClO_4$ and in 0.01 M $KClO_4$, respectively, are plotted again as three-dimensional (3-D) plots of pH, C_{eq}, and q in Figure 5.21. On the pH-q plane, the lines are very similar to that of pH-sorption % (in Figure 5.19); on the pH-C_{eq} plane, the concentration of Pb(II) remaining in the solution decreases with increasing pH. The projection on the pH-C_{eq} plane is just the inverted image of the projection on the pH-q plane; on the C_{eq}-q plane, the projection is a straight line containing all experimental data. The slope and the intercept calculated from the C_{eq}-q line are -1.0 and 4.83×10^{-5}, respectively, which are in good agreement with the values of $V/m = 1.0$ (L/g) and C_0 $V/m = 4.83 \times 10^{-5}$ (mol/L $\cdot$ L/g) (i.e., the values calculated from $V/m = 1.0$ L/g and $C_0 = 4.83 \times 10^{-5}$ mol/L). The complexity of the sorption edge relative to the sorption isotherm is demonstrated. The 3-D plots show the relationship of pH, C_{eq}, and q very clearly; that is, all the data of C_{eq}-q lie in a straight line with slope $-V/m$ and intercept $C_0 V/m$ for the same initial concentration of Pb(II) and the same solid content.

Figure 5.22 shows that the Pb^{2+} sorption rates increase dramatically in the first 10 minutes for various initial concentrations and reach equilibrium gradually at 20, 50, and 60 minutes, corresponding

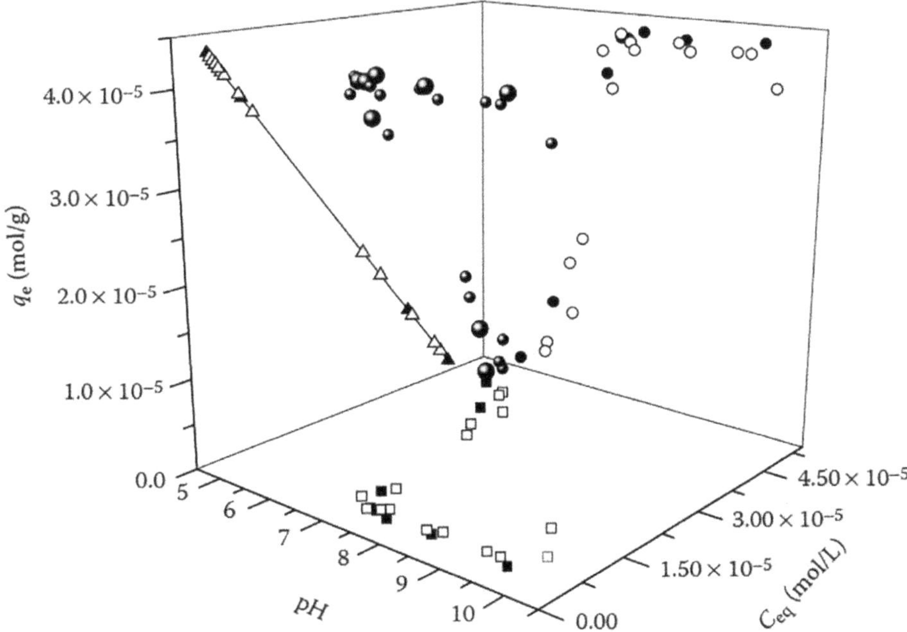

FIGURE 5.21 3-D plots of pH, Ceq, and q of Pb(II) sorption to oxidized MWCNTs. C[Pb(II)]initial= 4.83×10^{-5} mol/L, $m/V = 1.0$ g/L, $T = 293$ K. Solid points: $I = 0.01$ M NaClO4; open points: $I = 0.01$ M KClO4. (From Xu, D. et al., Removal of Pb(II) from aqueous solution by oxidized multiwalled carbon nanotubes. *J. Hazard. Mater.*, 154, 407–416, 2008. With permission.)

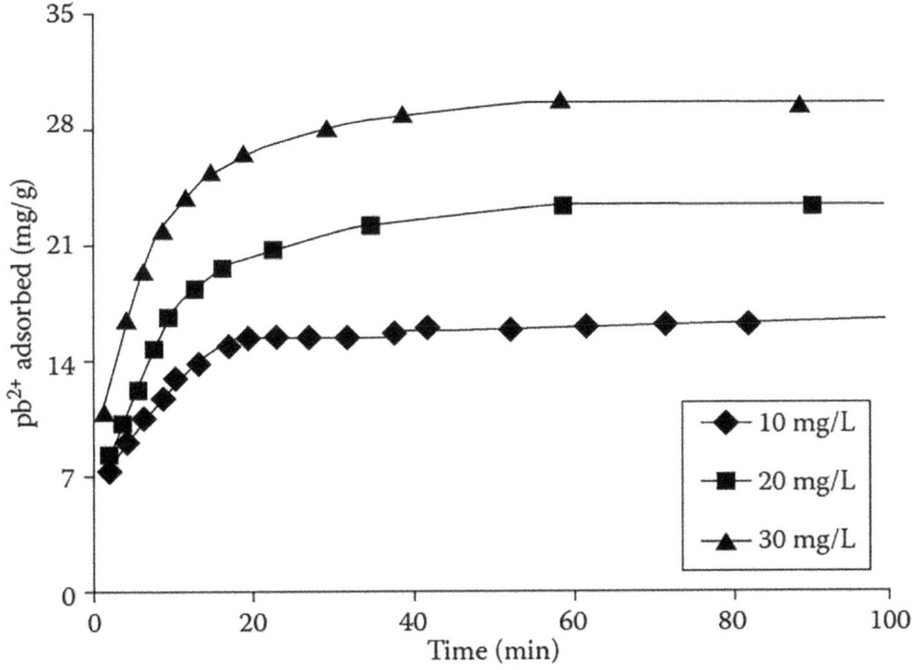

FIGURE 5.22 Effect of contact time on Pb^{2+} sorption rate for different concentrations (pH = 5.0, $T = 298$ K). (From Li, Y. et al., *Water Res.*, 39, 605–609, 2005 [10]. With permission.)

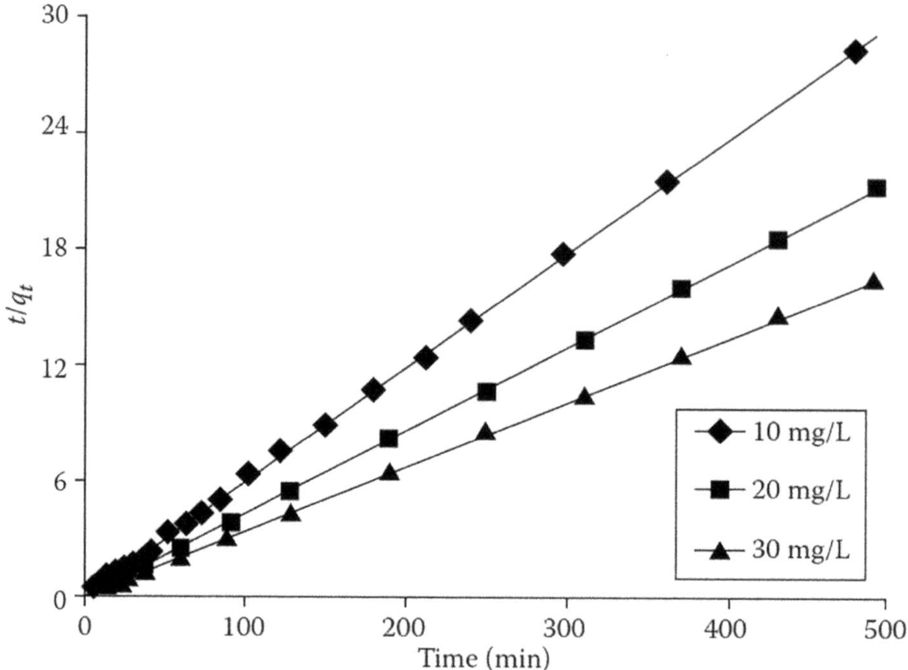

FIGURE 5.23 Test of pseudo-second-order rate equation for sorption of different concentrations of Pb^{2+} on CNTs (pH 5.0, $T = 298$ K). (From Li, Y. et al., *Water Res,* 39, 605–609, 2005 [10]. With permission.)

to Pb^{2+} initial concentrations of 10, 20, and 30 mg/L, respectively. The pseudo-second-order rate equation, $t/q_t = 1/(2K' q2) + t/q_e$, was evaluated based on the experimental data. A linear plot feature of t/q_t versus t (Figure 5.23) was achieved, and the K values calculated from the slopes and intercepts are summarized in Table 5.5. The correlation coefficients of the pseudo-second-order rate model for the linear plots are very close to 1, thus suggesting that kinetic adsorption can be described by the pseudo-second-order rate equation [10].

Figure 5.24 shows the sorption isotherms of Pb^{2+} on CNTs at 280, 298, and 321 K. The Pb^{2+} sorption capacity increases with an increase in temperature. Figure 5.25 shows linearized Freundlich isotherms for Pb^{2+} sorption on CNTs at different temperatures, and Freundlich parameters are listed in Table 5.6. The results indicate that the Freundlich isotherm model fits the Pb^{2+} sorption well. The thermodynamic parameters of Pb^{2+} sorption on CNTs are listed in Table 5.7. A positive standard enthalpy change suggests that the interaction of Pb^{2+} sorbed on CNTs is endothermic, which is supported by the

TABLE 5.5

Kinetic Parameters of Pb^{2+} Sorbed on CNTs at Different Initial Pb^{2+} Concentrations

Initial Pb^{2+} Concentration (mg/L)	Pseudo-Second-Order		
	q_e	K^r	R^2
10	17.09	0.0092	0.9989
20	23.41	0.0116	0.9999
30	30.32	0.0053	0.9987

Source: Li. Y., Di, Z., Ding, J., Wu, D., Luan, Z., and Zhu, Y., *Water Res.,* 39, 605–609, 2005. With permission.

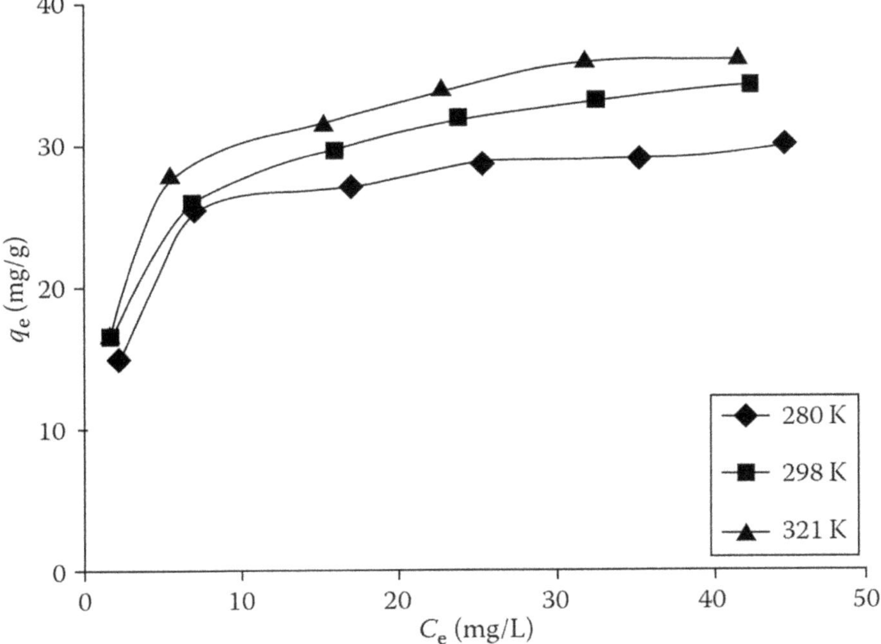

FIGURE 5.24 Sorption isotherms of Pb^{2+} onto CNTs at different temperatures. (From Li, Y. et al., *Water Res.*, 39, 605–609, 2005 [10]. With permission.)

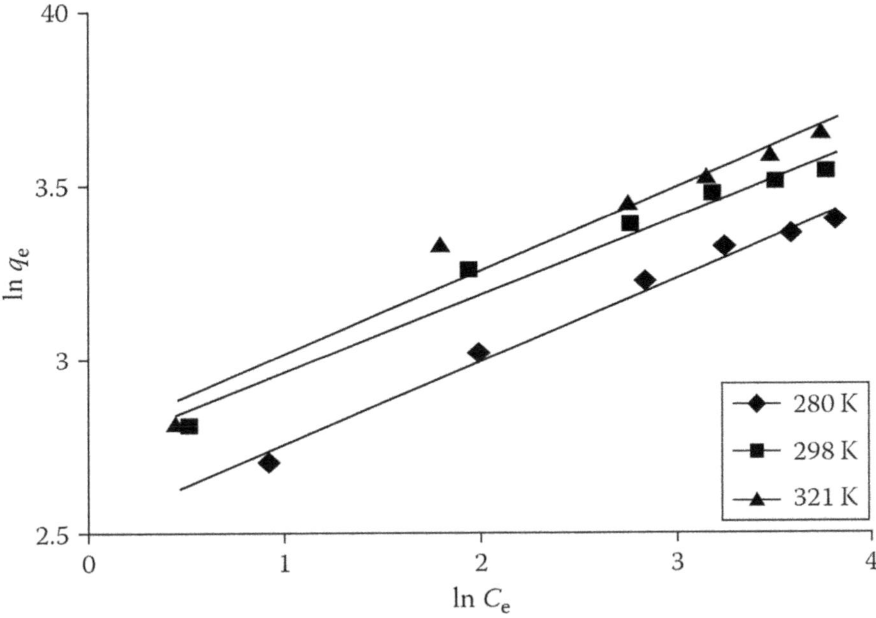

FIGURE 5.25 Linearized Freundlich isotherms for Pb^{2+} adsorption by CNTs at different temperatures. (From Li, Y. et al., *Water Res.*, 39, 605–609, 2005 [10]. With permission.)

increasing sorption of Pb^{2+} with an increase in temperature; a negative sorption standard free energy change and a positive standard entropy change indicates that the sorption reaction is a spontaneous process. The positive standard entropy change may be due to the release of a water molecule produced by the ion exchange reaction between the sorbate and the functional groups on the surfaces of CNTs.

TABLE 5.6

Parameters of Freundlich Adsorption Isotherm Models for Pb^{2+} on CNTs

Temperature (K)	K_F	N	R^2
280	12.4100	4.5269	0.9869
298	15.5646	4.4802	0.9659
321	16.1448	4.3440	0.9755

Source: Li, Y. et al., *Water Res.*, 39, 605–609, 2005 [10]. With permission.

TABLE 5.7

Values of Various Thermodynamic Parameters for Adsorption of Pb^{2+} on CNTs

Temperature (K)	Thermodynamic Constants		
	ΔG° (kcal/mol)	ΔH° (kcal/mol)	ΔS° (J/mol K)
280	−1.329	0.441	6.321
298	−1.453	0.441	6.356
321	−1.588	0.441	6.321

Source: Li, Y. et al., *Water Res.*, 39, 605–609, 2005 [10]. With permission.

To achieve molecular level information on Pb(II) sorption on oxidized MWCNTs at different pH values, the XPS technique is performed to identify the local structures of Pb(II) sorption on oxidized MWCNTs. Typical XPS spectra obtained after Pb(II) sorption on oxidized MWCNTs at pH 6.05 and 8.86 are shown in Figure 5.26a–c. Figure 5.26a shows the high-resolution XPS C1s spectra. The peak of typical graphitic carbon at 284.7 eV represents the C1s binding energy of MWCNTs. There is no difference in the C1s spectra of the two samples, which suggests that the species of carbon is not influenced by pH values. Figure 5.26b shows the high-resolution XPS O1s spectra of the sample around 532.5 eV. With reference to the XPS studies of CNTs, the experimental data show the functional groups present on the surface of MWCNTs: carboxyl oxygen [−O−C=O(H), 533.6 eV] and carbonyl oxygen [=C=O, 530.7 eV] [17]. Figure 5.26b shows that there is a significant difference between the oxygen peaks at pH 6.05 and 8.86. Pb(II) sorption is accompanied by a change in oxygen binding, providing evidence that the oxygen-containing functional groups on the surface of oxidized MWCNTs take part in Pb(II) sorption. The oxygen peak is shifted by 0.55 eV between pH 8.86 and 6.05. This shift is due to the fact that Pb(OH)$_2$ pellets begin to form and cover the adsorbent surface at pH 8.86. The characteristic low-binding energy XPS feature is present in the Pb4f XPS signal at pH 6.05 and 8.86 (Figure 5.26c). This XPS feature is associated with an MWCNT-OPb complex at pH 6.05 and MWCNT-Pb(OH)$_2$ pellet formation at pH 8.86. Figure 5.26c shows that doublets characteristic of Pb at pH 6.05 appear at 139.0 eV (assigned to Pb4f$_{7/2}$) and 143.85 eV (assigned to Pb4f$_{5/2}$), and doublets characteristic of Pb at pH 8.86 appear at 138.6 eV (assigned to Pb4f$_{7/2}$) and 143.45 eV (assigned to Pb4f$_{5/2}$). The peaks observed at 139.0 and 138.6 eV agree with the 139.0 and 138.4 eV values reported for PbC$_2$O$_4$ and Pb(OH)$_2$, respectively [48]. This indicates further complexation of Pb onto oxidized MWCNTs at pH 6.05, and precipitation of Pb occurs at high pH.

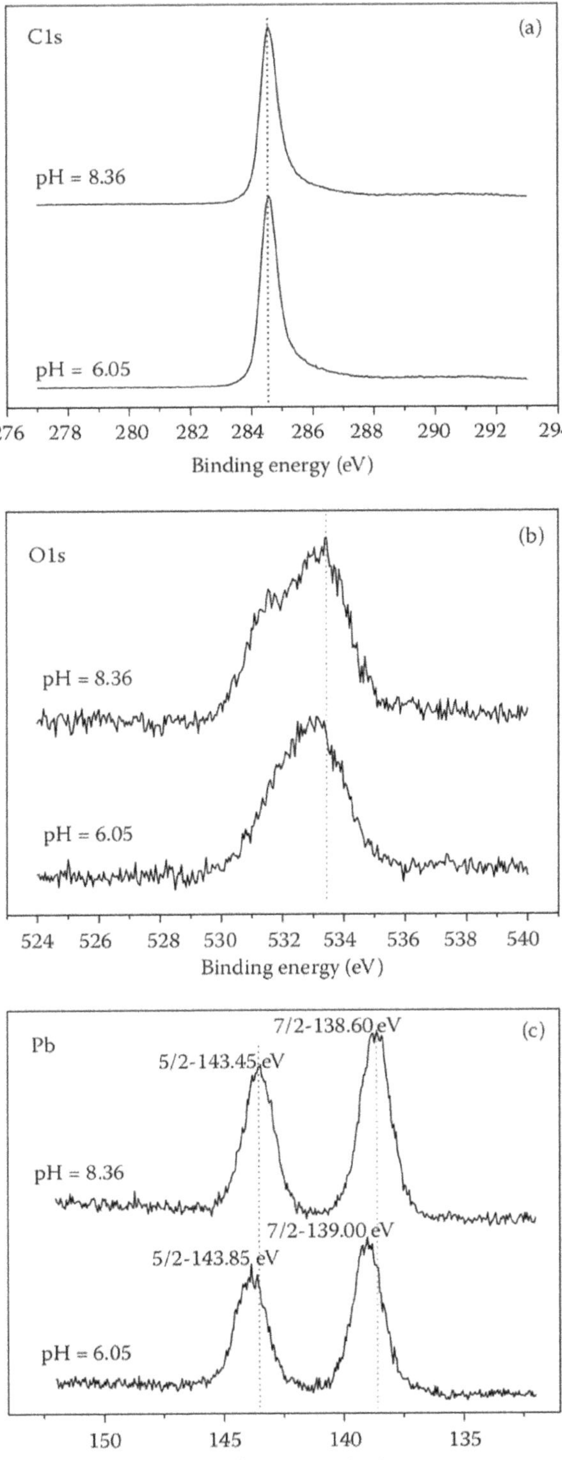

FIGURE 5.26 XPS spectra for free-dried samples of Pb(II) adsorbed on MWCNTs: (a) C1s; (b) O1s; and (c) Pb4f. (From Xu, D. et al., Removal of Pb(II) from aqueous solution by oxidized multiwalled carbon nanotubes. *J. Hazard. Mater.*, 154, 407–416, 2008. With permission.)

5.3.1.2.3   Removal of Cd(II)

Cadmium is considered to be one of the most toxic metals and causes severe health problems to animals and humans [49,50]. The tolerance limits of metals for drinking water are different, depending on the standards in different countries: for example, in the United States, these values are 0.005 mg/L for Cd, 0.1 mg/L for Cr, 1.3 mg/L for Cu, 0.002 mg/L for Hg, and 0.01 mg/L for As and Pb [51]. Cadmium could be introduced into water bodies through wastewaters from metal plating, cadmium-nickel battery industries, phosphate fertilizers, mining residues, and pigments [52].

The removal of Cd(II) from an aqueous solution using activated carbon, CNTs, and other sorbents has been studied extensively. Generally, the adsorption of Cd(II) is strongly dependent on pH values. Because many metal ions are present in the solution, the simultaneous removal of these ions from the solution to nanomaterials is inevitable. As mentioned earlier, the species of metal ions in solution is essential in understanding the physicochemical behavior in the natural environment, and therefore, it is helpful in evaluating the adsorption of metal ions on nanomaterials. The relative species of different metal ions in aqueous solution as a function of pH is shown in Figure 5.27, which, under ambient conditions, shows that the species of metal ions is strongly dependent on pH values and is obviously influenced by the presence of CO_2.

5.3.1.2.4   Removal of Cu(II)

Copper is one of the most widespread heavy metal contaminants in the environment. It is essential to human life and health but, like all heavy metals, is potentially toxic. The major sources of copper in industrial effluents are metal cleaning and electroplating. Figure 5.28 presents the adsorption isotherms of Cu^{2+} on CNTs at pH 6 as a relationship between the amount of Cu^{2+} adsorbed per unit mass of CNTs (q_e) and the equilibrium concentration in solution (C_e) at different temperatures. The adsorption capacities increased with concentration, reaching a plateau, which represents the maximum adsorption capacity. Table 5.8 summarizes the coefficients of the Langmuir and Freundlich isotherms for different temperatures and CNTs. Comparing the R^2 values given in Table 5.8 indicates that the Langmuir isotherm better fits the experimental data on the adsorption of Cu^{2+} onto CNTs than does the Freundlich isotherm. The validity of the Langmuir isotherm suggests that the adsorption is a monolayer process and the adsorptions of all molecules have equal activation energies. The Langmuir constant, K_L, increased with temperature, indicating that the adsorption of Cu^{2+} onto CNTs increased with temperature. The results implied that the affinity of the binding sites for Cu^{2+} increased with temperature. Moreover, K_L values for various CNTs followed the order NaOCl⁻ modified CNTs > HNO_3^- modified CNTs > as-produced CNTs, suggesting that the affinity of the binding sites for Cu^{2+} also followed this order.

The Freundlich isotherm does not describe the saturation behavior of the adsorbent. With regard to the coefficients of the Freundlich model, the Freundlich constant K_F increased with temperature, revealing that the adsorption capacity increased with temperature. Like K_F, the Freundlich constant n increased with temperature. Since the n values obtained from the isotherms all exceeded unity, Cu^{2+} adsorption is favorably adsorbed onto as-produced and modified CNTs at all temperatures. The highest value of n, 7.716 at 320 K for NaOCl-modified CNTs, represents favorable adsorption at high temperatures. Both the Langmuir and the Freundlich isotherms suggest that increasing the temperature increased adsorption capacity, suggesting that the adsorption is endothermic.

Table 5.9 presents the thermodynamic parameters at various temperatures for different CNTs. Positive $DH°$ values indicate that the adsorption of Cu^{2+} onto CNTs is endothermic, which is supported by an increase in the adsorption of Cu^{2+} with temperature. The increasing adsorption capacity of the adsorbent with temperature is attributable to the enlargement of pores and/or the activation of the adsorbent surface [54]. Furthermore, the positive $\Delta S°$ revealed that the degrees of freedom increased at the solid-liquid interface during the adsorption of Cu^{2+} onto CNTs. The $\Delta G°$ values were negative at all of the tested temperatures (280–320 K), verifying that the adsorption of Cu^{2+} onto CNTs was spontaneous and thermodynamically favorable. Restated, a more negative $\Delta G°$ implies

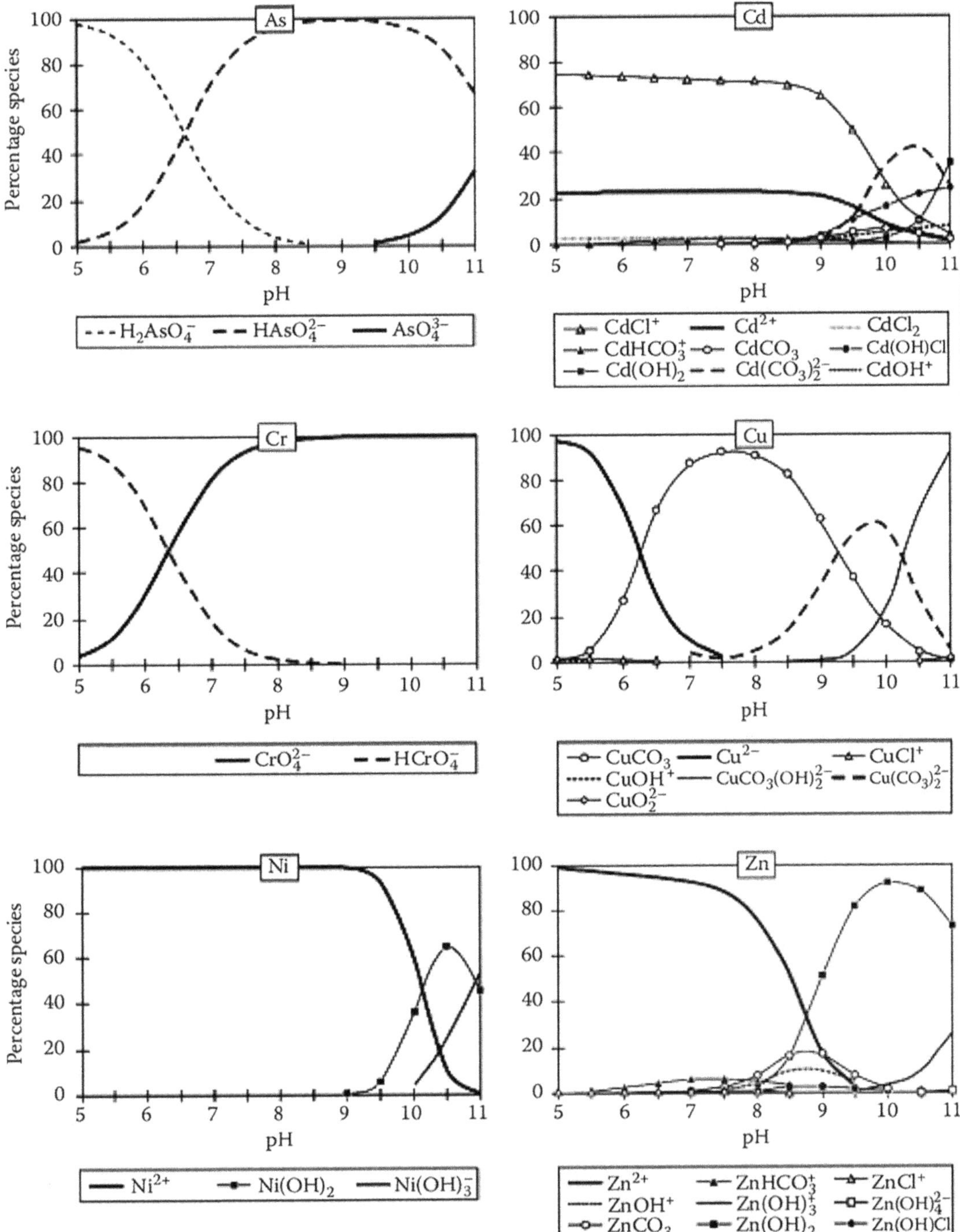

FIGURE 5.27   Speciation of As, Cd, Cr, Cu, Ni, and Zn in water obtained using the PHREEQ-C model with ionic strength of 0.01 M NaCl and 0.003 M NaHCO$_3$ buffer. (From Genc-Fuhrman, H., Mikkelsen, P.S., and Ledin, A., *Water Res.*, 41, 591–602, 2007 [56]. With permission.)

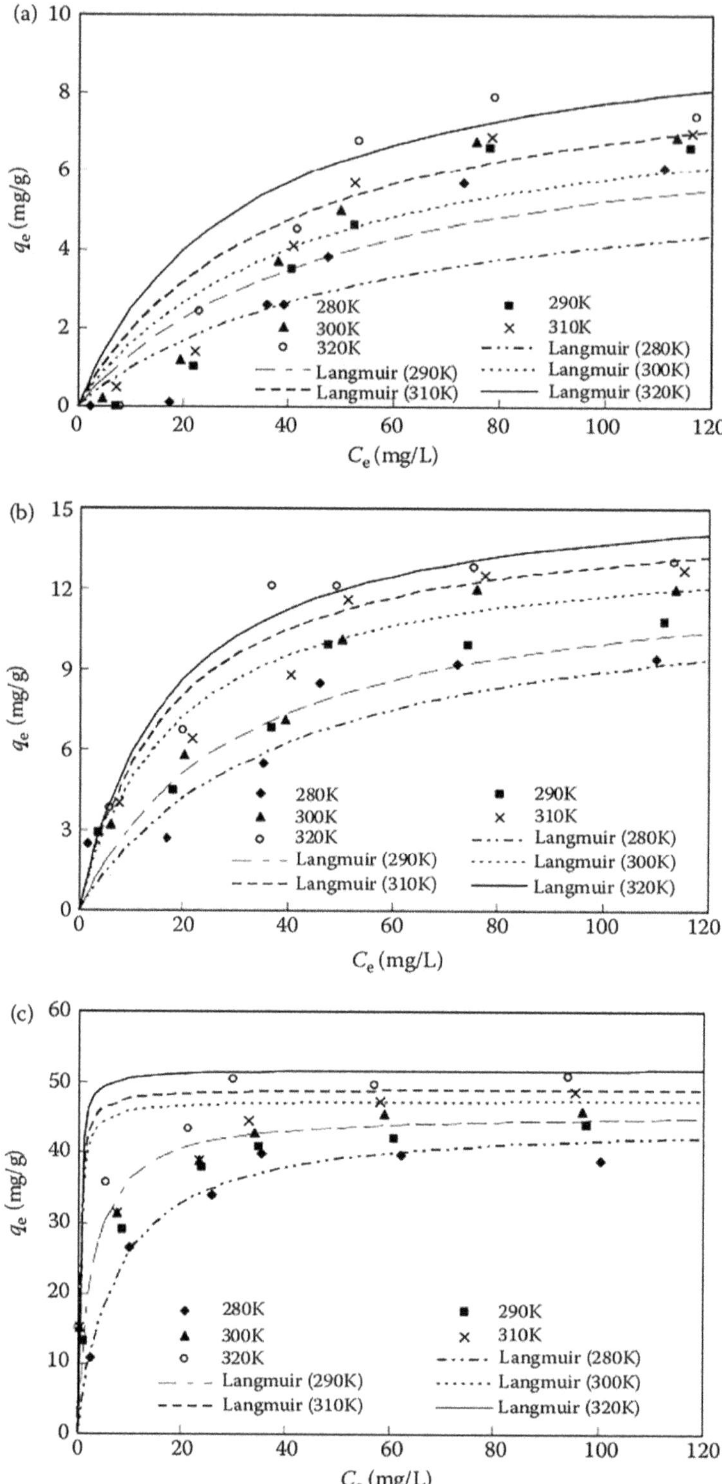

FIGURE 5.28 Effects of temperature for (a) as-produced CNTs, (b) HNO$_3$-modified CNTs, and (c) NaOCl-modified CNTs (pH 6, ionic strength = 0.01 M, CNTs = 0.5 g/L, and contact time = 24 hours). (From Wu, C.H., *J. Colloid Interface Sci.*, 311, 338–346, 2007 [53]. With permission.)

TABLE 5.8

Coefficients of Langmuir and Freundlich Isotherms

Temperature	Langmuir Isotherm			Freundlich Isotherm		
	K_L (L/mg)	q_m (mg/g)	R^2	K_L	n	R^2
As-produced CNTs						
280 K	0.0175	6.39	0.887	0.190	1.317	0.903
290 K	0.0198	7.87	0.910	0.374	1.600	0.862
300 K	0.0234	8.25	0.842	0.498	1.739	0.863
310 K	0.0255	9.34	0.801	0.753	2.058	0.798
320 K	0.0320	10.17	0.645	1.035	2.291	0.611
HNO_3-modified CNTs						
280 K	0.0250	12.46	0.901	0.455	1.451	0.858
290 K	0.0321	13.10	0.925	1.608	2.404	0.937
300 K	0.0542	13.87	0.960	1.542	2.330	0.986
310 K	0.0565	15.11	0.972	3.098	3.238	0.698
320 K	0.0584	16.04	0.945	2.784	2.841	0.699
NaOCl-modified CNTs						
280 K	0.1386	44.64	0.998	9.897	2.893	0.870
290 K	0.3854	45.87	0.997	14.160	3.610	0.937
300 K	2.9515	47.39	0.974	21.020	5.331	0.989
310 K	3.3800	49.02	0.952	22.300	5.540	0.951
320 K	4.1925	51.81	0.933	29.540	7.716	0.872

Source: Wu, C.H., *J. Colloid Interface Sci.*, 311, 338–346, 2007 [53]. With permission.

TABLE 5.9

Thermodynamic Parameters of CNTs at Various Temperatures

T (K)	$\Delta G°$ (kJ/mol)	$\Delta H°$ (kJ/mol)	$\Delta S°$ (J/mol K)
As-produced CNTs			
280	−16.34	10.84	96.89
290	−17.23		
300	−18.24		
310	−19.07		
320	−20.29		
HNO_3-modified CNTs			
280	−17.17	17.08	122.88
290	−18.38		
300	−20.33		
310	−21.12		
320	−21.89		
NaOCl-modified CNTs			
280	−21.16	67.77	319.76
290	−24.38		
300	−30.30		
310	−31.66		
320	−33.26		

Source: Wu, C.H., *J. Colloid Interface Sci.*, 311, 338–346, 2007 [53]. With permission.

a greater driving force of adsorption, resulting in higher adsorption capacity. As the temperature increased from 280 to 320 K, $\Delta G°$ became highly negative, suggesting that adsorption was more spontaneous at high temperatures. Additionally, adsorption by modified CNTs was more spontaneous than that by as-produced CNTs. Hence, the $\Delta G°$ results implied that the adsorption affinity of Cu^{2+} onto modified CNTs was stronger than that onto as-produced CNTs.

5.3.1.2.5 Removal of Zn(II)

Figure 5.29a and b shows the effect of pH on the sorption of Zn^{2+} onto oxidized SWCNTs and MWCNTs, respectively [5,6]. With a C_0 of 10 mg/L, the sorption of Zn^{2+} onto CNTs increased with an increase in pH in the range of 1–8, fluctuated very little and reached a maximum in the pH range of 8–11, and decreased at a pH of 12. It is known that zinc species can be present in deionized water as Zn^{2+}, $Zn(OH)^{+1}$, $Zn(OH)_2^0$, $Zn(OH)_3^-$, and $Zn(OH)_4^{2-}$ [55]. At pH<8, the predominant zinc species is always Zn^{2+} and its removal is mainly accomplished by adsorption reaction. Therefore, the low Zn^{2+} sorption that took place at low pH can be attributed in part to competition between H^+ and Zn^{2+} ions on the same sites [56,57]. Furthermore, the zeta potential of CNTs becomes more negative with increasing pH, which causes electrostatic attraction and thus results in the adsorption of more Zn^{2+} onto CNTs. In the pH range of 8–11, the removal of Zn remained constant and reached a maximum. The main species are $Zn(OH)^+$, $Zn(OH)_2^0$, and $Zn(OH)_3^-$, and thus the removal of Zn is possibly accomplished by the simultaneous precipitation of $Zn(OH)_2(s)$ and sorption of $Zn(OH)^+$ and $Zn(OH)_3^-$. At a pH of 12, the predominant zinc species are the negative species $Zn(OH)_3^-$ and $Zn(OH)_4^{2-}$. Therefore, the decrease in Zn removal that took place at a pH of 12 can be attributed in part to competition among OH-, $Zn(OH)_3^-$, and $Zn(OH)_4^{2-}$ ions on the same sites. With a C_0 of 80 mg/L, the sorption of Zn^{2+} onto CNTs increased with the increase in pH in the pH range of 1–7. Above a pH of 7, a large amount of $Zn(OH)_2(s)$ was formed and suspended in the solutions.

Figure 5.30a and b shows the effect of temperature on the sorption rate of Zn^{2+} by SWCNTs and MWCNTs, respectively. The initial Zn^{2+} concentration is 60 mg/L. For all the experiments, the sorption of Zn^{2+} increased quickly with time and then slowly reached equilibrium in 60 minutes, irrespective of temperature. Similar results have been reported for the sorption of Zn^{2+} onto activated carbon [58]. However, the times required to uptake 50% of the maximum sorption capacity (t_{50}) at 5°C, 15°C, 25°C, 35°C, and 45°C are 3.34, 2.90, 2.75, 2.38, and 2.28 minutes for SWCNTs and 5.04, 4.59, 4.28, 4.12, and 3.84 minutes for MWCNTs. It appears that the half-sorption capacity would be reached faster at a higher temperature. The equilibrium capacities of Zn^{2+} at 5°C, 15°C, 25°C, 35°C, and 45°C are 31.2, 35.6, 36.8, 39.6, and 41.4 mg/g for SWCNTs and 25.5, 27.5, 27.8, 28.5, and 29.5 mg/g for MWCNTs. The solution pH dropped to 0.5 and 0.25 pH units during the sorption of Zn^{2+} onto SWCNTs and MWCNTs, respectively. This could be explained by the release of H^+ ions from the surface site of CNTs, where Zn^{2+} ions are adsorbed and thus lead to a decrease in solution pH. As the temperature increased from 15°C to 25°C, the equilibrium pH value slightly dropped by 0.06 and 0.05 pH units for the SWCNTs and MWCNTs, respectively. This was because, as more Zn^{2+} ions are adsorbed onto CNTs at higher temperatures, more H^+ ions are desorbed from the surface site of CNTs into the solution, which therefore results in a lower equilibrium pH.

The Zn^{2+} adsorption isotherms data are fitted using the Langmuir and Freundlich models. The Langmuir and Freundlich constants were obtained from fitting the isotherm model to the adsorption equilibrium data and are listed in Table 5.10. For Zn^{2+} adsorption onto SWCNTs, the R^2 values of the Langmuir model are higher than those of the Freundlich model. For Zn^{2+} adsorption onto MWCNTs, the R^2 values of both models are very close. The constants a and K_f, which represent Zn^{2+} adsorption capacity, are greater for SWCNTs than for MWCNTs and increase with a rise in temperature. The constant b, which reflects the free energy of adsorption, presents generally the same trend. The slope $1/n$, which is related to the intensity of adsorption, is greater for SWCNTs, indicating the more favorable sorption of Zn^{2+} onto SWCNTs. The Langmuir isotherms of Zn^{2+}

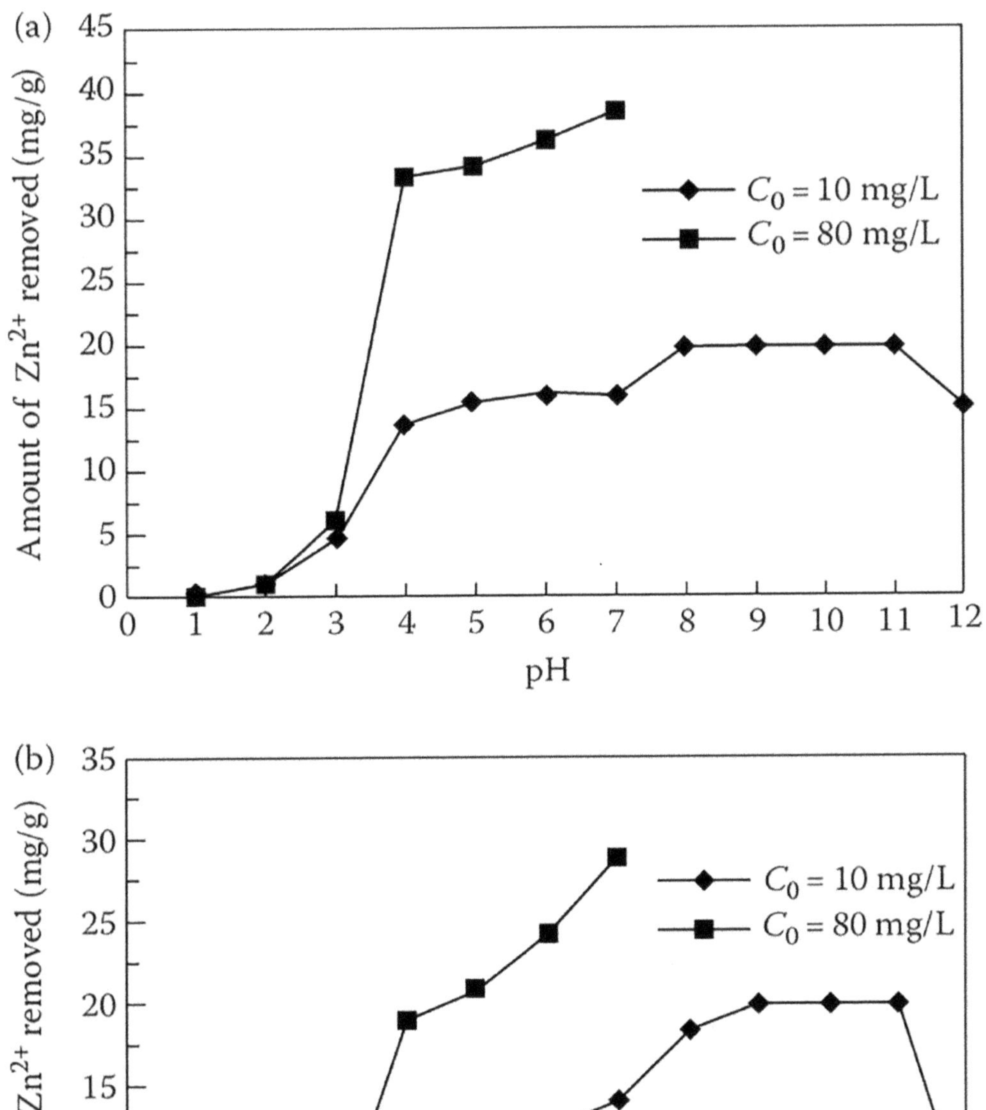

FIGURE 5.29 Effect of pH on the sorption of Zn^{2+} with purified CNTs: (a) SWCNTs and (b) MWCNTs. (From Lu, C. and Chiu, H., *Chem. Eng. Sci.*, 61, 1138–1145, 2006 [5]. With permission.)

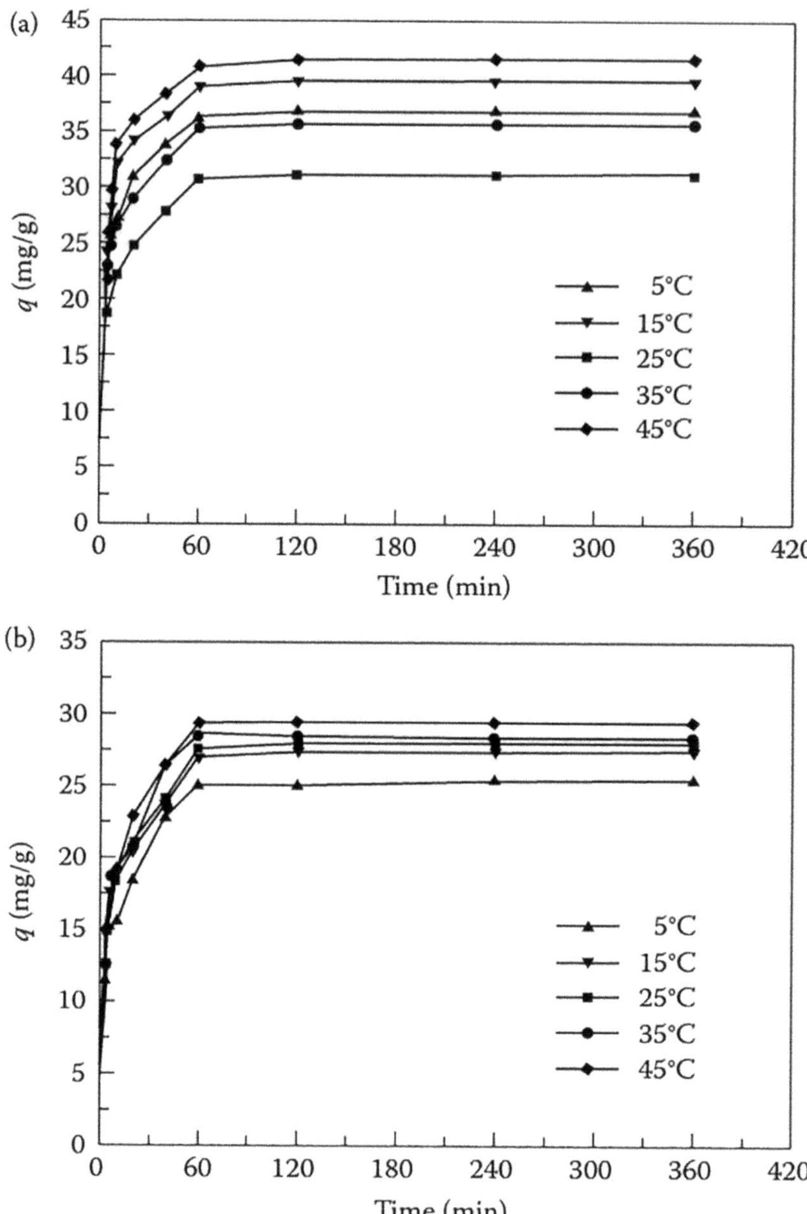

FIGURE 5.30 Effect of temperature on the sorption rate of Zn^{2+} by CNTs: (a) SWCNTs and (b) MWCNTs. (From Lu, C. et al., *Ind. Eng. Chem. Res.*, 45, 2850–2855, 2006 [6]. With permission.)

adsorption onto SWCNTs and MWCNTs at various temperatures are presented in Figure 5.31a and b, respectively. It is obvious that the Zn^{2+} adsorption capacity of CNTs increased with a rise in temperature, and high capacity was observed at 35°C and 45°C. The isotherm curves of 15°C and 25°C are relatively close, probably because of the experimental deviation in the determination of the amount of adsorbed Zn^{2+}. Zn^{2+} adsorption capacity of SWCNTs is greater than that of MWCNTs, which could be explained from the BET measurements, which show that the surface area of SWCNTs (423 m²/g) available for liquid phase mass transfer is higher than that of MWCNTs (297 m²/g). As the temperature increased from 5°C to 45°C, the maximum Zn^{2+} sorption capacity of SWCNTs and MWCNTs calculated by the Langmuir model increased from 37.03 to 46.94 mg/g

TABLE 5.10

Constants of Langmuir and Freundlich Models of Zn^{2+} Adsorption by CNTs at Various Temperatures

CNTs	Temperature (°C)	Langmuir Model			Freundlich Model		
		a	b	R^2	K_f	$1/n$	R^2
SWCNTs	5	37.03	0.132	0.997	9.97	0.303	0.979
	15	40.65	0.159	0.997	12.01	0.288	0.984
	25	41.84	0.161	0.996	12.37	0.289	0.984
	35	45.45	0.164	0.998	12.75	0.305	0.976
	45	46.94	0.178	0.997	13.58	0.302	0.971
MWCNTs	5	30.30	0.101	0.983	8.23	0.283	0.979
	15	31.74	0.124	0.987	10.43	0.242	0.983
	25	33.33	0.118	0.982	10.62	0.248	0.984
	35	33.78	0.129	0.979	12.15	0.219	0.972
	45	34.36	0.149	0.985	12.87	0.215	0.984

Source: Lu, C. et al., *Ind. Eng. Chem. Res.*, 45, 2850–2855, 2006 [6]. With permission.

Notes: Units: $a = mg/g$; $b = L/mg$, $K_f = (mg/g) (L/mg)^{1/n}$; $n = $ dimensionless; and $R = $ dimensionless.

and from 30.3 to 34.36 mg/g, respectively. These values are much greater than that of commercially available PAC (13.5 mg/g; San Ying Enterprises Co., Taipei, Taiwan) measured at 25°C in this study. This suggests that both MWCNTs and SWCNTs are efficient adsorbents for removing Zn^{2+} from an aqueous solution.

Figure 5.32 shows the Zn^{2+} recoveries of CNTs under various pH values of the regeneration solution ranging from 1 to 5. Zn^{2+} recovery is defined as the percentage ratio of the Zn^{2+} adsorption capacity of regenerated sorbents to that of virgin sorbents. The Zn^{2+} recovery of SWCNTs and MWCNTs reached 91.1% and 90.7% at a solution pH of 1 and decreased to 62.28% and 43.26% at a solution pH of 5. This could be explained by the fact that the surface charge of CNTs becomes more negative with an increase in the pH value of the solution, which causes more electrostatic attractions between the Zn^{2+} ions and the surface of CNTs and thus results in lower Zn^{2+} recovery.

Figure 5.33 shows the Zn^{2+} sorption capacities of SWCNTs, MWCNTs, and PAC, whereas Figure 5.34 displays the Zn^{2+} recoveries of SWCNTs, MWCNTs, and PAC under various regeneration cycles (RCs) of 0, 1, 5, and 10. As the RC increased, the amount of adsorbed Zn^{2+} and the Zn^{2+} recovery of SWCNTs and MWCNTs slightly decreased, but those of PAC sharply decreased. This could be explained by the fact that CNTs have no porous structure like PAC, in which Zn^{2+} ions have to move from the inner surface to the exterior surface of the pores on PAC, which makes desorption of Zn^{2+} ions from the surface site of CNTs much easier. The amounts of adsorbed Zn^{2+} under RCs of 0, 1, 5, and 10, respectively, are 34.1, 30.9, 28.8, and 27.25 mg/g for SWCNTs; 26.8, 24.2, 22.4, and 20.59 mg/g for MWCNTs; and 11.8, 4.8, 2.18, and 1.55 mg/g for PAC.

The Zn^{2+} recoveries under RCs of 1, 5, and 10, respectively, are 90.61%, 84.46%, and 79.91% for SWCNTs; 90.30%, 83.58%, and 76.83% for MWCNTs; and 40.68%, 18.47%, and 13.10% for PAC. It is evident that the Zn^{2+} ions would be easily removed from the surface sites of SWCNTs and MWCNTs by a 0.1-mol/L HNO_3 solution, and the adsorption capacity is maintained after ten cycles of the adsorption-desorption process. This indicates that SWCNTs and MWCNTs can be reused through many cycles of water treatment and regeneration for Zn^{2+} removal from an aqueous solution. The cost for the replacement of sorbents is, thus, greatly reduced. This is the key factor for whether a novel but expensive adsorbent can be accepted by the field or not. It is expected that the

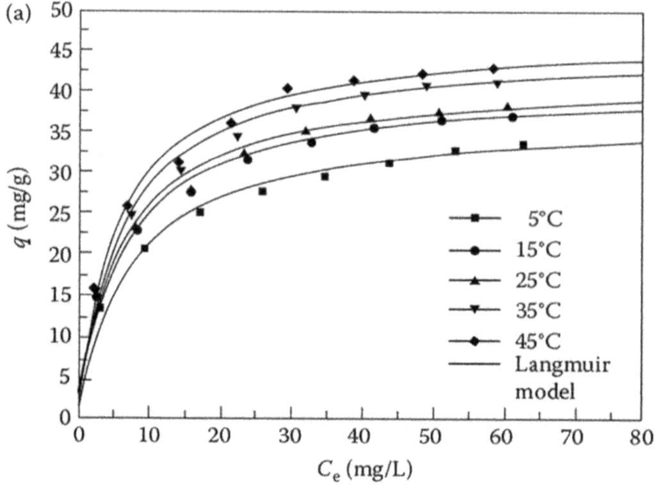

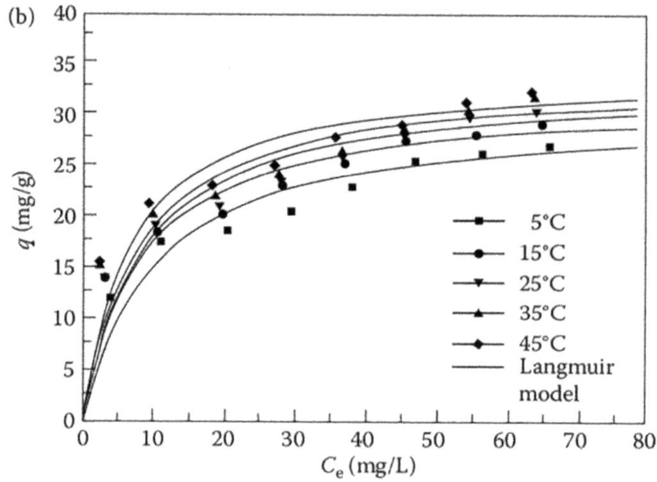

FIGURE 5.31 Langmuir isotherms of Zn^{2+} sorption by CNTs at various temperatures: (a) SWCNTs and (b) MWCNTs. (From Lu, C. et al., *Ind. Eng. Chem. Res.*, 45, 2850–2855, 2006 [6]. With permission.)

unit cost of CNTs can be further reduced in the future so that CNTs can possibly be cost-effective sorbents.

Table 5.11 lists comparisons of CNT characterizations and their sorption capacities of various divalent metal ions (Cd^{2+}, Cu^{2+}, Ni^{2+}, Pb^{2+}, Zn^{2+}). It is evident that the metal ion sorption capacity of CNTs does not have a direct correlation with their specific surface area, pore specific volume, and mean pore diameter but strongly depends on their surface total acidity. The metal ion sorption capacity of CNTs increased as a rise in the amount of surface total acidity (including carboxyls, lactones, and phenols) present on the surface site of CNTs. This reflects that the sorption of metal ions onto CNTs is by the chemisorption process rather than the physisorption process. Oxidized CNTs have more surface total basicities, which are responsible for the sorption of anions from aqueous solutions, than raw CNTs.

Competitive sorption is important in water and wastewater treatment because most metal ions to be sorbed exist in solution with other sorbable metal ions. Figure 5.35 shows the effect of CNT dosage on the competitive sorption of Pb^{2+}, Cu^{2+}, and Cd^{2+} by CNTs. The sorption capacities of all three heavy metal ions increase with increasing CNT dosages. The sorption percentages of Pb^{2+},

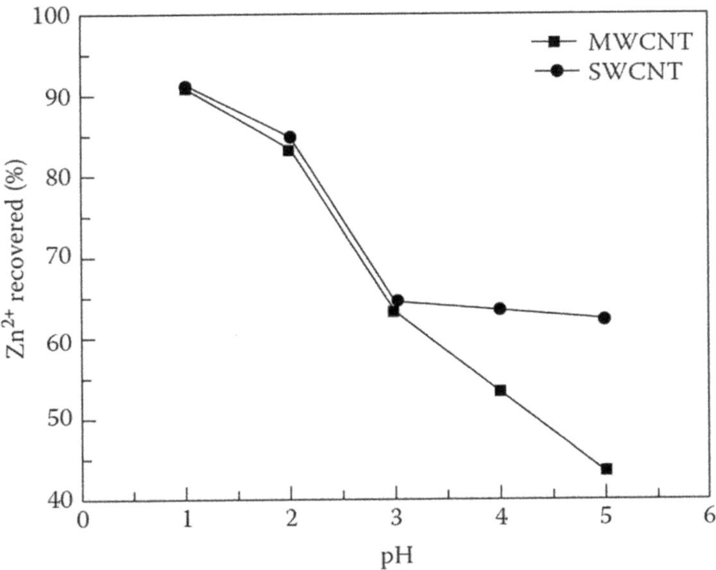

FIGURE 5.32 Effect of pH of the solution on the Zn^{2+} recovery of CNTs. (From Lu, C. et al., *Ind. Eng. Chem. Res.*, 45, 2850–2855, 2006 [6]. With permission.)

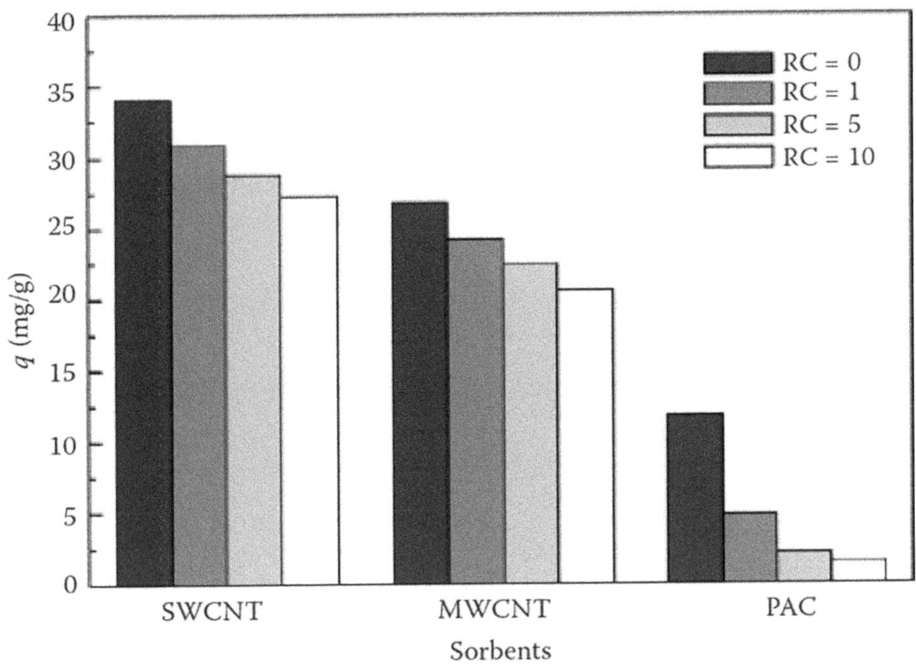

FIGURE 5.33 Zn^{2+} sorption capacities of CNTs and PAC under various RCs of 0, 1, 5, and 10. (From Lu, C. et al., *Ind. Eng. Chem. Res.*, 45, 2850–2855, 2006 [6]. With permission.)

Cu^{2+}, and Cd^{2+} are 56.1%, 23.7%, and 1.3%, respectively, at a CNT dosage of 0.05 g. The former two reach almost 100% sorption at a CNT dosage of 0.3 g/L. The increase of Cd^{2+} sorption percentage was slow at first and attained 75.4% at a CNT dosage of 0.3 g/L. At that point, Pb^{2+} and Cu^{2+} ions are almost adsorbed completely and there are more vacant active sorption sites used for Cd^{2+} sorption. In other words, the affinities of CNTs for Pb^{2+}, Cu^{2+}, and Cd^{2+} follow the order $Pb^{2+} > Cu^{2+} > Cd^{2+}$.

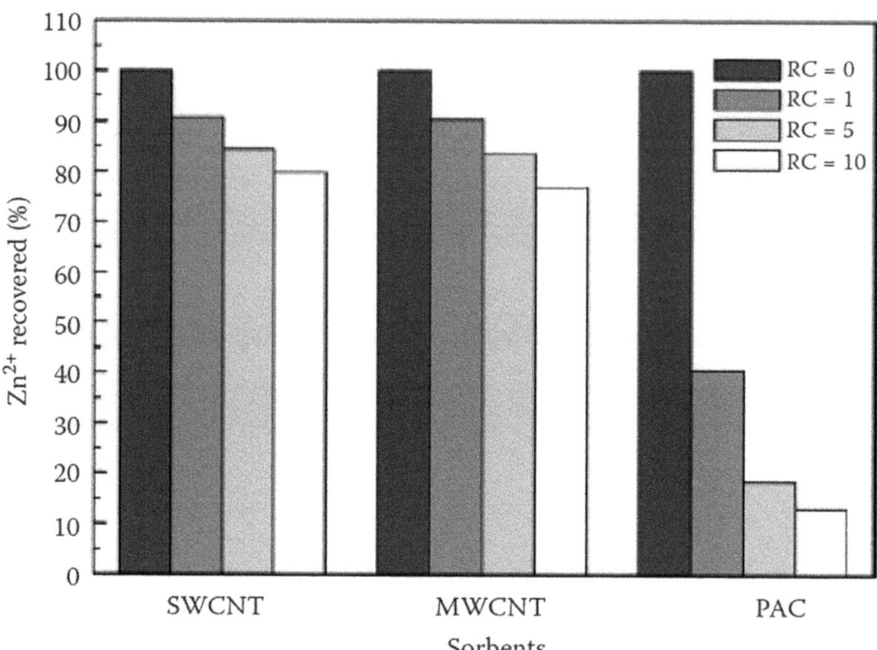

FIGURE 5.34 Zn^{2+} recoveries of CNTs and PAC under various RCs of 0, 1, 5, and 10. (From Lu, C. et al., *Ind. Eng. Chem. Res.*, 45, 2850–2855, 2006 [6]. With permission.)

TABLE 5.11

Comparisons of CNT Characterizations and Maximum Sorption Capacities of Various Divalent Metal Ions

Metal Ions	Sorbents	SA	PV	MPD	STA	STB	q_m
Ni^{2+}	SWCNTs	577	1.15	7.98	0.54	0.23	9.22
	SWCNTs (NaOCl)	397	0.46	4.62	4.42	0.35	47.85
	MWCNTs	448	1.10	8.26	0.44	0.19	7.53
	MWCNTs (NaOCl)	307	0.39	5.21	3.06	0.31	38.46
Pb^{2+}	CNTs (HNO_3)						
	Xylene-Fe	47	0.18	3.40	1.63		14.80
	Benzene-Fe	62	0.26	2.4–3.2	1.65		11.20
	Propylene-Ni	154	0.58	3.60	4.04		59.80
	Methane-Ni	145	0.54	3.60	4.31		82.60
Zn^{2+}	SWCNTs	590	1.12	7.60			11.23
	SWCNTs (NaOCl)	423	0.43	4.12			43.66
	MWCNTs	435	091	8.35			10.21
	MWCNTs (NaOCl)	297	0.38	5.17			32.68
Cd^{2+}	As-grown CNTs	122	0.28	3.60			1.10
	CNTs (H_2O_2)	130	0.36	3.60	2.52		2.60
	CNTs ($KMnO_4$)	128	0.32	3.60	3.36		11.00
	CNTs (HNO_3)	154	0.58	3.60	4.04		5.10

Source: Lu, C. and Liu, C., *J. Chem. Technol. Biotechnol.*, 81, 1932, 2006 [3]; Li, Y.H. et al., *Carbon*, 41, 1057, 2003 [4]; Lu, C. and Chiu, H., *Chem. Eng. Sci.*, 61, 1138, 2006 [5]; Li, Y.H. et al., D*iamond Relat. Mater.*, 15, 90, 2006 [7].

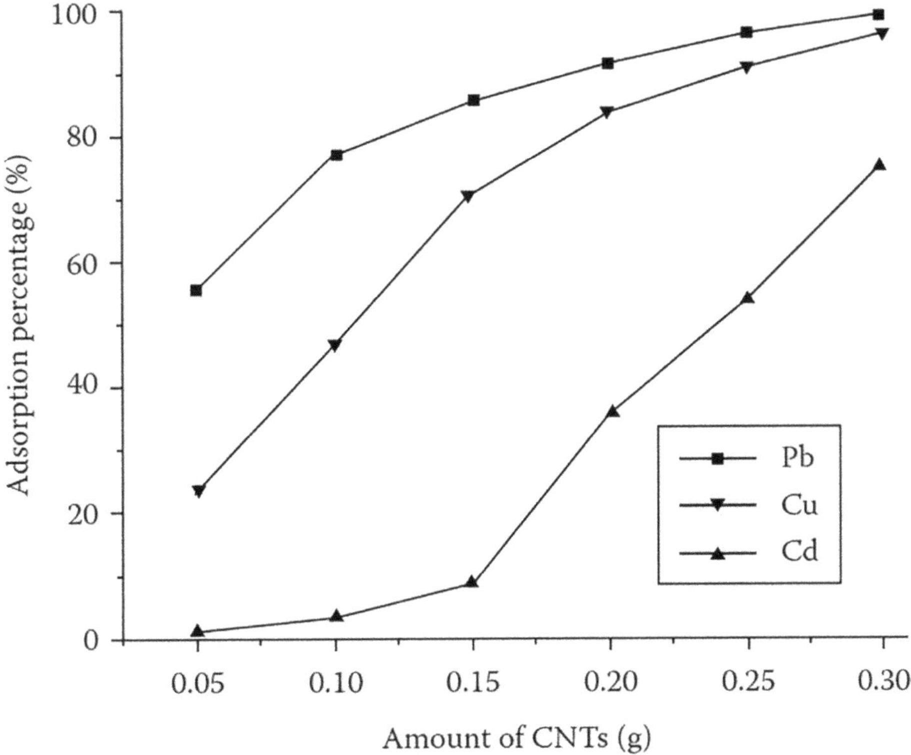

FIGURE 5.35 Effect of CNT dosage on the competitive sorption of Pb^{2+}, Cu^{2+}, and Cd^{2+} ions onto CNTs at room temperature and initial ion concentration of 30 mg/L. (From Li, Y. et al., *Carbon*, 41, 2787–2792, 2003 [11]. With permission.)

The simultaneous removal of Cd, Cu, Ni, and Zn using the sorbents are shown in Figure 5.36 on a double logarithmic scale. The amount of heavy metal removal increased with increasing initial heavy metal concentration. The simultaneous sorption of heavy metal ions depends not only on the sorbent properties but also on the heavy metal ions' properties. Generally, similar species and similar radii of metal ions have very similar physicochemical behaviors.

5.3.2 KINETIC SORPTION AND DESORPTION OF $^{152+154}$EU(III)

With the development of nuclear power, large volumes of high-level radioactive nuclear waste are being produced. The removal of long-lived radionuclides from nuclear waste solutions is an important environmental concern in nuclear waste management.

Figure 5.37 shows the adsorption of Eu(III) as a function of Eu(III) concentration at two different ionic strengths, 0.1 and 0.01 M $NaClO_4$. The ideal linear adsorption isotherms suggest that the adsorption of $^{152+154}$Eu(III) on MWCNTs is far from saturation, although the concentration of MWCNTs is quite low (0.5 g nanotubes/L solution), and the initial $^{152+154}$Eu(III) concentration is quite high (the maximum initial concentration 2.8×10^{-5} mol/L). The high adsorption capacity of MWCNTs indicates that MWCNTs are very suitable materials for the preconcentration of lanthanides/actinides from large volumes of aqueous solutions.

To study the reversibility-irreversibility of Eu(III) adsorption on MWCNTs, kinetic desorption analysis of Eu(III) from MWCNTs was investigated after different contact times of Eu(III) with MWCNTs (0.5 g/L MWCNTs, initial Eu(III) concentration 1.2×10^{-5} mol/L, pH 6.3 ± 0.2, and in 0.1 M $NaClO_4$). A purified filter membrane coated with a resin containing imminodiacetic

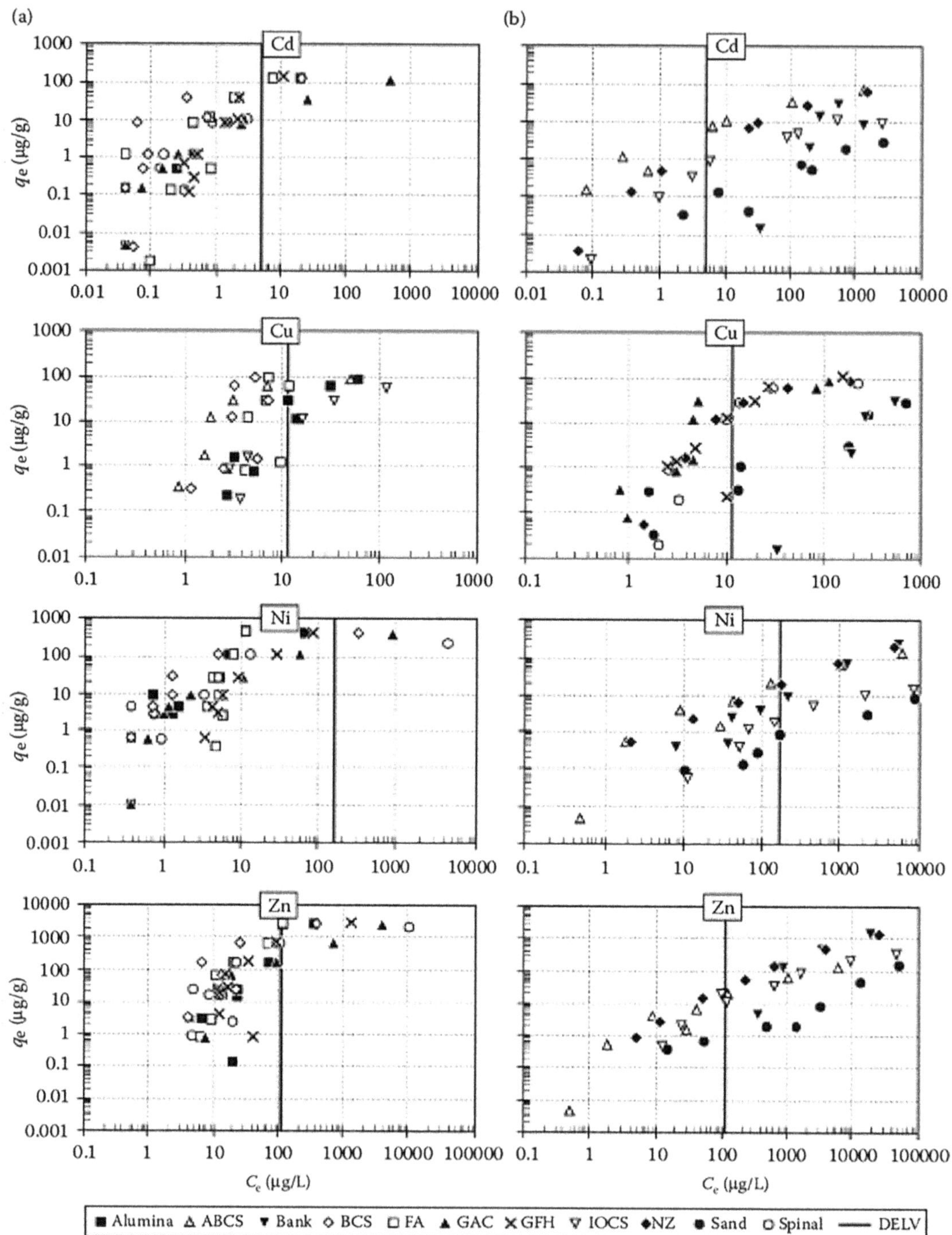

FIGURE 5.36 Cd, Cu, Ni, and Zn removal using (a) sorbents with high heavy metal removal efficiency and (b) sorbents with moderate or low heavy metal removal efficiency, with a starting pH of 6.5, a sorbent dosage of 20 g/L, an ionic strength of 0.01 M NaCl, and 0.003 M NaHCO$_3$ buffer. DELV is Danish emission limit values for fresh surface waters. (From Genc-Fuhrman, H. et al., *Water Res.*, 41, 591–602, 2007 [56]. With permission.)

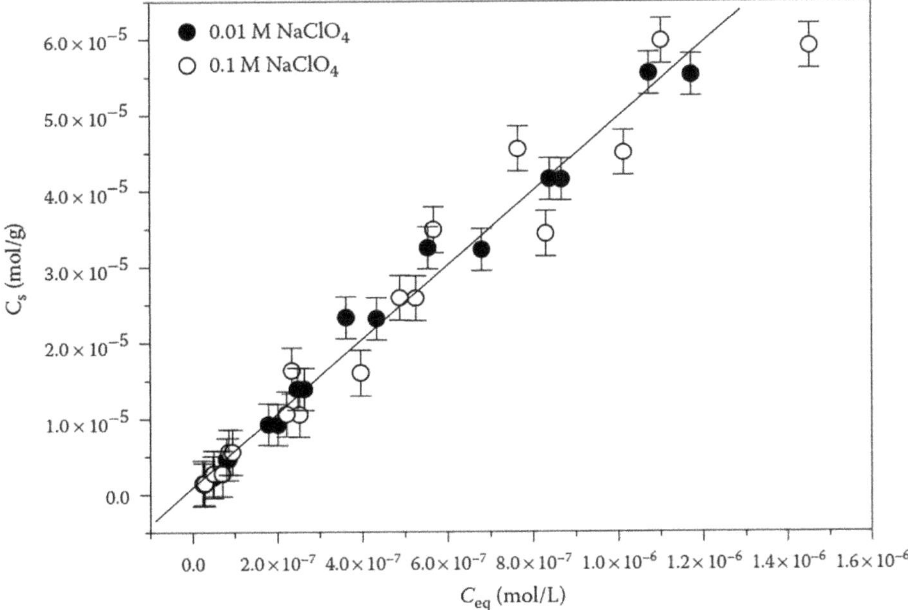

FIGURE 5.37 Sorption isotherms of Eu(III) on MWCNTs as a function of ionic strength and Eu(III) solution concentrations. 25°C, pH = 6.3 ± 0.2, t = 4 days, 0.5 g MWCNTs/L solution. (From Tan, X.L. et al., *Radiochim. Acta*, 96, 23–29, 2008 [20]. With permission.)

acid groups (Chelating Extraction Disk, 3M Empore) was added to the solution after previous equilibration with the electrolyte, and the pH was adjusted to 6.3 (Figure 5.38b) and 3.2 (Figure 5.38c). A fast reaction was found for the adsorption of Eu(III) to the resin in the absence of nanotubes (Figure 5.38a). More than 99.99% of the Eu(III) can form very strong complexes with the chelating resin. Adsorption of Eu(III) on the chelating resin was about 100% for 1.2×10^{-5} mol/L Eu(III). Chelating resin can form strong complexes with Eu(III), and therefore can be used to investigate the kinetic desorption of Eu(III) from MWCNTs. The adsorption/complexation reaction of Eu(III) with the chelating resin in the absence of MWCNTs, according to [20,59], is rather fast.

$$Eu^{3+} + Na/H - Resin \rightarrow Eu - Resin + 3Na^{+}/H^{+}, \tag{5.8}$$

The desorption reaction of Eu(III) from MWCNTs is observed in the following equation:

$$Eu - MWCNTs \xrightarrow{k_1} Eu^{3+} + MWCNTs + \cdots + \frac{Na}{H} - Resin$$
$$\xrightarrow{k_2} Eu - Resin + 3Na^{+}/H^{+}. \tag{5.9}$$

After the chelating resin is added to the MWCNT suspension, the free Eu(III) ion in solution is firstly adsorbed by chelating resin and then the adsorbed Eu(III) on the MWCNT surface is desorbed from MWCNTs and quickly adsorbed by chelating resin. Equation (5.9) is the reaction of Eu(III) from MWCNTs to chelating resin. In Equation (5.9), k_2 is rather fast, whereas k_1 is very slow compared to the reaction of free Eu(III) with chelating resin.

To quantify the differences in Eu(III) desorption behavior, we described the kinetic desorption of Eu(III) from MWCNTs using pseudo-first-order kinetics [20,59]. At least two different Eu(III)

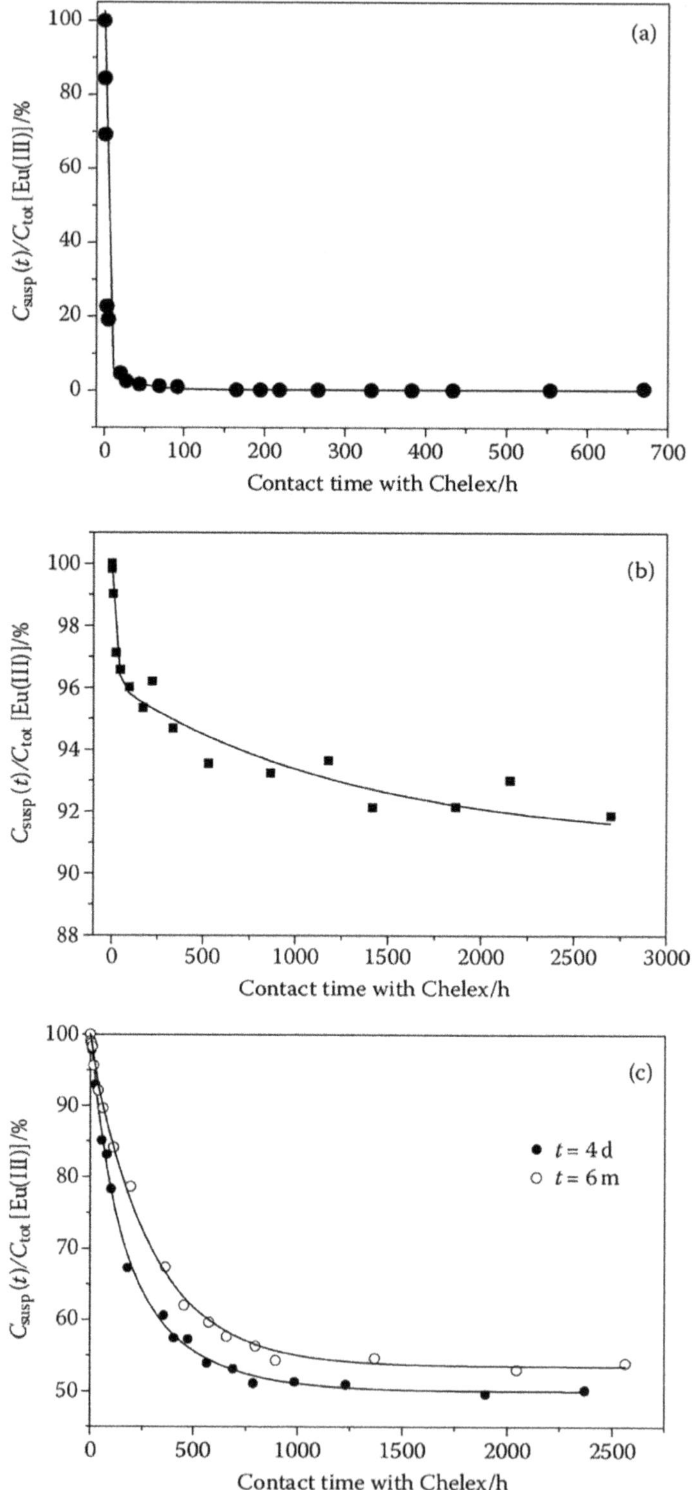

FIGURE 5.38 Fractions of [152+154]Eu(III) species in nanotube suspensions as a function of contact time with a chelating resin in 0.1M NaClO$_4$. (a) Free [152+154]Eu(III) solution, pH 5.0±0.2; (b) presence of CNTs, pH 6.3±0.2, t=4 days; (c) presence of CNTs, pH 3.2±0.1, the pH of the solution before the addition of chelating resin was 6.3±0.2, and then the pH was adjusted to 3.2±0.1 after the addition of chelating resin. (From Tan, X.L. et al., *Radiochim. Acta*, 96, 23–29, 2008 [20]. With permission.)

complexation species, showing "fast" and "slow" desorption kinetics, are used to simulate the kinetic desorption data:

$$\frac{C_{\text{susp}}(t)}{C_{\text{tot}}} = \frac{C_{\text{final}}}{C_{\text{tot}}} + A_1 \cdot \exp\left(\frac{-t}{\tau_1}\right) + A_2 \cdot \exp\left(\frac{-t}{\tau_2}\right), \tag{5.10}$$

where $C_{\text{susp}}(t)$ is the concentration of $^{152+154}$Eu(III) in the MWCNT suspension at time t (h); C_{final} is the final concentration of $^{152+154}$Eu(III) in the MWCNT suspension after equilibration with the chelating resin, which is considered as the portion of the irreversible Eu(III) adsorption on MWCNTs; and C_{tot} is the initial $^{152+154}$Eu(III) concentration in the system ($t=0$). Generally, the fraction of A_1 is considered as a "weak" binding fraction, and the fraction of A_2 is considered as a "strong" binding fraction. The adsorbed Eu(III) on "weak" sites can be "fast" desorbed from MWCNTs with a time constant τ_1 (%), whereas that adsorbed on "strong" sites can be "slow" desorbed from MWCNTs with a time constant τ_1 (%). τ_1 and τ_2 are desorption time constants (h).

The different systems were studied in the kinetic desorption of Eu(III) from MWCNTs and named as follows: aging time $t=4$ days, pH was adjusted to 6.3 ± 0.2 in the desorption study (Figure 5.38b); aging time $t=4$ days, pH was adjusted to 3.2 ± 0.1 in the desorption study (Figure 5.38c, solid points); and aging time $t=6$ months, pH was adjusted to 3.2 ± 0.1 in the desorption study (Figure 5.38c, open points). In the presence of MWCNTs, the initial Eu(III) concentration first drops rapidly at the initial time and then decreases very slowly. The rapid decrease at the beginning is related to the free Eu(III) not complexed with the MWCNTs in solution. The kinetic parameters obtained by fitting the experimental data to Equation (5.10) are listed in Table 5.12. The slow desorption kinetic of Eu(III) is attributed to the very slow desorption kinetics of Eu(III) from the MWCNTs. Even after more than 100 days of contact with the chelating resin, $(91.0\pm0.5)\%$ (pH 6.3 ± 0.2), $(50.0\pm0.4)\%$ ($t=4$ days, pH 3.2 ± 0.1), and $(53.5\pm0.5)\%$ ($t=6$ m, pH 3.2 ± 0.1) of Eu(III) still remains bound to MWCNTs. The experimental results indicate that Eu(III) forms kinetically stabilized complexes with MWCNTs and is not easily desorbed from MWCNTs. This finding clearly proves the existence of strong chemical binding of Eu(III) with nanotubes, and europium cannot easily be released from the surface of CNTs to solution.

Figure 5.38c shows that for the two different aging times (4 days and 6 months), the final concentration of Eu(III) remaining on MWCNTs (C_{final}) for 6 months is higher than that for 4 days. This indicates that Eu(III) is bound to MWCNTs in at least two forms, and the irreversible fractions increase with increasing aging time. The results are very similar to water conduction through the channels of CNTs [19]. The microstructure of adsorbed Eu(III) on the surface of MWCNTs perhaps changes with increasing aging time, the fractions of Eu(III) in the center channels of CNTs increase with increasing aging time, and the Eu(III) remaining in the channels of CNTs is irreversible to be desorbed from CNTs. This is a reasonable interpretation of the different final fractions of Eu(III) remaining on MWCNTs for 4 days and 6 months of aging time.

TABLE 5.12

Kinetic Parameters Obtained for Desorption Experiments by Fitting Experimental Data to the Kinetic Equation (6.10)

System	$C_{\text{final}}/C_{\text{tot}}$ (%)	τ_1 (h)	A_1 (%)	τ_2 (h)	A_2 (%)
Eu-Resin	0.3±0.8	27.6±7.3	8.1±0.4	2.0±0.3	94.2±0.2
Eu-MWCNTs-Resin (pH=6.3, $t=4$ days)	91.0±0.5	20±8	3.8±0.5	1296±210	5.1±0.4
Eu-MWCNTs-Resin (pH=3.2, $t=4$ days)	50.0±0.4	99±34	22.4±1.3	312±72	27.5±1.4
Eu-MWCNTs-Resin (pH=3.2, $t=6$ months)	53.5±0.5	10±13	1.8±1.0	297±10	44.7±0.9

Source: Tan, X.L. et al., *Radiochim. Acta*, 96, 23–29, 2008 [20]. With permission.

The fractions of A_1 are $(22.4 \pm 1.3)\%$ and $(1.8 \pm 1.0)\%$ for 4 days and 6 months of aging time, respectively; those of A_2 are $(27.5 \pm 1.4)\%$ and $(44.7 \pm 0.9)\%$ for 4 days and 6 months of aging time, respectively. The fraction of Eu(III) on "weak" sites decreases, whereas that of Eu(III) on "strong" sites increases with increasing aging time. This indicates that the surface-adsorbed Eu(III) on MWCNTs can be changed from "weak" to "strong" sites with increasing aging time. The results indicate the two changes in the microstructure on the surface and Eu(III) fraction with aging time.

The kinetic results suggest that metal ions adsorbed on MWCNTs are difficult to be desorbed from the CNTs at similar conditions to the sorption experiments. However, if the material is placed in solutions of low pH, the adsorbed metal ions on CNTs can be desorbed from solid surfaces. The fraction of metal ions that can be desorbed from solid surfaces is dependent on pH values and increases with decreasing pH.

5.3.3 EFFECT OF ORGANIC MATERIAL

In the natural environment, the presence of natural organic matters (such as humic acid or fulvic acid) influences the physicochemical behavior of metal ions because of the strong complexation capacity of organic matters with metal ions. The presence of humic substances obviously influences the species of metal ions. In this study, we present the results of Eu(III) species in the natural environment in the absence and presence of humic substances as a comparison. Table 5.13 lists the relative thermodynamic constants for the calculation of Eu(III) species, and the relative species of Eu(III) are shown in Figure 5.39.

Figure 5.39 clearly shows that the species of Eu(III) in the natural environment are strongly dependent on pH values and humic substances. The sorption behavior of metal ions is affected by the species of metal ions in the solution. Therefore, the presence of humic substances affects the sorption of metal ions on nanomaterials. Generally, the presence of humic substances enhances metal ion sorption at low pH values and decreases metal ion sorption at high pH values. The increase of sorption is explained by the sorption of humic substances onto the solid surface followed by the

---

TABLE 5.13

Thermodynamic Constants Used for Speciation Calculations

Reaction	log K
$H_2O \Leftrightarrow OH^- + H^+$	-13.79
$H_2CO_3 \Leftrightarrow 2H^+ + CO_3^{2-}$	-17.43
$Eu^{3+} + CO_3^{2-} + HA \Leftrightarrow \left[Eu(CO_3)HA\right]$	12.4
$Eu^{3+} + 2\,H_2O + HA \Leftrightarrow [Eu(OH)_2HA] + 2H^+$	-10.03
$Eu^{3+} + H_2O + HA \Leftrightarrow [Eu(OH)HA] + H^+$	-0.96
$Eu^{3+} + HA \Leftrightarrow [EuHA]^{3+}$	6.4
$Eu^{3+} + 3\,CO_3^{2-} \Leftrightarrow \left[Eu(CO_3)_3\right]^{3-}$	13.53
$Eu^{3+} + 2\,CO_3^{2-} \Leftrightarrow \left[Eu(CO_3)_2\right]^-$	10.81
$Eu^{3+} + CO_3^{2-} \Leftrightarrow \left[Eu(CO_3)\right]^+$	6.38
$Eu^{3+} + NO^- \Leftrightarrow [Eu(NO_3)]^{2+}$	1.22
$Eu^{3+} + 2\,H_2O \Leftrightarrow [Eu(OH)_2]^+ + 2H^+$	-16.15
$Eu^{3+} + H_2O \Leftrightarrow [Eu(OH)]^{2+} + H^+$	-8.08

Source: Geckeis, H. et al., *Environ. Sci. Technol.*, 36, 2946–2952, 2002 [60]. With permission.

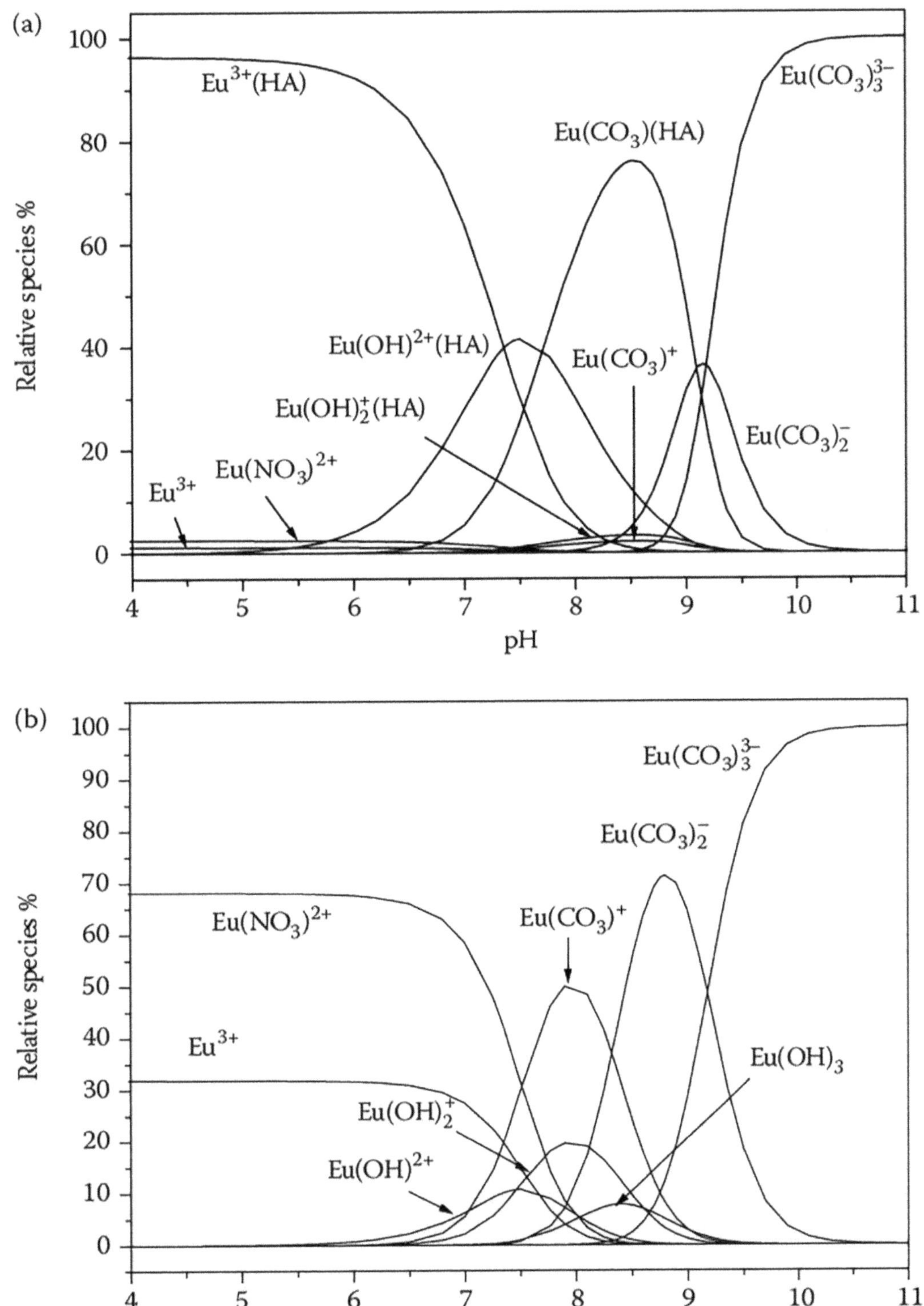

FIGURE 5.39 Relative species distribution of Eu(III) in solution in the presence of humic acid (HA) (a) and the absence of HA (b). I=0.1 mol/L KNO$_3$, T=20±1°C under ambient conditions. (From Tan, X.L. et al., *Environ. Sci. Technol.*, 42, 6532–6537, 2008 [61]. With permission.)

interaction of metal ions with surface-adsorbed humic substances, whereas the reduction of sorption is explained by the formation of soluble HS-metal ion complexes, which stabilize the metal ions in aqueous solution. Generally, humic substances are negatively charged and are easily adsorbed on solid surfaces at low pH values because of the positive surface charge at $pH < pH_{pzc}$ (point of zero charge). At $pH > pH_{pzc}$, the sorption of humic substances on solid surfaces decreases with increasing pH because of the negative surface charge at $pH > pH_{pzc}$.

To the best of our knowledge, the simultaneous sorption of heavy metal ions and organic pollutants on CNTs is still unavailable. The removal of organic pollutants or heavy metal ions from aqueous solutions using all kinds of CNTs has been extensively studied; the removal of heavy metal ions to CNTs in the presence of organic pollutants is still not reported. It is well known that organic pollutants and heavy metal ions coexist in wastewater; it is therefore inevitable that the sorption of organic pollutants and heavy metal ions should be investigated. From the results of heavy metal ion sorption on nanosized clay minerals and oxides [62–65], one can conclude that the presence of organic matter increases the sorption of heavy metal ions at low pH and decreases the sorption of heavy metal ions at high pH.

5.3.4 Composites of CNTs and Other Nanomaterials

5.3.4.1 MWCNT-TiO$_2$ Composites

MWCNTs are suitable materials in the preconcentration and solidification of heavy metal ions from large-volume solutions. In a natural environment, TiO$_2$-MWCNT composites are considered better than MWCNTs in the sorption of Cr(VI) and better than TiO$_2$ in the photocatalytic reduction of Cr(VI). The synthesis and characterization of TiO$_2$-MWCNT composites in view of their photocatalytic application has been reported [66–69]. TiO$_2$-MWCNT composites showed remarkable enhancement in photocatalytic properties compared with pure TiO$_2$ particles. The large sorption ability of MWCNTs and the remarkable photocatalytic reduction property of TiO$_2$ make TiO$_2$-MWCNTs a novel material for the reduction of organic and inorganic components in polluted environments.

The XRD patterns of MWCNTs, TiO$_2$, and treated TiO$_2$-MWCNTs are given in Figure 5.40. The most intense peaks of MWCNTs correspond to the (002) and (100) reflections. The peaks in TiO$_2$ correspond to anatase TiO$_2$ (JCPDS card of 894921, space group of P4$_2$ mnm^{-1}, $a_0 = 3.777$ Å, $c_0 = 9.501$ Å). In the case of TiO$_2$-MWCNT composites after thermal treatment at 450°C, it is worth noting that the (002) reflection due to MWCNTs overlaps the anatase TiO$_2$ (101) reflection. The treated TiO$_2$-MWCNT composites seem quite well crystallized, and only TiO$_2$ in the anatase phase can be identified.

The adsorption and photocatalytic reduction of Cr(VI) on TiO$_2$, MWCNTs, and TiO$_2$-MWCNT composites at different Cr(VI) concentrations are shown in Figure 5.41. For comparison, the activities of TiO$_2$, MWCNTs, and TiO$_2$-MWCNT composites were tested under the same conditions. It shows that different samples have apparently different capacities of Cr(VI) removal. As expected, the amount of Cr(VI) ions removed by the solid phases increased with increasing Cr(VI) concentration. Due to their large specific area, MWCNTs exhibit more of a superadsorption capacity than TiO$_2$ and TiO$_2$- MWCNTs in the dark. The photocatalytic activity of TiO$_2$-MWCNT composites is a little higher than that of pure TiO$_2$ under UV irradiation. Firstly, the introduction of MWCNTs into the composite catalysts remarkably induces not only a synergetic effect on Cr(VI) adsorption capacities but also high photocatalytic reduction activity. MWCNTs can adsorb Cr(VI) from solution, centralize them on the surface of TiO$_2$, and thereby favor phototcatalytic activity. Secondly, MWCNTs can conduct electrons and reduce the electronic accumulation of TiO$_2$; they can decrease the recombination of electron-hole pairs. The attachment of MWCNTs to TiO$_2$ surfaces possibly results in increased adsorption of cations and in turn its photocatalysis rate [68,69].

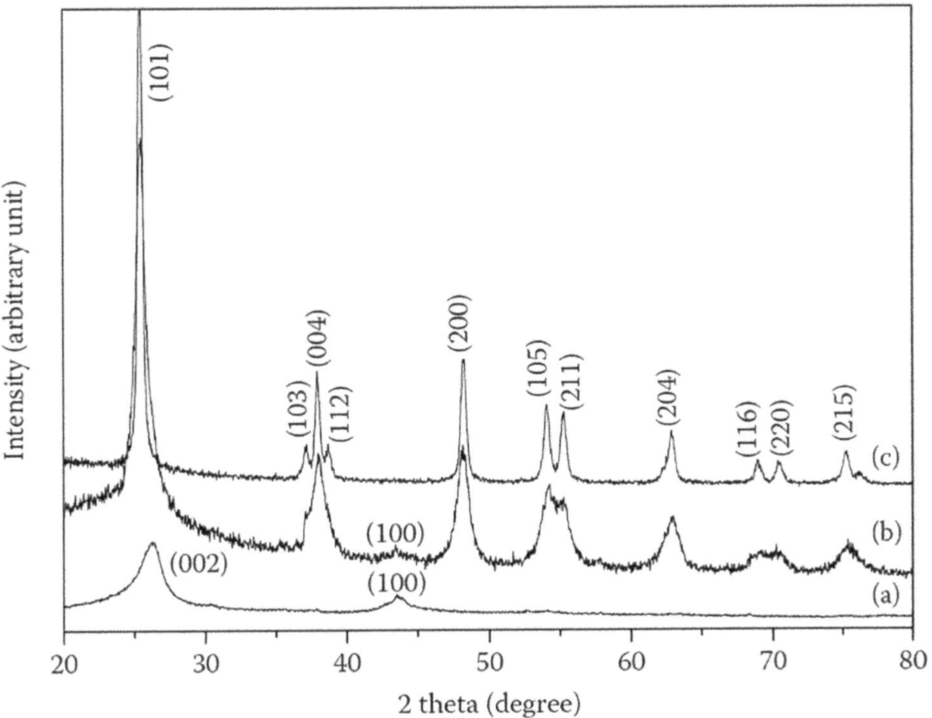

FIGURE 5.40 XRD patterns of samples: (a) MWCNTs, (b) TiO$_2$/MWCNTs, and (c) anatase TiO$_2$. (From Tan, X.L., Fang, M., and Wang, X.K., *J. Nanosci. Nanotech.*, 8, 5624, 2008 [70]. With permission.)

Figure 5.41 clearly shows that MWCNT-TiO$_2$ composites have much higher sorption and photocatalytic reduction abilities than pure TiO$_2$ under UV irradiation. Although MWCNT-TiO$_2$ composites have lower sorption capacities than pure MWCNTs in the dark, the photocatalytic reduction capacity of TiO$_2$ makes the composites have much higher capacities in the removal of Cr(VI) under UV-irradiation conditions. The MWCNT-TiO$_2$ composites may not be good material in the removal of heavy metal ions as compared with MWCNTs; however, if the photocatalytic reduction ability of TiO$_2$ is taken into account, the preconcentration property of MWCNTs and the photocatalytic reduction property of TiO$_2$ make the MWCNT-TiO$_2$ composites very good material in the removal and reduction of some special heavy metal ions such as Cr(VI).

5.3.4.2 MWCNT–Iron Oxide Magnetic Composites

The high sorption capacity of CNTs suggests that CNTs may be a promising adsorbent for treating wastewater containing organic and inorganic pollutants. However, it is difficult to separate CNTs from an aquatic phase because of their small size. The centrifugation method needs a very high rate, and the traditional filtration method may cause the blockage of filters. Compared with centrifugation and filtration methods, the magnetic separation method is considered a rapid and effective technique for separating magnetic particles from solutions. Magnetic separation methods, which represent a group of techniques based on the use of magnetic or magnetizable adsorbents, carriers, and cells, have been used for many applications in biochemistry, microbiology, cell biology, analytical chemistry, mining ores, and environmental technologies [71–73]. To facilitate the separation and recovery of CNTs from solution, the incorporation of magnetite with CNTs may be a promising method.

The preparation process of MWCNT–iron oxide magnetic composites is illustrated in Figure 5.42. Oxidized MWCNTs are functionalized with negative carboxylic and hydroxylic groups, which have

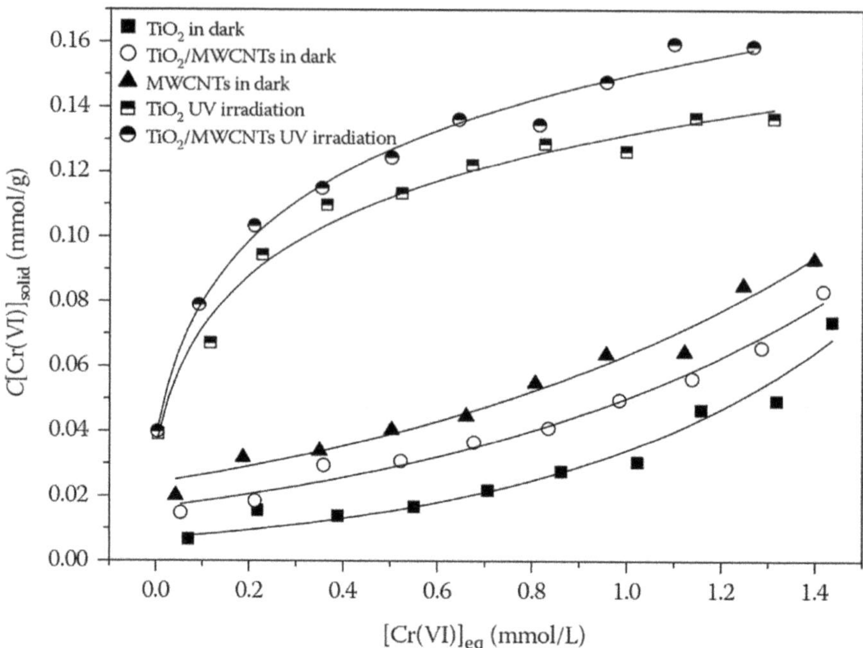

FIGURE 5.41 Effect of Cr(VI) concentration on removing Cr(VI) by TiO2, MWCNTs, and TiO2/MWCNTs in the dark and under UV irradiation. pH = 3.0 ± 0.1, C[TiO2/MWCNTs] = 0.2 g/L, C[K2SO4] = 7.5 × 10⁻⁵ mol/L, irradiation time = 6 hours. (From Tan, X.L., Fang, M., and Wang, X.K., *J. Nanosci. Nanotech.*, 8, 5624, 2008 [70]. With permission.)

the potential ability to bind metal ions. Positive ferrous and ferric ions are attached to oxidized MWCNTs due to the coordination reaction between ferrous and ferric ions and carboxylic and hydroxylic groups in the wet impregnation process.

We have applied synthesized MWCNT–iron oxide magnetic composites to adsorb metal ions and to separate magnetic composites from aqueous solutions by a magnetic process using a permanent magnet made of Nd-Fe-B. The results indicate that the magnetic method is as good as the centrifugation method in separating magnetic composites from aqueous solutions. As mentioned earlier, the application of magnetic composites in the removal of heavy metal ions from aqueous solutions is a very promising method. Further investigation, however, is still necessary.

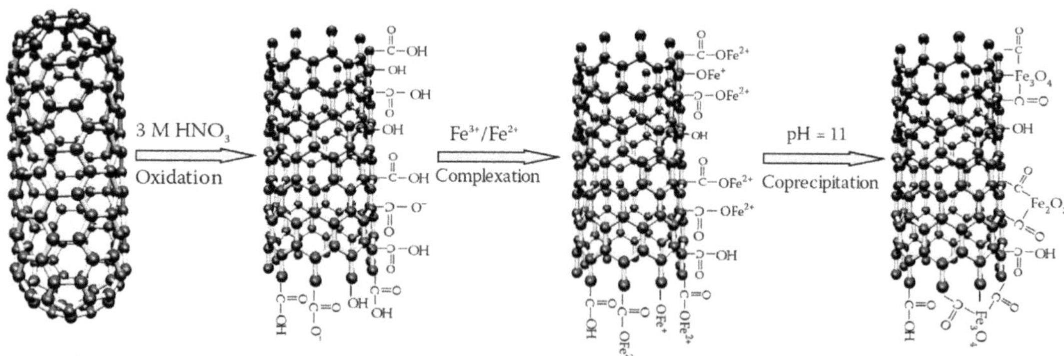

FIGURE 5.42 Preparation process of MWCNT–iron oxide magnetic composites.

5.4　CONCLUSIONS

The application of CNTs in the removal of heavy metal ions is summarized in this chapter. The special physicochemical properties of CNTs and their high sorption capacity and stability make them very suitable materials for the removal of heavy metal ions from large volumes of aqueous solutions. The organic pollutants in wastewater can also be removed easily by CNTs because of the strong interaction between CNTs and organic pollutants.

MWCNT-TiO$_2$ composites and MWCNT–iron oxide magnetic composites are briefly discussed at the end of the chapter. According to the application of composites in the removal of heavy metal ions from aqueous solution, the composites synthesized from MWCNTs and other oxides have special properties, and this feature enables the composites to have special applications. Recent advances in application of nanotechnology for removal of heavy metals from water can be found from the literature [74–95].

GLOSSARY OF NANOTECHNOLOGY

Carbon nanotube A carbon nanotube (CNT) is a cylindrical structure made of carbon atoms. CNTs exhibit remarkable properties, such as exceptional tensile strength and thermal conductivity, due to their nanostructure and strong carbon bonds. These properties make CNTs valuable in various areas of technology, including electronics, optics, composite materials, and nanomedicine. Here are some key points about CNTs: (a) Single-walled carbon nanotubes (SWCNTs) have diameters around 0.5–2.0 nanometers, which is about 100,000 times smaller than the width of a human hair. They can be idealized as cutouts from a two-dimensional graphene sheet rolled up to form a hollow cylinder; (b) Multiwalled carbon nanotubes (MWCNTs) consist of nested single-wall carbon nanotubes in a tube-in-tube structure.

Nanomaterials They are extremely small materials no thicker than 100 nanometers that are created by by manipulating atoms and molecules. These materials find applications in various industries, including healthcare, sports, and electronics.

Nanotechnology (a) It is a new technology involves designing, manufacturing or utilizing extremely small devices and structures. These technological creations typically operate on a scale of fewer than 100 nanometers (nm), in comparison with a sheet of paper that is about 100,000 nanometers thick; (b) Nanotechnology allows us to develop materials, devices, and systems with unique properties. Due to their small size, nanomaterials exhibit different physical and chemical characteristics than their larger counterparts. They have a large surface area-to-volume ratio, which enhances reactivity, strength, and conductivity. Plus, they're versatile and can be used in electronics, medical treatments, energy production, and environmental remediation.

REFERENCES

1. Iijima, S. Helical microtubules of graphic carbon. *Nature*, 354, 56, 1991.
2. Yang, K. and Xing, B.S. Desorption of polycyclic aromatic hydrocarbons from carbon nanomaterials in water. *Environ. Pollut.*, 145, 529, 2007.
3. Lu, C. and Liu, C. Removal of nickel(II) from aqueous solution by carbon nanotubes. *J. Chem. Technol. Biotechnol.*, 81, 1932, 2006.
4. Li, Y.H., Wang, S., Luan, Z., Ding, J., Xu, C., and Wu, D. Adsorption of cadmium(II) from aqueous solution by surface oxidized carbon nanotubes. *Carbon*, 41, 1057, 2003.
5. Lu, C. and Chiu, H. Adsorption of zinc(II) from water with purified carbon nanotubes. *Chem. Eng. Sci.*, 61, 1138, 2006.

6. Lu, C., Chiu, H., and Liu, C. Removal of zinc(II) from aqueous solution by purified carbon nanotubes: Kinetics and equilibrium studies. *Ind. Eng. Chem. Res.*, 45, 2850, 2006.

7. Li, Y.H., Zhu, Y., Zhao, Y., Wu, D., and Luan, Z. Different morphologies of carbon nanotubes effect on the lead removal from aqueous solution. *Diamond Relat. Mater.*, 15, 90, 2006.

8. Wang, X.K., Chen, C.L., Hu, W.P., Ding, A.P., Xu, D., and Zhou, X. Sorption of [243]Am(III) to multiwall carbon nanotubes. *Environ. Sci. Technol.*, 39, 2856, 2005.

9. Wang, Y.G., Shi, L., Gao, L., Wei, Q., Cui, L.M., Hu, L.H., Yan, L.G., and Du, B. The removal of lead ions from aqueous solution by using magnetic hydroxypropyl chitosan/oxidized multiwalled carbon nanotubes composites. *J. Colloid Interf. Sci.*, 451, 7–14, 2015.

10. Li, Y., Di, Z., Ding, J., Wu, D., Luan, Z., and Zhu, Y. Adsorption thermodynamic, kinetic and desorption studies of Pb^{2+} on carbon nanotubes. *Water Res.*, 39, 605, 2005.

11. Li, Y., Ding, J., Lun, Z., Di, Z., Zhu, Y., Xu, C., Wu, D., and Wei, B. Competitive adsorption of Pb^{2+}, Cu^{2+}, and Cd^{2+} ions from aqueous solutions by multiwalled carbon naotubes. *Carbon*, 41, 2787, 2003.

12. Chen, C.L. and Wang, X.K. Adsorption of Ni(II) from aqueous solution using oxidized multiwall carbon nanotubes. *Ind. Eng. Chem. Res.*, 45, 9144, 2006.

13. Konczyk, J., Zarska, S., and Ciesielski, W. Adsorptive removal of Pb(II) ions from aqueous solutions by multi-walled carbon nanotubes functionalised by selenophosphoryl groups: Kinetic, mechanism, and thermodynamic studies. *Colloids Surf. A: Physicochem. Eng. Asp.*, 575, 271–282, 2019.

14. Di, Z.C., Ding, J., Peng, X.J., Li, Y.H., Luan, Z.K., and Liang, J. Chromium adsorption by aligned carbon nanotubes supported ceria nanoparticles. *Chemosphere*, 62, 861, 2006.

15. Di Natale, F., Lancia, A., Molino, A., and Musmarra, D. Removal of chromium ions form aqueous solutions by adsorption on activated carbon and char. *J. Hazard. Mater.*, 145, 381, 2007.

16. Kratochvil, D., Pimentel, P., and Volesky, B., Removal of trivalent and hexavalent chromium by seaweed biosorbent. *Environ. Sci. Technol.*, 32, 2693, 1998.

17. Nowack, B., and Bucheli, T.D. Occurrence, behavior and effects of nanoparticles in the environment. *Environ. Pollut.*, 150, 5, 2007.

18. Rinzler, A.G., Liu, J., Dai, H., Nikolaev, P., Huffman, C.B., Rodriguez-Macias, F.J., Boul, P.J., Lu, A.H., Heymann, D., Colbert, D.T., Lee, R.S., Fischer, J.E., Rao, A.M., Eklund, P.C., and Smalley, R.E. Large-scale purification of single-wall carbon nanotubes: Process, product, and characterization. *Appl. Phys. A*, 67, 29, 1998.

19. Hummer, G., Rasaiah, J.C., and Noworyta, J.P. Water conduction through the hydrophobic channel of a carbon nanotube. *Nature*, 414, 188, 2001.

20. Tan, X.L., Xu, D., Chen, C.L., Wang, X.K., and Hu, W.P. Adsorption and kinetic desorption study of [152+154]Eu(III) on multiwall carbon nanotubes from aqueous solution by using chelating resin and XPS methods. *Radiochim. Acta*, 96, 23, 2008.

21. Alvarez-Puebla, R.A., Valenzuela-Calahorro, C., and Garrido, J.J. Retention of Co(II), Ni(II), and Cu(II) on a purified brown humic acid. Modeling and characterization of the sorption process. *Langmuir*, 20, 3657, 2004.

22. Chen, C.L., Hu, J., Xu, D., Tan, X.L., Meng, Y.D., and Wang, X.K. Surface complexation modeling of Sr(II) and Eu(III) adsorption onto oxidized multiwall carbon nanotubes. *J. Colloid Interface Sci.*, 323, 33, 2008.

23. Wang, H., Zhou, A., Peng, F., Yu, H., and Yang, J. Mechanism study on adsorption of acidified multi-walled carbon nanotubes to Pb(II). *J. Colloid Interface Sci.*, 316, 277, 2007.

24. Yang, C.M., Kaneko, K., Yudasaka, M., and Iijima, S. Surface chemistry and pore structure of purified HiPco single-walled carbon nanotube aggregates. *Appl. Phys. B*, 323, 140, 2002.

25. Gardner, S.D., Singamsetty, C.S.K., Booth, G.L., and He, G.R. Surface characterization of carbon fibers using angle-resolved XPS and ISS. *Carbon*, 33, 587, 1995.

26. Swiatkowski, A., Pakula, M., Biniak, S., and Walczyk, M. Influence of the surface chemistry of modified activated carbon on its electrochemical behaviour in the presence of lead(II) ions. *Carbon*, 42, 3057, 2004.

27. Deng, S.B. and Ting, Y.P. Characterization of PEI-modified biomass and biosorption of Cu(II), Pb(II), and Ni(II). *Water Res.*, 39, 2167, 2005.

28. Qurie, M., Khamis, M., Manassra, A., Ayyad, I., Nir, S., Scrano, L., Bufo, S.A., and Karaman, R. Removal of Cr(VI) from aqueous environments using micelle-clay adsorption. *Sci. World J.*, 942703, 2013.

29. Fang, J. and Gu, Z.M. Cr(VI) Removal from aqueous solution by activated carbon coated with quaternized poly(4-vinylpyridine). *Environ. Sci. Technol.*, 41, 4748, 2007.

30. Khezami, L. and Capart, R. Removal of chromium(VI) from aqueous solution by activated carbons: Kinetic and equilibrium studies. *J. Hazard. Mater.*, 123, 223, 2005.
31. Hu, J., Chen, C.L., Zhu, X.X., and Wang, X.K. Removal of chromium from aqueous solution by using oxidized multiwalled carbon nanotubes. *J. Hazard. Mater.*, 162, 1542, 2009.
32. Saleem, M.H., Afzal, J., Rizwan, M., Shah, Z.U.H., Depar, N., and Usman, K., Chromium toxicity in plants: consequences on growth, chromosomal behavior and mineral nutrient status. *Turk. J. Agric. For.*, 46, 371–389, 2022.
33. Stumm, W. and Morgan, J.J. *Aquatic Chemistry*, 3rd edition, John Wiley and Sons, New York, 1996.
34. Hu, J., Chen, G., and Lo, I.M.C. Removal and recovery of Cr(VI) from wastewater by maghemite nanoparticles. *Water Res.*, 39, 4528, 2005.
35. Bikerman, J.J., *Surface Chemistry: Theory and Applications*, 2nd ed., p. 294. Academic Press, New York, 1958.
36. Genc-Fuhrman, H., Tjell, J.C., and Mcconchie, D. Adsorption of arsenic from water using activated neutralized red mud. *Environ. Sci. Technol.*, 38, 2428, 2004.
37. Altundogan, H.S., Altundogan, S., Tumen, F., and Bildik, M. Arsenic removal from aqueous solutions by adsorption on red mud. *Waste Manag.*, 20, 761, 2000.
38. Rosen, M.J. and Li, F., The relationship between the interfacial properties of surfactants and their toxicity to aquatic organisms. *Environ. Sci. Technol.*, 35, 954, 2001.
39. Cserhati, T., Forgacs, E., and Oros, G. Biological activity and environmental impact of anionic surfactants. *Environ. Int.*, 28, 337, 2002.
40. Adak, A., Bandyopadhyay, M., and Pal, A. Removal of anionic surfactant from wastewater by alumina: A case study. *Colloid Surf. A*, 254, 165, 2005.
41. Vaism, L., Wagner, H.D., and Marom, G. The role of surfactants in dispersion of carbon nanotubes. *Adv. Colloid Interface Sci.*, 128–130, 37, 2006.
42. Islam, M.F., Rojas, E., Bergey, D.M., Johnson, A.T., and Yodh, A.G. High weight fraction surfactant solubilization of single-wall carbon nanotubes in water. *Nano Lett.*, 3, 269, 2003.
43. Tan, X.L., Fang, M., Chen, C.L., Yu, S.M., and Wang, X.K. Counterion effects of nickel and sodium dodecylbenzene sulfonate adsorption to multiwalled carbon nanotubes in aqueous solution. *Carbon*, 46, 1741, 2008.
44. Yang, K., Zhu, L., and Xing, B. Sorption of sodium dodecylbenzene sulfonate by montmorillonite. *Environ. Pollut.*, 145, 571, 2007.
45. Jones-Hughes, T. and Turner, A. Sorption of ionic surfactants to estuarine sediment and their influence on the sequestration of phenanthrene. *Environ. Sci. Technol.*, 39, 1688, 2005.
46. Weng, C.H. Modeling Pb(II) adsorption onto sandy loam soil. *J. Colloid Interface Sci.*, 272, 262, 2004.
47. Esmadi, F. and Simm, J. Sorption of cobalt(II) by amorphous ferric hydroxide. *Colloid Surf. A*, 104, 265, 1995.
48. Lee, S., Dyer, J.A., Sparks, D.L., Scrivner, N.C., and Elzinga, E.J. A multi-scale assessment of Pb(II) sorption on dolomite, *J. Colloid Interface Sci.*, 298, 20, 2006.
49. Nordberg, G.F., Jin, T., Hong, F., Zhang, A., Buchet, J.P., and Bernard, A. Biomarkers of cadmium and arsenic interactions. *Toxicol. Appl. Pharmacol.*, 206, 191, 2005.
50. Renu, Agarwal, M. and Singh, K., Heavy metal removal from wastewater using various adsorbents: A review. *J Water Reuse. Desal.*, 7, 387–419, 2017.
51. Environmental Protection Agency (EPA). Mexican Official Rule NOM-127-ssal–1994. Mexico.
52. Namasivayam, C. and Ranganathan, K. Removal of Cd(II) from wastewater by adsorption on "waste" Fe(III)/Cr(III) hydroxide. *Water Res.*, 29, 1737, 1995.
53. Wu, C.H. Studies of the equilibrium and thermodynamics of the adsorption of Cu^{2+} onto as-produced and modified carbon nanotubes. *J. Colloid Interface Sci.*, 311, 338, 2007.
54. Han, R., Lu, Z., Zou, W., Daotong, W., Shi, J., and Jiujun, Y. Removal of copper(II) and lead(II) from aqueous solution by manganese oxide coated sand: II. Equilibrium study and competitive adsorption. *J. Hazard. Mater.*, 137, 480, 2006.
55. Leyva, R.R., Bernal, J.L.A., Mendoza, B.J., Fuentes, R.L., and Guerrero, C.R.M. Adsorption of zinc(II) from an aqueous solution onto activated carbon. *J. Hazard. Mater.*, 90, 27, 2002.
56. Genc-Fuhrman, H., Mikkelsen, P.S., and Ledin, A. Simultaneous removal of As, Cd, Cr, Cu, Ni, and Zn from stormwater: Experimental comparison of 11 different sorbents. *Water Res.*, 41, 591, 2007.
57. Weng, C.H. and Huang, C.P. Adsorption characteristics of Zn(II) from dilute aqueous solution by fly ash. *Colloid Surf. A*, 247, 137, 2004.
58. Mohan, D. and Singh, K.P. Single- and multi-component adsorption of cadmium and zinc using activated carbon derived from bagasse—an agricultural waste. *Water Res.*, 36, 2304, 2002.

59. Wang, X.K., Zhou, X., Du, J.Z., Hu, W.P., Chen, C.L., and Chen, Y.X. Using of chelating resin to study the kinetic desorption of Eu(III) from humic acid-Al_2O_3 colloid surfaces. *Surf. Sci.*, 600, 478, 2006.

60. Geckeis, H., Rabung, Th., Ngo Manh, T., Kim, J.I., and Beck, H.P. Humic colloid-borne natural polyvalent metal ions: Dissociation experiment. *Environ. Sci. Technol.*, 36, 2946, 2002.

61. Tan, X.L., Wang, X.K., Geckeis, H., and Rabung, Th. Sorption of Eu(III) on humic acid or fulvic acid bound to hydrous alumina studied by SEM-EDS, XPS, TRLFS, and batch techniques. *Environ. Sci. Technol.*, 42, 6532, 2008.

62. Montavon, G., Markai, S., Andres, Y., and Grambow, B. Complexation studies of Eu(III) with aluminabound polymaleic acid: Effect of organic polymer loading and metal ion concentration. *Environ. Sci. Technol.*, 36, 3303, 2002.

63. Takahashi, Y., Kimura, T., Kato, Y., and Minai, Y. Speciation of europium(III) sorbed on a montmorillonite surface in the presence of polycarboxylic acid by laser-induced fluorescence spectroscopy. *Environ. Sci. Technol.*, 33, 4016, 1999.

64. Takahashi, Y., Kimura, T., and Minai, Y. Direct observation of Cm(III)-fulvate species on fulvic acid-montmorillonite hybrid by laser-induced fluorescence spectroscopy. *Geochim. Cosmochim. Acta*, 66, 1, 2002.

65. Xu, D., Shao, D.D., Chen, C.L., Ren, A.P., and Wang, X.K. Effect of pH and fulvic acid on sorption and complexation of cobalt onto bare and FA bound MX-80 bentonite. *Radiochim. Acta*, 94, 97, 2006.

66. Jitianu, A., Cacciaguerra, T., Benoit, R., Delpeux, S., Begum, F., and Bonnamy, S. Synthesis and characterization of carbon nanotubes-TiO_2 nanocomposites. *Carbon*, 42, 1147, 2004.

67. Yan, X.B., Tay, B.K., and Yang, Y. Dispersing and functionalizing multiwalled carbon nanotubes in TiO_2. *J. Phys. Chem. B*, 110, 25844, 2006.

68. Jang, S.R., Vittal, R., and Kim, K.J. Incorporation of functionalized single-wall carbon nanotubes in dye- sensitized TiO_2 solar cells. *Langmuir*, 20, 9807, 2004.

69. Yu, Y., Yu, J.C., Yu, J.G., Kwok, Y.C., Che, Y.K., Zhao, J.C., Ding, L., Ge, W. K., and Wong, P.K. Enhancement of photocatalytic activity of mesoporous TiO_2 by using carbon nanotubes. *Appl. Catal. A*, 289, 186, 2005.

70. Tan, X.L., Fang, M., and Wang, X.K. Preparation of TiO_2/multiwalled carbon nanotube composites and its application in photocatalytic reduction of Cr(VI) study. *J. Nanosci. Nanotech.*, 8, 5624, 2008.

71. Frehland, S., Kaegi, R., Hufenus, R., and Mitrano, D.M. Long-term assessment of nanoplastic particle and microplastic fiber flux through a pilot wastewater treatment plant using metal-doped plastics. *Water Res.*, 182, 2020.

72. Wu, R.C., Qu, J.H., and He, H. Removal of azo-dye acid red B (ARB) by adsorption and catalytic combustion using magnetic CuFe2O4 powder. *Appl. Catal. B*, 48, 49, 2004.

73. Wu, R.C., Qu, J.H., and Chen, Y.S. Magnetic powder MnO-Fe_2O_3 composite—A novel material for the removal of azo-dye from water. *Water Res.*, 39, 630, 2005.

74. Gong, J., Chen, L., and Zeng, G. Shellac-coated iron oxide nanoparticles for removal of cadmium (II) ions from aqueous solution. *J. Environ. Sci.*, 24(7), 1165–1173, 2012. https://doi.org/10.1016/S1001-0742(11)60934-0

75. Sheet, I., Kabbani, A., and Holail, H. Removal of heavy metals using nanostructured graphite oxide, silica nanoparticles and silica/graphite oxide composite. *Energy Procedia*, 50, 130–138, 2014. https://doi.org/10.1016/j.egypro.2014.06.016

76. Dargahi, A., Golestanifar, H., Darvishi, P., Karami, A., Hasan, S., Poormahammadi, A., and Behzadnia, A. An investigation and comparison of removing heavy metals (lead and chromium) from aqueous solutions using magnesium oxide nanoparticles. *Pol. J. Environ. Stud.*, 25(2), 557–562, 2016. https://doi.org/10.15244/pjoes/60281

77. Dubey, R., Bajpai, J., and Bajpai, A. K. Chitosan-alginate nanoparticles (CANPs) as potential nanosorbent for removal of Hg (II) ions. *Environ. Nanotechnol. Monit. Manag.*, 6, 32–34, 2016.

78. Ponnusamy, S., Anbalagan, S., and Yashwanthraj, S. Nanoscale zero-valent iron-impregnated agricultural waste as an effective biosorbent for the removal of heavy metal ions from wastewater. *Text. Cloth. Sustain.*, 2(1), 2017. https://doi.org/10.1186/s40689-016-0014-5

79. Rahbar, N., Ramezani, Z., and Mashhadizadeh, Z. One step in-situ formed magnetic chitosan nanoparticles as an efficient sorbent for removal of mercury ions from petrochemical waste water: Batch and column study. *Jundishapur J. Health Sci.*, 7(4), 2015. https://doi.org/10.17795/jjhs-30174

80. Thekkudan, V., Vaidvannathan, V., Ponnusamy, S., and Subramanian, S. Review on nanoadsorbents: A solution for heavy metal removal from wastewater. *IET Nanobiotechnol.*, 11(3), 213–224, 2017. https://doi.org/10.1049/iet-nbt.2015.0114

81. Hossein, V., Alireza, B., Shahriya, B., Ghosi, Z., Farnoush, F., and Re, M. A new nano-sorbent for fast and efficient removal of heavy metals from aqueous solutions based on modification of magnetic mesoporous silica nanospheres. *J. Magn. Magn. Mater.*, 441, 193–203, 2017. https://doi.org/10.1016/j.jmmm.2017.05.065

82. Mohamadiun, M., Dahrazma, B., Saghravani, F., and Darban, A. Removal of cadmium from contaminated soil using iron (III) oxide nanoparticles stabilized with polyacrylic acid. *J. Environ. Eng. Landsc. Manag.*, 26(2), 98–106, 2018. https://doi.org/10.3846/16486897.2017.1364645

83. Baysal, A., Kuznek, C., and Ozcan, M. Starch coated titanium dioxide nanoparticles as a challenging sorbent to separate and preconcentrate some heavy metals using graphite furnace atomic absorption spectrometry. *Int. J. Environ. Anal. Chem.*, 98, 45–55, 2018. https://doi.org/10.1080/03067319.2018.1427741

84. Cao, Y., Zhang, S., Zhong, Q., Wang, G., Xu, X., Li, T., Wang, L., Jia, Y., and Li, Y. Feasibility of nanoscale zero-valent iron to enhance the removal efficiencies of heavy metals from polluted soils by organic acids. *Ecotoxicol. Environ. Saf.*, 162, 464–473, 2018.

85. Ciotta, E., Prosposito, P., Moscone, D., Colozza, N., and Pizzoferrato, R. Detection and removal of heavy-metal ions in water by unfolded-fullerene nanoparticles. *AIP Conference Proceedings*, 2145, 2019. https://doi.org/10.1063/1.5123569Published

86. Parvin, F., Rikta, S., and Tareq, S. Application of nanomaterials for the removal of heavy metal from wastewater. In: Amimul Ahsan and Ahmad Fauzi Ismail (Eds.), *Nanotechnology in Water and Wastewater Treatment* (pp. 137–157). Springer, Cham, 2019. https://doi.org/10.1016/B978-0-12-813902-8.00008-3

87. Bankole, M. T., Abdulkareem, A. S., Mohammed, I. A., Ochigbo, S. S., Tijani, J. O., Abubakre, O. K., and Roos, W. D. Selected heavy metals removal from electroplating wastewater by purified and polyhydroxylbutyrate functionalized carbon nanotubes adsorbents. *Sci. Rep.*, 9, 4475, 2019. https://doi.org/10.1038/s41598-018-37899-4

88. Bassyouni, M., Mansi, A. E., Elgabry, A., Ibrahim, B. A., Kassem, O. A., and Alhebeshy, R. Utilization of carbon nanotubes in removal of heavy metals from wastewater: A review of the CNTs' potential and current challenges. *Appl. Phys. A*, 126, 38, 2020. https://doi.org/10.1007/s00339-019-3211-7

89. Nodeh, H. R., Shakiba, M., Gabris, M. A., Bidhendid, M., Shhabuddine, S., and Khanamf, R. Spherical iron oxide methyltrimethoxysilane nanocomposite for the efficient removal of lead(II) ions from wastewater: Kinetic and equilibrium studies. *Desalin. Water Treat.*, 192, 297–305, 2020. https://doi.org/10.5004/dwt.2020.25767192

90. Manyangadze, M., Chikuruwo, N., Narsaiah, T., Chakra, S., Charis, G., Danha, G., and Mamvura, T. A. Adsorption of lead ions from wastewater using nano silica spheres synthesized on calcium carbonate templates. *Heliyon*, 6, e05309, 2020.

91. Asadollahzadeh, H., Ghazizadeh, M., and Manzari, M. Developing a magnetic nanocomposite adsorbent based on carbon quantum dots prepared from pomegranate peel for the removal of Pb(II) and Cd(II) ions from aqueous solution. *Anal. Methods Environ. Chem. J.*, 4(3), 33–46, 2021. https://doi.org/10.24200/amecj.v4.i03.149

92. Baby, R., Hussein, M., Abdullah, A., and Zaina, Z. Nanomaterials for the treatment of heavy metal contaminated water. *Polymers*, 14(3), 583, 2022. https://doi.org/10.3390/polym14030583

93. Znad, H., Awual, M. R., and Martini, S. The utilization of algae and seaweed biomass for bioremediation of heavy metal-contaminated wastewater. *Molecules*, 27, 1275, 2022. https://doi.org/10.3390/molecules27041275

94. Inobeme, A., Mathew, J. T., Adetunji, C. O., Ajai, A. I., Inobeme, J., Maliki, M., Okonkwo, S., Adekoya, M. A., Bamigboye, M. O., Jacob, J. O., and Eziukwu, C. A. Recent advances in nanotechnology for remediation of heavy metals. *Environ. Monit. Assess.*, 195, 111, 2023. https://doi.org/10.1007/s10661-022-10614-7

95. Nworie, F. S. Development of nano-adsorbent for heavy metals removal from wastewater. In: Nitish Kumar (Ed.), *Heavy Metal Toxicity. Environmental Science and Engineering* (pp. 265–292). Springer, Cham, 2024. https://doi.org/10.1007/978-3-031-56642-4_9

6 Biosorption of Metals onto Granular Sludge

Shu Guang Wang, Xue Fei Sun, Wen Xin Gong, Yue Ma, ZiYi Yang, Xiao Liu, and Lawrence K. Wang

ACRONYMS

EDX energy dispersive X-ray
EPS exocellular polysaccharide
FTIR Fourier transform infrared
SEM Scanning electron microscopy
SBR sequencing batch reactors
XPS X-ray photoelectron spectroscopy
XRD X-ray diffraction

6.1 INTRODUCTION

Effluents from textile, leather, tannery, electroplating, galvanizing, pigment and dyes, metallurgical and paint industries, and other metal-processing and refining operations contain considerable amounts of toxic metal ions. Of the important metals, mercury, lead, copper, nickel, cobalt, cadmium, and chromium (VI) are regarded as highly toxic [1]. Radionuclides, such as uranium, possess high toxicity and radioactivity and exhibit a serious threat, even at small concentrations, to the health of human beings [2].

The toxic characteristics of heavy meals are as follows: (a) the toxicity can last for a long time; (b) some heavy metals can even be transformed from relevant low-toxic species into more toxic forms in a certain environment (e.g., mercury); (c) the bioaccumulation and bioaugmentation of heavy metals by food chain can damage normal physiological activity and eventually endanger human life; (d) metals can only be transformed and changed in valence and species, but they cannot be degraded by any method including biotreatment; and (e) the toxicity of heavy metals occurs in low concentrations of 1.0–10 mg/L. Some toxic metal ions, such as Hg and Cd, are very toxic even at trace levels of 0.001–0.1 mg/L [1,3].

Considerable researches have been carried out in developing cost-effective heavy metal removal techniques: chemical precipitation, chemical oxidation or reduction, evaporation, adsorption, ion exchange, and membrane technologies [4,5]. These techniques may be ineffective or expensive, especially when the concentrations are in the order of 1–100 mg/L [6–8]. It is important to find alternative, effective, economical, and practical techniques. Biosorption is the term given to the passive sorption and/or complexation of metal ions by biomass. The mechanisms of biosorption are generally based on physico-chemical interactions between metal ions and the functional groups present on the cell surface, such as electrostatic interactions, ion exchange, and metal ion chelation or complexation [9,10]. Because of low cost and good performance, biosorption is promising for removing heavy metals from aqueous solutions in recent years. A number of biomaterials have been used as biosorbents in the literature, including fungus, algal, biosludge, and microalgae [5,8,11–16]. However, most of the biosorbents currently used are in the form of suspended biomass; hence, one of the major operational problems associated is postseparation of biosorbent from the treated

 DOI: 10.1201/9781003541615-6

effluent. To overcome this drawback, a cell immobilization technique has been developed, but the immobilization procedure is often expensive and complex.

A great body of research showed that granular sludge is a microbial aggregate with a compact structure and excellent settleability. Granular sludge was first described for strictly anaerobic systems in 1980 [17], and by the late 1990s, the formation and application of aerobic granules had been reported [18–20]. Anaerobic granules possessed a compact, porous structure and excellent settling ability. The particulate biomass was found to possess high mechanical strength. Even under aggressive chemical environments (acidic or basic conditions), the biomass demonstrated excellent stability with no visible structural damage [21]. Although the aerobic granules were densely packed, microbial aggregates and their densities were much higher than that of conventional activated sludge. In addition, the aerobic granules were known to exhibit advantages of (a) regular, smooth, and nearly round in shape; (b) excellent settleability; (c) dense and strong microbial structure; (d) high biomass retention; (e) ability to withstand at high organic loading; and (f) tolerance to toxicity [22]. The gross morphology of anaerobic and aerobic granules is shown in Figure 6.1. The schematic diagrams of the upflow anaerobic sludge blanket (UASB) and the sequencing batch reactor (SBR) reactor are shown in Figures 6.2 and 6.3.

The average diameter of granular sludge is in the range of 0.2–5 mm. The settling velocity ranges from 30 to 100 m/h, whereas that of suspended flocs is usually lower than 9 m/h. The advantages of

FIGURE 6.1   Gross morphology of granular sludge: (a) anaerobic granules and (b) aerobic granules.

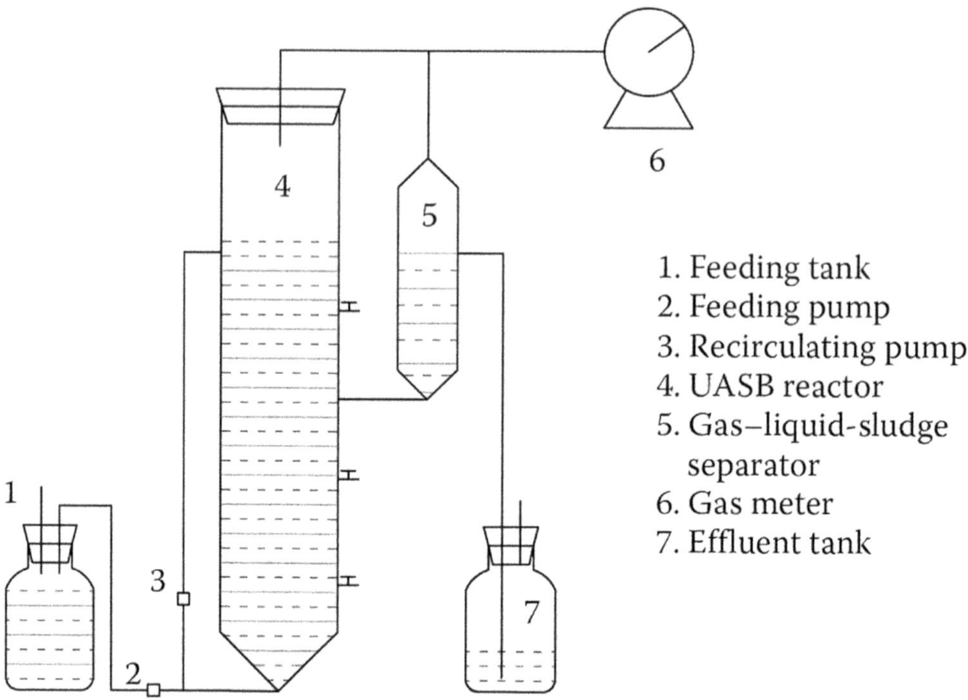

FIGURE 6.2 Schematic diagram of the UASB reactor. (Adapted from Shen, D.S. et al, Adsorption of phenol and copper onto ordered mesoporous carbon. *J. Hazard. Mater.*, 125, 231–236, 2005.)

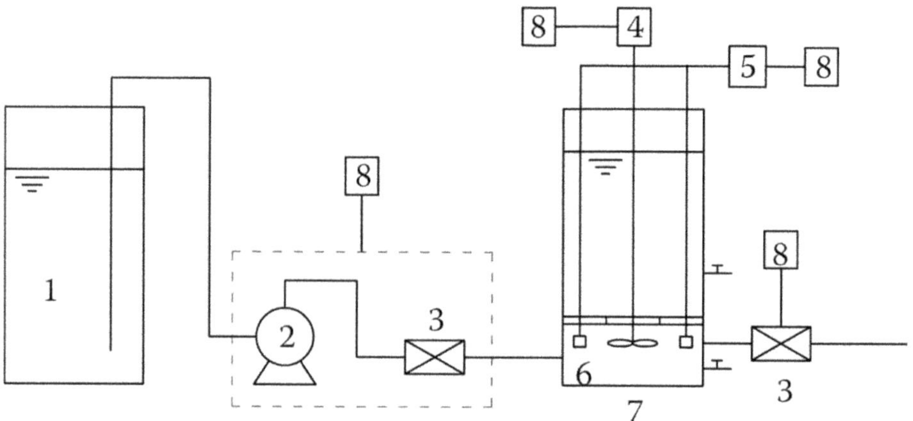

1. Water tank   2. Influent pump   3. Valve   4. Stirrer
5. Air pump   6. Air diffuser   7. Reactor   8. Timer

FIGURE 6.3 Schematic diagram of the SBR reactor.

compact structure and excellent settling property have been well recognized. They also exhibited good performance in heavy metals uptake and dyes uptake [23–31]. Granular sludge could be considered as a potential candidate for metal removal, such as cadmium(II), copper(II), nickel(II), and zinc(II). This chapter focuses on metal removal from aqueous solutions by granular sludge based on a substantial number of references on metal biosorption and our previous work.

6.2 BIOSORPTION CAPACITY OF GRANULAR SLUDGE

6.2.1 BIOSORPTION CAPACITY

The determination of the metal uptake rate by the biosorbent is often based on the equilibrium state of the sorption system. The sorption uptake is expressed in milligrams of metal sorbed per gram of the (dry) sorbent (the basis for engineering process-mass balance calculations) or mmol/g or meq/g (when stoichiometry and/or mechanism are considered) [32–34]. Metal ion uptake by granular sludge has been reported in a substantial number of references. Table 6.1 presents some data on the biosorptive capacities of different metal ions.

The biosorption capacity is strongly dependent on parameters such as the concentration of metals, pH, and adsorption time. The biosorption capacity of each metal is comparable with those of

TABLE 6.1
Biosorption Capacity of Different Metals

Metals	Sorbent	Uptake Capacity (mg/g)	T (°C)	pH	References
Copper(II)	Anaerobic granules	55	—	5	[23]
	Aerobic granules	59.6	26	4	[24]
		40.65	20	5	[28]
		44.90	25	6	[48]
		246.1	26	7	[54]
Cadmium(II)	Anaerobic granules	60	—	5	[23]
	Aerobic granules	172.7	26	4	[24]
		566	26	7	[25]
		90.26	25	6	[48]
Zinc(II)	Aerobic granules	164.5	26	4	[24]
		270	26	6	[41]
		180	26	7	[54]
		62.50	20	5	[26]
Uranium(VI)	Aerobic granules	218±2	30	4	[30]
Cerium(IV)	Aerobic granules	357			[52]
Lead(II)	Anaerobic granules	255	—	5	[23]
	Aerobic granules	87.72			[53]
Cobalt(II)	Anaerobic granules	12.34	—	7	[31]
		11.71	—	7	[31]
		9.40	30	7	[36]
		1.69	30	—	[55]
	Aerobic granules	55.25	20	7	[26]
Nickle(II)	Anaerobic granules	12.02	—	7	[31]
		13.33	—	7	[31]
		8.92	30	7	[36]
		26	—	5	[23]
	Aerobic granules	33.5	21	7	[50]
		17.5	25	6	[39]
		22.42	25	6	[48]

conventional suspended biosorbents [24,35,36]. As far as metal species are concerned, the investigation is narrow and not systemic. Arthrospira platensis (Spirulina) has been found successful in removing Cr in the range of 100%–73.5% from tannery wastewater [25]. Kafil et al. [26] indicated that heterotrophic cultivation of microalgae is beneficial for removing Cr(VI). For arsenic, recent studies have identified that microalgae are more efficient than other organisms in arsenic uptake [27]. Regulating the expression of transport protein DsPht1 in microalgae Chlamydomonas reinhardtii is beneficial for the absorption and transformation of arsenates [28].

6.2.2 GRANULAR SLUDGE VERSUS OTHER BIOMATERIALS

Table 6.2 presents a part of the comparative results of metal biosorption capacity between granular sludge and other biomaterials. By comparing the maximum sorption capacity (q_{max} of the Langmuir equation) with various types of biomass for the removal of zinc and cobalt ions, Sun et al. [30] indicated that aerobic granular sludge has a higher capacity for single- and binary-metal biosorption systems. Hawari and Mulligan [23] investigated various types of waste biomass including brown algae bacteria, fungus as well as activated sludge for biosorption of metals. They found that the metal removal capacity of anaerobic granular sludge was higher than that of granular and powder-activated carbons. Anaerobic granular sludge appeared more efficient in metal uptake than sugar beet pulp, activated sludge, *Penicillium chrysogenum*, and *Rhizopus arrhizus* fungus. The considerably low cost of the granular sludge, its physical characteristics, and the high uptake capacity of the heavy metals make it a very attractive biosorbent.

6.3 INFLUENTIAL FACTORS

6.3.1 INITIAL METAL CONCENTRATION

The uptake rate of the metal ion will increase with increasing the initial concentration [29,31]. At lower initial solute concentrations, the ratio of the initial moles of solute to the available surface area is low; the fractional sorption thus becomes independent of the initial concentration. However, at higher concentrations, the sites available for sorption become fewer compared to the moles of solute present, and hence, the removal of solute is strongly dependent on the initial solute concentration. It is always necessary to identify the maximum saturation potential of a biosorbent, for which experiments should be conducted at the highest possible initial solute concentration.

TABLE 6.2

A Comparative Study of Different Biomaterials with Granular Sludge

Metal	Biosorptive Capacity (mg metal/g Dry Weight Biomass)	References
Zn(II)	*Sargassum* sp. (118.5) > *Bacillus jeotgali* (105.2) > Powdered waste sludge (82.0) > Aerobic granule (62.5) > *Ascophyllum nodosum* (25.6) > *Penicillium chrysogenum* (19.2) > *F. vesiculosus* (17.3) > Activated sludge (9.7) > *S. rimosus* (6.63)	[3,26]
Co(II)	PFB1 (190.0) > *Mortierella* SPS 403 (61.2) > Aerobic granule (52.4) > Anaerobic granule (12.3) > Oscillatoria angustissima (5.3) > Myriophyllum spicatum L. (2.3)	[26,31]
Pb(II)	*Ascophyllum nodosum* (271.1) > Anaerobic granule (258.7) > *Sargassum natans* (252.5) > *Penicillium chrysogenum* (122.13) > *Rhizopus arrhizus* (89.8) > Activated sludge (89.0) > Aerobic granule (87.8)	[23,53]
Cu(II)	Aerobic granule (59.6) > Anaerobic granule (55.3) > Activated sludge (49.0) > Sugar beet pulp (21.0) > *Rhizopus arrhizus* (10.2) > *Penicillium chrysogenum* (8.9)	[23,24]
Cd(II)	Aerobic granule (172.7) > *Ascophyllum nodosum* (132.2) > *Sargassum natans* (131.0) > Activated sludge (68.3) > Anaerobic granule (59.4) > *Penicillium chrysogenum* (56.0) > Rhizopus arrhizus (26.9)	[23,24]
Ni(II)	*Ascophyllum nodosum* (40.5) > Aerobic granule (33.5) > Anaerobic granule (25.8) > Sargassum *natans* (24.1) > Activated sludge (18.2) > Sugar beet pulp (12.9)	[23,36]

6.3.2   pH

For the biosorption of heavy metal ions, pH is the most important environmental factor. The pH value of a solution strongly influences not only the site dissociation of the biomass surface, but also the solution chemistry of the heavy metals: hydrolysis, complexation by organic and/or inorganic ligands, redox reactions, precipitation, the speciation, and the biosorption availability of the heavy metals [37].

The biosorptive capacity of metal cations increases with increasing the pH of the sorption system, but not in a linear relationship (Figure 6.4). On the other hand, too high a pH value can cause precipitation of metal complexes, so it should be avoided during experiments.

Xu et al. [38] found that the amount of nickel uptake by granules tended to increase and a sharp increase in the maximum specific uptake (q_{max}) was observed in pH between 3 and 6. This led them to suggest that at low pH, the cell surface-binding sites should be protonized, thereby making them unavailable for the other cations. However, with an increase in pH, there was an increase in ligands with negative charges, which resulted in increased binding of cations. Hawari and Mulligan [23] proved that the biosorption capacity of Pb^{2+}, Cd^{2+}, Cu^{2+}, and Ni^{2+} is dependent on pH. For all the metal ions they studied, the optimal pH values are in the range 4.0–5.5. A further possible explanation of increasing sorption with increasing pH is that hydrolyzed species have a lower degree of hydration, that is, less energy is necessary for the removal or reorientation of the hydrated water molecules upon binding. With a further increase in pH (6–9), the solubility of metals decreases enough for precipitation to occur. This should be avoided during sorption experiments, as distinguishing between sorption and precipitation metal removal becomes difficult [39]. van Hullebusch [40] and Gai et al. [32] obtained a similar conclusion.

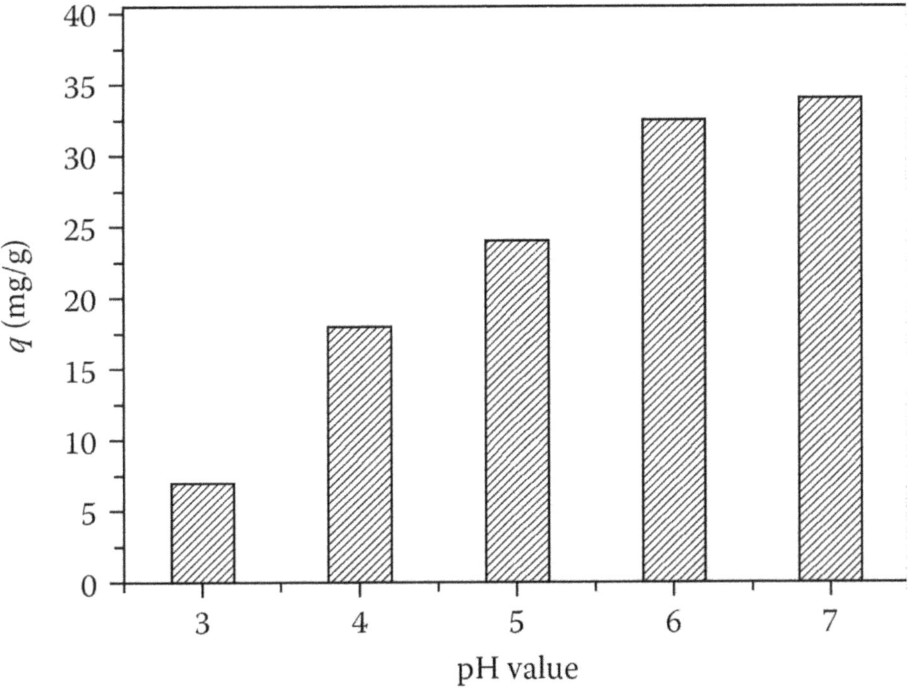

FIGURE 6.4 q_{max} at various initial pH values. (Adapted from Xu, H. et al., *Bioresour. Technol.*, 97, 359–363, 2006 [38.])

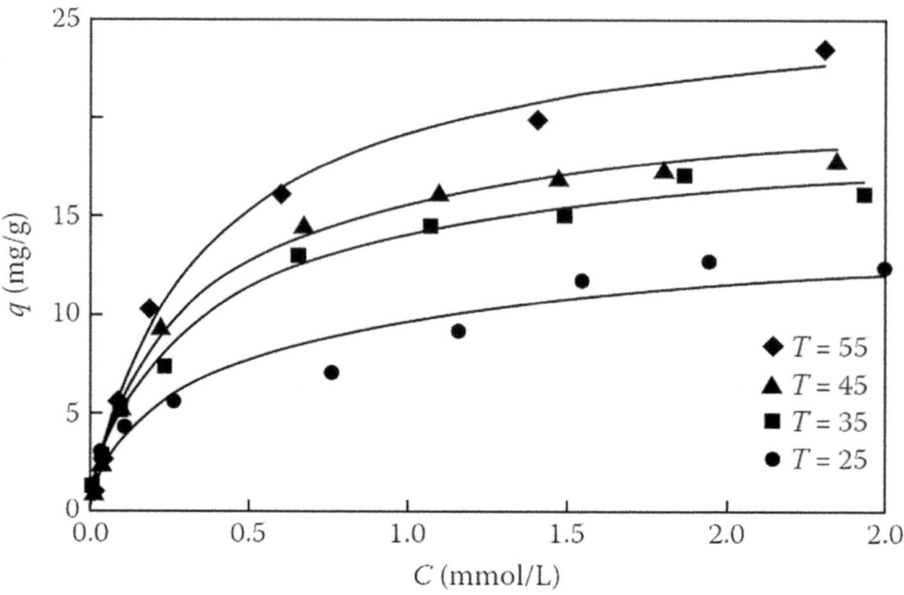

FIGURE 6.5 Biosorption isotherm of Ni^{2+} by aerobic granules at different temperatures. (Adapted from Xu, H. et al., *Bioresour. Technol.*, 97, 359–363, 2006 [38].)

6.3.3 TEMPERATURE

The biosorption process is usually not operated at high temperatures because it will increase the operational cost. Temperature affects biosorption only to a lesser extent within the range from 20°C to 35°C [2]. Higher temperatures usually enhance sorption due to the increased surface activity and kinetic energy of the solute [41,42]. Liu and Xu [43] found that nickel adsorption was enhanced by increasing the incubation temperature (Figure 6.5). This was based on the fact that q_{max} increased with an increase in temperature to a plateau at 55°C for 26.9 mg Ni(II)/g and 1.29 mmol Pd(II)/g, thereby implying the need for energy in the form of heat for maximal adsorption of metal ions.

6.3.4 CONTACT TIME

The biosorption process of heavy metal by granular sludge is usually completed rapidly. The biosorption of metals such as copper, cobalt, nickel, zinc, and uranium is a rapid process and often reaches equilibrium within several hours. Table 6.3 shows the equilibrium time for different metal-sludge systems. The biosorption of Pb, Cd, Cu, and Ni with anaerobic granules reached equilibrium within 30 minutes [23]. In the case of Cu by aerobic granules, equilibrium was attained in 30 minutes [32].

Generally speaking, metal biosorption consists of two phases: a very rapid initial sorption followed by a long period of much slower uptake. During the initial stage of sorption, a large number of vacant surface sites are available for biosorption. After some time, the remaining vacant surface sites are difficult to occupy due to repulsive forces between the solute molecules adsorbed on the solid surface and the bulk phase. Besides, the metal ions are adsorbed onto the mesopores that get almost saturated with metal ions during the initial stage of biosorption. Thereafter, the metal ions have to traverse farther and deeper into the pores, encountering much larger resistance. This results in the slowing down of the biosorption during the later period of biosorption [30].

TABLE 6.3
Equilibrium Time for the Metal-Sludge System

Metal	Biosorbents	Equilibrium Time (h)	References
Cu(II)	Aerobic granule	0.5	[28]
		0.5	[23]
Zn(II)	Aerobic granule	0.5	[26]
Pb(II)	Aerobic granule	0.5	[23]
Cd(II)	Aerobic granule	0.5	[23]
		2	[24]
		1.7	[25]
Uranium(VI)	Aerobic granules	1	[30]
Co(II)	Anaerobic granule	96	[36]
	Aerobic granule	0.5	[26]
Ni(II)	Anaerobic granule	96	[36]
	Aerobic granule	0.5	[23]

6.3.5 Presence of Competing Ions

Competitive sorption is important in water and wastewater treatment, because most metal ions to be sorbed exist in solution with other metal ions. Usually, the biosorption capacity of one metal ion is often interfered with and reduced by co-ions, including other metal ions and anions presenting in solution; however, the gross uptake capacity of all metals in solutions remains almost unchangeable. Sun et al. [30] studied the individual and competitive sorption of Zn^{2+} and Co^{2+} ions onto aerobic granules. The sorption capacities of granules for the two metal ions are in the order of $Zn^{2+} > Co^{2+}$ and the competitive sorption of metal ions also follows the same order. The sorption capabilities of granular sludge without competition are much better than those with competition because the granule surface is utilized by competing metal ions. Hawari and Mulligan [44] observed that the single-metal sorption uptake capacity of the biomass for Pb was slightly inhibited by the presence of Cu and Cd cations (by 6%) and by the presence of nickel (by 11%). Ni and Huang found that Pb (II) and Cd (II) have identical adsorption sites, but Pb (II) has a greater affinity than Cd (II) [45]. The affinity order of anaerobic biomass for the four metals was established as follows: Pb > Cu > Ni > Cd.

6.3.6 Thermodynamics

Both energy and entropy are the key factors to be considered in any process design. In this study, the thermodynamic parameters, such as standard free energy ($DG°$), enthalpy change ($DH°$), and entropy change ($DS°$), were estimated to evaluate the feasibility and endothermic nature of the adsorption process. These were calculated by using the following equations:

$$\Delta G^o = -RT \ln K_e \tag{6.1}$$

$$\ln K_e = \frac{\Delta S^o}{R} - \frac{\Delta H^o}{RT} \tag{6.2}$$

The Ke value can be obtained from the linearized form of the Langmuir equation.

The thermodynamic parameters for Ni^{2+} ions onto aerobic granules, which provide useful information concerning the inherent energetic changes of the sorption process, revealed that the negative free energy change ($DG°$) suggests that the sorption process is spontaneous with a high preference for Ni^{2+} ions for aerobic granules. The value of $DH°$ was estimated as 63.8 and 0.26 kJ/mol K for

$DS°$. The enthalpy change ($DH°$) is positive, indicating the endothermic nature of the sorption process [45]. The positive entropy change ($DS°$) reflects the increased randomness at the solid-solution interface during biosorption; it also indicates that ion replacement reactions occurred.

6.4 MODELS

The assessment of a solid-liquid sorption system is usually based on two types of investigations: equilibrium batch sorption tests and dynamic continuous-flow sorption studies [46].

6.4.1 Biosorption Isotherms

A biosorption isotherm, the plot of uptake (q) versus the equilibrium solute concentration in the solution (C_f), is often used to evaluate the sorption performance. Isotherm curves can be evaluated by varying the initial solute concentrations while fixing the environmental parameters, such as pH, temperature, and ionic strength. In general, the uptake increases with increase in concentration, and will reach saturation at higher concentrations. Figure 6.6 shows the outline of the general experimental procedure to obtain data points for the sorption isotherm [47].

Equilibrium isotherm models are usually classified into the empirical equations and mechanistic models, based on the mechanism of metal ion biosorption. Mechanistic models can be used not only to represent but also to explain and predict the experimental behavior [3].

Empirical models are simple mathematical relationships characterized by a limited number of adjustable parameters, which give a good description of the experimental behavior over a large range of operating conditions [2]. Within the literature, the Langmuir [48] and Freundlich [49] models (two-parameter models) have been used to describe the biosorption isotherm. The models are simple, well established, have physical meaning, and are easily interpretable, which are some of the important reasons for their frequent and extensive use.

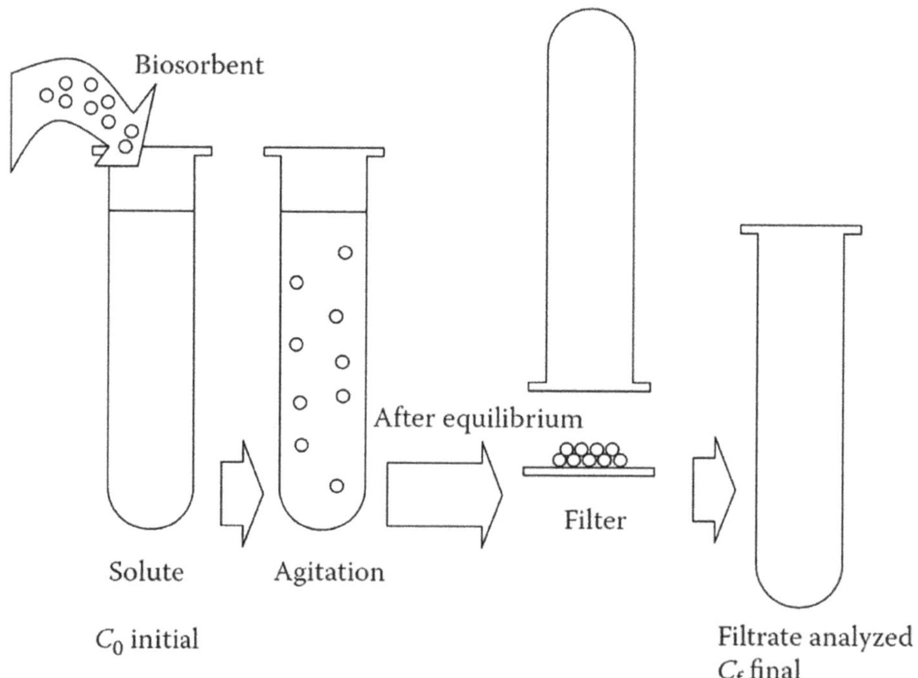

FIGURE 6.6 Outline of the experimental procedure to obtain data points for the sorption isotherm. (Adapted from Vijayaraghavan, K. and Yun, Y.S., *Biotechnol. Adv.*, 26, 266–291, 2008 [2].)

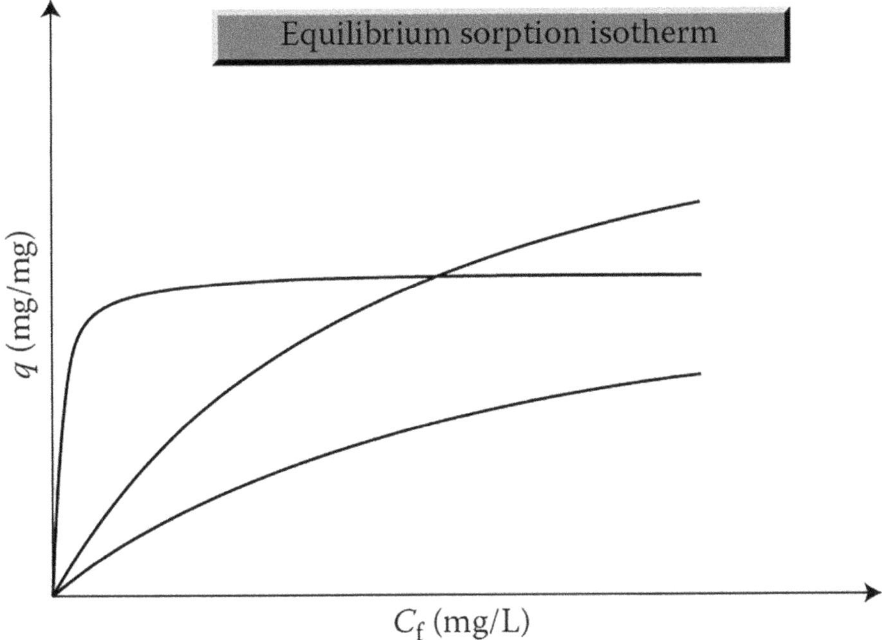

FIGURE 6.7 Biosorption-Langmuir isotherm relationship curves.

The Langmuir isotherm is valid for monolayer adsorption onto a surface with a finite number of identical sites. It can be expressed as

$$q = \frac{q_{max}bC_f}{1+bC_f} \tag{6.3}$$

This classical model incorporates two easily interpretable constants: q_{max}, which corresponds to the maximum achievable uptake by a system; and b, which is related to the affinity between the sorbate and sorbent [31]. A typical Langmuir adsorption isotherm is shown in Figure 6.7. The Langmuir constant "q_{max}" is often used to compare the performance of biosorbents; the other constant "b" characterizes the initial slope of the isotherm. Thus, for a good biosorbent, a high q_{max} and a steep initial isotherm slope (i.e., high b) are generally desirable [50].

The Freundlich isotherm is given as

$$q = K_F C_f^{1/n} \tag{6.4}$$

The Freundlich isotherm was originally empirical in nature but was later interpreted as the sorption to heterogeneous surfaces or surfaces supporting sites with various affinities. It is assumed that the stronger binding sites are initially occupied, with the binding strength decreasing with increasing degree of site occupation [49]. It incorporates two constants: K_F, which corresponds to the binding capacity; and n_F, which characterize the affinity between the sorbent and sorbate.

For the biosorption of metal ions by granular sludge, the classical Langmuir model and the Freundlich model have been employed to describe the single-metal biosorption system, and in most cases, both models fitted the experimental data very well (the correlation coefficient was usually larger than 0.9). Table 6.4 presents a part of biosorption isothermal constants for the Langmuir and Freundlich models reported in the references. Table 6.4 shows that, in most cases, the classical Langmuir model and the Freundlich model can represent single-metal biosorption behavior.

TABLE 6.4

Isothermal Constants of the Langmuir and Freundlich Models for the Metal-Sludge System

Metal	Sorbent	q_{max}	b	R^2	K_F	n	R^2	References
Zn(II)	Aerobic granule	56.50	0.004	0.934	0.681	1.496	0.993	[26]
Co(II)	Aerobic granule	51.28	0.003	0.988	0.462	1.396	0.987	[26]
	Anaerobic granule	8.42	0.031	0.944	13.31	0.628	0.991	[36]
Cu(II)	Aerobic granule	40.65	0.043	0.988	10.06	3.83	0.965	[28]
	Anaerobic granule	60	0.024	0.95				[23]
Pb(II)								
Cd(II)	Anaerobic granule	286	0.006	0.92				[23]
	Anaerobic granule	64	0.013	0.95				[23]
Ni(II)	Anaerobic granule	25	0.03	0.97				[23]
	Anaerobic granule	7.90	0.034	0.956	16.69	0.572	0.989	[36]

6.4.2 Biosorption Kinetics

For any practical applications, the process design, operation control, and sorption kinetics are very important [51]. Adsorption kinetics modeling explains how fast the sorption process occurs and the factors affecting the reaction rate. Information on the kinetics of solute uptake is required to select optimum operating conditions for the full-scale batch process. Also, it is important to establish the time dependence of adsorption systems under various process conditions. The nature of the sorption process will depend on the physical or chemical characteristics of the adsorbent system and also on the system conditions [52,53].

The kinetic modeling of metal sorption by granular sludge has been carried out using the Lagergren pseudo-first-order (Equation 6.5) and pseudo-second-order (Equation 6.6) kinetic models:

$$\log \frac{q_e - q_t}{q_e} = \frac{-K_{1,ad}t}{2.3} \tag{6.5}$$

$$\frac{1}{q_t} = \frac{1}{K_{2,ad}q_e^2} + \frac{t}{q_e} \tag{6.6}$$

where q_e is the amount of solute sorbed at equilibrium (mg/g), q_t is the amount of solute sorbed at time t (mg/g), and $K_{1,ad}$ is the first-order equilibrium rate constant (min^{-1}) and $K_{2,ad}$ is the second-order equilibrium rate constant (g/mgmin) [5,31].

Gai et al. [32] investigated the kinetics of biosorption of copper(II) ions by aerobic granules at varying initial solution pH and initial concentration C_0 (Table 6.5). The results showed that correlation coefficients of the pseudo-first-order kinetic model ($R_1^2 = 0.54 - 0.83$) were low compared with those of the pseudo-second-order kinetic model ($R_2^2 > 0.99$) at each pH studied. Lagergren's first- order rate equation describes the adsorption rate based on the adsorption capacity, whereas the pseudo-second-order rate expression is used to describe chemisorption involving valency forces and ion exchange. This suggests that the biosorption of the Cu(II) process may take place through ion exchange and chemisorption involving valency forces.

Sun et al. [30] reported similar results in the study of Co(II) and Zn(II) sorption kinetics by aerobic granules in a single- and binary-metal system. In the study, the values of the initial biosorption rate decreased as follows: Co(II)>Zn(II)>Co(II) (Co-Zn)>Zn(II) (Co-Zn), indicating that aerobic granules can adsorb Co(II) alone more rapidly than Zn(II) alone from aqueous solutions.

TABLE 6.5
Comparison of Pseudo-First-Order and Pseudo-Second-Order Adsorption Rate Constants at Different pH and C0 Values

C_0 (mg/L)	pH	$K_{1,ads}$ (min^{-1})	q_e (mg/g)	R_1^2	$K_{2,ads}$ (10^{-3}g/mg min)	q_e (mg/g)	R_2^2
50	3	0.015	6.812	0.5735	11.28	16.56	0.9966
	4	0.0221	10.817	0.6659	6.51	22.57	0.9973
	5	0.0297	13.744	0.8295	5.62	26.67	0.9994
125	3	0.0074	4.5394	0.1086	15.70	19.31	0.9923
	4	0.0205	11.593	0.7362	8.23	31.45	0.9979
	5	0.0168	14.067	0.7019	6.51	36.23	0.9963
250	3	0.0272	8.902	0.5367	11.21	28.17	0.9992
	4	0.0205	14.138	0.7057	6.24	34.72	0.9985
	5	0.0267	14.312	0.7296	5.10	39.84	0.9994

Source: Gai, L.H. et al., *J. Chem. Technol. Biotechnol.*, 83, 806–813, 2008.

6.5 BIOSORPTION MECHANISM

Generally speaking, biosorption is the interaction of metals in an aqueous solution and the organic interface of granular sludge. The analyses of elemental composition, X-ray diffraction (XRD), and Fourier transform infrared (FTIR) could reveal some interaction. For granular sludge, the multilayer structure and the role of exocellular polysaccharide (EPS) in the formation and stability of granules have focused attention on biosorption. The immobilization of heavy metal ions or radionuclides can be divided into three parts: (a) adsorption to cell walls, (b) adsorption to the EPS matrix, and (c) interaction with minerals present in the granular sludge [54]. To make clear the mechanism, further characterization of energy dispersive X-ray (EDX) and X-ray photoelectron spectroscopy (XPS) has been done, and encouraging results have been obtained [30,32,55].

6.5.1 SCANNING ELECTRON MICROSCOPY/EDX

Scanning electron microscopy (SEM) is an extremely useful method for visual confirmation of surface morphology and the physical state of the surface. Figure 6.8 shows the microbial structure of the granular sludge, and it can be seen that the surface of anaerobic granular sludge is rough, uneven, and porous [41], whereas aerobic granules have a compact structure with a regular, rigid, and stable surface as well as lots of cavities. SEM coupled with energy dispersive analysis of EDX is used to determine the metal uptake mechanism on granules. Liu and Xu [44] utilized SEM equipped with EDX to analyze fresh and Ni^{2+}-contaminated aerobic granules (Figures 6.9 and 6.10). The results proved that the adsorbed Ni^{2++} ions are uniformly distributed from the surface to the center of aerobic granules.

6.5.2 XRD

The crystal phases of aerobic granules before and after the biosorption experiments were analyzed by an X-ray powder diffraction analyzer [44,55]. According to Liu and Xu [44], chemical precipitation of Ni^{2+} would not be involved in Ni^{2+} biosorption by aerobic granules under the studied conditions. Xu and Liu [55] found a similar phenomenon; however, for Cd^{2+}- and Cu^{2+}-contaminated aerobic granules, their peak positions and ratios are different from those of fresh aerobic granules (Figures 6.11–6.13). The appearance of the new peaks indicates that there were different crystals

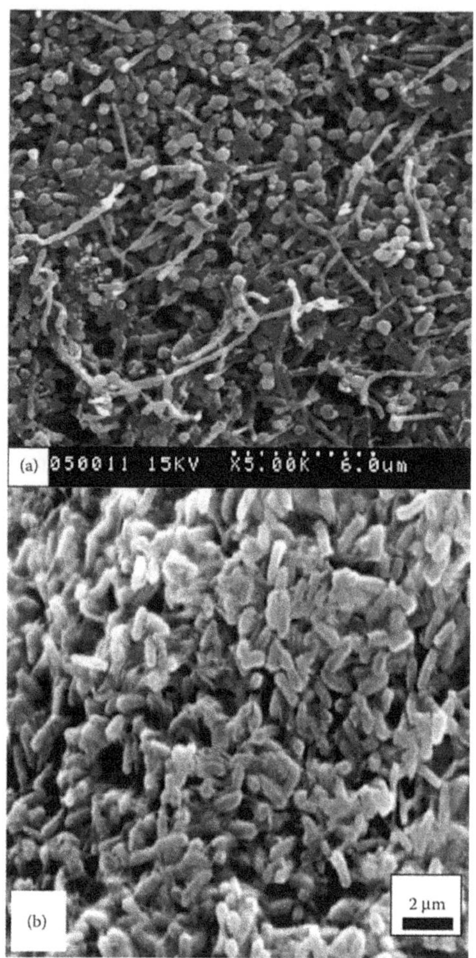

FIGURE 6.8 Typical SEM micrograph: (a) anaerobic granules and (b) aerobic granules. (Adapted from Cheng, W. et al., *Biochem. Eng. J.*, 39, 538–546, 2008 [41].)

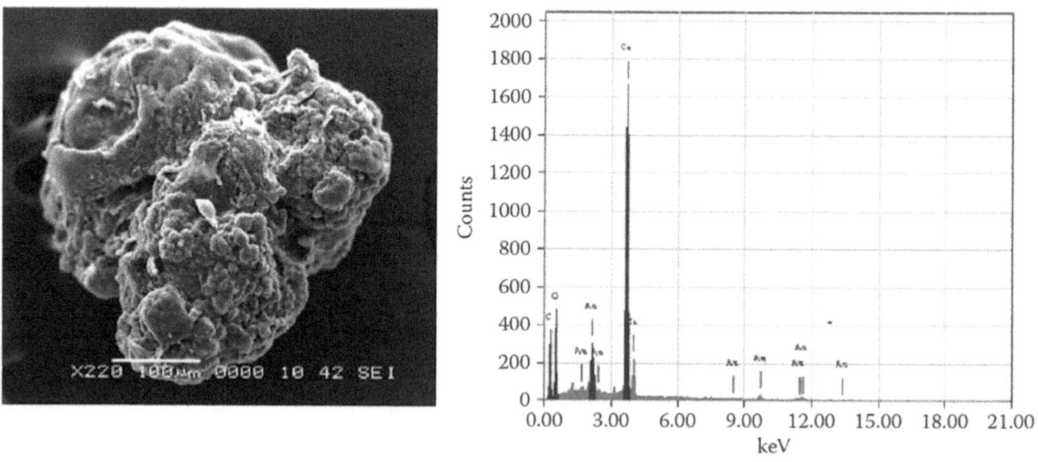

FIGURE 6.9 SEM image and EDX spectrum of fresh aerobic granules. (Adapted from Liu, Y. and Xu, H., *Biochem. Eng. J.*, 35, 174–182, 2007 [43].)

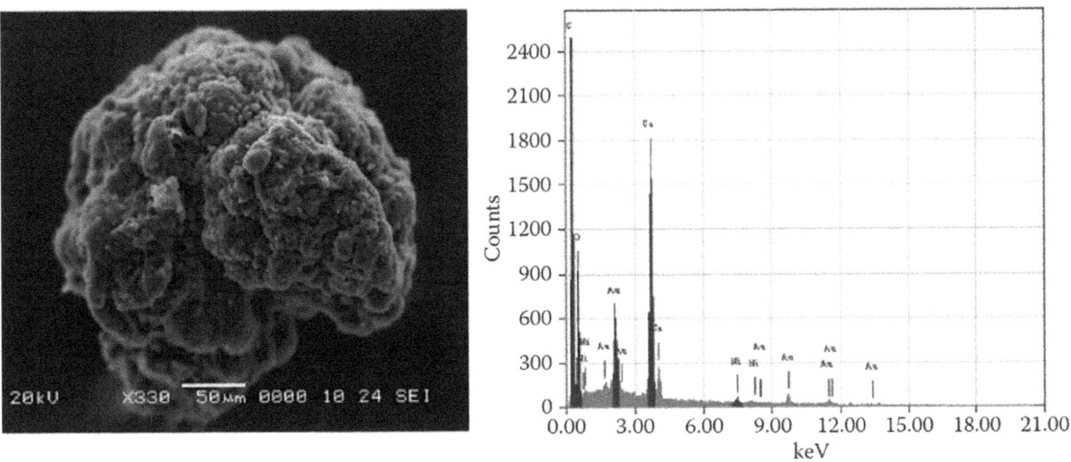

FIGURE 6.10 SEM image and EDX spectrum of the Ni^{2+}-contaminated aerobic granules. (Adapted from Liu, Y. and Xu, H., *Biochem. Eng. J.*, 35, 174–182, 2007 [43].)

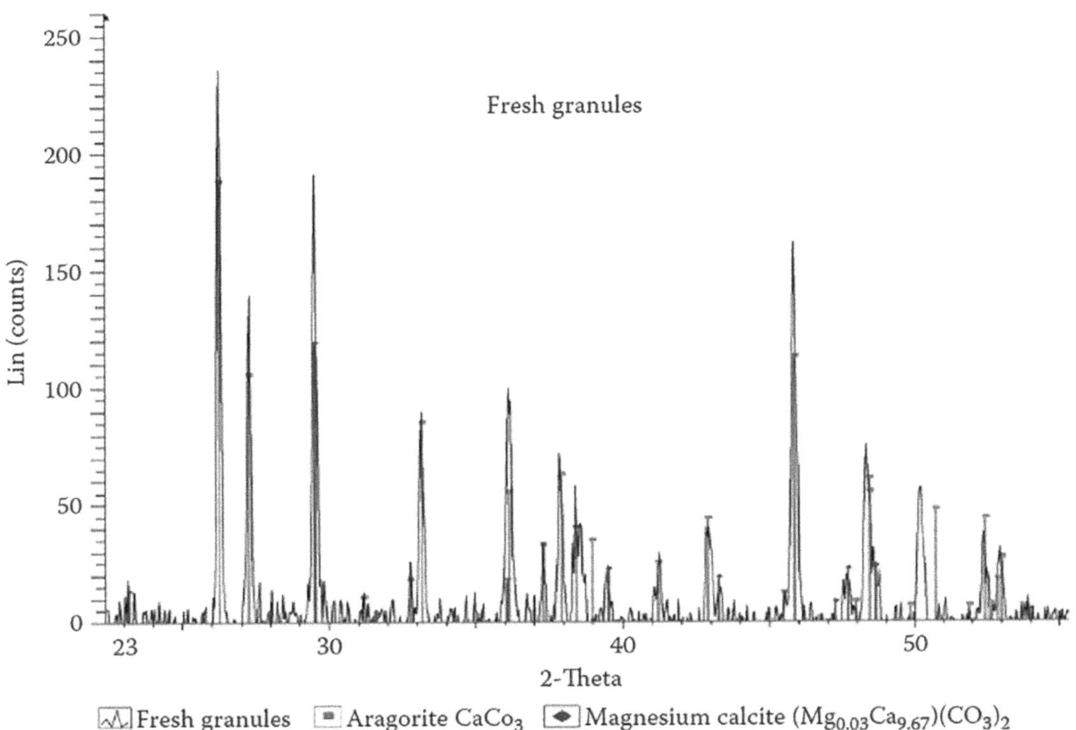

FIGURE 6.11 XRD analysis of fresh aerobic granules. (Adapted from Xu, H. and Liu, Y., *Sep. Purifi. Technol,* 58, 400–411, 2008 [55].)

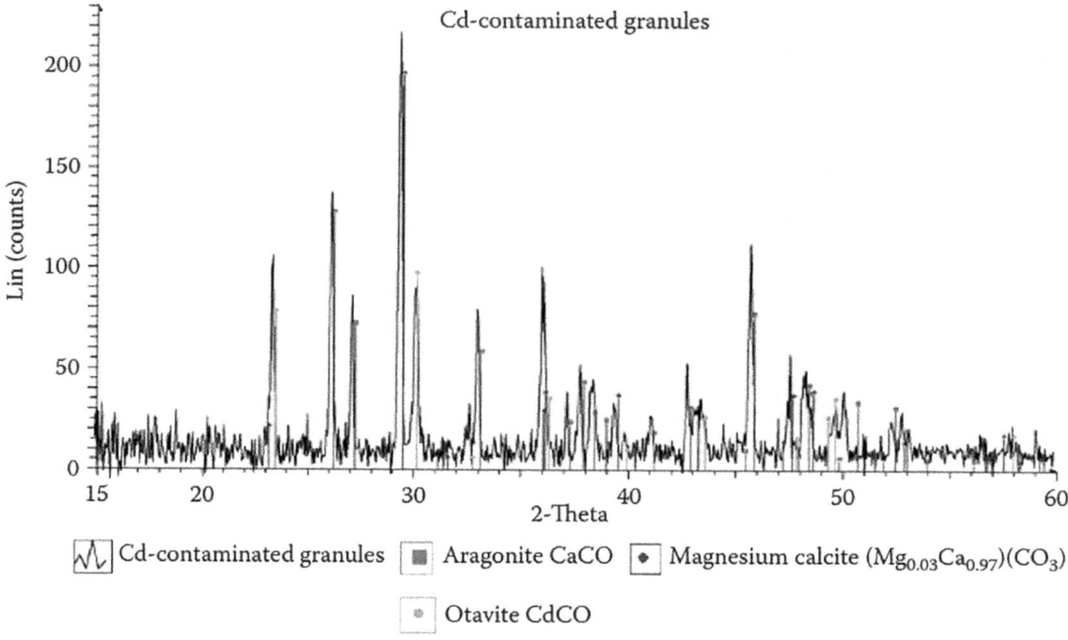

FIGURE 6.12   XRD analysis of Cd^{2+}-contaminated aerobic granules. (Adapted from Xu, H. and Liu, Y., *Sep. Purifi. Technol.*, 58, 400–411, 2008 [55].)

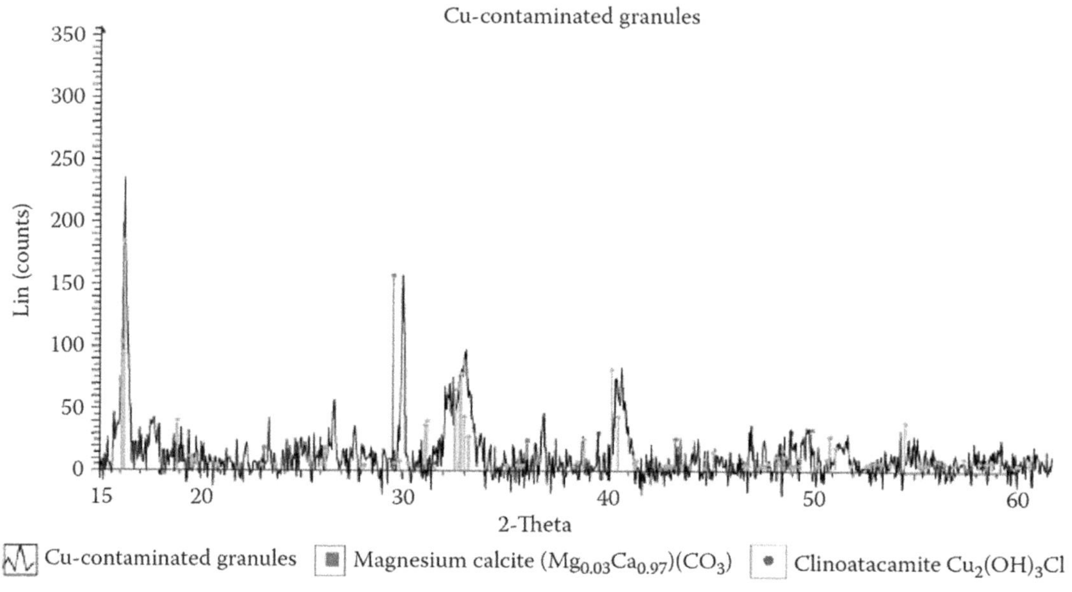

FIGURE 6.13   XRD analysis of Cu^{2+}-contaminated aerobic granules. (Adapted from Xu, H. and Liu, Y., *Sep. Purifi. Technol.*, 58, 400–411, 2008 [55].)

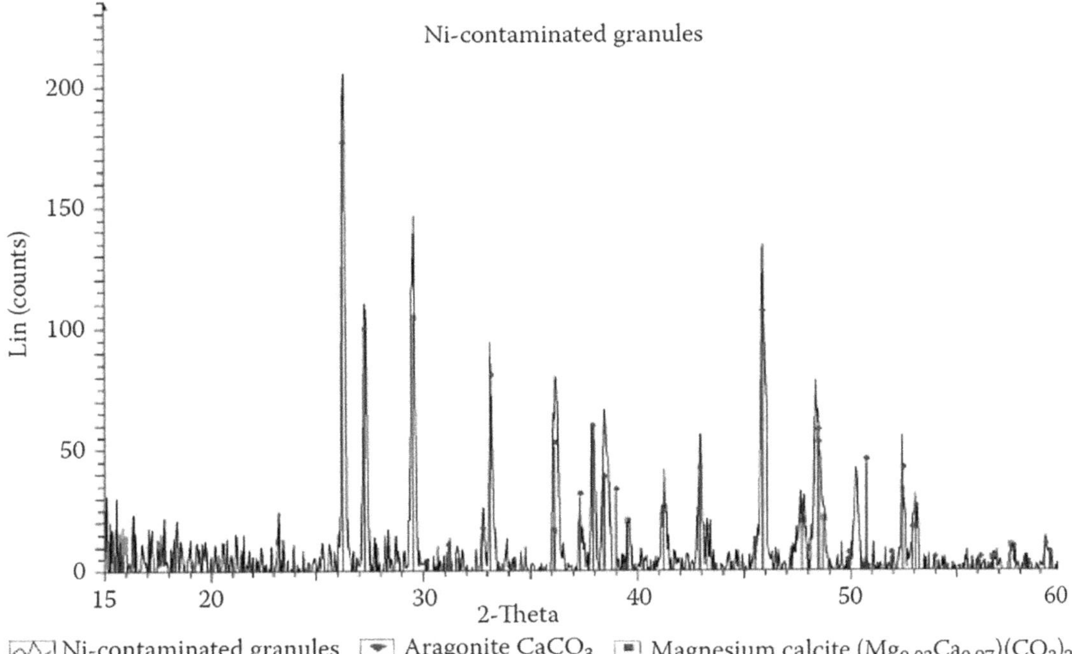

FIGURE 6.14 XRD analysis of Ni²⁺-contaminated aerobic granules. (Adapted from Xu, H. and Liu, Y., *Sep. Purifi. Technol.*, 58, 400–411, 2008 [55].)

precipitated in Cd²⁺- and Cu²⁺-contaminated aerobic granules. One of the mechanisms is believed to be chemical precipitation for the biosorption of Cu²⁺ and Cd²⁺ by aerobic granules; however, there is no evidence for the involvement of chemical precipitation in the removal of Ni²⁺ (Figure 6.14).

6.5.3 FTIR/XPS

Table 6.6 lists the main functional groups corresponding to the peaks observed on FTIR spectra of fresh aerobic granules. The FTIR spectra of raw and metal-loaded granules in the range of 400–4,000 cm⁻¹ were taken to obtain information on the nature of possible interactions between the functional groups of aerobic granules and the metal ions, as presented in Figure 6.15. The peak at 1,650–1,660 cm⁻¹ of granules, corresponding to the superimposition of different amide I bands (C-O stretching coupled with N-H deformation mode) from different materials (including proteins and amide-bearing materials), was significantly decreased after metal loading, indicating the complexation of metal ions with the functional groups from the protein. When the granules were loaded with the metal ions, the peaks at 1,455, 1,306, and 1,249 cm⁻¹ disappeared, leading to the formation of a complex between the metal and the C-O-C group of polysaccharides, the CH₂ group of lipids, and the carboxyl group of the protein. Compared to the intensities of the C-O stretching of the hydroxyl group (C-O-H) from saccharides at 1,053 cm⁻¹, the peak shifted to lower wave numbers 1,049 and 1,047 cm⁻¹ for Co(II)- and Zn(II)-loaded granules, respectively, which demonstrated that hydroxyl groups were involved in the metal biosorption [30].

Table 6.7 shows the elemental composition of different kinds of aerobic granules. It also shows that fresh aerobic granules mainly comprised seven major elements, namely, C, H, O, N, S, P, and Ca. As for heavy metals, the element Cd was not detected in fresh aerobic granules, whereas very small amounts of Cu, Ni, and Zn were found in fresh aerobic granules. These are simply because Cu, Ni, and Zn were found in fresh aerobic granules. These are simply because Cu, Ni, and that

TABLE 6.6

Main Functional Groups on Fresh Aerobic Granule

Wave Number (cm⁻¹)	Vibrational Type	Functional Type
≈3,200~3,600	Overlapping of stretching vibration of OH and NH	OH into polymeric compounds and amine
≈2,927	CH stretching (CH_2 and CH_3 groups)	
≈1,739~1,725	Stretching vibration of C=O	Membrane lipids, fatty acids, and carboxylic acids
≈1,648	Stretching vibration of C=O and C-N (amide I)	Protein (peptidic bond)
≈1,520~1,540	N-H bending and C-N stretching in amides (amide II)	Protein (peptidic bond)
≈1,370~1,410	Symmetric stretching for deprotonated COO⁻ group	
≈1,240~1,260	Deformation vibration of C=O	Carboxylic acids
≈1,130~1,160	Stretching vibration C-O-C	Polysaccharides
≈1,082	Bending vibration of C-O	Polysaccharides
≈1,050~1,060	Stretching vibration of OH	Polysaccharides
<1,000	"Fingerprint" zone	Phosphate or sulfur functional groups

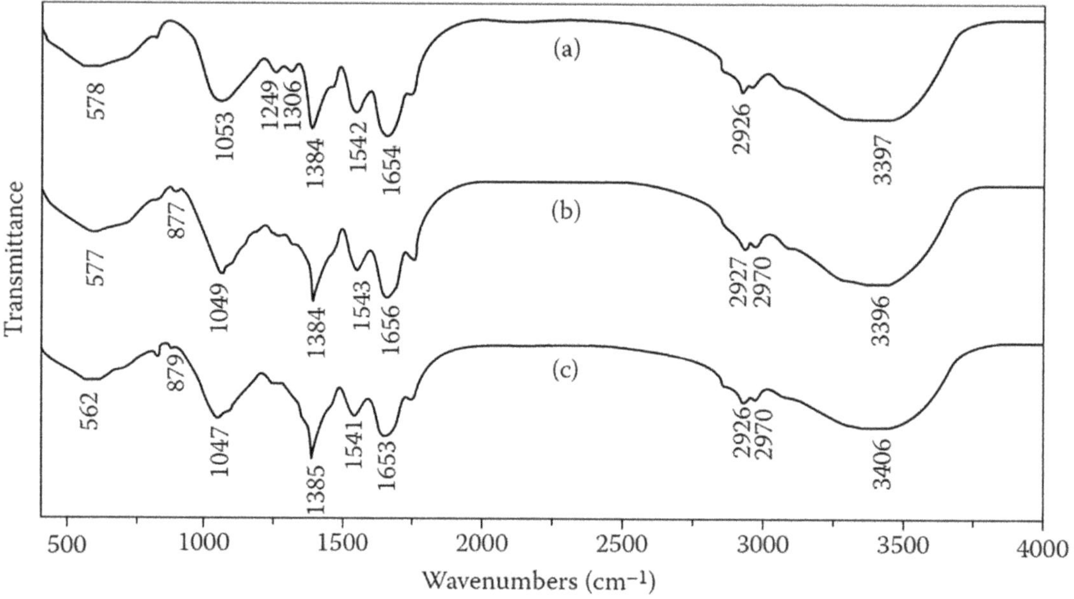

FIGURE 6.15 FTIR spectra of (a) fresh, (b) Co(II)-loaded, and (c) Zn(II)-loaded granules. (Adapted from Sun, X.F. et al., *J. Colloid Interface Sci.*, 324, 1–8, 2008 [30].)

Cu, Ni, and Zn were added to synthetic wastewater as the trace elements required for microbial growth. The Ca content in fresh aerobic granules was as high as 150.36 mg/g, whereas it was 1.37 and 3.47 mg/g for Mg and K, respectively. It seems that aerobic granules would contain a significant amount of light metal ions.

Moreover, Figures 6.15–6.18 showed the XPS survey scanning spectra of aerobic granules before and after Cd^{2+}, Cu^{2+}, and Ni^{2+} biosorption, respectively. The changes in binding energy (BE) of the coordination carbon atom (C1s) in aerobic granules before and after metal adsorption are presented in Figure 6.19. It is clear that the C1s spectra of all the samples comprised three peaks with BE of

TABLE 6.7
Elemental Compositions of Different Kinds of Aerobic Granules (mg/g Dry Weight of Granules)

| | Types of AG | | | |
Element	Fresh AG	Cd-AG	Cu-AG	Ni-AG
C	315.830	314.730	298.330	320.200
H	44.470	43.100	40.030	44.300
N	49.930	57.100	53.200	56.900
S	4.770	4.370	4.930	4.730
O	421.390	358.740	419.060	404.100
P	4.344	3.920	3.790	4.100
Cd	0.000	84.520	0.000	0.000
Cu	0.250	0.229	44.140	0.263
Ni	0.027	0.005	0.005	21.720
Zn	0.239	0.140	0.162	0.217
Ca	150.360	129.860	134.380	140.280
Mg	1.370	0.780	0.312	0.804
K	3.470	1.365	0.390	1.131
Others	3.550	1.141	1.270	1.255

Source: Xu, H. and Liu, Y., *Sep. Purifi. Technol.*, 58, 400–411, 2008 [55].
AG, aerobic granules.

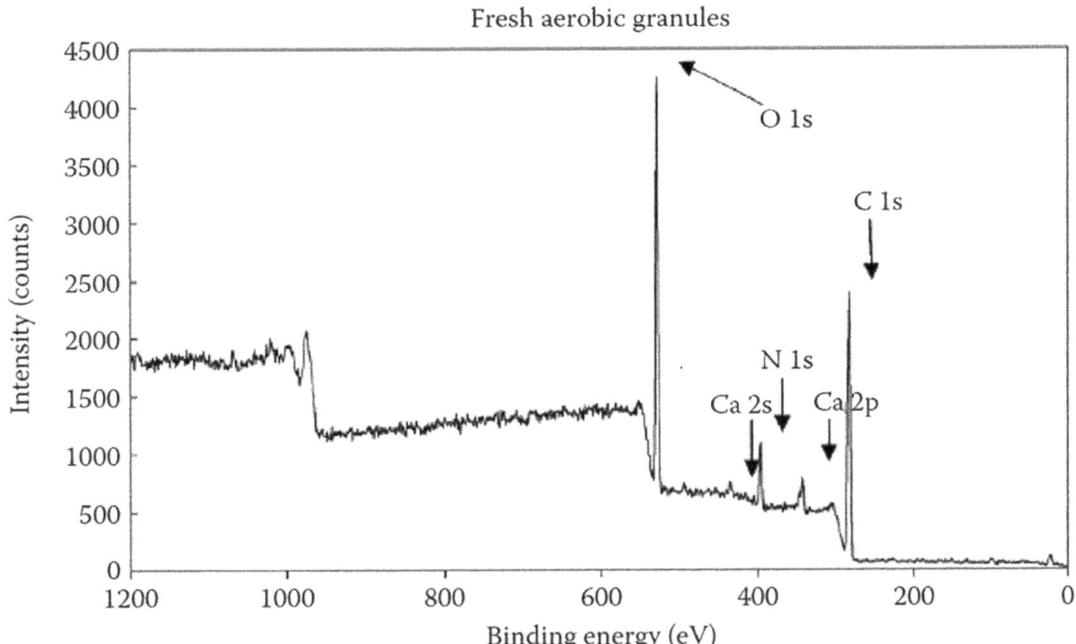

FIGURE 6.16 XPS survey scanning spectrum of Cd^{2+}-contaminated aerobic granules. (Adapted from Xu, H. and Liu, Y., *Sep. Purifi. Technol.*, 58, 400–411, 2008 [55].)

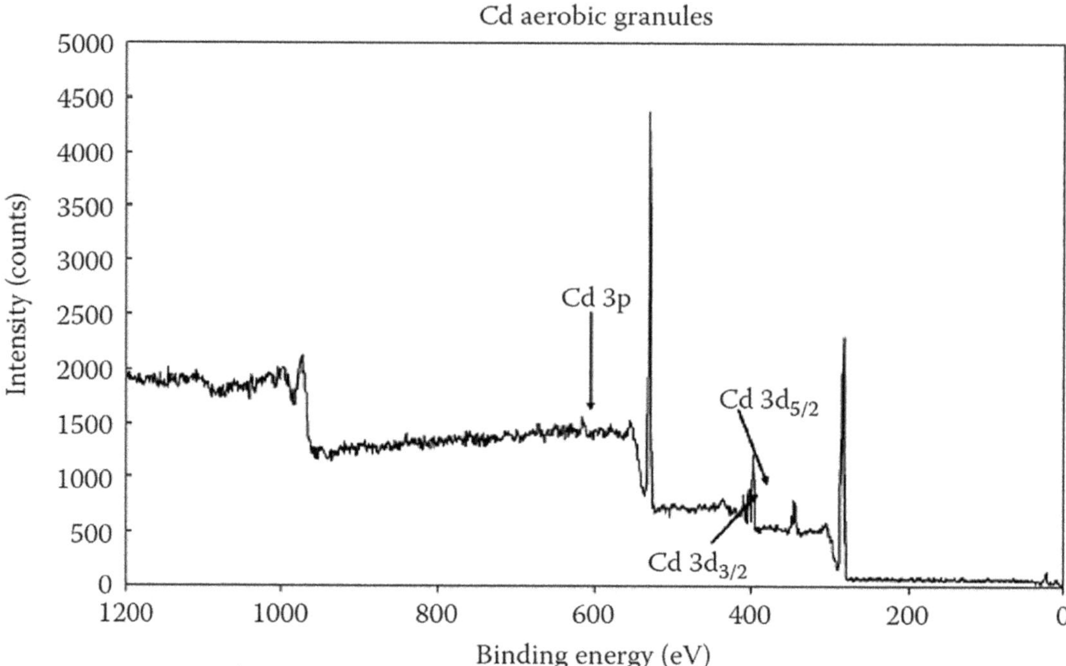

FIGURE 6.17 XPS survey scanning spectrum of Cd^{2+}-contaminated aerobic granules. (Adapted from Xu, H. and Liu, Y., *Sep. Purifi. Technol.*, 58, 400–411, 2008 [55].)

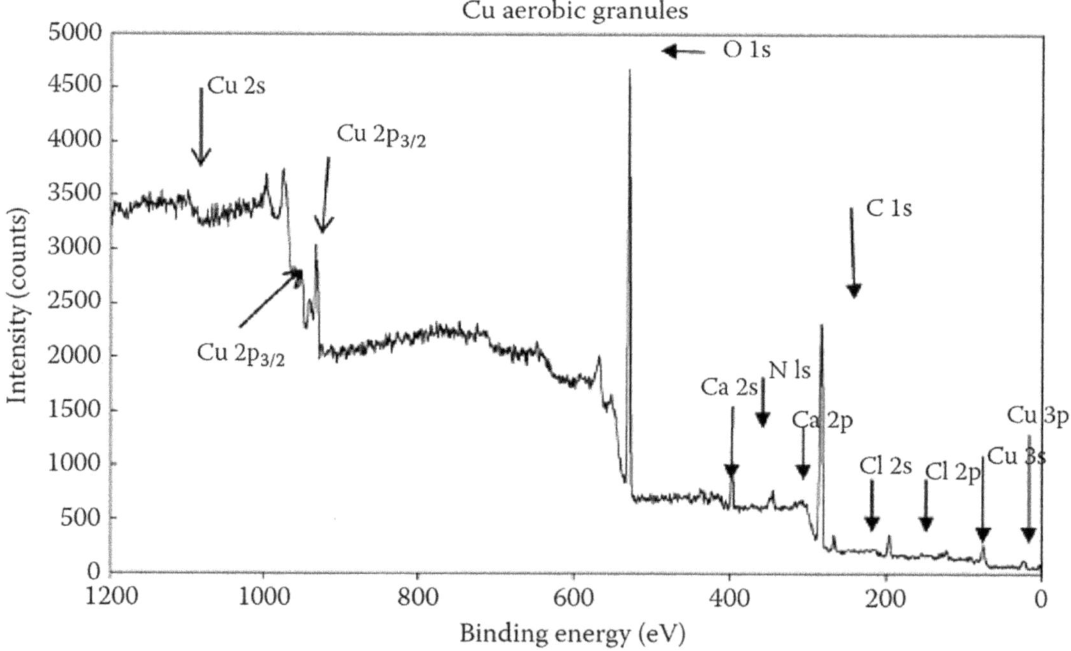

FIGURE 6.18 XPS survey scanning spectrum of Cu^{2+}-contaminated aerobic granules. (Adapted from Xu, H. and Liu, Y., *Sep. Purifi. Technol.*, 58, 400–411, 2008 [55].)

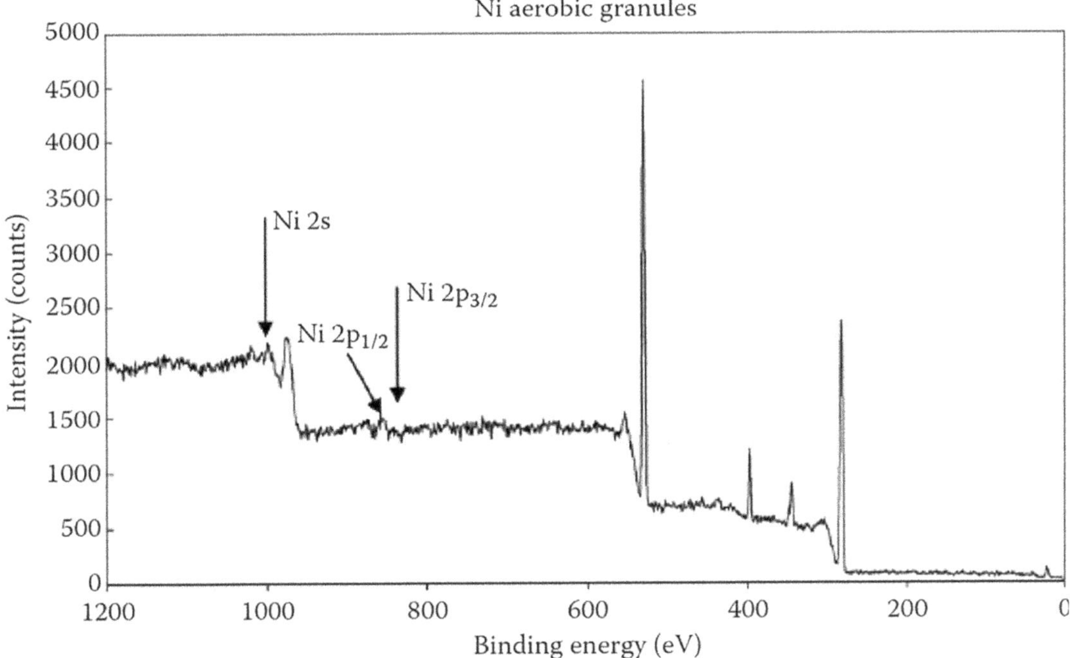

FIGURE 6.19 XPS survey scanning spectrum of Ni^{2+}-contaminated aerobic granules. (Adapted from Xu, H. and Liu, Y., *Sep. Purifi. Technol.*, 58, 400–411, 2008 [55].)

284.1, 285.2, and 287.2 eV, which are attributed to the carbon in C-C, C-OH, and O=C-O, respectively. The result verified that a high concentration of hydroxyl groups and carboxylate groups were present on the aerobic granule surface [30]. The analyses by FTIR and XPS showed that functional groups on aerobic granules, such as alcoholic, carboxylate, and ether, would be the main binding sites for biosorption of the studied heavy metals by aerobic granules (Figure 6.20).

6.6 ECONOMY

Most research about granular sludge has been limited to laboratory experiments: cultivation, formation mechanism, and characteristics. Little literature exists containing full costs and pilot-plant-scale successful application. Studies on metal biosorption evaluated adsorption capacity and mechanism, and cost evaluation remains to be explored [56–59].

For metals uptake, the advantages of granular sludge are high microbial density and excellent settling ability, so are the adsorption capacity and easy postseparation. In the pilot-scale granular sludge treatment system, adsorbents can be harvested as excess sludge, which is treated as waste. Thus, the cost of granular sludge is lower than that of chemically synthesized adsorbents or biosorbents developed from an expensive culture medium. Considering local availability, processing required, and treatment conditions, different biosorbents are difficult to compare. Therefore, much work is needed to demonstrate application costs at the industrial scale. The first step would be the design and operation of a continuous or column test. Potential problems are whether additional solid-liquid separation is needed to satisfy effluent quality.

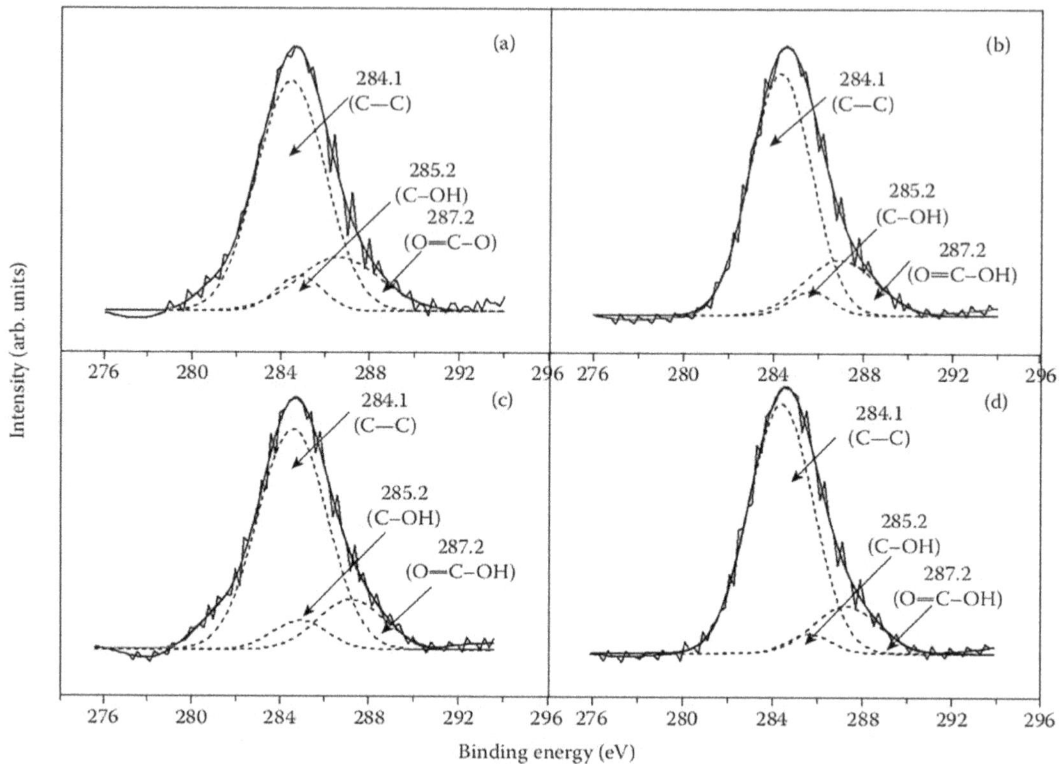

FIGURE 6.20 XPS spectra of aerobic granular sludge: (a) fresh, (b) Co(II)-loaded, (c) Zn(II)-loaded, and (d) Co(II)/Zn(II)-loaded granules. (Adapted from Sun, X.F. et al., *J. Colloid Interface Sci.*, 324, 1–8, 2008 [30].)

6.7 COMMENTS ON FUTURE DIRECTIONS

6.7.1 MECHANISM RESEARCH

1. The mechanisms involved in biosorption or metal-microbe interactions should be further studied with great effort by utilizing various techniques and their combinations [33,50]. Important influencing factors, such as pH and selectivity of co-ions for biosorption systems, have not been fully understood.
2. Mathematical models of equilibrium and kinetics, in particular mechanical models such as the surface complexation models, to simulate the multimetal ion or co-ion biosorption system, are important aspects of future biosorption studies [3] and should be developed.
3. The development of dynamic models to simulate the biosorption process and offer useful information for its application should receive more attention [60].
4. Molecular biotechnology, a powerful tool to elucidate the mechanism at the molecular level, should be considered more closely in the future to construct an engineered organism with higher sorption capacity and specificity for target metal ions.

6.7.2 APPLICATION OF BIOSORPTION TECHNOLOGY

Despite the failure of an attempt to commercialize biosorption in wastewater treatment, it is necessary to continue to explore the various aspects relevant to the application.

1. The physicochemical conditions, for example, pH and multi-ionic composition, should be chosen to simulate the real wastewater on the basis of thermodynamics and reaction kinetics studies.
2. Optimization of the parameters of biosorption process, including reuse and recycling by studying diffusion resistance and fluid dynamics on a sorption column or chemical engineering reactors, such as a fluidized bed reactor.
3. Immobilization of biomaterials is another key aspect for the purpose of biosorption application. It is important for decreasing the cost of immobilization and consequently distribution, regeneration, and reuse of biosorbents [60]. Although the continuous process of immobilized cells has been realized at the laboratory scale, there is still a long way to go for biosorption commercialization. Selection of good and cheap support materials for biomaterial immobilization, improvement of reuse methods, and enhancement of the properties of immobilized biosorbents such as pore ratio, mechanical intensity, and chemical stability are also important factors for applying the full-scale biosorption process.
4. The readers are referred to the recent literature for more technical information on the biosorption of heavy metals onto granular sludge [61–71].

GLOSSARY

Bioaccumulation It is an active metabolic process requiring energy to support the bioaccumulation process. Biosorption is a metabolically passive biochemical process that does not require energy and occurs naturally in all living organisms. Therefore, biosorption is different from bioaccumulation.

Biosorption (heavy metal) It is a biochemical process that occurs naturally in certain biomass, allowing the biomass to passively concentrate and bind contaminants onto its cellular structure without the support of energy. Biosorption is used for removing toxic heavy metals from industrial wastewater and aiding in environmental remediation. Common contaminants include heavy metals, pesticides, and other organic compounds.

Biosorption biomass Various waste materials, such as granular sludge, eggshells, bones, fungi, seaweed, and crab shells, can efficiently remove toxic heavy metal ions from contaminated water through biosorption. Biosorption offers an economical alternative for environmental cleanup by utilizing abundant biomass to sequester harmful contaminants.

Biosorption capability (heavy metals) It refers to the ability of biological materials to accumulate heavy metals from water or waste through either metabolically mediated or biochemical pathways of uptake.

Granular sludge It refers to a self-immobilized microbial consortium with a high density, commonly used in wastewater treatment reactors. Unlike simple flocs, granular sludge forms aggregates of microbial origin that do not coagulate under reduced hydrodynamic shear. These granules settle significantly faster than activated sludge flocs. They exhibit excellent settleability, high biomass retention, and simultaneous nutrient removal. The technology for cultivating aerobic granular sludge is often applied in SBR and has advantages such as tolerance to toxicity and the ability to treat high-strength wastewaters. Although it's not yet widely established, it shows promise as an alternative to conventional activated sludge processes.

REFERENCES

1. Volesky, B. *Biosorption and Biosorbents*. CRC Press, Boca Raton, FL, pp. 3–5, 1990.
2. Vijayaraghavan, K. and Yun, Y.S. Bacterial biosorbents and biosorption. *Biotechnol. Adv.*, 26, 266–291, 2008.
3. Wang, J.L. and Chen, C. Biosorption of heavy metals by *Saccharomyces cerevisiae*: A review. *Biotechnol. Adv.*, 24, 427–451, 2006.

4. Southichak, B., Nakano, K., Nomura, M., Chiba, N., and Nishimura, O. *Phragmites australis*: A novel biosorbent for the removal of heavy metals from aqueous solution. *Water Res.*, 40, 2295–2302, 2006.

5. Vilar, V.J.P., Botelho, C.M.S., and Boaventura, R.A.R. Equilibrium and kinetic modelling of Cd(II) biosorption by algae gelidium and agar extraction algal waste. *Water Res.*, 40, 291–302, 2006.

6. Suhasini, I.P., Sriram, G., Asolekar, S.R., and Sureshkumar, G.K. Biosorptive removal and recovery of cobalt from aqueous systems. *Process Biochem.*, 34, 239–247, 1999.

7. Kim, D.W., Cha, D.K., Wang, J., and Huang, C.P. Heavy metal removal by activated sludge: Influence of *Nocardia amarae*. *Chemosphere*, 46, 137–142, 2002.

8. Lodeiro, P., Rey-Castro, C., Barriada, J.L., Sastre de Vicente, M.E., and Herrero, R. Biosorption of cadmium by the protonated macroalga *Sargassum muticum*: Binding analysis with a nonideal, competitive, and thermodynamically consistent adsorption (NICCA) model. *J. Colloid and Interface Sci.*, 289, 352–358, 2005.

9. Haytoglu, B., Demirer, G.N., and Yetis, U. Effectiveness of anaerobic biomass in absorbing heavy metals. *Water Sci. Technol.*, 44, 245–252, 2001.

10. Liu, Y., Lam, M.C., and Fang, H.H. Adsorption of heavy metals by EPS of activated sludge. *Water Sci. Technol.*, 43, 59–66, 2001.

11. Kadukova, J. and Vircikova, E. Comparison of differences between copper bioaccumulation and biosorption. *Environ. Int.*, 31(2), 227–232, 2005.

12. Davis, T.A., Volesky, B., and Mucci, A. A review of the biochemistry of heavy metal biosorption by brown algae. *Water Res.*, 37(18), 4311–4330, 2003.

13. Niu, C.H., Volesky, B., and Cleiman, D. Biosorption of arsenic (V) with acid-washed crab shells. *Water Res.*, 41(11), 2473–2478, 2007.

14. Sheng, P.X., Ting, Y.P., Chen, J.P., and Hong, L. Sorption of lead, copper, cadmium, zinc, and nickel by marine algal biomass: Characterization of biosorptive capacity and investigation of mechanisms. *J. Colloid Interface Sci.*, 275, 131–141, 2004.

15. Lim, S.F., Zheng, Y.M., Zou, S.W., and Chen, J.P. Characterization of copper adsorption onto an alginate encapsulated magnetic sorbent by a combined FT-IR, XPS, and mathematical modeling study. *Environ. Sci. Technol.*, 42, 2551–2556, 2008.

16. Das, S.K., Das, A.R., and Guha, A.K. A study on the adsorption mechanism of mercury on *Aspergillus versicolor* biomass. *Environmental Sci. Technol.*, 41, 8281–8287, 2007.

17. Lettinga, G., van Velsen, A.F.M., Hobma, S.W., de Zeeuw, W., and Klapwijk, A. Use of the upflow sludge blanket (USB) reactor concept for biological waste water treatment especially for anaerobic treatment. *Biotechnol. Bioengin.*, 22, 699–634, 1980.

18. Morgenroth, E., Sherden, T., van Loosdrecht, M.C.M., Heijnen, J.J., and Wilderer, P.A. Aerobic granular sludge in a sequencing batch reactor. *Water Res.*, 31, 3191–3194, 1997.

19. Beun, J.J., Hendriks, A., van Loosdrecht, M.C.M., Morgenroth, E., Wilderer, P.A., and Heijnen, J.J. Aerobic granulation in a sequencing batch reactor. *Water Res.*, 33, 2283–2290, 1999.

20. Dangcong, P., Bernet, N., Delgenes, J.P., and Moletta, R. Aerobic granular sludge—A case report. *Water Res.*, 33, 890–893, 1999.

21. Hawari, A.H. and Mulligan, C.N. Heavy metals uptake mechanisms in a fixed-bed column by calcium-treated anaerobic biomass. *Process Biochem.*, 41, 187–198, 2006.

22. Adav, S.S., Lee, D.J., Show, K.Y., and Tay, J.H. Aerobic granular sludge: Recent advance. *Biotechnol. Adv.*, 26, 411–423, 2008.

23. Hawari, A.H. and Mulligan, C.N. Biosorption of lead(II), cadmium(II), copper(II) and nickel(II) by anaerobic granular biomas. *Bioresource Technol.*, 97, 692–700, 2006.

24. Liu, Y., Xu, H., Yang, S.F., and Tay, J.H. A general model for biosorption of Cd^{2+}, Cu^{2+} and Zn^{2+} by aerobic granules. *J. Biotechnol.*, 102, 233–239, 2003.

25. Sarker, S.S., Akter, T., Parveen, S., Uddin, M.T., Mondal, A.K., and Sujan, S.M.A. Microalgae-based green approach for effective chromium removal from tannery effluent: A review. *Arab. J. Chem.*, 16(10), 105085, 2023.

26. Kafil, M., Berninger, F., Koutra, E., and Kornaros, M. utilization of the microalga *Scenedesmus quadricauda* for hexavalent chromium bioremediation and biodiesel production. *Bioresource Technol.*, 346, 126665, 2022.

27. Duncan, E.G., Maher, W.A., and Foster, S.D. Contribution of arsenic species in unicellular algae to the cycling of arsenic in marine ecosystems. *Environ. Sci. Technol.*, 49(1), 2015.

28. Xi, Y., Han, B., Kong, F., You, T., Bi, R., Zeng, X., Wang, S., and Jia, Y. Enhancement of arsenic uptake and accumulation in green microalga *Chlamydomonas reinhardtii* through heterologous expression of the phosphate transporter DsPht1. *J. Hazard. Mater.*, 459, 132130, 2023.

29. Liu, Y., Yang, S.F., Xu, H., Woon, K.H., Lin, Y.M., and Tay, J.H. Biosorption kinetics of cadmium(II) on aerobic granular sludge. *Process Biochem.*, 38, 997–1001, 2003.
30. Sun, X.F., Wang, S.G., Liu, X.W., Gong, W.X., Bao, N., and Gao, B.Y. Competitive biosorption of zinc(II) and cobalt(II) in single- and binary-metal systems by aerobic granules. *J. Colloid Interface Sci.*, 324, 1–8, 2008.
31. Sun, X.F., Wang, S.G., Liu, X.W., Gong, W.X., Bao, N., Gao, B.Y., and Zhang, H.Y. Biosorption of malachite green from aqueous solutions onto aerobic granules: Kinetic and equilibrium studies. *Bioresource Technol.*, 99, 3475–3483, 2008.
32. Gai, L.H., Wang, S.G., Gong, W.X., Liu, X.W., Gao, B.Y., and Zhang, H.Y. Influence of pH, ionic strength on Cu^{2+} biosorption by aerobic granular sludge and biosorption mechanism. *J. Chemical Technol. Biotechnol.*, 83, 806–813, 2008.
33. van Hullebusch, E.D., Peerbolte, A., Zandvoort, M.H., and Lens, P.N.L. Sorption of cobalt and nickel on anaerobic granular sludges: Isotherms and sequential extraction. *Chemosphere*, 58, 493–505, 2005.
34. Sağ, Y. and Kutsal, T. Determination of the biosorption heats of heavy metal ions on *Zoogloea ramigera* and *Rhizopus arrhizus*. *Biochem. Eng. J.*, 6, 145–151, 2000.
35. Liu, Y., Yang, S.F., Tan, S.F., Lin, Y.M., and Tay, J.H. Aerobic granules: A novel zinc biosorbent. *Lett. Appl. Microbiol.*, 35(6), 548–551, 2002.
36. Xu, H., Tay, J.H., Foo, S.K., Yang, S.F., and Liu, Y. Removal of dissolved copper(II) and zinc(II) by aerobic granular sludge. *Water Sci. Technol.*, 50(9), 155–160, 2004.
37. Esposito, A., Pagnanelli, F., and Veglio, F. pH-related equilibria models for biosorption in singlemetal systems. *Chem. Eng. Sci.*, 57, 307–313, 2002.
38. Xu, H., Liu, Y., and Tay, J.H. Effect of pH on nickel biosorption by aerobic granular sludge. *Bioresource Technol.*, 97, 359–363, 2006.
39. Schiewer, S. and Volesky, B. Modeling of the proton-metal ion exchange in biosorption. *Environ. Sci. Technol.*, 29, 3049–3058, 1995.
40. van Hullebusch, E.D., Zandvoort, M.H., and Lens, P.N.L. Nickel and cobalt sorption on anaerobic granular sludges: Kinetic and equilibrium studies. *J. Chem. Technol. Biotechnol.*, 79, 1219–1227, 2004.
41. Cheng, W., Wang, S.G., Lu, L., Gong, W.X., Liu, X.W., Gao, B.Y., and Zhang, H.Y. Removal of malachite green (MG) from aqueous solutions by native and heat-treated anaerobic granular sludge. *Biochem. Eng. J.*, 39, 538–546, 2008.
42. Vijayaraghavan, K. and Yun, Y.S. Utilization of fermentation waste (*Corynebacterium glutamicum*) for biosorption of reactive black five from aqueous solution. *J. Hazard. Mater.*, 141, 45–52, 2007.
43. Liu, Y. and Xu, H. Equilibrium, thermodynamics and mechanisms of Ni^{2+} biosorption by aerobic granules. *Biochem. Eng. J.*, 35, 174–182, 2007.
44. Hawari, A.H. and Mulligan, C.N. Effect of the presence of lead on the biosorption of copper, cadmium and nickel by anaerobic biomass. *Process Biochem.*, 42, 1546–1552, 2007.
45. Ni, B.-J., Huang, Q.-S., Wang, C., Ni, T.-Y., Sun, J., and Wei, W. Competitive adsorption of heavy metals in aqueous solution onto biochar derived from anaerobically digested sludge. *Chemosphere*, 219, 351–357, 2019.
46. Volesky, B. and Holan, Z.R. Biosorption of heavy metals, *Biotechnol. Progress*, 11, 235–250, 1995.
47. Volesky, B. *Sorption Biosorption*. BV Sorbex, Inc., Montreal, Canada, p. 316, 2003.
48. Langmuir, I. The adsorption of gases on plane surfaces of glass, mica and platinum. *J. Am. Chem. Soc.*, 40, 1361–1403, 1918.
49. Freundlich, H. Ueber die adsorption in loesungen. *Z. Phys. Chem.*, 57, 385–470, 1907.
50. Kratochvil, D. and Volesky, B. Advances in the biosorption of heavy metals. *Trends Biotechnol.*, 16(7), 291–300, 1998.
51. Azizian, S. Kinetic models of sorption: A theoretical analysis, *J. Colloid Interface Sci.*, 276, 47–52, 2004.
52. Ho, Y.S. and McKay, G. Pseudo-second order model for sorption processes. *Process Biochem.*, 34, 451–465, 1999.
53. Ho, Y.S. and McKay, G. The kinetics of sorption of divalent metal ions onto sphagnum moss peat. *Water Res.*, 34, 735–742, 2000.
54. Nancharaiah, Y.V., Joshi, H.M., Mohan, T.V.K., Venugopalan, V.P., and Narasimhan, S.V. Aerobic granular biomass: A novel biomaterial for efficient uranium removal. *Current Sci.*, 91(4), 503–509, 2006.
55. Xu, H. and Liu, Y. Mechanisms of Cd^{2+}, Cu^{2+} and Ni^{2+} biosorption by aerobic granules. *Sep. Purif. Technol.*, 58, 400–411, 2008.

56. Chen, J.P., Lie, D., Wang, L., Wu, S., and Zhang, B. Dried waste activated sludge as biosorbents for metal removal: Adsorptive characterization and prevention of organic leaching. *J. Chem. Technol. Biotechnol.*, 77(6), 657–662, 2002.

57. Wei, Y.T., Zheng, Y.M., and Chen, J.P. Functionalization of regenerated cellulose membrane via surface initiated atom transfer radical polymerization for boron removal from aqueous solution. *Langmuir* 27(10), 6018–6025, 2011.

58. Lim, S.F., Zheng, Y.M., Zou, S.W., and Chen, J.P. Uptake of arsenate by an alginate-encapsulated magnetic sorbent: Process performance and characterization of adsorption chemistry. *J. Colloid Interface Sci.*, 333(1), 33–39, 2009.

59. He, J and Chen, J.P. Cu (II)-imprinted poly (vinyl alcohol)/poly (acrylic acid) membrane for greater enhancement in sequestration of copper ion in the presence of competitive heavy metal ions: Material development, process demonstration, and study of mechanisms. *Ind. Eng. Chem. Res.*, 53(52), 20223–20233, 2014.

60. Tsezos, M. Biosorption of metals. The experience accumulated and outlook for technology development. *Hydrometallurgy*, 59, 241–243, 2001.

61. Wang, X., Gai, L., Sun, X., Xie, H., Gao, M., and Wang S. Effects of long-term addition of Cu (II) and Ni (II) on the biochemical properties of aerobic granules in sequencing batch reactors. *Appl. Microbiol. Biotech.*, 86, 1967–1975, 2010. https://doi.org/10.1007/s00253-010-2467-9,

62. Kończak, B., Karcz, J., and Miksch, K. Influence of calcium, magnesium, and iron ions on aerobic granulation. *Appl. Biochem. Biotech.*, 174, 2910–2918, 2014. https://doi.org/10.1007/s12010-014-1236-0

63. Ahn, K. H. and Hong, S. W. Characteristics of the adsorbed heavy metals onto aerobic granules: isotherms and distributions. *Desalin. Water Treat.*, 53, 2388–2402, 2015. https://doi.org/10.1080/19443994.2014.927125

64. Li, N., Wei, D., Wang, S., Hu, L., Xu, W., and Du, B., and Wei, Q. Comparative study of the role of extracellular polymeric substances in biosorption of Ni(II) onto aerobic/anaerobic granular sludge. *J. Colloid Interf. Sci.*, 490, 754–761, 2017. https://doi.org/10.1016/j.jcis.2016.12.006

65. Wang, L., Liu, X., Lee, D., Tay, J., Zhang, Y., Wan, C., and Chen, X. Recent advances on biosorption by aerobic granular sludge. *J. Hazard. Mater.*, 357, 253–270, 2018. https://doi.org/10.1016/j.jhazmat.2018.06.010

66. Ajao, V., Nam, K., Chatzopoulos, P., Spruijt, E., Bruning, H., Rijnaarts, H., and Temmink, H. Regeneration and reuse of microbial extracellular polymers immobilised on a bed column for heavy metal recovery. *Water Res.*, 171, 115472, 2020. https://doi.org/10.1016/j.watres.2020.115472

67. Purba, L. D. A., Ibiyeye, H. T., Yuzir, A., Mohamad, S. E., Iwamoto, K., Zamyadi, A., and Abdullah, N. Various applications of aerobic granular sludge: A review. *Environ. Technol. Innovat.*, 20, 101045, 2020. https://doi.org/10.1016/j.eti.2020.101045

68. Pagliaccia, B., Carretti, E., Severi, M., Berti, D., Lubello, C., and Lotti, T. Heavy metal biosorption by extracellular polymeric substances (EPS) recovered from anammox granular sludge. *J. Hazard. Mater.*, 424, 126661, 2022. https://doi.org/10.1016/j.jhazmat.2021.126661

69. Wan, C., Fu, L., Li, Z., Liu, X., Lin, L., and Wu, C. Formation, application, and storage-reactivation of aerobic granular sludge: A review. *J. Environ. Manag.*, 323, 116302, 2022. https://doi.org/10.1016/j.jenvman.2022.116302

70. Fei, Y. and Hu, Y. H. Recent progress in removal of heavy metals from wastewater: A comprehensive review. *Chemosphere*, 335, 139077, 2023. https://doi.org/10.1016/j.chemosphere.2023.139077

71. Liu, C., Shen, Y., Li, Y., Huang, F., Wang, S., and Li, J. Aerobic granular sludge for complex heavy metal-containing wastewater treatment: Characterization, performance, and mechanisms analysis. *Front Microbiol.*, 15, 1356386, 2024. https://doi.org/10.3389/fmicb.2024.1356386

7 Arsenic Pollution

Occurrence, Distribution, and Technologies

Huijuan Liu, Ruiping Liu, Jiuhui Qu, Gaosheng Zhang, Yi Yang, and Lawrence K. Wang

ACRONYMS AND NOMENCLATURES

$(CH_3)_2As(OH)$	dimethyl arsenic acid
$CH_3AsO(OH)_2$	methyl arsenic acid
CMC	critical micelle concentration
CPC	cetylpyridinium chloride
DBPs	disinfection byproducts
DAF	dissolved air flotation
DMA	dimethylarsinate
$dMnO_2$	freshly prepared manganese dioxide
$FeSO_4 \cdot 7H_2O$	iron(II) sulfate heptahydrate
HAAs	haloacetate acids
IX	ion exchange
$KMnO_4$	potassium permanganate
MF	microfiltration
MFM	microfiltration membrane
MMA	monomethylarsonate
NF	nanofiltration
NOM	natural organic matter
PES	polyethersulfone
RAs/Fe	ratio of residual arsenic to iron
RC	regenerated cellulose
RO	reverse osmosis
TDS	total dissolved solids
THM	trihalomethanes
UF	ultrafiltration
UFM	ultrafiltration membranes
USEPA	U.S. Environmental Protection Agency
WHO	World Health Organization

7.1 ARSENIC CHEMISTRY AND ARSENIC POLLUTION SOURCES

7.1.1 ARSENIC CHEMISTRY

Arsenic is a ubiquitous element found in the atmosphere, soils and rocks, natural waters, and organisms [1]. It is classified as a nonmetal or a metalloid, but it is a gray-like metal material usually present in the environment in hexagonal crystalline form [2]. Arsenic is usually combined with other elements such as oxygen, chlorine, or sulfur. Arsenic generally includes organic arsenic and inorganic arsenic.

DOI: 10.1201/9781003541615-7

In the natural circulation of arsenic, organic arsenic, for example, monomethylarsonate (MMA) and dimethylarsinate (DMA), mainly results from the arsenic methylation effects of different organisms such as bacteria, fungi, and mammals [3], mostly in surface waters [1]. In the past several decades, the worldwide use of As-contained pesticides and herbicides has resulted in the increase of organic arsenic in environments. Organic arsenic is generally less harmful than the inorganic form.

Comparatively, the environmental behaviors, distribution, and treatment technologies of organic arsenic have resulted in fewer studies than those of inorganic arsenic. This may be ascribed to the lower prevalence of organic arsenic pollution in water, and to its lower toxicity effects, when compared with inorganic arsenic species. Interestingly, the methylation of inorganic arsenic is demonstrated to reduce toxicity, which is contrary to mercury (Hg).

Soluble, inorganic arsenic exists in either one of two valence states, oxyanions of trivalent arsenite [As(III)] or pentavalent arsenate [As(V)], depending on local oxidation-reduction conditions [4]. Arsenic trioxide, As_2O_3, is also an important species of inorganic arsenic and has a long history of use in Chinese and Western medicine for the treatment of diseases such as acute promyelocytic leukemia, ascribed to its being active against malignancies and showing some activity in patients with accelerated phase chronic myelogenous leukemia and multiple myeloma [5]. Nevertheless, this chapter mainly focuses on As(III) and As(V), owing to their dominant presence in water environments and their contributive effects on worldwide arsenicism.

Arsenite exists in four different species, such as H_3AsO_3, $H_2AsO_3^-$, $HAsO_3^{2-}$, and AsO_3^{3-}; similarly, arsenate exists as H_3AsO_4, $H_2AsO_4^-$, $HAsO_4^{2-}$, and AsO_4^{3-}. Redox potential (Eh) and pH are the most important factors controlling As speciation. Under oxidizing conditions, $H_2AsO_4^-$ is dominant at low pH (pH less than about 6.9), whereas at higher pH, $HAsO_4^{2}$ becomes dominant. However, H_3AsO_4 and AsO_4^{3-} may be present under extremely acidic and alkaline conditions, respectively. Under reducing conditions at pH less than about 9.2, the uncharged arsenite species H_3AsO_3 will predominate [6]. The dissociation and species transformation of these species as a function of pH have been reported earlier [7].

As(III) mainly exists under anoxic conditions in underground water and sediment, and As(V) is often found in aerobic surface waters. In a typical water body, however, the simultaneous presence of As(III) and As(V) is often observed.

The existing forms of arsenic may also be classified as soluble arsenic and particulate arsenic, depending on its adsorption onto colloids or not. Particulate arsenic may generally be removed through filtration units such as media filtration and membrane filtration. This kind of arsenic species classification is of importance in practice, because most technologies are devoted to the transformation of soluble arsenic to particulate arsenic, which will be discussed in detail later.

7.1.2 Sources of Arsenic Pollution

Anthropogenic activities and natural geochemical processes are the two main sources contributing to arsenic pollution in the water environment.

As for anthropogenic contribution, the most important point-pollution source is the discharge of wastewater and waste foils from industries such as mining and smelting. Additionally, the use of arsenic pesticides, herbicides, and crop desiccants and the use of arsenic as an additive to livestock feed, particularly for poultry [1], contribute to nonpoint arsenic pollution all over the world.

In natural geochemical cycles, aqueous arsenic mainly comes from As-containing minerals, rocks, sediments, and soils through processes such as weathering, desorption [8], and reductive dissolution [9]. The occurrence of aqueous arsenic concentration increase is often observed to be with the changes in pH [10], oxidation-reduction conditions, and the reduction in the surface areas of oxide minerals or binding strength between arsenic and mineral surfaces. Atmospheric precipitation also contributes to arsenic pollution, especially in surface waters; however, there is still a lack of sufficient data to quantitatively evaluate its impact on drinking water sources [1].

Arsenic is a major constituent in more than 200 minerals, including elemental As, arsenides, sulfides, oxides, arsenates, and arsenites [1]. The most abundant As ore mineral is arsenopyrite, FeAsS, which is mainly formed under high-temperature conditions in the Earth's crust but is also reported in sediments and orpiment by the microbial precipitation effect [11,12].

Arsenic concentrations in rocks, sediments, and soils generally vary from several mg/kg to several decade mg/kg, and the high-arsenic concentrations in sediments are usually positively correlated with the more pyrite or Fe oxides present. Datta and Subramanian [13] found concentrations averaging 2.0 mg/kg (range 1.2–2.6 mg/kg) in sediments from the River Ganges, 2.8 mg/kg (range 1.4–5.9 mg/kg) in sediments from the Brahmaputra River, and 3.5 mg/kg (range 1.3–5.6 mg/kg) in sediments from the Meghna River [13]. However, in contaminated sediments and soils (e.g., tailing piles and tailing-contaminated soils), arsenic concentrations as high as several decades thousand mg/kg are reported [14,15].

In unpolluted areas, concentrations of As in the atmosphere are usually as low as $10^{-5} - 10^{-3}$ mg/m^3 but increase to 0.003–0.18 mg/m^3 in urban areas and to higher than 1 mg/m^3 close to industrial plants [16]. It is estimated that anthropogenic sources of atmospheric arsenic (around 18,800 tons per year), mainly from industrial processes (e.g., smelting) and fossil fuel combustion, amounted to around 70% of the global atmospheric As flux [17]. Although the health effect of atmospheric As on drinking water sources has not been suitably evaluated, it has been demonstrated to be a major health threat to residents in parts of Guizhou Province, China, who consume foods that are dried over domestic coal fires and directly inhale domestic coal-fire smoke [18].

7.2 OCCURRENCE AND DISTRIBUTION OF UNDERGROUND ARSENIC PROBLEMS

7.2.1 Overview of Worldwide Arsenic Pollution

Quite a few countries (e.g., Argentina, Bangladesh, Chile, China, Hungary, India (West Bengal), Mexico, Romania, Vietnam, Nepal, Myanmar, Cambodia, and the United States) have large aquifers that have been identified to suffer from arsenic occurrence at concentrations above 0.05 mg/L [1]. Arsenic associated with geothermal waters has also been reported in several areas, including hot springs from parts of Argentina, Japan, New Zealand, Chile, Kamchatka, Iceland, France, Dominica, and the United States [1]. Most of the groundwater is contaminated with arsenic at levels from 100 to over 2,000 ppb. Typical As concentrations in different kinds of natural waters have been well determined previously [1]. Generally, arsenic levels in seas and estuaries are low, with average concentrations of several ppb. In underground water, the arsenic concentration is dependent on local geochemical conditions, varying from <0.5 ppb to as high as 5,000 ppb. Anthropogenic activities, such as mining, oil extraction, and arsenical herbicide production, significantly increase arsenic levels in related water bodies such as underground water, geothermal water, sediment porewater, and brine [1].

It is noteworthy that the occurrence of arsenic is the result of complicated geochemical and hydro-geological processes. The presence of arsenic in different As-rich aquifers is ascribed to different geochemical effects. Smedley and Kinniburgh [1] systematically classified the As-rich aquifers in the world into reducing environments, arid oxidizing environments, mixed oxidizing and reducing environments, and sulfide mineralization and mining-related arsenic problems [1].

7.2.2 Arsenic Pollution in Bangladesh

The groundwater contamination problem in Bangladesh is the worst in the world. Bangladesh extends between longitudes 88°01′ and 92°40′ east and latitudes 20°25′ and 26°38′ north. In Bangladesh, arsenopyrite has been identified as the prime source of As pollution [19].

The British Geological Survey (BGS) reported that the groundwater problem in Bangladesh is the unfortunate combined effect of three factors: source of As (As is present in aquifer sediments), mobilization (As is released from the sediments to groundwater), and transport (As is flushed in the natural groundwater circulation) [20]. There are two prevailing hypotheses to describe the mobilization of As into the groundwater in Bangladesh: (a) pyrite oxidation and (b) oxyhydroxide reduction [21–23]. The distribution of As in the groundwater of Bangladesh has been generalized in Ref. [21]. Of the 36 districts with sufficient data on the arsenic levels in groundwater from the shallow aquifer (less than 150 m deep), 14 indicate that more than 50% of their groundwater exceeds the Bangladesh standard of 0.05 mg/L for arsenic.

Arsenic in Bangladesh's underground water mainly exists as the following four species: $H_2AsO_3^-$, $H_2AsO_4^-$, methyl arsenic acid [$CH_3AsO(OH)_2$], and dimethyl arsenic acid [$(CH_3)_2As(OH)$] [22]. BGS [20] has analyzed the 2022 samples collected in deepened shallow tube wells. It is indicated that 51% of the samples are above 0.010 mg/L [the World Health Organization (WHO) Guideline Value], 35% are above 0.050 mg/L (the Bangladesh Drinking Water Standard), 25% are above 0.10 mg/L, 8.4% are above 0.30 mg/L, and 0.1% are above 1.0 mg/L, and the maximum As concentration found in this survey is as high as 1.67 mg/L [20]. The typical ranges of arsenic concentrations in underground water and tube well water, and the analytical report of 41 districts of Bangladesh where As found in groundwater was >0.05 mg/L have been reported in Ref. [23].

There are also many investigations on arsenic contamination in surface waters and soils [24] and arsenic content in food, forage plants, plant tissues, and plants grown in contaminated sites or irrigated by arsenic-polluted water [25].

7.3 ARSENIC TOXICOLOGY AND ARSENIC-RELATED DISEASES

7.3.1 ARSENIC TOXICOLOGY

Arsenic has a long and nefarious history; its very name has become synonymous with poison. In the fifteenth and sixteenth centuries, the Italian family of Borgias used arsenic as their favorite poison for political assassinations. Some have even suggested that Napoleon was poisoned by arsenic-tainted wine served to him while in exile.

Arsenic interferes with a number of essential physiological activities, including the actions of enzymes, essential cations, and transcriptional events in cells [26], and damages the nerves, stomach, intestines, and skin. The toxicology of arsenic is generally classified into acute and chronic toxicity. The high levels of inorganic arsenic (greater than 60 ppm) in food or water can be fatal. The main early manifestations of acute arsenic poisoning include burning and dryness of the mouth and throat, dysphasia, colicky abnormal pain, projectile vomiting, profuse diarrhea, and hematuria. Muscular cramps, facial edema, cardiac abnormalities, and shock can develop rapidly as a result of dehydration [27]. Long-term exposure to arsenic via drinking water results in chronic health risks, including cancer of the skin, kidney, lung, and bladder, as well as other diseases of the skin, the neurological system, and the cardiovascular system [28]. Chronic arsenic toxicity due to drinking arsenic-contaminated water has been one of the worst environmental health hazards affecting eight districts of West Bengal since the early 1980s [29].

Generally, there are four recognized stages of arsenicosis (or chronic arsenic poisoning) as follows.

Preclinical: The patient shows no symptoms, but arsenic can be detected in urine or body tissue samples.
Clinical: Various effects can be seen on the skin at this stage. Darkening of the skin (melanosis) is the most common symptom, often observed on the palms. Dark spots on the chest, back, limbs, or gums have also been reported. Edema (swelling of hands and feet) is often seen. A more serious symptom is keratosis, or hardening of skin into nodules, often on palms and soles. The WHO estimates that this stage requires 5–10 years of exposure to arsenic [30–32]:

Complications: Clinical symptoms become more pronounced, and internal organs are affected. Enlargement of the liver, kidneys, and spleen have been reported. Some research indicates that conjunctivitis (pinkeye), bronchitis, and diabetes may be linked to arsenic exposure at this stage.

Malignancy: Tumors or cancers (carcinoma) affect the skin or other organs. The affected person may develop gangrene or skin, lung, or bladder cancer.

Arsenicosis is greatly associated with the arsenic concentrations in drinking water. Exposure to high levels of arsenic in drinking water leads to excess cancer risk (e.g., skin and lung). However, exposure to lower levels (<100–200 mg/L) does not significantly increase the risks of arsenic-related cancers. Recently, more and more studies have focused on the effects of long-term exposure to low levels of arsenic [33,34].

Long-term exposure to arsenic would lead to a build-up of arsenic in ectodermic tissues (e.g., hair and nails) and an increase in arsenic concentration in the urine of a patient. Generally, the higher arsenic content in these tissues is positively related to the higher arsenic concentrations and the longer exposure to arsenic-polluted water, as reported previously by Eisler [35].

7.3.2 Distribution of Arsenic-Related Diseases in the World

The first chronic endemic arsenic via drinking water in the world was found in Taiwan in 1968, and the largest population influenced by arsenic in drinking water was in Bangladesh. It has been estimated that globally, tens of millions of people are at risk due to exposure to excessive levels of arsenic [28]. To date, exposure to high concentrations of arsenic via drinking water is reported to be associated with chronic symptoms such as skin lesions [35], peripheral vascular disease [36], hypertension [37], Blackfoot disease [38,39], and a high risk of cancers [35,38].

West Bengal and India have long been known to suffer from the problem of arsenic-contaminated groundwater, which is claimed to be the most enormous calamity in the world [32]. Only in the Bengal Basin do more than 40 million people drink water containing excessive As. After the independence of Bangladesh, the government endeavored to provide safe drinking water to the people (by installing tube wells that extract water from subsurface alluvial aquifers) in order to reduce the risks of waterborne diseases. Unfortunately, another serious problem of arsenicosis, which is induced by arsenic-contaminated groundwater, affected this poor agricultural country.

It was reported that about 25 million people in Bangladesh are at risk of As poisoning, and 3,695 (20.6%) out of 17,896 people examined suffer from arsenicosis [1,25,31]. Chowdhury et al. [23] reported that 22 out of 64 districts in Bangladesh had arsenicosis patients, and 21 districts had people with arsenical skin manifestations [23]. They surveyed only 98 villages in these districts and found As patients in 95 villages. From these 95 villages, they surveyed at random 6,973 people, and 2,309 people (33.1%) were found with arsenical skin lesions [23].

Additionally, a survey from the School of Environmental Studies and Dhaka Community Hospital source said that 47 districts are contaminated with As, which represents a total of 241 villages where As patients are suspected, thousands are currently As patients, and a total of 40 deaths are due to As-related diseases. According to some estimates, arsenic in drinking water will cause 200,000–270,000 deaths from cancer in Bangladesh alone [32,40].

In the mainland of China, the first chronic endemic arsenism via drinking water was found in Xinjiang Autonomy Region in 1983. Up to 2004, chronic endemic arsenic via drinking water was found in Taiwan, Xingjiang, Inner Mongolia, Shanxi, Ningxia, Jilin, Qinghai, Anhui province, and certain suburbs of Beijing, and involved eight provinces, including 40 counties and 1,047 villages, with a 2,343,238 population exposed to high-arsenic water [41]. As for the distribution of high-arsenic areas and exposed at-risk populations in China, the Inner Mongolia, Shanxi, Xingjiang, and Jilin regions have the highest populations exposed to high-arsenic water, which accounts for about 50% of the total As-exposed population in China. The highest arsenic concentration detected is 1.86 mg/L in Inner Mongolia and 0.783 mg/L in Shanxi. According to the latest survey [42], the average prevalence of arsenism patients is 9.65%, which includes 0.57% serious patients, 1.47% medium patients, 5.74% mild patients, and 1.86% suspected patients. Skin cancer cases due to arsenic exposure have also been reported [43].

It has been reported that people of different ages are susceptible to arsenism if they are exposed to high-arsenic drinking water, and the prevalence of arsenism increases with increasing age and extended years of residence. The youngest patient that has been reported is only 3 years old [44].

Taiwan district has also been facing serious underground arsenic pollution. Based on the large-scale arsenicosis cases who were exposed to high levels of arsenic concentration in well water (higher than 1,000 mg/L), the United States Environmental Protection Agency (U.S. EPA) assessed the risks of skin cancer [45,46] and bladder and lung cancer [47–49] due to the arsenic in drinking water [50].

7.3.3 DRINKING WATER STANDARD ON ARSENIC

The WHO first revised the guideline for arsenic from 0.05 to 0.01 mg/L in 1993 [51]. Then, Germany lowered its permissible limit of arsenic to 0.01 mg/L in 1996, and the Australian drinking water limit was also reduced from 0.05 to 0.007 mg/L [32]. The current standard in France, Vietnam, and Mexico is 0.05 mg/L for arsenic in drinking water [32]. In the European Union, the arsenic standard level is now set to 0.01 mg/L [32]. The U.S. EPA also implemented the reduction of permissible values of arsenic in drinking water from 0.05 to 0.01 mg/L in 2002 [4]. In China, the newly issued drinking water standard decreased the permissible arsenic level to 0.01 mg/L in 2006 [52]. The stricter standard requires more feasible technologies in engineering for arsenic pollution control.

7.4 TECHNOLOGIES FOR ARSENIC REMOVAL

Arsenic pollution, arsenic-related diseases, and strict standards have generated numerous studies on the technologies and processes for arsenic removal. Generally, the removal of arsenic from water mainly includes (a) transformation of arsenic from aqueous to particulate arsenic through adsorption, precipitation/coprecipitation, coagulation, biological process, and so on, and then removal of particulate arsenic by the liquid-solid separation units (e.g., sedimentation, media filtration, and membrane filtration); (b) removal of arsenic species by the exchange of surface active sites on ion exchange resin; and (c) removal of arsenic by physical pressure-driven processes (i.e., membrane filtration).

On the other hand, As(III), which is commonly present in underground water, is neutral under natural pH conditions and is difficult to remove by adsorption, precipitation/coprecipitation, and ion exchange. The oxidation of As(III) to negatively charged As(V) is important for facilitating As(III) removal and is usually used in engineering [53].

In the following subsections, the different techniques, that is, coagulation/filtration, lime softening (LS), adsorption, ion exchange, membrane, and biological treatment, for arsenic removal are discussed. Generally, each technique shows its advantages and disadvantages in practice, which should be carefully evaluated before choosing a process in engineering.

7.4.1 OXIDATION

Frankly, oxidation can only transform As(III) to As(V) rather than remove arsenic from the aqueous phase, and oxidation should be coupled with a removal process such as coagulation, adsorption, or ion exchange. Nevertheless, oxidation is of critical importance for the achievement of optimal performance of arsenic removal for the processes noted above [4].

The conversion of As(III) to As(V) is often accomplished by providing an oxidizing agent (e.g., chlorine, permanganate, ozone, and hydrogen peroxide) at the head of any proposed arsenic removal process. However, other kinds of oxidants in the solid phase, such as Filox-R™, manganese ore, freshly formed manganese dioxide, and ferric and manganese binary oxide (FMBO), could also contribute to As(III) oxidation, through the effects of interfacial oxidation and/or catalytic oxidation. When these solid-phase oxidants are used, As(III) oxidation and arsenic adsorption often occur simultaneously and contribute to arsenic removal.

Most of the oxidants that are commonly employed in drinking water treatment, that is, chlorine, permanganate, and ozone, are highly effective for this purpose. However, chlorine dioxide and monochloramine are ineffective in oxidizing As(III) [4]. Ultraviolet (UV) radiation alone is also ineffective for As(III) oxidation, but UV oxidation of As(III) may be catalyzed by the presence of sulfite, ferric iron, or citrate. Other kinds of oxidants, such as ferrate, hydrogen peroxide, and potassium peroxydisulfate peroxides, can also preferably oxidize As(III) but will not be discussed in this chapter due to the difficulty in using them in engineering.

7.4.1.1 Chlorine

Chlorine is a very effective oxidant and can steadily oxidize As(III) to As(V) in less than 1 minute. It can be easily used in the gas phase and large-scale water treatment plants; however, in rural areas, liquid sodium hypochlorite is preferred because of its safety during operation. The stoichiometric oxidant demand is 0.95 mg of chlorine (as Cl_2) per mg of arsenite. The ability of chlorine to oxidize arsenite to arsenate is relatively independent of pH in the range of 6.3–8.3 [4]. Dissolved ferrous ion (Fe^{2+}), manganese ion (Mn^{2+}), and organic matter show no significant side effects on the oxidation rate of As(III). However, a higher dosage of chlorine is required for the removal of both Fe^{2+} and Mn^{2+} in practice. Chlorine may also act as a secondary oxidant while being added at the head and can facilitate the regeneration and polishing of media surfaces in media filtration units. Chlorine dosage should be carefully calculated to obtain As(III) oxidation, reductive species oxidation, chlorine consumption in treatment units, disinfection purpose, and residual chlorine demand.

There is an important issue that should be evaluated before using chlorine in the formation of disinfection byproducts (DBPs) such as trihalomethanes (THMs) and haloacetate acids (HAAs). It is a less serious problem than that in surface water due to the lower DBP precursor concentration in underground water. Other issues associated with chlorine are the sensitivity of the treatment process to chlorine and operator safety because it is corrosive.

7.4.1.2 Permanganate

As a powerful oxidizing agent, permanganate is first used for the oxidation of Fe^{2+} and Mn^{2+} and the removal of iron and manganese in underground water. It is also used in surface water treatment for taste and odor control, DBP formation control, coagulation and/or filtration aiding, and so on. In the treatment of urgent source water pollutants that require oxidation, permanganate is also used in some cases due to its being easy to dose as compared with other oxidants. Similarly, permanganate can oxidize arsenite to arsenate effectively and rapidly. Freshly prepared manganese dioxide ($dMnO_2$), which is formed after permanganate reduction, could also oxidize As(III) [54,55] and will be discussed later. Subsequent removal of $dMnO_2$ is required when permanganate is used. Permanganate is effective in refreshing the arsenic removal activity of a filter through the in situ coating of $dMnO_2$ on filter media but will inevitably decrease the running period at cost. Permanganate oxidation produces hardly any DBP and is valuable for a drinking water source whose DBP precursor concentration is high. However, the disinfection potential should be evaluated, and a secondary disinfectant is required under this condition. Permanganate is widely available and is relatively stable with a long shelf life. Consequently, it is easy to transport to and conserve in rural villages.

The use of permanganate has several disadvantages. First, it is difficult to handle. It is very corrosive and stains almost everything purple. The second drawback is the formation of $dMnO_2$, the removal of which requires a solid-liquid separation such as filtration. Additionally, the dosage of permanganate should be carefully controlled to avoid the pink color in the effluent. It is considered an expensive oxidant in some countries, although not in China.

7.4.1.3 Ozone

Ozone is the most powerful and rapid-acting oxidizer produced. It is created by exposing oxygen to high energy, such as an electric discharge field or UV light. The only byproduct from oxidation with ozone is oxygen, which is dissolved in aqueous systems. The stoichiometric oxidant demand is 0.64 mg of ozone per mg of As(III), and As(III) oxidation may be easily finished within 1 minute at this dosage. In the pH range from 6.3 to 8.3, the ability of ozone to convert As(III) to As(V) is relatively independent of pH [4]. Ozone shows promising capabilities of inactivating most microorganisms, but the residual disinfectant concentration is not satisfactorily assured, especially in urban cities. However, in villages where the distribution system is small scale, secondary ozone might be feasible, although evaluation of this is necessary.

In Europe and increasingly in the United States, ozone is more and more widely used as an oxidant and disinfectant in large-scale drinking water treatment plants. However, in rural areas of developing and underdeveloped countries, the use of ozone is inhibited by its high cost. That ozone must be produced on site is another disadvantageous factor for its application in rural areas for As(III) oxidation. Additionally, the UV and the ozone formed by the exposure of oxygen to UV may be used as primary and secondary oxidants in small systems for arsenic removal, owing to its possibility of being installed as an instrument in engineering.

7.4.1.4 Oxidants in Solid Phase

Manganese oxides are interesting metal oxides that exhibit oxidative, adsorptive, and catalytic activities. Commercial Filox-R™, a MnO_2-based medium, is a kind of solid-phase oxidant and has been used in engineering to oxidize Fe^{2+}, Mn^{2+}, sulfite, As(III), and so on [4]. Precisely, Filox-R™ is a catalyst rather than an oxidant because the oxidation of As(III) to As(V) is accomplished by the oxygen in water, which is facilitated by the catalysis effect of Filox-R™. Consequently, the stoichiometric oxidant demand in the Filox-R™ oxidation process depends on the reactions between oxygen and reductive species such as As(III) and Fe^{2+}. The conversion rate of As(III) by Filox-R™ increases with lower pH conditions, higher dissolved oxygen (DO) concentrations, and elevated empty bed contact time (EBCT) values [4].

There are other kinds of manganese oxides, that is, synthetic MnO_2, freshly prepared $dMnO_2$, and $dMnO_2$-coated quartz (which may be produced in the processes of removing Mn(II) and Fe(II) [54,56]), that are valuable for As(III) oxidation. Compared with Filox-R™, these manganese oxides could convert As(III) through heterogeneous oxidation effects, and the capabilities of adsorbing arsenic are satisfactory [54,55,57]. However, the Mn^{2+} that results from reductive dissolution reactions and its concentrations should be well evaluated to meet the guidelines of drinking water standards.

7.4.2 Coagulation/Flocculation Followed by Filtration

7.4.2.1 Coagulation/Flocculation

Coagulation using metal salts with subsequent filtration is the most commonly used treatment method for arsenic removal, especially in large-scale water treatment plants, and is considered one of the best available technologies (BATs) by the U.S. EPA for arsenic removal [4].

Coagulation is the process of destabilizing the surface charges of colloidal and suspended matter to allow for the agglomeration of particles. This process results in the formation of large, dense floc, which is amenable to removal by clarification and/or filtration. Coagulation is typically described as a process consisting of three steps: coagulant formation, particle destabilization, and interparticle collisions. The first two steps, coagulant formation and particle destabilization, occur during rapid mixing, and the third step occurs during flocculation. Comparatively, coagulation is the destabilization of colloids by neutralizing the forces that keep them apart, and flocculation is the action of polymers to form bridges between the larger mass particles or flocs and bind the particles into large agglomerates or clumps. In this chapter, the authors would like to express "coagulation" in brief. Similarly to oxidation, coagulation processes could not remove arsenic from solution, but transform aqueous arsenic to particulate arsenic, which should be removed by subsequent sedimentation and/or filtration processes.

During coagulation, arsenic is transformed into particulate arsenic through three main mechanisms: precipitation, coprecipitation, and adsorption. Precipitation refers to forming insoluble compounds of $Fe(AsO_4)$ or $Al(AsO_4)$. Coprecipitation is defined as the incorporation of soluble arsenic species into a growing hydroxide phase via inclusion, occlusion, or adsorption. Adsorption refers to the electrostatic interaction and/or formation of surface complexes between soluble arsenic and the solid oxyhydroxide surface sites. All three mechanisms can independently contribute to contaminant removal. In the case of arsenic removal, direct precipitation has not been shown to play an important role, whereas coprecipitation and adsorption are both active in arsenic removal. Several factors affect the coagulation process, including coagulant dosage, pH, turbidity, natural organic matter (NOM), anions and cations in solution, zeta potential, and temperature. Additionally, the addition of permanganate is beneficial to arsenic removal by ferric precipitation, mainly through the effect of oxidizing As(III) to As(V) rather than the adsorption capability of in situ formed $dMnO_2$ (Figure 7.1) [55].

Arsenic removal with metal salts has been used since at least 1934 [57]. The most widely used coagulants for arsenic removal are aluminum salts such as aluminum sulfate, and ferric salts such as ferric chloride or ferric sulfate, which hydrolyze to form aluminum and iron hydroxide particulates, respectively. Ferrous sulfate has also been used but is less effective [58]. Generally, iron-based coagulants, including ferric sulfate and ferric chloride, are more effective in removing As(V) than aluminum-based counterparts. For example, Gulledge and O'Connor [59] spiked water with As(V) and tested 10–50 mg/L alum and ferric sulfate at pH 5–8 in a bench-scale study. The ferric sulfate achieved higher removal than the alum (88.6%–99.0% versus 18.5%–93.6%). This is because iron hydroxides have a higher affinity toward As(V) species and are more stable than aluminum hydroxides in the pH range of 5.5–8.5 [59]. A fraction of the aluminum remains a soluble complex,

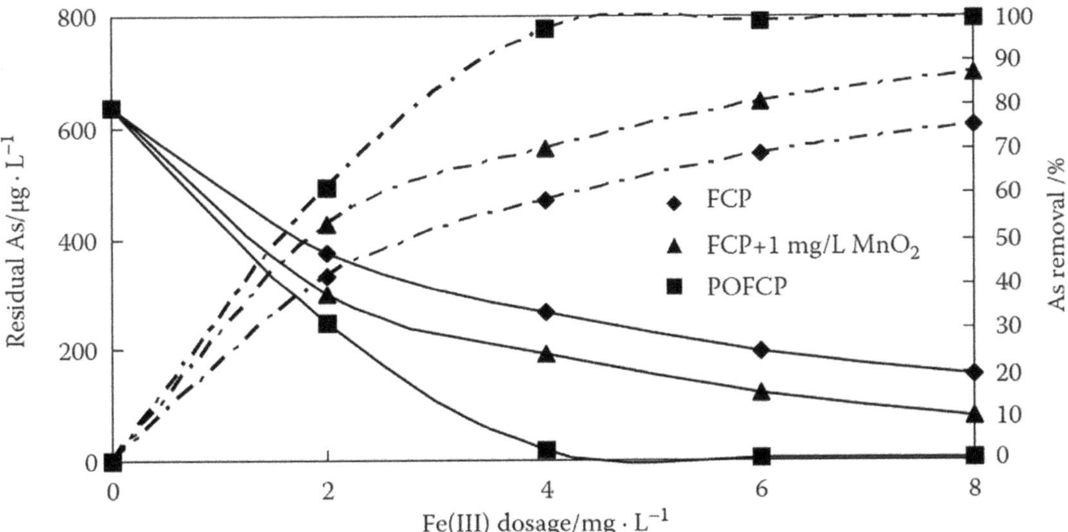

FIGURE 7.1 Comparison of As removal between permanganate-facilitated ferric coprecipitation (POFCP) and combined effects of FCP and $MnO_2(s)$ adsorption. (Adapted from Liu, R.P. et al., *Environ. Sci.* 26(1), 2005, 72–75 (in Chinese).)

incapable of adsorbing As(V) and can pass through the filtration stage. The optimal pH ranges for coagulation with ferric and aluminum salts are 5–8 and 5–7, respectively. At pH values above 7, the removal efficiency of aluminum-based coagulants decreases significantly [59].

1. Excellent arsenic removal is possible with either ferric or aluminum coagulants, with laboratories reporting over 99% removal under optimal conditions and residual arsenic concentrations of less than 1 mg/L [60]. Numerous bench- and pilot-scale studies and short-term full-scale evaluations have been performed to evaluate arsenic removal using conventional coagulation/filtration. Full-scale plants typically reported a somewhat lower efficiency, from 50% to over 90% removal. Scott et al. [61] found that arsenic removal was 81%–96% using 3–10 mg/L of ferric chloride in a full-scale conventional treatment plant [61].

The presence of coexistent ions and their effects on arsenic removal by coagulation with iron-based coagulants have also been investigated. It has been reported that chloride, sulfate, carbonate, and nitrate show little adverse effect on arsenic removal [62]; however, silicate and phosphate significantly reduce arsenic removal.

The effects of silicate on arsenic removal with ferric hydroxide precipitation at different silicate concentrations are strongly pH-dependent, as indicated in Figure 7.2 [63]. The presence of 1.4 mg/L silicate as Si decreased arsenic removal from 99.3% to 62.8% at pH 9.2 and from 98.5% to 67.4% at pH 7.4. Silicate showed a lower extent of inhibiting arsenic removal at lower pH levels of 6.6, which correspondingly decreased arsenic removal from 92.5% to 84.7%. Davis et al. [64] also reported that the higher pH level was normally related to silicate's higher degree of inhibiting arsenic removal with preformed ferric hydroxide precipitates [64].

The negative effect of silicate on arsenic removal by ferric chloride precipitation may be ascribed to (a) the decreased z potential of ferric precipitates and the increased repulsive forces

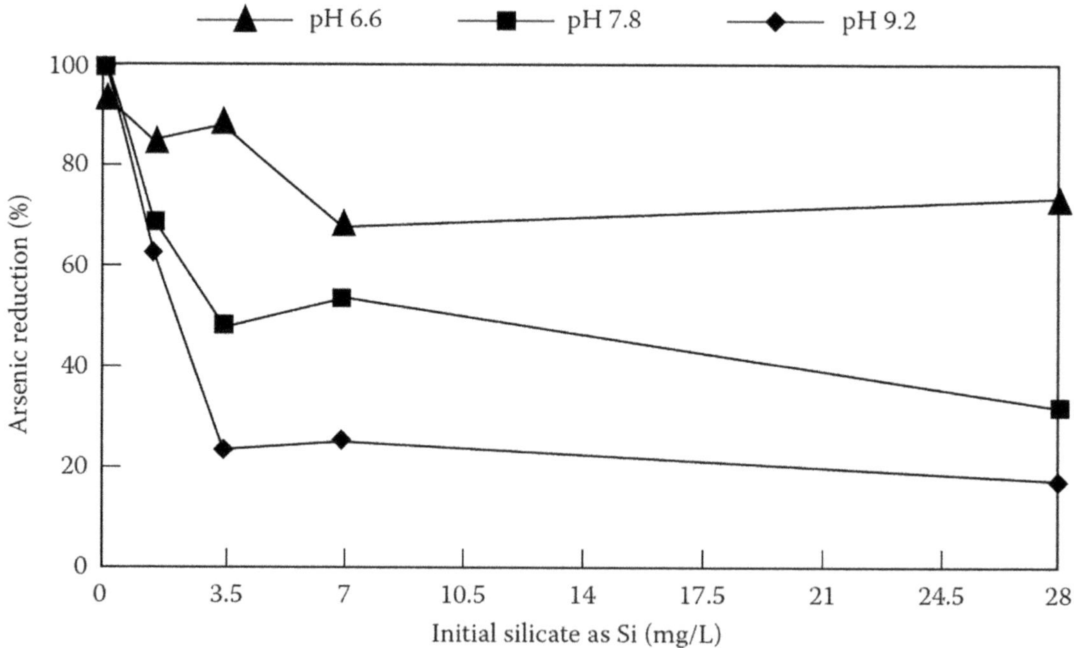

FIGURE 7.2 Effects of silicate on arsenic coprecipitation with ferric hydroxide at different pH levels. (Adapted from Liu, R.P. et al., *Environ. Eng. Sci.* 24(5), 2007, 707–715 [63].)

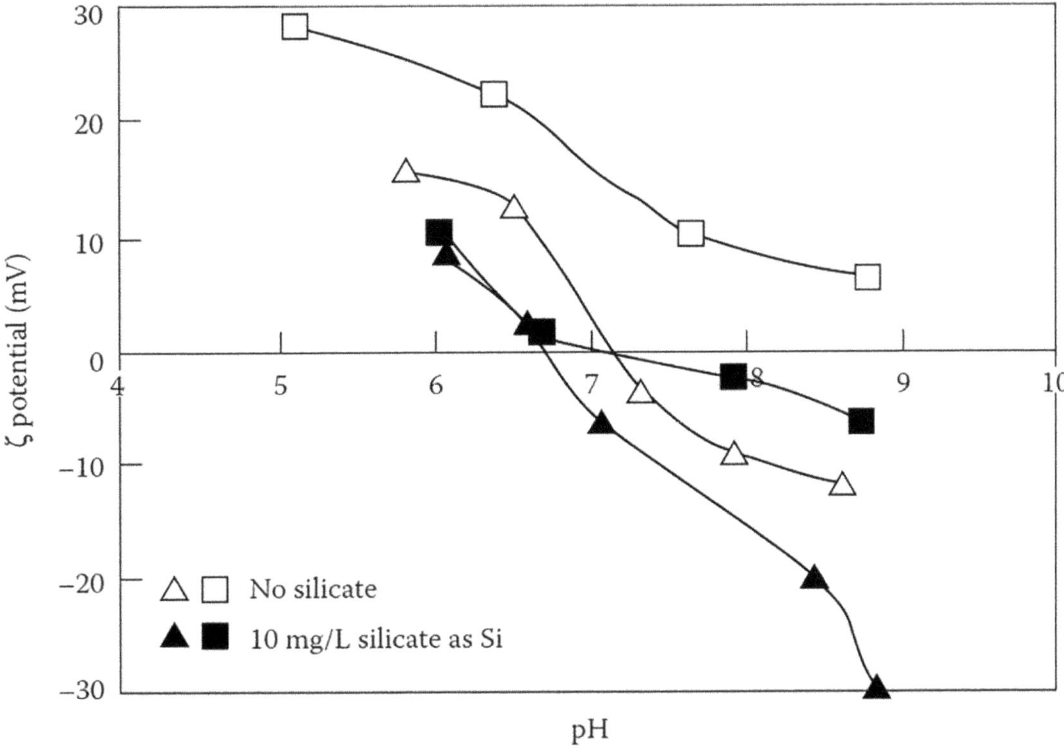

FIGURE 7.3 Effects of silicate on z potential of ferric hydroxide in the absence and presence of 40 mg/L Ca^{2+}. (Adapted from Liu R.P. et al., *Environ. Eng. Sci.* 24(5), 2007, 707–715 [63].)

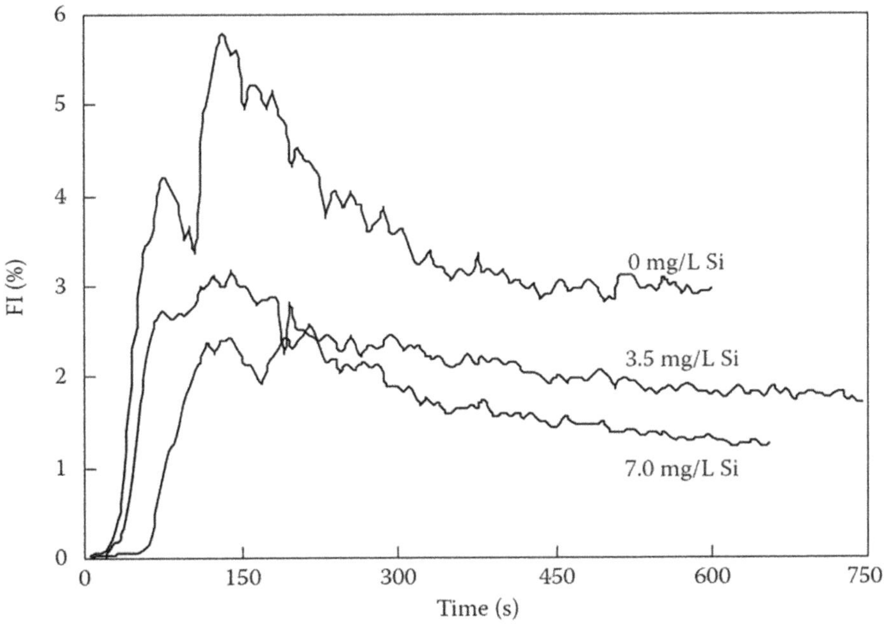

FIGURE 7.4 Effects of silicate on the dynamic online flocculating index (FI) value of ferric hydroxide precipitates. (Adapted from Liu, R.P. et al., *Environ. Eng. Sci.* 24(5), 2007, 707–715 [63].)

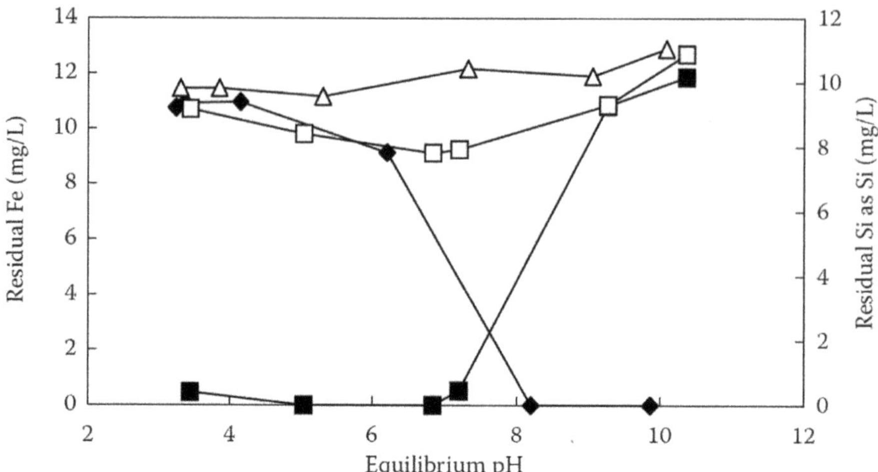

FIGURE 7.5 Residual silicate and filtered iron concentration variation at different pH levels. (Adapted from Liu, R.P. et al., *Environ. Eng. Sci.* 24(5), 2007, 707–715 [63].)

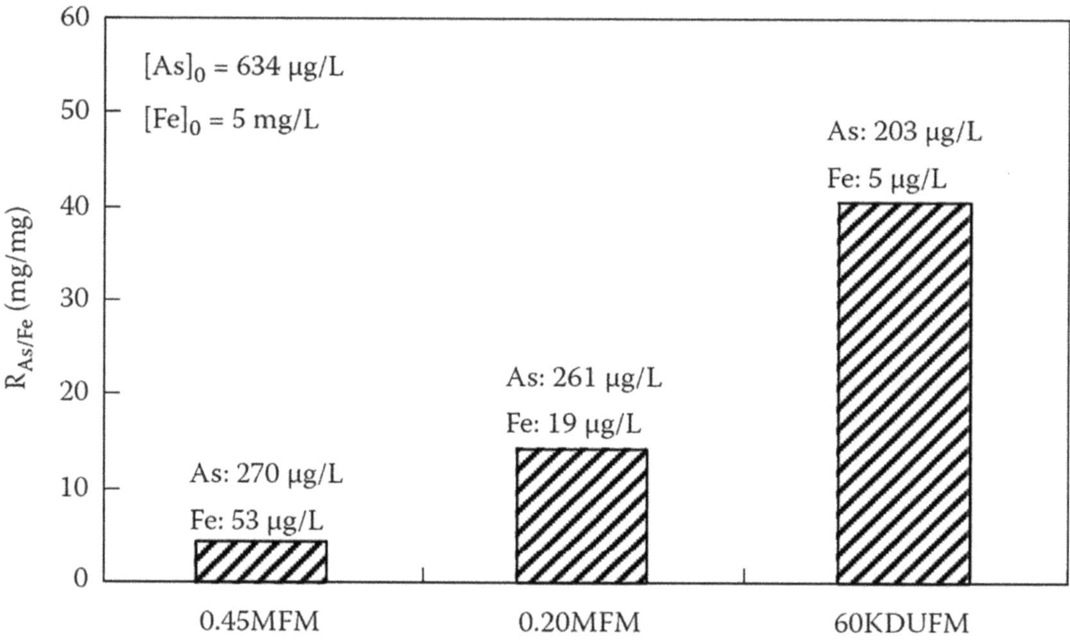

FIGURE 7.6 Ratio of residual arsenic to iron ($R_{As/Fe}$) in filtrate after sequential filtering through 0.45 MFM, 0.20MFM, and 60KDUFM. (Adapted from Liu, R.P. et al., *Environ. Eng. Sci.* 24(5), 2007, 707–715 [63].)

between arsenic and ferric precipitates due to the presence of silicate (Figure 7.3), (b) the inhibited precipitation, aggregation, and flocculation of ferric hydroxides (Figure 7.4), and (c) the increased filterable iron concentration and subsequent decreased content of precipitates for arsenic adsorption (Figure 7.5) [23].

To investigate the dominant mechanism involved in silicate hindering arsenic removal, the ratio of residual arsenic to iron (RAs/Fe) was compared for filtrates passing through the different membrane filters: 0.45 mm microfiltration membrane (0.45MFM), 0.20 mm microfiltration membrane

(0.20MFM), and 60 kDa pore size ultrafiltration membranes (60KDUFM). Although the concentrations of arsenic and iron decreased steadily after sequential filtering with 0.45MFM, 0.20MFM, and 60KDUFM, the RAs/Fe values increased from 4.43 to 40.51 mg/mg (Figure 7.6), indicating the existence of high levels of aqueous ionic soluble arsenic in the filtrate. Consequently, the effects of silicate increasing repulsive charges and decreasing surface sites available for arsenic, rather than facilitating the formation of tiny ferric colloids with adsorbed arsenic, are more pronounced in silicate, hindering arsenic removal [23].

Fortunately, the widely presented calcium ion (Ca^{2+}) reduces the adverse effects of silicate and facilitates arsenic removal to a certain extent [62,65,66], and Ca^{2+} favoring ferric aggregation is more significant than Ca^{2+} improving arsenic removal, in comparison with the increased z potential of ferric (hydro)oxides due to the presence of Ca^{2+} [66].

7.4.2.2 Filtration

Filtration units aim to remove particulate arsenic, destabilized colloids, flocs, and so on. The process of filtration involves the flow of water through a fixed bed of granular media, which is called media filtration, or forcedly through a semipermeable membrane by a pressure differential at low speed, which is called membrane filtration. Units of sedimentation or clarification are required when turbidity in the source water is higher than 10 NTU; most underground water fortunately would not exceed this value.

As for the design and optimization of a filter for arsenic removal, the parameters of media materials, namely, grain size and size distribution, filtration velocity, and filter depth, should be well evaluated. The common materials are sand, anthracite coal, garnet, and ilmenite. If direct filtration or contact filtration is employed, relatively low velocity is preferred except that the floc is satisfactorily formed before filtration.

Membrane filtration includes microfiltration (MF), ultrafiltration (UF), nanofiltration (NF), and reverse osmosis (R/O). Generally, NF and RO remove components through chemical diffusion.

UF and MF remove components through physical sieving. MF and UF show little potential of removing aqueous arsenic, but could promisingly remove particulate arsenic species. Consequently, MF and UF may be chosen as the membrane type following coagulation, which is more cost-effective than NF and R/O.

As for the coagulation/MF process, the actual efficiency of arsenic removal is highly dependent on the quality of the raw water. pH adjustment might be required prior to coagulation. The dosing of polyelectrolytes as coagulant aids may lead to enhanced permeate fluxes of MF; however, the polyelectrolyte has no effect on residual arsenic concentration [67].

Conclusively, coagulation followed by filtration can effectively remove many suspended and dissolved constituents from water besides arsenic, notably iron, manganese, phosphate, and fluoride. Great reductions are also possible in turbidity, odor, and color. Therefore, coagulation/filtration to remove arsenic will improve other water quality parameters, which is valuable in engineering.

7.4.3 ADSORPTION

Adsorption is one of the most widely used technologies in the treatment of both drinking water and wastewater. Numerous adsorbents, such as active carbon, metal oxides, mineral oxides, hydrotalcites, agriculture products, industrial wastes, adsorbents, and polymer resins, have been proposed to remove different pollutants, including arsenic [68]. The surface modification of adsorbents, that is, active carbon, gravel sand, and zeolite, is also an important strategy for the design and application of adsorbents. There are also some commercial adsorbents available for arsenic removal, such as Kanchan™ Arsenic Filter, ArsenXnp, and Electromedia® IX [69].

Most adsorbents mainly focus on the removal of arsenic from the aqueous phase and cannot contribute to the conversion of As(III) to more easily adsorbed As(V). The addition of oxidants is usually inevitable for achieving high-arsenic removal potential [4]; however, this unfortunately increases the operational complexity of the treatment system and is disadvantageous for its application in rural areas. This chapter introduces two adsorbents that readily oxidize and adsorb arsenic.

7.4.3.1 Manganese Oxides

Manganese oxides, which mainly include synthetic MnO_2, natural manganese oxide minerals, and biogenic manganese oxide, have been widely used as adsorbents, solid oxidants, and catalysts due to their important environmental interfacial characteristics.

The mechanism involved in the heterogeneous oxidation of As(III) by birnessite has been well investigated. Nesbitt et al. [70] demonstrated by X-ray photoelectron spectroscopy (XPS) that the oxidation of As(III) by the synthetic 7°A birnessite surface proceeded by a two-step pathway, involving the reduction of Mn(IV) to Mn((III) [70]:

$$2MnO_2 + H_3AsO_3 = 2MnOOH * + H_3AsO_4 \tag{7.1}$$

where MnOOH* is an Mn(III) intermediate reaction product.

This reaction is followed by the reaction of As(III) with MnOOH*:[71]

$$2MnOOH * + H_3AsO_3 + 4H^+ = 2Mn^{2+} + H_3AsO_4 + 3H_2O \tag{7.2}$$

where Mn-OH represents a reactive hydroxyl group on the MnO_2 surface and $(MnO)_2AsOOH$ represents the As(V) surface complex.

An additional reaction could include the adsorption of As(V) by the MnO_2 surface:

$$2Mn\text{-}OH + H_3AsO_3 = (MnO)_2 AsOOH + 2H_2O \tag{7.3}$$

Manning et al. [54] investigated arsenic removal using synthetic birnessite (MnO_2), indicating that As(III) is oxidized by MnO_2 followed by the adsorption of the As(V) reaction product onto the MnO_2 solid phase [54]. The most likely As(V)-MnO_2 complex is a bidentate binuclear corner-sharing (bridged) complex occurring at MnO_2 crystallite edges and interlayer domains. The As(V)-Mn interatomic distance determined by Extended X-ray Absorption Fine Structure analysis was 3.22°A for both As(III)- and As(V)-treated MnO_2 [54]. Interestingly, the reductive dissolution of MnO_2 by As(III) causes a surface alteration and promotes more active sites for As(V), which is demonstrated to be beneficial to As(V) adsorption [54].

In the As(III)-$dMnO_2$ system, the addition of permanganate slightly increases arsenic removal and shows a much less positive effect on As(III) removal by ferric hydroxide precipitation [56]. NOM decreases, and Ca^{2+} increases arsenic removal by $dMnO_2$. The mechanisms of Ca^{2+} favoring arsenic removal include the increased z potential and the facilitated destabilization of freshly prepared $dMnO_2$ due to the presence of Ca^{2+} [56].

Column experiments were conducted to evaluate the As removal potential for a natural manganese oxide using different particle sizes and flow rates [71]. Generally, total adsorption capacity varies with flow rate and particle size, which were interpreted using the effective diffusivity of arsenate in the grain as a single adjustable parameter by a transport model including Langmuir adsorption and mass transfer. Transport was influenced by nonlinear adsorption and intraparticle diffusion, and diffusivities between 0.6 and $7.0 \times 10^{-11}\,m^2/s$ that included intraparticle diffusion were calculated.

7.4.3.2 FMBO

As noted above, traditional adsorbents are generally effective for As(V) removal but fail in the case of As(III). To remove As(III) effectively, a novel FMBO material, which combines the oxidation property of manganese dioxide and the high adsorption features of iron oxides to As(V), was developed by our group [72,73]. The FMBO adsorbent was prepared from low-cost materials, such as potassium permanganate ($KMnO_4$) and iron(II) sulfate heptahydrate ($FeSO_4 \cdot 7H_2O$), using a simultaneous oxidation and coprecipitation method [73]. The prepared FMBO with a high surface area ($265\,m^2/g$) was amorphous. The Fe/Mn molar ratio of this adsorbent was about 2.86 on the surface, a little lower than that of bulk, which was in the range of 2.93–3.02. Iron and manganese were evenly distributed on the surface and existed mainly in the oxidation states +III and +IV, respectively. The presence of abundant surface hydroxyl groups (Fe-OH) was determined [73].

The FMBO adsorbent was effective for both As(V) and As(III) removal, particularly As(III). The maximal adsorption capacities of As(V) and As(III) were 0.93 and 1.77 mmol/g, respectively [73]. Figure 7.7 shows the adsorption isotherms for As(III) and As(V) by Fe-Mn adsorbent. The adsorbent could effectively oxidize As(III) to As(V) within 16 hours under tested conditions. The kinetics of arsenite removal and change in concentrations of arsenic species in the aqueous phase with time are demonstrated in Figure 7.8. No significant Mn release was observed during As(III) removal when the solution pH value was over 6.5. Arsenic removal was enhanced by the presence of metal cations (Mg^{2+} and Ca^{2+}), whereas it was inhibited by coexisting oxyanions such as CO_3^{2-}, SiO_3^{2-}, and PO_3^{4-}, especially at high concentrations. The presence of F^-, Cl^-, SO_2^{4-}, and humic acid did not significantly affect arsenic removal [73].

FMBO had a much higher adsorption capacity toward As(III) than As(V). This was interesting because if only oxidation occurred, as mentioned above, maximal adsorption capacities toward As(V) and As(III) oxidized and then adsorbed should be the same. The higher maximal As(III) adsorption capacity than that of As(V) indicated that something else must take place on the removal

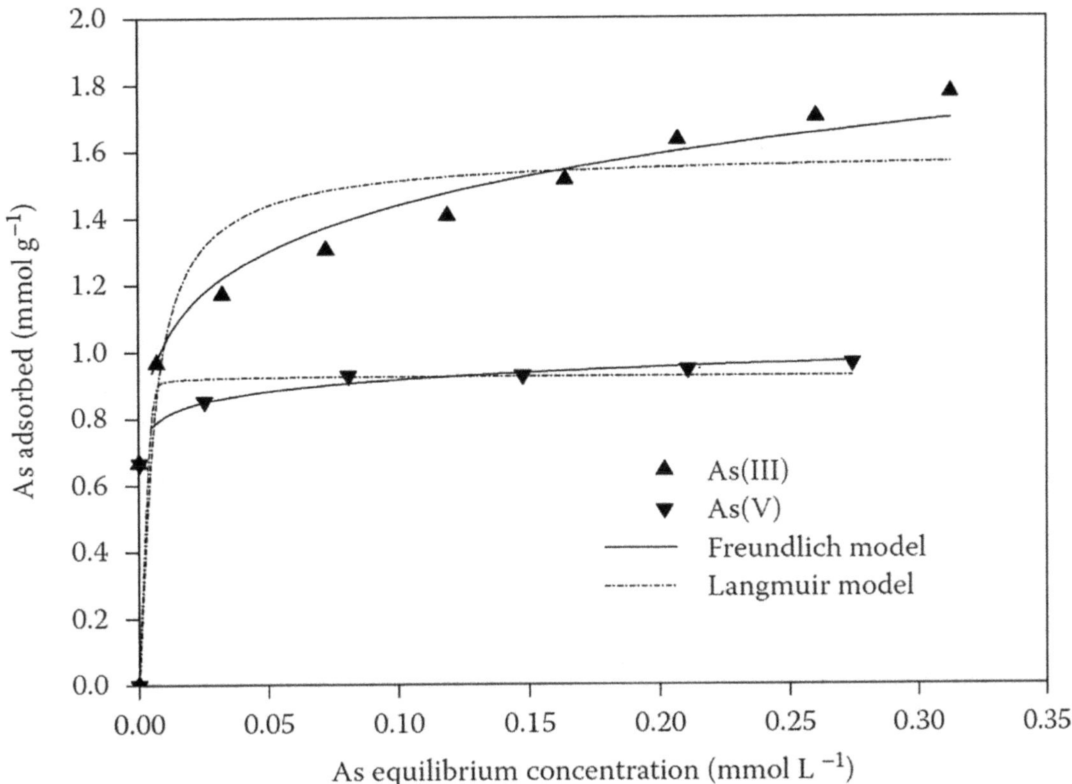

FIGURE 7.7 Adsorption isotherms for As(III) and As(V) by Fe-Mn adsorbent in a 200 mg/L suspension at pH 5.0±0.1 at low equilibrium As solution concentrations. (Adapted from Zhang, G.S. et al., *Water Res.*, 41, 2007, 1921–1928 [73].)

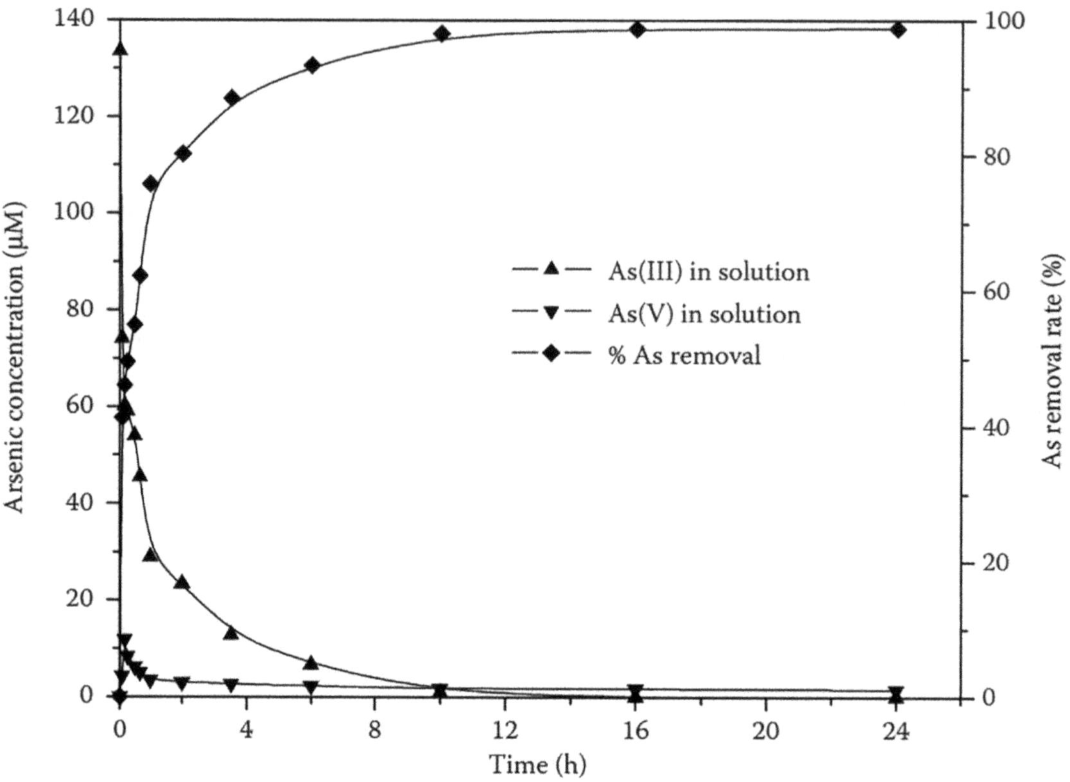

FIGURE 7.8   Kinetics of arsenite removal and change in concentration of arsenic species in aqueous phase with time. (Adapted from Zhang, G.S. et al., *Water Res.*, 41, 2007, 1921–1928 [73].)

of As(III). It was suggested that fresh adsorption sites for arsenic adsorption were created at the solid surface during As(III) oxidation, resulting in an increase of formed As(V) removal.

The removal of As(V) was achieved by forming an inner-sphere complex on the surface of the adsorbent. However, the removal mechanism of As(III) is distinct from that of As(V) and much more complicated since both oxidation and sorption reactions are involved. To understand the process of As(III) removal, a control experiment was conducted to investigate the effect of Na_2SO_3 treatment on arsenic removal, which can provide useful information on the As(III) removal mechanism. The adsorbent was firstly treated with Na_2SO_3, which can lower its oxidizing capacity by reductive dissolution of the Mn oxide, and then reacted with As(V) or As(III). From the experimental results along with FTIR and XPS analyses, an oxidation and sorption mechanism was established and the processes of As(III) removal could be depicted as follows [72].

As(III) was firstly transported to the solid-water interface by convection or diffusion from bulk solution. Then, As(III) was adsorbed onto the surface by forming a surface complex. Adsorbed As(III) near the Mn atoms was oxidized to As(V) by MnO_2 and the formed As(V) was released into the solution with the reductive dissolution of MnO_2. During this process, fresh active adsorption sites were formed at the solid surface. As(V) was then transported to the solid-water interface and adsorbed onto the surface of the Fe-Mn adsorbent, occupying empty adsorption sites or replacing sorptive As(III). The whole process can be briefly represented by reaction/reactions (7.4–7.6) [72]:

$$As(III)\ (aq) + (-S_{Fe\text{-}Mn}) = As(III) - S_{Fe\text{-}Mn} \tag{7.4}$$

$$As(III) - S_{Fe\text{-}Mn} + MnO_2 + 2H^+ = As(V) \ (aq) + Mn^{2+} + H_2O, \tag{7.5}$$

$$As(V) \ (aq) + As(III) - S_{Fe\text{-}Mn} = As(V) - S_{Fe\text{-}Mn}.+ As(III) \ (aq), \tag{7.6}$$

where $(-S_{Fe\text{-}Mn})$ represents an adsorption site on the Fe-Mn adsorbent surface, As(III)-$S_{Fe\text{-}Mn}$ represents the As(III) surface species, and As(V)-$S_{Fe\text{-}Mn}$ represents the As(V) surface species. This process proceeded until As(III) or the available MnO_2 was depleted.

During As(III) oxidation, fresh adsorption sites for arsenic adsorption were created at the solid surface due to the reductive dissolution of manganese dioxide. This was responsible for the higher As(III) uptake.

FMBO can be easily loaded on the surface of porous granular materials such as diatomite, which allows it to be conveniently used in a packed column. In summary, the novel FMBO adsorbent is highly efficient, of low cost, and environmentally friendly and has a high potential of being used in the removal of As(III) from water [72].

7.4.4 Lime Softening

LS is a chemical-physical treatment process commonly used to remove calcium and magnesium cations from water. The addition of lime to water raises the pH, thereby causing a shift in the carbon phosphate equilibrium and the formation of calcium carbonate and magnesium hydroxide precipitates. These formed precipitates can be amenably removed by the filtration process. The arsenic in raw water would not only adsorb onto the calcium and magnesium precipitates but would also form weak complexes with Ca^{2+} and Mg^{2+}. Quantitatively, the interactions between arsenate and Ca^{2+} lead to the formation of calcium-arsenate complexes, and $CaHAsO_4$, $CaHAsO_4 \cdot H_2O$, and $CaHAsO_4 \cdot 2H_2O$ are the dominant species whose water solubilities vary from 2.9×10^{-2} to 5.8×10^{-2}M [74]. As for the calcium-arsenite complexes, Qiu and Liu [75] reported that the water solubility of $Ca(AsO_2)_2$ was 3.5×10^{-3}M [75]. In drinking water treatment, arsenic levels are in the range of several decades ppb to several ppm. The formation of calcium-arsenic complexes and its contribution to arsenic removal is relatively minor. However, as in the case of industrial wastewater with high-arsenic levels (several thousand ppm), lime precipitation may be an important strategy for arsenic removal.

To remove As(V) in underground water, lime is added to increase the solution pH above 10.5. In this range, magnesium hydroxide precipitates, and As(V) is removed by coprecipitation with it. As(V) removal by coprecipitation with calcium carbonate is poor (less than 10%) [4]. LS solely for arsenic removal is uneconomical and is generally considered cost-prohibitive. Only when LS is used to remove hardness from water, this process can be used to simultaneously remove arsenic. There are several drawbacks to LS methods. Firstly, it is difficult to reduce arsenic concentration in water to less than 10 mg/L by LS alone [4]. Sludge disposal is another problem in this treatment method [76] because the total volume of waste produced from LS is typically higher than that produced by coagulation/filtration and coprecipitation processes. Prior to disposal, this waste residual will require thickening and dewatering, most likely using mechanical devices [76].

7.4.5 Ion Exchange

Ion exchange (IX) is a physical-chemical process in which ions are exchanged between a solution phase and a solid phase. Synthetic ion exchange resins are widely used in water treatment to remove many undesirable ions from water. These resins are based on a three-dimensional cross-linked polymer skeleton called the "matrix," which is commonly composed of polystyrene and divinylbenzene. A large number of ionizable functional groups are attached to the matrix through covalent bonding.

These groups are exchanged for ions of similar charge in solution that have a strong exchange affinity for the resins. Charged functional groups fall into four types, namely strongly acidic (e.g., sulfonate, -SO$_3$-), weakly acidic (e.g., carboxylate, -COO-), strongly basic [e.g., quaternary amine, −N+ (CH$_3$)$_3$], and weakly basic [e.g., tertiary amine, −N(CH$_3$)$_2$] [77].

Arsenic removal is accomplished by continuously passing water through one or more columns packed with exchange resins under pressure. In this process, arsenate anion species in the water were exchanged with the negatively charged groups of the resins and were immobilized in the solid phase. As(V) can be removed using strong-base anion exchange resin in either chloride or hydroxide form. These resins are insensitive to pH in the range of 6.5–9.0 [78]. Arsenate can be effectively removed from solution, producing effluent with less than 1 mg/L arsenic by various strong-base anion exchange resins that are commercially available, whereas arsenite, being uncharged, cannot be removed in this way. Therefore, unless arsenic is present exclusively as an arsenate species, an oxidation pretreatment step will be necessary for arsenic removal.

Different ions have different exchange affinities toward the same resin. For example, some common anions compete for sites on a strong-base anions resin according to the following selectivity sequence [4,78]:

$$SO_4^{2-} > NO_3^- > HAsO_4^{2-} > NO_2^- > Cl^-.$$

Traditional sulfate-selective resins are suited for arsenate removal. Nitrate-selective resins can also remove arsenic but are not as effective as sulfate-selective resins. Arsenate removal is relatively independent of solution pH and influent arsenic concentrations. However, competing anions, most notably sulfate and nitrate, have a strong effect. For low-sulfate waters, ion exchange resin can easily remove over 95% of arsenate and treat over a thousand bed volumes before breakthrough. In addition, high levels of total dissolved solids (TDSs) may adversely influence the efficiency of arsenate removal. The ion exchange process is not an economical treatment technology if the source water contains over 500 mg/L of TDS or over 50 mg/L of sulfate [4,78].

In summary, the ion exchange technique is an effective and simple procedure for arsenic removal. But some problems should be considered. One is the phenomenon known as chromatographic peak damping, which can cause the As(V) level in the treatment effluent to exceed that in the influent stream. Another problem is resin fouling. Resin fouling occurs when mica or mineral scale coats the resin or when ions bond the active sites and are not removable by standard regeneration methods. This may have a great influence on the resin's capacity as the media becomes older.

7.4.6 MEMBRANE SEPARATION

Membrane separation processes are receiving much attention due to their provision of extremely low arsenic levels in treated water without solid byproducts [79]. MF and UF alone could not effectively remove arsenic. Quite a few studies investigated the potential of NF membranes (aromatic polyamide and composite polyamide) and RO membranes (polyamide, cellulose acetate, polyether urea, and polyvinyl alcohol) for the removal of arsenic at different pH and arsenic levels from 10 to 600 ppb [80–82]. NF and RO were found to remove arsenate ions over 90% and 95%, respectively [81,82]. However, both use relatively dense membranes requiring high operating pressures (240 kPa [81] and 550–690 kPa [83] for NF membranes and 750 kPa [82] and 138–1,345 kPa [84] for RO membranes).

NF and RO produce very low fluxes (water recovery is 15% for NF and between 10% and 50% for RO). Furthermore, the removal efficiency of NF drops to 65% when water recovery is 65% and to 16% when water recovery is 90%. Additionally, avoiding membrane fouling risks often requires high-quality influent for NF and RO processes. Consequently, in engineering, NF and RO are not recommended for arsenic removal in rural villages in developing and underdeveloped countries.

There are some emerging studies on the pretreatment procedures before MF and UF (which transform aqueous arsenic to species that are unable to pass through MF or UF membranes) for arsenic removal by cost-effective and energy-saving techniques. Coagulation and adsorption are feasible for this strategy, which is discussed above. Recently, Gecol et al. [85] proposed the addition of surfactants (cetylpyridinium chloride, CPC) to the contaminated water stream to achieve high-arsenic rejection by UF [regenerated cellulose (RC) and polyethersulfone (PES)] [85].

In detail, when ionic surfactant (e.g., cationic surfactant) is added to contaminated drinking water above its critical micelle concentration (CMC), it forms micelles (aggregates of 20–200 surfactant molecules) that can effectively bind oppositely charged anionic arsenic species. When a water stream containing these surfactant aggregates is passed through a suitable UF membrane, these surfactant aggregates with bound inorganic arsenic ions are physically too large to pass through membrane pores [85].

The addition of surfactant was observed to significantly increase arsenic removal, which was found to be between 90.9% and 100%. It is clearly indicated that the cationic surfactant micelles can effectively bind the oppositely charged As(V) anions, replacing the counterion Cl- and forming CP[As(V)] micelles. The resultant CP[As(V)] micelles are physically too large to pass through the PES and RC membrane pores and are retained on the retentate side. The removal of arsenic is lower at decreased pH values, and this is attributed to the difference in arsenic dissociation under different pH conditions. The surface micelles in feed water reduce the permeate flux significantly since micelles, being larger than the pore size of the membranes, accumulate on the membrane surface. Comparatively, at pH 5.5, the clean water permeate fluxes of 10 kDa PES, 10 kDa RC, and 5 kDa PES decrease by 77.8%, 24.5%, and 37.1%, respectively. At pH 8, the clean water permeate fluxes of 10 kDa PES, 10 kDa RC, and 5 kDa PES decrease by 75.1%, 21.7%, and 32.7%, respectively. This indicates that the molecular interaction between the surfactant and PES membrane is more pronounced than the RC membrane [85].

However, surfactant concentrations in the permeates are high under different experimental conditions, which should be well controlled before the application of these technologies in engineering.

7.4.7 Dissolved Air Flotation and Filtration

In a research conducted by Wang and Wang [86], over 4,500 gallons of arsenic-contaminated storm run-off water was trucked from a petrochemical chemical company in New Jersey, USA, to the Lenox Institute in Massachusetts, USA, for a feasibility study using a dissolved air flotation (DAF) and sand filtration process system which was continuously operated at 9 gpm (34 L/min) flow for removal of arsenic, oil and grease (O&G), color, turbidity, and chemical oxygen demand (COD) from the contaminated stormwater. Ferric chloride/ferric sulfate (25 mg/L as Fe) and sodium aluminate (15 mg/L as Al_2O_3) were used as the coagulants in DAF, and the sand filter depth was 11 in. (27.94 cm). The combined DAF-filtration (DAFF) system removed arsenic 100%, 90.3%, and 90.6% when influent arsenic concentrations were 0.015, 0.518, and 1.010 mg/L, respectively. The same DAFF system removed over 93% turbidity, over 98% color, over 61% O&G, and over 51% COD at all time. The process water pH was controlled in the range of 6.9–7.2. If the storm run-off water contains 1.010 mg/L (1010 ppb) or below, and the effluent arsenic standard is equal to or less than 200 ppb, DAF alone would be sufficient for stormwater treatment. If the influent arsenic concentration can occasionally be higher than 1010 ppb, and/or the effluent arsenic standard is more stringent (equal to or less than 100 ppb As), using the DAFF system for stormwater treatment is highly recommended.

7.4.8 Technologies Selection Strategy

The advantages and disadvantages of different technologies have been compared and discussed above. It is inferred that there is no single technique that could resolve all the arsenic-related problems in practice, and each technique has its advantages and disadvantages, as concluded earlier [68].

On the contrary, these technologies should be well compared and evaluated before choosing a process for a specified engineering. Generally, the following factors should be carefully considered while selecting a technique in engineering: arsenic concentration in polluted water, coexistent ions and their effect on arsenic removal, the local standard for arsenic in drinking water, treatment capacity of the system, gross investment, acceptable operating fees, educational background of the operator of the system, available field areas, the existence of distribution systems or not, requirements of auto-control systems, and the population that is being served.

GLOSSARY [86–90]

Arsenic water pollution Arsenic contamination in surface and groundwater is a significant environmental concern worldwide. While efforts have primarily focused on eliminating inorganic arsenic As(V), the toxicity of **organic arsenic**, such as dimethyl arsenic acid (DMA), has been somewhat overlooked. DMA is discharged through nonpoint sources, such as storm run-off water, into water bodies, including both groundwater and surface water, posing severe health risks.

Dissolved air flotation (DAF) DAF is a water and wastewater treatment process that clarifies wastewaters by removing suspended matter such as chemical flocs, biological flocs, O&G. DAF system includes the following process steps: (a) Coagulation and Flocculation: The feed water to the DAF float tank is often dosed with a coagulant (such as ferric chloride or, sodium aluminate, aluminum sulfate, and magnesium bicarbonate) to coagulate the colloidal particles and/or a flocculant to conglomerate the particles into bigger clusters; (b) Air Saturation: A portion of the clarified effluent water leaving the DAF tank is pumped into a small pressure vessel (called the air drum) into which compressed air is introduced. This saturates the pressurized effluent water with air; (c) Bubble Formation The air-saturated water stream is recycled to the front of the float tank and flows through a pressure reduction valve. As it enters the front of the float tank, the air is released in the form of tiny bubbles. These bubbles form at nucleation sites on the surface of the suspended particles, adhering to them; and d) Floating and Skimming: As more bubbles form, the lift from the bubbles eventually overcomes the force of gravity. This causes the suspended matter to float to the surface, where it forms a scum layer. The scum layer (called float) is then removed by a skimmer. DAF is widely used in treating industrial wastewater effluents from oil refineries, petrochemical and chemical plants, natural gas processing plants, paper mills, and other industrial facilities and is newly improved for potable water treatment. It's a cost-effective method for separating solids from liquids.

Ferric manganese binary oxide (FMBO) FMBO is developed by researchers as an effective adsorbent for combating organic-arsenic pollution, particularly dimethyl arsenic acid (DMA). The following are some key points: (a) Synthesis: FMBO was synthesized using the coprecipitation method and compared with manganese dioxide (MnO_2) and ferric oxide (FeOOH); (b) Characterization: Various methods, including X-ray powder diffraction (XRD), specific surface area (SBET), zeta (ζ)-potential, scanning electron microscopy (SEM) with energy dispersive X-ray (EDX), and Fourier transform infrared spectroscopy (FTIR), were used to evaluate the produced adsorbents; (c) Adsorption Capacity: At pH 4.0, FMBO exhibited a maximum adsorption capacity of 665.28 µg/g for DMA, surpassing FeOOH (581.32 µg/g) and MnO_2 (510.56 µg/g); (d) Adsorption Models: The Elovich model (heterosphere diffusion reactions) and the pseudo-second-order model best explained the adsorption data; (e) Adsorption Isotherms: The Langmuir and Sips models described DMA adsorption onto FMBO, FeOOH, and MnO_2; (f) Role of FMBO: Systematic characterization highlighted FMBO's predominant role in DMA adsorption at lower pH; (g) **Regeneration** FMBO maintained substantial capability even after five consecutive cycles, with a slight decrease in DMA adsorption (from 99.04% to 94.90%).

REFERENCES

1. P.L. Smedley, D.G. Kinniburgh, A review of the source, behaviour and distribution of arsenic in natural waters, *Appl. Geochem.* 17, 2002, 517–568. https://doi.org/10.1016/S0883-2927(02)00018-5.

2. M. Grayson, D. Eckroth, R. Kromer, The making of an encyclopedia: the 3rd edition of the "Kirk-Othmer encyclopedia of chemical technology", *J. Chem. Inf. Comput. Sci.* 19, 1979, 117–122. https://doi.org/10.1021/ci60019a001.

3. W.R. Cullen, K.J. Reimer, Arsenic speciation in the environment, *Chem. Rev.* 89, 1989, 713–764. https://doi.org/10.1021/cr00094a002.

4. U.S. EPA, *Arsenic Treatment Technology Evaluation Handbook for Small Systems*, US Environmental Protection Agency, Washington DC, USA, 2003.

5. M. Abele, S. Müller, S. Schleicher, U. Hartmann, M. Döring, M. Queudeville, P. Lang, R. Handgretinger, M. Ebinger, Arsenic trioxide in pediatric cancer – a case series and review of literature, *Pediatr. Hematol. Oncol.* 38, 2021, 471–485. https://doi.org/10.1080/08880018.2021.1872748.

6. D.G. Brookins, *Eh-pH Diagrams for Geochemistry*, Springer, New York. 1988.

7. M. Guo, J. Zhou, S. Fan, L. Li, H. Chen, L. Lin, Q. Zhao, X. Wang, W. Liu, Z. Wu, X. Hai, Characteristics and clinical influence factors of arsenic species in plasma and their role of arsenic species as predictors for clinical efficacy in acute promyelocytic leukemia (APL) patients treated with arsenic trioxide, *Expert Rev. Clin. Pharmacol.* 14, 2021, 503–512. https://doi.org/10.1080/17512433.2021.1893940.

8. S. Saha, A.H.M.S. Reza, M.K. Roy, Arsenic geochemistry of the sediments of the shallow aquifer and its correlation with the groundwater, Rangpur, Bangladesh, *Appl. Water Sci.* 11, 2021. https://doi.org/10.1007/s13201-021-01495-1.

9. D.G. Kinniburgh and P.L. Smedley, *Arsenic Contamination of Groundwater in Bangladesh*, British Geological Survey (BGS), 2001.

10. Z. Gao, H. Guo, W. Qiao, T. Ke, Z. Zhu, Y. Cao, X. Su, L. Wan, Abundant Fe(III) oxide-bound arsenic and depleted Mn oxides facilitate arsenic enrichment in groundwater from a sand-gravel confined aquifer, *J. Geophys. Res. Biogeosci.* 127 2022. https://doi.org/10.1029/2022JG006942.

11. E.E. Rios-Valenciana, T. Moreno-Perlin, R. Briones-Gallardo, R. Sierra-Alvarez, L.B. Celis, The key role of biogenic arsenic sulfides in the removal of soluble arsenic and propagation of arsenic mineralizing communities, *Environ. Res.* 220, 2023, 115124. https://doi.org/10.1016/j.envres.2022.115124.

12. P. Mazumder, S.K. Sharma, K. Taki, A.S. Kalamdhad, M. Kumar, Microbes involved in arsenic mobilization and respiration: a review on isolation, identification, isolates and implications, *Environ. Geochem. Health.* 42, 2020, 3443–3469. https://doi.org/10.1007/s10653-020-00549-8.

13. D.K. Datta, V. Subramanian, Texture and mineralogy of sediments from the Ganges-Brahmaputra-Meghna river system in the Bengal Basin, Bangladesh and their environmental implications, *Environ. Geol.* 30, 1997, 181–188. https://doi.org/10.1007/s002540050145.

14. G. Huang, J. Xie, T. Li, P. Zhang, Worker health risk of heavy metals in pellets of recycled plastic: a skin exposure model, *Int. Arch. Occup. Environ. Health.* 94, 2021, 1581–1589. https://doi.org/10.1007/s00420-021-01727-6.

15. K.J. Jacob-Tatapu, S. Albert, A. Grinham, Sediment arsenic hotspots in an abandoned tailings storage facility, gold ridge mine, Solomon islands, *Chemosphere.* 269, 2021, 128756. https://doi.org/10.1016/j.chemosphere.2020.128756.

16. WHO, *Environmental Health Criteria 224: Arsenic Compounds*, World Health Organization, Geneva, Switzerland, 2001.

17. J.O. Nriagu, J.M. Pacyna, Quantitative assessment of worldwide contamination of air, water and soils by trace metals, *Nature.* 333, 1988, 134–139. https://doi.org/10.1038/333134a0.

18. R.B. Finkelman, H.E. Belkin, B. Zheng, Health impacts of domestic coal use in China, In: *Proceedings of the National Academy of Sciences,* Vol. 96, 1999, pp. 3427–3431. https://doi.org/10.1073/pnas.96.7.3427.

19. S.B. Adeloju, S. Khan, A.F. Patti, Arsenic contamination of groundwater and its implications for drinking water quality and human health in under-developed countries and remote communities—a review, *Appl. Sci.* 11, 2021, 1926. https://doi.org/10.3390/app11041926.

20. B.G.S., Executive summary of the main report of Phase I. Groundwater studies of as contamination in Bangladesh, by British geological survey and Mott Macdonald (UK) for the government of Bangladesh, Rural Development and Cooperatives DPHE and DFID (UK), 2000.

21. T. Biswas, S.C. Pal, A. Saha, D. Ruidas, A.R.M.T. Islam, M. Shit, Hydro-chemical assessment of groundwater pollutant and corresponding health risk in the Ganges delta, indo-Bangladesh region, *J. Clean. Prod.* 382, 2023, 135229. https://doi.org/10.1016/j.jclepro.2022.135229.

22. M.A. Fazal, T. Kawachi, E. Ichion, Extent and severity of groundwater arsenic contamination in Bangladesh, *Water Int.* 26, 2001, 370–379. https://doi.org/10.1080/02508060108686929.

23. T.R. Chowdhury, G.K. Basu, B.K. Mandal, B.K. Biswas, G. Samanta, U.K. Chowdhury, C.R. Chanda, D. Lodh, S.L. Roy, K.C. Saha, S. Roy, S. Kabir, Q. Quamruzzaman, D. Chakraborti, Arsenic poisoning in the Ganges delta, *Nature.* 401, 1999, 545–546. https://doi.org/10.1038/44056.

24. B.K. Mandal, K.T. Suzuki, Arsenic round the world: a review, *Talanta.* 58, 2002, 201–235. https://doi.org/10.1016/S0039-9140(02)00268-0.

25. A.V. Sindireva, Local biogeochemical cycles of trace elements in agroecosystems of Western Siberia, *Geochem. Int.* 61, 2023, 1048–1060. https://doi.org/10.1134/S0016702923100129.

26. N.R.C., *Arsenic in Drinking Water*, National Academy Press, 1999. https://doi.org/10.17226/6444.

27. A.K. Done, A.J. Peart, Acute toxicities of arsenical herbicides, *Clin. Toxicol.* 4, 2008, 343–355. https://doi.org/10.3109/15563657108990485.

28. P.M. Nguyen, C.N. Mulligan, Study on the influence of water composition on iron nail corrosion and arsenic removal performance of the Kanchan arsenic filter (KAF), *J. Hazard. Mater. Adv.* 10, 2023, 100285. https://doi.org/10.1016/j.hazadv.2023.100285.

29. M.F. Hossain, Arsenic contamination in Bangladesh—an overview, *Agri. Ecosyst. Environ.* 113, 2006, 1–16. https://doi.org/10.1016/j.agee.2005.08.034.

30. H. Romero-Schmidt, A. Naranjo-Pulido, L. Mendez-Roclriguez, B. Acosta-Vargas, A. Ortega-Rubio, Environmental health risks by arsenic consumption in water wells in the cape region, Mexico, *WIT Trans. Biomed. Health.* 5, 2001, 8. https://doi.org/10.2495/EHR010131.

31. WHO (2017) *Guidelines for Drinking-Water Quality*, 4th Edition, incorporating the 1st addendum, World Health Organization, Geneva, Switzerland, ISBN: 978-92-4-154995.

32. T.S.Y. Choong, T.G. Chuah, Y. Robiah, F.L. Gregory Koay, I. Azni, Arsenic toxicity, health hazards and removal techniques from water: an overview, *Desalination.* 217, 2007, 139–166. https://doi.org/10.1016/j.desal.2007.01.015.

33. M.R. Karagas, T.D. Tosteson, J.S. Morris, E. Demidenko, L.A. Mott, J. Heaney, A. Schned, Incidence of transitional cell carcinoma of the bladder and arsenic exposure in new hampshire, *Cancer Causes Control.* 15, 2004, 465–472. https://doi.org/10.1023/B:CACO.0000036452.55199.a3.

34. H. Ferdosi, E.K. Dissen, N.A. Afari-Dwamena, J. Li, R. Chen, M. Feinleib, S.H. Lamm, Arsenic in drinking water and lung cancer mortality in the United States: an analysis based on us counties and 30 years of observation (1950–1979), *J. Environ. Public Health.* 2016, 2016, 1–13. https://doi.org/10.1155/2016/1602929.

35. J. Prieto, Effective teaching on infection prevention and control: it's all about what learning is going on, *J. Infect. Prev.* 10, 2009, 211–213. https://doi.org/10.1177/1757177409350824.

36. J. Ahn, S.H. Lamm, H. Ferdosi, I.J. Boroje, Aortic elasticity and arsenic exposure: a step function rather than a linear function, *Risk Anal.* 41, 2021, 2293–2300. https://doi.org/10.1111/risa.13756.

37. C. Chen, Y. Hsueh, M. Lai, M. Shyu, S. Chen, M. Wu, T. Kuo, T. Tai, Increased prevalence of hypertension and long-term arsenic exposure, *Hypertension.* 25, 1995, 53–60. https://doi.org/10.1161/01.HYP.25.1.53.

38. W. Tseng, Effects and dose-response relationships of skin cancer and blackfoot disease with arsenic, *Environ. Health Perspect.* 19, 1977, 109–119. https://doi.org/10.1289/ehp.7719109.

39. C. Chen, Blackfoot disease, *The Lancet.* 336, 1990, 442. https://doi.org/10.1016/0140-6736(90)91990-R.

40. R.L. Kane, R.A. Kane, Assessment in long-term care, *Annu. Rev. Public Health.* 21, 2000, 659–686. https://doi.org/10.1146/annurev.publhealth.21.1.659.

41. Y. Xia, J. Liu, An overview on chronic arsenism via drinking water in PR China, *Toxicology.* 198, 2004, 25–29. https://doi.org/10.1016/j.tox.2004.01.016.

42. C. Tian, F. Hu, S. Lu, G. Tan, L. Wang, Iatrogenic arsenism characterized by palmoplantar hyperkeratosis and diffused skin cancers for over decades, *Am. J. Dermatopathol.* 43, 2021, 373–376. https://doi.org/10.1097/DAD.0000000000001903.

43. Y.L.C.K. Jin, The investigation of endemic arsenism distribution in China, In: *Proceedings of China CDC Endemic Arsenism Conference,* 2003.

44. H. Nemoto, M. Otake, T. Matsumoto, R. Izutsu, J.P. Jehung, K. Goto, M. Osaki, M. Mayama, M. Shikanai, H. Kobayashi, T. Watanabe, F. Okada, Prevention of tumor progression in inflammation-related carcinogenesis by anti-inflammatory and anti-mutagenic effects brought about by ingesting fermented brown rice and rice bran with aspergillus oryzae (FBRA), *J. Funct. Food.* 88, 2022, 104907. https://doi.org/10.1016/j.jff.2021.104907.

45. A.V. Gillespie, K.E.K. Rowson, The influence of sex upon the development of friend virus leukaemia, *Int. J. Cancer.* 3, 1968, 867–875. https://doi.org/10.1002/ijc.2910030620.

46. W. Tseng, Effects and dose-response relationships of skin cancer and blackfoot disease with arsenic, *Environ. Health Perspect.* 19, 1977, 109–119. https://doi.org/10.1289/ehp.7719109.

47. S. Amin, N. Hussain, G. Balanikas, K. Huie, S.S. Hecht, Mutagenicity and tumor initiating activity of methylated benzofluoranthenes, *Cancer Lett.* 26, 1985, 343–347. https://doi.org/10.1016/0304-3835(85)90059-X.

48. C. Chen, C.W. Chen, M. Wu, T. Kuo, Cancer potential in liver, lung, bladder and kidney due to ingested inorganic arsenic in drinking water, *Br. J. Cancer.* 66, 1992, 888–892. https://doi.org/10.1038/bjc.1992.380.

49. New products, *Chem. Eng. News Arch.* 67, 1989, 27. https://doi.org/10.1021/cen-v067n034.p027.

50. P.J. Mink, D.D. Alexander, L.M. Barraj, M.A. Kelsh, J.S. Tsuji, Low-level arsenic exposure in drinking water and bladder cancer: a review and meta-analysis, *Regul. Toxicol. Pharmacol.* 52, 2008, 299–310. https://doi.org/10.1016/j.yrtph.2008.08.010.

51. WHO (1993) *Guidelines for Drinking-Water Quality*, World Health Organization, Geneva, Switzerland.

52. SAC (2007) *Sanitary Standards for Drinking Water,* Standardization Administration of the People's Republic of China (SAC), Beijing, China, April 2007.

53. X. Long, J. Li, F. Gao, H. Wu, J. Deng, Bioinspired synthesis of spirochensilide A from lanosterol, *J. Am. Chem. Soc.* 144, 2022, 16292–16297. https://doi.org/10.1021/jacs.2c07198.

54. B.A. Manning, S.E. Fendorf, B. Bostick, D.L. Suarez, Arsenic(III) oxidation and arsenic(V) adsorption reactions on synthetic birnessite, *Environ. Sci. Technol.* 36, 2002, 976–981. https://doi.org/10.1021/es0110170.

55. Y. Zhang, D. Wu, Y. Sun, S. Peng, Sol–gel synthesis of methyl modified optical silica coatings and gels from DDS and TEOS, *J. Sol-Gel Sci. Technol.* 33, 2005, 19–24. https://doi.org/10.1007/s10971-005-6694-y.

56. R.P. Liu, B.L. Yuan, X. Li, S.J. Xia, Y.L. Yang, G.B. Li, The oxidative and adsorptive effectiveness of hydrous manganese dioxide for arsenite removal from aqueous solution, *High Technol. Lett.* 12, 2006, 30–34.

57. A.M. Buswell, R.C. Gore, H.E.H. Jr., A.C. Wiese, T.E. Larson, War problems in analysis and treatment, *J. Am. Water Works Assoc..* 35, 1943, 1303–1311. https://doi.org/10.1002/j.1551-8833.1943.tb19890.x.

58. J.G. Hering, P. Chen, J.A. Wilkie, M. Elimelech, Arsenic removal from drinking water during coagulation, *J. Environ. Eng.* 123, 1997, 800–807. https://doi.org/10.1061/(ASCE)0733-9372(1997)123:8(800).

59. J.H. Gulledge, J.T. O'Connor, Removal of arsenic (V) from water by adsorption on aluminum and ferric hydroxides, *J. Am. Water Works Assoc.* 65, 1973, 548–552. https://doi.org/10.1002/j.1551-8833.1973.tb01893.x.

60. R.C. Cheng, S. Liang, H.C. Wang, M.D. Beuhler, Enhanced coagulation for arsenic removal, *J. Am. Water Works Assoc.* 86, 1994, 79–90. https://doi.org/10.1002/j.1551-8833.1994.tb06248.x.

61. K.N. Scott, J.F. Green, H.D. Do, S.J. McLean, Arsenic removal by coagulation, *J. Am. Water Works Assoc.* 87, 1995, 114–126. https://doi.org/10.1002/j.1551-8833.1995.tb06347.x.

62. X. Meng, Effects of silicate, sulfate, and carbonate on arsenic removal by ferric chloride, *Water Res.* 34, 1255–1261. https://doi.org/10.1016/S0043-1354(99)00272-9.

63. R. Liu, J. Qu, S. Xia, G. Zhang, G. Li, Silicate hindering *in situ* formed ferric hydroxide precipitation: inhibiting arsenic removal from water, *Environ. Eng. Sci.* 24, 2007, 707–715. https://doi.org/10.1089/ees.2006.0074.

64. C.C. Davis, W.R. Knocke, M. Edwards, Implications of aqueous silica sorption to iron hydroxide: mobilization of iron colloids and interference with sorption of arsenate and humic substances, *Environ. Sci. Technol.* 35, 2001, 3158–3162. https://doi.org/10.1021/es0018421.

65. J.A. Wilkie, J.G. Hering, Adsorption of arsenic onto hydrous ferric oxide: effects of adsorbate/adsorbent ratios and co-occurring solutes, *Colloids Surf. A: Physicochem. Eng.* 107, 1996, 97–110. https://doi.org/10.1016/0927-7757(95)03368-8.

66. L. Ruiping, L. Xing, X. Shengji, Y. Yanling, W. Rongcheng, L. Guibai, Calcium-enhanced ferric hydroxide co-precipitation of arsenic in the presence of silicate, *Water Environ. Res.* 79, 2007, 2260–2264. https://doi.org/10.2175/106143007X199324.

67. S.R. Wickramasinghe, B. Han, J. Zimbron, Z. Shen, M.N. Karim, Arsenic removal by coagulation and filtration: comparison of groundwaters from the united states and Bangladesh, *Desalination.* 169, 2004, 231–244. https://doi.org/10.1016/j.desal.2004.03.013.

68. D. Mohan, C.U. Pittman, Arsenic removal from water/wastewater using adsorbents—a critical review, *J. Hazard. Mater.* 142, 2007, 1–53. https://doi.org/10.1016/j.jhazmat.2007.01.006.

69. S. Zou, K.Y. Koh, Z. Chen, Y. Wang, J.P. Chen, Y. Zheng, Adsorption of organic and inorganic arsenic from aqueous solution: optimization, characterization and performance of Fe–Mn–Zr ternary magnetic sorbent, *Chemosphere.* 288, 2022, 132634. https://doi.org/10.1016/j.chemosphere.2021.132634.

70. H.W. Nesbitt, G.W. Canning, G.M. Bancroft, XPS study of reductive dissolution of 7Å-birnessite by H_3AsO_3, with constraints on reaction mechanism, *Geochim. Cosmochim. Acta.* 62, 1998, 2097–2110. https://doi.org/10.1016/S0016-7037(98)00146-X.

71. S. Ouvrard, M. Simonnot, P. de Donato, M. Sardin, Diffusion-controlled adsorption of arsenate on a natural manganese oxide, *Ind. Eng. Chem. Res.* 41, 2002, 6194–6199. https://doi.org/10.1021/ie020269m.

72. G. Zhang, J. Qu, H. Liu, R. Liu, G. Li, Removal mechanism of As(III) by a novel Fe–Mn binary oxide adsorbent: oxidation and sorption, *Environ. Sci. Technol.* 41, 2007, 4613–4619. https://doi.org/10.1021/es063010u.

73. G. Zhang, J. Qu, H. Liu, R. Liu, R. Wu, Preparation and evaluation of a novel Fe–Mn binary oxide adsorbent for effective arsenite removal, *Water Res.* 41, 2007, 1921–1928. https://doi.org/10.1016/j.watres.2007.02.009.

74. P.M. Swash, A.J. Monhemius, Synthesis, characterisation and solubility testing of solids in the Ca-Fe-AsO_4 system, In: *Conference on Mining and the Environment,* 1995.

75. Q.L.P.A.R., *Bench Scale Study on the Treatment of Acidic Wastewater with High Concentration Arsenic,* Chinese Water Wastewater, Beijing, China, 2000.

76. USEPA, *Regulations on the Disposal of Arsenic Residuals from Drinking Water Treatment Plants,* US Environmental Protection Agency, Washington DC, USA, 2000.

77. A.O. Regis, J. Vanneste, S. Acker, G. Martínez, J. Ticona, V. García, F.D. Alejo, J. Zea, R. Krahenbuhl, G. Vanzin, J.O. Sharp, Pressure-driven membrane processes for boron and arsenic removal: ph and synergistic effects, *Desalination.* 522, 2022, 115441. https://doi.org/10.1016/j.desal.2021.115441.

78. D. Clifford, *Ion Exchange and Inorganic Adsorption in Water Quality and Treatment,* American Water Works Association, Denver, CO, USA, 1999.

79. E.O. Kartinen, C.J. Martin, An overview of arsenic removal processes, *Desalination.* 103, 1995, 79–88. https://doi.org/10.1016/0011-9164(95)00089-5.

80. P. Brandhuber, G. Amy, Alternative methods for membrane filtration of arsenic from drinking water, *Desalination.* 117, 1998, 1–10. https://doi.org/10.1016/S0011-9164(98)00061-7.

81. T. Urase, J. Oh, K. Yamamoto, Effect of pH on rejection of different species of arsenic by nanofiltration, *Desalination.* 117, 1998, 11–18. https://doi.org/10.1016/S0011-9164(98)00062-9.

82. M. Kang, M. Kawasaki, S. Tamada, T. Kamei, Y. Magara, Effect of pH on the removal of arsenic and antimony using reverse osmosis membranes, *Desalination.* 131, 2000, 293–298. https://doi.org/10.1016/S0011-9164(00)90027-4.

83. E.M. Vrijenhoek, J.J. Waypa, Arsenic removal from drinking water by a "loose" nanofiltration membrane, *Desalination.* 130, 2000, 265–277. https://doi.org/10.1016/S0011-9164(00)00091-6.

84. USEPA, *Technologies and Costs for Removal of Arsenic from Drinking Water,* Standards and Risk Management Division, US Environmental Protection Agency, Washington DC, USA, 2000.

85. H. Gecol, E. Ergican, A. Fuchs, Molecular level separation of arsenic (V) from water using cationic surfactant micelles and ultrafiltration membrane, *J. Membr. Sci.* 241, 2004, 105–119. https://doi.org/10.1016/j.memsci.2004.04.026.

86. L. K. Wang, M. H. S. Wang and N. K. Shammas, Treatment of wastewater, storm runoff and combined sewer overflow by dissolved air flotation and filtration, In: L.K. Wang, M. H. S. Wang, Y.T. Hung, N. K. Shammas and J. P. Chen (eds), *Handbook of Advanced Industrial and Hazardous Wastes Management,* CRC Press, FL, USA. pp. 577–610, 2018.

87. L.K. Wang, N.K. Shammas, W.A. Selke, D.B. Aulenbach, *Flotation Technology,* Humana Press, Totowa, NJ, p. 680, 2010.

88. L.K. Wang, M.H.S. Wang, N.K. Shammas, D.B. Aulenbach, *Environmental Flotation Engineering,* Springer Nature, Cham, Switzerland, p. 433, 2021.

89. Wang, Y., Deng, Y., Huang, K., Liu, X., and Wang, L. K. Environmental geochemistry of high-arsenic aquifer systems. In: *Control of Heavy Metals in the Environment* Volume 3, Wang, L. K., Wang, M. H. S., Hung, Y. T. and Chen, J. P. (editors). CRC Press, Florida, USA, p. 140–173, 2025.

90. Naja, G. M., Volesky, B., Yang, Y. and Wang, L. K. Treatment of Metal-bearing effluents: removal and recovery. In: *Control of Heavy Metals in the Environment* Volume 3, Wang, L. K., Wang, M. H. S., Hung, Y. T. and Chen, J. P. (editors). CRC Press, Florida, USA, p. 273–318, 2025.

8 Treatment of Metal-Bearing Effluents: Removal and Recovery

Ghinwa M. Naja, Bohumil Volesky,
Yi Yang, and Lawrence K. Wang

ACRONYMS AND NOMENCLATURES

AMD:	acid mine drainage
BOD:	biochemical oxygen demand
Ca$_5$(OH)(PO$_4$)$_3$:	calcium hydroxyapatite
CaCO$_3$:	calcium carbonate
CFW:	subsurface flow wetlands
COD:	chemical oxygen demand
CW:	`constructed wetland
FWS:	free water surface
GAC:	granular activated carbon
PAC:	powdered activated carbon
Pb(OH)$_2$:	lead hydroxide
PbCO$_3$:	lead carbonate
USEPA:	U.S. Environmental Protection Agency

8.1 PHYSICAL-CHEMICAL TREATMENT

8.1.1 CHEMICAL PRECIPITATION

The most common method for heavy metal removal from wastewater is chemical precipitation. As shown in Figure 8.1, the basic treatment process for precipitating heavy metals includes pretreatment, pH adjustment, flocculation/clarification, sludge thickening, sludge dewatering, and effluent polishing [1].

The general principle of heavy metal removal by chemical precipitation is based on the low solubility of heavy metal hydroxides. The treatment reduces the heavy metal content of the water dramatically, thus the residual concentrations of heavy metals in the overflow from the settling tank are usually below the environmental limits. Furthermore, the process can handle relatively high flow rates of wastewater containing high concentrations of heavy metals, and so it is suitable for the treatment of many industrial and mining effluents. However, the efficiency of the treatment hinges on the rates of formation and settling of the solid metal hydroxides. The fact that these rates are usually low in water with low metal content leads to increased consumption of lime and/or caustic soda in the process and the design of large mixing and settling tanks. Furthermore, the treatment generates toxic sludge that has to be dewatered, stabilized, and disposed of.

Three theories have been put forward to explain the effects and efficiency of chemical precipitation. The first theory uses the fact that heavy metal salts, when treated with alkaline substances, form heavy voluminous precipitates that carry down colloidal suspensions through mechanical entrapment principles. It has been demonstrated that colloidal particles possess an electric charge.

DOI: 10.1201/9781003541615-8

"

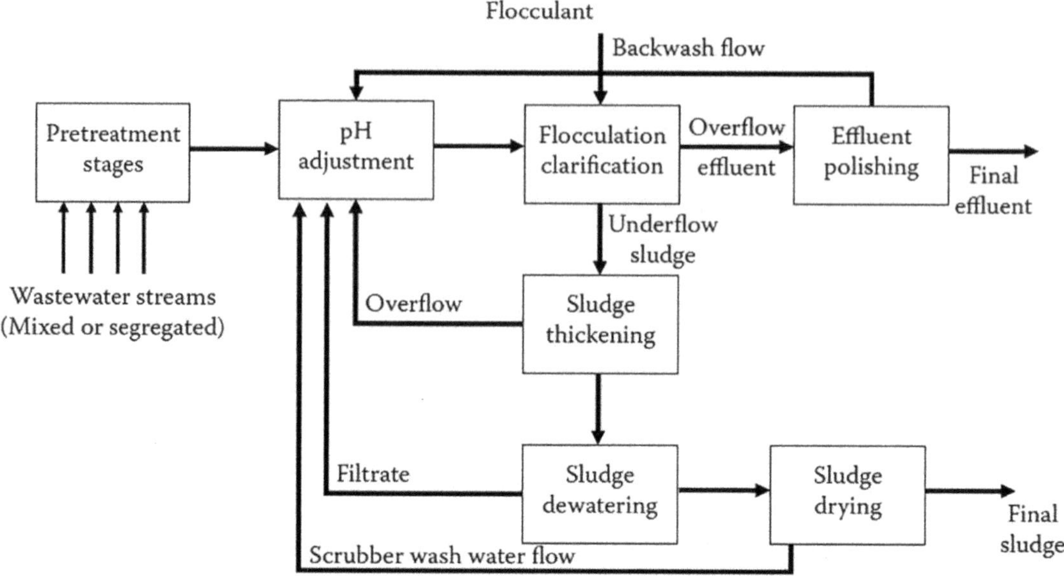

FIGURE 8.1   Basic waste treatment process for heavy metals.

Since these charges are alike, the particles repel each other and thus tend to remain in suspension. If a colloidal particle with an opposite charge is added, the charges neutralize, and the settling of the particles is affected. This explains the efficiency of the multivalent ions and why ferric salt is more efficient than ferrous salts. Clay suspensions are also claimed to exert a charge-neutralizing effect. The third theory pertains to physical behavior. Insoluble substances that have a large particle surface area can effectively sorb colloids. They can also act as nuclei for the initiation of precipitation. Activated charcoal is a material that uses this type of action. Of the three theories mentioned above, the second one is generally predominantly accepted.

8.1.1.1   Pretreatment

Prior to precipitation, the wastewater will be subjected to a pretreatment stage. Pretreatment is used to remove materials such as grease and scum before sedimentation to improve process feasibility. Common pretreatment stages include oil removal and chromate reduction [Cr(VI) to Cr(III)].

If significant levels of oil are present in metal-bearing wastewater streams, the oil must be removed before clarification to prevent interference with the settling of the precipitated solids. The oil is usually removed by skimming, and emulsified oil can be removed by ultrafiltration. Chromium will only precipitate in the trivalent form, and thus it must be reduced from its hexavalent form prior to precipitation. Hexavalent chromium reduction is achieved at low pH levels (~2–3) with a reducing agent—usually sodium metabisulfite. Owing to the low pH level involved, chromium- bearing wastewater is usually segregated for pretreatment before being mixed with other metal-bearing wastewater streams [2].

8.1.1.2   pH Adjustment

The precipitation of metals occurs at various pH levels depending on a number of factors. For wastewater streams that contain different heavy metals, the pH level for precipitation must be carefully chosen so that all of the metals have an acceptable level of insolubility. If this is not possible, the stream will have to be segregated to treat the particular component metal at an appropriate pH level [3]. Typical minimum pH values for precipitation are given in Table 8.1, depending on the solubility product K_{sp} of the metal hydroxide [4]. To precipitate metals in the hydroxide form, caustic soda and lime are

TABLE 8.1
Minimum pH Values for Complete Precipitation of Metal Ions as Hydroxide

Metal	Sn^{2+}	Fe^{3+}	Al^{3+}	Pb^{2+}	Cu^{2+}	Zn^{2+}	Ni^{2+}	Fe^{2+}	Co^{2+}	Mn^{2+}
pH	4.2	4.3	5.2	6.3	7.2	8.4	9.3	9.5	9.7	10.6

Source: U.S. Environmental Protection Agency (U.S. EPA). Processes, procedures and methods to control pollution from mining Activities. U.S. EPA # 430/9-73-011, 1973 [4].

frequently used for pH adjustment. In large wastewater treatment systems, lime is preferred due to its lower cost, whereas in small wastewater treatment systems, caustic soda is used due to ease of handling. In the case where fluoride or phosphate removal from the wastewater is required, lime is used for pH adjustment since it will generally cause precipitation of the fluorides and the phosphates. Precipitate particle size (and filterability) appears to be greater when lime is used. The use of lime or caustic soda seems to have no great effect on the rate of sedimentation, but the volume of sludge will be twice as great when using lime as when using caustic [5].

8.1.1.3 Coagulation/Flocculation/Clarification

As the metal hydroxides come out of the solution due to pH adjustment, chemicals are often added to promote coagulation and flocculation. The inorganic coagulants are often trivalent cations that neutralize the negative charge of the colloids. The higher the valency, the more effective the coagulating action will be (Schultz-Hardy theory: a trivalent ion is ten times more effective than a divalent ion). When choosing a coagulant, its harmlessness and its cost must be taken into account. Thus, trivalent iron or aluminum salts have been and continue to be widely used in all water coagulation treatments. Organic coagulants may also be used. These are cationic polyelectrolytes that directly neutralize the negative colloids. Inorganic polymers (activated silica) and natural polymers (starches, alginate) were the first to be used as common flocculants. The appearance of the widely varying synthetic polymers has changed flocculation results considerably. Of chief importance is the timing of the introduction of the coagulant and that of the flocculant. In fact, a flocculant usually does not take effect until the coagulation stage is over. The use of synthetic flocculants often results in a minimum amount of sludge. Combined with modern separation techniques, this can lead to the production of very dense sludge that can be directly treated in a dewatering unit.

High-molecular-weight polymers are fed to the neutralized waste as it enters the clarifier at a dosage of 10–100 ppm [6]. The exact flocculant dosage is usually determined by individual bench tests. Prior to the clarification stage, flash mixing and flocculating chambers allow the flocculant to be well mixed into the wastewater and provide gentle particle contact to aid the formation of larger, heavier particles, which will settle well in the clarification stage.

The clarification stage involves the removal of the solids (the flocs) from the wastewater stream. This is typically a gravity-settling process that occurs in a sedimentation tank or an inclined plate clarifier, also called a Lamella clarifier. Under proper conditions, precipitated solids can be concentrated about 10:1 in a clarifier. Clarifier performance is largely a function of the settling surface that increases with the utilization of the Lamella clarifier. In summary, a clarifier usually has to perform three different functions to do its task well [7]. The settling tank, thus must

Provide for effective removal of suspended solids from the effluent.

Have an adequate sludge removal capacity.

Thicken the sludge satisfactorily.

Any failure in one of these functions will impair the performance of the settler and, if serious, destroy the effectiveness of the process almost completely. A poor design results in the propagation of the problem to successive units within the plant, and inevitably in a decrease in the overall performance of the treatment.

8.1.1.4　Solid-Liquid Separation

If further reduction in the level of suspended solids is required, the effluent can be polished. Filtration using backwash sand in-depth filters is the most popular method for suspended solids removal in the polishing stage. The overflow from the settler/clarifier enters the filtration process. After passing downward through the granular medium, it can be discharged. During backwashing, wash water passing upward through the filter (fluidizing the sand filter medium) carries out the impurities that accumulate in the filter bed. A continuous backwash sand filter can provide continuous streams of polished effluent and reject flow so that the filter never needs to be shut down for backwashing. The reject flow would be returned to the front end of the treatment system for further processing. The effluent from this type of sand filter is usually quite clear, containing only 3–5 ppm of total suspended solids [8].

Apart from gravity settling, the dissolved air flotation (DAF) process offers a cost-effective solution for solid-water separation due to its short process detention time and small footprint [9–11]. Membrane filtration processes provide another approach to concentrating the solids from the water and wastewater streams and producing clear effluent. Ultrafiltration could be quoted as an example. It is a physical membrane separation process whereby the membrane acts as a barrier to precipitate particles and prevents their passage into the discharge stream. No flocculants are needed, and the total suspended solids in the effluent are essentially zero [12,13]. Thus, tighter effluent limits are met, and even high flow rates can nowadays be handled [14]. Ultrafiltration systems are based on the cross-flow membrane technology. A bundle of parallel hollow fiber membranes is sealed into a shell to form a cartridge. Each cartridge has a process inlet, outlet, and a pair of permeate outlets. Inside each fiber, waste is separated at the membrane surface. Cleaned effluent passes through the membrane, whereas contaminants are rejected and exit at the opposite end. Turbulent flow across the membrane surface reduces waste build-up and minimizes cleaning.

8.1.1.5　Sludge Thickening

The flocculation/clarification stage is usually followed by the sludge-thickening stage. A sludge thickener is typically a conical bottom tank that receives the underflow from the clarifier and provides storage where further gravity settling of solids can occur. The sludge concentrated at the bottom of the tank contains about 4%–6% solids [6]. The advantage of operating with a high solids content sludge is that it generally improves the operation and performance of the dewatering equipment.

8.1.1.6　Sludge Dewatering

Dewatering of the concentrated sludge can be accomplished using a wide variety of equipment such as centrifuges, rotary vacuum filters, belt presses, and filter presses [15]. Rotary vacuum filters produce a relatively dry cake (20%–25% solids) and can operate continuously so that they are suitable for applications involving large volumes of sludge. For most industrial applications, filter presses (plate and frame pressure filters) are the most economical method of producing a dry cake (20%–30% solids). Filter presses have wide applicability and are probably the most common type of dewatering equipment in use [6]; however, they do not operate on a continuous-flow basis. A filter press consists of a series of plates lined with polypropylene filter cloths. The plates form chambers into which the sludge is pumped. The solids are retained by the filter media, whereas the filtrate flows through the porous filter cloths. The filtrate flow is discharged from the press to the front end of the treatment process (prior to the clarification stage). As the filter cycle progresses, the chambers become completely full of solids, and at this point, no more sludge can be pumped into the press. The press must then be opened, and the dry cakes are discharged. Therefore, the term "press" is a misnomer because no mechanical squeeze is involved. It should be noted that only the thickening of neutralized metal hydrates using a clarifier and subsequent dewatering of sludge has been discussed to this point. The neutralized waste can also be pressure filtered directly, depending on its concentration (about 300–500 ppm of suspended solids).

8.1.1.7 Sludge Disposal

With the increased production of industrial wastes, sludge management is becoming increasingly important [16]. There exist various methods of disposing of sludge (landfill 25%, ocean dumping 10%, incineration 40%, and land application 25%). Metal-bearing sludges are difficult to handle and, due to their toxicity, their final disposal is often troublesome and expensive. If incorrectly disposed of, these sludges could be a potential source of pollution of surface and ground waters. There appears to be little problem concerning the incineration of waste sludges with respect to heavy metals, but this depends on where and how the pollutant is released. Ocean dumping is being less used compared to land applications and sludge landfilling. In determining the sludge disposal site, many factors need to be considered, such as type of soils, ground water table, hydrology, composition, and pH of the sludge. With a sludge pH of less than 6.5, the potential for release of cadmium, chromium, lead, mercury, and selenium increases. Suitable disposal sites are areas where natural (clay, rock) or artificial means (plastic liner) can prevent excessive amounts of leachate from getting into the groundwater. Where these conditions do not exist, it may be necessary to install a collection system at the bottom of the pond so that leachate can be collected and piped to a treatment system to remove soluble metal salts. The U.S. Environmental Protection Agency (USEPA) has suggested limiting sludge land application as a function of specific metal content and soil cation exchange capacity [17]. Chemical treatments have been developed to reduce the leaching possibilities of metal-bearing sludges and may be used in conjunction with a landfill program to further minimize potential leachate contamination. One other solution could be to remove metals from these sludges (Section 8.1.4).

8.1.2 TREATMENT METHODS

Chemical precipitation of metal may be accomplished in either batch (Figure 8.2) or continuous-treatment systems (Figure 8.3). Batch treatment is usually preferred when the volumes to be treated are small or where the waste may be variable from day-to-day and require modification of the treatment as characteristics change [7].

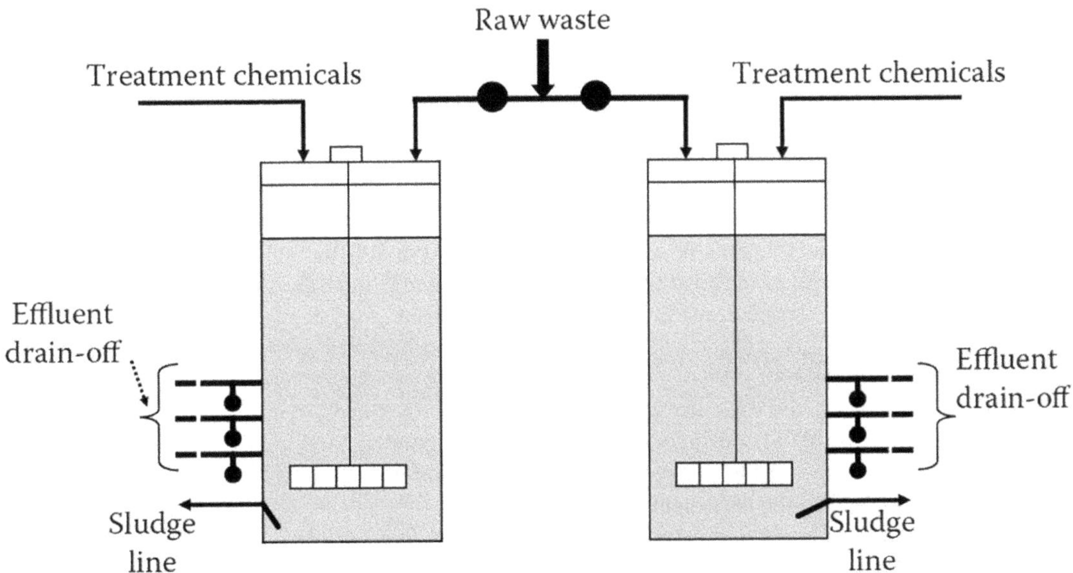

FIGURE 8.2 Typical batch treatment system.

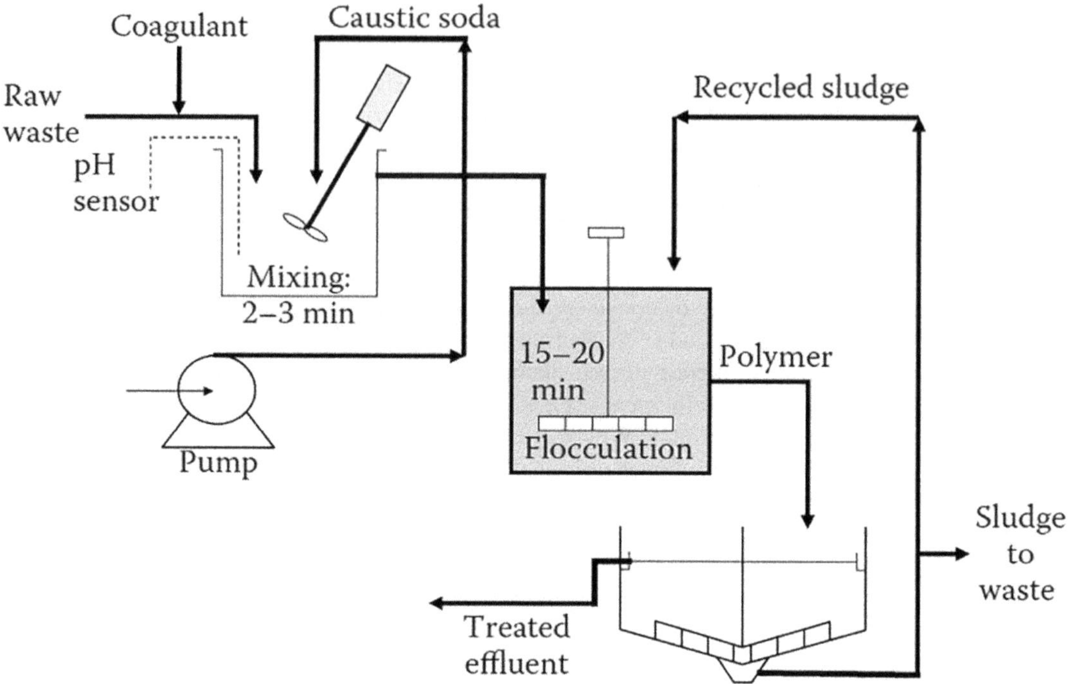

FIGURE 8.3 Typical continuous-treatment system for heavy metals.

8.1.2.1 Batch Treatment

Batch treatment systems can be economically designed for flows as high as 50,000 gpd. A batch system is usually designed with two tanks, each one of sufficient volume to handle the waste generated in a specific time. When one tank is full, a mixer is used to provide a homogeneous mixture, and a sample is taken and analyzed to determine the amount of metal contaminants present. Chemical addition, based on the metal contaminants present and pH of the waste, is then calculated and the required amount of chemicals added. The tank contents are then mixed, using pH for controlling purposes and metal removal, and allowed to settle for 2–4 hours.

When treatment is complete, a second sample can be taken and analyzed to ensure that all contaminants have been removed. If, for any reason, contaminants are still present, treatment can be repeated, or alternative treatment can be applied. When the operator is satisfied that the treated waste is suitable for discharge, the clear liquor is decanted. The settled sludge is drawn off periodically for disposal. The advantages of a batch treatment system are that nothing is discharged from the plant until the operator is satisfied that it meets effluent requirements. The system is also simple in its design and is easy to operate (Figure 8.2).

8.1.2.2 Continuous Treatment

When wastewater characteristics are uniform or when volumes are large, a continuous-treatment system is applicable. A usual continuous-flow treatment system has an equalization tank of several hours to a day of detention time to even out any fluctuations in the wastewater characteristics and provide a uniform feed to the treatment system.

The first process step is the adjustment of the pH by the addition of acid or alkali to the proper level of optimum precipitation. This chemical addition is controlled by a pH probe in the reaction tank, which activates the speed control of the chemical feed pump. A polymer is usually added to aid coagulation. Reaction times are in the range of 15–60 minutes.

The waste stream then flows to a sedimentation basin where the metal sediment settles out of the solution, leaving a clear treated overflow for discharge to the receiving water body (Figure 8.3).

8.1.2.3 Sludge Recirculation

Recirculation of precipitated sludge to be mixed with the raw waste at the time of chemical addition can have beneficial effects. The presence of precipitated particles provides a seed for the newly formed precipitate to agglomerate. In a batch treatment, the settled sludge is collected at the bottom of the tank. When a new batch is put in the tank, the mixer is turned on to resuspend the sludge and mix it with the tank contents.

In continuous-treatment systems, the sludge can be recycled either externally to the clarifier or internally within the clarifier. With external recirculation, the sludge is pumped out of the sludge hopper in the clarifier and introduced to the raw feed in the chemical mixing tanks. With internal recirculation, the clarifier is designed with an internal mixer and baffles that provide recirculation within the clarifier.

8.1.3 APPLICATIONS

8.1.3.1 Zinc

The precipitation process most frequently involves adjustment of pH with either lime or caustic to achieve alkaline conditions, and precipitation of zinc hydroxide. Lime addition has been the widely accepted method for pH adjustment despite the concurrent precipitation of calcium sulfate at elevated sulfate concentrations in some waters. The precipitation of calcium sulfate along with zinc hydroxide increases the total amount of sludge to be disposed of. Table 8.2 summarizes some results of zinc precipitation to its hydroxide. These values reflect a wide range of industrial systems; the treatment is usually not just for zinc removal alone. In cases where cyanide or chromate is also present in the waste, as frequently occurs in zinc and brass plating, cyanide removal and chromate reduction must precede metal hydroxide precipitation. Settling efficiency affects effluent concentration, as seen in Table 8.2, by improvements in effluent zinc levels resulting from the filtration of settled effluent. Incomplete cyanide treatment will increase effluent zinc levels due to complexation, as will improper control of the treatment process pH.

TABLE 8.2

Summary of Hydroxide Precipitation Treatment Results for Zinc in Wastewaters Zinc Concentration (mg/L)

Industrial Source	Initial	Final	Comments
Zinc plating	—	0.2–0.5	pH 8.7–9.3
General plating	4.1–120	0.39–2.9	pH 7.5–10.5
Vulcanized fiber	100–300	1.0	pH 8.5–9.5
Brass wire mill	36–374	0.08–1.60	Integrated treatment for copper recovery
Tableware plant	16.1	0.02–0.23	Sand filtration
Viscose rayon	20–120	0.88–5	pH 5
Metal fabrication	—	0.5–1.2	Sedimentation
	—	0.1–0.5	Sand filtration
Blast furnace gas scrubber water	50	0.2	pH 8.8
Zinc smelter	744–1,500	26–50	
Ferroalloy wastes	3–89	0.29–7.9	

8.1.3.2 Lead

In the precipitation process, lead is normally precipitated as lead carbonate ($PbCO_3$) or lead hydroxide $Pb(OH)_2$. The lead form precipitated depends on the amount of carbonate in or added to the wastewater, and the treatment pH. Initial acidic wastes are typically low in carbonate, and precipitation treatment of these waters would normally yield lead hydroxide unless supplemental carbonate were added. Lead carbonate precipitate is more crystalline than lead hydroxide, resulting in desirable settling and sludge-dewatering characteristics. A large excess of carbonate, or treatment above pH 9.0, may yield less effective precipitation, however. The optimum pH range for lead carbonate precipitation is between 7.5 and 9.0.

In forming insoluble lead hydroxide, lime is the treatment chemical of choice, although caustic has also been used. The results with caustic or lime treatment are equivalent but there is an interference with lead hydroxide precipitation as calcium ion concentration increases.

8.1.4 Metal Recovery from Sludges

The settled sludge from a clarification basin is frequently in the range of 1%–2% solids. Hydroxide precipitation of the metals produces sludge that is usually gelatinous in character, thereby increasing the difficulty of dewatering. Lime will make a considerably greater quantity of sludge than caustic, but that kind of sludge is easier to dewater. Similar to the case with wastewater sludges, the composition of these water treatment sludges varies from plant to plant, necessitating individual attention [5]. Some sludge compositions for North American plants are available in Table 8.3.

TABLE 8.3

Metal Composition in the Sludges and Their Recommended Levels Composition (mg/kg of Dry Sludge)

	Al	Cd	Cr	Cu	Mn	Ni	Pb	Zn
Plant #1[a]	27,640	5.0	87	215	933	28	110	419
Plant #1[b]	26,320	2.6	66	200	1,053	42	234	392
Plant #2[a]	43,630	9.2	401	1,070	445	141	278	413
Plant #2[b]	22,120	10.0	1,719	1,827	395	177	336	596
Plant #3[a]	74,097	1.4	50	178	323	17	23	359
Plant #3[b]	77,734	0.7	26	147	365	13	15	285
Plant #4[a]	30,907	4.5	124	737	4,613	30	177	379
Plant #4[b]	18,589	4.0	87	625	5,696	26	129	343
Plant #5[a]	32,484	2.3	99	1,211	2,914	142	266	181
Plant #5[b]	18,039	1.9	97	1,282	2,450	151	225	151
Plant #6[a]	28,786	0.8	349	1,017	1,458	50	43	1,430
Plant #6[b]	27,430	7.7	321	603	1,519	45	118	1,205
Plant #7[c]	21,705	11.2	116	3,689	166	23	447	1,024
Plant #8[c]	19,340	7.9	98	2,279	444	13	646	646
Plant #9[a]	13,520	2.0	155	391	418	222	106	1,456
Plant #6[c]	27,009	3.7	288	462	1,013	274	155	1,926
Recommended levels	—	15	1,000	1,000	1,500	180	500	2,500

[a] Secondary activated sludge.

[b] Aerobically digested sludge.

[c] Anaerobically digested sludge.

Metal recovery from sludges has been studied [18]. Digesting the sludge in an acid medium, neutralization, and electrolytic recovery have been investigated [19,20]. The cost estimate for the recovery of copper, nickel, and chromium in a small plant was $13.25 kg^{-1}, which was quite high compared to the current market prices. However, since these metal values are steadily increasing, one approach is to stockpile these metal-bearing sludges, either separately or in a regional disposal site, so that they are available for economical metal recovery in the future.

Nevertheless, digestion of the sludge can be done biologically. The further paragraphs will compare acid and microbial leaching for metal removal from municipal sludge.

A number of chemical methods for toxic metal solubilization from sludges have been studied [21,22]: ion exchange [23,24], utilization of chelating agents (EDTA and similar) [25–27], aerobic digestion coupled with or without hydrochloric acid [28–30], or oxidative acid hydrolysis [31]. The relatively high operating costs and sometimes insufficient yield of metal solubilization are obstacles to their practical applicability. The acid leaching (H_2SO_4, HCl, HNO_3, CH_3COOH) with or without heating are the processes that have been given more attention. The requirement of large amounts of acid to adjust pH and large amounts of alkali for residual sludge neutralization after the leaching makes these processes

unattractive from a practical standpoint. Consequently, interests have also focused on developing new microbiological methods. Two microbial leaching processes have been studied to remove toxic metals from sewage sludge. The bioleaching process with iron-oxidizing bacteria requires a lowering of the initial sludge pH to 4.0 and the addition of ferrous sulfate as a substrate, whereas the microbial leaching process with elemental sulfur as a substrate does not require an initial addition of acid. The principal advantage of microbiological processes is a considerable reduction in using significant quantities of acid to solubilize metals.

When comparing the microbial processes with acid-leaching treatment [18,19,32], some conclusions could be drawn:

The indigenous adapted sulfur- or iron-oxidizing bacteria can be utilized for toxic metal removal from sludges.

The use of a microbial leaching process with elemental sulfur and ferrous sulfate as substrates permits to considerably reduce the quantity of acid required for metal extraction with a reduction of 100% and 83%, respectively.

The bioleaching process with sulfur as a substrate for sulfur-oxidizing bacteria was revealed to be better than the acid treatment process and microbial leaching with ferrous sulfate and iron-oxidizing bacteria for solubilization of all metals examined.

The microbial leaching process with ferrous sulfate as a substrate permits a better solubilization of cadmium, copper, manganese, and zinc than the acid treatment with sulfuric acid. However, the solubilization of aluminum, chromium, nickel, and lead was less than that for acid leaching.

8.1.4.1 Alum Sludges

These sludges are usually fairly high in moisture content—98% or more. The biochemical oxygen demand (BOD) is low, about 50 mg/L, but the chemical oxygen demand (COD) is fairly high, from 500 to 1,500 mg/L. The pH of waste alum sludge is about 6, and ~40% of the solids are volatile. The major impact of these waste alum discharges is the formation of mudbanks along the stream.

One of the main problems of waste alum sludge is its very high moisture content. Thickeners have been used in some places to reduce this and make the sludge more efficient to handle. Tube settlers have been introduced for waste alum sludge and appear to be quite effective in increasing the solids concentration.

Dewatering waste alum sludge is difficult. The specific resistance of alum sludges is about 10–40×10^{12} m/kg, which is approximately the same as activated sludge. Interestingly, the specific resistance of alum sludge decreases with increasing solids concentration. Alum sludges at even high solids concentrations behave as a liquid, with Newtonian flow characteristics. Centrifugal dewatering is possible with high polymer dosages. It has been found that a 2lb polymer/ton of solids had almost no effect on solids recovery, but the addition of 1lb polymer/ton more polymer resulted in a jump to better than 90% solids recovery. In other words, the recovery-polymer dose curves were very steep. Cake solids of about 15% were obtained, which was considered acceptable since a drying system followed centrifugation. Pressure filters are used for dewatering alum sludges in a number of cities, with lime conditioning to aid the dewatering.

8.1.4.2　Lime Sludges

Lime CaO can be used to remove many impurities in wastewater. By adding sufficient quantities of lime, the pH can be raised to about 11.5, and calcium carbonate, metal hydroxides, and phosphates are precipitated. The phosphorus present is precipitated mostly as calcium hydroxyapatite $Ca_5(OH)(PO_4)_3$. The small quantities of aluminum, magnesium, and manganese oxides aid in the removal of silt and other impurities.

It is possible to recover lime from the sludge produced in lime precipitation. One method is to centrifuge the slurry at low solids recoveries to remove only the heavy calcium carbonate $CaCO_3$. The magnesium hydroxide and other light solids are then dewatered in a second centrifuge (Figure 8.4).

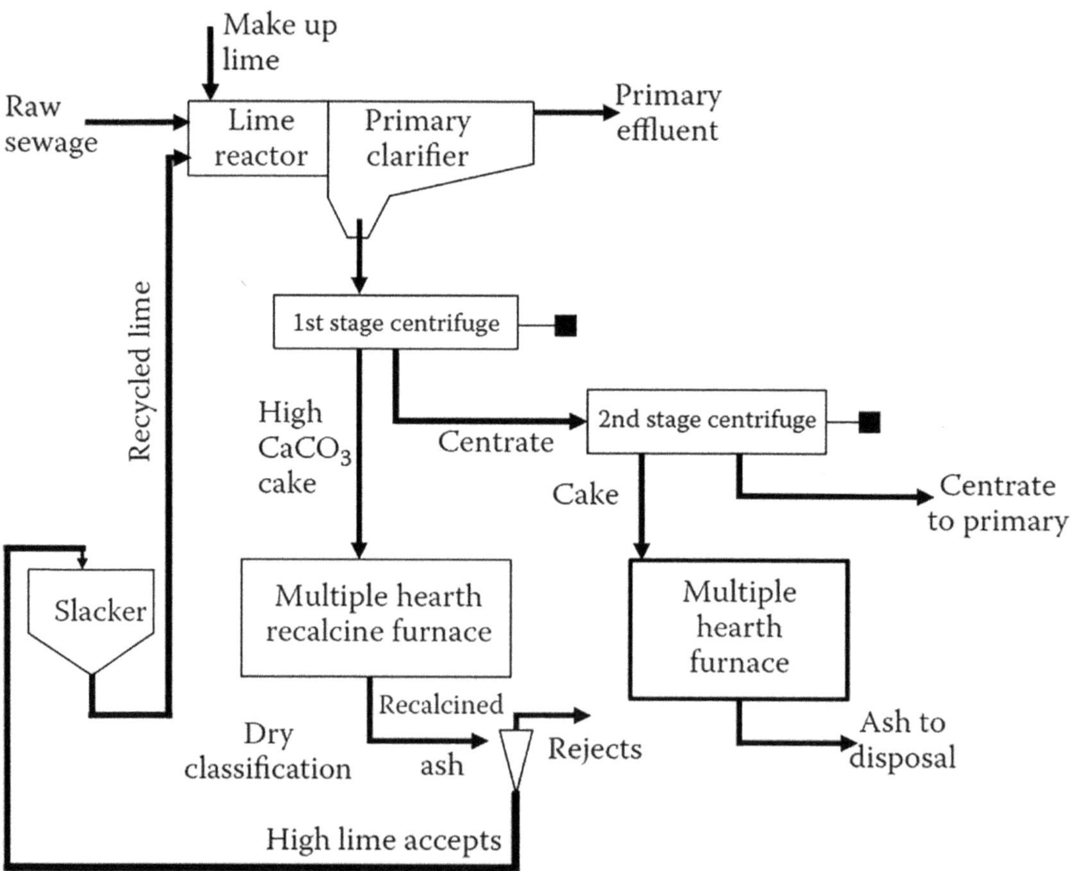

FIGURE 8.4　Sludge treatment by centrifugation.

The $CaCO_3$ cake is then recalcined with the addition of heat according to the following reaction:

$$CaCO_3 = CaO + CO_2$$

The recalcination process is fairly simple because the cakes produced in centrifuges and vacuum filters are quite dry, with solids ranging from 40% to 50%. Lime sludges with a high pH, however, have proved difficult to dewater. Recalcination can be conducted in a rotary kiln or a fluidized bed furnace. In both cases, the CO_2 produced can be used to dissolve some of the hydroxides or for recarbonizing the finished water to bring the pH down. A recently completed study showed that the addition of lime to the primary clarifier produced a thick sludge that centrifuges and could easily process. Approximately 60% of the calcium carbonate fed to the first centrifuge was recovered in the cake, whereas 50%–75% of the other solids were rejected as the centrate. The calcium carbonate slurry was subsequently dewatered to 50% solids and incinerated. The lime produced in recalcining is CaO or quicklime, a dangerous compound. It is often slaked by adding water following the reaction below, and the resulting hydrated product $Ca(OH)_2$ is much safer to handle:

$$CaO + H_2O = Ca(OH)_2.$$

Quicklime can also be used in the dewatering or drying of biological sludge by mixing the lime and sludge in a common concrete mixer. The above reaction is exothermic, and thus sludge is dried and disinfected as a result of the high temperatures produced. This process, used in some European treatment plants, yields a product that is marketable as a soil conditioner, especially where the soil is acidic; however, it lacks permeability and has poor water-holding capacity.

8.1.4.3 Iron Sludges

The sludges formed by both ferric and ferrous compounds are surprisingly soft and fluffy and difficult to dewater to more than 10% or 12% solids. Such sludges still behave as liquids. The recovery of iron from such operations is theoretically possible but not economically feasible.

8.2 ION EXCHANGE MATERIALS

Ion exchangers are insoluble granular substances that have in their molecular structure acidic or basic radicals that can exchange, without any apparent modification in their physical appearance and without deterioration or solubilization, the positive or negative ions fixed on these radicals for ions of the same sign in solution in the liquid upon contact with them [33]. This process, known as ion exchange, enables the ionic composition of the liquid being treated to be modified without changing the total number of ions in the liquid before the exchange.

For each reaction involving two ions A and B, the equilibrium between the respective concentrations A and B in the liquid and the ion exchange substance can be shown graphically (Figure 8.5).

Under conditions of equilibrium, and for a concentration B of X% in the solution, the exchange material becomes saturated up to a concentration of Y%. When the two ions A and B have the same affinity for the exchange material, the equilibrium curve corresponds to the diagonal of the square. The more marked the exchange material's preference for ion B, the further the curve moves in the direction of the arrows. The form of the curve for a given system of two ions depends on a number of factors: nature and valence of the ions, concentration of ions in the liquid, and the type of exchange material. To achieve substantially complete exchange, it is necessary to create successive equilibrium stages by percolating the water through superimposed layers of exchange material (Figure 8.6).

If we take a layer of exchange material entirely in form A, and if a liquid containing ions B or B' (the exchange material has a much greater affinity for ion B' than for ion B) is passed through it,

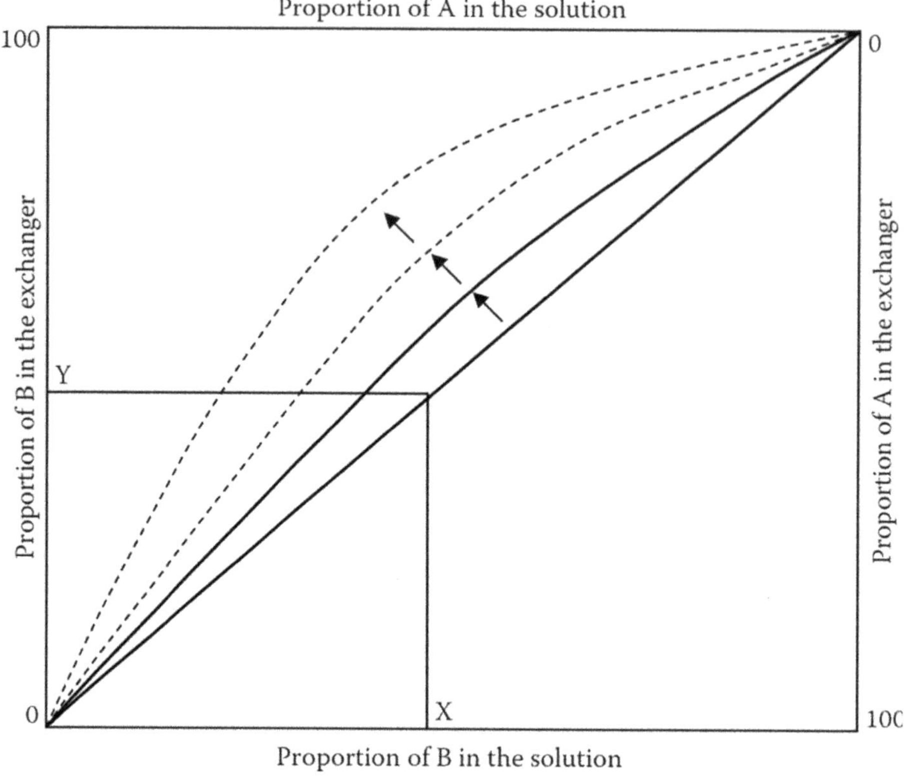

FIGURE 8.5   Ion exchange equilibrium curves.

successive equilibrium points between (A and B) and (A and B') give different series of concentration curves. The possible breakthrough or exhaustion curves (Figure 8.7) depend not only on the static equilibrium curve mentioned above but also on the exchange kinetics between the liquid and the exchange material: this type of kinetics involves the penetration of solutes into the exchanger and is governed by laws known as the "Donnan equilibrium laws."

The first ion exchange substances were natural earths (zeolites); they were followed by synthetic inorganic compounds (aluminosilicates) and organic compounds; the latter materials are used today almost exclusively and, derived from hydrocarbon feedstocks, they are called resins. This term has often been wrongly extended to cover just about any kind of ion exchanger.

Ion exchange resins have proven to be an efficient means of controlling the concentration of heavy metals in wastewaters. There is a noticeable and expanding activity in the application of ion exchange to the recovery and recycling of water and heavy metals from waste. There are various types of resins available for the removal of different metals from effluents. Each resin has certain advantages and limitations, and a proper choice, depending on effluent composition, should be made for its application.

8.2.1   Types of Resins

Ion exchange resins are insoluble polymers with chemical active groups that, when ionized, bond with opposite-charged metal ions. Those resins capable of exchanging cations are called cation exchangers. Resins capable of exchanging anions are called anion exchangers. The ion exchange function of a resin is generally limited by pH levels, flow rate, turbidity, type of regenerant, and complexity of wastewater. Organic resins generally feature a complex matrix, a three-dimensional

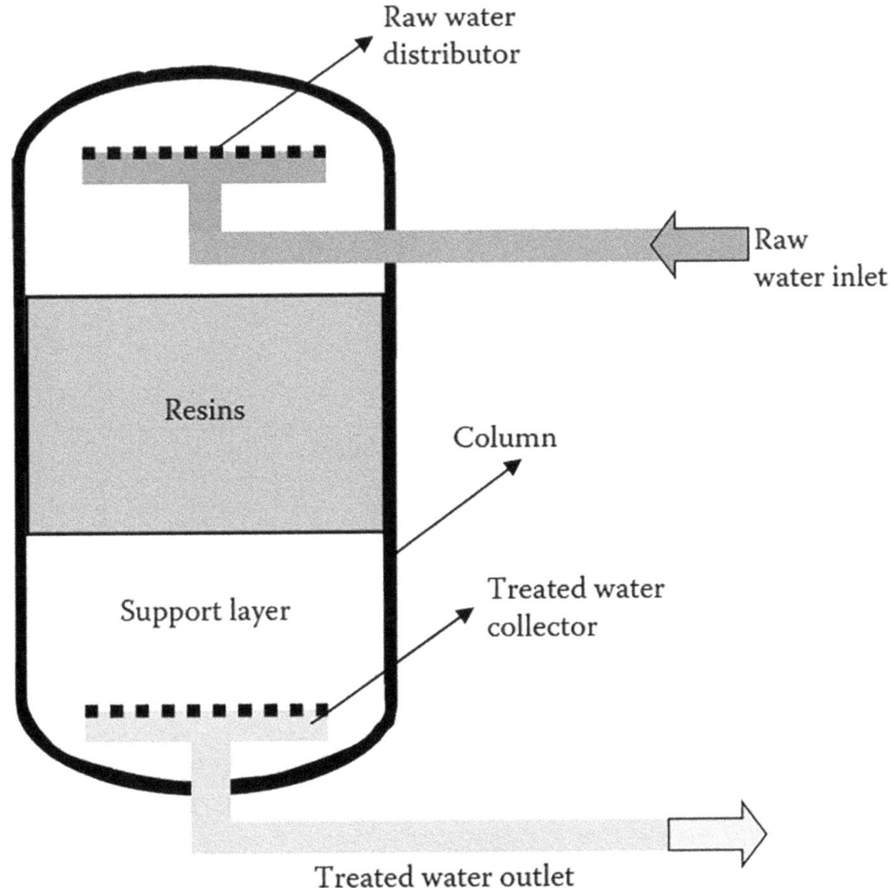

FIGURE 8.6 Common column filled with ion exchange material to treat raw polluted water.

network of hydrocarbon chains [34]. According to the structure, there are two categories: the resins of the gel type and those of the macroporous or loosely cross-linked type. Their basic macromolecular structure is identical, obtained in both cases by copolymerization of, for example, styrene and divinylbenzene. The difference between them lies in their porosity. Their high cross-linking degree increases their mechanical strength to both physical (pressure—negative pressure) and chemical (change in the ionic saturation or exhaustion state) stresses. Gel-type resins have a natural porosity that results from the polymerization process and are limited to intermolecular distances. It is a microporous-type structure. Macroporous-type resins have an additional artificial porosity that is obtained by adding a substance designed for this purpose. Thus, a network of large canals known as macropores is created in the matrix. These products have a better capacity for adsorption and desorption of organic substances.

The cation exchangers can be categorized into two groups: strong acid and weak acid. Anion exchangers can also be divided into strong base and weak base groups. Those resins that remove a specific metal ion are known as chelating exchangers. Tables 8.4 and 8.5 summarize the physical and chemical characteristics of the macroporous- and gel-type resins as well as the main suppliers of these types of ion exchangers.

8.2.1.1 Strong Acid Cations

They are characterized by having HSO_3 sulfonic radicals and acidities close to that of sulfuric acid. In current use, these are sulfonated polystyrenes obtained by

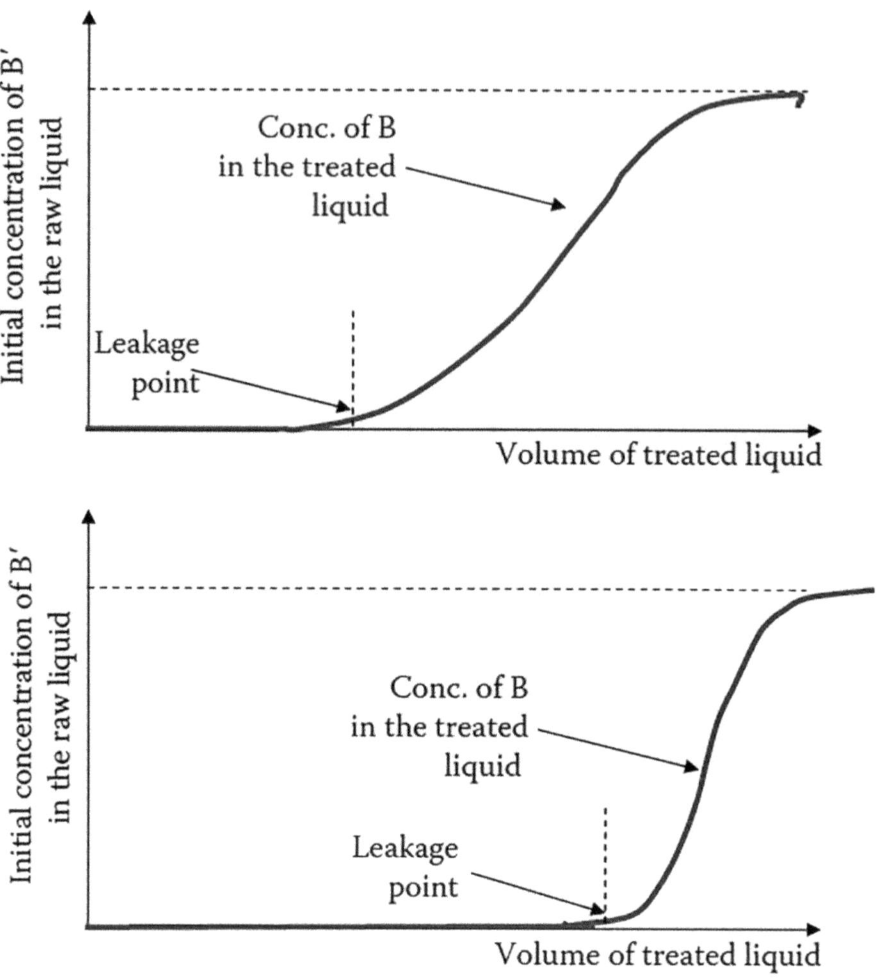

FIGURE 8.7 Exhaustion curves for an ion exchange material entirely in form A. The exchange material has a much greater affinity for ion B' than for ion B.

Copolymerization of styrene and divinylbenzene in emulsion form to obtain perfect spheres on solidification

Sulfonation of the beads was thus obtained.

The products obtained by this process are virtually monofunctional. Their physical and chemical properties vary depending on the percentage of divinylbenzene to styrene, which, in turn, determines the degree of cross-linking, generally ranging from 6% to 16%.

The ion exchange process follows the general reaction of this type:

$$RSO^-H^+ = RSO^-Na^+ + H^+.$$

The selectivity of these resins is usually as follows:

$$Fe^{3+} > Al^{3+} > Ca^{2+}$$

$$La^{3+} > Y^{3+} > Ba^{2+}$$

$$Th^{4+} > Hf^{4+} > Zn^{2+}$$

TABLE 8.4

Physical and Chemical Characteristics of Gel Resins (Strong Acid, Weak Acid, Strong Base, and Weak Base) as Well as the Main Suppliers of These Types of Ion Exchangers

Resins/Gel	Strong Acid	Weak Acid	Strong Base	Weak Base
Particle diameter (mm)	0.3–1.2	0.3–1.2	0.3–1.2	0.3–1.2
Moisture content (%)	45–48	46–53	45–48	46–53
pH range	0–14	1–14	0–14	1–14
T maximum (°C)	120	120	120	120
Turbidity tolerance (NTU)	5	5	5	5
Tolerance (g/m^3)	Chlorine 1.0	Iron 0.5	Chlorine 1.0	Iron 0.5
Total capacity (eq/L)	1.4–2.2	3.5–4.2	1.2–1.4 (type I) 1.3–1.5 (type II)	1.4–2.0
Regeneration	NaCl 80–300 H2SO4 80–250 HCl 40–200	110% of the capacity used	NaOH 40–100 NH$_3$ 30–60 Na$_2$CO$_3$ 60–130 NaOH 40–200	—
Supplier				
Bayer	Lewatit S100		Lewatit M 500 (I) 600 (II)	
Duolite	Duloite C20	Duloite C433	Duloite A101 (I) 102 (II)	
Dow Chemical	Dowex HCR-S	Dowex CCR-2	Dowex SBR (I) SAR (II)	Dowex WGR-2
Rohm&Haas	Amberlite IR120	Amberlite (IRC50, IRC84)	Amberlite IRA 400 (I) 410 (II)	Amberlite IRA68

Notes: The total exchange capacities of various categories of exchange materials are expressed in gram equivalents per liter of resin. The values of regeneration levels listed above are expressed in grams of pure product per liter of resin.

TABLE 8.5

Physical and Chemical Characteristics of Macroporous Resins (Strong Acid, Weak Acid, Strong Base, and Weak Base) as Well as the Main Suppliers of These Types of Ion Exchangers

Resins/Macroporous	Strong Acid	Weak Acid	Strong Base	Weak Base
Particle diameter (mm)	0.3–1.2	0.3–1.2	0.3–1.2	0.3–1.2
Moisture content (%)	40–46	52–57	40–46	52–57
pH range	0–14	5–14	0–14	5–14
T maximum (°C)	150	150	150	150
Turbidity tolerance (g/m^3)	5	5	5	5
Tolerance (g/m^3)	Chlorine 1.0	Iron 0.5	Chlorine 1.0	Iron 0.5
Total capacity (eq/L)	1.7–1.9	2.7–4.8	1.0–1.1 (type I) 1.1–1.2 (type II)	1.2–1.5
Regeneration	NaCl 80–300 H$_2$SO$_4$ 80–250 HCl 40–200	110% of the capacity used	NaOH 40–100 NH$_3$ 30–60 Na$_2$CO$_3$ 60–130/ NaOH 40–200	
Supplier				
Bayer	Lewatit SP112	Lewatit CNP80	Lewatit M 500 (I) MP 600 (II)	Lewatit MP64
Duolite	Duloite C26	Duloite C464	Duloite A 161 (I) 162 (II)	Duloite A378

(Continued)

TABLE 8.5 (*Continued*)

Physical and Chemical Characteristics of Macroporous Resins (Strong Acid, Weak Acid, Strong Base, and Weak Base) as Well as the Main Suppliers of These Types of Ion Exchangers

Resins/Macroporous	Strong Acid	Weak Acid	Strong Base	Weak Base
Dow Chemical	Dowex MSC-1		Dowex MSA1 (I) MSA2 (II)	Dowex MWA-1
Rohm&Haas	Amberlite IR200		Amberlite IRA 900 (I) 910 (II)	Amberlite IRA93

Notes: The total exchange capacities of various categories of exchange materials are expressed in gram equivalents per liter of resin. The values of regeneration levels listed above are expressed in grams of pure product per liter of resin.

$$Ac^{3+} > La^{3+}$$

$$Th^{4+} > La^{3+} > Ce^{2+} > Na^+$$

$$Mg^{2+} > Be^{2+}$$

8.2.1.2 Weak Acid Cations

These are polyacrylic resins characterized by the presence of HCO_2 carboxyl radicals that can be likened to organic acids such as formic or acetic acid. They differ from strong acid exchangers in two respects:

They retain only the Ca, Mg, Na, and so on cations that are bound to bicarbonates, but they cannot exchange cations at equilibrium with strong anions (SO_4^{2-}, Cl^-, and NO_3^-).

They can be regenerated more easily, and their regeneration rates are close to those of stoichiometric efficiency.

The ion exchange process follows the general reactions of this type:

$$RCO_2H = RCO_2^- + H^+,$$

$$RCO_2H + HCO_3^- + Na^+ = RCO_2^- Na^+ + H_2O + CO_2.$$

The selectivity of these resins is usually as follows:

$$H^+ > Ca^{2+} > Mg^{2+} > K^+ > Na^+.$$

8.2.1.3 Strong and Weak Base Anions

Anion exchangers can be divided into weak or intermediate and strong base anion exchangers. These two types can be distinguished in practice as follows:

The weak base types do not retain very weak acids, such as carbonic acid or silica, but the strong base types retain them completely.

The strong base types alone are able to release the bases from their salts following the typical reaction:

$$R - OH + NaCl = R - Cl + NaOH.$$

The weak base types are more or less sensitive to hydrolysis, in the form of the displacement, by pure water, of the anions previously attached to the resin:

$$R - Cl + H_2O = R - OH + HCl.$$

The strong base types are practically unaffected by this phenomenon.

The weak base types are regenerated more easily. The existence of quaternary ammoniums in the molecule is typical of the strong base anion exchangers. All the strong base resins used for demineralization purposes belong to two main groups commonly known as type I and type II. The former consists of simple quaternary ammonium radicals, and the latter consists of quaternary ammonium radicals with alcohol function. Each type has its own field of application, depending on the nature of the water to be treated and the conditions applied to the regeneration cycle. The two types differ in the following respects:

In type I, the basicity is strong and the capacity low; the regeneration efficiency is poor.
In type II, the basicity is weaker and the capacity higher; the regeneration efficiency is also better.

The weak anion exchangers consist of a mixture of primary, secondary, tertiary, and sometimes quaternary amines. The nucleus of the molecule is highly varied in nature and may be aliphatic, aromatic, or heterocyclic.

The exchange process of the strong base type follows the reaction:

$$RN^+OH^- + H^+ + A^- \Leftrightarrow RN^+A^- + H_2O.$$

The selectivity is as follows:

$$NO_3^- > CrO_4^{2-} > Br^- > SCN^- > Cl^-.$$

The exchange process of the weak base type follows the reaction:

$$R_3N + H^+ + A^- \Leftrightarrow R_3NH^+A^-.$$

The selectivity is as follows:

$$OH^- > SO_4^{2-} > CrO_4^{2-} > NO_3^- > PO_4^{3-} > MoO_4^{2-} > HCO_3^- > Br^-.$$

8.2.1.4 Adsorbent Resins

These are products that are designed to retain nonionic compounds (basically organic molecules) in solution in polar and nonpolar solvents by means other than ion exchange and by a reversible technique. This process of adsorption on solids is very complex and involves various types of interaction between the adsorbent surface and the adsorbed molecules. For this reason, the adsorptive capacity of the resins depends on numerous factors, such as the chemical composition of the skeleton (polystyrenic, polyacrylic, and formophenolic), the type of functional groups of polar adsorbents (secondary and tertiary amines, quaternary ammonium), the degree of polarity, the porosity (usually macroporous materials with pore sizes up to 130 nm), the specific surface area: up to 750 m²/g, the hydrophilic nature, and the shape of the grains.

Their possible uses include the following:

Protection of the ion exchange system by retaining the pollutants present in feed water (humic acids, detergents, etc.)

Decolorization of sugar syrups, glycerin, grape musts, whey, and so on

Separation, purification, and concentration processes in the pharmaceutical industry and synthetic chemistry.

The regeneration method of adsorbent resins basically depends on the product adsorbed. The traditional eluants are acids, bases, sodium chloride, methanol, adapted organic solvents, and, in certain cases, pure water or steam.

The choice of the correct adsorbent presents some difficulty; it must be guided by the properties of each adsorbent and the products to be retained. Therefore, laboratory or pilot studies are indispensable in the majority of cases.

8.2.1.5 Special Resins

Polyfunctional resins: These are products that combine the properties of strong resins with those of weak resins. This is the case with anion resins that are able to remove all the anions including silica and CO_2 while ensuring a high exchange capacity and an excellent regeneration efficiency due to their weak base function.

Nonionic polymeric adsorbent is a man-made porous resin with extremely high surface area for adsorption of heavy metals and/or organic substances similar to granular activated carbon (GAC) [35].

Chelate resins: These comprise special functional groups (aminophosphoric, aminodiacetic, aminodioxime, mercaptan) that permit the selective retention of heavy metals from various effluents (zinc, lead, mercury, etc.), gas chromatographic separations of metals, and also the final softening of brine from the electrolysis process.

Ferric manganese binary oxide (FMBO) is developed by researchers as an effective adsorbent for combating organic-arsenic pollution, particularly **dimethyl** arsenic acid (DMA). FMBO was synthesized using the co-precipitation method and compared with manganese dioxide (MnO_2) and ferric oxide (FeOOH).

Resins for nuclear use: These involve products with a higher degree of purity than that of resins used in common operations. Among these are strong acid cation resins in H^+ form that are regenerated to 99%, and strong base anion resins in OH form with less than 0.1% of Cl^-.

Catalyzing resins: These conventional resins are used in a basic or acidic catalyst process (e.g., the inversion of glucose in the manufacture of liquid sugar). They could also be used with a metallic catalyst (e.g., a palladium resin for deoxygenation of demineralized water or sea water).

8.2.2 SPECIFIC APPLICATIONS

It is important to emphasize that the techniques related to ion exchange processes should not be used unless the raw water has been subjected to a form of preliminary treatment suited to its type, which must include the removal of suspended solids, organic matter, residual chlorine, chloramines, and so on. The continuous removal of heavy metals by ion exchange takes place in fixed-bed columns that are packed with cationic and/or anionic resins. The metals sorb onto the resins in exchange of hydroxide anions, protons, and/or light metal cations that are released into the solution. An ion exchange system can perform both heavy metal removal and neutralization of acidic water. Furthermore, ion exchange systems open the possibility of recovering heavy metals in the form of liquid concentrates. These concentrates can be either returned to the manufacturing process, or they can be

Efficiently precipitated yielding small volumes of sludges for disposal or metal recovery

Further processed directly to recover the metal(s) in solid and resaleble (reusable) form.

The higher the selectivity of a resin, the more strongly the metals are bound to it, and the more difficult it usually is to desorb them. This increases the consumption of a regenerant and hence the operating cost. Moreover, the resins are prone to fouling (poisoning) by organic substances [36]. Finally, as ion exchange resins are hydrocarbon based, their price is coupled with the price of crude oil and they are thus relatively expensive, with the price per kilogram ranging from USD10 to USD50 (prior to the year 2004). Therefore, in order to keep the operating cost sufficiently low, the ion exchange processes are usually applied only to effluents containing medium or low levels of heavy metals.

8.2.2.1　Nickel

The most commonly applied process for the removal of nickel from wastewater is the use of a strong acid cation resin. Unfortunately, this type of resin can only be applied if nickel is the only polyvalent metal ion in the wastewater. Consider a wastewater composed of nickel in the presence of ammonium molybdate. The nickel is most effectively recovered by the use of an aminophosforic acid resin. The Russian-made chelating resin ANKF-80 used on a wastewater at pH 2 is approximately 19 times more effective than a conventional resin. Also available, but not as effective, is the Amberlite XE-318, having an ion capacity of 34 g/L of resin. Another type of resin effective in nickel removal is a weak acid resin. This type of resin shows high selectivity for nickel ions even in wastewater polluted with organic carbon. A Wofatit CA-20 resin is the most effective weak acid resin manufactured and available. Another weak acid resin available is the Zerolit 236 possessing an ion capacity of 108 g/L of resin. The regenerant used for these resins is ammonium carbonate.

Strong acid resins such as Zerolit 525, Amberlite IR-20, and Amberlite 200 have ion capacities of 48, 31, and 30 g/L of resin, respectively. The regenerant used is again ammonium carbonate.

8.2.2.2　Copper

In the industrial production of copper, large amounts of acidic wastewaters are produced. There are mainly two specific types of resins applied for copper recovery. One resin is Dowex XFS-4195. This N-(hydroxyalkyl) picolyamines-based resin can be applied for very acidic wastewaters. Another resin, having the same base, is Dowex XFS-4196, which performs well for wastewaters of higher pH. The regenerant used in both cases is sulfuric acid.

Unfortunately, not all wastewaters can be purified of copper with ordinary ion exchangers. Wastewaters that contain organic ligand, for example, cannot be treated in this way because they form a coordination complex. Complex compounds having carboxylic acid groups such as tartaric, citric, and lactic acids will interfere with the efficiency of a standard ion exchanger. In these cases the stability for the copper complexes is lower. The chelating resin will successfully remove copper ions from such wastewaters.

8.2.2.3　Zinc

Zinc can be extracted from wastewaters as a Zn^{2+} cation or $ZnCl_4^-$ anion. The zinc cation may be sorbed by a cation exchanger or a chelating resin and the anion by an anion exchanger. Zinc salts are present in wastewaters from, for example, the kaolin industry and the blowdown of cooling towers. For the kaolin industry, a strong acid resin may be employed for the removal of the Zn^{2+} cation, using sodium chloride as a regenerant. In the case of the cooling tower blowdown water, phosphoric acid resins such as Duolite ES-63 or Duolite TSAP-40 are commonly employed.

8.2.2.4　Mercury

As a metal, mercury is probably one of the worst water pollutants. The source of most mercury pollution is the wastewater produced from chlorine and alkali manufacturers. Resins containing the thio group possess a high affinity for mercury ions. The Imac-TMR resin is an example. Their operational capacity depends on the concentration of mercury present in the wastewater (see Table 8.6).

TABLE 8.6

Operational Capacity of a Resin to Uptake Mercury

Hg Concentration (mg/L)	Capacity for Hg (g/L of Resin)
0.6	48
0.8	57
1.0	66
4.0	80
6.0	88
8.0	95
10.0	100

Other types of molecules that exhibit high affinity for mercury ions contain the R-S-C(NH)-NH$_2$ (isothiouronium) group, an example of which could be resin Srafion-NMRR. Another highly active chemical group is R-NH-C(S)-SH, the dithiocarbonate group, which is found in the resin Nisso-ALM-125. Regeneration using sulfuric acid is applied to recover mercury from the resin.

Another resin available is the Nisso-ALM-126 whose capacity can be increased by heating the resin to a maximum of 50°C. Regeneration is not feasible in this case. However, the mercury can be removed from the resin by roasting it. Strong base resins can also be applied for the removal of mercury from wastewaters. For acidic wastewaters within a pH range of 2.5–3.5, Wofatit SDW or Varion AD resins can be used. They possess the ability to remove up to 85% of the mercury. The use of a strong acid resin such as the Amberlite IR-120 could produce an effluent containing levels of 0.05 mg/L of Hg^{2+} from the initial solution containing 10 mg of Hg^{2+} per liter of wastewater.

8.2.2.5　A Case Study

A typical and ubiquitous industrial operation known to have a metal-pollution problem is metal plating. The amount of electroplating process effluent encountered in an average size operation generally varies between 25,000 US gal day^{-1} and 100,000 US gal day^{-1}. For the case design calculations, the wastewater flow of 48,000 US gal day^{-1} (182.4 m^3 day^{-1}) has been chosen. It is assumed to contain 20 mg Zn^{2+} L^{-1}, 30 mg Ni^{2+} L^{-1}, 40 mg CuSO$_4$ L^{-1}, and 130 mg CrO$_3$ L^{-1}. The latter chromium content eventually complicates the treatment system considerably as will be demonstrated in the design of the treatment facility. Removal of the basic metals from the given solution is considered as a major objective. Chromium is a valuable element and its recovery would be highly desirable. However, since it cannot be retained in its chromate form on the same cationic ion exchanger as all the other metals, a special sequence of additional two ion exchange operations would have to be added to the basic cationic one. A typical treatment system is outlined in Figure 8.8. The pH of the solution to be treated has a crucial effect on the uptake of metals from acidic solutions. The resin IRC-718 selected for the treatment process is more selective for metal ions and calcium interference is minimized at the optimum operation pH 4. Requirement for this pH level dictates the necessity of a pH adjustment in a tank that has to be placed prior to the sorption contact stage. Following the pH adjustment, the wastewater solution should be filtered to remove precipitates, particulate matter, and insoluble salts from the solution. The filtration step decreases the possibility of resin fouling.

The solution to be treated is pumped into two columns operating in a sequence. The whole arrangement could operate for five days until saturation of the bed. For practical reasons, it may be more convenient to regenerate the first column approximately half-way through this period, whereas the second one becomes first in their sequence. The newly regenerated column would be always phased in as a polishing second one in that sequence. This arrangement would require three columns, two operating whereas the third one would be on the regeneration cycle. The columns are designed to be backwashed to move any entrained solids under up-flow conditions. The partly fluidized bed

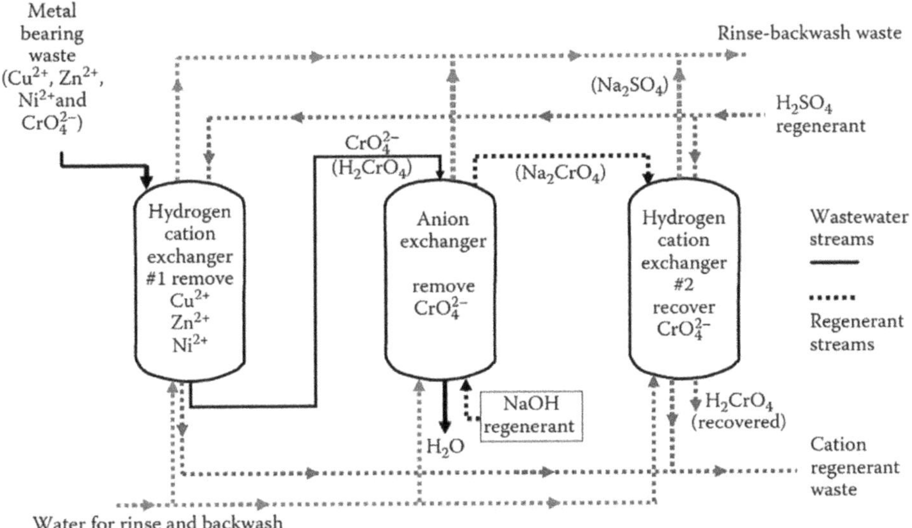

FIGURE 8.8 Ion exchange process: Schematic diagram of a case study.

expansion of 50% has been estimated for this operation. Regeneration is done with 5% of sulfuric acid. The operations of backwashing, regeneration, and rinsing are done in situ, in the same columns. The spent regenerant and rinse water from this first cationic exchange contain metal sulfates, $M_2(SO_4)n$, and some sulfuric acid. They must therefore be neutralized in a separate operation and the metals precipitated prior to discharge to any receiving body.

8.3 REVERSE OSMOSIS

Reverse osmosis is based on the separation of the solvent from the influent waste stream by a pressure in excess of the osmotic pressure of the solution. The wastewater flows under high pressure through an inner tube made of a semipermeable membrane material. The purified solvent is removed from the outer tube, which is at atmospheric pressure (Figure 8.9) [37].

The disadvantages associated with reverse osmosis involve the sensitivity of the membrane. Organics as well as other impurities precipitate, which leads to membrane fouling. It is therefore necessary to have a consistent composition of the influent waste stream. Apart from the membrane sensitivity, the process also requires elevated pressures that drive up the operating costs of pumping.

8.3.1 PRINCIPLE OF REVERSE OSMOSIS

Reverse osmosis makes use of the properties of semipermeable membranes that allow water to pass through while solutes are retained except for certain organic molecules very similar to water (with a low molecular weight and strong polarity). If a concentrated saline solution is separated from a more dilute solution by such a membrane, the difference in chemical potential tends to promote the passage of water from a compartment with a low potential to that with a higher potential in order to dilute it (natural osmosis). To stop this diffusion, a pressure must be exerted on the "filtered" fluid. At equilibrium, the pressure difference established in this way is known as the osmotic pressure of the system (Figure 8.10).

A simple Equation (8.1) relates osmotic pressure to concentration:

$$P = \Delta C * R * T \tag{8.1}$$

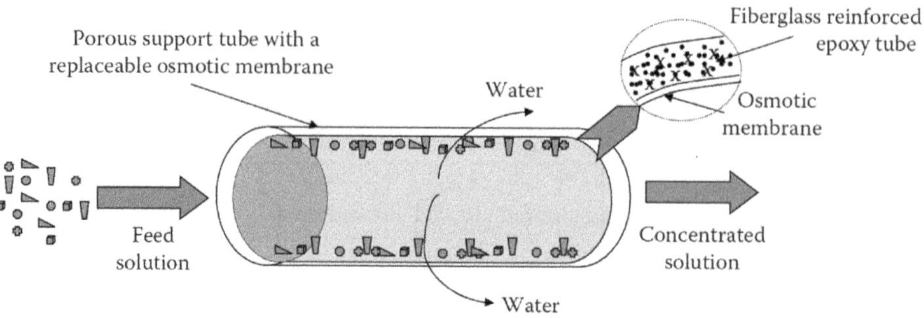

FIGURE 8.9   Tubular module section of a reverse osmosis operation.

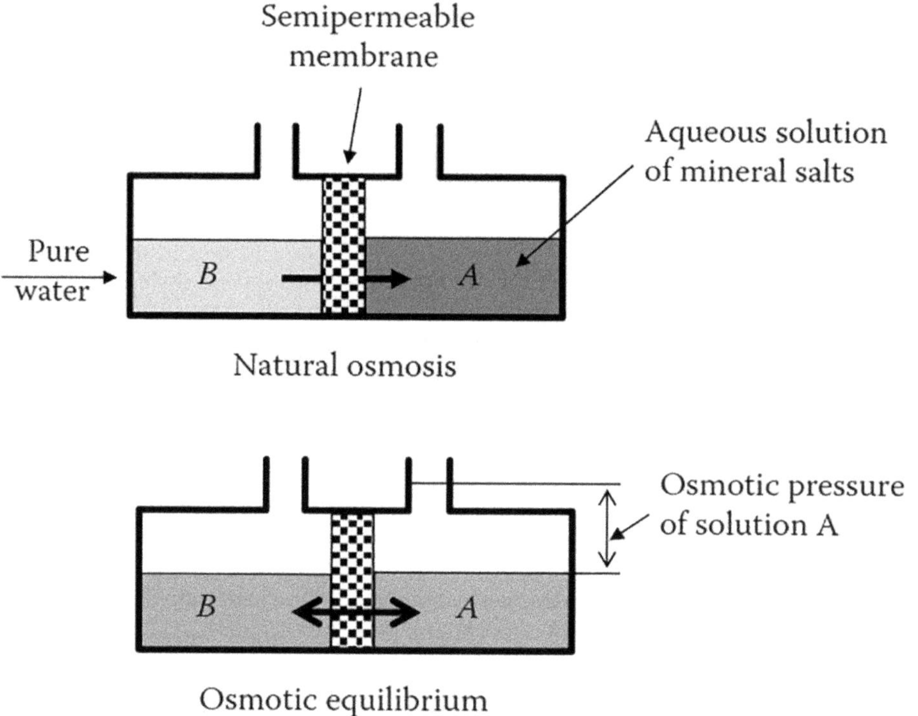

FIGURE 8.10   Osmosis phenomenon.

where P is the osmotic pressure (Pa), DC is the difference in concentration in mol/m^3, R is the constant of an ideal gas = 8.314 (J/(mol K)), and T is the temperature in degrees kelvin.

Clearly, the smaller the molecule (i.e., the lower the molecular weight), the greater the osmotic pressure set up by the same difference in concentration. This explains why ultrafiltration leads to an osmotic backpressure that is much lower than that experienced with reverse osmosis.

In fact, to produce "pure" water from a saline solution, the osmotic pressure of the solution must be exceeded. In the same way, in order to obtain economically viable flows, at least twice the osmotic pressure must be exerted. For instance, for a brine containing several grams of salt per liter, pressures of 5–30 bar would be needed, and for sea water, pressures of 50–80 bar would be needed. A second phenomenon can amplify this effect.

As Figure 8.11 illustrates, when water is transferred, the molecules and ions retained by the membrane tend to accumulate along its entire surface, thereby increasing both the salinity actually "treated" by the membrane and the osmotic pressure that must be "overcome" in order to desalinate

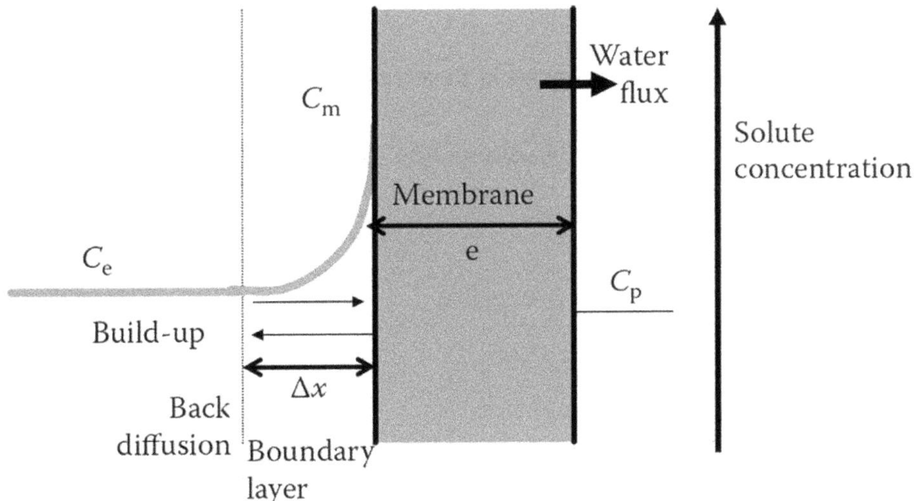

FIGURE 8.11 Concentration polarization during a reverse osmosis process.

the solution. This results in higher energy costs, as well as in the risk of causing precipitation if the solubility product of one of the cation-anion couples is exceeded in the boundary layer all along the membrane.

This phenomenon is known as concentration polarization of the membrane and it is defined by the coefficient Y (Equation 8.2):

$$\varphi = \frac{C_m}{C_e} \tag{8.2}$$

where C_m is the concentration of the liquid in contact with the membrane and C_e is the concentration of the liquid to be treated.

The concentration polarization phenomenon can be reduced to a minimum by maintaining a circulation flow across the upstream surface of the membrane. That limits the thickness of the boundary layer and facilitates the reverse diffusion of the rejected solutes; however, this also limits the fraction of desalinated water. This technique is used in industrial systems to maintain the coefficient Y between 1 and 1.4.

To describe the phenomena observed, best models call upon the laws of diffusion, water being considered dissolved by the polymer making up the membrane (water used for swelling the polymer); this water moves under the effect of the pressure gradient, whereas the salts move under the effect of their concentration gradient alone.

For a saline solution, the water and salt flux rates may be obtained by Fick's and Henry's laws for water (Equation 8.3) and for salts (Equation 8.4):

$$Q_p = K_p - S\,e - \left(\Delta P - \Delta_p\right) K_t \tag{8.3}$$

where Q_p is the flow of water through the membranes, K_p is the membrane permeability coefficient for water, S is the membrane surface area, e is the thickness of the membrane, DP is the hydraulic pressure differential across the membrane, D_p is the osmotic pressure differential across the membrane, and K_t is the temperature coefficient.

Thus, the flow of water through the membrane is directly proportional to the effective pressure gradient, represented by the difference between the hydraulic and the osmotic pressure.

TABLE 8.7

General Tendencies in Reverse Osmosis Systems

	Product Flow Q_p	Product Salinity C_p
Pressure ↑	↑	↓
Temperature ↑	↑	=
Salinity ↑	↑	↑
φ ↑	↑	↑

The coefficient K_t takes the viscosity of water into account. The latter decreases when the temperature rises. Therefore, the flow is greater when the temperature rises (2.5%–3% difference per degree at about 15°C).

$$Q_p = K_s - S\,e - (\text{DC})\,K_t \qquad (8.4)$$

where Q_s is the flow of salt through the membrane, K_s is the membrane permeability coefficient for solutes, S is the membrane surface area, e is the thickness of the membrane, DC is the ion concentration differential across the membrane ($C_m - C_p$ or $C_e * Y - C_p$), and K_t is the temperature coefficient.

The flow of salt is directly proportional to the gradient of concentration through the membrane; for a given membrane and a given solution, its value is independent of the applied pressure.

In practice, Table 8.7 summarizes the general tendencies in all reverse osmosis systems.

8.4 ELECTROCHEMICAL PROCESSES, ELECTROWINNING

Public concern over water contamination and the steep rise in heavy metal prices triggered both governmental and industrial activity in the removal and recuperation of heavy metals from domestic and particularly from industrial wastewater. As the world consumption of resources increases, reserves will exhaust themselves even faster and the need for recycling is becoming more and more urgent. Traditionally, heavy metals are removed using precipitation (Section 8.1) or ion exchange methods (Section 8.2). Particularly, the most widely practiced precipitation, however, produces highly toxic sludge for which disposal is difficult and expensive. Electrolytic techniques can recuperate these metals from wastewater solutions either directly or when preconcentrated by ion exchange. Metals can then be profitably salvaged for resale and reuse. Six major electrochemical metal recovery methods will be discussed in detail: electrodialysis, forced flow, rotating cathode, mesh, packed bed, and fluidized bed recovery cells [32]. Generally, the capital costs for electrochemical processes range between two and three times the capital cost for a physico-chemical removal system. The operating costs are lower than physico-chemical systems due to the salvage value of the metal and the absence of chemical reagents [38,39].

8.4.1 BACKGROUND

In electrochemical processes, an electric potential is used to move charged ionic particles in solution from one medium to another [40] (Figure 8.12). For example, positively charged metal ions can be plated out on an electronegative cathode by applying a potential in an electrolyte solution. By varying the electric potential, metals can be plated out selectively. Problems in removing metals even in concentrations below 200 mg/L using electrolytic methods have been overcome, thus providing industries with the tools to reduce concentrations below government guidelines. The problems of dilute electroplating are due to the low mass transfer rate of migrating ions in solution. As Fick's law states, the mass transfer rate decreases with decreasing concentration gradient, resulting in lower rate at which

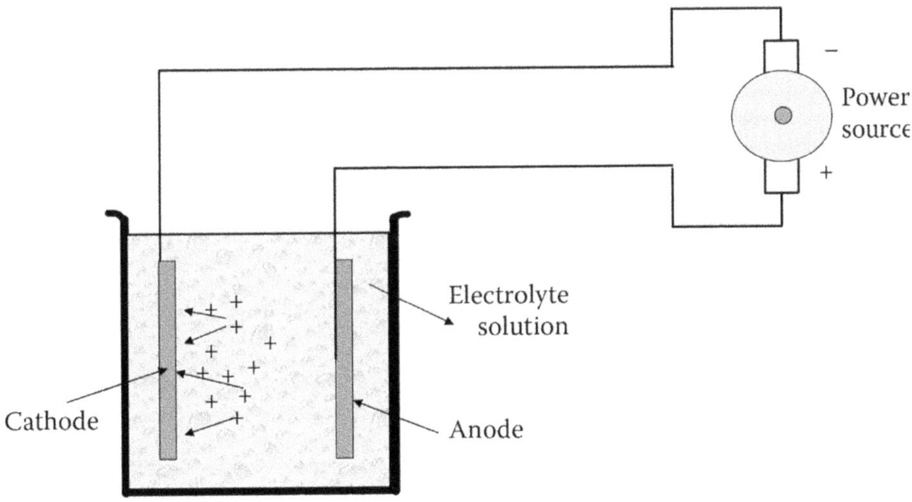

FIGURE 8.12 Electrochemical cell. Direction of travel of cations in an electrolyte solution applying a potential.

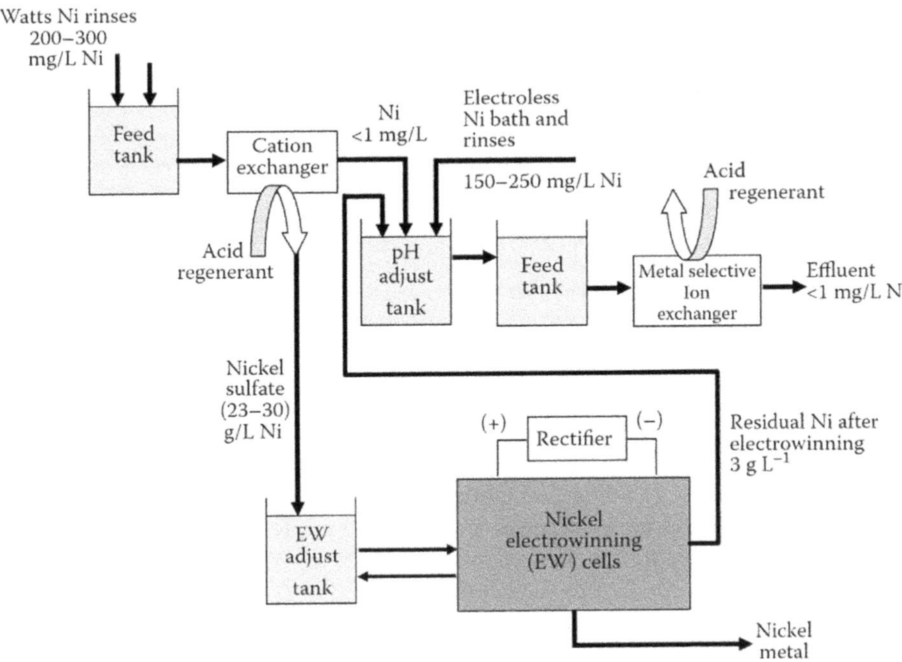

FIGURE 8.13 Nickel recovery treatment using an electrochemical cell process in conjunction with cation- and metal-selective exchangers.

metal plates on the cathode. As the plating-out rate decreases, so does the efficiency of the process. Furthermore, creation of hydrogen gas at the cathode surface due to the redox reaction creates an additional barrier for metal ions to plate on the cathode. These two problems can be solved by either having very large cathode surfaces (flow-through processes) or by increasing turbulence in the solution (flow-by processes). Mesh, packed bed, and fluidized bed cathodes are examples of flow-through cathodes, whereas forced flow and rotating cathodes are examples of flow-by processes [39,41]. Figure 8.13 represents a general metal (nickel) recovery treatment using an electrochemical cell process in conjunction with cation- and metal-selective exchangers.

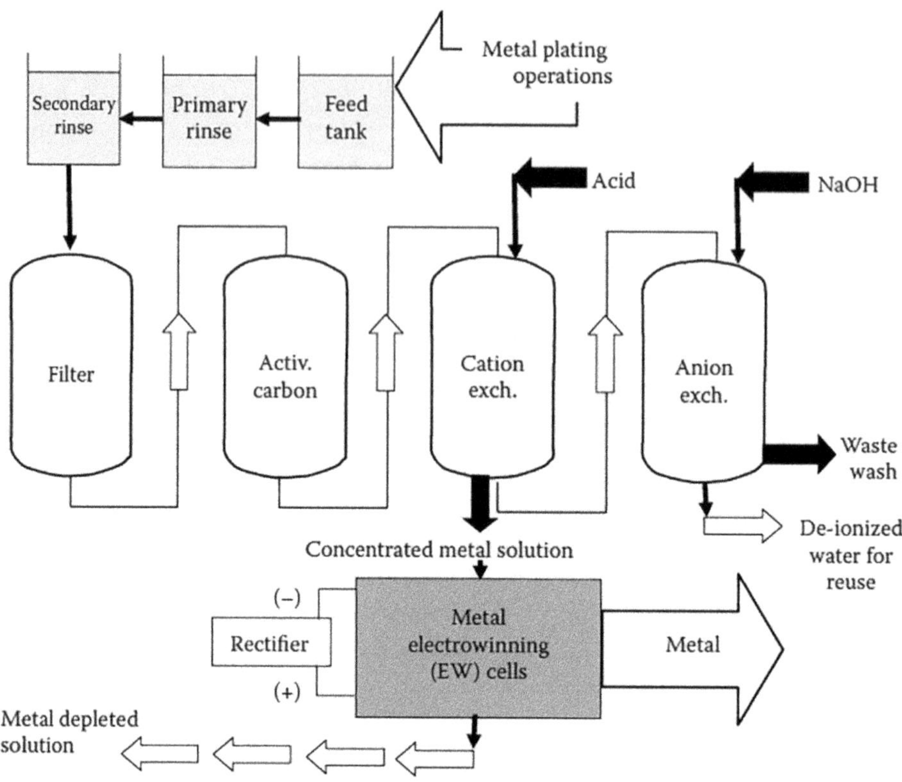

FIGURE 8.14 Ion exchange/physico-chemical/electrochemical combination.

8.4.2 Forced Flow Cell (Flow-by Cathode)

A forced flow cell is characterized by a mechanical or physical agitation of the contaminated solution around the cathode. The agitation can be created either by intensive solution circulation by using pumps

or by ventures that introduce air bubbles at the bottom of the cell or by using static mixers. Since the process mechanism is simple, operating and capital costs are low. The metal is also easily recoverable since it plates out directly on flat immobile cathodes. However, the process suffers from low efficiencies when heavy metal concentrations are below 50 mg/L and cannot be made continuous due to the periodic removal of the cathodes for stripping. Owing to its low operating cost and good efficiencies above 50 mg/L, forced flow cells are often used in conjunction with preconcentration ion exchange or physico-chemical processes that are inexpensive and very efficient at concentrations below 50 mg/L. Figure 8.14 represents a general process scheme combining electrochemical, physico-chemical, and ion exchange treatments.

8.4.3 Rotating Cathodes (Flow-by Cathode)

As in forced flow cells, the solution is agitated in order to increase the efficiency of the process. The difference resides in the fact that the agitation of the solution is accomplished by the brisk movement of the cathode. The most common arrangement is the rotation and impaction of cylindrical cathodes. Due to the process simplicity, the operating and capital costs are low. Furthermore, the process can be made continuous by impacting the rods together or by scraping the rods while in movement, and by collecting the metals as powder at the bottom of the cell. The process, however, still suffers from low efficiencies when concentration decreases below 50 mg/L and from possible

FIGURE 8.15   An electrowinning system using a rotating electrode to improve metal yields and allow economic metals recovery from solutions with lower metals concentrations.

breakdown problems due to the many moving parts. Figure 8.15 represents an electrowinning system using a rotating electrode to improve metal yields and allow economic metal recovery from solutions with lower metal concentrations.

8.4.4 Mesh Cathode (Flow-through Cathode)

Although flow-by cathodes are essentially two-dimensional cathodes, since their thickness is irrelevant, mesh cathodes and other flow-through cathodes are three dimensional due to the particulate or fibrous nature of the cathode. Flow-through cathodes can also be distinguished by their very high real-to-apparent surface area ratio that can reach 10,000. The mesh cathode consists of an intertwining matrix of conducting fibers through which the contaminated solution flows and on which the metal deposits. The heavy metal can be recovered by using the mesh cathode as the anode in a conventional plating cell and by stripping the metal from the flat cathode. Mesh cathode cells have low removal limits (0.5 mg/L) and low capital costs due to the simplicity of the plating mechanism. As an example, Figure 8.16 represents the kinetics of copper and cyanide electrochemical removal (using a mesh electrode) from a heavy-metal-laden wastewater. It appears that after 30 hours, the removal of 98% and 99% could be reached for cyanide and copper, respectively. However, some

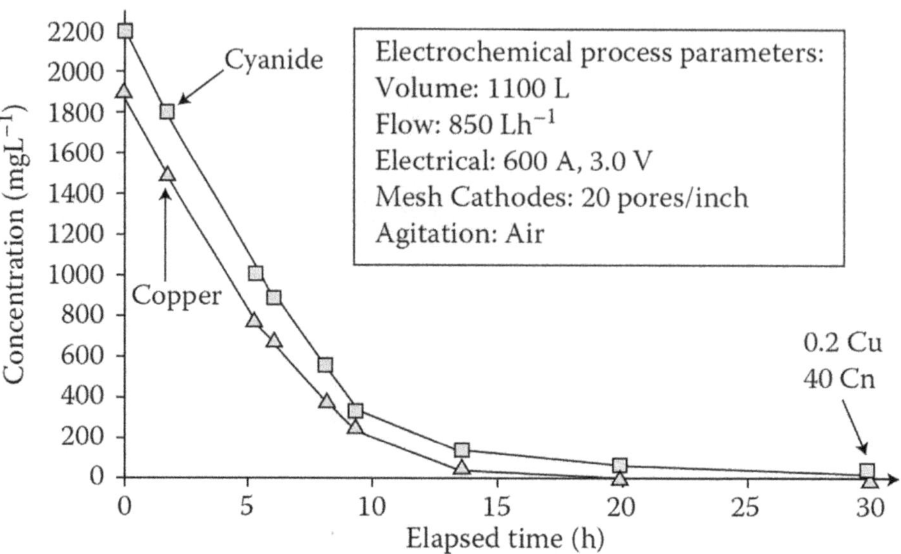

FIGURE 8.16 Kinetics of copper and cyanide electrochemical removal using a mesh electrode. The process parameters are indicated as a legend.

problems do exist in the operation of mesh cathodes: blockage in the mesh due to large particles being caught in the matrix or by unequal depositing of metals in the mesh, batch processing of the solution, and high operating costs due to the complicated process of recovering the heavy metals.

a conventional plating cell and by stripping the metal from the flat cathode. Mesh cathode cells have low removal limits (0.5 mg/L) and low capital costs due to the simplicity of the plating mechanism. As an example, Figure 8.16 represents the kinetics of copper and cyanide electrochemical removal (using a mesh electrode) from a heavy-metal-laden wastewater. It appears that after 30 hours, the removal of 98% and 99% could be reached for cyanide and copper, respectively. However, some problems do exist in the operation of mesh cathodes: blockage in the mesh due to large particles being caught in the matrix or by unequal depositing of metals in the mesh, batch processing of the solution, and high operating costs due to the complicated process of recovering the heavy metals.

8.4.5 PACKED BED CELLS (FLOW-THROUGH CATHODE)

As in mesh cathodes, packed bed cells excel due to the high surface area of the cathode [42]. However, packed bed cathodes are made of packed metallic particles and not fibers such as in mesh electrodes. The particles are made of the same metal as the one being recuperated. The heavy metals plate out on the particle surface and are recuperated by either scraping off the metal from their surface or by melting them. Packed bed cells have the advantages in giving low residual metal concentrations (0.1 mg/L) and in the easy recuperation of heavy metals from the cathode. However, some inherent problems have made this process inaccessible to large industrial applications: high capital and operating costs due to the complicated mechanism, channeling in the bed, difficulties in scaling up the process, high-power consumption, and the batch processing of the solution.

8.4.6 FLUIDIZED BED CELLS (FLOW-THROUGH CATHODE)

Fluidized bed cells are being examined very closely due to their low removal limit (0.1 mg/L) [43]. Furthermore, some problems associated with packed bed cells are eliminated in fluidized flow. Fluidized bed cells differ from packed bed cells due to the circulating motion given to the particles by fluidizing the particle bed with the contaminated solution. Apart from low metal removal limits,

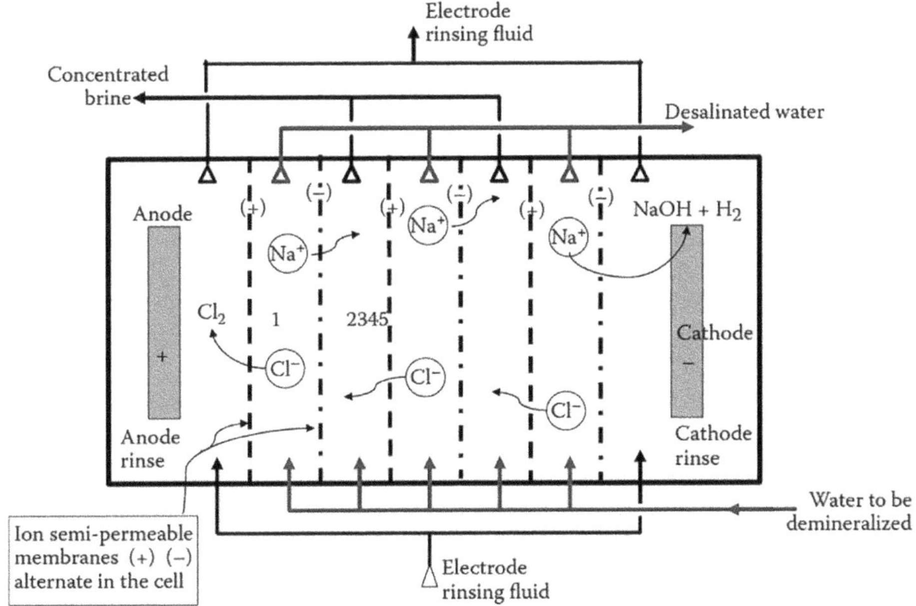

FIGURE 8.17 The principle of electrodialysis.

fluidized bed cells also take advantage of easy metal recovery from the particulate cathode and, unlike packed bed cells, the possibility of making the process continuous by bleeding the particles in a recycle stream. Once again, the complexity of the process makes fluidized bed cells rather expensive. Problems with channeling and scale up and the high-power consumption have also left the process at the experimental stage.

8.4.7 ELECTRODIALYSIS

Although electrodialysis does not remove heavy metals from the solution, this process has proved very effective in concentrating heavy metals in the "brine" solution while simultaneously purifying the contaminated stream [44]. This is accomplished by separating a contaminated solution with an ion-selective membrane and by applying an electrical potential across the system. As the potential is applied, cations (e.g., heavy metals in solution) migrate through semipermeable membranes toward the cathode, thus becoming concentrated in one solution compartment while the in-flow solution becomes purified. The concentrated solution can then be returned to an electroplating cell. Figure 8.17 represents the principle of the electrodialysis process. In the past, electrodialysis was limited by the strength, high cost, and efficiency of the cation-selective membrane. However, continued improvements in membrane properties have resulted in day-to-day improvements of the electrodialysis process. The most attractive aspect of the process is the selective extraction performance of the membrane. For the present, its high removal limits (30 mg/L) and its relatively high-power consumption make its industrial applications still rare.

8.5 ADSORPTION

Adsorption refers to the ability of certain materials to retain molecules (gas, metallic ions, organic molecules, etc.) on their surface in a more or less reversible manner. There is a mass transfer of sorbate from the bulk of liquid or gas phase to the surface of the solid. The solid sorbent thus acquires superficial (hydrophobic or hydrophilic) properties liable to modify the state of equilibrium of the medium (diffusion, flocculation) [45].

The adsorptive capacity of the solid depends on [46]:

- The developed surface area or specific surface area of the material. Natural adsorbents (clays, silica, etc.) possess specific surface areas that vary with the physico-chemical state of the liquid medium (pH value, nature of the bound cations, surface saturation by organic molecules, etc.). Thus, certain clays such as bentonites (montmorillonite for instance) have a surface area that is accessible to most molecules and ranges from 40 to 800 m²/g. Their adsorptive capacity is quite variable but constitutes the main parameter in the regulation of transfers and in the mobility of elements in the natural environment. Industrial adsorbents (mainly activated carbon) can feature extensive surface areas (roughly between 600 and 1,200 m²/g) that are characteristic of a very strong microporosity. Other adsorbents such as metallic hydroxides that are formed in the course of the coagulation-flocculation process also develop very large surface areas whose expanse is closely dependent on the pH value.

The nature of the adsorbate-adsorbent bond, in other words, on the free energy of interaction (*G*) between the adsorption sites and that part of the molecule which is in contact with the surface. This energy is directly measurable in the case of gas adsorption. However, in a liquid medium, the calorimetric methods only record the differential enthalpy of adsorption that corresponds to the difference between the adsorption energy of adsorbed molecules and the desorption energy of bound water at the interface.

The contact time between the solid and the solutes. At equilibrium, there is a dynamic exchange between the molecules of the adsorbed phase and those that remain in solution. Many theories have attempted to model the relation that exists between the number of molecules adsorbed (g/g, g/m², etc.) and the number at equilibrium. One of the most commonly employed theories in the field of adsorption on activated carbon is reflected in the Freundlich equation [47] (Equation 8.5) depicted in Figure 8.18:

$$\frac{X}{m} = K\,C_{eq}^{1/n} \tag{8.5}$$

where *X/m* is the weight of pollutant retained per unit weight of the adsorbent, and C_{eq} is the equilibrium concentration of pollutant molecules in the aqueous phase. *K* and *n* are energy constants depending on the adsorbate-adsorbent couple at a given temperature, which is kept constant during the operation (isotherm).

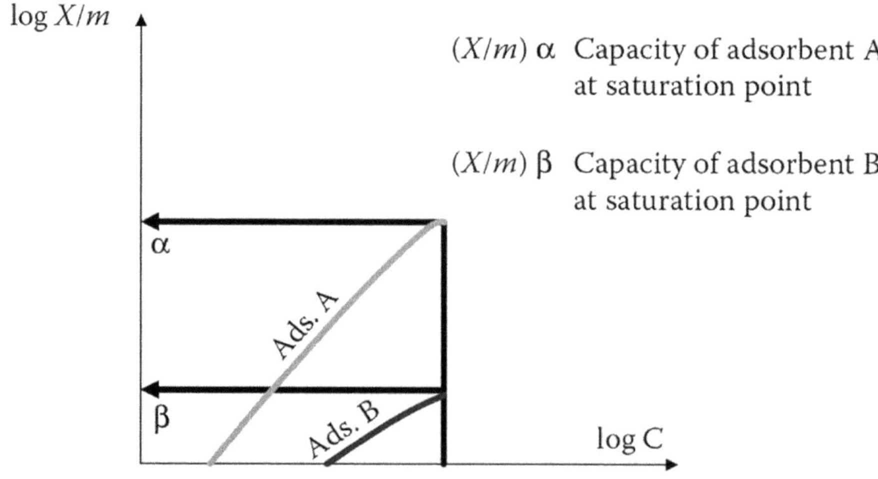

FIGURE 8.18 Freundlich isotherms.

In fact, hardly any modeling, no matter how "complex," can adequately reflect the structure of the isotherm, and a fortiori explain the mechanisms of adsorption. The basic reason for this is that any surface is heterogeneous both as regards physical aspects and energy.

It is mainly the van der Waals-type attraction and the Coulomb electrostatic-type attraction that provided the basis for adsorption that is ultimately based in thermodynamics and reflected in the resulting affinity between the sorbate moiety and the sorption active site. For instance, it can be seen that there is a strong affinity of aromatic molecules for the graphitic structure of carbon and a repulsion of the nonaromatic polar molecules.

8.5.1 Main Adsorbents

8.5.1.1 Activated Carbon

Experience shows that activated carbon has a broad spectrum of adsorptive activity, as most organic molecules are retained on its surface. The hardest to retain are the molecules that are the most polar and the linear ones with a very low molecular weight (simple alcohols, primary organic acids, etc.) [48]. Molecules that are slightly polar, generating taste and smell, and molecules with a relatively high molecular weight are for various reasons well adsorbed on carbon. It is interesting to note that, apart from these adsorbent properties, activated carbon can also provide a fine solid support for the growth of bacteria that are capable of breaking down a fraction of the adsorbed phase. In this combination, a part of the support is continuously being regenerated and capable of freeing sites, allowing new molecules to be retained. This combined action has successfully been used to enhance purification of certain types of wastewaters.

Activated carbon is available in two forms: powdered carbon and granular carbon. Powdered activated carbon (PAC) takes the form of grains between 10 and 50 mg and its use is generally combine calcium hydroxyapatite $Ca_5(OH)(PO_4)_3$. The small quantities of aluminum, magnesium, and manganese oxides aid in the removal of silt and other impurities.

It is possible to recover lime from the sludge produced in lime precipitation. One method is to centrifuge the slurry at low solids recoveries so as to remove only the heavy calcium carbonate $CaCO_3$ with a clarification treatment. If it is added continuously to the water together with flocculating reagents, it enters the floc and is then extracted from the water with it. The PAC is about 2–3 times less expensive than GAC. The investment costs are low when the treatment involves only a flocculation/settling PAC takes the form of grains between 10 and 50 mg and its use is generally combine calcium hydroxyapatite $Ca_5(OH)(PO_4)_3$.g stage (an activated carbon feeder is all that is needed).

The physical characteristics of GAC vary considerably depending on the raw materials used to prepare the GAC (Table 8.8).

TABLE 8.8
Physical Characteristics of GAC

Raw Material Form	Peat, Wood, Coconut, Anthracite, Crushed, Extruded
Grain size ES (mm)	0.25–3
UC	1.4–2.2
Friability 750 strokes (%)	10–50
1,500 strokes (%)	20–100
Bulk density (compacted)	0.25–0.55
Specific surface area(m^2g^{-1})	500–1,300
Ash content (%)	4–12

Note: ES: effective size.

8.5.1.2 Main Applications

Activated carbon is used

In the polishing treatment of drinking water or very pure industrial process water. In this case, the activated carbon will retain the dissolved organic compounds not broken down by natural biological means (self-purification of waterways): micropollutants and substances determining the taste and flavor of the water. The carbon will also adsorb traces of certain heavy metals.

In the treatment of industrial wastewater, when the effluent is not biodegradable or when it contains certain organic toxic elements [49,50] that rule out the use of biological techniques. In this case, the use of activated carbon often allows the selective retention of toxic elements and the resultant liquid can thus be degraded by normal biological means.

In the "tertiary" treatment of municipal and industrial wastewater. The carbon retains dissolved organic compounds that have resisted upstream biological treatment, and thus removes a large part of the residual COD.

8.5.1.2.1 Adsorptive Capacity of Carbon

GAC is used as a filter bed through which the water to be treated passes, leaving behind its impurities that are thus extracted methodically: the water, as it progressively loses its pollutants, encounters zones of activated carbon that are less and less saturated and therefore more and more active. Whether treatment using activated carbon is economical or not largely depends on the adsorptive capacity of the carbon, expressed in grams of retained COD per kilogram (or volume) of activated carbon, which characterizes the "carbon requirements" for a given result [51]. For a given polluted water-carbon system, this capacity depends on the following [52,53]:

The depth of the bed: The deeper the bed, the easier it deals with extended adsorptive fronts inside the bed without excessive breakthrough leakage—while still ensuring thorough saturation of the upper layers.

The exchange rate: The experience shows that three volumes of water per volume of carbon per hour can seldom be exceeded when treating high levels of pollution. In the case of drinking water, in which the content of adsorbable products is very low, any decision as to the economic optimum has to take the high investment costs into account, with the result that higher bed volumes are used: 5–10 vol/(vol h), with a smaller degree of carbon saturation.

The theory only gives an indication of the trend of the laws of adsorption. It still remains indispensable to call upon the experience of the expert and to carry out dynamic tests on columns of sufficient size so that results can be extrapolated.

A compact bed has three functions:

Filtration: This must often be reduced to a minimum in order to avoid clogging of the bed which is unavoidable without efficient washing systems to break up the layers completely after each cycle. In addition, the carbon tends to extract adsorbable products from the floc with which it is in contact, causing premature saturation. That is why it is often advisable to use sand filtration as a preliminary clean-up step to remove suspended matter particles.

Medium for supporting microbial growth: This phenomenon can contribute to the process of purification, but can also be very dangerous if not properly controlled (anaerobic fermentation gives off odors, clogging of the bed, etc.).

Adsorption: This must remain the basic role of the carbon.

There are three possible arrangements:

Simple fixed beds: This technique is widely used in drinking water treatment.

Fixed beds in series: A series of several columns, which are regenerated by permutation, are used (Figure 8.19). Thus, a countercurrent extraction system is organized.

Moving beds: These usually make use of the countercurrent principle (Figure 8.20). The base of the bed can even be fluidized.

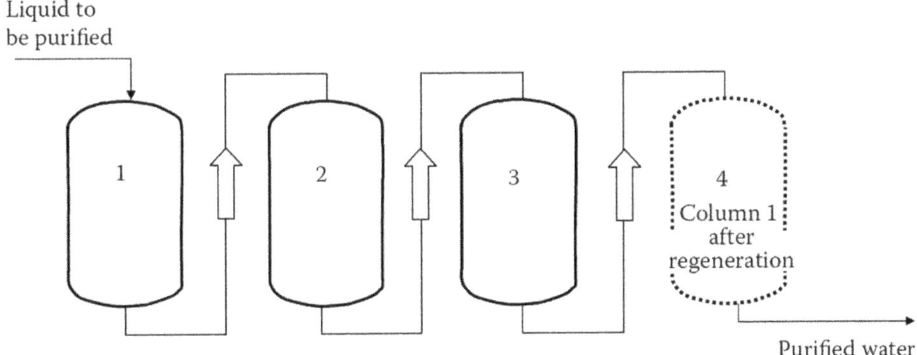

FIGURE 8.19 Diagram of fixed bed in series.

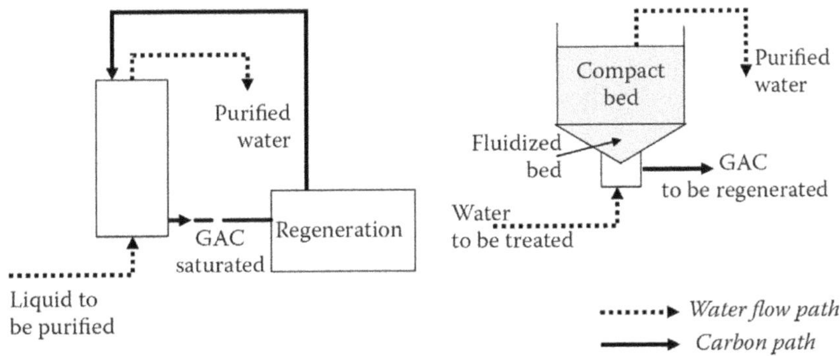

FIGURE 8.20 Diagrams of moving beds.

8.5.1.2.2 Regeneration of the Activated Carbon

Activated carbon (such as artificial adsorbents) is an expensive product. In most cases the cost of replacing the saturated carbon would be prohibitive [54]. It should therefore be regenerated, and three methods have been developed for this purpose:

Steam regeneration: This method is restricted to regenerating carbon which has only retained a few very volatile products. However, steam treatment can be useful in unclogging the surface of the grains and disinfecting the carbon.

Thermal regeneration: By pyrolysis and burning off of adsorbed organic substances. To avoid igniting the carbon, it is heated to about 800°C in a controlled anoxic atmosphere. This is the most widely used method and regenerates the carbon very well, but it has two disadvantages: (a) It requires considerable investment in a multiple-hearth furnace, a fluidized bed furnace, or a rotary kiln. The furnace must have monitoring devices for its gaseous atmosphere and temperature, a dewatering system at the inlet, and a carbon quenching system at the outlet. (b) It causes high carbon losses (7%–10% per regeneration), so that after 10–14 regenerations, the GAC volume will, on average, have been entirely replaced. The use of electrical heating (infrared furnace, induction furnace) reduces these losses. However, these methods, which are expensive, are only used for the recovery of costly metals.

Chemical regeneration: The advantage of this process is that for the same capital outlay, only minimum carbon loss occurs (about 1% of the quantity treated). However, the use of chemical reagents for regeneration (alkaline reagents, solvents) leads to the formation of eluates from which the solvent must be separated by distillation. The pollutants are then destroyed by incineration unless they can be recovered. The process is less widely used than thermal regeneration.

Biological regeneration: This method of regeneration is somewhat complicated due to the necessity of controlling the microbial action and thus it is rarely applied on an industrial scale.

8.5.1.3 Other Adsorbents

Apart from the activated carbon, new adsorbents have been developed:

Inorganic adsorbents: Alumina [55,56] and other metallic oxides [57]: they can have a very large specific surface area, but these solids adsorb more selectively than carbon. Their capacity depends very much on the pH value and their mesoporosity. Below the isoelectric point only negatively charged molecules are adsorbed on positively charge sites. In the current state of their development, these adsorbents are not competitive with activated carbon. However, some of these solids such as the alumina or the ferric oxyhydroxides have the advantage of removing arsenic, fluoride, phosphates, nitrates, and so on.

Organic adsorbents: Macromolecular resins with specific surface areas of between 300 and 750 m^2/g [58,59]: their adsorptive capacity is generally lower when compared with that of activated carbon. However, these resins can have special adsorptive properties and are often easier to regenerate (low binding energy). Here the "scavengers" should also be mentioned, which are highly porous anionic resins. However, these resins have a smaller specific surface area and their action on polar substances (such as humic acids and anionic detergents) is partly due to their ionic charge, which distinguishes them from other adsorbents.

8.6 OXIDATION-REDUCTION

Some substances are found either in oxidized or in reduced form, and are converted from one to the other by gaining electrons (reduction) or by losing electrons (oxidation). A system comprising an acceptor and a donor of electrons is known as an "oxidation-reduction" system.

Oxidation-reduction reactions are used in the treatment of water for disinfection and to convert an element from its dissolved state to a state in which it may be precipitated (Fe, Mn, sulfur removal, etc.) [60].

The definition and monitoring of the pH value in a reaction is very important. Eh-pH diagrams exist in handbooks to represent the state of various forms of elements and their evolution depending on the pH and the redox potential, in order to

Convert an element from its dissolved state to its gaseous state (e.g., denitrification)

Break down a substance into several simpler substances, the presence of which is acceptable in water (e.g., phenols, etc.)

Break down a nonbiodegradable substance into several simpler substances, which can be removed by bacterial assimilation during a later treatment phase (e.g., micropollutants).

Oxidation can take place by means of chemotrophic bacteria such as in the oxidation of iron and manganese, the oxidation of sulfur compounds, the oxidation-reduction of nitrogen compounds, and methane-forming reduction [61–64].

8.6.1 Main Oxidation Techniques

This mainly concerns the following industrial waters and effluents:

Using oxidizing reagents:

Cyanide-laden waters from electroplating or gas scrubbing

Hydrazine-laden condensates: oxidation by H_2O_2 catalyzed on specific resins Nitrite baths from electroplating: oxidation by $H_2O_2 + Cu^{2+}$ (Fenton reagent), NaClO, or H_2SO_5 (Caro's acid)

Solutions of thiosulfates oxidizable from H_2O_2.

Using air or oxygen [30,62]:
Spent caustic soda, rich in S^{2-}
Waters from pickling, loaded with Fe^{2+}
Uranium leachates U^{4+}.
Most of these reactions present a high enough potential and rapid enough kinetics to permit regulation except in the case of thiosulfates. If other less dangerous reducing agents coexist, a posteriori monitoring to limit the over consumption of a costly oxidizing agent, as in the case of cyanide-laden effluents from gas scrubbing, is considered adequate. The use of air and oxygen in the equipment known as "oxidizers" requires high temperatures and pressures in order to achieve adequate kinetics and efficiency.

Ozone can also be used for oxidation reactions, especially for
Effluents containing low CN or phenol concentrations
Effluents from methionine units or those containing refractory compounds.

8.6.1.1 Application: Treatment of Cyanides

The oxidation of cyanides in an alkaline environment theoretically comprises two successive stages. The first one, in which there is practically no toxicity, is the cyanate state and then the nitrogen and bicarbonate state. Powerful oxidizing agents employed are sodium hypochlorite, chlorine, and permonosulfuric acid (Caro's acid). In practice, for economic reasons, only the first stage is employed.

First stage (cyanates):
The overall reactions that come into play are
Using sodium hypochlorite: $NaCN + NaClO \Leftrightarrow NaCNO + NaCl$
Using chlorine gas: $NaCN + Cl_2 + 2NaOH \Leftrightarrow NaCNO + 2NaCl + H_2O$
Using Caro's acid: $NaCN + H_2SO_5 \Leftrightarrow NaCNO + H_2SO_4$.
The first two reactions occur almost instantaneously when the pH level is above 12, but the reaction speed drops rapidly if the pH level falls (critical threshold: pH 10.5). Whatever the pH level, an intermediate compound that is formed is cyanogen chloride CNCI, which is just as dangerous as hydrocyanic acid:

$$NaCN + NaClO + H_2O \Leftrightarrow CNCl + 2NaOH.$$

With a pH level starting at 10.5, however, cyanogen chloride is hydrolyzed the moment it is formed according to the reaction:

$$CNCl + 2NaOH \Leftrightarrow NaCl + NaCNO + H_2O.$$

With Caro's acid, an adequate reaction speed is observed for pH level above 9.5.
Second stage (nitrogen):
The breakdown of cyanate into nitrogen occurs according to the reaction:

$$2NaCNO + 3Cl_2 + 6NaOH \Leftrightarrow 2NaHCO_3 + N_2 + 6NaCl + 2H_2O.$$

It also takes place at the pH level of 12, but requires three times the amount of reagent and a reaction time of about 1 hour as it is impossible to regulate the potential.

8.6.2 MAIN REDUCTION TECHNIQUES

The most common examples involve the reduction of oxygen, that of hexavalent chromium [65], as well as the destruction of residual oxidizing agents employed in disinfection. It is also necessary to mention the reduction of nitrites in the process of surface treatment (sulfamic acid or $NaHSO_3$).

8.6.2.1　Application: Reduction of Hexavalent Chromium

The reduction of toxic hexavalent chromium into trivalent chromium which is less toxic and can be precipitated in the form of hydroxide occurs in an acid medium through the action of sodium bisulfate or ferrous sulfite.

Using sodium bisulfite:

$$H_2Cr_2O_7 + 3NaHSO_3 + 3H_2SO_4 \Leftrightarrow Cr_2(SO_4)_3 + 3NaHSO_4 + 4H_2O.$$

Using ferrous sulfate:

$$H_2Cr_2O_7 + 6FeSO_4 + 6H_2SO_4 \Leftrightarrow Cr_2(SO_4)_3 + 3Fe_2(SO_4)_3 + 7H_2O.$$

The first of these reactions occurs almost instantaneously when the pH level is below 2.5, but the reaction speed falls rapidly when the pH level rises (the critical threshold is at pH 3.5).

The reduction of ferrous iron has fewer restrictions and may occur with a pH level below 6, with monitoring. It is less used because a significant amount of hydroxide sludge is produced during the final neutralization stage.

8.7　LIQUID-LIQUID EXTRACTION

The liquid-liquid extraction process is a basic operation that allows a component (solute) to be extracted from an inert liquid by another liquid known as a solvent [66]. The liquid phase 1 is a homogeneous mixture. The solvent must not be miscible with one of the two initial compounds. The inert compound and the solvent are usually not miscible. Liquid-liquid extraction is also governed by the laws of mass transfer and it is necessary to determine features that favor exchange, such as maximum interfacial area, wide concentration difference, and a notable transfer (or extraction) coefficient. There are two main types of industrial equipment employed in the liquid-liquid extraction process [67]:

Contactors with several separate stages in series. At each stage, the functions of dispersion followed by separation of the two phases take place in two successive units: the mixing settling tank and the hydrocyclone-settling tank.

Differential contactors in which one phase is dispersed into the other on a countercurrent basis. Following this, the phases are separated in the two ends of a vertical column (Figure 8.21). The method of dispersion of the two phases may be by gravity, mechanical stirring, pulsation, etc. Spinning mechanical countercurrent contactors (Podbielniak) proved to be extremely efficient—for the price and elevated energy consumption.

Liquid-liquid extraction is used for phenol removal [68] from spent caustic soda from refineries using gas oil as a solvent. The efficiency of phenol removal reported was high (90%–95%) when pulsed columns with perforated trays were used.

8.8　WETLAND METAL MINERALIZATION

A broadly accepted definition of a wetland is "an area that meets one or more of the following conditions: (a) areas supporting predominately hydrophytes, (b) areas with predominately undrained hydritic soil producing anaerobic conditions, and (c) areas with a nonsoil substrate" [69]. Figure 8.22 pictures a constructed wetland (CW). Wetlands are also ephemeral in that they depend on perturbations to their

environment Application in order to exist [70]. Without these disturbances, wetlands will eventually dry up and become stable soil substrate for other terrestrial ecosystems. It is this dynamic property of wetlands that makes them suitable for the treatment of polluted wastewaters because they have the capacity to absorb and smooth variations in hydrological input and reduce substantial concentrations

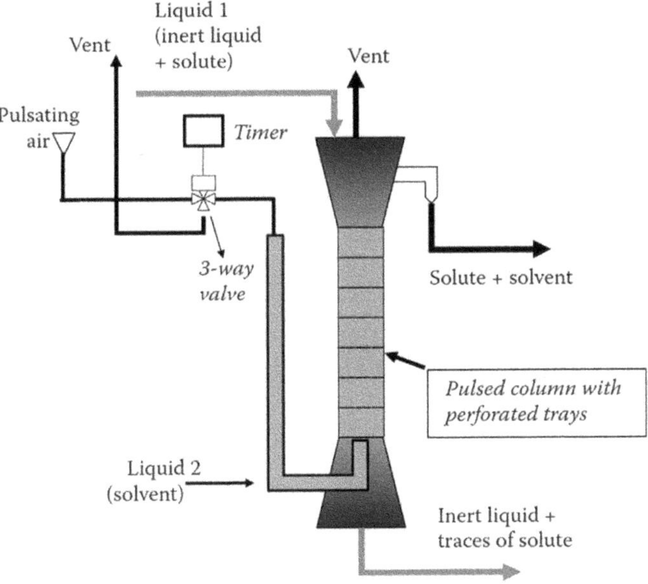

FIGURE 8.21   Diagram of a pulsed column liquid-liquid extraction.

FIGURE 8.22   A constructed wetland.

of pollutants. Wetlands have also shown incredible tolerance to metals [71,72], and their ability to accumulate metals such as iron and manganese has been known for centuries. There are generally two types of CWs: (a) free water surface (FWS) and (b) subsurface flow (also referred to as vegetated submerged bed types, root zone method, microbial rock filters, and hydrobotanical systems). In engineering terms, both can be referred to as attached growth biological reactors. The FWS wetlands are constructed so that an exposed area of water will always exist and also so that mixing will lead to oxygenation of the surface sediments. These wetlands have the disadvantage that they usually require some form of primary treatment to their influent waters [73]. The subsurface flow wetlands (SFWs) are the most common for treating metal-rich wastewaters. They are designed with an organic, porous medium and gradual slope so that no FWS exists and so that the saturated conditions and organic matter decomposition lead to anoxic sediments. The lack of a FWS also has the advantage of avoiding odor and insect vector problems. If effluent acidity is a problem (as it always is in an acid amine drainage (AMD) and can sometimes be for industrial processes), then the SFW is very advantageous because it can be underlain with crushed limestone which, under the anoxic conditions, can slowly release alkalinity without becoming armored by metal precipitate coatings which would otherwise occur under oxic conditions [74]. Furthermore, SFWs are also advantageous compared with FWSs because they usually require less preliminary treatment [73].

8.8.1 ELEMENTS OF DESIGN

CWs are attempts at replicating and optimizing natural ecosystems. They are essentially systems that incorporate microbial, planktonic, invertebrate, soil, and hydrophytic components into a dynamic ecosystem [75,76]. For the purpose of CWs though, designers are usually only interested in three components: (a) vegetation; (b) microbes, algae, and microbially mediated processes; and (c) substrate conditions.

8.8.2 VEGETATION

Many plant species have been used in CWs. These include (most common to least) *Typha latifolia* (cattails), *Scirpus validus*, *Scirpus robustus* (bulrush), *Phragmites communis* (common reed), *Leersia oryzoides* (cutgrass), and *Lemna minor* (duckweed). Plants in CWs generally play two roles: (a) their roots provide increased surface area for the attachment of microbes and epiphytes, and (b) by their hydrophytic nature, wetland plants have the capability to transport oxygen down into their roots where some leakage occurs through radial oxygen loss, thus oxygenating the surrounding anoxic substrate. Although it may seem deceiving due to their dominant presence in wetlands, the plants themselves only account for at most ~1% of the total heavy metal removal via uptake [77]. Most of this removal occurs in the plant root tips, which is not advantageous if plant harvesting is a desired method for permanent metal removal, as roots remain in the substrate after harvesting. Another important function of emergent plants is that they accelerate the establishment of the necessary microbial population to less than a year, compared with almost two years without vegetation [78]. The most important role of emergent vegetation in the CW is their ability to transport oxygen into the anoxic sediment. This function is necessary to support aerobic microorganisms and remove dissolved metals through oxidization. Transport results from thermoosmosis of gases that requires a special plant morphological structure called aerenchyma, but does not actually require physiological activities. The oxygen-transporting capabilities of wetland plants vary among species. Studies have shown that *Typha* can transport the most with mixed assemblages following in the sequence [79]:

Typha latofolia > Juncus effuses > Solanum americanum > Eleocharis quadrangulata.

Not only do *Typha* exhibit the greatest capacity to transport oxygen, but they also concentrate most metals in their roots. Meiorin et al. [80] found *Typha* to accumulate 447–1,220 mg/kg dry weight of Mn in much larger concentrations than other species studied. Furthermore, the rhizospheres of *Typha* were found to be extremely conducive environments for aerobic heterotrophs such as *Thiobacillus* [81], which are essential for metal removal. Overall then, the literature indicates that Typha species might be the most appropriate wetland plants for heavy metal removal.

Some studies have also focused on the metal removal by Sphagnum species which, at times, can be substantial [72]. However, since the removal capacities of this plant are directly linked to its cation exchange capacity [82], which eventually becomes saturated [83], and *Sphagnum* species have a very low net productivity and so are dependent on outside sources for nutrients [82], their use in CWs does not promise sustainable heavy metal removal.

8.8.3 Algae, Microbes, and Microbially Mediated Processes

Algae and microbes are by far the most abundant living organisms in a wetland. The microbial population can sometimes reach 10^{10} m/L in organic sediments that are ubiquitous in wetland ecosystems [84]. The processes by which algae remove metals are bioaccumulation, biosorption, and either aerobic or anaerobic biologically mediated deposition by mineralization. The capacity of microbes and algae to sequester heavy metals or to cause their removal from solution is measured not only by the percent of metals retained, but also by the bioavailability of these metals in the marsh system after being removed from solution. Therefore, the ideal mechanism for heavy metal removal would be one where the removed metal could not re-enter the solution phase or the biota.

The metal accumulation capacity of certain algae is a remarkable phenomenon in nature [85]. Their ability to sequester metals from solution was first noticed in the early 1970s when it was observed that lagoon effluents from lead/zinc mining and milling operations were causing dense mat-like growths of algae in receiving streams. These algal growths were irregularly dense partly because the receiving waters were so polluted with heavy metals that the normal competitors of algae were practically eliminated.

The microbially mediated reactions can be broken down into those that occur in aerobic sediments and those that occur in anaerobic sediments. In aerobic regions, heterotrophic microbes such as *Arthrobacter*, *Pseudomonas*, and *Citrobacter* can mediate the following reaction [80]:

$$2MnSO_4 + O_2 + 2H_2O \Leftrightarrow 2MnO_2 + 2H_2SO_4,$$

whereas *Metallogenium* mediates another reaction:

$$2MnSO_4 + H_2O_2 \Leftrightarrow MnO_2 + H_2SO_4.$$

The overall effect is the precipitation of the insoluble Mn oxide into the sediments where it will hopefully remain. Another metal removal reaction results from the microbial decay of organic matter that promotes the formation of NH_3 and HCO_3^-, which increase the pH and cause hydroxide and Mn and Fe oxyhydroxide formation and consequent precipitation in aerobic zones [86]. These microbially mediated reactions are further enhanced by the fact that other dissolved metals adsorb onto the metal oxyhydroxides and are also precipitated from solution. In the latter adsorption reaction, Mn oxyhydroxides are more important than Fe oxyhydroxides [86] and there are many different bacteria that can oxidize Mn [84], making the removal of Mn by this process very important. Moreover, the formation of oxyhydroxides, especially those of Al [87], is not favored under acidic conditions, which also emphasizes the importance of the buffering limestone underlayer of SFWs to this oxidation reaction. Finally, Giblin et al. [88] found that Fe, Pb, and Mn were trapped in the sediments in a form that was unavailable for reuptake into the biota or resuspension back into solution, so these metals are, for the most part, permanently removed by the above processes.

In SFWs, the most important microbially mediated process to remove heavy metals is the anaerobic formation of H_2S. In anoxic regions, bacteria such as *Desulfovibrio* oxidize organic matter using sulfate as the electron acceptor and reducing it to H_2S. Some of the important chemical reactions include [89]:

Sulfate reduction: $SO_4^{2-} + 2CH_2O \Leftrightarrow H_2S + 2HO_3^-$.
Iron monosulfate formation: $Fe^{2+} + HS^- \Rightarrow FeS + H^+$
Pyrite formation: $FeS + S^0 \Rightarrow FeS_2$
Pyrite formation: $Fe^{2+} + H_2S + S^0 \Rightarrow FeS_2 + 2H^+$
Pyrite formation: $Fe^{2+} + 2HS^- \Rightarrow FeS_2 + 2H^+$.

The formation of pyrite is better than Fe oxyhydroxides because pyrite accumulates within the organic substrates and is less likely to form surface deposits or be washed out during storm events [89]. Two of the biggest advantages of sulfate reduction are that it produces alkalinity and also that it is not constrained by accumulation limits or toxic effects. Pyrite is also negligibly soluble in acid of neutral water, so the formation of this mineral represents a long-term, sustainable mechanism for removing Fe from solution. However, the importance of sulfate reduction in heavy metal removal depends on the availability of sulfate in the influent waters and also on the type of metal sulfide complexes formed. The latter issue is unclear in the literature as Fe sulfides are the only complexes that have been extensively studied, whereas other metal complexes have been mentioned but their formation mechanisms were not included.

8.8.4 Substrate Conditions

Many different substrates have been used in CWs. The overall goal of the substrate in an SFW is to provide nutrients and support for vegetation, while at the same time it must be porous so as not to inhibit lateral hydrological flow. The substrate can also have certain functions in metal removal because of its sorption properties. The best-known example of this is the use of *Sphagnum* as a substrate [73,82]. However, the performance of *Sphagnum* is limited because it does not have a very high nutrient status and because its sorption sites eventually become saturated. Another study, which investigated the sorption of Pb onto particles [90], reported that the amount of organic matter in the particles was the key factor limiting sorption. Therefore, it appears that the most appropriate substrate for metal removal in CWs requires a high percent of organic matter content. This is advantageous because wetlands themselves can contribute up to 67% of their annual net primary productivity to the sediments [77], replacing thus consumed sorption sites. The most frequently used substrate for treating AMD has been mushroom compost, which has a very high organic matter content (1%–3%), combined with crushed limestone. Mushroom compost is a byproduct of the mushroom-growing industry and is formed by the anaerobic composting of a mixture of hay, manure, and agricultural wastes. Brodie et al. [91] evaluated the performance of several substrates but did not include mushroom compost. They found that there were no significant differences between the metal removal efficiencies of different substrates (soils included natural wetland, acid wetland, clay, mine spoil, and pea gravel). Except for the two natural wetland soils, none of the substrates investigated by Brodie et al. [91] could be used for long-term projects because of their low nutrient status. Furthermore, these results were difficult to interpret due to inconsistent experimental hydrologic flow rates. Therefore, further work regarding the effect of substrate type on metal removal should consider more commonly used substrates, especially mushroom compost.

8.8.5 Metal Removal Efficiencies

There are many publications that cite metal removal efficiencies of wetlands (Table 8.9). Unfortunately, these citations cannot contribute to further understanding of the complex process of heavy metal removal because very few concurrently give information on the respective wetland parameters. The removal efficiencies for the wetlands studied here range from 100% for Fe, Cu, or Pb to 0.6% for Cd.

The relevant literature [91–94] seems to agree that in terms of substrate and plant species selection, high organic matter is important in the substrate and that *Typha* species are associated with high metal removal rates. Furthermore, the metal removal efficiencies decrease with increasing flushing rate, wetland depth, and for some metals, with wetland area, experiment duration, and temperature. On the other hand, Mn removal increased with longer experiment durations. Some other wetland characteristics that are important to metal removal are length-to-width ratio, which increases with increasing Fe removal, and limestone, which also significantly improved Fe removal. The results also suggest that costs can be reduced by building shallower wetlands with natural substrate conditions (except for a little limestone). These financial implications might also enhance the capabilities of CWs as a low-tech alternative to heavy metal abatement of wastewaters.

TABLE 8.9
Heavy Metal Removal Efficiencies of Various Wetlands and CWs from the Literature
Removal Efficiency (%)

References	Cd	Cr	Cu	Pb	Hg	Zn	Fe	Mn	Ni
[69]	50–90	50–90	50–90	80–95	50–90	50–90			
[88]	20–35	20–50	60–100	55–100		20–45	60–100	55–60	
[80]		40–53	5–32	30–83		6–51			12–32
[95]			99	94		98	86		
[86]			92						84
[86]	0.3		14			5.8	70	0.7	

Note: The removal efficiencies for each metal (calculation of removal efficiency assumed to be (([in] − [out])/[in])*100%).

TABLE 8.10
Metal-Binding Capacities of Selected Biomass

Biomass Species	Biosorbent Capacity (m eq/g)	References
Sargassum natans	2–2.3	[99]
Ascophyllum nodosum	2–2.5	[99]
Rhizopus arrhizus	1.1	[97]
Eclonia radiata	1.8–2.4	[100]
Peat moss	4.5–5	[101]
Commercial resins	0.35–5	[102]

8.9 BIOSORPTION

Metal accumulation has been demonstrated in a significant number of microorganisms (bacteria, algae, yeasts, and fungi) and this phenomenon is associated with a wide spectrum of microbial activities. Leaving aside the metabolically mediated metal sequestration by living microbial cells (*bioaccumulation*), which has been studied particularly from toxicological point of view, it has been observed that some microbial cells tend to bind metals even when they are dead and metabolically nonactive (Table 8.10). This type of metal uptake, termed *biosorption*, is usually rapid and sometimes very high, which makes it particularly interesting from the application point of view [96]. The metals may be bound in the biomass through a variety of mechanisms: adsorption, ion exchange, coordination, complexation, chelation, and microprecipitation seem to occur to a varying degree [97].

The metal-binding chemical groups of the biological materials include carboxylate, phosphate, hydroxyl, sulfhydryl, and amines, present in durable and decomposition-resistant cell walls [98].

The pH of the metal-bearing solution can play a critical role in influencing the metal-sequestering ability of the biomass materials [103,104].

Earlier patents awarded for the application of biomaterials in metal concentration indicate the technological potential of the biosorption phenomenon [105–107]. The packed bed contacting column appears to be the most effective mode of bringing together the metal-bearing solution and biosobent material [108]. Most often it is possible to wash the resulting metal-saturated biosorbent in the same column, releasing the deposited metal in a small volume of "desorption" solution which then contains the metal in high concentrations [109,110]. This makes the subsequent recovery of the metal possible by routine commercial methods (e.g., electrowinning). The regenerated biosorbent can be then used in many sequential metal sorption cycles. Multiple uptake desorption cycles further decrease already quite cost-effective potential of the biosorption process, making its applications economically very attractive. The published book on *Sorption and Biosorption* [111] summarizes the current knowledge in the field of biosorption [112].

GLOSSARY

Constructed wetlands (CW): These are the engineered ecosystems that play a crucial role in treating wastewater and enhancing water quality using wetlands. There are two types of CWs: (a) free water surface (FWS) wetlands and (b) SFW. FWS wetlands offer natural habitat but have risks, while SFW wetlands focus exclusively on treatment function

Free water surface (FWS) wetlands: (a) In FWS wetlands, the water surface is exposed to the atmosphere. These natural wetlands include bogs (with primary vegetation like mosses), wamps (with trees), and marshes (with grasses and emergent macrophytes); (b) FWS wetlands provide habitat for various wildlife but also come with risks like mosquitoes and public access.

Subsurface flow (SF) wetlands: (a) SF wetlands are specifically designed for wastewater treatment; (b) They are constructed as beds or channels containing appropriate media (such as gravel); (c) The water surface in SFWs remains below the top surface of the medium, preventing mosquitoes and odors; (d) Emergent vegetation (similar to marshes) is planted in the medium; (e) Microorganisms attached to the submerged substrate (plant roots and media surfaces) contribute to water quality improvement; (f) SFWs can be more efficient than FWS wetlands for most contaminants due to their larger surface area for microbial reactions.

REFERENCES

1. Saleh, T.A., M. Mustaqeem and M. Khaled, Water treatment technologies in removing heavy metal ions from wastewater: A review. *Environmental Nanotechnology, Monitoring & Management*, 2022. 17: p. 100617.
2. Acharyya, S., A. Das and T.P. Thaker, Remediation processes of hexavalent chromium from groundwater: a short review. AQUA - Water Infrastructure. *Ecosystems and Society*, 2023. 72(5): pp. 648–662.
3. Kremser, K., et al., Bioleaching and selective precipitation for metal recovery from basic oxygen furnace slag. *Processes*, 2022. 10: p. 576.
4. U.S. Environmental Protection Agency (U.S. EPA). Processes, procedures, and methods to control pollution from mining activities. [Coal mining in eastern USA]. 1973: United States.
5. Madeira, L., et al., Reuse of lime sludge from immediate one-step lime precipitation process as a coagulant (aid) in slaughterhouse wastewater treatment. *Journal of Environmental Management*, 2023. 342: p. 118278.
6. Eckenfelder, W.W. and J.L. Musterman. *Activated Sludge: Treatment of Industrial Wastewater*. CRC Press, Boca Raton, FL, USA, ISBN: 978-1566763028, 281 pages, 1998.
7. Droste, R.L. *Theory and Practice of Water and Wastewater Treatment*, John Wiley and Sons, New York, USA, ISBN: 978-0471124443, 800 pages, 1996.
8. Fitriani, N., et al., Performance of modified slow sand filter to reduce turbidity, total suspended solids, and iron in river water as water treatment in disaster areas. *Journal of Ecological Engineering*, 2023. 24(1): pp. 1–18.
9. L. K. Wang, M. H. S. Wang and N. K. Shammas, Treatment of wastewater, storm runoff and combined sewer overflow by dissolved air flotation and filtration, In: L.K. Wang, M. H. S. Wang, Y.T. Hung, N. K. Shammas and J. P. Chen (Eds), *Handbook of Advanced Industrial and Hazardous Wastes Management*, CRC Press, FL, USA, pp. 577–610, 2018.
10. Wang, L.K., N.K. Shammas, W.A. Selke, and D.B. Aulenbach, *Flotation Technology*. 2010: Humana Press, Totowa, NJ. 680 pages.
11. Wang, L.K., M.H.S. Wang, N.K. Shammas, and D.B. Aulenbach, *Environmental Flotation Engineering*. 2021: Springer Nature, Cham, Switzerland. 433 pages.
12. Wang, L.K., J.P. Chen, Y.T. Hung, and N.K. Shammas, *Membrane and Desalination Technologies*. 2011: Humana Press, Totowa, NJ. 716 pages.
13. Shammas, N.K. and L.K. Wang. *Water Engineering: Hydraulics, Distribution and Treatment*. 2016: Wiley, New York. 806 pages.

14. Noble, R.D. and S.A. Stern, *Membrane Separations Technology Principles and Applications*. 1995: Elsevier, Amsterdam, Netherlands, ISBN: 978-0444816337, 738 pages, January 17, 1995.

15. Orris E. Albertson, *Dewatering Municipal Wastewater Sludges*. 1991: Noyes Data Corp, Park Ridge, New Jersey, USA, ISBN: 978-0815512660, 189 pages.

16. Guo, D., et al., A system engineering perspective for net zero carbon emission in wastewater and sludge treatment industry: A review. *Sustainable Production and Consumption*, 2024. 46: pp. 369–381.

17. Zenz, D. R., Lue-Hing, C., and Kuchenrither, R. *Municipal Sewage Sludge Management: Processing, Utilization, and Disposal*. 1992: Technomic Pub. Co., Lancaster, Pennsylvania, USA. ISBN: 978-0877629306, 663 pages.

18. Krishnan, R., et al., *Recovery of Metals from Sludges and Wastewaters*. 1993: Noyes Data Corp., Park Ridge, New Jersey, USA.

19. Wang, Z., et al., Acidic aerobic digestion of anaerobically-digested sludge enabled by a novel ammonia-oxidizing bacterium. *Water Research*, 2021. 194: p. 116962.

20. Nayeri, D. and S.A. Mousavi, A comprehensive review on the coagulant recovery and reuse from drinking water treatment sludge. *Journal of Environmental Management*, 2022. 319: p. 115649.

21. Dobson, R.S. and J.E. Burgess, Biological treatment of precious metal refinery wastewater: A review. *Minerals Engineering*, 2007. 20(6): pp. 519–532.

22. Al-Mutair, M., et al., An overview of metals extraction and recovery from industrial wastewater sludge. *The Canadian Journal of Chemical Engineering*, 2024. 102: pp. 2892–2908.

23. Lee, I.H., Y. Kuan and J. Chern, Factorial experimental design for recovering heavy metals from sludge with ion-exchange resin. *Journal of Hazardous Materials*, 2006. 138(3): pp. 549–559.

24. Pang, H., et al., Innovative cation exchange-driven carbon migration and recovery patterns in anaerobic fermentation of waste activated sludge. *Bioresource Technology*, 2024. 394: p. 130168.

25. Sirianuntapiboon, S. and T. Hongsrisuwan, Removal of Zn^{2+} and Cu^{2+} by a sequencing batch reactor (SBR) system. *Bioresource Technology*, 2007. 98(4): pp. 808–818.

26. Geng, H., et al., An overview of removing heavy metals from sewage sludge: Achievements and perspectives. *Environmental Pollution*, 2020. 266: p. 115375.

27. Vandevivere, P.C., et al., Metal decontamination of soil, sediment, and sewage sludge by means of transition metal chelant [S,S]-EDDS. *Journal of Environmental Engineering*, 2001. 127: pp. 802–811.

28. Blais, J., et al., Pilot plant study of simultaneous sewage sludge digestion and metal leaching. *Journal of Environmental Engineering*, 2004. 130(5): p. 516–525.

29. Benmoussa, H., R.D. Tyagi and P.G.C. Campbell, Simultaneous sewage sludge digestion and metal leaching using an internal loop reactor. *Water Research*, 1997. 31(10): pp. 2638–2654.

30. Filali-Meknassi, Y., R.D. Tyagi and K.S. Narasiah, Simultaneous sewage sludge digestion and metal leaching: effect of aeration. *Process Biochemistry*, 2000. 36(3): pp. 263–273.

31. Wang, L., et al., Carbon-rich and low-ash hydrochar formation from sewage sludge by alkali-thermal hydrolysis coupled with acid-assisted hydrothermal carbonization. *Waste Management*, 2024. 177: pp. 182–195.

32. Krishnan, S., et al., Current technologies for recovery of metals from industrial wastes: An overview. *Environmental Technology & Innovation*, 2021. 22: p. 101525.

33. Zagorodni, A.A., *Ion Exchange Materials*. 2007: Elsevier, Amsterdam, Netherlands.

34. Korkisch, J., *Handbook of Ion Exchange Resins*. 1989: CRC Press, FL, USA.

35. Wang, L.K., Y.T. Hung and N.K. Shammas, *Physicochemical Treatment Processes*. 2005, NJ, USA: Humana Press. 723 pages.

36. Arden, T., *Ion Exchange: Theory and Practice*. Royal Society of Chemistry, Cambridge, UK, 1994, ISBN: 978-0-85186-484-6, 302 pages.

37. Wang, W., et al., Development of novel high anti-pollution polyamide/polysulfate disk tubular reverse osmosis membrane modules and their application in simulated space bathing wastewater. *Journal of Water Process Engineering*, 2024. 60: p. 105119.

38. Tran, T., et al., Electrochemical treatment of heavy metal-containing wastewater with the removal of COD and heavy metal ions. *Journal of the Chinese Chemical Society*, 2017. 64(5): pp. 493–502.

39. Kim, B.M. and J.L. Weininger, Electrolytic removal of heavy metals from wastewaters. Elimination of end-of-the-line treatment is the key to important savings of capital and operating costs. *Environmental Progress*, 1982. 1: p. 121–125.

40. Wang, J., et al., Evaluation of electrokinetic removal of heavy metals from sewage sludge. *Journal of Hazardous Materials*, 2005. 124(1): pp. 139–146.

41. McLay Dedietrich, W.J. and F.P. Reinhard, Waste minimization and recovery technologies. *Metal Finishing*, 2007. 105(10): pp. 715–742.

42. Zhang, J., et al., Enhanced capacitive deionization of saline water using N-doped rod-like porous carbon derived from dual-ligand metal–organic frameworks. *Environmental Science: Nano*, 2020. 7(3): pp. 926–937.

43. Scott, K., A consideration of circulating bed electrodes for the recovery of metal from dilute solutions. *Journal of Applied Electrochemistry*, 1988. 18(4): pp. 504–510.

44. Kim, J., et al., Metal ion recovery from electrodialysis-concentrated plating wastewater via pilot-scale sequential electrowinning/chemical precipitation. *Journal of Cleaner Production*, 2022. 330: p. 129879.

45. Karim, A.R., et al., A review of pre- and post-surface-modified neem (Azadirachta indica) biomass adsorbent: Surface functionalization mechanism and application. *Chemosphere*, 2024. 351: p. 141180.

46. Cheremisinoff, P.N. and A.C. Morresi, *Carbon Adsorption Handbook*. 1978: Science Publishers, Ann Arbor, MI, USA.

47. Holmes, H.N., Colloid and Capillary Chemistry. By Herbert Freundlich. Translated from the third German edition by H. Stafford Hatfield. New York, E. P. Dutton and Company, 1926. 886 pages, 156 figures. *Science*, 1927. 65(1672): pp. 40–41.

48. Chowdhury, Z.K., *Activated Carbon: Solutions for Improving Water Quality*. 2013: American Water Works Association, Denver, CO, USA..

49. Bhatt, P., et al., Nanobioremediation: A sustainable approach for the removal of toxic pollutants from the environment. *Journal of Hazardous Materials*, 2022. 427: p. 128033.

50. Shang, J., et al., Kinetics of catalytic oxidation of methylene blue with La/Cu Co-doped in attapulgite. *Materials*, 2023. 16(5): p. 2087.

51. Bergmann, C.P. and F.M. Machado, *Carbon Nanomaterials as Adsorbents for Environmental and Biological Applications*. 2015: Springer, NY, USA.

52. Paul Chen, J. and M. Lin, Equilibrium and kinetics of metal ion adsorption onto a commercial H-type granular activated carbon: Experimental and modeling studies. *Water Research*, 2001. 35(10): pp. 2385–2394.

53. Netzer, A. and D.E. Hughes, Adsorption of copper, lead and cobalt by activated carbon. *Water Research*, 1984. 18(8): pp. 927–933.

54. Bankole, D.T., A.P. Oluyori and A.A. Inyinbor, The removal of pharmaceutical pollutants from aqueous solution by Agro-waste. *Arabian Journal of Chemistry*, 2023. 16(5): pp. 104699.

55. Habila, M.A., et al., Influence of synthesis-heating conditions on the porosity and performance of a carbon nanotube/SDS-alumina nanocomposite for effective wastewater treatment: Fabrication, characterization, and adsorption modeling. *Industrial & Engineering Chemistry Research*, 2023. 62(32): pp. 12571–12588.

56. Ragab, A.H., et al., Exploring the sustainable synthesis pathway and comprehensive characterization of magnetic hybrid alumina nanoparticles phase (MHAl-NPsP) as highly efficient adsorbents and selective copper ions removal. *Environmental Technology & Innovation*, 2024. 34: p. 103628.

57. Ghosh, N., et al., Review on some metal oxide nanoparticles as effective adsorbent in wastewater treatment. *Water Science and Technology*, 2022. 85(12): pp. 3370–3395.

58. Balaji, T. and H. Matsunaga, Adsorption characteristics of As(III) and As(V) with titanium dioxide loaded amberlite XAD-7 resin. *Analytical Sciences*, 2002. 18(12): pp. 1345–1349.

59. Els, E.R., L. Lorenzen and C. Aldrich, The adsorption of precious metals and base metals on a quaternary ammonium group ion exchange resin. *Minerals Engineering*, 2000. 13(4): pp. 401–414.

60. Zhang, Y., et al., Application of Fe(II)/peroxymonosulfate for efficient alkali lignin wastewater treatment: Insight into the synergistic interactions between redox reactions and coagulation. *Separation and Purification Technology*, 2024. 328: p. 125037.

61. Lloyd, J.R., C.L. Harding and L.E. Macaskie, Tc(VII) reduction and accumulation by immobilized cells of Escherichia coli. *Biotechnology and Bioengineering*, 1997. 55(3): pp. 505–510.

62. Ayala-Muñoz, D., et al., Microbial carbon, sulfur, iron, and nitrogen cycling linked to the potential remediation of a meromictic acidic pit lake. *The ISME Journal*, 2022. 16(12): pp. 2666–2679.

63. Sun, H., et al., Preparation of microbial agent immobilized composites for Cr(VI) removal from wastewater. *Environmental Technology*: Volume 45, pp. 1–13.

64. Park, D., et al., Effects of ionic strength, background electrolytes, heavy metals, and redox-active species on the reduction of hexavalent chromium by Ecklonia biomass. *Journal of Microbiology and Biotechnology*, 2005. 15: pp. 780–786.

65. Sun, Y., et al., Microbial communities outperform extracellular polymeric substances in the reduction of hexavalent chromium by anaerobic granular sludge. *Environmental Technology & Innovation*, 2024. 34: p. 103616.

66. Pramanik, S., et al., Supramolecular chemistry of liquid–liquid extraction. *Chemical Science*, 2024. 15: 7824–7847.

67. Godfrey, J.C. and Slater, M.J. *Liquid-Liquid Extraction Equipment*. 1994: John Wiley & Sons, Hoboken, NJ, USA.

68. Yaseen, S.M., et al., Enhancement of phenol recovery from wastewater by nanofluid utilizing liquid-liquid extraction method in a membrane-based contactor. *Chemical Engineering Research and Design*, 2023. 196: pp. 404–412.

69. Hammer, D.A., *Constructed Wetlands for Wastewater Treatment*. 1989: Lewis Publishers and CRC Press, Boca Raton, FL, USA.

70. Xu, Z., et al., Geographical and environmental distance differ in shaping biogeographic patterns of microbe diversity and network stability in lakeshore wetlands. *Ecological Indicators*, 2024. 158: p. 111575.

71. Ieviņa, S., et al., Coastal wetland species Rumex hydrolapathum: Tolerance against flooding, salinity, and heavy metals for its potential use in phytoremediation and environmental restoration technologies. *Life*. 2023. 13: p. 1604.

72. Schück, M. and M. Greger, Salinity and temperature influence removal levels of heavy metals and chloride from water by wetland plants. *Environmental Science and Pollution Research*, 2023. 30(20): pp. 58030–58040.

73. Kelman Wieder, R., Metal cation binding to Sphagnum peat and sawdust: Relation to wetland treatment of metal-polluted waters. *Water, Air, and Soil Pollution*, 1990. 53(3): pp. 391–400.

74. Henrot, J. and R.K. Wieder, Processes of iron and manganese retention in laboratory peat microcosms subjected to acid mine drainage. *Journal of Environmental Quality*, 1990. 19(2): pp. 312–320.

75. Kushwaha, A., et al., Chapter 10 - Bio-treatment of the swine wastewater and resource recovery: A sustainable approach towards circular bioeconomy, in *Bio-Based Materials and Waste for Energy Generation and Resource Management*, C. M. Hussain, et al., Editors. 2023: Elsevier, Amsterdam, Netherlands, pp. 299–329.

76. Karametos, I., et al., Mathematical modeling of constructed wetlands for hexavalent chromium removal. *Science of the Total Environment*, 2024. 917: p. 170088.

77. Nguyen, H.T.H., et al., Pilot-scale removal of arsenic and heavy metals from mining wastewater using adsorption combined with constructed wetland. *Minerals*. 2019. 9(6): p. 379.

78. Kalske, A., J.D. Blande and S. Ramula, Soil microbiota explain differences in herbivore resistance between native and invasive populations of a perennial herb. *Journal of Ecology*, 2022. 110(11): pp. 2649–2660.

79. Rehman, F., et al., Constructed wetlands: Perspectives of the oxygen released in the rhizosphere of macrophytes. *CLEAN – Soil, Air, Water*, 2017. 45(1). 1500988, DOI: 10.1002/clen.201500988

80. Meiorin, E.C. Urban runoff treatment in a fresh/brackish water marsh in Fremont, California. In *Constructed Wetlands for Wastewater Treatment* Alexandros, S (editor), 2020. pp. 677–685.

81. Nyer, S.C., et al., Nitrogen transformations in constructed wetlands: A closer look at plant-soil interactions using chemical imaging. *Science of the Total Environment*, 2022. 816: p. 151560.

82. Kelman Wieder, R. and G.E. Lang, Fe, Al, Mn, and S chemistry of Sphagnum peat in four peatlands with different metal and sulfur input. *Water, Air, and Soil Pollution*, 1986. 29(3): pp. 309–320.

83. Spratt, A.K. and R.K. Wieder, Growth responses and iron uptake in sphagnum plants and their relation to acid mine drainage treatment. *Journal of the American Society of Mining and Reclamation*, 1988. 1988: p. 279–285.

84. Gregory, E. and J.T. Staley, Widespread distribution of ability to oxidize manganese among freshwater bacteria. *Applied and Environmental Microbiology*, 1982. 44(2): pp. 509–511.

85. Le, V., et al., Ecotoxicological response of algae to contaminants in aquatic environments: a review. *Environmental Chemistry Letters*, 2024. 22(2): pp. 919–939.

86. Wildeman, T.R. and L.S. Laudon. *Use of Wetlands for Treatment of Environmental Problems in Mining: Non-Coal-Mining Applications*. CRC Press, Boca Raton, FL, USA, ISBN: 978-1003069850, 11 pages, 2020.

87. Li, W., et al., Selective separation of aluminum, silicon, and titanium from red mud using oxalic acid leaching, iron precipitation and pH adjustments with calcium carbonate. *Hydrometallurgy*, 2024. 223: p. 106221.

88. Giblin, A.E., et al., Uptake and losses of heavy metals in sewage sludge by a New England salt marsh. *American Journal of Botany*, 1980. 67: pp. 1059–1068.

89. Hedin, R., D.M. Hyman and R.W. Hammack, Implications of sulfate-reduction and pyrite formation processes for water quality in a constructed wetland: Preliminary observation. *Journal of the American Society of Mining and Reclamation*, 1988. 1988: pp. 382–388.

90. Salim, R., Adsorption of lead on the suspended particles of river water. *Water Research*, 1983. 17(4): pp. 423–429.

91. Brodie, G.A., D.A. Hammer and D.A. Tomljanovich, An evaluation of substrate types in constructed wetlands acid drainage treatment systems. *Journal of the American Society of Mining and Reclamation*, 1988. 1988: pp. 389–398.

92. Stark, L.R., et al., The Simco 4Wetland: Biological patterns and performance of a wetland receiving mine drainage. *Journal of the American Society of Mining and Reclamation*, 1988. 1988: pp. 332–344.

93. Hiel, M. and F.J. Kerins, The Tracy wetlands: A case study of two passive mine drainage treatment systems in Montana. *Journal of the American Society of Mining and Reclamation*, 1988. 1988: pp. 352–358.

94. Kepler, D.A., An overview of the role of algae in the treatment of acid mine drainage. *Journal of the American Society of Mining and Reclamation*, 1988. 1988: pp. 286–290.

95. Pelletier, T., End of the pipe? 1991: *Industry Week*. 38–57.

96. Volesky, B. and Z.R. Holan, Biosorption of heavy metals. *Biotechnology Progress*, 1990. 11: pp. 235–250.

97. Naja, G., et al., Lead biosorption study with Rhizopus arrhizus using a metal-based titration technique. *Journal of Colloid and Interface Science*, 2005. 292(2): pp. 537–543.

98. Dong, B., et al., Evidence of weak interaction between ferric iron and extracellular polymeric substances of Acidithiobacillus ferrooxidans. *Hydrometallurgy*, 2022. 209: p. 105817.

99. Holan, Z.R., B. Volesky and I.M. Prasetyo, Biosorption of cadmium by biomass of marine algae. *Biotechnology and Bioengineering*, 1993. 41: 819–825.

100. Matheickal, J.T. and Q. Yu, Biosorption of lead from aqueous solutions by marine algae Ecklonia radiata. *Water Science and Technology*, 1996. 34: pp. 1–7.

101. Smith, R.V. and Misra, M. *Mineral Bioprocessing*. 1992: Minerals,Metals & Materials Society, Warrendale, Pennsylvania, USA. ISBN: 978-0873391757, 495 pages.

102. Mijangos, F. and M. Díaz, Kinetic of copper ion exchange onto iminodiacetic resin. *Canadian Journal of Chemical Engineering*, 1994. 72: pp. 1028–1035.

103. Garg, R., et al., Biosynthesized silica-based zinc oxide nanocomposites for the sequestration of heavy metal ions from aqueous solutions. *Journal of King Saud University - Science*, 2022. 34(4): p. 101996.

104. Liu, Q., et al., A lignin-derived material improves plant nutrient bioavailability and growth through its metal chelating capacity. *Nature Communications*, 2023. 14(1): p. 4866.

105. Volesky, B. and Kuyucak, N. Biosorbent for gold. U.S. Patent no. 4 769 233, 1988.

106. Volesky, B. and Tsezos, M. Separation of uranium by biosorption. U.S. Patent no. 4 320 093, 1981.

107. Stamberg, K., Katzer, J., Prochazka, H., Jilek, R., Nemec, P., and Hulak, P. Canadian Patent no. 1 009 600, 1977.

108. Volesky, B. and I. Prasetyo, Cadmium removal in a biosorption column. *Biotechnology and Bioengineering*, 1994. 43(11): p. 1010–1015.

109. Ravindiran, G., et al., Conversion of seaweed waste to biochar for the removal of heavy metal ions from aqueous solution: A sustainable method to address eutrophication problem in water bodies. *Environmental Research*, 2024. 241: p. 117551.

110. Dong, Z., et al., Microbe-encapsulated silica gel biosorbents for selective extraction of scandium from coal byproducts. *Environmental Science & Technology*, 2021. 55(9): pp. 6320–6328.

111. Volesky, B., *Sorption and Biosorption*. 2003, Montreal, Canada: BV Sorbex, Inc.

112. Wang, L.K., J.P. Chen, Y.T. Hung and N.K. Shammas, *Heavy Metals in the Environment*. 2009, Boca Raton, FL: CRC Press. 514 pages.

9 Advances in the Management and Treatment of Acid Pickling Wastes Containing Heavy Metals

Mu-Hao Sung Wang, Lawrence K. Wang, Veysel Eroglu, and Ferruh Erturk

ACRONYMS

BOD$_5$	Biochemical oxygen demand, 5-day.
Cd^{2+}	Cadmium ion
COD	Chemical oxygen demand
CPL	Concentrated pickling liquor
Cr^{6+}	Chromium (VI)
Cu^{2+}	Copper ion
DAF	Dissolved air flotation
DAFF	Dissolved air flotation-filtration
Fe	Iron
Fe^{2+}	Ferrous ion
Fe$_2$O$_3$	Ferric oxide
Fe^{3+}	Ferric ion
Fe$_3$O$_4$	Fe$_2$O$_3$FeO
FeCl$_2$	Ferrous chloride
FeO	Ferrous oxide
H$_2$	Hydrogen gas
H$_2$O	Water
H$_2$SO$_4$	Sulfuric acid
HCl	Hydrochloric acid
NESHAP	National Emission Standard for Hazardous Air Pollutants
NH$_4^+$-N	Ammonium nitrogen
Ni^{2+}	Nickel ion
O&G	Oil and grease
SPL	Spent pickling liquor
TSS	Total suspended solids
U.S. EPA	US Environmental Protection Agency
WPL	Waste pickling liquor
Z$_n^{2+}$	Zinc ion

DOI: 10.1201/9781003541615-9

9.1 INTRODUCTION

9.1.1 Metal Finishing Industry

Metal industries use substantial quantities of water in processes such as metal finishing and galvanized pipe manufacturing in order to produce corrosion-resistant products. Effluent wastewaters from such processes contain toxic substances, metal acids, alkalis, and other substances that must be treated, such as detergents, oil, and grease. These effluents may interfere with biological treatment processes in sewage treatment plants. In case the effluents are to be discharged directly to a watercourse, treatment requirements will be more stringent and costly [1–9].

9.1.2 Acid Pickling and Acid Cleaning of Metal Surfaces

Laser cutting, welding, and hot working leave a discolored oxidized layer or scale on the surface of the worked steel. This must be removed to perform many of the surface finishing processes. The acid pickling process, in which acids or acid mixtures are used, removes the oxide or scale of metals and corrosion products.

Acid cleaning removes inorganic contaminants that are not removable by other primary cleaning solutions. Acid cleaning has limitations because it is difficult to handle because of its corrosiveness and does not apply to all steels. Hydrogen embrittlement is problematic for some alloys and high-carbon steels [5]. The hydrogen from the acid reacts with the surface and makes it brittle and crack. Because of its high reactivity to treatable steels, acid concentrations and solution temperatures must be controlled to assure desired pickling rates [5].

Technically speaking, acid pickling is the treatment of metallic surfaces to remove impurities, stains, rust, or scale with a solution called pickle liquor, containing strong mineral acids, before subsequent processing (i.e., extrusion, rolling, painting, galvanizing, or plating) with tin or chromium. The two acids commonly used are hydrochloric acid and sulfuric acid.

9.1.3 Pickling Liquor and WPL

The most common acid used for pickling is sulfuric acid. Other acids such as hydrochloric, phosphoric, hydrofluoric, or nitric acids are also used individually or as mixtures. Sulfuric or hydrochloric acids are used for pickling carbon steels, and phosphoric, nitric, and hydrofluoric acids are used together with sulfuric acid for stainless steel. Water is used in pickling and rinsing. The quantity of water used can vary from less than 100 to –3,000 L/ton, depending on whether once-through or recycle systems are used [1,2].

Carbon steel is usually pickled by either sulfuric acid or hydrochloric acid. At one time, sulfuric acid was the pickling agent of choice for picklers running integrated steel works [1]. Hydrochloric acid is chosen in more modern lines when bright surfaces, low energy consumption, reduced overpickling, and the total recovery of the pickling agent from the WPL are desired [5,6].

The spent pickling liquor (SPL) is called waste pickling liquor (WPL), which must be properly treated for disposal or reuse. Wastewaters from pickling include acidic rinsewaters, metallic salts, and waste acid. WPL is considered a hazardous waste by the U.S. Environmental Protection Agency (U.S. EPA).

9.1.4 Acid Pickling Operation

Pickle solutions that are used in the removal of metal oxides or scales and corrosion products are acids or acid mixtures. Depending on the product being pickled, the acid pickling operation process can be batch or continuous. In continuous strip pickling, more water is required for the uncoilers, looping pit, and coilers. In the case of pickling hot rolled coils, the coils are transported to the pickling line. In the uncoiler section, the coil is fed through a pit containing water for washing off the surface dirt and then fed through the pickling line.

9.1.5 WPL Treatment and Recycling

Lime or alkaline substances are used to neutralize the WPL. In addition, 5-day biochemical oxygen demand (BOD$_5$), chemical oxygen demand (COD), total suspended solids (TSS), oil and grease (O&G), ammonium nitrogen (NH$_4^+$-N), pH, cyanides, fish toxicity, and several relevant metal ions such as cadmium (Cd^{2+}), iron (Fe^{2+}), zinc (Zn^{2+}), nickel (Ni^{2+}), copper (Cu^{2+}), and chromate (Cr^{6+}) have to be reduced below the maximum allowable limits.

Some acid pickling plants, particularly those using hydrochloric acid, operate acid recovery plants where the mineral acid is boiled away from the iron salts, but there still remains a large volume of highly acid ferrous sulfate or ferrous chloride to be disposed of. Since the 1960s, total hydrochloric acid regeneration processes have become widely accepted [6]. The by-product of nitric acid pickling is marketable to a couple of secondary industries, including the fertilizer industry.

9.2 PICKLING PROCESS REACTIONS AND WPL CHARACTERISTICS

9.2.1 WPL Generation

During the application of the pickling process in the finishing of steel, in which steel sheets are immersed in a heated bath of acid (sulfuric, hydrochloric, phosphoric, etc.), scale (metallic oxides) is chemically removed from the metal surface. The process can be batch or continuous. In these processes, water is used in pickling and rinsing operations. In continuous pickling, wet fume scrubbing systems are also used. Effluent water from the pickling tanks, which is called WPL, consists of spent acid and iron salts. Waste hydrochloric liquor contains 0.5%–1% free hydrochloric acid and 10% dissolved iron, and the production of WPL is approximately 1 kg free hydrochloric acid and 10 kg dissolved iron per ton of steel pickled [2]. In waste sulfuric acid pickle liquor, the free acid and dissolved iron content are approximately 8% each, resulting in 10 kg each of free sulfuric acid and dissolved iron per ton of steel pickled. WPL may also contain other metal ions, sulfates, chlorides, lubricants, and hydrocarbons. Rinsewater, which includes smaller concentrations of the above contaminants, ranges in quantities from 200 to 2,000 L/ton. Fume scrubber water requirements range from 10 to 200 L/ton [2].

In hot rolling processes, pickling is used for further processing to obtain the surface finish and proper mechanical properties of a product. In the case of pickling hot rolled coils, the coil is fed through a pit containing water for washing off surface dirt and then fed through the pickling line. In the pickling section, the coil strip comes in contact with the pickle liquor (sulfuric or hydrochloric acid). Wastewater sources are processor water, WPL, and rinse water.

In the case of batch pickling, the product is dipped into a pickling tank and then rinsed in a series of tanks. The quantity of wastewater discharged from a batch process is less than that from continuous operation. The wastewater is usually treated by neutralization and sedimentation.

9.2.2 Sulfuric Acid Pickling Reaction

In sulfuric acid pickling, ferrous sulfate is formed from the reaction of iron oxides with sulfuric acid:

$$FeO + H_2SO_4 \rightarrow FeSO_4 + H_2O. \tag{9.1}$$

The ferrous sulfate formed in the above reaction is either monohydrate or heptahydrate (FeSO$_4 \cdot$ 7H$_2$O).

9.2.3 Hydrochloric Acid Pickling Reactions

During the hot forming or heat treating of steel, oxygen from the air reacts with iron to form iron oxides or scale on the surface of the steel. This scale must be removed before the iron is subsequently shaped or coated. One method of removing this scale is pickling with hydrochloric acid [6].

Pickling is conducted by continuous, semicontinuous, or batch modes depending on the form of metal processed. In developing a National Emission Standard for the Steel Pickling industry, the U.S. EPA recently surveyed the industry and produced a background information document containing detailed information on the various processes in the industry, pollution control devices, and emissions [6].

When iron oxides dissolve in hydrochloric acid, ferrous chloride is formed according to the following reactions:

$$Fe_2O_3 + Fe + 6HCl \rightarrow 3FeCl_2 + 3H_2O, \tag{9.2}$$

$$FeO + 2HCl \rightarrow FeCl_2 + H_2O. \tag{9.3}$$

Since Fe_3O_4 is Fe_2O_3FeO, the reaction for Fe_3O_4 is the sum of the two reactions. Some of the base metal is consumed in the above reaction as well as in the following reaction:

$$Fe + 2HCl \rightarrow FeCl_2 + H_2. \tag{9.4}$$

An inhibitor is usually added to reduce the acid attack on the base metal while permitting its action on iron oxides. The rate of pickling increases with temperature and concentration of HCl. As pickling continues, HCl is depleted and ferrous chloride builds up in the pickling liquid to a point where pickling is no longer effective. At this point, the old liquid is discharged, and the pickling tank is replenished with fresh acid. Typical HCl concentrations in a batch pickling process are 12 wt% for a fresh solution and 4 wt% before acid replenishment. At these concentrations, the concentration of HCl in the vapor phase increases rapidly with temperature [6].

9.3   TREATMENT OF WPLS AND CLEANING WASTES

9.3.1   TREATMENT, DISPOSAL, OR RECYCLE

Throughout the late 1980s, SPL was traditionally land disposed of by steel manufacturers after lime neutralization. The lime neutralization process raises the pH of spent acid and makes heavy metals in the sludge less likely to leach into the environment. Today however, some of the SPL can be recycled or regenerated onsite by steel manufacturers [5,6].

Treated wastewater effluents, in general, can be discharged to either a watercourse or a public sewer system. In the former case, the treatment requirements will be more stringent.

The WPL, rinse water discharges, and fume scrubber effluent can be combined in an equalization tank for subsequent treatment. Basically, three methods are used to treat WPLs:

1. Neutralization and clarification [sedimentation or dissolved air flotation (DAF)].
2. Crystallization of ferric sulfates and regeneration of the acid.
3. Deep well disposal.

The most commonly used methods are the first two.

9.3.2   NEUTRALIZATION AND CLARIFICATION (SEDIMENTATION OR DAF)

In old plants, neutralization and sedimentation are applied to treating wastewaters in general, including WPLs. A typical treatment system for continuous pickling water is shown in Figure 9.1 [1,3].

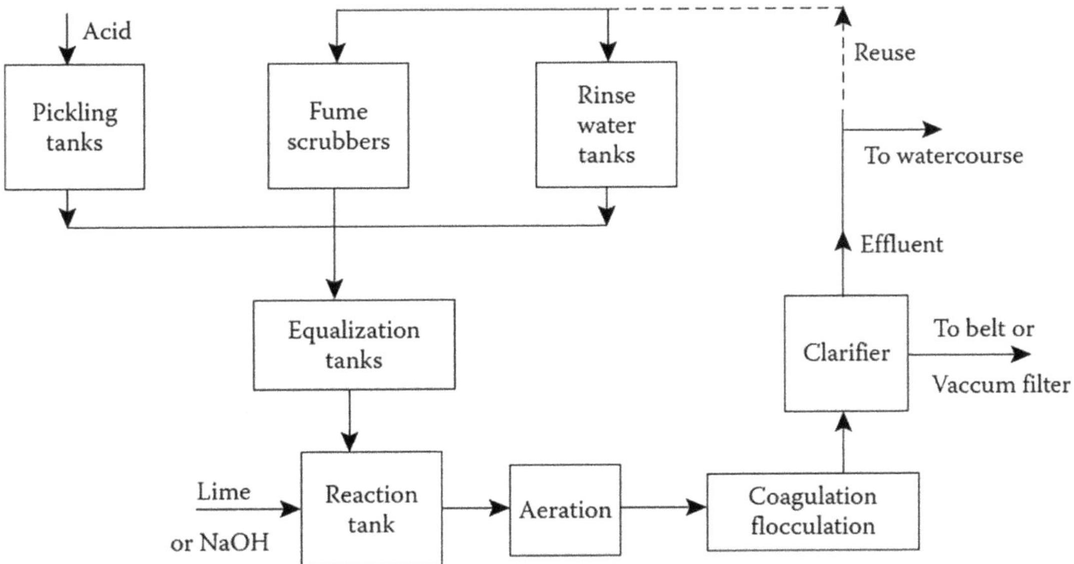

FIGURE 9.1 Typical treatment system for pickling. (Adapted from Eroglu V, Erturk F. In: Wang LK, Wang MHS (Eds), *Handbook of Industrial Waste Treatment.* Marcel Dekker, Inc., New York, pp. 293–306, 1991 [1]; Eroglu V, Topacik, D, Ozturk, I. *Wastewater Treatment Plant for Cayirova Pipe Factory.* Environmental Engineering Department, Istanbul Technical University, Turkey, 1989 [3].)

In an integrated steel mill, a central wastewater treatment system is used to treat wastewater from pickling lines, cold rolling mills, and coating lines.

Pickling wastewater has a low pH and contains dissolved iron and other metals. The blowdown and dumps from cold rolling mill solutions, which may contain up to 8% oil, are collected in emulsion-breaking tanks in which the emulsions are broken by heat and acid. The oil is then skimmed, and the water phase containing 200–300 mg/L of oils is treated together with wastewaters from pickling, cold rolling, and coating lines. The combined wastewater flows to a settling and skimming tank where solids and oil are removed. Effluent from the settling/skimming tank is then treated in a series of settling tanks where chemicals (coagulants and/or lime) and air are added to oxidize the remaining iron to ferric ions (Fe^{3+}), to further break the oil emulsions and neutralize the excess acid in wastewater. Effluent from the mixing tanks then enters a flocculator/clarifier system, the overflow from the clarifier is discharged, and the settled sludge is pumped to a dewatering system consisting of centrifuges, belt, or vacuum filters. The dewatered sludge is disposed of, and the water phase is returned to the clarifier effluent.

The clarifier shown in Figure 9.1 can be a sedimentation clarifier, a DAF clarifier, or a dissolved air flotation-filtration (DAFF) clarifier, depending on space availability, pretreatment requirements, effluent limitations, and costs [7–12]. Modern pickling plants use DAF or DAFF for more cost-effective clarification or more efficient clarification, respectively.

9.3.3 Crystallization and Regeneration

Using lime or other alkaline substances to neutralize acid is quite costly, especially when large capacities are involved. There is also potential resale value in the acids and ferrous ions. Therefore, recovery of these substances will reduce the pollution load, and their sale or reuse will represent a profit to the industry.

Crystallization is one of the treatment methods for sulfuric acid WPL. Thus, it is possible to decrease the pollution load and, at the same time, recover various hydrates of $FeSO_4$.

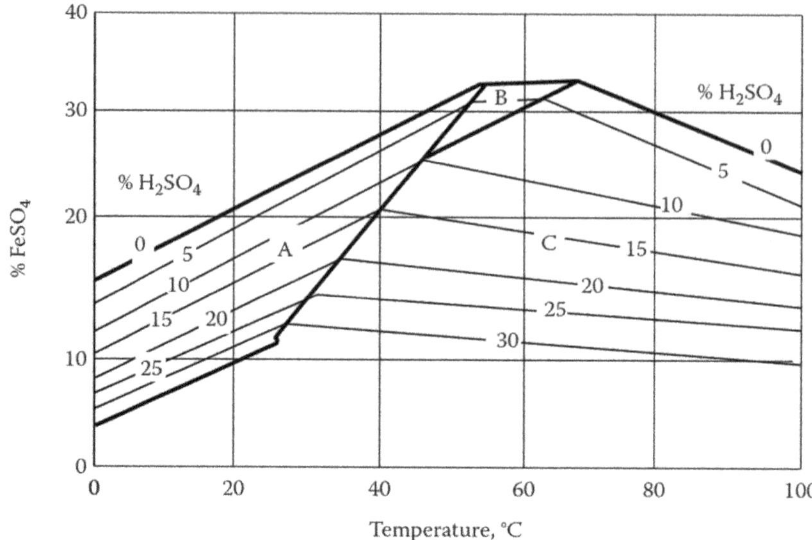

FIGURE 9.2   Solubility of ferrous sulfate FeSO$_4$ as a function of temperature and sulfuric acid concentration. (Adapted from Eroglu V, Erturk F. In: Wang LK, Wang MHS (Eds), *Handbook of Industrial Waste Treatment*. Marcel Dekker, Inc., New York, pp. 293–306, 1991 [1]; Eroglu V, Topacik, D, Ozturk, I. *Wastewater Treatment Plant for Cayirova Pipe Factory*. Environmental Engineering Department, Istanbul Technical University, Turkey, 1989 [3].)

The crystallization of FeSO$_4$ depends on the characteristics of the water and acid, and the solubility of FeSO$_4$. The solubility of ferrous sulfate as a function of temperature and sulfuric acid concentration is shown in Figure 9.2 [4]. In this figure, FeSO$_4$ • 7H$_2$O is dominant in region A, FeSO$_4$ • 4H$_2$O in region B, and FeSO$_4$ • H$_2$O in region C.

The crystallization of ferrous sulfate as heptahydrate is commonly used today. The concentration of iron in the acid bath is approximately 80 g/L as Fe^{3+}. The crystallization of FeSO$_4$ • 7H$_2$O is achieved by cooling the acid waters in heat exchangers or evaporation under vacuum after pickling. Make-up acid must be added to the bath. During countercurrent cooling, the acid bath waste passes through two to three crystallization tanks and is cooled to between 0°C and 5°C. The crystallized ferric sulfates are recovered by centrifuging. A typical flow diagram for FeSO$_4$ • 7H$_2$O crystallization is shown in Figure 9.3.

The WPL is sprayed above a cyclone crystallizer, and air is blown from the bottom countercurrent to the liquid. Packing material is also present to increase the area of contact between the air and the liquid. Acid wastewaters are then cooled, and the FeSO$_4$ • 7H$_2$O crystals are recovered by centrifuging.

In the Ruthner process [1], WPL is first concentrated in an evaporator. The concentrate is then pumped to a reactor where it is combined with hydrochloric acid gas, in which ferrous chloride and sulfuric acid are formed. The sulfuric acid is then separated by centrifuging. Ferrous chloride goes to a roaster in which it is converted to ferric oxide. The gases liberated from the roaster and the acid from the centrifuge go to a degassing chamber; the sulfuric acid is removed and returned to the pickling process or can be sold. The remaining gases from the degasser are passed through an absorption system and then reused in the reaction chamber.

In the Lurgi process [1] that was developed in Germany, hydrochloric acid is recovered from the WPL. The acid is regenerated in a fluidized bed. During pickling with HCl, the acid circulates between a pickling tank and a storage tank, and the acid reacts with the iron oxide scale from the steel producing ferric chloride, resulting in increasing concentrations of dissolved iron and decreasing concentrations of acid.

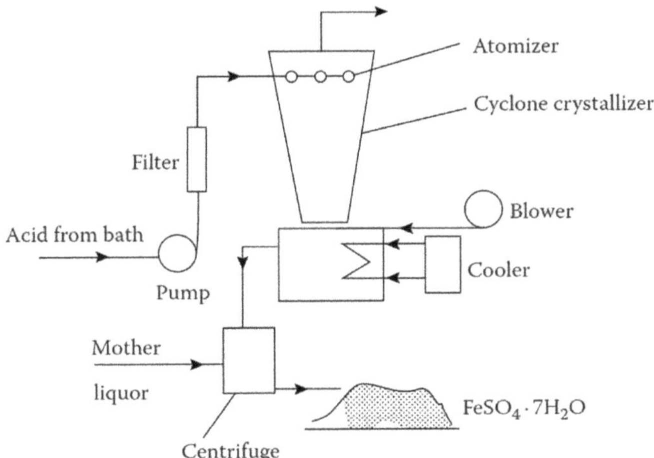

FIGURE 9.3 Flow diagram for ferrous sulfate FeSO$_4$ • 7H$_2$O crystallization. (Adapted from Eroglu V, Erturk F. In: Wang LK, Wang MHS (Eds), *Handbook of Industrial Waste Treatment.* Marcel Dekker, Inc., New York, pp. 293–306, 1991 [1]; Eroglu V, Topacik, D, Ozturk, I. *Wastewater Treatment Plant for Cayirova Pipe Factory.* Environmental Engineering Department, Istanbul Technical University, Turkey, 1989 [3].)

In the Lurgi system [1], the acid level in the pickling liquor remains constant at about 10%. A continuous bleed stream is removed from the system at the same rate as it is pickled. The bleed stream, or spent pickle, is fed to a preevaporator and heated with gases from the regeneration reactor. Concentrated liquor from the preevaporator then enters the lower part of the reactor containing 13% acid and 20% ferrous chloride. The reactor contains a fluidized bed of sand and is fired by oil or gas to maintain an operating temperature of about 800°C. The reaction products leave from the top of the reactor. The ferric oxide is removed by a cyclone, and the hot gases enter the preevaporator. The overhead from the evaporator, which is at a temperature of about 120°C, contains water vapor, HCl, combustion products, and also some HCl that vaporizes directly from the plant liquor that enters the system. The gas mixture from the preevaporator enters the bottom of the adiabatic absorption tower, where HCl is absorbed by another bleed stream of the pickle liquor, and thus the regenerated acid is placed back into the pickle liquor circuit. The regenerated acid contains 12% acid and about 70 g/L of iron. The unabsorbed gases go to a condenser.

9.4 TREATMENT OF WASTEWATER FROM ACID PICKLING TANKS IN A GALVANIZED PIPE MANUFACTURING FACTORY USING SULFURIC ACID

9.4.1 GENERAL DESCRIPTION

This study was conducted at Cayirova Boru Sanayii AS (a galvanized pipe manufacturing factory) in Gebze, Kocaeli, Turkey [1,3]. At this plant, batch pickling is applied. During the manufacturing process, the pipes are immersed in an acid bath that contains 25% sulfuric acid at 80°C and then prepared for the galvanization process by passing through cold water, hot water, and flux baths. The purpose of a cold water bath is to clean the acid from the surface of the pipes following pickling. A hot bath is applied in order to dry and prevent water and acid from entering the flux bath. The purpose of the flux bath, in which ammonium zinc chloride (NH$_4$ZnCl$_3$) is used, is to prepare a suitable surface for galvanization and prevent oxidation of the pipe. The flow diagram of the baths is shown in Figure 9.4.

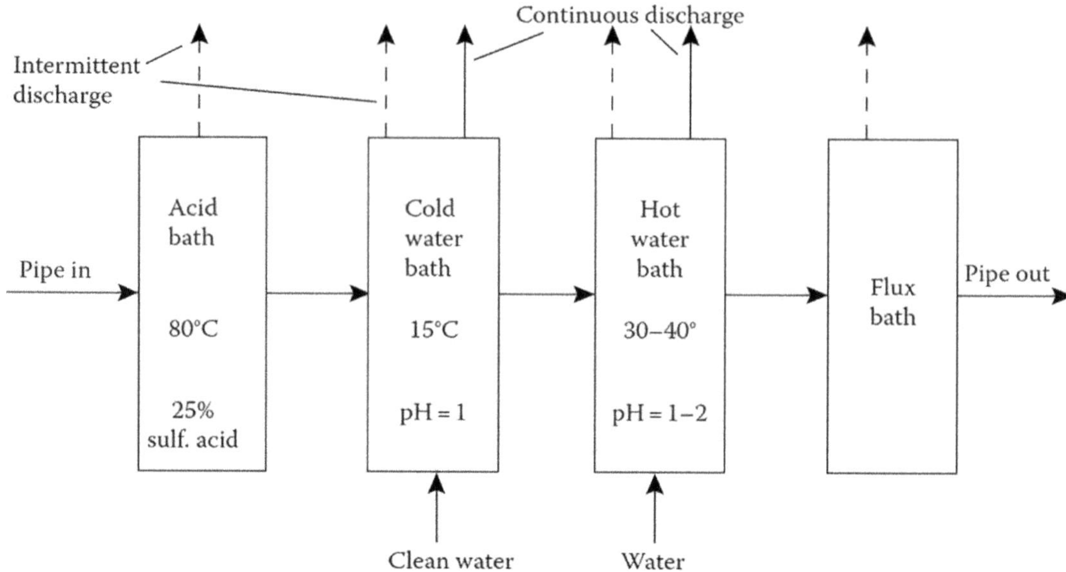

FIGURE 9.4 Flow diagram showing sources of wastewaters in the galvanized pipe manufacturing process. (Adapted from Eroglu V, Erturk F. In: Wang LK, Wang MHS (Eds), *Handbook of Industrial Waste Treatment.* Marcel Dekker, Inc., New York, pp. 293–306, 1991 [1]; Eroglu V, Topacik, D, Ozturk, I. *Wastewater Treatment Plant for Cayirova Pipe Factory.* Environmental Engineering Department, Istanbul Technical University, Turkey, 1989 [3].)

TABLE 9.1

Types and Quantities of Wastewaters in Acid and Flux Baths

Wastewater Source	Average	Maximum
Continuous discharge from hot and cold water baths	$4\,m^3/h$	$8\,m^3/h$
Intermittent discharge (once every 7 days)	$15\,m^3$	$15\,m^3$
Cold water bath (once every 15 days)	$15\,m^3$	$15\,m^3$
Hot water bath (once every 15 days)	$15\,m^3$	$15\,m^3$
Flux bath	$5\,m^3$	$5\,m^3$

Source: Eroglu V, Erturk F. In: Wang LK, Wang MHS (Eds), *Handbook of Industrial Waste Treatment.* Marcel Dekker, Inc., New York, pp. 293–306, 1991 [1]; Eroglu V, Topacik, D, Ozturk, I. *Wastewater Treatment Plant for Cayirova Pipe Factory.* Environmental Engineering Department, Istanbul Technical University, Turkey, 1989 [3].

Acid bath wastewaters are usually discharged once a week. The average flow rate of these wastewaters is $4\,m^3/hr$, with a maximum of $8\,m^3/hr$. The hot and cold water baths are discharged once every 15 days. The quantities and flow rates of these wastewaters are shown in Table 9.1 [1,3].

9.4.2 CHARACTERISTICS OF WASTEWATERS

Wastewater characteristics must be known in order to select a suitable treatment system. For this purpose, wastewater samples from the sources were analyzed to determine various parameters.

Also, the quantities of chemicals (NaOH) required for neutralization and settling characteristics were determined. These were made separately for continuous and batch discharges. Since the system is to be designed according to the continuous discharge of wastewaters from the batch system to the treatment plant, "mixed wastewater" was prepared in quantities proportional to the flow rates. The quantity of NaOH required for 1,000 mL of mixed wastewater is shown in Table 9.2 [1,3].

Since the continuous discharge quantities are much larger than the batch discharges, they were analyzed separately. Wastewaters from continuous discharge were neutralized with 2 N NaOH. The results are given in Table 9.3 [1,3]. The settling characteristics of continuous discharge wastewaters are shown in Table 9.4 [1,3]. The experimental results for "mixed wastewaters," the quantities of which are shown in Table 9.2, are given in Table 9.5 [1,3]. The settling characteristics of mixed wastewaters are shown in Table 9.6 [3].

Neutralization can also be carried out by a combination of NaOH and lime. Experiments were conducted in order to determine the optimum combination of NaOH and lime. For this purpose, various quantities of lime were added to 1 L of mixed wastewater, and then the amount of NaOH required was determined to obtain a pH of 8.5. The results are shown in Table 9.7 [3].

As can be seen from Table 9.7, the required dosage of NaOH does not increase significantly when the limb dosage is more than 20 g/1,000 mL. The mixed wastewater, which was treated with the dosages of lime and NaOH shown in Table 9.7, was then aerated for 15 minutes after a pH of 8.5

TABLE 9.2

Quantities of NaOH Required for 1,000 mL "Mixed Wastewater"

Source	Flow Rate (m³/2 Months)	Quantity of NaOH Required for 1,000 mL
Continuous discharge	8,640	971
Acid bath	129	14.5
Cold water bath	60	6.8
Hot water bath	60	6.8
Flux bath	50	0.9
Total	8,894	1,000

Source: Eroglu V, Erturk F. In: Wang LK, Wang MHS (Eds), *Handbook of Industrial Waste Treatment.* Marcel Dekker, Inc., New York, pp. 293–306, 1991 [1]; Eroglu V, Topacik, D, Ozturk, I. *Wastewater Treatment Plant for Cayirova Pipe Factory.* Environmental Engineering Department, Istanbul Technical University, Turkey, 1989 [3].

TABLE 9.3

Experimental Results for Continuous Discharge

Parameter	Unit	Original Sample	After Neutralization and Separation
Total iron	mg/L	5,980	350
Chromate	mg/L	0	0
Lead	mg/L	0	0
COD	mg/L	350	20
Zinc	mg/L	0	0
pH	—	1.6	8.0
Color	—	—	Greenish

Source: Eroglu V, Erturk F. In: Wang LK, Wang MHS (Eds), *Handbook of Industrial Waste Treatment.* Marcel Dekker, Inc., New York, pp. 293–306, 1991 [1]; Eroglu V, Topacik, D, Ozturk, I. *Wastewater Treatment Plant for Cayirova Pipe Factory.* Environmental Engineering Department, Istanbul Technical University, Turkey, 1989 [3].

TABLE 9.4

Settling Characteristics of Continuous Discharge Wastewater

Time	Volume of Clear Phase, h (mL/L)
15 minutes	20
30 minutes	50
1.0 hour	90
2.5 hours	240
3.5 hours	400
4.5 hours	460
5.5 hours	500
20 hours	720

Source: Eroglu V, Erturk F. In: Wang LK, Wang MHS (Eds), *Handbook of Industrial Waste Treatment.* Marcel Dekker, Inc., New York, pp. 293–306, 1991 [1]; Eroglu V, Topacik, D, Ozturk, I. *Wastewater Treatment Plant for Cayirova Pipe Factory.* Environmental Engineering Department, Istanbul Technical University, Turkey, 1989 [3].

TABLE 9.5

Experimental Results for Mixed Wastewater Samples

Parameter	Original Sample	After Neutralization and Separation in Clear Phase
Total iron (mg/L)	6,100	300
Sulfate (mg/L)	19,000	16,000
Chromate (mg/L)	0	0
Lead (mg/L)	0	0
Zinc (mg/L)	15	0
COD (mg/L)	360	15
pH	0.7	8.5

TABLE 9.6

Settling Characteristics of Mixed Wastewaters

Time	Volume of Clear Phase (mL/L)
30 minutes	40
1 hour	100
2.5 hours	220
3.5 hours	350
4.5 hours	410
5.5 hours	460
20 hours	700

Source: Eroglu V, Erturk F. In: Wang LK, Wang MHS (Eds), *Handbook of Industrial Waste Treatment.* Marcel Dekker, Inc., New York, pp. 293–306, 1991 [1]; Eroglu V, Topacik, D, Ozturk, I. *Wastewater Treatment Plant for Cayirova Pipe Factory.* Environmental Engineering Department, Istanbul Technical University, Turkey, 1989 [3].

TABLE 9.7

Quantities of Sodium Hydroxide Required for Different Quantities of Lime to Obtain a pH of 8.5

10 g Lime NaOH Added		20 g Lime NaOH Added		26 g Lime NaOH Added		32 g Lime NaOH Added	
I		II		III		IV	
pH	(mL)	pH	(mL)	pH	(mL)	pH	(mL)
3.4	0	6.8	0	7.1	0	7.4	0
6.5	20	7.4	8	7.6	4	7.6	4
7.4	28	8.4	16	7.9	12	8.6	12
8.0	30	8.5	17.2	8.6	15.2		
8.5	32						

Source: Eroglu V, Erturk F. In: Wang LK, Wang MHS (Eds), *Handbook of Industrial Waste Treatment.* Marcel Dekker, Inc., New York, pp. 293–306, 1991 [1]; Eroglu V, Topacik, D, Ozturk, I. *Wastewater Treatment Plant for Cayirova Pipe Factory.* Environmental Engineering Department, Istanbul Technical University, Turkey, 1989 [3].

TABLE 9.8

Analysis of Mixed Wastewater after Neutralization, Aeration, and Clarification

Parameter (mg/L)	Settling Time (min)	Lime+NaOH (g)			
		10	20	26	32
Iron	30	125	30	5	0
	120	0	0	0	0
Sulfate	30	5,750	5,759	5,000	3,000
	120	5,750	5,750	5,000	2,750
Settleable matter	30	120	280	320	440
	120	400	520	410	480

Source: Eroglu V, Erturk F. In: Wang LK, Wang MHS (Eds), *Handbook of Industrial Waste Treatment.* Marcel Dekker, Inc., New York, pp. 293–306, 1991 [1]; Eroglu V, Topacik, D, Ozturk, I. *Wastewater Treatment Plant for Cayirova Pipe Factory.* Environmental Engineering Department, Istanbul Technical University, Turkey, 1989 [3].

TABLE 9.9

Analysis of Wastewater after Neutralization, Aeration, and Clarification

Parameter	Concentration (mg/L)
COD	0
Total iron	0
Zinc	0
Sulfate	2,100

was reached. After aeration, it was allowed to settle for a period of 30–120 minutes. An analysis of the clear phase after settling is shown in Table 9.8 [3]. The wastewater was treated with 15 g/L of lime and NaOH to attain a pH of 8.5, aerated for 1 hour, mixed for 23 hours, and one additional hour was allowed for clarification. The analysis of the clear clarifier effluent is shown in Table 9.9 [1,3].

TABLE 9.10

Effluent Standards for Metal Industry Wastewaters in Turkey

Parameter	*2h* Composite Sample (mg/L)
COD	200
Suspended solids	125
Oil and grease	20
Ammonium nitrogen	400
Cd	0.1
Fe	3
Fluoride	50
Zn	5
Fish toxicity	10
pH, unH	6.9

Source: Eroglu V, Erturk F. In: Wang LK, Wang MHS (Eds), *Handbook of Industrial Waste Treatment.* Marcel Dekker, Inc., New York, pp. 293–306, 1991 [1].

9.4.3 TREATMENT METHODS

As indicated in the previous section, the concentration of iron in mixed wastewaters ranged from 5,980 to 6,100 mg/L; its pH was 0.7 and the zinc concentration was 15 mg/L. Since these wastewaters come only from acid baths and not from other processes of the plant, parameters like cadmium and fluoride are not encountered. The discharge standards for metal industry effluents set by the Turkish Water Pollution Control Regulation (Official Gazette, Table 15.7, September 4, 1988) are shown in Table 9.10 [1].

The experiments conducted on wastewaters, the results of which were shown in the previous section, indicated that neutralization/aeration/settling gave satisfactory results. The sludge formed must be disposed of after dewatering in a filter-press, a horizontal belt filter, or a centrifuge. An equalization tank is required to compensate for the effects of intermittent discharges. The treated wastewater can then be recycled for use in the process or discharged to the river. The flow diagram of the selected system is shown in Figure 9.5.

9.5 MANAGEMENT AND TREATMENT OF WASTEWATER AND AIR EMISSIONS FROM ACID PICKLING TANKS USING HYDROCHLORIC ACID

9.5.1 ENVIRONMENTAL MANAGEMENT AT STEEL/IRON HYDROCHLORIC ACID PICKLING PLANTS

Hydrochloric acid aerosols are produced and released into the air during the pickling process as HCl volatilizes, and steam and hydrogen gas with entrained acid fumes rise from the surface of the pickling tank and the pickled material as it is transferred from the pickling tank to the rinse tank. Pickling and rinse tanks are covered, and the acid fumes are generally collected and treated by control devices (e.g., packed tower scrubbers) to remove HCl. Emissions from many batch operations are uncontrolled. Pickling is sometimes accomplished in vertical spray towers. In this process, all the HCl in the pickling solution produces hydrochloric acid aerosols that are also used. Acid storage tanks and loading and unloading operations are also potential sources of HCl emissions.

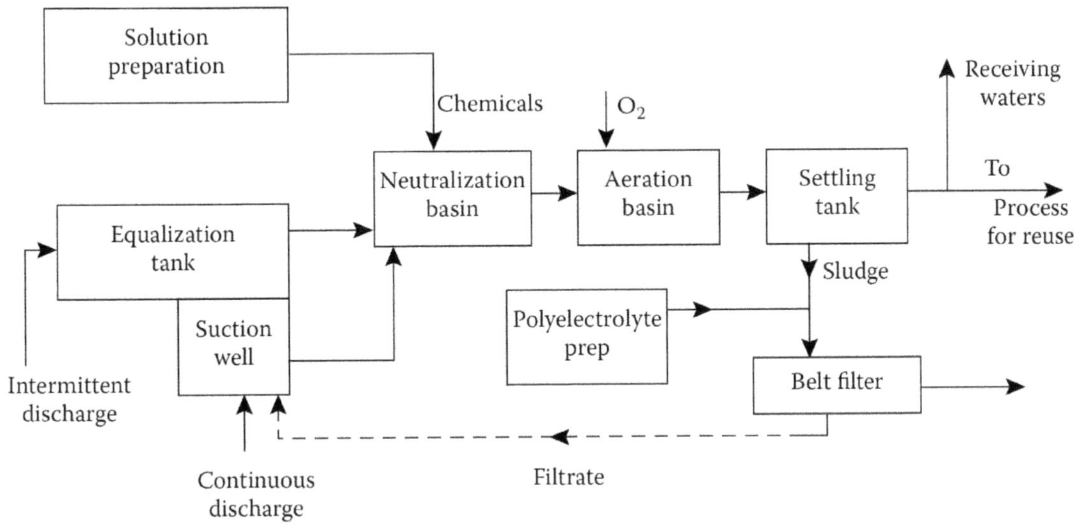

FIGURE 9.5 Flow diagram of the selected treatment system. (Adapted from Eroglu V, Erturk F. In: Wang LK, Wang MHS (Eds), *Handbook of Industrial Waste Treatment*. Marcel Dekker, Inc., New York, pp. 293–306, 1991 [1]; Eroglu V, Topacik, D, Ozturk, I. *Wastewater Treatment Plant for Cayirova Pipe Factory.* Environmental Engineering Department, Istanbul Technical University, Turkey, 1989 [3].)

Uncontrolled HCl emissions from a storage tank may be on the order of 0.07–0.4 tons per year (tpy) of HCl per tank, depending on tank size and usage. For each million ton of steel processed at continuous coil or push-pull coil model facilities, storage tank losses are estimated to amount to 0.39 tpy. For other types of pickling facilities, storage tank losses are estimated to be about 11.19 tpy of HCl per million ton of steel processed.

The U.S. EPA guidance for acid storage tanks can be applicable to storage tanks used in conjunction with the pickling process and may be extended to apply to the pickling process itself [6]. For storage tanks, one applies the amount of hydrochloric acid aerosol generated from a tank under average capacity and other conditions to the manufacturing threshold and multiplies that by the number of times the tank has been drawn down and refilled. The amount of acid aerosol manufactured during the picking process can be similarly determined by the amount of HCl generated from the pickling tanks during the processing of a certain amount of material and scaling that figure up to apply to all the materials processed by the same process and under the same conditions. The amount of hydrochloric acid aerosols lost from the pickling tanks is counted toward the material released to air unless the aerosol is collected and removed before exiting the stack. The hydrochloric acid aerosol collected in a scrubber is converted to the nonaerosol form, not reportable; the hydrochloric acid aerosol removed by the scrubber is considered to have been treated for destruction.

Hydrochloric acid may be recovered from the WPL in an acid regeneration process. This process has the potential of emitting significant amounts of hydrochloric acid aerosols. In ten acid regeneration plants surveyed by the U.S. EPA [6], annual capacities ranged from 3.2 to 39.8 million gallons (MG) per year for a single facility. The spray roasting acid regeneration process is the dominant one presently employed. One older facility used a fluidized bed roasting process.

In the spray roasting acid regeneration process, WPL at 2%–4% HCl comes in contact with hot flue gas from the spray roaster, which vaporizes some of the water in the WPL. WPL then becomes concentrated pickling liquor (CPL). CPL is sprayed on the spray roaster where ferrous chloride in the droplets falling through the rising hot gases reacts with oxygen and water to form ferric oxide and HCl,

$$2FeCl_2 + 0.5O_2 + 2H_2O \rightarrow Fe_2O_3 + 4\,HCl. \tag{9.5}$$

Flue gas containing HCl goes to a venturi preconcentrator and an absorption column. There generated acid contains approximately 18% HCl by weight. Emissions from acid regeneration plants range from about 1 to more than 10 tpy from existing facilities with and without pollution control devices (controlled and uncontrolled facilities).

Acid regeneration plants have storage tanks for spent and regenerated acid and these tanks are potential sources of HCl emissions. Emission estimates for uncontrolled and controlled storage tanks at acid regeneration facilities are 0.0126 and 0.008 tpy per 1,000 gallons of storage capacity, respectively.

Acid recovery systems are used to recover the free acid in the WPL. They are not employed in larger facilities because they recover only 2%–4% free HCl in the spent acid, but leave $FeCl_2$ in the solution, which must be processed or disposed of separately. These acid recovery systems are generally closed-loop processes that do not emit HCl. In their survey, the U.S. EPA compiled data from different types of pickling operations and their estimated emissions [6]. This information is reproduced in Table 9.11.

To estimate emissions from pickling facilities, the U.S. EPA developed 17 model plants to represent five types of pickling operations and one acid regeneration process [6]. The model plants include one or more size variations for each process model and were developed from information obtained from a survey of steel pickling operations and control technologies. The U.S. EPA estimated emission rates for model facilities. By using these emission rates and the production and hours of operation for the model pickling plants, emission factors were calculated. These are shown in Table 9.12.

A National Emission Standard for Hazardous Air Pollutants (NESHAP) for new and existing hydrochloric acid process steel pickling lines and HCl regeneration plants pursuant to Section 112 of the Clean Air Act as amended in November 1990 has been proposed (62 FR 49051, September 18, 1997). The purpose of this rulemaking is to reduce emissions of HCl by about 8,360 Mg/yr.

9.5.2 A STAINLESS STEEL PIPES AND FITTINGS MANUFACTURER: CASE HISTORY

9.5.2.1 Manufacturing Process

This stainless steel pipes and fittings manufacturing plant is located in the United States and produces stainless steel pipes of various diameters and lengths and custom-made pipe fittings [13]. It operates over 6,240 hr/yr to produce nearly 30 million pounds of pipe annually.

The raw materials used by the plant include coil and sheet metal stock, solvent-based marking ink, and protective plastic end caps. The two major operations in this plant, pipe and fitting formation and acid pickling, are described in the following section.

TABLE 9.11

Annual Emission Estimates from Steel Pickling Operations

Type of Facility	Number of Facilities	Number of Operations	Uncontrolled Emission (mg/yr)	Controlled Emissions (mg/yr)
Continuous coil	36	64	22,820	2,640
Push-pull coil	19	22	815	29
Continuous tube	20	55	6,524	4,252
Batch	4	11	100	52
Acid regeneration	26	59	2,632	1,943
Storage tanks	10	13	5,662	393
	99	369 (est.)	41	24

Source: U.S. EPA. *Steel Pickling.* The U.S. Environmental Protection Agency, Washington, DC, June 2008. Available at: http://www.epa.gov/tri/TWebHelp/WebHelp/hcl_section_3_1_4_ steel_pickling.htm [6].

TABLE 9.12

Air Emissions and Emission Factors for Model Pickling Plants

	Production	Hours of Operation	HCl Emissions	Control Efficiency	Emission Factor lb HCl/Tons Processed[b]	
			Uncontrolled			
Type Facility	(tpy)[a]	(hr)	(lb/hr)	(%)	(U)	(C)
Continuous coil (S)	450,000	6,300	111	93	1.6	0.1
Continuous coil (M)	1,000,000	6,300	179	92	1.1	0.1
Continuous coil (L)	2,700,000	7,000	347	92	0.9	0.1
Push-pull coil (S)	300,000	5,000	12	98	0.2	0.0
Push-pull coil (M)	550,000	4,400	27	98	0.2	0.0
Push-pull coil (L)	1,300,000	8,760	42	95	0.3	0.0
Continuous rod/wire (S)	10,000	5,100	46	98	23.5	0.5
Continuous rod/wire (M)	55,000	7,800	119	84	16.9	2.7
Continuous rod/wire (L)	215,000	7,200	413		13.8	
Continuous tube (S)	80,000	6,400	73	95	5.8	0.3
Continuous tube (L)	420,000	6,700	312	95	5.0	0.2
Batch (S)	15,000	4,400	16	94	4.7	0.3
Batch (M)	75,000	4,600	65	90	4.0	0.4
Batch (L)	170,000	5,700	147	81	4.9	0.9
Acid regeneration (S)	4	8,200	7	98	14,350.0	287.0
Acid regeneration (M)	13.5	7,700	28	98	15,970.4	319.4
Acid regeneration (L)	30	8,760	1,064	98.5	310,688.0	4,660.3

Source: U.S. EPA. *Steel Pickling.* The U.S. Environmental Protection Agency, Washington, DC, June 2008. Available at: http://www.epa.gov/tri/TWebHelp/WebHelp/hcl_section_3_1_4_steel_pickling.htm [6].

C, controlled; L, large; M, medium; S, small; U, uncontrolled.

[a] Units of production for acid regeneration facilities are in millions of gallons/yr.

[b] Emission factor units for acid regeneration facilities are in lb of HCl per MG of HCl produced.

9.5.2.1.1 Pipe and Fitting Formation

Stainless steel coil and sheet stock is unloaded and stored outdoors under protective cover. When it is needed, the coil stock is moved indoors by a forklift to one of six automatic tube mills where the sides of unrolled metal strips are curled up to form a continuous, cylindrical pipe. The seam of the resulting pipe is fused in an electric in-line welding operation. An abrasive saw is used to cut the continuously formed pipe to specified lengths; sections of poorly welded pipe are cut away.

Stainless steel sheet stock is used to form custom products such as tees, elbows, and reducers. The sheets are cut with a band saw or plasma torch into smaller pieces and custom-formed into final product shapes using various forming and bending equipment.

All pipes and fittings are hardened in electric induction or gas annealing furnaces. After annealing, the pipes are water spray-quenched or quenched in a water-filled tank outdoors, depending on their size.

The roughened ends of the pipe are manually deburred with an air grinder. The pipes are then straightened as necessary and transported to the acid pickling process.

9.5.2.1.2 Acid Pickling

All pipes and fittings are transported to the pickling process. An overhead crane is used to lower them into an acidic pickle liquor solution, which chemically cleans and etches the black oxide surface layer, resulting in a clean, rust-resistant pipe.

Each pipe is rinsed with water in one of two rinse tanks and is then mounted on a wash rack and manually sprayed with water in a second rinsing operation. After the pipes dry, they are labeled with a solvent-based ink spray jet and protective plastic caps are hammered onto the ends. The finished products are stored outdoors until they are shipped to customers.

9.5.2.2 Waste Minimization and Pollution Prevention

This plant has already implemented the following techniques to manage and minimize its wastes:

1. The polymer previously used by this plant as a flocculent in the onsite wastewater treatment system has been replaced by magnesium hydroxide in order to reduce the volume of sludge generated and shipped offsite.
2. An acid regeneration system has been installed to regenerate SPL for reuse onsite.

The type of waste currently generated by the plant, the source of the waste, the waste management method, the quantity of the waste, and the annual waste management cost for each waste stream identified are given in Figure 9.6 and Table 9.13. Acid pickling appears to be one of many stainless steel product manufacturing operations according to Figure 9.6. However, from Table 9.13,

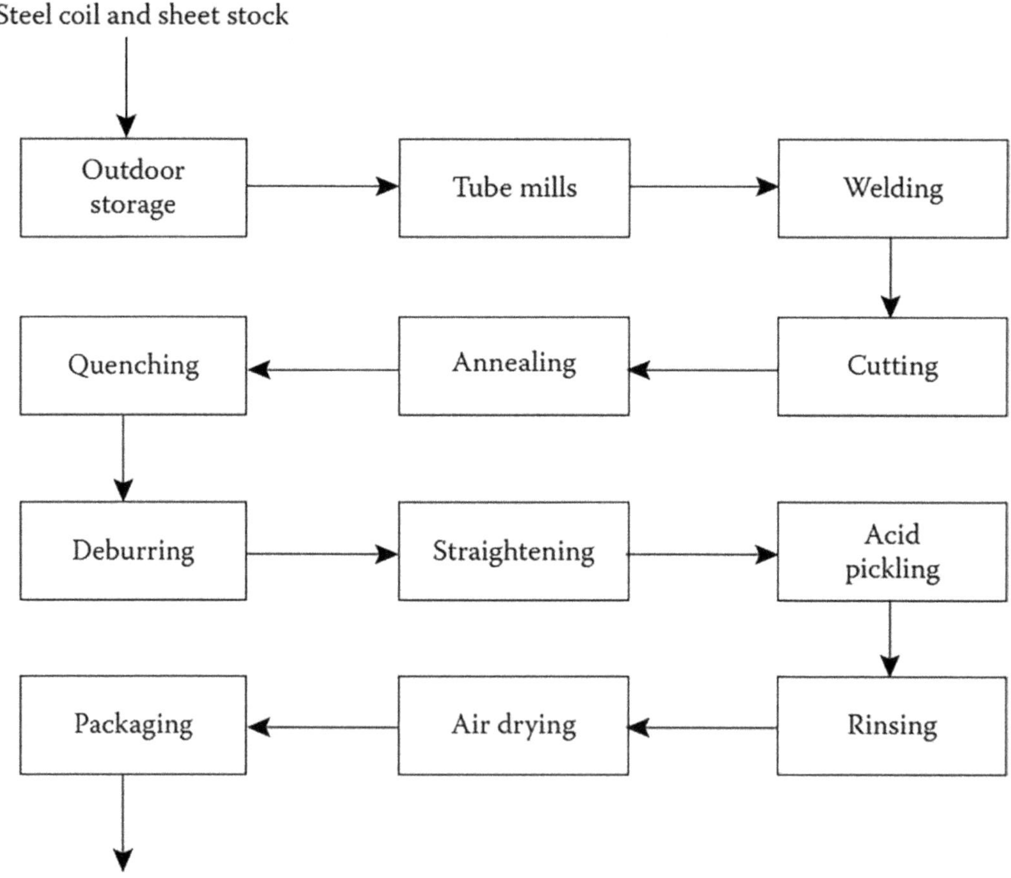

FIGURE 9.6 Flow diagram of a typical stainless steel product manufacturing plant involving acid pickling operation. (U.S. EPA.)

TABLE 9.13

Summary of Current Waste Generation

Waste Generated	Source of Waste	Waste Management Method	Annual Quantity Generated (lbs/yr)	Annual Waste Management Cost
Packaging and protective barrier waste	Receipt and storage of raw materials	Shipped to municipal landfill	7,500	$0[a]
Leaked and spent lubricating oil	Machining	Shipped to fuels blending program	8,540	5,980
Spent abrasive saw blades	Cutting of pipe	Shipped to municipal landfill	5,200	0[a]
Stainless steel scrap	Machining and cutting of pipe	Sold to scrap recycler	700,000	−164,300 (net revenue received)
Oxidized metal fakes and metal dust	Annealing, deburring, and cutting	Shipped to special landfill	30,000	15,810
Quench water	Quenching of pipes following annealing	Sewered to POTW	49,800	40
Damaged plastic end caps	Packaging of finished product	Shipped to municipal landfill	130	0[a]
Pickling rinse water	Acid pickling of product	Treated in onsite WWTP; Sewered to POTW	84,598,000	89,100
Wastewater treatment sludge	Onsite treatment of wastewater	Shipped to hazardous waste landfill	1,560,000	265,370
Miscellaneous solid waste landfill	Various plant operations	Shipped to municipal	135,000 cu ft[b]	26,990

Source: Jendrucko RJ, Myers JA, Thomas TM, Looby GP. *Pollution Prevention Assessment for Manufacturer of Stainless Steel Pipes and Fittings.* Report no. EPA-600/S-95/017. The U.S. Environmental Protection Agency, Cincinnati, OH, August 1995 [13].

[a] Included in annual waste management cost for miscellaneous solid waste.

[b] Includes specific quantities given for packaging and protective barrier waste, spend abrasive saw blades, and damaged plastic end caps. The majority of this waste stream is cardboard waste.

the importance of pickling rinsewater treatment is clearly identified [13]. The toxic waste sludge disposal is the highest, while the miscellaneous solid waste disposal cost is ranked third.

Table 9.14 shows the opportunities for pollution prevention that the U.S. EPA recommended for the plant [13]. The opportunity, type of waste, possible waste reduction and associated savings, and implementation cost along with simple payback time are given in the table. The quantities of waste currently generated by the plant and possible waste reduction depend on the production level of the plant. All values should be considered in that context.

It should be noted that the financial savings of the opportunities result from the need for less raw material and from the reduced present and future costs associated with waste management. Other savings not quantifiable by this study include a wide variety of possible future costs related to changing emission standards, liability, and employee health. It should also be noted that the savings given for each pollution prevention opportunity reflect the savings achievable when implementing each opportunity independently and do not reflect duplication of savings that would result when the opportunities are implemented in a package.

9.5.2.3 Regeneration of WPL and Bright Dipping Liquors

Acid cleaning, or pickling, is often used to remove contaminants from the workpiece using an acid. Acid pickling is used to remove oxides (rust), scale, or tarnish as well as to neutralize any base remaining on the parts. Acid pickling uses aqueous solutions of sulfuric, hydrochloric, phosphoric,

TABLE 9.14

Summary of Recommended Pollution Prevention Opportunities

		ANNUAL WASTE REDUCTION				
Pollution Prevention Opportunity	**Waste Reduced**	**Quantity (lb/yr)**	**Percent**	**Net Annual Savings**	**Implementation Cost**	**Simple Payback (yr)**
Install a propane-fired sludge drying oven to reduce the volume and weight of the sludge that is generated in the onsite wastewater treatment system and shipped offsite	Wastewater treatment sludge	928,200	60	$141,150	$66,200	0.5
Utilize a trash compactor to reduce the volume of municipal trash shipped offsite thereby reducing disposal costs.	Miscellaneous solid waste	0[a]	0	12,810	15,000	1.2
Remove the poor quality length of each coil of raw material prior to forming in the mills. Current practice is for the entire length of raw material to undergo the normal forming and welding operations, regardless of the quality. The current procedure leads to unnecessary expenditures of welding gases, worker labor, and energy.	n/a	—	—	9,300	0	0
Automate the addition of caustic to the wastewater treated in the onsite wastewater treatment plant in order to reduce caustic purchases and reduce labor costs.	n/a			12,620	12,600	1.0

Source: Jendrucko RJ, Myers JA, Thomas TM, Looby GP. *Pollution Prevention Assessment for Manufacturer of Stainless Steel Pipes and Fittings.* Report no. EPA-600/S-95/017. The U.S. Environmental Protection Agency, Cincinnati, OH, August 1995 [13].

[a] A significant volume reduction would occur.

and/or nitric acids. For instance, most carbon steel is pickled in sulfuric or hydrochloric acids, although hydrochloric acid can embrittle certain types of steel and is used only in specific applications. In the pickling process, the workpiece generally passes from the pickling bath through a series of rinses and then onto plating. Acid pickling is similar to acid cleaning, but is more commonly used to remove the scale from semifinished mill products whereas acid cleaning is usually used for nearfinal preparation of metal surfaces prior to finishing.

9.5.2.3.1 *Copper and Alloys*

Straight electrolytic recovery as described earlier is highly effective in many copper pickling and milling solutions, including sulfuric acid, cupric chloride, and ammonium chloride solutions. Solutions based on hydrogen peroxide are generally regenerated best by crystallization and removal of copper sulfate with the crystals sold as by-products or redissolved for further treatment by electrolytic metal recovery [14,15].

Highly concentrated bright dipping nitric/sulfuric acids are a difficult challenge for regeneration because of the small quantities (5–25 gallons) used and the high drag-out losses. Regeneration is possible by distillation of nitric acid and removal of copper salts; however, the economics are usually not favorable.

9.5.2.3.2 *Sulfuric and Hydrochloric Acids*

Both sulfuric and hydrochloric acids are commonly used for cleaning steel. Sulfuric acid can be regenerated by crystallizing ferrous sulfate. Hydrochloric acid can be recovered by distilling off the acid and leaving behind iron oxide. These techniques have been used for many years in large facilities. The economics of these processes, however, are usually not favorable for smaller facilities [14,15].

WPLs from these operations can often be of use in sanitary waste treatment systems for phosphate control and sludge conditioning. Some industrial firms can use spent process waste from the pickling operation. Iron in the waste is used as a coagulant in wastewater treatment systems [14,15].

9.5.2.4 Engineering Calculations for the Determination of Hydrochloric Acid Requirements

On submerging steel in acid, two main reactions take place. Both use hydrogen chloride and both produce ferrous chloride:

$$FeO + 2HCl \rightarrow FeCl_2 + H_2O, \tag{9.3}$$

$$Fe + 2HCl \rightarrow FeCl_2 + H_2. \tag{9.4}$$

Equation (9.4) shows the chemical reaction where the acid reacts with the base metal (the steel under the scale). This reaction (Equation 9.4) is quite slow and produces hydrogen gas as a by-product, which accounts for the bubbling and foaming in the tank. The ferrous chloride produced is the by-product, which can be sold for cost recovery.

Another reaction shown by Equation (9.3) is the reaction of acid reacts with the scale (FeO) itself. This reaction is faster than the first and produces both ferrous chloride and water as by-products.

It is in the plant's interest to discourage reaction (9.4) from taking place, since the plant is only interested in removing the scale. This is done by leaving the steel in the acid only as long as it is absolutely necessary to remove the scale and by adding a chemical (inhibitor) to the acid, which inhibits this reaction to the minimum.

By applying the chemistry indicated by Equations (9.3 and 9.4), an environmental engineer can calculate the acid requirement to pickle 1 ton of steel, assuming an average iron loss of 0.35%, which is typical for pickling strip steel.

$$FeO + 2HCl \rightarrow FeCl_2 + H_2O,$$

$$(72) + (73) \rightarrow (127) + (18).$$

(9.3)

73 lb of hydrogen chloride (not acid) is required to react with 56 lb of iron in the scale. Since 1 ton of steel has 0.35% of iron to be removed as scale, which is

$$(2,000 \times 0.35)/100 = 7 \text{ lb}$$

it follows that if 56 lb Fe requires 73 lb HCl, then 7 lb Fe requires

$$(73 \times 7)/56 = 9.13 \text{ lb HCl (pure chemical).}$$

The hydrochloric acid commercially available is shipped at 32 lb HCl (gas) per 100 lb of aqueous (in water) solution (32% concentration); it follows then that if 32 lb HCl is dissolved in 100 lb of solution, then 9 lb HCl is dissolved in

$$(100 \times 9.13)/32 = 28.5 \text{ lb of solution.}$$

One gallon of 32% HCl acid weighs approximately 9.7 lb; therefore, the required volume of 32% hydrochloric acid solution is calculated by the following equation:

$$(28.5)/9.7 = 2.94 \text{ gallons of 32% hydrochloric acid.}$$

The above constitutes the theoretical or stoichiometric amount of liquid hydrochloric acid needed to pickle 1 ton of steel. Stoichiometric means, as per chemical formula, using absolutely pure materials with no losses [16]. In practice, it is not possible to use up all the acid in the pickle tank if pickling is to be complete in any acceptable time. Depending on the pickling equipment, between 70% and 80% of the free acid will be used up in dissolving the scale, and 20%–30% will remain as "free" acid in the SPL [16].

Accordingly, between 3.6 and 4.2 gallons of 32% hydrochloric acid are needed to pickle 1 ton of steel (assuming that the iron loss is 0.35%).

9.5.2.5 Engineering Calculations for the Determination of Ferrous Chloride Recovered

The amount of ferrous chloride produced as a by-product during the pickling of 1 ton of steel can also be determined in the same manner [16]. Since a molecular weight of 56 is equal to 7 lb in the formula ratio, a molecular weight of 127 is equal to $(7 \times 127)/56 = 15.88$ lb. 5.88 lb $FeCl_2$ is produced per ton of steel pickled. The molecular weight of water is 18. Therefore, $(7 \times 18)/18 = 2.25$ lb of water is produced per ton of steel pickled.

9.6 CASE STUDIES

9.6.1 Closed-Loop System for the Treatment of WPL

Peterson [17] has demonstrated the feasibility of converting ferrous sulfate in the waste pickle liquor (WPA) to marketable ferric oxide (Fe_2O_3) when ferrous sulfate is obtained by low-temperature crystallization from sulfuric acid WPL generated by the acid cleaning of steel surfaces. A closed-loop system consisting of a crystallizer, ion exchange unit, oxidizer, and hydrolyzer has been used by Peterson [17]. All acids are recycled, and the net effect is that ferrous sulfate from sulfuric acid pickling WPL is consumed, and marketable ferric oxide is produced. The ferrous sulfate solution was contacted with hydrogen ion exchange resin in a continuous ion exchange unit. Removal of ferrous ion was 90%. About 11%–15% by weight sulfuric acid was generated for recycle to pickling. The resin was regenerated with 4M nitric acid, and a ferrous-ferric nitrate solution was produced. This product was heated to 180°C and contacted with air to achieve complete oxidation to the ferric state and to oxidize any by-product nitrogen oxide to nitric acid. The nitrate solution was then hydrolyzed to ferric oxide and nitric acid in a continuous coil autoclave at 205°C. The nitric acid was about 20% by weight and can be recycled to the ion exchange unit. After drying, the ferric oxide was about 99% pure.

9.6.2 A Promising Process System Consisting of Limestone-Packed Bed Filtration, Physicochemical Sequencing Batch Reactor (PC-SBR) Neutralization, and Membrane Filtration

Zhang et al. [18] have developed a new technique for treating pickling wastewater discharged from the steel industry using limestone-packed bed filtration, physicochemical sequencing batch reactor (PC-SBR) neutralization, and membrane filtration (MF) in a full scale process system. The morphological and chemical properties were examined through scanning electron microscopy (SEM), X-ray diffraction (XRD), energy dispersive X-ray spectroscopy (EDS) and inductively coupled plasma optical emission spectroscopy (ICP-OES) etc. The results showed that by increasing the Zeta potential of the pickling wastewater, the membrane fouling could be controlled. Their process system shown in Figure 9.7 [18] could assist in understanding of the unit processes, and conducting further research for improvement of pickling wastewater treatment.

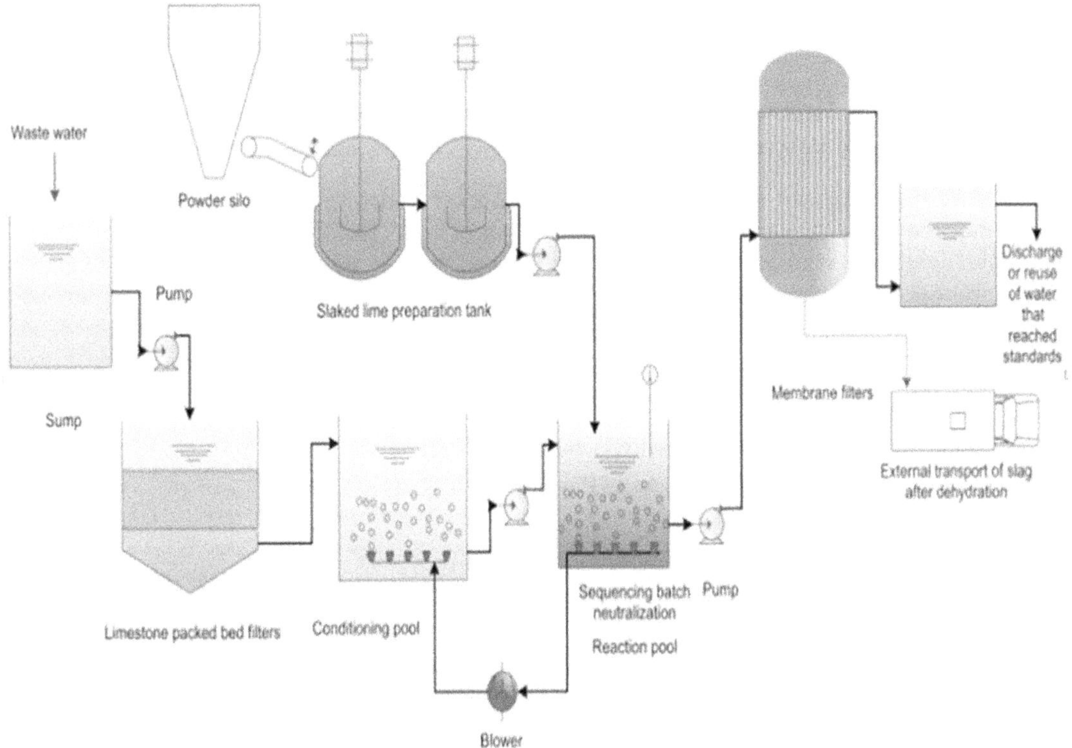

FIGURE 9.7 A process system for treatment of waste pickle liquor from steel industry [18].

9.6.3 TREATMENT OF BATTERY RECYCLING WASTEWATER CONTAINING SULFURIC ACID AND NICKEL SULFATE BY MEMBRANE PROCESSES

Recycling waste lithium batteries generates a large amount of wastewater which contains high concentration of heavy metals and acids, and causes severe environmental hazards. Wu et al. [19] have separated, recovered, and utilized Ni^{2+} and H_2SO_4 in the wastewater using a combined process of diffusion dialysis (DD) and electrodialysis (ED). In their DD process, the acid recovery rate and Ni^{2+} rejection rate could reach 75.96% and 97.31%, respectively, with a flow rate of 300 L/h and a W/A flow rate ratio of 1:1. In the ED process, the recovered acid from DD is concentrated from 43.1 to 150.2 g/L H_2SO_4 by the two-stage ED, which could be used in the front-end procedure of battery recycling process. Zhu et al. [20] have concluded that the DD-ED treatment of battery wastewater may have industrial application prospects for the recycling and utilization of Ni^{2+} and H_2SO_4 present in the acidic wastewater.

9.6.4 SUSTAINABLE FLAT- MEMBRANE PLATE-AND-FRAME MODULE FOR SPENT ACID REGENERATION AND METAL ION RECOVERY

Perveen et al. [21] have provided techno-economical insights for acid regeneration and metal recovery from spent acidic wastewater by a DD plate-and-frame module using Quaternized Polyepichlorohydrin - Polyacrylonitrile (QPECH-PAN) membranes. Mechanical characterization reveals that the QPECH-PAN membranes possess mechanical strengths (tensile strength: 329.56 MPa), sheet resistivity (6.11 Ω cm^2) and conductivity (0.16 S/cm^2). Percent acid recovery and metal ion recovery ratios were found to be 72% and 48% respectively, and separation factors were 126.8 and 84.57 respectively. The QPECH-PAN membrane's techno-economic feasibility was also reported within the context of a textile industry processing up to 5,500 kg/d of acidic wastewater. Saving on H_2SO_4 and NaOH, as well as an OPEX against a semicontinuous acid neutralizer was reported.

9.6.5 SUSTAINABLE SELECTIVE RECOVERY OF SULFURIC ACID AND VANADIUM FROM ACIDIC WASTEWATER WITH TWO-STEP SOLVENT EXTRACTION

Zhu et al. [20] investigated a two-stage extraction process to recover sulfuric acid and vanadium from simulated acid solution and titanium dioxide waste acid (TDWA). The stripping efficiencies of sulfuric acid and vanadium were more than 90% and 98%, respectively. The real waste acid of TDWA was operated to extract and separate sulfuric acid and vanadium with the same parameters, in which the characteristics of high extraction efficiency and good selectivity were obtained. The technique may separate and recover valuable components from TDWA.

9.6.6 SELECTIVE EXTRACTION OF ZINC (II) OVER IRON (II) FROM SPENT HYDROCHLORIC ACID PICKLING EFFLUENTS BY LIQUID-LIQUID EXTRACTION

Mansur et al. [22] studied the selective removal of zinc (II) over iron (II) by liquid-liquid extraction from spent hydrochloric acid pickling effluents produced by the zinc hot-dip galvanizing industry at room temperature. The following extraction results were documented using two extractants: (a) TBP (tri-n-butyl phosphate), and (b) Cyanex 301 [bis(2,4,4-trimethylpentyl) dithiophosphinic acid]. Excellent separation results were obtained for extractants TBP and Cyanex 301. Around 92.5% of zinc and 11.2% of iron were extracted from effluent 1 (containing 70.2 g/L of Zn, 92.2 g/L of Fe and pH 0.6) in one single contact using 100% (v/v) of TBP. With Cyanex 301, around 80%–95% of zinc and less than 10% of iron were extracted from effluent 2 (containing 33.9 g/L of Zn, 203.9 g/L of

Fe and 2M HCl) at pH 0.3–1.0. Reactions for the extraction of zinc with TBP and Cyanex 301 from hydrochloric acid solution were proposed by the researchers [22].

9.6.7 NEUTRALIZATION OF RED MUD WITH PICKLING WASTE LIQUOR

Alumina refinery generates 'red mud' or 'bauxite residue' which is an alkaline waste with a pH of 10.5–12.5. Red mud poses serious environmental problems such as alkali seepage in groundwater and alkaline dust in air. Rai et al. [23] investigate the feasibility of using highly acidic pickling waste liquor (WPL) to neutralize the hazardous highly alkaline red mud, hoping to make both the wastes less hazardous and safe for disposal. Under the optimized operational conditions, neutralization to pH value of 7 can be achieved by mixing the red mud and WPL. About 25%–30% of the total soda from the red mud is being neutralized, and alkalinity is getting reduced by 80%–85%. Their research data presented will be useful in view of environmental concern of red mud disposal [23].

9.6.8 MINIMIZING THE SPL CREATION IN A HYDROCHLORIC ACID PICKLING PROCESS THROUGH THE SYNERGISTIC CORROSION INHIBITION OF OP-10 AND POTASSIUM IODIDE

Tang et al. [24] propose a strategy for minimizing the SPLs generation through the synergistic corrosion inhibition of OP-10 and potassium iodide, in turn, facilitating a cleaner production process for acid pickling of metals with a high-concentration solution (6.0 mol/L) of hydrochloric acid. A synergistic effect was identified by Tang et al. [24] when KI and OP-10 were present in suitable proportions to form a complex mixed-type inhibitor with a compact film on the metal surface, thus providing an effective protection for the metal in the acidic solutions. The end results are an increase in the usage life of pickling liquor and a decrease in the production of SPL.

9.7 REVIEW OF RECENT ADVANCES

9.7.1 A REVIEW ON METHODS OF REGENERATION OF SPENT PICKLING SOLUTIONS FROM STEEL PROCESSING

Regel-Rosocka [25] reviews various techniques of regeneration of SPL, including (a) the methods with acid recovery, such as DD, electrodialysis, membrane electrolysis and membrane distillation, evaporation, precipitation, and spray roasting; and (b) the methods with acid and metal recovery, such as ion exchange, retardation, crystallization solvent, and membrane extraction. The advantages and disadvantages of the techniques are presented, discussed, and confronted with the best available techniques (BAT) requirements. The BAT are electrodialysis, DD, and crystallization; however, in practice spray roasting and retardation/ion exchange are applied most frequently for SPL regeneration. Solvent extraction, non-dispersive solvent extraction and membrane distillation are well investigated and developed.

9.7.2 A REVIEW ON METHODS OF RECOVERY OF ACID(S) FROM SPL OF STEEL INDUSTRY AND ELECTROPLATING INDUSTRIES

Ghare et al. [26] present an overview on different aspects of generation and treatment of WPL with an aim to recovery of acid for recycling. Their investigation involves the use of hydrofluoric acid-Nitric acid mixture for stainless steel pickling. Pickling solutions are spent when acid concentration in pickling solutions decreases by 75%–85%, which also has metal content up to 150–250 g per cubic dm. WPL from steel pickling in hot-dip galvanizing plants contains zinc (II), iron, traces of lead, chromium, and other heavy metals (max. 500 mg per cubic dm) and hydrochloric acid. Zinc (II) passes to the spent solution after dissolution of this metal from zinc (II)-covered racks, chains and baskets used for transportation of galvanized elements. Unevenly covered zinc layers are usually removed in another pickling bath. Due to this, zinc (II) concentration increases even up to 110 g per cubic dm, while iron content may reach or exceed even 80 g per cubic dm in the same solution.

Agrawal and Sahu [27] provide a critical review on the acidic waste streams generated from steel and electroplating industries particularly from WPL and spent bleed streams (SBS). Various aspects on the WPL and SBS generation and their treatment either for the acid recovery and reuse or for waste disposal are discussed. Major stress is laid upon the hydrometallurgical methods such as solvent extraction.

9.8 SUMMARY

1. The basic unit operations/processes required for treating acid pickling wastewater are (a) neutralization with NaOH and/or lime to increase the pH and (b) physicochemical methods, such as chemical coagulation, precipitation, clarification (sedimentation or DAF), and filtration; to remove BOD_5, COD, and iron.
2. The iron present in the wastewater appears in the form of ferrous ion (Fe^{2+}), which is soluble in water, and can be recovered as a by-product.
3. Ferrous ion can be removed either by oxidation to ferric (Fe^{3+}) or by crystallization.
4. Sulfuric acid, hydrochloric acid, and other acids, individually or in combination, can be used for acid pickling of metals, although sulfuric and hydrochloric acids are commonly used for cleaning steel.
5. Sulfuric acid can be regenerated by crystallizing ferrous sulfate.
6. Hydrochloric acid can be recovered by distilling off the acid and leaving behind iron oxide.
7. In the hydrochloric acid pickling process, ferrous chloride can also be recovered as a by-product.
8. Waste minimization and pollution prevention are very important for saving overall manufacturing cost in a steel product manufacturing plant.
9. Treatment of pickling wastewater by neutralization/aeration/clarification gave satisfactory results. The sludge formed must be disposed of after dewatering in a filter-press, a horizontal belt filter, or a centrifuge. An equalization tank is required in order to compensate the effects of intermittent discharges. The treated wastewater can be recycled for use in the process or discharged to the river.
10. A sedimentation clarifier, a DAF clarifier, or a DAFF can be used for the clarification of pickling wastewater [28].
11. Both air emission control and sludge disposal are extremely important in a steel acid pickling plant [29–31].
12. In a steel product manufacturing plant involving acid pickling operation, disposal of hazardous metal sludge is the most expensive engineering task, and treatment of pickling liquor and rinsewater is the second most expensive engineering task.

GLOSSARY OF ACID PICKLING

Acid cleaning (metal work) It is a metal cleaning process involves the use of acid to remove inorganic contaminant not removable by other primary cleaning solutions from metal surface. Acid cleaning has its limitations in that it is difficult to handle because of its corrosiveness and is not applicable to all steels. The hydrogen from the acid reacts with the metal surface and makes it brittle and crack. Because of its high reactivity to treatable steels, acid concentrations and solution temperatures must be kept under control to assure desired pickling rates.

Acid pickling Technically speaking, acid pickling is the treatment of metallic surfaces with acid in order to remove impurities, stains, rust, or scale with a solution called pickle liquor, containing strong mineral acids, before subsequent processing (i.e., extrusion, rolling, painting, galvanizing, or plating) with tin or chromium. The two acids commonly used are hydrochloric acid and sulfuric acid.

Concentrated pickling liquor (CPL) In the spray roasting acid regeneration process, or similar, the WPL at 2%–4% HCl in contact with the hot flue gas from the spray roaster, which vaporizes some of the water in the WPL, the WPL then becomes CPL.

Hydrochloric acid pickling It is a chemical acid pickling process involving the use of hydrochloric acid as the main ingredient of a pickle liquor for removing impurities, stains, rust, or scale from a metal surface. If the metal surface is iron and the rust is iron oxide, ferrous chloride is formed from the reaction of iron oxides with hydrochloric acid.

Pickle liquor, or pickling solution (metal work) (a) It is usually an acid solution, such as sulfuric acid, hydrochloric acid, or nitric acid, etc. which is used in a metal surface treatment process (known as pickling process) for removal of superficial impurities, such as rust or scale, from metal; (b) Sometimes an alkaline solution can also be used as a pickle liquor for cleaning the metal surface; (c) Pickle liquor may contain additives such as wetting agents and corrosion inhibitors; (d) Pickling is sometimes called acid cleaning if descaling is not needed; (e) Pickling solutions are spent when acid concentration in pickling solutions decreases by 75%–85%, and/or its metal content increases up to 150–250 g per cubic dm.

Pickling process (metal work) Pickling is a chemical metal surface treatment process that removes superficial impurities (such as rust, scale, corrosion products) from metal (such as, iron, copper, precious metals, aluminum alloys) using usually a strong acidic solution such as sulfuric acid, hydrochloric acid, or nitric acid, etc. although an alkaline solution can also be used for cleaning the metal surface.

Pickling sludge Pickling sludge is a solid waste product from the pickling process, and includes dust, undissolved rust or scale, or the waste sludge generated from a WPL treatment process, such as lime precipitation, neutralization, or filtration.

Red mud or bauxite residue It is a waste generated from alumina refinery which is highly alkaline in nature with a pH of 10.5–12.5. Red mud may become alkali seepage in ground water or alkaline dust in air posing serious environmental problems. Neutralization with acid or an acidic waste may make red mud less hazardous.

Sulfuric acid pickling It is a chemical acid pickling process involving the use of sulfuric acid as the main ingredient of a pickle liquor for removing impurities, stains, rust, or scale from a metal surface. If the metal surface is iron and the rust is iron oxide, ferrous sulfate is formed from the reaction of iron oxides with sulfuric acid.

Waste pickle liquor (WPL) or spent pickle liquor (SPL) It is a wastewater product from the pickling process. In steel industries, pickling process removes superficial impurities from metal surface using an acid solution (pickle liquor) and then generates WPL (or SPL) as a waste product when the pickle liquor is exhausted and unwanted. The composition of WPL varies depending on the types of metal treated, and the type of acid solution used (For instance, if hydrochloric acid is used to treat the iron surface, WPL will contain iron chlorides). WPL contains the dissolved metals, residual free acid, and some additives, such as wetting agents, corrosion inhibitors, and is considered a hazardous waste by the USEPA.

REFERENCES

1. Eroglu V, Erturk F. Treatment of acid pickling wastes of metals. In: Wang LK, Wang MHS (Eds), *Handbook of Industrial Waste Treatment*. Marcel Dekker, Inc., New York, pp. 293–306, 1991.
2. Barnes D, Forster CF, Hudrey SE (Eds). Surveys in industrial wastewater treatment. In *Manufacturing and Chemical Industries. The Treatment of Wastewaters from Steel Plants*, vol. 3. Longman Scientific and Technical, Essex, UK, 1987.

3. Eroglu V, Topacik, D, Ozturk, I. *Wastewater Treatment Plant for Cayirova Pipe Factory.* Environmental Engineering Department, Istanbul Technical University, Turkey, 1989.

4. Abwasser Technik, A.T.V. *Behandling von Beizereiabwassern und Abbeizen bei der Halbzeugfertigung,* pp. 4–7, 149–54, 1986.

5. Wikimedia Foundation, Inc. *Metal's Acid Pickling, Cleaning and Waste Products,* June 2008. Available at: https://en.wikipedia.org/wiki/Pickling_(metal)

6. U.S. EPA. *Steel Pickling.* The U.S. Environmental Protection Agency, Washington, DC, June 2008. Available at: https://www.epa.gov/tri/TWebHelp/WebHelp/hcl_section_3_1_4_steel_pickling.htm

7. Wang LK, Hung YT, Lo HH, Yapijakis C (Eds). *Hazardous Industrial Waste Treatment.* CRC Press/Taylor and Francis Group, Boca Raton, FL, p. 516, 2007.

8. Wang LK, Shammas, NK, Hung YT (Eds). *Advances in Hazardous Industrial Waste Treatment.* CRC Press/Taylor and Francis Group, Boca Raton, FL, 2009.

9. Wang LK, Shammas NK, Hung YT (Eds). *Waste Treatment in the Metal Manufacturing, Forming, Coating and Finishing Industries.* CRC Press/Taylor and Francis Group, Boca Raton, FL, 2009.

10. Wang LK, Hung, YT, Shammas NK (Eds). *Physicochemical Treatment Processes.* Humana Press, Totowa, NJ, p. 723, 2005.

11. Wang LK, Hung, YT, Shammas NK (Eds). *Advanced Physicochemical Treatment Processes.* Humana Press, Totowa, NJ, p. 690, 2006.

12. Wang LK, Hung YT, Shammas NK (Eds). *Advanced Physicochemical Treatment Technologies.* Humana Press, Totowa, NJ, p. 710, 2007.

13. Jendrucko RJ, Myers JA, Thomas TM, Looby GP. *Pollution Prevention Assessment for Manufacturer of Stainless Steel Pipes and Fittings.* Report no. EPA-600/S-95/017. The U.S. Environmental Protection Agency, Cincinnati, OH, August 1995.

14. UIUC. *Metal Finishing Industry—Pre-Finishing Operations.* University of Illinois, Urbana, IL, June 2008. Available at: www.wmrc.uiuc.edu/info/1ibrary_docs/manuals/finishing/prefinop.htm

15. Steward FA, McLay WJ. *Waste Minimization and Alternate Recovery Technologies.* Alcoa Separations Technology, Inc., Warrendale, PA, 1985.

16. Hasler F, Stone N. *Hydrochloric Acid Pickling.* Esco Engineering, Kingsville, ON, Canada. pp. 21–25, 1997.

17. Wang LK, Kurylkop L, Wang MHS. *Sequencing Batch Liquid Treatment.* U.S. patent no. 5354458. U.S. Department of Commerce, Office of Patents and Trademarks, Washington, DC, October 11, 1994.

18. Wang LK, Kurylkop, L. *Liquid Treatment System with Air Emission Control.* U.S. patent no. 5399267. U.S. Department of Commerce, Office of Patents and Trademarks, Washington, DC, March 21, 1995.

19. University of *Massachusetts. Sulfuric Acid and Fuming Sulfuric Acid.* University of Massachusetts, Toxics Use Reduction Institute, Lowell, MA, 2008. Available at: www.turi.org

20. Roth-Evans W. *Facing the New Emission Standards for Hydrochloric Acid Pickling.* Trinity Consultants, Inc., Chesterfield, MS, 2008.

21. Peterson J. *Closed Loop System for the Treatment of Waste Pickle Liquor.* Technical Report No. EPA/600/2-77/127. U.S. Environmental Protection Agency, Washington, DC, 2004. Available at: https://cfpub.epa.gov/si/si_public_comments.cfm

22. Zhang J, Chen G, Ma Y, Xu M, Qin S, Liu X, Feng H, Hou L. Purification of pickling wastewater from the steel industry using membrane filters: Performance and membrane fouling. *Environ Eng Res.* 2022;27(1). https://doi.org/10.4491/eer.2020.486

23. Wu S, Zhu H., Wu Y, Li S, Zhang G, Zhiwei Miao Z. Resourceful treatment of battery recycling wastewater containing H_2SO_4 and $NiSO_4$ by diffusion dialysis and electrodialysis. *Membranes.* 2023 May 31;13(6):570. doi: 10.3390/membranes13060570.

24. Perveen S, Hussain SG, Ahmed MJ, Khawar R, Siraj TB, Saleem, M. A Viable and sustainable flat-membrane plate-and-frame module for spent acid regeneration and metal ion recovery. *Heliyon.* 2023 July 17;9(8):e18344. doi: 10.1016/j.heliyon.2023.e18344

25. Zhu X, Ma C, Li W. Sustainable selective recovery of sulfuric acid and vanadium from acidic wastewater with two-step solvent extraction. *ACS Omega.* 2023 July 19;8(30):27127–38. doi: 10.1021/acsomega.3c02180

26. Mansur MB, Rocha SD, Magalhães FS, dos Santos Benedetto S. Selective extraction of zinc (II) over iron (II) from spent hydrochloric acid pickling effluents by liquid-liquid extraction. *J Hazard Mater.* 2008 Feb 11;150(3):669–78. doi: 10.1016/j.jhazmat.2007.05.019.

27. Rai S, Wasewar KL, Lataye DH, Mishra RS, Puttewar SP, Chaddha MJ, Mahindiran P, Mukhopadhyay J. Neutralization of red mud with pickling waste liquor using Taguchi's design of experimental methodology. *Waste Manag Res.* 2012 Sep;30(9):922–30. doi: 10.1177/0734242X12448518.

28. Tang B, Su W, Wang J, Fu F, Yu G, Zhang J. Minimizing the creation of spent pickling liquors in a pickling process with high-concentration hydrochloric acid solutions: mechanism and evaluation method. *J Environ Manage.* 2012 May 15;98:147–54. doi: 10.1016/j.jenvman.2011.12.027.
29. Regel-Rosocka M. A review on methods of regeneration of spent pickling solutions from steel processing. *J Hazard Mater.* 2010 May 15;177(1–3):57–69. doi: 10.1016/j.jhazmat.2009.12.043.
30. Ghare NY, Wani KS, Patil VS. A review on methods of recovery of acid(s) from spent pickle liquor of steel industry. *J Environ Sci Eng.* 2013 Apr;55(2):253–266.
31. Agrawal A, Sahu KK. An overview of the recovery of acid from spent acidic solutions from steel and electroplating industries. *J Hazard Mater.* 2009 Nov 15;171(1–3):61–75. doi: 10.1016/j.jhazmat.2009.06.099.

10 Advances in the Treatment and Management of Metal Finishing Industry Wastes

Mu-Hao Sung Wang, Nazih K. Shammas, and Lawrence K. Wang

ACRONYMS

BAT	Best available technology
BDL	Below detection limit
BPT	Best practicable control technology
CFR	Code of Federal
CWA	Clean Water
DAF	Dissolved air flotation
HSWA	Hazardous and Solid Wastes Amendments
MF	membrane filtration
NPDES	National Pollutant Discharge Elimination System
NSPS	New source performance standards
PFAS	Per- and polyfluoroalkyl substances
POTW	Publicly-owned treatment works
PSES	Pretreatment standards for existing sources
PSNS	Pretreatment standards for new sources
RCRA	Resource Conservation and Recovery Act
SIC	Standard industrial classification
TTO	Total toxic organics
U.S. EPA	U.S. Environmental Protection Agency
UF	Ultrafiltration
USD	US dollars

10.1 INDUSTRY DESCRIPTION

The metal finishing industry is one of many industries subject to regulation under the Resource Conservation and Recovery Act (RCRA) [1,2] and the Hazardous and Solid Wastes Amendments (HSWA) [3]. The metal finishing industry has also been subject to extensive regulation under the Clean Water Act (CWA) [4]. Compliance with these regulations requires highly coordinated regulatory, scientific, and engineering analyses to minimize costs [5].

10.1.1 GENERAL DESCRIPTION

The metal finishing industry consists of 44 unit operations involving the machining, fabrication, and finishing of metal products (SIC groups 34–39). There are approximately 160,000 manufacturing facilities in the United States that are classified as being part of the metal finishing industry [6]. These facilities are engaged in the manufacture of a variety of products that are constructed primarily by

DOI: 10.1201/9781003541615-10"

using metals. The operations performed usually begin with a raw stock in the form of rods, bars, sheets, castings, forgings, and so on, and can progress to sophisticated surface finishing operations. The facilities vary in size from small job shops employing fewer than ten people to large plants employing thousands of production workers. Wide variations also exist in the age of the facilities and the number and type of operations performed within facilities. Because of the differences in size and processes, production facilities are custom-tailored to the specific needs of each plant. The possible variations in unit operations within the metal finishing industry are extensive. Some complex products could require the use of nearly all of the 44 possible unit operations, whereas a simple product might require only a single operation. Each of the 44 individual unit operations is listed below with a brief description [7].

1. *Electroplating* produces a thin coating of one metal on another by electrodeposition.
2. *Electroless plating* is a chemical reduction process that depends on the catalytic reduction of a metallic ion in an aqueous solution containing a reducing agent and the subsequent deposition of metal without the use of external electric energy.
3. *Anodizing* is an electrolytic oxidation process that converts the surface of the metal to an insoluble oxide.
4. Chemical conversion coatings are applied to previously deposited metal or basis material for increased corrosion protection, lubricity, preparation of the surface for additional coatings, or formulation of a special surface appearance. This operation includes chromating, phosphating, metal coloring, and passivating.
5. *Etching and chemical milling* produce specific design configurations and tolerances on parts by controlled dissolution with chemical reagents or etchants.
6. *Cleaning* involves the removal of oil, grease, and dirt from the surface of the basis material using water with or without a detergent or other dispersing materials.
7. *Machining* is the general process of removing stock from a workpiece by forcing a cutting tool through the workpiece, removing a chip of basis material. Machining operations such as turning, milling, drilling, boring, tapping, planning, broaching, sawing and cutoff, shaving, threading, reaming, shaping, slotting, hobbing, filing, and chamfering are included in this definition.
8. *Grinding* is the process of removing stock from a workpiece using a tool consisting of abrasive grains held by a rigid or semirigid binder. The processes included in this unit operation are sanding (or cleaning to remove rough edges or excess material), surface finishing, and separating (as in cutoff or slicing operations).
9. *Polishing* is an abrading operation used to remove or smooth out surface defects (scratches, pits, tool marks, etc.) that adversely affect the appearance or function of a part. The operation usually referred to as buffing is included in the polishing operation.
10. *Barrel finishing* or tumbling is a controlled method of processing parts to remove burrs, scale, flash, and oxides, as well as to improve surface finish.
11. *Burnishing* is the process of finish sizing or smooth finishing a workpiece (previously machined or ground) by displacement, rather than removal, of minute surface irregularities. It is accomplished with a smooth point or line contact and fixed or rotating tools.
12. *Impact deformation* is the process of applying an impact force to a workpiece such that the workpiece is permanently deformed or shaped. Impact deformation operations include shot peening, forging, high-energy forming, heading, and stamping.
13. *Pressure deformation* is the process of applying force (at a slower rate than an impact force) to permanently deform or shape a workpiece. Pressure deformation includes operations such as roiling, drawing, bending, embossing, coining, swaging, sizing, extruding, squeezing, spinning, seaming, staking, piercing, necking, reducing, forming, crimping, coiling, twisting, winding, flaring, or weaving.

14. *Shearing* is the process of severing or cutting a workpiece by forcing a sharp edge or opposed sharp edges into the workpiece, stressing the material to the point of shear failure and separation.

15. *Heat treating* is the modification of the physical properties of a workpiece through the application of controlled heating and cooling cycles. Such operations as tempering, carburizing, cyaniding, nitriding, annealing, normalizing, austenizing, quenching, austempering, siliconizing, martempering, and malleablizing are included in this definition.

16. *Thermal cutting* is the process of cutting, slotting, or piercing a workpiece using an oxy-acetylene oxygen lance or electric arc cutting tool.

17. *Welding* is the process of joining two or more pieces of material by applying heat, pressure, or both, with or without filler material, to produce a localized union through fusion or recrystallization across the interface. Included in this process are gas welding, resistance welding, arc welding, cold welding, electron beam welding, and laser beam welding.

18. *Brazing* is the process of joining metals by flowing a thin, capillary thickness layer of nonferrous filler metal into the space between them. Bonding results from the intimate contact produced by the dissolution of a small amount of base metal in the molten filler metal without the fusion of the base metal. The term "brazing" is used where the temperature exceeds 425°C (800°F).

19. *Soldering* is the process of joining metals by flowing a thin, capillary thickness layer of nonferrous filler metal into the space between them. Bonding results from the intimate contact produced by the dissolution of a small amount of base metal in the molten filler metal without the fusion of the base metal. The term "soldering" is used where the temperature range falls below 425°C (800°F).

20. *Flame spraying* is the process of applying a metallic coating to a workpiece using finely powdered fragments of wire and suitable fluxes, which are projected together through a cone of flame onto the workpiece.

21. *Sand blasting* is the process of removing stock, including surface films, from a workpiece by the use of abrasive grains pneumatically impinged against the workpiece. The abrasive grains used include sand, metal shot, slag, silica, pumice, or natural materials such as walnut shells.

22. *Abrasive jet machining* is a mechanical process for cutting hard, brittle materials. It is similar to sand blasting but uses much finer abrasives carried at high velocities (150–910 m/s [500–3,000 ft/s]) by a liquid or gas stream. Uses include frosting glass, removing metal oxides, deburring, and drilling and cutting thin sections of metal.

23. *Electrical discharge machining* is a process that can remove metal with good dimensional control from any metal. It cannot be used for machining glass, ceramics, or other nonconducting materials. Electrical discharge machining is also known as spark machining or electronic erosion. The operation was developed primarily for machining carbides, hard nonferrous alloys, and other hard-to-machine materials.

24. *Electrochemical machining* is a process based on the same principles used in electroplating, except that the workpiece is the anode, and the tool is the cathode. Electrolyte is pumped between the electrodes and a potential applied, resulting in rapid removal of metal.

25. *Electron beam machining* is a thermoelectric process in which heat is generated by high-velocity electrons impinging the workpiece, converting the beam into thermal energy. At the point where the energy of the electrons is focused, the beam has sufficient thermal energy to vaporize the material locally. The process is generally carried out in a vacuum. The process results in X-ray emission, which requires that the work area be shielded to absorb radiation. At present, the process is used for drilling holes as small as 0.05 mm (0.002 in.) in any known material, cutting slots, shaping small parts, and machining sapphire jewel bearings.

26. *Laser beam machining* is the process of using a highly focused, monochromatic-collimated beam of light to remove material at the point of impingement on a workpiece. Laser

beam machining is a thermoelectric process, and material removal is largely accomplished by evaporation, although some material is removed in the liquid state at high velocity. Since the metal removal rate is very small, this process is used for such jobs as drilling microscopic holes in carbides or diamond wire drawing dies and for removing metal in the balancing of high-speed rotating machinery.

27. *Plasma arc machining* is the process of material removal or shaping of a workpiece by a high-velocity jet of high-temperature ionized gas. A gas (nitrogen, argon, or hydrogen) is passed through an electric arc causing it to become ionized and raising its temperatures in excess of 16,000°C (30,000°F). The relatively narrow plasma jet melts and displaces the workpiece material in its path.

28. *Ultrasonic machining* is a mechanical process designed to remove material using abrasive grains, which are carried in a liquid between the tool and the work and bombard the work surface at high velocity. This action gradually chips away minute particles of material in a pattern controlled by the tool shape and contour. Operations that can be performed include drilling, tapping, coining, and making openings in all types of dies.

29. *Sintering* is the process of forming a mechanical part from a powdered metal by fusing the particles together under pressure and heat. The temperature is maintained below the melting point of the basis metal.

30. *Laminating* is the process of adhesive bonding of layers of metal, plastic, or wood to form a part.

31. *Hot dip coating* is the process of coating a metallic workpiece with another metal by immersion in a molten bath to provide a protective film. Galvanizing (hot dip zinc) is the most common hot dip coating.

32. *Sputtering* is the process of covering a metallic or nonmetallic workpiece with thin films of metal. The surface to be coated is bombarded with positive ions in a gas discharge tube, which is evacuated to a low pressure.

33. *Vapor plating* is the process of decomposition of a metal or compound upon a heated surface by reduction or decomposition of a volatile compound at a temperature below the melting point of either the deposit or the basis material.

34. *Thermal infusion* is the process of applying fused zinc, cadmium, or other metal coating to a ferrous workpiece by imbuing the surface of the workpiece with metal powder or dust in the presence of heat.

35. *Salt bath descaling* is the process of removing surface oxides or scale from a workpiece by immersion of the workpiece in a molten salt bath or a hot salt solution. The work is immersed in the molten salt (temperatures range from 400°C to 540°C [750°F–1,000°F]), quenched with water, and then dipped in acid. Oxidizing, reducing, and electrolytic baths are available, and the particular type needed depends on the oxide to be removed.

36. *Solvent degreasing* is a process for removing oils and grease from the surfaces of a workpiece by the use of organic solvents, such as aliphatic petroleum, aromatics, oxygenated hydrocarbons, halogenated hydrocarbons, and combinations of these classes of solvents. However, ultrasonic vibration is sometimes used with liquid solvent to decrease the required immersion time with complex shapes. Solvent cleaning is often used as a precleaning operation, such as prior to the alkaline cleaning that precedes plating, as a final cleaning of precision parts, or as a surface preparation for some painting operations.

37. *Paint stripping* is the process of removing an organic coating from a workpiece. The stripping of such coatings is usually performed with caustic, acid, solvent, or molten salt.

38. *Painting* is the process of applying an organic coating to a workpiece. This process includes the application of coatings such as paint, varnish, lacquer, shellac, and plastics by methods such as spraying, dipping, brushing, roll coating, lithographing, and wiping. Other processes included in this unit operation are printing, silk screening, and stenciling.

39. *Electrostatic painting* is the application of electrostatically charged paint particles to an oppositely charged workpiece followed by thermal fusing of the paint particles to form a cohesive paint film. Both waterborne and solvent-borne coatings can be sprayed electrostatically.
40. *Electropainting* is the process of coating a workpiece by either making it anodic or cathodic in a bath that is generally an aqueous emulsion of the coating material. The electrodeposition bath contains stabilized resin, dispersed pigment, surfactants, and sometimes organic solvents in water.
41. *Vacuum metalizing* is the process of coating a workpiece with metal by flash-heating metal vapor in a high-vacuum chamber containing the workpiece. The vapor condenses on all exposed surfaces.
42. *Assembly* is the fitting together of previously manufactured parts or components into a complete machine, unit of a machine, or structure.
43. *Calibration* is the application of thermal, electrical, or mechanical energy to set or establish reference points for a component or complete assembly.
44. *Testing* is the application of thermal, electrical, or mechanical energy to determine the suitability or functionality of a component or complete assembly.

Table 10.1 presents an industry summary for the metal finishing industry including the total number of subcategories, number of subcategories studied, and the type and number of dischargers.

10.1.2 SUBCATEGORY DESCRIPTIONS

The primary purpose of subcategorization is to establish groupings within the metal finishing industry such that each subcategory has a uniform set of quantifiable effluent limitations. Several bases were considered in establishing subcategories within the metal finishing industry. These included the following:

1. Raw waste characteristics
2. Manufacturing processes
    1. Raw materials
    2. Product type or production volume
    3. Size and age of facility
    4. Number of employees
    5. Water usage
    6. Individual plant characteristics

TABLE 10.1
Metal Finishing Industry Summary

Item	Number
Total subcategories	51
Subcategories studied	28
Discharges in industry	98,418
Direct	20,632
Indirect	77,586
Zero discharge	200

Source: U.S. EPA. *Treatability Manual, Volume II. Industrial Descriptions.* Report no. EPA-600/2-82-001b. U.S. Environmental Protection Agency, Washington, DC, September 1981 [7].

After these subcategorization bases were evaluated, raw waste characterization was selected as the basis for subcategorization. The raw waste characterization is divided into two components: inorganic and organic wastes. These components are further subdivided into the specific types of wastes that occur within the components. Inorganics include common metals, precious metals, complex metals, hexavalent chromium, and cyanide. Organics include oils and solvents.

Table 10.2 lists the unit operations associated with each of the seven industry subcategories (raw waste characteristics). Common metals are found in the raw waste of all 44 unit operations. Precious metals are found in only seven unit operations; complexed metals are found in three unit operations; hexavalent chromium is found in seven unit operations; and cyanide is found in eight unit operations. Within the organics, oils are found in 22 unit operations, and solvents are found in nine unit operations. A unit operation will often be found in more than one subcategory.

TABLE 10.2
Subcharacterization of Unit Operations

Industry Subcategory (Raw Waste Characteristics)	Unit Operations	
Common metals		
All 44 unit operations		
Precious metals		
Electroplating	Etching	Burnishing
Electroless plating	Cleaning	
Conversion coating	Polishing	
Complexed metals		
Electroless plating		
Etching		
Cleaning		
Hexavalent chromium		
Electroplating	Etching	Electrostatic painting
Anodizing	Cleaning	
Conversion coating	Tumbling	
Cyanide		
Electroplating	Cleaning	Heat treating
Electroless plating	Tumbling	Electrochemical machining
Conversion coating	Burnishing	
Oils		
Cleaning	Pressure deformation	Solvent degreasing
Machining	Shearing	Paint stripping
Grinding	Heat treating	Painting
Polishing	Other abrasive jet machining	Assembly
Tumbling	Electrostatic painting	Calibration
Burnishing	Electrical Discharge machining	Testing
Impact deformation	Electrochemical machining	
Solvents		
Cleaning	Solvent degreasing	Electrostatic painting
Heat treating	Paint stripping	Electropainting
Electrochemical machining	Painting	Assembly

Source: U.S. EPA. *Treatability Manual, Volume II. Industrial Descriptions.* Report no. EPA-600/2-82-001b. U.S. Environmental Protection Agency, Washington, DC, September 1981 [7].

10.2 WASTEWATER CHARACTERIZATION

In this section, the water uses in the metal finishing industry are presented, and the waste constituents are identified and quantified.

Water is used for rinsing workpieces, washing away spills, air scrubbing, processing fluid replenishment, cooling and lubrication, washing equipment and workpieces, quenching, spray booths, and assembly and testing. Unit operations with significant water usage include electroplating, electroless plating, anodizing, conversion coating, etching, cleaning, machining, grinding, tumbling, heat treating, welding, sandblasting, salt bath descaling, paint stripping, painting, electrostatic painting, electroplating, and testing. Unit operations with zero discharge are electron beam machining, laser beam machining, plasma arc machining, ultrasonic machining, sintering, sputtering, vapor plating, thermal infusion, vacuum metalizing, and calibration [7].

Table 10.3 displays the ranges of flows in the metal finishing industry. Approximately 81% of the plants have flows of between 1.9 and 57 m^3/h (67–2,000 ft^3/h). For those plants with common metals waste streams, the average contribution of these streams to the total wastewater flow within a particular plant is 62.4% (range of 0.007%–100%). All of the plants have a waste stream requiring common metals treatment.

Of the plants, 4.8% have production processes that generate precious metals wastewater. The average precious metals wastewater flow is 21.5% of total plant flow.

The average contribution of the complexed metal streams to total plant flow is 22.2%. The percentage was computed from data for plants whose complex metal streams could be segregated from the total stream.

Of the plants, 42.5% have segregated hexavalent chromium waste streams. The average flow contribution of these waste streams to the total wastewater stream is 28.7%. At those plants with cyanide wastes, the average contribution of the cyanide-bearing stream to the total wastewater generated is 28.8% (range of 0.1%–100%). Of the plants, 31.2% have segregated cyanide-bearing wastes.

TABLE 10.3

Wastewater Flow Characterization of the Metal Finishing Industry

Flow of Plants (m³/h)	Percentage of Plants Represented by this Flow
<0.38	2.8
0.38–1.9	5.0
1.9–3.8	13
3.8–9.5	17
9.5–19	20.7
19–28	10.7
28–38	10.7
38–57	9.1
57–95	5.0
95–190	3.8
190–380	0.7
>380	1.5

Source: U.S. EPA. *Treatability Manual, Volume II. Industrial Descriptions.* Report no. EPA-600/2-82-001b. U.S. Environmental Protection Agency, Washington, DC, September 1981 [7].

Segregated oily wastewater is defined as oil waste collected from machine sumps and process tanks. The water is segregated from other wastewaters until it has been treated by an oily waste removal system. Of the plants, 12.4% are known to segregate their oily wastes. The average contribution of these wastes to the total plant wastewater flow is 6.6% (range of approximately 0.0%–55.4%).

To characterize the waste streams in each subcategory, raw waste data were collected. Discrete samples of raw wastes were taken for each subcategory, and analyses of the samples were performed. The results of these analyses are presented for each subcategory in Tables 10.4–10.9. In each table, data are presented on the number of detections of a pollutant, the number of samples analyzed, the median concentration, the range in concentrations, and the mean concentration of those samples detected. The minimum detection limit for the toxic pollutants in the sampling program was 1 ppb, and any value below this is listed in the six tables as below detection limit (BDL).

10.2.1 COMMON METALS SUBCATEGORY

Pollutant parameters found in the common metals subcategory raw waste stream from sampled plants are shown in Table 10.4. The major constituents shown are parameters that originate from process solutions (such as from plating or galvanizing) and then enter wastewaters by drag-out to rinses. These metals appear in waste streams in widely varying concentrations.

10.2.2 PRECIOUS METALS SUBCATEGORY

Table 10.5 shows the concentrations of pollutant parameters found in the precious metals subcategory raw waste streams. The major constituents are silver and gold, which are much more commonly used in metal finishing industry operations than palladium and rhodium. Because of their high cost, precious metals are of special interest to metal finishers.

10.2.3 COMPLEXED METALS SUBCATEGORY

The concentrations of metals found in complexed metals subcategory raw waste streams are presented in Table 10.6. Complexed metals may occur in a number of unit operations but come primarily from electroless and immersion plating. The most commonly used metals in these operations are copper, nickel, and tin. Wastewaters containing complexing agents must be segregated and treated independently of other wastes in order to prevent further complexing of free metals in the other streams.

10.2.4 CYANIDE SUBCATEGORY

Cyanide has been used extensively in the surface finishing industry for many years; however, it is a hazardous substance that must be handled with caution. The use of cyanide in plating and stripping solutions stems from its ability to weakly complex many metals typically used in plating. Metal deposits produced from cyanide plating solutions are finer grained than those plated from an acidic solution. In addition, cyanide-based plating solutions tend to be more tolerant of impurities than other solutions, offering preferred finishes over a wide range of conditions:

1. Cyanide-based strippers are used to selectively remove plated deposits from the base metal without attacking the substrate.
2. Cyanide-based electrolytic alkaline descalers are used to remove heavy scale from steel.
3. Cyanide-based dips are often used before plating or after stripping processes to remove metallic smuts on the surface of parts.

TABLE 10.4

Concentrations of Pollutants Found in the Common Metals Subcategory Raw Wastewater

Pollutant	Number of Samples	Number of Detections	Range of Detections	Median of Detections	Mean of Detections
		Toxic Pollutants (ppb)			
Metals and Inorganics					
Antimony	106	22	1–430	6	34
Arsenic	105	31	2–64	10	16
Beryllium	27	23	1–44	5	9
Cadmium	108	60	BDL–19,000	8	1,000
Chromium	105	89	3–35,000	180	16,000
Copper	108	105	3–500,000	180	16,000
Lead	108	73	3–42,000	120	1,400
Mercury	99	32	BDL–400	10	18
Nickel	108	88	4–420,000	200	24,000
Selenium	26	21	1–60	5	9
Thallium	26	21	1–62	3	10
Zinc	108	107	9–330,000	290	19,000
Phthalates					
Bis(2-ethylhexyl) phthalate	93	91	BDL–1,900	6	57
Butyl benzyl phthalate	65	38	BDL–10	BDL	1
Di-*n*-butyl phthalate	89	79	BDL–10	BDL	BDL
Di-*n*-octyl phthalate	65	25	BDL–10	BDL	BDL
Diethyl phthalate	83	66	BDL–240	5	31
Dimethyl phthalate	65	7	BDL–10	BDL	2
Nitrogen Compounds					
3,3-Dichlorobenzidene	4	1	BDL		
N-nitroso-di-*n*-propylamine	4	1	570		
Phenols					
2-Nitrophenol	4	1	24		
Phenol	23	15	BDL–1,000	45	240
Aromatics					
Benzene	6	4	BDL–16	7	8
Ethylbenzene	37	9	BDL–1,200	250	340
Toluene	39	17	2–690	77	140
Polycyclic Aromatic Hydrocarbons					
Fluoranthene	4	1	74		
Isophorone	4	4	13–310	180	170
Naphthalene	89	61	BDL–2,000	1	83
Anthracene	82	56	BDL–30	1	2
Fluorene	2	2	BDL–160		80
Phenanthrene	71	55	BDL–30	1	2
Pyrene	4	1	190		

(Continued)

TABLE 10.4 (*Continued*)

Concentrations of Pollutants Found in the Common Metals Subcategory Raw Wastewater

Pollutant	Number of Samples	Number of Detections	Range of Detections	Median of Detections	Mean of Detections
Toxic Pollutants (ppb)					
Halogenated Aliphatics					
Carbon tetrachloride	57	37	BDL–1	BDL	BDL
1,2-Dichloroethane	4	1	3		
1,1,1-Trichloroethane	57	43	BDL–550	BDL	18
1,1,2-Trichloroethane	57	21	BDL–3	BDL	BDL
Chloroform	65	48	BDL–140	BDL	5
1,1-Dichloroethylene	58	4	BDL–110	BDL	20
1,2-*Trans*-dichloroethylene	5	3	1–5	2	3
1,2-Dichloropropylene	4	1	2		
Methylene chloride	80	27	BDL–570	BDL	53
Methyl chloride	74	3	BDL–60	3	21
Methyl bromide	4	1	2		
Dichlorobromomethane	5	2	3–8		6
Chlorodibromomethane	4	1	8		
Tetrachloroethylene	59	23	BDL–66	BDL	6
Trichloroethylene	77	49	BDL–480	BDL	22
Pesticides and Metabolites					
Dieldrin	4	1	BDL		
Alpha-endosulfan	4	1	9		
Endrin aldehyde	4	1	BDL		
Alpha-BHC	4	1	BDL		
Beta-BHC	4	1	4		
Delta-BHC	4	1	BDL		
Concentration (mg/L)					
Classical Pollutants					
TSS	107	104	0.56–11,000	63	520
Aluminum	8	6	0.03–200	0.29	62
Barium	4	3	0.027–0.071	0.03	0.043
Calcium	3	3	25–76	52	51
Cobalt	4	4	0.009–0.023	0.02	0.017
Fluorides	7	3	0.021–36	1.1	5.3
Iron	85	76	0.035–490	1.9	28
Magnesium	88	87	5.6–31	14	16
Manganese	4	4	0.059–0.5	0.085	0.22
Molybdenum	7	7	0.031–0.3	0.27	0.2
Phosphorous	4	3	0.007–77	3	7.9
Sodium	4	3	17–310	140	160
Tin	4	4	0.002–15	0.86	3.7
Titanium	5	2	0.006–0.08	0.03	0.039
Vanadium	7	3	0.01–0.22	0.036	0.087
Yttrium	4	3	0.002–0.02	0.018	0.013

Source: U.S. EPA. *Development Document for Effluent Limitations Guidelines and Standards for the Metal Finishing Point Source Category*. Report no. EPA-440/1-80/091. U.S. Environmental Protection Agency, Washington, DC, 1980 [6].

TABLE 10.5

Concentrations of Pollutants Found in the Precious Metals Subcategory Raw Wastewater

Pollutant	Number of Samples	Number of Detections	Range of Detections	Median of Detections	Mean of Detections
			Concentration (mg/L)		
Classical Pollutants					
Silver	15	12	0.033–600	0.38	86
Gold	15	9	0.56–43	0.86	15
Palladium	13	3	0.09–0.12	0.09	0.10
Rhodium	12	1	0.22		

Source: U.S. EPA. *Development Document for Effluent Limitations Guidelines and Standards for the Metal Finishing Point Source Category.* Report no. EPA-440/1-80/091. U.S. Environmental Protection Agency, Washington, DC, 1980 [6].

TABLE 10.6

Concentrations of Pollutants Found in the Complexed Metals Subcategory Raw Wastewater

Pollutant	Number of Samples	Number of Detections	Range of Detections	Median of Detections	Mean of Detections
			Concentration (ppb)		
Toxic Pollutants					
Cadmium	31	9	1–3,600	67	850
Copper	31	28	10–63,000	6,700	11,000
Lead	31	10	2–3600	420	1,200
Nickel	31	25	26–290,000	3,200	28,000
Zinc	31	31	23–18,000	210	3,000
			Concentration (mg/L)		
Classical Pollutants					
Aluminum	1	1	0.1		
Calcium	1	1	17		
Iron	31	31	0.038–99	0.74	9.9
Magnesium	1	1	2		
Manganese	1	1	0.1		
Phosphorus	31	31	0.023–100	8.2	23
Sodium	1	1	110		
Tin	31	10	0.013–6	0.68	1.6

Source: U.S. EPA. *Development Document for Effluent Limitations Guidelines and Standards for the Metal Finishing Point Source Category.* Report no. EPA-440/1-80/091. U.S. Environmental Protection Agency, Washington, DC, 1980 [6].

TABLE 10.7

Concentrations of Pollutants Found in the Cyanide Subcategory Raw Wastewater

Pollutant	Number of Samples	Number of Detections	Range of Detections	Median of Detections	Mean of Detections
		Concentration (ppb)			
Toxic Pollutants					
Cyanide	20	20	45–500,000	45,000	110,000
Cyanide, amenable to chlorination	19	18	5–460,000	4,500	86,000

Source: U.S. EPA. *Development Document for Effluent Limitations Guidelines and Standards for the Metal Finishing Point Source Category.* Report no. EPA-440/1-80/091. U.S. Environmental Protection Agency, Washington, DC, 1980 [6].

TABLE 10.8

Concentrations of Pollutants Found in the Hexavalent Chromium Subcategory Raw Wastewater

Pollutant	Number of Samples	Number of Detections	Range of Detections	Median of Detections	Mean of Detections
		Concentration (ppb)			
Toxic Pollutants					
Chromium, hexavalent	49	41	5–13,000,000	20,000	420,000

Source: U.S. EPA. *Development Document for Effluent Limitations Guidelines and Standards for the Metal Finishing Point Source Category.* Report no. EPA-440/1-80/091. U.S. Environmental Protection Agency, Washington, DC, 1980 [6].

TABLE 10.9

Concentrations of Pollutants Found in the Oils Subcategory Raw Wastewater

Pollutant	Number of Samples	Number of Detections	Range of Detections	Median of Detections	Mean of Detections
		Concentration (ppb)			
Toxic Pollutants					
Phthalates					
Bis(2-ethylhexyl)phthalate	37	20	2–9,300	73	820
Butyl benzyl phthalate	37	9	1–10,000	130	1,600
Di-*n*-butyl phthalate	37	13	1–3,100	16	270
Di-*n*-octyl phthalate	37	2	4–120		62
Diethyl phthalate	37	9	1–1,900	48	420
Dimethyl phthalate	37	34	1–1,200	1	400
Ethers					
Bis(chloroethyl)ether	37	1	9		
Bis(2-chloroethyl)ether	37	2	4–10		7
Bis(2-chloroisopropyl)ether	37	1	4		

(Continued)

TABLE 10.9 (*Continued*)

Concentrations of Pollutants Found in the Oils Subcategory Raw Wastewater

Pollutant	Number of Samples	Number of Detections	Range of Detections	Median of Detections	Mean of Detections
Bis(2-chloroethoxy)methane	37	1	3		
Nitrogen compounds					
1,2-Diphenylhydrazine	37	2	3-12		8
Phenols					
2,4,6-Trichlorophenol	37	3	10–1,800	30	610
Parachlorometa cresol	37	8	4–8,00,000	2,300	100,000
2-Chlorophenol	37	2	76–620		350
2,4-Dichlorophenol	37	2	10–68		39
2,4-Dimethylphenol	37	6	1–31,000	10	5,200
2-Nitrophenol	37	3	10–120	15	120
4-Nitrophenol	37	1	10		
2,4-Dinitrophenol	37	3	10–10,000	13	3,300
N-nitrosodiphenylamine	37	5	4–900	750	490
Pentachlorophenol	37	3	10–50,000	5,200	18,000
Phenol	37	3	3–6,600	440	1,700
4,6-Dinitro-o-cresol	37	2	10–5,700		2,800
Aromatics					
Benzene	37	18	1–110	8	12
Chlorobenzene	37	2	11–610		310
Nitrobenzene	37	2	1–10		5
Toluene	37	25	1–37,000	33	1,800
Ethylbenzene	37	16	1–9,500	12	380
Polynuclear Aromatic Hydrocarbons					
Acenaphthene	37	2	57–5,700		2,900
2-Chloronaphthalene	37	1	130		
Fluoranthene	37	8	1–55,000	110	6,300
Naphthalene	37	10	1–260	100	36
Benzo(a)pyrene	37	1	10		
Chrysene	37	3	1–73	2	25
Acenaphthylene	37	3	77–1,000	140	410
Anthracene	43	7	3–2,000	34	360
Fluorene	37	7	1–760	75	180
Phenanthrene	37	8	8–8,000	80	400
Pyrene	37	5	31–150	75	79
Halogenated Hydrocarbons					
Carbon tetrachloride	37	5	1–10,000	97	2,600
1,2-Dichloroethane	37	6	9–2,100	1,400	1,100
1,1,1-Trichloroethane	37	18	1–1,300,000	260	75,000
1,1-Dichloroethane	37	11	2–1,100	600	460
1,1,2-Trichloroethane	37	6	6–1,300	10	330
1,1,2,2-Tetrachloroethane	37	8	6–370		290
Chloroform	37	19	8–690	10	50

(Continued)

TABLE 10.9 (*Continued*)
Concentrations of Pollutants Found in the Oils Subcategory Raw Wastewater

Pollutant	Number of Samples	Number of Detections	Range of Detections	Median of Detections	Mean of Detections
1,1-Dichloroethylene	37	18	8–10,000	800	1,500
1,2-*trans*-dichloroethylene	37	9	8–1,700	88	510
Methylene chloride	37	89	5–7,600	92	600
Methyl chloride	37	4	1–4,700	9	1,800
Bromoform	37	1	10		
Dichlorobromomethane	37	2	1–10		5
Trichlorofluoromethane	37	8	860–890		800
Chlorodibromomethane	37	3	1–10	8	4
Tetrachloroethylene	37	10	1–110,000	10	8,900
Trichloroethylene	37	11	1–130,000	110	23,000
Pesticides and Metabolites					
Aldrin	37	2	4–11		7
Dieldrin	37	1	3		
Chlordane	37	8	1–13		7
4,4-DDT	37	8	8–10		6
4,4-DDE	37	4	801–33	8	14
4,4-DDD	37	3	1–10	4	5
Alpha-endosulfan	37	2	6–88		18
Beta-endosulfan	37	8	801–6		3
Endosulfan sulfate	37	4	1–16	11	10
Endrin	37	8	7–10	13	8
Endrin aldehyde	37	8	10–14	7	18
Heptachlor	37	1	801		
Heptachlor epoxide	37	1	801		
Alpha-BHC	37	3	4–18	13	12
Gamma-BHC	37	3	1–9	7	6
Delta-BHC	37	8	4–11		7
Polychlorinated Biphenyls					
Aroclor 1234	37	2	76–1,100		590
Aroclor 1248	37	2	160–1,800		960
Concentration (mg/L)					
Classical Pollutants					
Ammonia	37	10	0.46–270	7.9	46
BOD	37	21	10–17,000	1,400	3,800
COD	37	16	310–1,900,000	18,000	180,000
Oil and grease	37	17	60–900,000	6,100	41,000
Phenols total	37	14	9,062–48	624	8.3
Total dissolved solids	37	9	850–4,960	1,600	8,000
Total organic carbon	37	17	1–560,000	1,600	88,000
TSS	37	35	35–18,000	600	2,700

Source: U.S. EPA. *Development Document for Effluent Limitations Guidelines and Standards for the Metal Finishing Point Source Category.* Report no. EPA-440/1-80/091. U.S. Environmental Protection Agency, Washington, DC, 1980 [6].

Cyanide-based metal finishing solutions usually operate at basic pH levels to avoid the decomposition of the complexed cyanide and the formation of highly toxic hydrogen cyanide gas. The cyanide concentrations found in the cyanide subcategory raw waste streams are shown in Table 10.7. The levels of cyanide range from 0.045 to 500 lig/L. Streams with high cyanide concentrations normally originate from electroplating and heat-treating processes. Cyanide-bearing waste streams should be segregated and treated before being combined with other raw waste streams.

10.2.5 Hexavalent Chromium Subcategory

Concentrations of hexavalent chromium from metal finishing raw wastes are shown in Table 10.8. Hexavalent chromium enters wastewater as a result of many unit operations and can be very concentrated. Because of its high toxicity, it requires separate treatment to be efficiently removed from wastewater.

10.2.6 Oils Subcategory

Pollutants and their concentrations found in the oily waste subcategory streams are shown in Table 10.9. The oily waste subcategory for the metal finishing industry is characterized by both concentrated and dilute oily waste streams that consist of a mixture of free oils, emulsified oils, greases, and other assorted organics. Applicable treatment of oily waste streams is dependent on the concentration levels of the wastes, but oily wastes normally receive specific treatment for oil removal prior to solids removal waste treatment.

The majority of the pollutants listed in Table 10.9 are priority organics that are used either as solvents or as oil additives to extend the useful life of the oils. Organic priority pollutants, such as solvents, should be segregated and disposed of or reclaimed separately. However, when they are present in wastewater streams, they are most often at the highest concentration in the oily waste stream because organics generally have a higher solubility in hydrocarbons than in water. Oily wastes will normally receive treatment for oil removal before being directed to waste treatment for solids removal.

10.2.7 Solvent Subcategory

The solvent subcategory raw wastes are generated in the metal finishing industry by the dumping of spent solvents from degreasing equipment (including sumps, water traps, and stills). These solvents are predominately comprised of compounds classified by the U.S. EPA as toxic pollutants. Spent solvents should be segregated, hauled for disposal or reclamation, or reclaimed on-site. Solvents that are mixed with other wastewaters tend to appear in the common metals or the oily waste stream.

10.3 SOURCE REDUCTION

It is not currently feasible to achieve a zero discharge of chemical pollutants from metal finishing operations. However, substantial reductions in the type and volume of hazardous chemicals wasted from most metal finishing operations are possible [8]. Because end-of-pipe waste detoxification is costly for small- and medium-sized metal finishers, and the cost and liability of residual disposal have increased for all metal finishers, management and production personnel may be more willing to consider production process modifications to reduce the amount of chemicals lost to waste.

This section provides guidance for reducing waterborne wastes from metal finishing operations in order to avoid or reduce the need for waste detoxification and the subsequent off-site disposal of detoxification residuals. Waste reduction practices may take the form of [5]

1. Chemical substitution
2. Waste segregation
3. Process modifications to reduce drag-out loss
4. Capture/concentration techniques.

10.3.1   Chemical Substitution

The incentive for substituting process chemicals containing nonpolluting materials has only been present in recent years with the advent of pollution control regulations. Chemical manufacturers are gradually introducing such substitutes. By eliminating polluting process materials such as hexavalent chromium and cyanide-bearing cleaners, and deoxidizers, the treatments required to detoxify these wastes are also eliminated. It is particularly desirable to eliminate processes employing hexavalent chromium and cyanide since special equipment is needed to detoxify both.

Substituting nonpolluting cleaners for cyanide cleaners can avoid cyanide treatment entirely. For a 7.6 L/min rinse water flow, this means a savings of about USD18,400 in equipment costs and USD10/kg of cyanide treatment chemical costs. In this case, treatment chemical costs are about four times the cost of the raw sodium cyanide cleaner.

There can be disadvantages to using nonpolluting chemicals. Before making a decision, the following questions should be asked of the chemical supplier [5]:

- Are substitutes available and practical?
- Will substitution solve one problem but create another?
- Will tighter chemical controls be required of the bath?
- Will product quality and/or production rate be affected?
- Will the change involve any cost increases or decreases?

Based on a survey of chemical suppliers and electroplaters who use nonpolluting chemicals, some commonly used chemical substitutes are summarized in Table 10.10.

TABLE 10.10
Chemical Substitutes

Polluting	Substitute	Comments
Fire dip (NaCN)	Muriatic acid with additives	Slower acting than $+H_2O_2$ traditional fire dip Excellent throwing power with a bright, smooth, rapid finish
Heavy copper cyanide plating bath	Copper sulfate	A copper cyanide strike may still be necessary for steel, zinc, or tin-lead base metals Requires good preplate cleaning Noncyanide process eliminates carbonate build-up in tanks
Chromic acid pickles, deoxidizers, and bright dips	Sulfuric acid and hydrogen peroxide	Nonchrome substitute
Chrome-based antitarnish	Benzotriazole (0.1%–1.0% solution in methanol) or water-based proprietaries	Nonfuming Nonchrome substitute Extremely reactive, requires ventilation Noncyanide cleaner Good degreasing when hot and in an ultrasonic bath
Cyanide cleaner	Trisodium phosphate or ammonia	Highly basic May complex with soluble metals if used as an intermediate rinse between plating baths where metal ion may be dragged into the cleaner and cause wastewater treatment problems
Tin cyanide	Acid tin chloride	Works faster and better

Source: U.S. EPA. *Meeting Hazardous Waste Requirements for Metal Finishers.* Report no. EPA/625/4-87/018. U.S. Environmental Protection Agency, Cincinnati, OH, 1987 [5].

TABLE 10.11

Cyanide and Noncyanide Plating Processes

Metal	Cyanide	Noncyanide
Brass	Proven	No
Bronze	Proven	No
Cadmium	Proven	Yes
Copper	Proven	Proven
Gold	Proven	Developing
Indium	Proven	Yes
Silver	Proven	Developing
Zinc	Proven	Proven

Source: U.S. EPA. *Managing Cyanide in Metal Finishing.* Capsule Report no. EPA 625/R-99/009. U.S. Environmental Protection Agency, Cincinnati, OH, December 2000 [21].

The chemical supplier can also identify any regulated pollutants in the facility's treatment chemicals and offer available substitutes. The federally regulated pollutants are cyanide, chrome, copper, nickel, zinc, lead, cadmium, and silver. Local and/or state authorities may regulate other substances, such as tin, ammonia, and phosphate. The current status of cyanide and noncyanide substitute plating processes is shown in Table 10.11.

10.3.2 WASTE SEGREGATION

After eliminating as many pollutants as possible, the next step is for polluting streams to be segregated from nonpolluting streams. Nonpolluting streams can go directly to the sewer, although pH adjustment may be necessary. The segregation process will likely require some physical relayout and/or repiping of the shop. These potentially nonpolluting rinse streams represent about one-third of all plating process water. Caution must be exercised to make certain that the so-called nonpolluting baths contain no dissolved metal. The cost savings in segregating polluting from nonpolluting streams is realized through wastewater treatment equipment and operating costs. The remaining polluting sources, which require some form of control, include all dumped spent solutions, including tumble finishing and burnishing washes, cyanide cleaner rinses, plating rinses, rinses after "bright dips," and aggressive cleaning solutions.

10.3.3 PROCESS MODIFICATIONS TO REDUCE DRAG-OUT LOSS

Plating solution wasted by being carried over into the rinse water as a workpiece emerges from the plating bath is defined as drag-out and is the largest volume source of chemical pollutants in the electroplating shop. Numerous techniques have been developed to control drag-out; the effectiveness of each method varies as a function of the plating process, operator cooperation, racking, barrel design, transfer dwell time, and plated part configuration.

Wetting agents and longer workpiece withdrawal/drainage times are two techniques that significantly control drag-out. These and other techniques are discussed below.

10.3.3.1 Wetting Agents

Wetting agents lower the surface tension of process baths. To remove the plating solution dragged out with the plated part, gravity-induced drainage must overcome the adhesive force between the solution and the metal surface. The drainage time required for racked parts is a function of the surface tension of the solution, part configuration, and orientation. Lowering the surface tension reduces the drainage time and minimizes the edge effect (the bead of liquid adhering to the part edge); thus, there is less drag-out. Plating baths such as nickel and heavy copper cyanide also use wetting agents to maintain grain quality and provide improved coverage. The chemical supplier should be asked if the baths he supplies contain wetting agents and, if not, whether wetting agents can be added. In some baths the use of wetting agents has the potential to reduce drag-out by 50%.

10.3.3.2 Longer Drain Times

With slower withdrawal rates and/or longer drain times, drag-out of process solutions can be reduced by up to 50%. Where high-temperature plating solutions are used, slow withdrawal of the rack may also be necessary to prevent evaporative "freezing," which can actually increase drag-out. In the extreme case, too rapid a withdrawal rate causes "sheeting," where huge volumes of drag-out are lost to waste. Figure 10.1 shows the drainage rates for plain and bent-shaped pieces. Drainage for all shapes is almost complete within 15 seconds after withdrawal, indicating that this is an optimum drain time for most pieces.

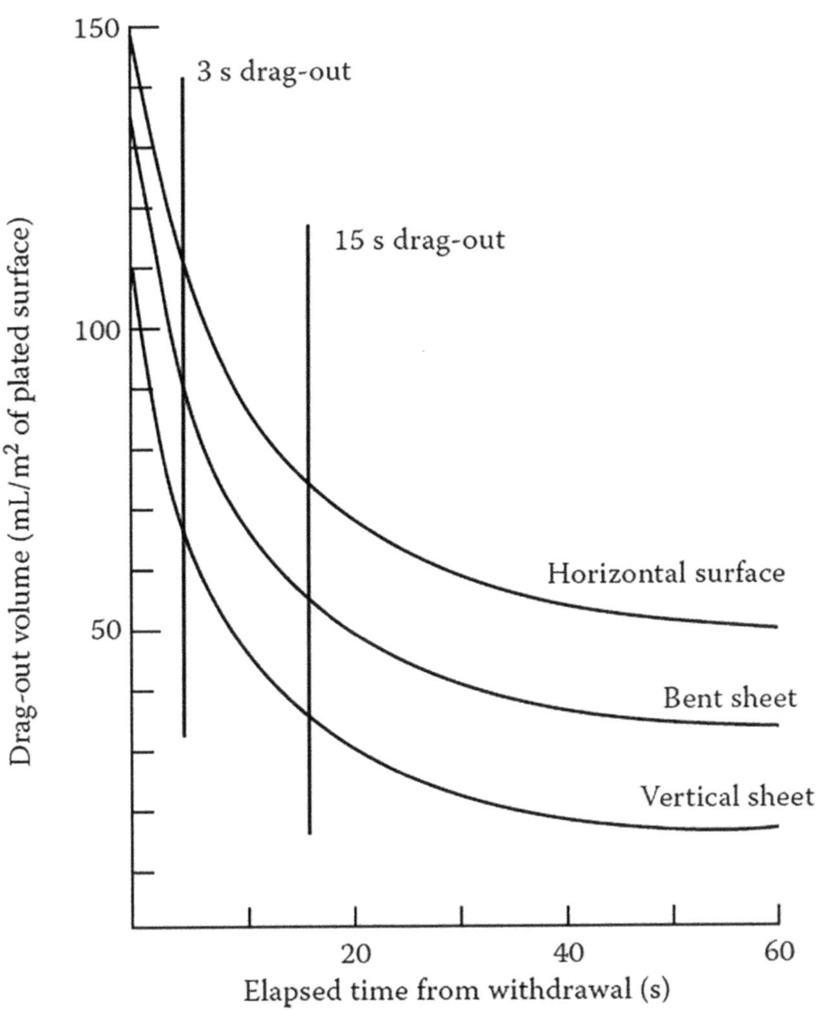

FIGURE 10.1 Typical drag-out drainage rates. (Adapted from U.S. EPA. *Meeting Hazardous Waste Requirements for Metal Finishers*. Report no. EPA/625/4-87/018. U.S. Environmental Protection Agency, Cincinnati, OH, 1987 [5].)

One of the best ways to control drag-out loss from rack plating on hand lines is to provide drain bars over the tank from which the rack can be hung to drain for a brief period. Hanging and removing the racks from the drain bars ensures an adequate drain time. Slightly jostling the racks helps shake off the adhering solution.

In barrel plating, the barrel should be rotated for a time just above the plating tank in order to reduce the volume of dragged-out chemicals. Holes in the barrels should be as large as possible to improve solution drainage while still containing the pieces. A fog spray directed at the barrel or its contents can also help drag-out drainage. Deionized water is recommended to minimize bath contamination.

The combined application of wetting agents and longer withdrawal/drainage times can significantly reduce the amount of drag-out for many cleaning or plating processes. For example, a typical nickel drag-out can be reduced from 1 to *4 L/h by these techniques.

10.3.3.3 Other Drag-Out Reduction Techniques

10.3.3.3.1 Rinse Elimination

The rinse between a soak cleaner and an electrocleaner may be eliminated if the two baths are compatible.

10.3.3.3.2 Low-Concentration Plating Solutions

Low-concentration plating solutions reduce the total mass of chemicals being dragged out. The mass of chemicals removed from a bath is a function of the solution concentration and the volume of solution carried from the bath. Traditionally, the bath concentration is maintained at a midpoint within a range of operating conditions. With the high cost of replacement, treatment, and disposal of dragged-out chemicals, the economics of low-concentration baths are favorable.

As an illustration, a typical nickel plating operation with five nickel tanks has an annual nickel drag-out of about 10,000 L. Assuming the nickel baths are maintained at the midpoint operating concentration, as shown in Table 10.12, the annual cost of chemical replacement, treatment, and disposal is about USD20,700 in terms of 2007 dollars. If the bath is converted to the modified operating condition as shown in the table, the annual cost of chemical replacement, treatment, and disposal is approximately USD18,700, a savings of about USD2,000 per year. Generally, any percent decrease in bath chemical concentration results in the same percent reduction in the mass of chemicals lost in the drag-out. The disadvantage of low-concentration baths may be lowered plating efficiencies, which may require higher current densities and closer process control. The reduction in plating chemical replacement, treatment, and disposal costs could be partially offset by the added labor and power costs associated with using the lower-concentration baths.

TABLE 10.12
Standard Nickel Solution Concentration Limits

Chemical	Concentration Range (g/L)	Midpoint Operating Condition (g/L)	Modified Operating Condition (g/L)
Nickel sulfate			
$NiSO_4\text{-}6H_2O$	300–375	338	308
as $NiSO_4$	—	200	182
Nickel chloride			
$NiCl_2\text{-}6H_2O$	60–90	75	64
as $NiCl_2$	—	41	35
Boric acid H_3BO_3	45–49	47	46

Source: U.S. EPA. *Meeting Hazardous Waste Requirements for Metal Finishers.* Report no. EPA/625/4-87/018. U.S. Environmental Protection Agency, Cincinnati, OH, 1987 [5].

10.3.3.3.3 Clean Plating Baths

Contaminated plating baths, for example carbonate build-up in cyanide baths, can increase drag-out as much as 50% by increasing the viscosity of the bath. Excessive impurities also make applying recovery technology difficult, if not impossible.

10.3.3.3.4 Low-Viscosity Conducting Salts

Bath viscosity indexes are available from chemical suppliers. As the bath viscosity increases, drag-out volume also increases.

10.3.3.3.5 High-Temperature Baths

High-temperature baths reduce surface tension and viscosity, thus decreasing drag-out volume. Disadvantages to be considered are more rapid solution decomposition, higher energy consumption, and possible dry-on pattern on the workpiece.

10.3.3.3.6 No Unnecessary Components

Additional bath components (chemicals) tend to increase both viscosity and drag-out.

10.3.3.3.7 Fog Sprays or Air Knives

Fog sprays or air knives may be used over the bath to remove drag-out from pieces as they are withdrawn. The spray of deionized water or air removes the plating solution from the part and returns as much as 75% of the drag-out to the plating tank. Fog sprays, located just above the plating bath surface, dilute and drain the adhering drag-out solution, thus reducing the concentration and mass of chemicals lost. Fog sprays are best when tank evaporation rates are sufficient to accommodate the added volume of spray water. Air knives, also located just above the plating bath surface, reduce the volume of drag-out by mechanically scouring the adhering liquid from the workpiece. The drag-out concentration remains constant, but the mass of chemicals lost is reduced. Air knives are best when the surface evaporation rates of the bath are too low to allow additional spray water. In some cases, the use of supplementary atmospheric evaporators may be justified by economic considerations. Air knives can be installed for about USD 750–800 per bath if an oil-free, compressed air source is available. Fog sprays can also be installed for about USD 750–800 per bath if a deionized water source is available. The spray should be actuated only when work is in the spraying position. Properly designed spray nozzles distribute the water evenly over the work, control the volume of water used, and avoid snagging workpieces as they are withdrawn from the tank.

10.3.3.3.8 Proper Racking

Every piece has at least one racking position in which drag-out will be at a minimum. In general, to minimize drag-out:

- Parts should be racked with major surfaces vertically oriented.
- Parts should not be racked directly over one another.
- Parts should be oriented so that the smallest surface area of the piece leaves the bath surface last.

The optimum orientation will provide faster drainage and less drag-out per piece. However, in some cases, this may reduce the number of pieces on a rack, or the optimum draining configuration may not be the optimum plating configuration. In addition, the user should maintain rack coatings, replace rack contacts when broken, strip racks before plating build-up becomes excessive, and ensure that all holes on racks are covered or filled.

10.3.3.4 Capture/Concentration Techniques

10.3.3.4.1 Capture/Concentration with Full Reuse of Drag-Out

The pioneer in simple, low-cost methods of reducing waste in the plating shop was Dr. Joseph B. Kushner. In *Water and Waste Control for the Plating Shop* (1972), he describes a "simple waste recovery system," which captures drag-out in a static tank or tanks for return to the plating bath. The drag-out tanks are followed by a rinse tank which flows to the sewer with only trace amounts of polluting salts and is often in compliance with sewer discharge standards. A simplified diagram of this reuse system is shown in Figure 10.2. It is not difficult to automate the direct drag-out recovery process, and commercial units are available.

The Kushner concept is easily applicable to hot plating baths where the bath evaporation rate equals or exceeds the pour-back rate, Q_2. The drag-out concentration depends on the bath drag-out rate, the number of drag-out tanks, the rinse water flow rate, Q_2, the plating bath evaporation rate, and the drag-out return rate. The number of drag-out tanks must be based on the available space. The higher the number of counterflowed drag-out tanks, the smaller will be the return rate necessary to obtain good rinsing. The Kushner multiple drag-outs are not feasible if there is no room for the required drag-out tanks. If there is little or no evaporation from the bath, supplementary evaporation should be considered. Bath contamination must be minimized by using purified (RO) water for Q_2.

10.3.3.4.2 Capture/Concentration with Partial Reuse of Drag-Out

By adding a trickling water supply and drain, Q_3, to the drag-out tank, the application of Kushner's concept can be extended to other metal finishing processes that may not be amenable to full reuse but can allow partial reuse. Figure 10.3 depicts the partial reuse scheme. The trickle concentrate can also be batch treated in a small volume on-site, recycled at a central facility, or mixed with Q_1, for discharge, if the combined metal content is below sewer discharge standards.

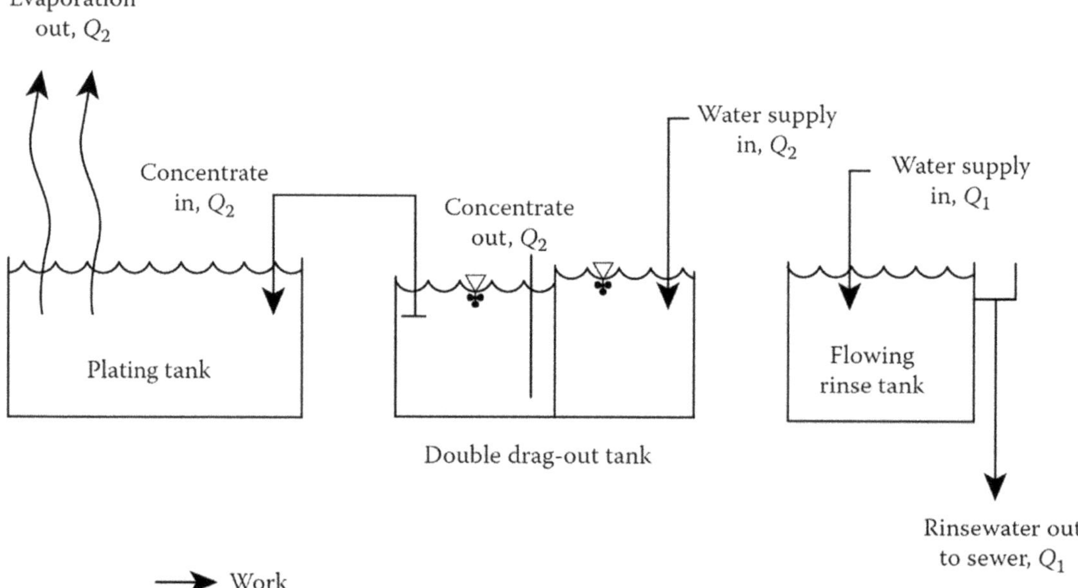

FIGURE 10.2 Kushner method of double drag-out for full reuse. (Adapted from U.S. EPA. *Meeting Hazardous Waste Requirements for Metal Finishers*. Report no. EPA/625/4-87/018. U.S. Environmental Protection Agency, Cincinnati, OH, 1987 [5].)

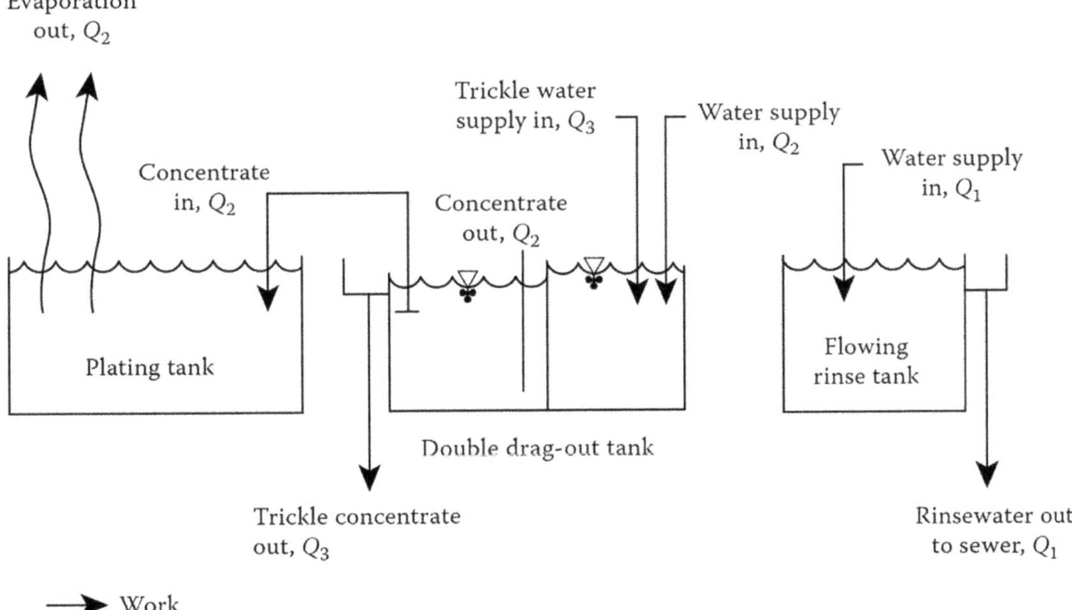

FIGURE 10.3 Modified method of double drag-out for partial reuse. (Adapted from U.S. EPA. *Meeting Hazardous Waste Requirements for Metal Finishers.* Report no. EPA/625/4-87/018. U.S. Environmental Protection Agency, Cincinnati, OH, 1987 [5].)

10.3.4 WASTE REDUCTION COSTS AND BENEFITS

Benefits of waste reduction in the metal finishing shop include

1. Reduced chemical cost
2. Reduced water cost
3. Reduced volume of "hazardous" residuals
4. Reduced pretreatment cost.

The benefits of saving valuable chemicals and water and reducing sludge disposal costs can best be illustrated by an example. An electroplating operation discharges 98,400 L/d of wastewater containing 0.91 kg of copper, 1.14 kg of nickel, and 0.91 kg of cyanide. The shop can reduce its generation of cyanide and copper waste by about 50% by eliminating cyanide cleaners and utilizing pour-back of copper cyanide solution; generation of nickel waste can be reduced by 90% by pour-back of the nickel solution. Reducing wasted salts also reduces rinse water flow rate, thus saving water and sewer use fees. The chemical costs of treatment are given in Table 10.13, and the annual replacement costs of chemicals are given in Figure 10.4. Calculations of the annual dollar savings are shown in Table 10.14. All costs have been converted into 2007 USD using the U.S. ACE Yearly Average Cost Index for Utilities [9].

10.4 POLLUTANT REMOVABILITY

This section reviews the technologies currently available and used to remove or recover pollutants from the wastewater generated in the metal finishing industry [5–7,10]. Treatment options are presented for each subcategory within the metal finishing industry. Table 10.15 lists the treatment techniques available for treating wastes from each subcategory.

TABLE 10.13

Chemical Costs of Treatment and Disposal in 2007 USD

Chemical Cost (2022 USD[a]/kg)

Pollutant	Treatment[b]	Disposal[c]
Nickel	3.68	9.03
Copper	3.68	9.03
Cyanide	23.76	NA

Source: U.S. EPA. *Meeting Hazardous Waste Requirements for Metal Finishers.* Report no. EPA/625/4-87/018. U.S. Environmental Protection Agency, Cincinnati, OH, 1987 [5].

[a] Costs were converted from 1979 USD to 2007 USD using U.S. ACE Yearly Average Cost Index for Utilities [32,50].

[b] Cost of NaOH at USD 1.35/kg and NaOCl at USD 3.17/kg.

[c] Cost of disposal at USD 2.48/kg of sludge (USD 539/drum) at 30% solids content.

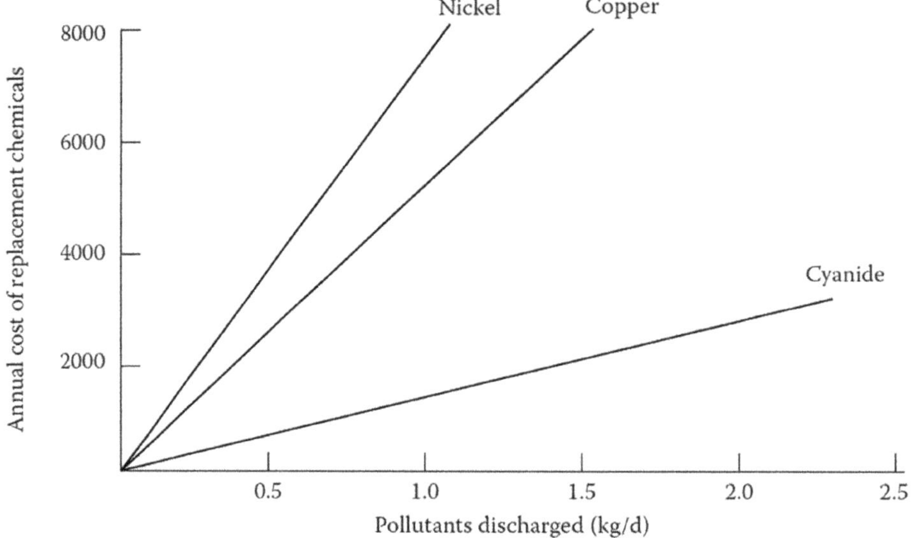

FIGURE 10.4 Annual replacement cost of chemicals in 2007 USD. (Adapted from U.S. EPA. *Meeting Hazardous Waste Requirements for Metal Finishers.* Report no. EPA/625/4-87/018. U.S. Environmental Protection Agency, Cincinnati, OH, 1987 [5].)

10.4.1 Common Metals

The treatment methods used to treat wastes within the common metals subcategory fall into two groups:

1. Recovery techniques
2. Solids removal techniques.

TABLE 10.14

Illustration of Annual Cost Savings for Waste Reduction

Item	Cost Saving[a] 2022 USD
Process Chemical Savings[b]	
Copper	3268
Cyanide	654
Nickel	10457
Treatment Chemical Saving[c]	
Copper	418
Cyanide	2695
Nickel	943
Reduced Treatment Sludge Disposal[c]	
Copper	1024
Cyanide	0
Nickel	2291
Water and sewer use fee reduction[d]	5876
Total annual savings	27626

Source: U.S. EPA. *Meeting Hazardous Waste Requirements for Metal Finishers.* Report no. EPA/625/4-87/018. U.S. Environmental Protection Agency, Cincinnati, OH, 1987 [5].

[a] Costs were converted from 1979 USD to 2022 USD using U.S. ACE. Yearly Average Cost Index for Utilities. (Adapted from U.S. ACE. Yearly Average Cost Index for Utilities. In: *Civil Works Construction Cost Index System Manual.* 110-2-1304. U.S. Army Corps of Engineers, Washington, DC, 44 pp. [32,40].

[b] From Figure 10.4.

[c] From Table 10.12 and Figure 10.4.

[d] USD 1.04/m^3.

Recovery techniques are treatment methods used for the purpose of recovering or regenerating process constituents that would otherwise be discarded. This group includes [5–7]

1. Evaporation
2. Ion exchange
3. Electrolytic recovery
4. Electrodialysis
5. Reverse osmosis.

Solids removal techniques are employed to remove metals and other pollutants from process wastewaters to make these waters suitable for reuse or discharge. These methods include [5–7]

1. Hydroxide and sulfide precipitation
2. Sedimentation
3. Diatomaceous earth filtration
4. Membrane filtration
5. Granular bed filtration
6. Peat adsorption
7. Insoluble starch xanthate treatment
8. Flotation.

TABLE 10.15

Treatment Methods in Current Use or Available for Use in the Metal Finishing Industry

Subcategory/Technology	Number of Plants
Common Metals	
Hydroxide followed by sedimentation	103
Hydroxide followed by sedimentation and filtration	30
Evaporation (metal recovery, bath concentrates, rinse waters)	41
Evaporation recovery	63
Ion exchange	11
Electrolysis	3
Reverse osmosis	8
Post adsorption	0
Insoluble starch xanthenes	2
Sulfide precipitation	3
Flotation	29
Membrane flotation	7
Precious Metals	
Evaporation	1
Ion exchange	NR
Electrolytic	NR
Complexed Metals	
High pH precipitation with sedimentation	NR
Membrene filtration	NR
Hexavalent Chromium	
Chemical chrome reduction	343
Electrochemical chromium reduction	2
Electrochemical chromium regeneration	0
Advanced electrodialysis	NR
Evaporation	1
Ion exchange	1
Cyanida	
Oxidation by chlorine	201
Oxidation by ozone	2
Oxidation by ozone with uv radiation	NR
Oxidation by hydrogen peroxide	3
Electrochemical cyanide oxidation	4
Chemical precipitation	3
Reverse osmosis	NR
Evaporation	NR
Oils (Segregated)	
Emulsion breaking	26
Skimming	94
Emulsion breaking and skimming	
Ultrafiltration	20

(Continued)

TABLE 10.15 (*Continued*)
Treatment Methods in Current Use or Available for Use in the Metal Finishing Industry

Subcategory/Technology	Number of Plants
Reverse osmosis	3
Carbon adsorption	10
Coalescing	3
Flotation	29
Centrifugation	5
Integrated adsorption	0
Resin adsorption	0
Ozonation	0
Chemical oxidation	0
Aerobic decomposition	14
Thermal emulsion breaking	0
Solvent Wastes	
Segregation	NR
Contract handling	NR
Sludges	
Gravity thickening	76
Pressure filtration	66
Vacuum filtration	68
Centrifugation	55
Sludge bed drying	77
In Process Control	
Flow reduction	NR

Source: U.S. EPA. *Treatability Manual, Volume II. Industrial Descriptions.* Report no. EPA-600/2-82-001b. U.S. Environmental Protection Agency, Washington, DC, September 1981 [7].

Note: NR, not reported.

Three treatment options are used in treating common metal wastes:

1. *The option 1* system consists of hydroxide precipitation [11] followed by sedimentation [12]. This system accomplishes the end-of-pipe metals removal from all common metal-bearing wastewater streams that are present at a facility. The recovery of precious metals, the reduction of hexavalent chromium, the removal of oily wastes, and the destruction of cyanide must be accomplished prior to common metals removal.
2. *The option 2* system is identical to the option 1 treatment system with the addition of filtration devices [13] after the primary solids removal devices. The purpose of these filtration units is to remove suspended solids such as metal hydroxides that do not settle out in the clarifiers. The filters also act as a safeguard against pollutant discharge should an upset occur in the sedimentation device. Filtration techniques applicable to option 2 systems are diatomaceous earth and granular bed filtration [14,15].
3. *The option 3* treatment system for common metal wastes consists of the option 2 end-of-pipe treatment system plus the addition of in-plant controls for lead and cadmium. In-plant controls would include evaporative recovery, ion exchange, and recovery rinses [15].

In addition to these three treatments, there are several alternative treatment technologies applicable to the treatment of common metal wastes. These technologies include electrolytic recovery, electrodialysis, reverse osmosis, peat adsorption, insoluble starch xanthate treatment, sulfide precipitation, flotation, and membrane filtration [14,15].

10.4.2   PRECIOUS METALS

Precious metal wastes can be treated using the same treatment alternatives as those described for the treatment of common metal wastes. However, due to the intrinsic value of precious metals, every effort should be made to recover them. The treatment alternatives recommended for precious metal wastes are the recovery techniques—evaporation, ion exchange, and electrolytic recovery.

10.4.3   COMPLEXED METAL WASTES

Complexed metal wastes within the metal finishing industry are a product of electroless plating, immersion plating, etching, and printed circuit board manufacture. The metals in these waste streams are tied up or complexed by particular complexing agents whose function prevents metals from coming out of the solution. This counteracts the technique employed by most conventional solids removal methods. Therefore, segregated treatment of these wastes is necessary. The treatment method well suited to treating complex metal wastes is high pH precipitation. An alternative method is membrane filtration [16] that is primarily used in place of sedimentation for solids removal.

10.4.4   HEXAVALENT CHROMIUM

Hexavalent chromium-bearing wastewaters are produced in the metal finishing industry in chromium electroplating, chromate conversion coatings, etching with chromic acid, and metal finishing operations carried out on chromium as a basis material.

The selected treatment option involves the reduction of hexavalent chromium to trivalent chromium either chemically or electrochemically. The reduced chromium can then be removed using a conventional precipitation-solids removal system. Alternative hexavalent chromium treatment techniques include chromium regeneration, electrodialysis, evaporation, and ion exchange [15].

10.4.5   CYANIDE

Cyanides are introduced as metal salts for plating and conversion coating or are active components in plating and cleaning baths. Cyanide is generally destroyed by oxidation. Chlorine, in either elemental or hypochlorate form, is the primary oxidation agent used in industrial waste treatment to destroy cyanide. Alternative treatment techniques for the destruction of cyanide include oxidation by ozone, ozone with ultraviolet (UV) radiation (oxyphotolysis), hydrogen peroxide, and electrolytic oxidation [17]. Treatment techniques, which remove cyanide but do not destroy it, include chemical precipitation, reverse osmosis, and evaporation [15,17].

10.4.6   OILS

Oily wastes and toxic organics that combine with the oils during manufacturing include process coolants and lubricants, wastes from cleaning operations, wastes from painting processes, and machinery lubricants. Oily wastes are generally of three types: free oils, emulsified or water-soluble oils, and greases. Oil removal techniques commonly employed in the metal finishing industry include skimming, coalescing, emulsion breaking, flotation, centrifugation, ultrafiltration, reverse osmosis, carbon adsorption, and aerobic decomposition [17–19].

Because emulsified oils and processes that emulsify oils are used extensively in the metal finishing industry, the exclusive occurrence of free oils is nearly nonexistent.

Treatment of oily wastes can be carried out most efficiently if oils are segregated from other wastes and treated separately. Segregated oily wastes originate in the manufacturing areas and are collected in holding tanks and sumps. Systems for treating segregated oily wastes consist of separating oily wastes from the water. If oily wastes are emulsified, techniques such as emulsion breaking or dissolved air flotation (DAF) [20] with the addition of chemicals are necessary to remove oil. Once the oil-water emulsion is broken, the oily waste is physically separated from the water by decantation or skimming. After the oil-water separation has been carried out, the water is sent to the precipitation/sedimentation unit used for metal removal. There are three options for oily waste removal:

- *The option 1* system incorporates the emulsion breaking process followed by surface skimming (gravity separation is adequate if only free oils are present).
- *The option 2* system consists of the option 1 system followed by ultrafiltration.
- *The option 3* treatment system consists of the option 2 system with the addition of either carbon adsorption or reverse osmosis.

In addition to these three treatment options, several alternative technologies are applicable to treating oily wastewater. These include coalescing, flotation, centrifugation, integrated adsorption, resin adsorption, ozonation, chemical oxidation, aerobic decomposition, and thermal emulsion breaking [17–19].

10.4.7 SOLVENTS

Spent degreasing solvents should be segregated from other process fluids to maximize the value of the solvents, to preclude contamination of other segregated wastes, and to prevent the discharge of priority pollutants to any wastewaters. This segregation may be accomplished by providing and identifying the necessary storage containers, establishing clear disposal procedures, training personnel in using these techniques, and checking periodically to ensure that proper segregation is occurring. Segregated waste solvents are appropriate for on-site solvent recovery or maybe contract hauled for disposal or reclamation.

Alkaline cleaning is the most feasible substitute for solvent degreasing. The major advantage of alkaline cleaning over solvent degreasing is the elimination or reduction in the quantity of priority pollutants being discharged. Major disadvantages include high-energy consumption and the tendency to dilute oils removed and to discharge these oils as well as the cleaning additive.

10.5 TREATMENT TECHNOLOGIES

10.5.1 NEUTRALIZATION

One technique, used in a number of facilities that utilize molten salt for metal surface treatment prior to pickling, is to take advantage of the alkaline values generated in the molten salt bath in treating other wastes generated in the plant. When the bath is determined to be spent, it is in many instances manifested, hauled off-site, and land disposed. Another technique is to take the solidified spent molten salt (molten salt is sold at ambient temperatures) and circulate acidic wastes generated in the facility over the material prior to entry into the waste treatment system. This, in effect, neutralizes the acid wastes and eliminates the requirements of manifesting and land disposal.

10.5.2 Cyanide-Containing Wastes

There are eight methods applicable to the treatment of cyanide wastes for metal finishing [5,21]:

1. Alkaline chlorination
2. Electrolytic decomposition
3. Ozonation
4. UV/ozonation
5. Hydrogen peroxide
6. Thermal oxidation
7. Acidification and acid hydrolysis
8. Ferrous sulfate precipitation.

Alkaline chlorination is the most widely used method in the metal finishing industry. A schematic diagram of cyanide reduction via alkaline chlorination is provided in Figure 10.5. This technology generally applies to wastes containing less than 1% cyanide, generally present as free cyanide. It is conducted in two stages: the first stage is operated at a pH greater than 10 and the second stage is operated at a pH range of 7.5–8. Alkaline chlorination is performed using sodium hypochlorite and chlorine.

Electrolytic decomposition technology was applied to cyanide-containing wastes in the early part of this century. It fell from favor as alkaline chlorination came into use at large-scale facilities. However, as wastes become more concentrated, this technology may find more widespread application in the future. The reason is that it applies to wastes containing cyanide over 1%. The basis of this technology is the electrolytic decomposition of the cyanide compounds at an elevated temperature (200°F) to yield nitrogen, CO_2, ammonia, and amines (Figure 10.6).

$$NaCN + Cl_2 \rightarrow CNCl + NaCl$$

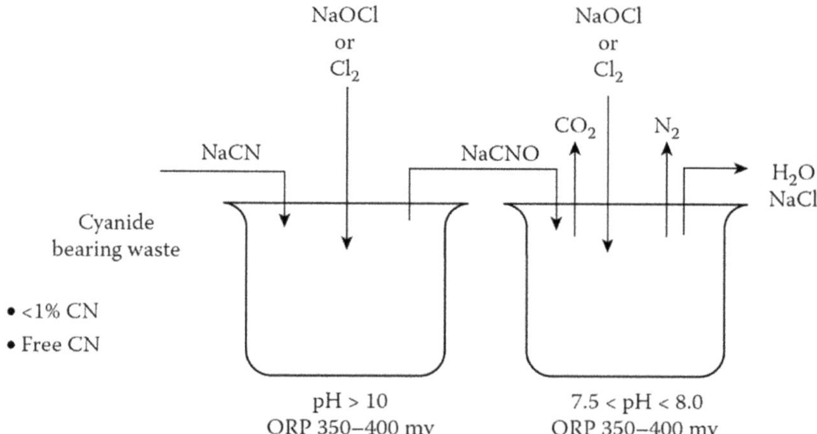

FIGURE 10.5 Cyanide reduction via alkaline chlorination. (Adapted from U.S. EPA. *Meeting Hazardous Waste Requirements for Metal Finishers.* Report no. EPA/625/4-87/018. U.S. Environmental Protection Agency, Cincinnati, OH, 1987 [5].)

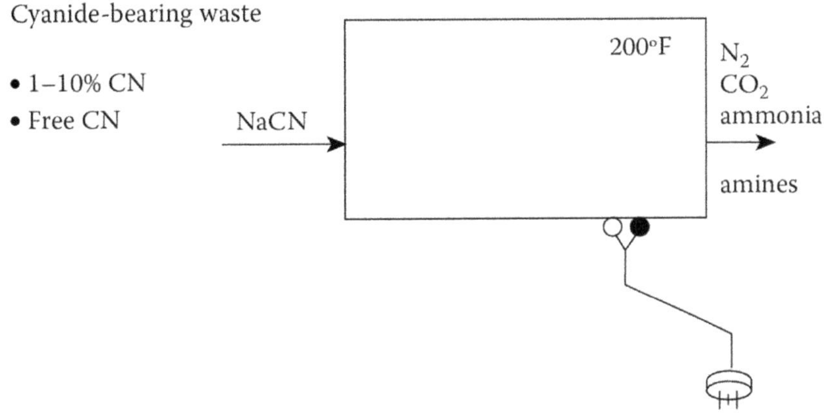

FIGURE 10.6 Cyanide reduction via electrolytic decomposition. (Adapted from U.S. EPA. *Meeting Hazardous Waste Requirements for Metal Finishers*. Report no. EPA/625/4-87/018. U.S. Environmental Protection Agency, Cincinnati, OH, 1987 [5].)

$$CNCl + 2NaOH \rightarrow NaCNO + H_2O + NaCl$$

$$2NaCNO + 3Cl_2 + 4NaOH \rightarrow 2CO_2 + N_2 + 6NaCl + 2H_2O$$

Ozonation treatment can be used to oxidize cyanide, thereby reducing the concentration of cyanide in wastewater. Ozone, with an electrode potential of +1.24 V in alkaline solutions, is one of the most powerful oxidizing agents known. Cyanide oxidation with ozone is a two-step reaction similar to alkaline chlorination [21]. Cyanide is oxidized to cyanate, with ozone reduced to oxygen as per the following equation:

$$CN + O_3 \rightarrow CNO + O_2 \tag{10.1}$$

Then cyanate is hydrolyzed, in the presence of excess ozone, to bicarbonate and nitrogen and oxidized as per the following reaction:

$$2CNO + 3O_3 + H_2O \rightarrow N_2 + 2HCO_3^- + 3O_2. \tag{10.2}$$

The reaction time for complete cyanide oxidation is rapid in a reactor system with 10–30 minutes retention time being typical. The second-stage reaction is much slower than the first-stage reaction. The reaction is typically carried out in the pH range of 10–12, where the reaction rate is relatively constant. Temperature does not influence the reaction rate significantly.

One interesting variation of ozonation technology is augmentation with UV radiation. This is a technology that has been applied to wastes in the coke byproduct manufacturing industry. A significant development has been made that has resulted in significantly less ozone consumption through the use of UV radiation. UV absorption has the following effects:

- Ozone and cyanide are raised to a higher energy status
- Free radicals are formed
- More rapid reaction
- Less ozone is required.

Cyanide reduction with hydrogen peroxide is effective in reducing cyanide. It has been applied on a less frequent basis within this industry, because there are high operating costs associated with hydrogen peroxide generation. The reduction of cyanide with peroxide occurs in two steps and yields CO_2 and ammonia:

$$NaCN + H_2O_2 \rightarrow NaCNO + H_2O, \tag{10.3}$$

$$NaCNO + 2H_2O \rightarrow CO_2 + NH_3 + NaOH. \tag{10.4}$$

Thermal oxidation is another alternative for destroying cyanide. Thermal destruction of cyanide can be accomplished through either high-temperature hydrolysis or combustion. At temperatures between 140°C and 200°C and a pH of 8, cyanide hydrolyzes quite rapidly to produce formate and ammonia [22]. Pressures up to 100 bar are required, but the process can effectively treat waste streams over a wide concentration range and applies to both rinse water and concentrated solutions [21].

$$CN^- + 2H_2O \rightarrow HCOO^- + NH_3. \tag{10.5}$$

In the presence of nitrates, formate and ammonia can be destroyed in another reactor at 150°C, according to the following equations:

$$NH_4^+ + NO_2^- \rightarrow N_2 + 2H_2O, \tag{10.6}$$

$$3HCOOH + 2NO_2^- + 2H^+ \rightarrow 3CO_2 + 4H_2O. \tag{10.7}$$

Direct acidification of cyanide waste streams was once a relatively common treatment. Cyanide is acidified in a sealed reactor that is vented to the atmosphere through an air emission control system. Cyanide is converted to gaseous hydrogen cyanide, treated, vented, and dispersed.

Acid hydrolysis of cyanates is still commonly used, following a first-stage cyanide oxidation process. At pH 2, the reaction proceeds rapidly, whereas at pH 7, cyanate may remain stable for weeks [23]. This treatment process requires specially designed reactors to assure that HCN is properly vented and controlled. The hydrolysis mechanisms are as follows [21]:

In acid medium,

$$HOCN + H^+ \rightarrow NH_4^+ + CO_2 \ \left(\text{rapid}\right), \tag{10.8}$$

$$HOCN + H_2O \rightarrow NH_3 + CO_2 \ \left(\text{slow}\right) \tag{10.9}$$

In a strongly alkaline medium,

$$NCO^- + 2H_2O \rightarrow NH_3 + HCO_3^- \ \left(\text{very slow}\right) \tag{10.10}$$

Each of the technologies described above is effective in treating wastes containing free cyanides; that is, cyanides present as CN in solution. There are instances in metal finishing facilities where complex cyanides are present in wastes. The most common are complexes of iron, nickel, and zinc.

A technology that has been applied to remove complex cyanides from aqueous wastes is ferrous sulfate precipitation. The technology involves a two-stage operation in which ferrous sulfate is first added at a pH of 9 to complex any trace amounts of free cyanide. In the second stage, the complex cyanides are precipitated by adding ferrous sulfate or ferric chloride at a pH range of 2–4 [5].

10.5.3 Chromium-Containing Wastes

There are three treatment methods applicable to wastes containing hexavalent chromium. Wastes containing trivalent chromium can be treated using chemical precipitation and sedimentation, as discussed below. The three methods applicable to the treatment of hexavalent chromium are

1. Sulfur dioxide
2. Sodium metabisulfite
3. Ferrous sulfate

Hexavalent chromium reduction using sulfur dioxide and sodium metabisulfite has found the widest application in the metal finishing industry. It is not truly a treatment step but a conversion process in which the hexavalent chromium is converted to trivalent chromium.

The hexavalent chromium is reduced by adding the reductant at a pH range of 2.5–3 with a retention time of approximately 30–40 minutes (Figure 10.7).

Ferrous sulfate has not been as widely applied. However, it is particularly applicable in facilities where ferrous sulfate is produced as part of the process or is readily available. The basis for this technology is that the hexavalent chromium is reduced to trivalent chromium, and the ferrous iron is oxidized to ferric iron.

10.5.4 Arsenic and Selenium-Containing Wastes

It may be necessary to segregate waste streams containing elevated concentrations of arsenic and selenium, especially waste streams with concentrations in excess of 1 mg/L for these pollutants. Arsenic and selenium form anionic acids in solution (most other metals act as cations) and require special preliminary treatment prior to conventional metals treatment. Lime, a source of calcium

$$SO_2 + H_2O \rightarrow H_2SO_3$$

$$2H_2CrO_4 + 3H_2SO_3 \rightarrow Cr_2(SO_4)_3 + 5H_2O$$

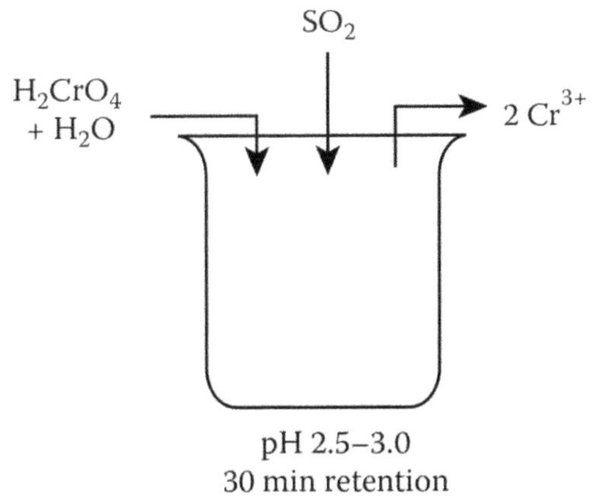

FIGURE 10.7 Hexavalent chromium reduction. (Adapted from U.S. EPA. *Meeting Hazardous Waste Requirements for Metal Finishers*. Report no. EPA/625/4-87/018. U.S. Environmental Protection Agency, Cincinnati, OH, 1987 [5].)

ions, is effective in reducing arsenic and selenium concentrations when the initial concentration is below 1 mg/L. However, preliminary treatment with sodium sulfide at a low pH (i.e., 1–3) may be required for waste streams with concentrations over 1 mg/L [21]. The sulfide reacts with the anionic acids to form insoluble sulfides, which are readily separated using filtration.

10.5.4.1 Chemical Precipitation and Sedimentation

The most important technology in metal treatment is chemical precipitation and sedimentation. It is accomplished by adding a chemical reagent to form metal precipitants, which are then removed as solids in a sedimentation step. The options available to a facility as precipitation reagents are lime $Ca(OH)_2$, caustic $NaOH$, carbonate $CaCO_3$ and Na_2CO_3, sulfide $NaHS$ and FeS, and sodium borohydride $NaBH_4$. The advantages and disadvantages of these reagents are summarized below [21]:

1. Lime
   - Least expensive precipitation reagent
   - Generates the highest sludge volume
   - Sludges generally cannot be sold to smelter/refiners
2. Caustic
   - More expensive than lime
   - Generates smaller volume of sludge
   - Sludges can be sold to smelter/refiners
3. Carbonates
   - Applicable for metals where solubility within a pH range is insufficient to meet treatment standards.

Lime is the least expensive reagent; however, it generates the highest volume of residue. It also generates a residue that cannot be resold to smelters and refiners for reclaiming because of the presence of the calcium ion. Caustic is more expensive than lime; however, it generates a smaller volume of residue. One key advantage of caustic is that the resulting residues can be readily reclaimed. Carbonates are particularly appropriate for metals where solubility within a pH range is insufficient to meet a given set of treatment standards. The sulfides offer the benefit of achieving effective treatment at lower concentrations due to the lower solubilities of the metal sulfides. Sodium borohydride has applications where small volumes of sludge that are suitable for reclamation are desired.

It is appropriate to look at reagent use in the context of the current regulatory framework under HSWA. Historically, lime has been the reagent of choice. It was relatively inexpensive and simple to handle. The phrase "Lime and Settle" refers to applying lime precipitation and sedimentation technology. In the 1970s, new designs used caustic as the precipitation reagent because of the reduction in residue volume realized and the ability for reclamation. In the 1980s, a return to lime and the use of combined reagent techniques came into use.

One obvious question is why return to lime as a treatment reagent, given that caustic results in a smaller residue volume and a waste that can undergo reclamation? The answer lies in the three points that result from the implementation of the HSWA hierarchy. As source reduction and material reuse and recovery techniques are applied, facilities will generating

- More concentrated wastes
- Wastes with a varied array of constituents
- Wastes with a greater degree of complexation.

10.5.4.2 Complexation

Complexation is a phenomenon that involves a coordinate bond between a central atom (the metal) and a ligand (the anions). In a coordinate bond, the electron pair is shared by the metal and the ligand. A complex containing one coordinate bond is referred to as a monodentate complex. Multiple coordinate bonds are characteristic of polydentate complexes. Polydentate complexes are also referred to as chelates. An example of a monodentate forming ligand is ammonia. Examples of chelates are oxylates (bidentates) and EDTA (hexadentates).

The reason for the return to lime is that calcium ions are present in lime. The calcium ions present in solution on the addition of lime are very effective in competing with the ligand for the metal ions. The sodium ions contributed by caustic is not effective. As such, lime dramatically reduces complexation and is more effective in treating complexed wastes. The term "high lime treatment" has been applied in cases where excess calcium ions are introduced into the solution. This is accomplished through the addition of lime to raise the pH to approximately 11.5 or through the addition of calcium chloride (which has a greater solubility than lime).

The use of combinations of precipitation reagents has been most effective in taking advantage of the attributes of caustic as well as the advantages of lime. As an example, a system may use caustic in the first stage to make a coarse pH adjustment followed by the addition of lime to make a fine adjustment. This achieves an overall reduction in the sludge volume through the use of the caustic and a more effective metal removal through the use of lime. Sulfide reagents are used in a similar fashion in combination with caustic or lime to provide additional metal removal due to the lower solubility of the metal sulfides. Sulfides also apply to wastes containing elevated concentrations (i.e., over 2 mg/L) of selenium and arsenic compounds [21].

10.5.5 OTHER METAL WASTES

There are three techniques applicable to managing solids generated in metal finishing. These are

1. Dewatering
2. Stabilization
3. Incineration.

There are four dewatering techniques that have been applied in metal processing. The most widely applied techniques are vacuum and belt filtration [24]. They have a higher relative capital cost but generally have a lower relative operating cost. Plate and frame filter presses have experienced less widespread application. Belt filters generally have a lower relative capital cost and higher relative operating costs. The higher operating costs are because the units are more labor intensive. Centrifuges [24] have been applied in specific instances but are more difficult to operate when a widely varying mix of wastes is treated.

Experience has shown that companies are most successful in applying a dewatering technique that they have successfully designed and operated in similar applications within the company. For example, many companies operate plate and frame filter presses as a part of metal manufacturing operations. The knowledge gained in metal processing has been successfully transferred to treating metal finishing wastes.

There are six stabilization techniques currently available; however, only two of them have found widespread application. These are cementation and stabilization through the addition of lime and fly ash [24,25]. There is currently developmental work being undertaken to make use of bitumen, paraffin, and polymeric materials to reduce the degree to which metals can be taken into solution. Encapsulation with inert materials is also under development.

10.6 COSTS

The investment cost, operating and maintenance [26,27] and energy costs for applying control technologies to the wastewaters of the metal finishing industry have been analyzed. These costs were developed to reflect the conventional use of technologies in this industry. The detailed presentation of the cost methodology and cost data is available in a U.S. EPA publication [6]. The available industry-specific cost information is characterized below.

10.6.1 TYPICAL TREATMENT OPTIONS

Many waste treatment options are available [27–31]. Only several unit operation/unit process configurations have been analyzed for the cost of application to the wastewater of this industry. The components included in these configurations are

- **Option 1:** Emulsion breaking and oil separation by skimming, cyanide oxidation, chromium reduction, chemical precipitation and sedimentation, and sludge drying beds.
- **Option 2:** All of option 1 plus multimedia filtration.
- **Option 3:** All of option 2 plus ultrafiltration and carbon adsorption for oily waste, zero discharge of any processes using either cadmium or lead by using the evaporative system.

The flow diagram for suggested option 1 is shown in Figure 10.8. The flow diagram for the other options would be similar.

10.6.2 COST ANALYSIS

The cost estimates prepared for the treatment technologies commonly used in this industry are briefly described below. More details of the factors considered in the cost analysis are available in the source [6].

1. Emulsion breaking and oil separation
   Method: Emulsion broken by mixing oily waste with alum and a chemical emulsion breaker, followed by gravity oil separation in a tank.
   System component: Small mixing tank, two chemical feed tanks, a mixer, and a large tank equipped with an oil skimmer and a sludge pump. The mixing tank has a retention time of 15 minutes and the oil skimming tank has a retention time of 2.5 hours.
2. Cyanide oxidation
   Method: Cyanide is destroyed by reaction with sodium hypochlorite under alkaline conditions.

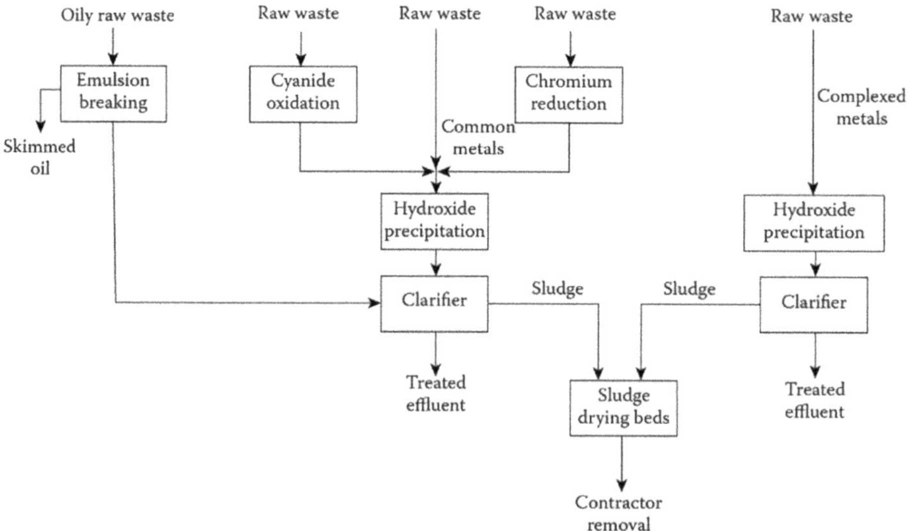

FIGURE 10.8 Metal finishing wastewater treatment flow diagram. (Adapted from U.S. EPA. *Treatability Manual, Volume II. Industrial Descriptions.* Report no. EPA-600/2–82-001b. U.S. Environmental Protection Agency, Washington, DC, September 1981 [7].)

System component: Reaction tanks, a reagent storage and feed system, mixers, sensors, and controls. Two identical reaction tanks sized as aboveground cylindrical tanks with a retention time of 4 hours are provided. Chemical storage consists of covered concrete tanks to store 60 days supply of sodium hypochlorite and 90 days supply of sodium hydroxide.

3. Chromium reduction

   Method: Chemical reduction of hexavalent chromium by sulfur dioxide under acid conditions for the continuous operating system and by sodium bisulfite under acid conditions for the batch operating system. The reduced trivalent form of chromium is subsequently removed by precipitation as the hydroxide.

   System component: Reaction tanks, a reagent storage and feed system, mixers, sensors, and controls for continuous chromium reduction. A single aboveground concrete tank with a retention time of 45 minutes is provided. For batch operation, dual aboveground concrete tanks with 4 hours retention time are provided.

4. Lime precipitation and sedimentation

   Method: Chemical precipitation of dissolved and complexed metals by reaction with lime and subsequent removal of the precipitated solids by gravity settling in a clarifier. Alum and polyelectrolytes are added for coagulation and flocculation.

   System component: The continuous treatment system includes reagent storage and feed equipment, a mix tank for reagent feed addition, sensors and controls, and a clarification basin with associated sludge rakes and pumps. Lime is fed as 30% lime slurry prepared by using hydrated lime. The mix tank is sized for a retention time of 45 minutes, and the clarifier is sized for hydraulic loading of 1,360 L/m^2 and a retention time of 4 hours. Batch treatment includes dual reaction-settling tanks sized for 8 hours of retention time and sludge pumps.

5. Sludge drying beds

   Method: Sludge dewatered using gravity drainage and natural evaporation.

   System component: Beds of highly permeable gravel and sand underlain by drain pipes [28].

6. Multimedia filter

   Method: Polishing treatment after chemical precipitation and sedimentation by filtration through a bed of particles of several distinct size ranges.

   System components: Filter beds, media, backwash mechanism, pumps, and controls. Filter beds sized for hydraulic loading of 81 $L/min/m^2$ (2 gpm/ft^2).

7. Ultrafiltration

   Method: Process used for oily waste stream after emulsion breaking-gravity oil separation.

   System component: Filter modules sized based on hydraulic loading of 1 $L/min/m^2$.

8. Carbon adsorption

   Method: A packed-bed throwaway system to remove organic pollutants from oily waste stream.

   System component: Contactor system and a pump station designed for a contact time of 30 minutes and a hydraulic loading of 162 $L/min/m^2$ (4 gpm/ft^2).

Unit costs shown in Table 10.16 are for the complete treatment options described previously. Unit costs are computed for a model plant where flows are contributed by several waste streams as follows:

- 30% oily waste stream,
- 4% cyanide waste stream,
- 9% chromium waste stream,
- 52.5% common metals stream, and
- 4.5% complex metal stream.

TABLE 10.16
Total Annual Unit Cost (USD/m³ in 2022 Dollars)[a]

	Option 1		Option 2		Option 3	
Flow (m³/h)	Continuous	Batch	Continuous	Batch	Continuous	Batch
2.36	—	19.24	—	32.86	—	38.20
11.81	8.20	6.74	17.54	11.32	15.28	13.87
59.07	3.40	—	6.23	—	7.07	—
118.16	2.83	2.83	4.81	5.09	5.66	5.94

Source: U.S. EPA. *Treatability Manual, Volume II. Industrial Descriptions.* Report no. EPA-600/2-82-001b. U.S. Environmental Protection Agency, Washington, DC, September 1981 [7].

[a] Costs were converted from 1979 USD to 2022 USD using U.S. ACE Yearly Average Cost Index for Utilities. (Adapted from U.S. ACE. Yearly Average Cost Index for Utilities. In: 110-2-1304. U.S. Army Corps of Engineers, Washington, DC. [32,50] See Table A1 for the ACE Cost Index.

10.7   U.S. CODE OF FEDERAL REGULATIONS FOR METAL FINISHING EFFLUENT DISCHARGE MANAGEMENT

This section introduces the U.S. Code of Federal Regulations (CFR) Title 40, Part 433 (40 CFR part 433) for effluent discharge management of the metal finishing point source category.

The topics introduced in this section include (a) the applicability and description of the metal finishing point source category; (b) the monitoring requirements of metal finishing effluent discharges; (c) the effluent limitations representing the degree of effluent reduction attainable by applying the best practicable control technology (BPT) available; (d) the effluent limitations representing the degree of effluent reduction attainable by applying the best available technology (BAT) economically achievable; (e) the pretreatment standards for existing sources (PSES); (f) the new source performance standards (NSPS); and (g) the pretreatment standards for new sources (PSNS).

10.7.1   APPLICABILITY AND DESCRIPTION OF THE METAL FINISHING POINT SOURCE CATEGORY

Except as noted in the following two paragraphs of this section, the provisions of this subpart apply to plants that perform any of the following six metal finishing operations on any basis material: electroplating, electroless plating, anodizing, coating (chromating, phosphating, and coloring), chemical etching and milling, and printed circuit board manufacture. If any of those six operations are present, then this part applies to discharges from those operations and also to discharges from any of the following 40 process operations: cleaning, machining, grinding, polishing, tumbling, burnishing, impact deformation, pressure deformation, shearing, heat treating, thermal cutting, welding, brazing, soldering, flame spraying, sandblasting, other abrasive jet machining, electric discharge machining, electrochemical machining, electron beam machining, laser beam machining, plasma arc machining, ultrasonic machining, sintering, laminating, hot dip coating, sputtering, vapor plating, thermal infusion, salt bath descaling, solvent degreasing, paint stripping, painting, electrostatic painting, electropainting, vacuum metalizing, assembly, calibration, testing, and mechanical plating.

In some cases, effluent limitations and standards for the following industrial categories may be effective and applicable to wastewater discharges from the metal finishing operations listed above. In such cases, the 40 CFR part 433 limits shall not apply, and the following regulations shall apply:

- Nonferrous metal smelting and refining (40 CFR part 421)
- Coil coating (40 CFR part 465)
- Porcelain enameling (40 CFR part 466)

- Battery manufacturing (40 CFR part 461)
- Iron and steel (40 CFR part 420)
- Metal casting foundries (40 CFR part 464)
- Aluminum forming (40 CFR part 467)
- Copper forming (40 CFR part 468)
- Plastic molding and forming (40 CFR part 463)
- Nonferrous forming (40 CFR part 471)
- Electrical and electronic components (40 CFR part 469).

The 40 CFR part 433 does not apply to the following: (a) metallic platemaking and gravure cylinder preparation conducted within or for printing and publishing facilities and (b) existing indirect discharging job shops and independent printed circuit board manufacturers that are covered by 40 CFR part 413.

10.7.2 Monitoring Requirements of Metal Finishing Effluent Discharges

In lieu of requiring monitoring for total toxic organics (TTO), the permitting authority (or, in the case of indirect dischargers, the control authority) may allow dischargers to make the following certification statement:

> Based on my inquiry of the person or persons directly responsible for managing compliance with the permit limitation [or pretreatment standard] for total toxic organics (TTO), I certify that, to the best of my knowledge and belief, no dumping of concentrated toxic organics into the wastewaters has occurred since filing of the last discharge monitoring report. I further certify that this facility is implementing the toxic organic management plan submitted to the permitting [or control] authority.

For direct dischargers, this statement is to be included as a "comment" on the Discharge Monitoring Report required by 40 CFR 122.44(i), formerly 40 CFR 122.62(i).

For indirect dischargers, the statement is to be included as a comment to the periodic reports required by 40 CFR 403.12(e). If monitoring is necessary to measure compliance with the TTO standard, the industrial discharger need to analyze for only those pollutants that would reasonably be expected to be present.

In requesting the certification alternative, a discharger shall submit a solvent management plan that specifies to the satisfaction of the permitting authority (or, in the case of indirect dischargers, the control authority) the toxic organic compounds used; the method of disposal used instead of dumping, such as reclamation, contract hauling, or incineration; and procedures for ensuring that toxic organics do not routinely spill or leak into the wastewater. For direct dischargers, the permitting authority shall incorporate the plan as a provision of the permit.

Self-monitoring for cyanide must be conducted after cyanide treatment and before dilution with other streams. Alternatively, samples may be taken of the final effluent, if the plant limitations are adjusted based on the dilution ratio of the cyanide waste stream flow to the effluent flow.

10.7.3 Effluent Limitations Based on the BPT Currently Available

Except as specifically provided in the U.S. CFR, any existing point source subject to the 40 CFR part 433 must achieve the effluent limitations shown in Table 10.17, which represents the degree of effluent reduction attainable by applying the best practicable control technology currently available (BPT). Alternatively, for metal finishing industrial facilities with cyanide treatment, and upon agreement between a source subject to those limits and the pollution control authority, the amenable cyanide limit shown in Table 10.18 may apply in place of the total cyanide limit specified in Table 10.17. No user subject to the provisions of these regulations shall augment the use of process wastewater or otherwise dilute the wastewater as a partial or total substitute for adequate treatment to achieve compliance with this limitation.

TABLE 10.17

U.S. Best Practicable Control Technology (BPT) Currently Available Effluent Limitations for the Metal Finishing Point Source Category

Pollutant or Pollutant Property	Maximum for Any 1 day	Monthly Average not Exceed
	(mg/L except for pH)	
Cadmium (T)	0.69	0.26
Chromium (T)	2.77	1.71
Copper (T)	3.38	2.07
Lead (T)	0.69	0.43
Nickel (T)	3.98	2.38
Silver (T)	0.43	0.24
Zinc (T)	2.61	1.48
Cyanide (T)	1.20	0.65
TTO	2.13	
Oil & grease	52	26
TSS	60	31
pH	6-9	6-9

Source: U.S. EPA. Code of Federal Regulations. *Metal Finishing Point Source Category.* Title 40, Vol. 27, Part 433. U.S. Environmental Protection Agency, Washington, DC, revised as of July 1, 2003 [33].

TABLE 10.18

Alternative U.S. Best Practicable Control Technology (BPT) Currently Available Effluent Limitations on Cyanide (A) for the Metal Finishing Point Source Category

Pollutant or Pollutant Property	Maximum for Any 1 day (mg/L)	Monthly Average Shall not Exceed (mg/L)
Cyanide (A)	0.86	0.32

Source: U.S. EPA. Code of Federal Regulations. *Metal Finishing Point Source Category.* Title 40, Vol. 27, Part 433. U.S. Environmental Protection Agency, Washington, DC, revised as of July 1, 2003 [33].

10.7.4 EFFLUENT LIMITATIONS BASED ON THE BAT

Except as specifically provided in the U.S. CFR, any existing point source subject to this subpart must achieve the effluent limitations shown in Table 10.19, which represents the degree of effluent reduction attainable by applying the BAT. Alternatively, for the metal finishing industrial facilities with cyanide treatment, and upon agreement between a source subject to those limits and the pollution control authority, the amenable cyanide limit shown in Table 10.20 may apply in place of the total cyanide limit specified in Table 10.19. No user subject to the provisions of these regulations shall augment the use of process wastewater or otherwise dilute the wastewater as a partial or total substitute for adequate treatment to achieve compliance with this limitation.

10.7.6 PSES

Except as specifically provided in the U.S. CFR, any existing source subject to this 40 CFR part 433 that introduces pollutants into POTW must also comply with 40 CFR part 403 and achieve the PSES. Table 10.21 indicates the PSES for all metal finishing plants except job shops and independent

TABLE 10.19

U.S. BAT Effluent Limitations for the Metal Finishing Point Source Category

Pollutant or Pollutant Property	Maximum for Any 1 day	Monthly Average Shall not Exceed
	(mg/L except for pH)	
Cadmium (T)	0.69	0.26
Chromium (T)	2.77	1.71
Copper (T)	3.38	2.07
Lead (T)	0.69	0.43
Nickel (T)	3.98	2.38
Silver (T)	0.43	0.24
Zinc (T)	2.61	1.48
Cyanide (T)	1.20	0.65
TTO	2.13	

Source: U.S. EPA. Code of Federal Regulations. *Metal Finishing Point Source Category.* Title 40, Vol. 27, Part 433. U.S. Environmental Protection Agency, Washington, DC, revised as of July 1, 2003 [33].

TABLE 10.20

Alternative U.S. BAT Effluent Limitations on Cyanide (A) for the Metal Finishing Point Source Category

Pollutant or Pollutant Property	Maximum for Any 1 day (mg/L)	Monthly Average Shall not Exceed (mg/L)
Cyanide (A)	0.86	0.32

Source: U.S. EPA. Code of Federal Regulations. *Metal Finishing Point Source Category.* Title 40, Vol. 27, Part 433. U.S. Environmental Protection Agency, Washington, DC, revised as of July 1, 2003 [33].

TABLE 10.21

U.S. PSES for All Metal Finishing Plants Except Job Shops and Independent Printed Circuit Board Manufacturers

Pollutant or Pollutant Property	Maximum for Any 1 day	Monthly Average Shall not Exceed
	(mg/L except for pH)	
Cadmium (T)	0.11	0.07
Chromium (T)	2.77	1.71
Copper (T)	3.38	2.07
Lead (T)	0.69	0.43
Nickel (T)	3.98	2.38
Silver (T)	0.43	0.24
Zinc (T)	2.61	1.48
Cyanide (T)	1.20	0.65
TTO	2.13	

Source: U.S. EPA. Code of Federal Regulations. *Metal Finishing Point Source Category.* Title 40, Vol. 27, Part 433. U.S. Environmental Protection Agency, Washington, DC, revised as of July 1, 2003 [33].

printed circuit board manufacturers. Alternatively, for industrial facilities with cyanide treatment, upon agreement between a source subject to those limits and the pollution control authority, the amenable cyanide limit shown in Table 10.22 may apply in place of the total cyanide limit specified in Table 10.21. No user introducing wastewater pollutants into a POTW under the provisions of this subpart shall augment the use of process wastewater as a partial or total substitute for adequate treatment to achieve compliance with this standard. An existing source submitting a certification in lieu of monitoring pursuant to this regulation must implement the toxic organic management plan approved by the control authority. An existing source subject to this subpart shall comply with a daily maximum pretreatment standard for TTO of 4.57 mg/L.

10.7.7 NSPS

Any new metal finishing point source subject to the 40 CFR part 433 regulations must achieve the NSPS shown in Table 10.23. Alternatively, for the metal finishing industrial facilities with cyanide treatment, and upon agreement between a source subject to those limits and the pollution control authority, the amenable cyanide limit shown in Table 10.24 may apply in place of the total cyanide limit specified in Table 10.23. No user subject to the provisions of this subpart shall augment the use of process wastewater or otherwise dilute the wastewater as a partial or total substitute for adequate treatment to achieve compliance with this limitation.

TABLE 10.22

Alternative U.S. PSES on Cyanide (A) for all Metal Finishing Plants Except Job Shops and Independent Printed Circuit Board Manufacturers

Pollutant or Pollutant Property	Maximum for Any 1 day (mg/L)	Monthly Average Shall not Exceed (mg/L)
Cyanide (A)	0.86	0.32

Source: U.S. EPA. Code of Federal Regulations. *Metal Finishing Point Source Category.* Title 40, Vol. 27, Part 433. U.S. Environmental Protection Agency, Washington, DC, revised as of July 1, 2003 [33].

TABLE 10.23
U.S. NSPS for the Metal Finishing Point Source Category

Pollutant or Pollutant Property	Maximum for Any 1 day	Monthly Average Shall not Exceed
	(mg/L except for pH)	
Cadmium (T)	0.11	0.07
Chromium (T)	2.77	1.71
Copper (T)	3.38	2.07
Lead (T)	0.69	0.43
Nickel (T)	3.98	2.38
Silver (T)	0.43	0.24
Zinc (T)	2.61	1.48
Cyanide (T)	1.20	0.65
TTO	2.13	
Oil & grease	52	26
TSS	60	31
pH	6-9	6-9

Source: U.S. EPA. Code of Federal Regulations. *Metal Finishing Point Source Category.* Title 40, Vol. 27, Part 433. U.S. Environmental Protection Agency, Washington, DC, revised as of July 1, 2003 [33].

TABLE 10.24

Alternative U.S. NSPS on Cyanide (A) for the Metal Finishing Point

Pollutant or Pollutant Property	Maximum for Any 1 day (mg/L)	Monthly Average Shall not Exceed (mg/L)
Cyanide (A)	0.86	0.32

Source Category

Source: U.S. EPA. Code of Federal Regulations. *Metal Finishing Point Source Category.* Title 40, Vol. 27, Part 433. U.S. Environmental Protection Agency, Washington, DC, revised as of July 1, 2003 [33].

TABLE 10.25

U.S. PSNS for the Metal Finishing Point Source Category

Pollutant or Pollutant Property	Maximum for Any 1 day	Monthly Average Shall not Exceed
	(mg/L except for pH)	
Cadmium (T)	0.69	0.26
Chromium (T)	2.77	1.71
Copper (T)	3.38	2.07
Lead (T)	0.69	0.43
Nickel (T)	3.98	2.38
Silver (T)	0.43	0.24
Zinc (T)	2.61	1.48
Cyanide (T)	1.20	0.65
TTO	2.13	

Source: U.S. EPA. Code of Federal Regulations. *Metal Finishing Point Source Category.* Title 40, Vol. 27, Part 433. U.S. Environmental Protection Agency, Washington, DC, revised as of July 1, 2003 [33].

TABLE 10.26

Alternative U.S. PSNS on Cyanide (A) for the Metal Finishing Point Source Category

Pollutant or Pollutant Property	Maximum for Any 1 day (mg/L)	Monthly Average Shall not Exceed(mg/L)
Cyanide (A)	0.86	0.32

Source: U.S. EPA. Code of Federal Regulations. *Metal Finishing Point Source Category.* Title 40, Vol. 27, Part 433. U.S. Environmental Protection Agency, Washington, DC, revised as of July 1, 2003 [33].

10.7.8 PSNS

Except as provided in the U.S. CFR, any new source subject to this subpart that introduces pollutants into a POTW must comply with 40 CFR part 403 and achieve the PSNS, shown in Table 10.25. Alternatively, for industrial facilities with cyanide treatment, and upon agreement between a source subject to these limits and the pollution control authority, the amenable cyanide limit shown in Table 10.26 may apply in place of the total cyanide limit specified in Table 10.25.

No user subject to the provisions of this subpart shall augment the use of process wastewater or otherwise dilute the wastewater as a partial or total substitute for adequate treatment to achieve compliance with this limitation. An existing source submitting a certification in lieu of monitoring pursuant to Section 433.12 (a) and (b) of this regulation must implement the toxic organic management plan approved by the control authority.

10.8　SPECIALIZED DEFINITIONS

The definitions set forth in the U.S. CFR for the metal finishing point source category are incorporated in this section for reference.

1. The term "T," as in "Cyanide T," shall mean total.
2. The term "A," as in "Cyanide A," shall mean amenable to alkaline chlorination.
3. The term "job shop" shall mean a facility which owns not more than 50% (annual area basis) of the materials undergoing metal finishing.
4. The term "independent" printed circuit board manufacturer shall mean a facility that manufactures printed circuit boards principally for sale to other companies.
5. The term "TTO" shall mean TTO, which is the summation of all quantifiable values greater than 0.01 mg/L for the following toxic organics:

Acenaphthene
Acrolein
Acrylonitrile
Benzene
Benzidine
Carbon tetrachloride (tetrachloromethane) Chlorobenzene
1,2,4-Trichlorobenzene Hexachlorobenzene
1,2,-Dichloroethane
1,1,1-Trichloroethane Hexachloroethane
1,1-Dichloroethane
1,1,2-Trichloroethane
1,1,2,2-Tetrachloroethane Chloroethane Bis(2-chloroethyl) ether
2-Chloroethyl vinyl ether (mixed)
2-Chloronaphthalene
2,4,6-Trichlorophenol Parachlorometa cresol Chloroform (trichloromethane)
2-Chlorophenol
1,2-Dichlorobenzene
1,3-Dichlorobenzene
1,4-Dichlorobenzene
3,3-Dichlorobenzidine
1,1-Dichloroethylene
1,2-*Trans*-dichloroethylene
2,4-Dichlorophenol
1,2-Dichloropropane
1,3-Dichloropropylene (1,3-dichloropropene)
2,4-Dimethylphenol
2,4-Dinitrotoluene
2,6-Dinitrotoluene
1,2-Diphenylhydrazine Ethylbenzene Fluoranthene
4-Chlorophenyl phenyl ether
4-Bromophenyl phenyl ether
Bis(2-chloroisopropyl) ether
Bis(2-chloroethoxy) methane
Methylene chloride (dichloromethane) Methyl chloride (chloromethane)
Methyl bromide (bromomethane) Bromoform (tribromomethane)
Dichlorobromomethane
Chlorodibromomethane

Hexachlorobutadiene
Hexachlorocyclopentadiene
Isophorone
Naphthalene
Nitrobenzene
2-Nitrophenol
4-Nitrophenol
2.4-Dinitrophenol
4.6-Dinitro-*o*-cresol
N-nitrosodimethylamine
N-nitrosodiphenylamine
N-nitrosodi-*n*-propylamine
Pentachlorophenol
Phenol
Bis(2-ethylhexyl) phthalate
Butyl benzyl phthalate
Di-*n*-butyl phthalate
Di-*n*-octyl phthalate
Diethyl phthalate
Dimethyl phthalate
I,2-Benzanthracene (Benzo(*a*) anthracene)
Benzo(*a*)pyrene (3,4-benzopyrene)
3.4-Benzofluoranthene (benzo(*b*) fluoranthene)
II,12-Benzofluoranthene (benzo(*k*) fluoranthene)
Chrysene
Acenaphthylene
Anthracene
1,12-Benzoperylene (benzo(*ghi*)perylene) Fluorene
Phenanthrene
1.2.5.6-Dibenzanthracene (dibenzo(*a,h*) anthracene)
Indeno(1,2,3-*cd*) pyrene (2,3-*o*-phenlene pyrene)
Pyrene
Tetrachloroethylene
Toluene
Trichloroethylene
Vinyl chloride (chloroethylene)
Aldrin
Dieldrin
Chlordane (technical mixture and metabolites)
4.4-DDT
4.4-DDE (*p,p*-DDX)
4.4-DDD (*p,p*-TDE)
Alpha-endosulfan
Beta-endosulfan
Endosulfan sulfate
Endrin
Endrin aldehyde
Heptachlor
Heptachlor epoxide (BHC-hexachloro-cyclohexane)
Alpha-BHC
Beta-BHC

Gamma-BHC
Delta-BHC (PCB-polychlorinated biphenyls)
PCB-1242 (Arochlor 1242)
PCB-1254 (Arochlor 1254)
PCB-1221 (Arochlor 1221)
PCB-1232 (Arochlor 1232)
PCB-1248 (Arochlor 1248)
PCB-1260 (Arochlor 1260)
PCB-1016 (Arochlor 1016)
Toxaphene
2,3,7,8-Tetrachlorodibenzo-*p*-dioxin (TCDD)

10.9 U.S. ENVIRONMENTAL PROTECTION AGENCY METAL FINISHING EFFLUENT GUIDELINES UPDATE

10.9.1 METAL FINISHING EFFLUENT GUIDELINES

As stated in previous section, metal finishing is the process of changing the surface of an object, for the purpose of improving its appearance and/or durability, while metal finishing is related to electroplating, which is the production of a thin surface coating of the metal upon another by electrode position. The U.S. Environmental Protection Agency (USEPA) promulgated the federal Metal Finishing Effluent Guidelines (40 CFR Part 433) in 1983, with technical amendments in 1984 and 1986. The USEPA regulations cover wastewater discharges from a wide variety of industries performing various metal finishing operations. About 44,000 facilities perform various metal finishing operations and discharge process wastewater directly to surface waters or indirectly to surface waters through public owned treatment works (POTWs). The Metal Finishing Effluent Guidelines are incorporated into National Pollutant Discharge Elimination System (NPDES) permits for direct dischargers, and permits or industrial effluent Pretreatment Program.

10.9.2 CURRENT RULEMAKING FOR CHROME FINISHING FACILITIES

Based on information and data, USEPA has collected since it began studying per- and polyfluoroalkyl substances (PFAS) in industrial wastewater, and has determined that PFAS have, and continue to be, used by metal finishing facilities in the United States. USEPA has identified chromium electroplating and chromium anodizing operations (collectively referred to as "chromium electroplating facilities" or "chrome finishing facilities,") as the most significant source of PFAS in the metal finishing point source category due to their use of PFAS-based mist/fume suppressants to control toxic hexavalent chromium emissions. USEPA has also identified PFAS as a human carcinogen and inhalation hazard. In conclusion, the "chrome finishing facilities" includes chromium plating, chromium anodizing, chromic acid etching, and chromate conversion coating operations, which are the predominant sources of PFAS discharges by the Metal Finishing and Electroplating point source categories [34,35].

10.9.3 FACILITIES AND THE CORE OPERATIONS COVERED BY THE CURRENT US ENVIRONMENTAL PROTECTION AGENCY METAL FINISHING EFFLUENT GUIDELINES

It is important to note that the USEPA Metal Finishing regulation is defined by manufacturing processes and not by industrial sectors. However, facilities regulated by the Metal Finishing Effluent Guidelines are often included in the following SIC Major Groups 34–39:

1. Fabricated Metal Products, except Machinery and Transportation
2. Machinery, except Electrical

TABLE 10.27
The U.S. National Levels of Water Pollution Control

Type of Sites Regulated	BPT	BCT	BAT	NSPS	PSES	PSNS
Existing Direct Dischargers	•	•	•			
New Direct Dischargers				•		
Existing Indirect Dischargers					•	
New Indirect Dischargers						•

Pollutants Regulated	BPT	BCT	BAT	NSPS	PSES	PSNS
Priority Pollutants	•		•	•	•	•
Conventional Pollutants	•	•		•		
Nonconventional Pollutants	•		•	•	•	•

3. Electrical and Electronic Machinery, Equipment and Supplies
4. Transportation Equipment
5. Measuring, Analyzing and Controlling Instruments: Photograph; Optical Goods; Watches and Clocks
6. Miscellaneous Manufacturing Industries

The category covers plants which perform one or more of the following six operations: (a) electroplating; electroless plating; (c) anodizing; (d) coating (phosphating, chromating, and coloring); (e) chemical etching and milling; and (f) printed circuit board manufacture [34,35].

10.9.4 LEVELS OF WATER POLLUTION CONTROL

The US national regulations for industrial wastewater discharges set technology-based numeric limitations for specific pollutants at several levels of control: Best Practicable Control Technology Currently Available (BPT), Best Available Technology Economically Achievable (BAT), Best Conventional Pollutant Control Technology (BCT), New Source Performance Standards (NSPS), Pretreatment Standards for New Sources (PSNS) **or** Pretreatment Standards for New Sources (PSNS. Each of these terms is defined below. Effluent limitations are based on performance of specific technologies, but the regulations do not require use of a specific control technology. The levels of water pollution control is illustrated by Table 10.27.

10.10 RECENT ADVANCES

The readers are referred to the literature [36–48] for the recent advances in treatment and management of metal finishing wastewater. Only a sustainable and cost-effective solution to manage hazardous shock loads from metal finishing and electro-coating industry is reported here [48]. About-Elela et al. [48] report that (a) The three main hazardous wastewater sources come from (a1) batch chemical cleaning of degreasing basin (CCDB) (pH 13), (a2) batch chemical cleaning of phosphating basin (CCPB) (pH 1.03); and (a3) wastewater containing high concentrations of iron (2,300 mg/L), zinc (2,400 mg/L) and degreasing basin contents (DBC); (b) Mixing CCDB with CCPB at their actual discharge time forms a self-coagulant of metal hydroxide chemical flocs which can be utilized to treat the DBC in clarification (such as sedimentation); (c) Removal efficiency of COD (87%), total suspended solids (TSS) (94%), and oil and grease (92%) has been achieved; (d) For the purpose of comparison, conventional chemical coagulation of DBC has been carried out using $FeCl_3$ but demonstrated to have high O&M costs; (e) Advantage of using self-coagulation to treat DBC is proven since it eliminates the use of external chemicals and provides an integrated solution for the three main sources of hazardous pollutants; (e1) An innovative and sustainable solution to the shock loads of hazardous wastewater generated from metal finishing and E-coating industry by utilizing iron-rich wastewater from CCPB and alkaline wastewater from CCDB to produce metal hydroxide; (e2) The metal hydroxide was cost-effective and technically effectively than external coagulant in treating highly polluted degreasing basin content (DBC) at due discharge time; (e3) Iron-rich wastewater could be used to produce self-coagulant of iron hydroxide; (e4) Mixing iron-rich wastewater and alkaline wastewater produce iron hydroxide which is cost-effective in treating hazardous wastewater of degreasing basin. The use one hazardous waste stream to treat another hazardous waste stream within an industrial facility has also been demonstrated by Wang and Wang [49].

GLOSSARY

Best Available Technology Economically Achievable (BAT) BAT is defined at CWA section 304(b)(2). In general, BAT represents the best available economically achievable performance of plants in the industrial subcategory or category. Factors considered in assessing BAT include: (a) cost of achieving BAT effluent reductions; (b) age of equipment and facilities involved; (c) the processes employed by the industry and potential process changes; (d) non-water quality environmental impacts, including energy requirements; and (e) other factors as EPA deems appropriate

Best Conventional Pollutant Control Technology (BCT) BCT **is** defined at CWA section 304(b)(4), addresses conventional pollutants from existing industrial point sources. In addition to considering the other factors specified in section 304(b)(4)(B), USEPA establishes BCT limitations after consideration of a two part "cost-reasonableness" test. The initial BCT methodology and some BCT effluent limitations were published in 1979. A revised methodology, with additional effluent limitations, and some revised limitations, were published in 1986.

Best Management Practice (BMP) (a) BMP is defined as a permit condition used in place of, or in conjunction with effluent limitations, to prevent or control the discharge of pollutants. BMPs may include a schedule of activities, prohibition of practices, maintenance procedure, or other management practice. Sections 304(e), 308(a), 402(a), and 501(a) of the CWA authorize USEPA to prescribe BMPs as part of Effluent Guidelines and Standards or as part of a permit; (b) BMP regulations: (b1) Direct dischargers: 40 CFR 122.2, 40 CFR 122.44(k); and (b2) Indirect dischargers: 40 CFR 403.3(e); and (c) CWA section 304(e) authorizes USEPA to include BMPs in Effluent Guidelines for certain toxic or hazardous pollutants for the purpose of controlling "plant site runoff, spillage or leaks, sludge or waste disposal, and drainage from raw material storage." CWA section 402(a)(1) and 40 CFR 122.44(k) also provide for BMPs to control or abate the discharge of pollutants when numeric limitations and standards are infeasible.

Best Practicable Control Technology Currently Available (BPT) BPT is defined at CWA section 304(b)(1). In specifying BPT, USEPA looks at a number of factors: (a) the total cost of applying the control technology in relation to the effluent reduction benefits; (b) age of the equipment and facilities; (c) processes employed by the industry and any required process changes; (d) engineering aspects of the control technologies; (e) non-water quality environmental impacts, including energy requirements; (f) other factors as USEPA deems appropriate.

Both direct and indirect dischargers (a) A facility can be both a direct and indirect discharger if some of its wastewater goes directly to a water body and some into a sewer. Most metal finishers are indirect dischargers; (b) Both direct and indirect dischargers must meet effluent limits and conduct periodic monitoring and reporting. Both direct and indirect dischargers will need to pretreat their wastewaters and/or implement pollution prevention measures to successfully meet the applicable effluent limits

Clean water act (CWA) CWA is the principal federal law protecting *surface water quality* in the United States. Originally passed in 1972 as the Water Pollution Control Act, the CWA has been amended several times, most notably in 1978 and 1982. The two principal goals of CWA are: (a) To restore and maintain the chemical, physical, and biological integrity of the nation's waters; and (b) Where attainable, to achieve water quality that promotes protection and propagation of fish, shellfish, and wildlife, and provides for recreation in and on the water.

Control authority permit or pretreatment standards They are the wastewater discharge requirements for indirect dischargers. **As** an indirect discharger, its wastewater must meet specific pretreatment standards that control its quality. Standards for indirect discharges are established by the "Control Authority" for its area and the standards re usually specified in a document issued by the Control Authority. This document may be referred to by a number of names, including "permit", "agreement," or "Control Authority permit.

Conventional pollutants Clean Water Act (CWA) section 304(a)(4) designates the following as "conventional" pollutants: biochemical oxygen demand (BOD_5), total suspended solids (TSS), fecal coliform, pH, and any additional pollutants USEPA defines as conventional. USEPA designated "oil and grease" (O&G) as an additional conventional pollutant on July 30, 1979 (refer to 44 FR 44501).

CWA (metal finishers) Many programs have been developed to achieve the federal Clean Water Act goals. Three of these programs affect metal finishers: (a) Control of wastewater discharges, which affects all metal finishers;(b); Permitting of storm water discharges, which affects some metal finishers; and (c) Preparation of oil spill control and countermeasure plans, which affects a select few metal finishers.

Direct discharger If a facility discharges its *process and/or nonprocess wastewater* directly to a *surface water body* (including rivers, streams, lakes, and oceans), it is a *direct discharger.* Direct discharges are controlled under the National Pollutant Discharge Elimination System (NPDES) program, established under the Clean Water Act. Direct dischargers must obtain an NPDES permit to discharge wastewater.

Effluent guidelines or effluent limitation guidelines (ELG) (a) Effluent guidelines are the US national standards for industrial and commercial wastewater discharges to surface waters and publicly owned treatment works (POTW); (b) USEPA publishes Effluent Guidelines, nationally-applicable water pollution regulations for industrial and commercial facilities under Title III of the *Clean Water Act* for the categories of existing sources and new sources.; and (c) The ELG standards are technology-based (i.e. they are based on the performance of treatment and control technologies); they are not based on risk or impacts upon receiving waters.

Effluent guidelines Effluent guidelines are national wastewater discharge standards that are developed by USEPA on an industry-by-industry basis. These are technology-based regulations, and are intended to represent the greatest pollutant reductions that are economically achievable for an industry. The standards for direct dischargers are incorporated into National Pollutant Discharge Elimination System (NPDES) permits issued by States and EPA regional offices, and permits or other control mechanisms for indirect dischargers (refer to Pretreatment Program).

Effluent limitation guidelines (ELG) and effluent guidelines program plan (EGPP) (a) US EPA periodically reviews the *existing Effluent Guideline regulations* (also called Effluent Limitation Guidelines or ELG), and updates them, as appropriate; (b) The Effluent Guidelines Program Plan (EGPP), published every 2 years, identifies existing industries selected for regulatory revisions and new industries identified for regulation. The EGPP provides a rulemaking schedule for any such activities; (c) The Effluent Guidelines Program Plan 15 announces the initiation of new rulemakings and detailed studies (January 20, 2023).

Indirect discharger If a facility discharges its *process wastewater* to a sewer that flows to a *publicly owned treatment works* (POTW), it is an *indirect discharger*. Indirect dischargers must meet pretreatment standards. (Nonprocess wastewaters from metal finishing [e.g., boiler blowdown, noncontact cooling water, and air compressor condensate] discharged to sewers are not regulated under the CWA. However, if discharged directly to a POTW, nonprocess wastewaters are subject to the local POTW's permitting or sewer use ordinance requirements and to state requirements.)

Levels of water pollution control The US national regulations for industrial wastewater discharges set technology-based numeric limitations for specific pollutants at several levels of control: Best Practicable Control Technology Currently Available (BPT), Best Available Technology Economically Achievable (BAT), Best Conventional Pollutant Control Technology (BCT), New Source Performance Standards (NSPS), Pretreatment Standards for New Sources (PSNS) **or** Pretreatment Standards for New Sources (PSNS. Each of these terms is defined below. Effluent limitations are based on performance of specific technologies, but the regulations do not require use of a specific control technology. (See Table 10.27)

Metal finisher Under the federal Clean Water Act regulations, a "metal finisher" is any facility that performs any of the following operations: (a) Electroplating; (b) Anodizing; (c) Chemical etching and milling; (d) Electroless plating; (e) Coating (chromating, phosphating, and coloring); (f) Printed circuit board manufacturing

Metal Finishing Effluent Guidelines (40 CFR Part 433) USEPA promulgated the Metal Finishing Effluent Guidelines (40 CFR Part 433) in 1983, with technical amendments in 1984 and 1986. The regulations cover wastewater discharges from a wide variety of industries performing various metal finishing operations. About 44,000 facilities perform various metal finishing operations and discharge process wastewater directly to surface waters or indirectly to surface waters through POTWs. The Metal Finishing Effluent Guidelines are incorporated into NPDES permits for direct dischargers, and permits or other control mechanisms for indirect dischargers (refer to Pretreatment Program).

Metal finishing Metal finishing is the process of changing the surface of an object, for the purpose of improving its appearance and/or durability. Metal finishing is related to electroplating, which is the production of a thin surface coating of the metal upon another by electrodeposition.

New Source Performance Standards (NSPS) NSPS is defined at CWA section 306, apply to direct dischargers. NSPS reflect effluent reductions that are achievable based on the "best available demonstrated control technology." New sources have the opportunity to install the best and most efficient production processes and wastewater treatment technologies. As a result, NSPS should represent the most stringent controls attainable through the application of the best available demonstrated control technology for all pollutants (i.e., conventional, nonconventional, and priority pollutants). In establishing NSPS, EPA is directed to take into consideration the cost of achieving the effluent reduction and any non-water quality environmental impacts and energy requirements.

Nonconventional pollutants All pollutants which are not classified as conventional pollutants (such as, biochemical oxygen demand, total suspended solids, fecal coliform, pH, and oil and grease), and not classified as toxic pollutants are nonconventional pollutants.

Pretreatment Standards for Existing Sources (PSES) PSES is defined at CWA section 307(b). Like PSNS, PSES are national, uniform, technology-based standards that apply to indirect dischargers. They are designed to prevent the discharge of pollutants that pass through, interfere with, or are otherwise incompatible with the operation of POTWs. Dischargers subject to PSES are required to comply with those standards by a specified date, typically no more than 3 years after the effective date of the categorical standard.

Pretreatment Standards for New Sources (PSNS) PSNS is defined at CWA section 307(c). PSNS are national, uniform, technology-based standards that apply to dischargers to publicly owned treatment works (POTWs) from specific industrial categories (i.e., indirect dischargers). They are designed to prevent the discharges of pollutants that pass through, interfere with, or are otherwise incompatible with the operation of POTWs. PSNS are to be issued at the same time as NSPS. The Agency considers the same factors in promulgating PSNS as it considers in promulgating NSPS.

REFERENCES

1. Federal Register. *Resource Conservation and Recovery Act (RCRA)*, 42 U.S. Code s/s 6901 *et seq.* (1976), U.S. Government, Public Laws. US Environmental Protection Agency, Washington DC, USA.
2. U.S. EPA. *Resource Conservation and Recovery Act (RCRA)—Orientation Manual.* Report no. EPA530-R-02-016. U.S. Environmental Protection Agency, Washington, DC, January 2003.
3. U.S. EPA. *Federal Hazardous and Solid Wastes Amendments (HSWA).* U.S. Environmental Protection Agency, Washington, DC, November 1984. 2007.
4. Federal Register. *Clean Water Act (CWA)*, 33 U.S. Code s/s 1251 *et seq.* (1977), U.S. Government, Public Laws. May 2002.
5. U.S. EPA. *Meeting Hazardous Waste Requirements for Metal Finishers.* Report no. EPA/625/4-87/018. U.S. Environmental Protection Agency, Cincinnati, OH, 1987.
6. U.S. EPA. *Development Document for Effl uent Limitations Guidelines and Standards for the Metal Finishing Point Source Category.* Report no. EPA-440/1-80/091. U.S. Environmental Protection Agency, Washington, DC, 1980.
7. U.S. EPA. *Treatability Manual, Volume II. Industrial Descriptions.* Report no. EPA-600/2-82-001b. U.S. Environmental Protection Agency, Washington, DC, September 1981.
8. PRC Environmental Management, Inc. *Hazardous Waste Reduction in the Metal Finishing Industry.* Noyes Data Corporation, Park Ridge, NJ, 1989.
9. U.S. ACE. Yearly Average Cost Index for Utilities. In: *Civil Works Construction Cost Index System Manual.* 110-2-1304. U.S. Army Corps of Engineers, Washington, DC, p. 44. A PDF file is available on the Internet at https://www.nww.usace.army.mil/cost, 2007.
10. Patterson, J.W. *Industrial Wastewater Treatment Technology*, 2nd edition. Butterworths, Scotland, 1985.
11. Wang, L.K., Vaccari, D.A., Li, Y., and Shammas, N.K. Chemical precipitation. In: Wang, L.K., Hung, Y.T., and Shammas, N.K. (Eds), *Physicochemical Treatment Processes.* Humana Press, Totowa, NJ, pp. 141–198, 2005.
12. Shammas, N.K., Kumar, I.J., and Chang, S.Y. Sedimentation. In: Wang, L.K., Hung, Y.T., and Shammas, N.K. (Eds), *Physicochemical Treatment Processes.* Humana Press, Totowa, NJ, pp. 379–430, 2005.
13. Chen, J.P., Chang, S.Y., Huang, J.Y.C., Baumann, E.R., and Hung, Y.T. Gravity filtration. In: Wang, L.K., Hung, Y.T. and Shammas, N.K. (Eds), *Physicochemical Treatment Processes.* Humana Press, Totowa, NJ, pp. 501–544, 2005.
14. Wang, L.K., Hung, Y.T., and Shammas, N.K. (Eds), *Physicochemical Treatment Processes.* Humana Press, Totowa, NJ, p. 723, 2005.
15. Wang, L.K., Hung, Y.T., and Shammas, N.K. (Eds), *Advanced Physicochemical Treatment Processes.* Humana Press, Totowa, NJ, p. 690, 2006.
16. Chen, J.P., Mou, H., Wang, L.K., and Matsuura, T. Membrane filtration. In: Wang, L.K., Hung, Y.T., and Shammas, N.K. (Eds), *Advanced Physicochemical Treatment Technologies.* Humana Press, Totowa, NJ, pp. 203–260, 2007.
17. Wang, L.K., Hung, Y.T., and Shammas, N.K. (Eds), *Advanced Physicochemical Treatment Technologies.* Humana Press, Totowa, NJ, p. 710, 2007.
18. Wang, L.K., Pereira, N., and Hung, Y.T. (Eds), and Shammas, N.K. (Consulting Editor). *Biological Treatment Processes.* Humana Press, Totowa, NJ, 2008.
19. Wang, L.K., Shammas, N.K., and Hung, Y.T. (Eds), *Advanced Biological Treatment Processes.* Humana Press, Totowa, NJ, 2008.
20. Wang, L.K., Fahey, E.M., and Wu, Z. Dissolved air flotation. In: Wang, L.K., Hung, Y.T., and Shammas, N.K. (Eds), *Physicochemical Treatment Processes.* Humana Press, Totowa, NJ, pp. 431–500, 2005.
21. U.S. EPA. *Managing Cyanide in Metal Finishing.* Capsule Report no. EPA 625/R-99/009. U.S. Environmental Protection Agency, Cincinnati, OH, December 2000.

22. Hartinger, L. *Handbook of Effl uent Treatment and Recycling for the Metal Finishing Industry*, 2nd edition. Finishing Publications, Warrington, 1994.

23. Eilbeck, W.J. and Mattock, G. *Chemical Processes in Wastewater Treatment*. Ellis Horwood Limited, Sussex, 1987.

24. Wang, L.K., Shammas, N.K., and Hung, Y.T. (Eds), *Biosolids Treatment Processes*. Humana Press, Totowa, NJ, p. 820, 2007.

25. Singh, I.B., Chaturvedi, K., Singh, D.R., and Yegneswaran, A.H. Thermal stabilization of metal finishing waste with clay. *Environmental Technology*, 26(8), 877–884, 2005.

26. Roy, C.H. *Operation and Maintenance of Surface Finishing Wastewater Treatment Systems*. American Electroplaters and Surface Finishers Society, Orlando, FL, 1988.

27. Altmayer, F. *Plating and Surface Finishing, Advice & Council*. AESF, Orlando, FL, 1997.

28. Wang, L.K., Li, Y. Shammas, N.K., and Sakellaropoulos, G.P. Drying beds. In: Wang, L.K., Shammas, N.K., and Hung, Y.T. (Eds), *Biosolids Treatment Processes*. Humana Press, Totowa, NJ, pp. 403–430, 2007.

29. Wang, L.K., Kurylko, L., and Wang, M.H.S. *Sequencing Batch Liquid Treatment*. U.S. Patent No. 5354458. U.S. Patent and Trademark Office, Washington, DC, October 1994.

30. Wang, L.K. and Wang, M.H.S. (Eds), *Handbook of Industrial Waste Treatment*. Marcel Dekker, New York, pp. 127–172, 1992.

31. Wang, L.K., Yung, Y.T., Lo, H.H., and Yapijakis, C. (Eds), *Hazardous Industrial Waste Treatment*. CRC Press, New York, pp. 289–360, 2007.

32. Wang, L.K., Wang, M.H.S., and Hung, Y.T. *Waste Treatment in the Biotechnology, Agricultural ad Food Industries, Volume 1*. Springer Nature, Cham, Switzerland, p. 485, 2022.

33. U.S. EPA. Code of Federal Regulations. *Metal Finishing Point Source Category*. Title 40, Vol. 27, Part 433. U.S. Environmental Protection Agency, Washington, DC, revised as of July 1, 2003.

34. USEPA. *Metal Finishing Guidance Manual: A Practical Guide to Environmental Compliance and Continuous Improvement*. US Environmental Protection Agency, Washington DC, 1996.

35. USEPA. *Preliminary Effluent Guidelines Program Plan*. US Environmental Protection Agency, Washington DC, 15 September 2021

36. Cavaco, S.A., Fernandes, S., Quina, M.M., Ferreira, L.M. Removal of chromium from electroplating industry effluents by ion exchange resins. *Journal of Hazardous Materials*, 144(3), 634–638, 2007. doi: 10.1016/j.jhazmat.2007.01.087.

37. Dutra, A.J.B., Rocha, G.P. and Pombo, F.R. Copper recovery and cyanide oxidation by electrowinning from a spent copper-cyanide electroplating electrolyte. *Journal of Hazardous Materials*, 152(2), 648–655, 2008. doi: 10.1016/j.jhazmat.2007.07.030.

38. Sousa, F.W., Sousa, M.J., Oliveira, I.R., Oliveira, A.G., Cavalcante, R.M., Fechine, P.B., Neto, V.O., de Keukeleire, D. and Nascimento, R.F. Evaluation of a low-cost adsorbent for removal of toxic metal ions from wastewater of an electroplating factory. *Journal of Environmental Management*, 90(11), 3340–3344, 2009. doi: 10.1016/j.jenvman.2009.05.016.

39. Babu, B.R., Bhanu, S.U., Meera, K.S. Waste minimization in electroplating industries: A review. *Journal of Environmental Science and Health Part C*, 27(3), 155–177, 2009. doi: 10.1080/10590500903124158.

40. Agrawal, A. and Sahu, K.K. An overview of the recovery of acid from spent acidic solutions from steel and electroplating industries. *Journal of Hazardous Materials*, 171(1–3), 61–75, 2009. doi: 10.1016/j.jhazmat.2009.06.099.

41. Gartiser, S., Hafner, C., Hercher, C., Kronenberger-Schäfer, K., and Paschke, A. Whole effluent assessment of industrial wastewater for determination of BAT compliance. Part 2: Metal surface treatment industry. *Environmental Science and Pollution Research*, 17(5), 1149–1157, 2010. doi: 10.1007/s11356-009-0290-6.

42. Zhang, Z. Treatment of oilfield wastewater by combined process of micro-electrolysis, Fenton oxidation and coagulation. *Water Science and Technology*, 76(11–12), 3278–3288, 2017. doi: 10.2166/wst.2017.486.

43. Badawy, M.I. and Ali, M.E. Fenton's peroxidation and coagulation processes for the treatment of combined industrial and domestic wastewater. *Journal of Hazardous Materials*, 136(3), 961–966, 2006. doi: 10.1016/j.jhazmat.2006.01.042.

44. Rajoria, S., Vashishtha, M., Sangal, V.K. Treatment of electroplating industry wastewater: A review on the various techniques. *Environmental Science and Pollution Research*, 29(48), 72196–72246, 2022. doi: 10.1007/s11356-022-18643-y.

45. Abou-Elela, S.I., El-Kamah, H.M., and Aly, H.I. Waste minimization in the electroplating industry (case study). *International Journal of Environmental Studies*, 55(4), 287–296, 1998.

46. Abou-Elela, S.I., El-Shafai, S.A., Fawzy, M.E., Hellal, M.S., and Kamal, O. Management of shock loads wastewater produced from water heaters industry. *International Journal of Environmental Science and Technology*, 15(4), 743–754, 2018.
47. Almeida, C.M.V.B., Sevegnani, F., Agostinho, F., Liu, G., Yang, Z., Coscieme, L., and Giannetti, B.F. Accounting for the benefits of technology change: Replacing a zinc-coating process by a water-based organo-metallic coating process. *Journal of Cleaner Production*, 174, 170–176, 2018.
48. Abou-Elela, S.I., Fawzy, M.E., and El-Shafai, S.A. Treatment of hazardous wastewater generated from metal finishing and electro-coating industry via self-coagulation: Case study, *Water Environment Research*, 93(9), 1476–1486, 2021. doi: 10.1002/wer.1552.
49. Wang, L.K. and Wang, M.H.S. Application of secondary flotation-filtration and coagulant recycle for improvement of a paper mill primary waste treatment facility. In: Wang, L.K., Wang, M.H.S., and Hung, Y.T. (Eds), *Waste Treatment in the Biotechnology, Agricultural ad Food Industries, Volume 2*. Springer Nature, Cham, Switzerland, pp. 1–45, 2024.
50. USAC. *Civil Works Construction Cost Index System Manual*. 110-2-1304, U.S. Army Corps of Engineers, Washington, DC, p. 44, (2020-Tables Revised 31 March), 2020.

TABLE A1
United States Yearly Average Cost Index for Utilities
[32,50]

Year	Index	Year	Index
1967	100	1995	439.72
1968	104.83	1996	445.58
1969	112.17	1997	454.99
1970	119.75	1998	459.40
1971	131.73	1999	460.16
1972	141.94	2000	468.05
1973	149.36	2001	472.18
1974	170.45	2002	486.16
1975	190.49	2003	497.40
1976	202.61	2004	563.78
1977	215.84	2005	605.47
1978	235.78	2006	645.52
1979	257.20	2007	681.88
1980	277.60	2008	741.36
1981	302.25	2009	699.70
1982	320.13	2010	720.80
1983	330.82	2011	758.79
1984	341.06	2012	769.30
1985	346.12	2013	776.44
1986	347.33	2014	791.59
1987	353.35	2015	786.32
1988	369.45	2016	782.46
1989	383.14	2017	803.93
1990	386.75	2018	841.84
1991	392.35	2019	866.18
1992	399.07	2020	867.71
1993	410.63	2021	893.02[a]
1994	424.91	2022	918.91[a]

[a] Projected future cost index values.

11 Use of Biosorption of Agro-Industrial Wastes and Microorganisms for Removal of Potentially Toxic Metals

Jéssica Mesquita do Nascimento,
Jonas Juliermerson Silva Otaviano,
Helayne Santos de Sousa,
Jorge Diniz de Oliveira, and
Yung-Tse Hung

ACRONYMS

AgNPs	Silver nanoparticles
AuNPs	Gold nanoparticles
BCM	Biologically Controlled Mineralization
BIM	Biological Induced Mineralization
BNPs	Biogenically synthesized nanoparticles
DLS	Dynamic Light Scattering
DNA	Deoxyribonucleic acid
EPS	Extracellular polymeric substances
FTIR	Fourier Transform Infrared Spectroscopy
LPS	Lipopolysaccharides
M	molar
MIC	Minimum Inhibitory Concentration
mm	Millimolar
nm:	nanometers
NPs	Nanoparticles
pH PZC	Point of Zero Charge
pH	Hydrogenic Potential
qe	Biosorption capacity
R%	Removal efficiency
rRNA	Ribosomal Ribonucleic Acid
RSM	Response Surface Methodology
TEM-EDX	Transmission Electron Microscopy–Energy Dispersive X-ray Spectroscopy

11.1 INTRODUCTION

The expansion of the industry connected to population growth, on the one hand, has expanded the processes of technological innovation and sophisticated products, on the other hand, environmental complications related to the contamination of terrestrial and aquatic ecosystems in much of the

DOI: 10.1201/9781003541615-11

"

planet have evolved. Thus, studies related to the mitigation of impacts caused on the environment by potentially toxic metals have become progressively necessary, which translates into developing sophisticated treatment methods that reduce this pollution [1].

Potentially toxic metals are highly soluble elements, existing in cationic and anionic forms with different oxidation states, which can be identified in the form of complexes present in mining activities, agricultural/industrial chemicals, and solid waste [2,3]. Among the elements responsible for contamination in soil and water are cadmium (Cd), chromium (Cr), lead (Pb), mercury (Hg), nickel (Ni), zinc (Zn), copper (Cu), and arsenic (As) [4,5].

Since they can occur naturally in the environment, potentially toxic metals pose risks to human health even at low concentrations [4,6]. However, a considerable portion of these elements comes from most industries, which in the large-scale production of materials, release numerous types of hazardous waste for which metals stand out for having high atomic densities and weights, and therefore, result in high rates of mobilization in the environment [7–9].

For the removal of potentially toxic metals in the environment, whether in aquatic or terrestrial environments, there are numerous amounts of methods used for this. But biosorption, when compared to conventional treatments (membrane separation, flocculation, coagulation, ion exchange), has proved to be a relevant parameter for researchers from a strategic, economic, simple, and ecologically adequate point of view [10,11].

Biosorption is an adsorption method related to the passive bond between organic and metallic ions by biomass, which through physicochemical parameters, ion exchanges, chelation, complexation, electrostatic interaction occur, allowing the desorption, regeneration and reuse of the biosorbent in constant cycles of sorption/biosorption [12–14].

The use of both agro-industrial residues and microorganisms is one of the most important biomass sources for biosorption under the observation of most researchers at a global level. The application of agricultural residues as biosorbents, for example, constitutes an ecologically correct methodology in the same way that it reduces waste, also making it economically efficient [15].

Regarding microorganisms, many species of microalgae, bacteria, and fungi can produce good results in removing potentially toxic metals. Some studies have shown that the use of microalgae for remediation purposes has several advantages such as the requirement of low investment and operating cost; significant amounts of biomass; sustainability; efficiency in the sensitivity and selectivity of potentially toxic metals; good ability to absorb and remove these metals [16–18].

Works related to biosorption showed that bacteria can be efficient in removing contaminating metals. As an example, studies have identified that, from the metabolism of anaerobic ammonium oxidizing bacteria, a biological process commonly called anammox, they are capable of generating granular residues that can act as biosorbents in the effluent treatment process, with cost-effectiveness and high performance [19].

Studies have found that fungi originating from shellfish and fungal biomass showed a high degree of correction in soils contaminated by potentially toxic metals, for example, the production of particles from fungal mycelia was effective in the remediation and biosorption of contaminating metals such as Cu^{2+} and Pb^{2+} [20].

The biosorption technique using live or dead material or biomass by-products, in addition to its proven effectiveness in the removal of potentially toxic metals, can also be efficient for the removal of radiotoxic sources. The effect, for example, of inactivated yeasts such as *Saccharomyces cerevisiae* and activated *S. cerevisiae* can help in the treatment of environments contaminated by Americium-241 [21].

Biosorption, therefore, becomes one of the most promising decontamination methodologies and, therefore, is widely studied worldwide due to its performance, easy implementation, low cost, and abundant biomass that act as biosorbents in the removal of inorganic and organic contaminants [22–24].

11.2 BIOSORPTION AND ENVIRONMENTAL DECONTAMINATION

11.2.1 ENVIRONMENTAL CONTAMINATION

The planet Earth on which we live is a source of fundamental resources for the maintenance of the entire ecosystem. The union of all these resources and consequently their balance guarantee the perpetuation of life. However, human and industrial activities consume natural resources and generate pollutants that, if released inappropriately, contaminate the environment and pose a health risk to various organisms such as animals, plants, and human beings [25–29].

The routes of exposure to these components can occur in several ways, indirect exposure occurs mainly through the food chain, including absorption of contaminants by plants and animals. Thus, the risk of exposing humans to environmental contaminants has become one of the main concerns of regulatory agencies. The main environmental contaminants are organic and inorganic components, such as potentially toxic metals [30–33].

The industrial activities carried out over time by humanity have caused new changes with a negative impact on the environment. The bioaccumulation of pollutants, whether organic or inorganic, makes it easier to concentrate and transport them to the most remote parts of planet Earth. In an era of exponential environmental changes caused by anthropogenic activities, the reestablishment of balance becomes evident [30,33,34].

Environmental contamination is one of the factors that alter the natural cycling of the ecosystem. Persistent toxic contaminants, such for example, metallic species are most often concentrated in the food chain, consequently, they can appear in more distant areas [30,31,33,34].

Metals are major contaminants because some species are extremely toxic at low concentrations [33,35]. Soil pollution by metals has worsened over the years, although certain metallic species occur naturally in the soil, additional contributions from anthropogenic activities promote an imbalance of these metallic species [30,35,36].

Soil pollution by metallic species can inhibit the activity of microbial enzymes and reduce population diversity. As for the mechanisms of toxicity, the most relevant is certainly the chemical inactivation of enzymes. All divalent transition metals readily react with protein groups such as amine, imine, and sulfhydryl. Some (Cd and Hg) can compete with essential elements like zinc and displace it in metalloenzymes. Furthermore, some metals can also harm cells, acting as antimetabolites, forming precipitates or chelates with essential metabolites [29–31,33].

11.2.2 CONVENTIONAL TECHNIQUES

The presence of inorganic pollutants such as countless metallic species in the ecosystem causes serious problems for the environment. Several metals are considered toxic to the environment and living beings when they exceed certain concentrations, such as Cr, Cu, Pb, Cd, Hg, Zn, Mn, and Ni because they are not biodegradable and highly persistent [30,37].

Therefore, there is a need to remove potentially toxic metals in effluents before they are released into the ecosystem. Numerous technologies have been developed to remove potentially toxic metals from water and industrial effluents, such as precipitation, evaporation, ion exchange, membrane separation, coagulation, and other techniques [9,38,39].

Removing potentially toxic metals by technologies based on physicochemical processes is widely used in several sectors because they are consolidated technologies. The selection of the most adequate technique depends mainly on the initial concentration of the metal, the components of the wastewater, the investment, and the operational cost [9,40]. In Figure 11.1, some techniques used in the decontamination of effluents containing potentially toxic metals are summarized.

Coagulation is one of the most used technologies for wastewater treatment. In coagulation, the charge of the particles is neutralized. This technique is capable of filtering potentially toxic metal

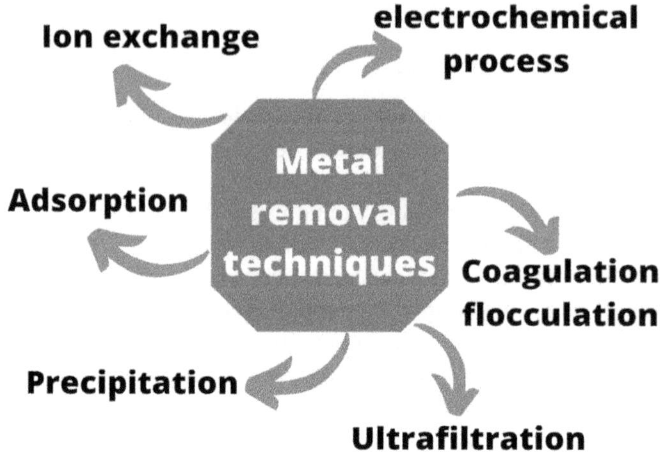

FIGURE 11.1 Some conventional techniques used in the treatment of effluents containing potentially toxic metals.

solids from the solution. This is a very practical way to manage metal contamination with or without water treatment tools. By introducing highly charged small molecules into wastewater, this mechanism destabilizes particle charges. Furthermore, this methodology is normally coupled to flocculation, where colloidal particles leave the suspension to flocculant and form sediments [9].

Many metallurgical industries opt for cheaper technologies such as precipitation. Precipitation of metals using lime, sulfide, and caustic soda is the most common method of metal removal. However, chemical precipitation is not selective and produces large amounts of solid sludge. Due to these restrictions, other treatment processes such as ion exchange, membrane filtration, and adsorption using activated carbon have become popular in wastewater treatment. These techniques are established and represent significant capital investments by the industry [41].

However, conventional technologies have some disadvantages such as secondary pollution, high cost, energy, large amounts of chemical reagents, and poor treatment efficiency at low metallic concentrations, some of these advantages and advantages of these methodologies are summarized in (Table 11.1) [9,38].

Ultrafiltration performed by membranes generates less solid waste, uses fewer chemicals, and is highly efficient. However, one of its disadvantages, in addition to the high cost, is the low flow rates. Although the flocculation/coagulation technique has the advantage of achieving sludge dehydration and sedimentation, the high cost and high consumption of chemicals are its disadvantages [9].

The ion exchange process has the advantage of not changing the pH of the effluent in addition to being safe due to the non-use of chemicals, stability, and high reliability of the process. However, the high costs of the membrane, as well as the need for resin renewal due to fouling, are its main disadvantages. Therefore, the decontamination technology selected is determined according to the initial concentration of pollutants, capital available for the treatment of the effluent, operational costs, and environmental impact of the generated effluent [9].

Adsorption is an economical technology that has several advantages over conventional methods, such as sludge reduction, high efficiency, minimal cost, the possibility of metal recovery, and adsorbent regeneration. The biosorption mechanism is the same used in adsorption, the difference is the nature of the adsorbent, which in this case is organic, making the amount of biomass used as biosorbents abundant. On the other hand, the main adsorbent used in the adsorption is activated carbon, which presents good performance but at considerable cost [9,41,42].

TABLE 11.1

Main Conventional Techniques of Environmental Decontamination

Technique	Definition	Advantage	Disadvantage
Chemical precipitation	Use of chemicals to precipitate metals	Low cost and easy implementation	Production of large volume of sludge
Ion exchange	Use of synthetic resins that will effect ion exchange	Effectiveness and promotion of metal recovery	Resins are expensive; need for resin regeneration with chemical reagents
Electrochemical process	Use of electric current to promote the destabilization of metal ions	Metal recovery	High cost
Coagulation/ flocculation	Neutralization of particle charges/ flocculation of suspended particles	Less sludge	High cost and consumption of chemicals
Ultrafiltration	Use of membranes with a diameter of 0.1–0.001 micrometers	Effective	High cost; membrane fouling; production of secondary contaminated effluent
Adsorption	Accumulation of metallic species on another surface called adsorbent	Effective	Need for adsorbent regeneration by chemical reagents

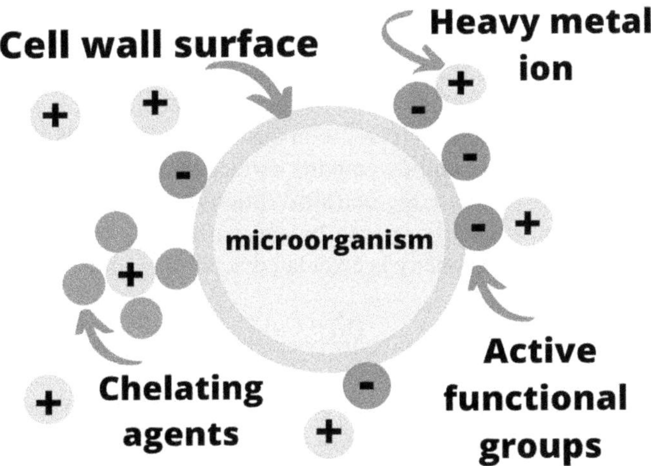

FIGURE 11.2   Schematic representation of some mechanisms involved in biosorption.

11.2.3 Fundamentals of Biosorption

The search and development of new technologies to remove metals more efficiently and at a low cost is extremely relevant. In this context, biosorption emerged as a sustainable proposal. Biosorption is an adsorption process that fundamentally uses low-cost biomaterials to separate environmental pollutants from aqueous solutions by a wide range of physicochemical mechanisms, such as ion exchange, chelation, complexation, physical adsorption, and surface microprecipitation. These biomaterials are called biosorbents [32,38].

Biosorption consists of a physical-chemical process of removing substances, whether organic or inorganic, by biological material. Living and dead organisms have components that facilitate the biosorptive process. The complex matrix of biosorbents contains several functional groups that attract and sequester metals acting as active binding sites (Figure 11.2). Many biosorbents contain functional groups such as phosphodiester, phosphate, phenolic, thioether, sulfhydryl, sulfonate, imidazole, imine, hydroxyl, carbonyl, carboxyl, amino, and amide [43]. Biosorption is recognized as a promising biotechnological process, due to its intelligibility, and operation similar to ion exchange technology, in addition to its evident efficiency and vast availability of biosorbents [9,38,44].

Biosorption is an innovative and growing technology in the removal of potentially toxic metals from effluents. This technique is considered an efficient process even at low concentrations, in addition to being economically viable [44–46]. In this process, you can live and/or dead biomasses are used to remove pollutants. Dead biomass is usually one of the most preferred because it is cheaper and does not require special conditioning and nutrient supply [1].

Biosorption is a process based on adsorption where the solid-state biomass, called biosorbent, will be in contact with the solvent that consists of the aqueous metallic solution called adsorbate. Adsorption is considered an efficient and admirable method compared to other wastewater treatment technologies, as it provides high-quality treated effluents. In adsorption, adsorbents can be regenerated and reused, in addition to being easy to implement [45,47], [48]. In biosorption, several biomasses have a promising character in metallic removals, such as fungi, bacteria, agro-industrial residues, and others.

The biosorption process is stimulated by the concentration gradient that exists between the solid biosorbent and the liquid part (metallic solution) and with the transport of solute molecules through the interphase mediated by molecular diffusion. In a batch mode of biosorption process (batch), this spontaneous molecular diffusion can be augmented with advection currents to form convective mass transfer at the solid-liquid interface [2].

These convective currents form the circular motions of the solute-enriched liquid phase and increase the average collision frequencies of the diffusing molecules. This results in a decrease in diffusion resistance and, subsequently, an increase in the mass transfer flux through the solid-liquid interface. When adsorbate molecules collide with the surface of the biosorbent, a change in molecular dynamics is expected at the interface region (biosorption) [2].

The biosorption potential of a metallic species by a biosorbent is evaluated by its removal capacity and efficiency. The biosorption capacity is calculated according to the following formula:

$$q_e = \frac{(C_i - C_0)\, V}{m}$$

Where (q_e) is the biosorption capacity (in mg of the metal per g of biosorbent), (C_i and C_0) are the initial concentration and the final concentration of the metal ion solution (in mg/L), (V) is the volume of the solution (in Liters), and (m) is the mass of the biosorbent used (in grams) [46,49]. The biosorption or removal efficiency represents the percentage of removal of the metallic species by the biosorbent ($R\%$). Efficiency is calculated according to the formula:

$$R\% = \frac{(C_i - C_0)}{Ci} \times 100$$

The results regarding the removal capacity and efficiency are usually plotted on graphs where the analyzed variable is located on the abscissa axis (metallic concentration, biosorbent dosage, contact time, pH and others) and the ordinate axis is the capacity or efficiency of removal (Figure 11.3a and b), [49,50].

Several factors can influence the biosorption capacity of biosorbents. The effects of these parameters on the process are usually evaluated and optimized during biosorption tests, as each biosorbent and metal investigated in the biosorption process generally has its optimal characteristics [51–53]. Some of the factors that are most commonly studied and that influence the biosorptive process are temperature, pH of the medium, contact time, biosorbent dosage, and initial concentration of metal ions [9,40,54].

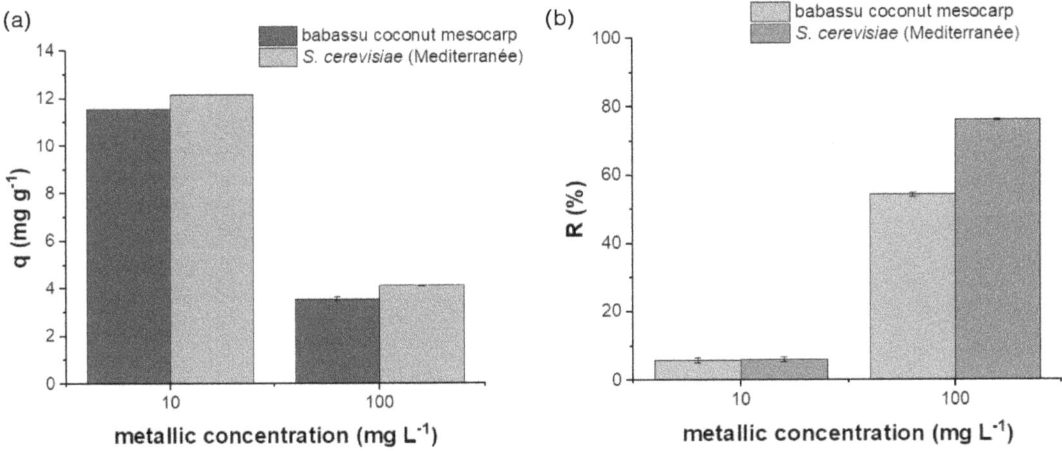

FIGURE 11.3 Graphic representation of Cu(II) biosorption potential by lignocellulosic biomass and yeast as a function of metal concentration. (a) biosorption capacity; (b) removal efficiency.

11.2.4 Factors that Influence the Biosorption Process

Biosorption is a process derived from physicochemical interactions and identified by the attraction of molecules/metallic ions on the surface of the biomass, where the unbalance of forces through surface contact of the solid biosorbent with the aqueous solution containing the adsorbent causes a solute layer is formed on the biosorbent [55,56].

Due to this fact, several factors can influence the biosorption of metal ions and, consequently, their removal efficiency. Among them, we can mention the contact time, the pH that affects the chemistry of the solution of the metallic species, the Point of Zero Charge (PZC), the nature of the solvent, the properties and concentration of the adsorbate, the temperature of the system, the characteristics and amount of the biosorbent, the activity of the functional groups of the biomass, the concentration of the metal in the aqueous solution and the competition of the metallic ions for the binding sites. In addition to the biomass conditions (live or dead), size (granulometry) and pretreatment performed [57–60].

The important factors that influence the biosorption process and that tend to be more investigated are pH, temperature, biosorbent dosage, initial metal concentration, and contact time. Batch experiments generally focus on these factors to assess the biosorption potential of the biosorbent by the metallic species (Figure 11.4). These factors are essential to optimize the conditions of the removal process [61,62].

11.2.4.1 Dosage of Biosorbent

Biosorbent dosage is related to the amount of biosorbent used according to the actual volume of contaminated effluent solution or synthetic metallic solution. As it is the biosorbent that will provide the active binding sites through the presence of functional groups present on the surface of the biomass for the biosorption of the metal, the dosage of the biosorbent is an important factor in the biosorptive process [2].

The biosorption of metal ions evolves as the dosage of the biosorbent is increased in a certain initial concentration of metal due to the enlargement of the surface area, increasing the number of accessible fixation sites [2].

However, if the amount of biosorbent is increased above the ideal dose, the biosorption capacity will decrease, as a large amount of biomass in an aqueous medium with metals will decrease the

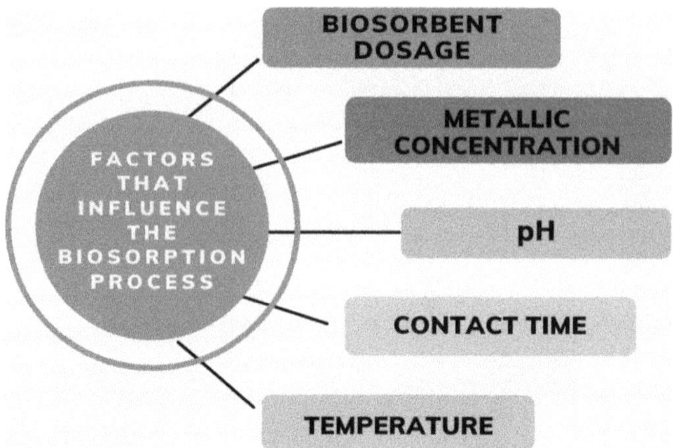

FIGURE 11.4   Summary of the main factors that affect the biosorptive process.

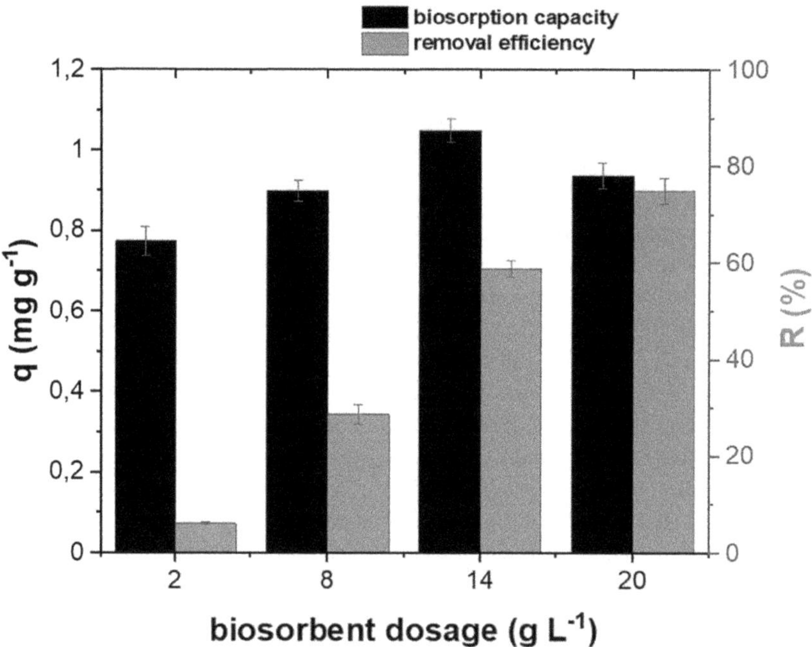

FIGURE 11.5   Effect of biosorbent dosage (commercial yeast *S. cerevisiae* Perlage® BB) on Cu(II) biosorption, for 240 minutes and metallic concentration of 25 mg/L.

ability of each particle to capture cations in solution. This behavior can be observed in Figure 11.5, since, from the biosorbent dosage of 14 g/L, there is a decrease in the biosorptive capacity of copper by the biomass of the yeast *Saccharomyces cerevisiae* [63].

This phenomenon occurs due to the repulsion of the active site between the surfaces of the biosorbent, and when the surfaces are close to each other, this effect is amplified, generating a loss of efficiency [64]. In Figure 11.5, the removal efficiency increased with the addition of more biomass in the system. With 20 g/L of biosorbent, the efficiency was the highest found, being above 70%. On the other hand, the lowest removal efficiency was observed at the concentration of 2 g/L of yeast.

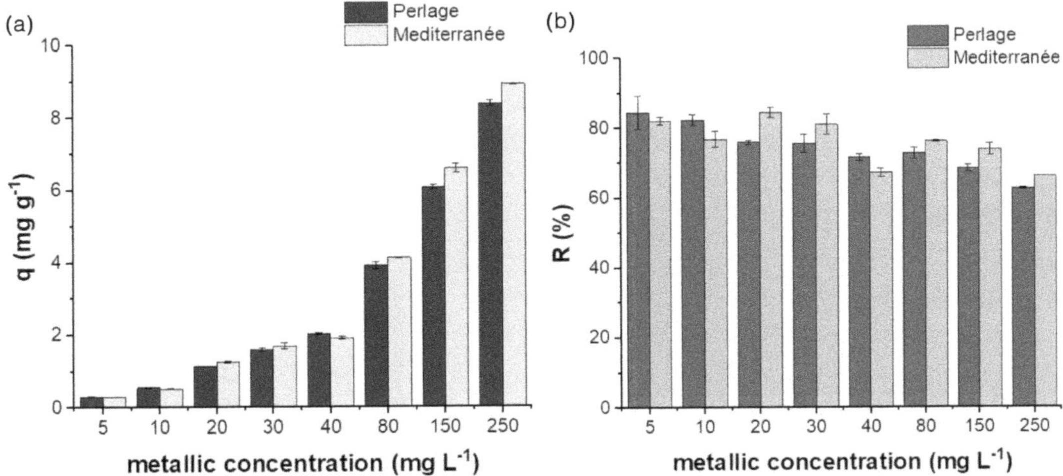

FIGURE 11.6 Effect of metallic concentration on Cu(II) biosorption by commercial yeasts *S. cerevisiae* Perlage® BB and Fermol Mediterranée. (a) biosorption capacity; (b) removal efficiency.

11.2.4.2 Metallic Concentrations

The initial concentration of metal ions in solution plays an important role, as it alters the existing driving force to overcome the resistance of mass transfer between the aqueous and solid phases [38]. In general, the solution with low metal concentration has a higher percentage of removal and lower biosorption capacity, while the solution with high metallic concentration can result in a low removal rate and high biosorption capacity (Figures 11.6 and 11.7) [65].

Figures 11.6 and 11.7 show the potential for copper biosorption by different biomasses and ranges of metallic concentration. In Figure 11.6, the biosorptive potential is related to the comparison of commercial yeasts *Saccharomyces cerevisiae* Perlage® BB and Fermol® Mediterranée which, according to the portfolio of the AEB® group, the Perlage® BB strain has alcoholic power and cryophilic character and Fermol® Mediterranée has to produce high amounts of polysaccharides and mannoproteins [65].

The analysis of Figure 11.6a shows that these strains destined for the wine industry did not present a different removal profile, that is, the commercial modifications did not affect the biosorption capacity since they presented the same behavior, proving that, according to the concentration metallic increases there is an increase in the biosorptive capacity. In Figure 11.6b, referring to the removal efficiency, it can be seen that the metallic concentration did not greatly influence the removal rate, being above 70% for metallic concentrations of (5–30 mg/L) and above 60% for concentrations of (40–250 mg/L) [65].

The biosorption capacity of the fungus *Trichoderma harzianum,* in comparison with the yeast *S. cerevisiae* Mediterranée (Figure 11.7a), shows that the biosorption capacity of biomass is proportional to the increase in metallic concentration, the same behavior shown in Figure 11.6a [65].

While the removal efficiency (Figure 11.7b) was not influenced by the metallic concentration, demonstrating that this factor did not change the removal rate. In the studied range (10–1,600 mg/L), copper removals by the studied biomass were above 50% [65].

This is because, at low concentrations, biosorption is independent, while at high concentrations, the process becomes dependent on the number of adsorbate molecules. The number of binding sites available on the biosorbents and their accessibility to the adsorbate molecules will determine the rate of removal from the biosorption process [2].

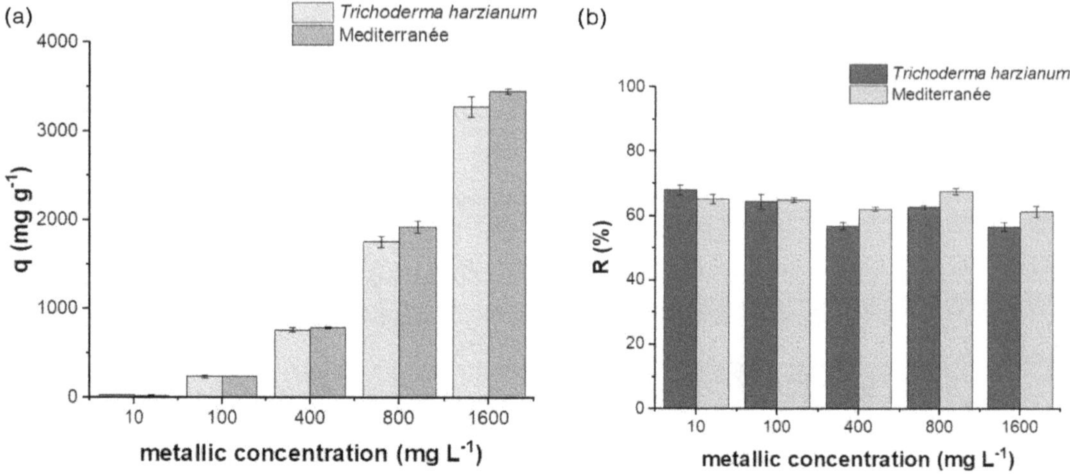

FIGURE 11.7 Effect of metal concentration on Cu(II) biosorption by a fungus and a yeast. (a) biosorption capacity; (b) removal efficiency.

Thus, the dosage of the biosorbent can strongly influence the efficiency of the process, as it determines the number of sites available for biosorption. However, beyond a certain biosorbent dosage, the potentially toxic metal removal process will become independent as the biosorbent surface reaches equilibrium with the adsorbate molecules (biosorption equilibrium) [2].

The higher the initial concentration of the metal in the solution, the faster the biosorbent will become saturated due to the occupation of active sites present on the surface of the biomass. The initial biosorption of the metal is greater at higher concentrations in the medium, which is why the use of rapid contact biomass systems is beneficial in significantly reducing contamination of highly enriched waters with potentially toxic metals [64].

According to Silva et al. [66], the influence of metal concentration is one of the most important factors in metal removal. With the increase of equilibrium metal concentration, the biosorption capacity (mg of the metal per gram of biomass) increases, and the efficiency of biosorption decreases [67].

11.2.4.3 pH

In the biosorption process, some variables can affect the performance of this environmental decontamination technology, such as pH, which mainly affects two aspects, the solubility of metal ions and the total charge of the biosorbent, since protons can be adsorbed or released. The acidity of the medium affects the competition of hydrogen ions for metal ions for active sites present in the biomass [38].

Alkaline pH favors the removal of positively charged (cationic) metals and basic dyes, while the removal of negatively charged (anionic) metals and acidic dyes is better in acidic media. Since H^+ ions and positively charged cations will be competing with each other for binding regions on the surface of the biosorbent and the formation of metal complexes or the formation of hydroxides at acidic pH will reduce the binding potential of cationic adsorbates [2].

Consequently, the negatively charged OH^- ions will be competing with the anions for binding to the surface, resulting in their low adsorption capacity at alkaline pH [2]. In addition, the increase in pH can cause deprotonation of the functional groups present on the surface of the biosorbent, causing them to behave negatively charged and consequently begin to attract positively charged metal ions. In contrast, as the pH is reduced, the overall surface rates will become positive, which inhibits the approach of positively charged metal cations [38].

The effect of pH on the biosorptive process can also be justified by the Point of Zero Charge ($_{pH}$ PZC) which is defined as the pH range in which the surface of the biosorbent is neutral. When biomass is kept in a solution with a pH lower than the ($_{pH}$ PZC) of the biomass, protonation of certain functional groups occurs, and the biomass behaves like a surface of positive charge. An increase in pH above this point causes deprotonation of the surface groups of the biomass, acting as negatively charged sites [38] [62].

As the pH rises above the ($_{pH}$ PZC) even though the surface of the adsorbent is negatively charged, the adsorption will still increase as long as the metallic species are positively or neutrally charged. However, when both the surface charges of the biosorbent and the metallic species become negative, the biosorption will decrease significantly [38].

Metallic speciation is another aspect that must be considered, which involves pH, the copper ion, for example, exists in the form Cu^{2+} at pH 3, above this pH, other species of Cu(II), such as Cu (OH)$^+$ at pH 4–5 and $Cu(OH)_2$ are found at pH above 6 [38].

Changes in the pH of the media influence the redox potential, dissociation state of binding sites, ionic conditions of metals, and competition for active sites. In the absence of complexing agents, the hydrolysis and precipitation of metal ions can be affected by the concentration of soluble metal species and their shape by the pH variable [64].

In turn, the pH also affects the specificity of the metal ion in the solution since there is a decrease in the solubility of the metal complexes with increasing pH. As adsorption depends not only on the attraction of the adsorbate to the surface of the solid but also on the lyophobic behavior (sorption increases with decreasing solubility), for most metals, this means that adsorption increases with increasing pH. On the other hand, very high pH values, which cause precipitation of metal complexes, should be avoided during sorption experiments, as the distinction between sorption and precipitation in metal removal would be difficult [68].

The solubility of metal ions is controlled by the concentration of hydroxides or carbonates that modify their speciation. The optimum pH of biosorption generally varies depending on the metal and the biomass, in general, the pH varies between 3.0 and 6.5 [39,69]. Values below this range cause competition of dissociated H$^+$ ions for active sites on the cell surface. Values above this range can lead to the precipitation of metals in their forms as hydroxides, making the use of biosorbents impossible and generating toxic sludge highly enriched with metals [64].

The evaluation of the influence of pH on copper biosorption by the commercial yeast *S. cerevisiae* Perlage® BB (Figure 11.8) showed that the pH that had the best removal capacity and

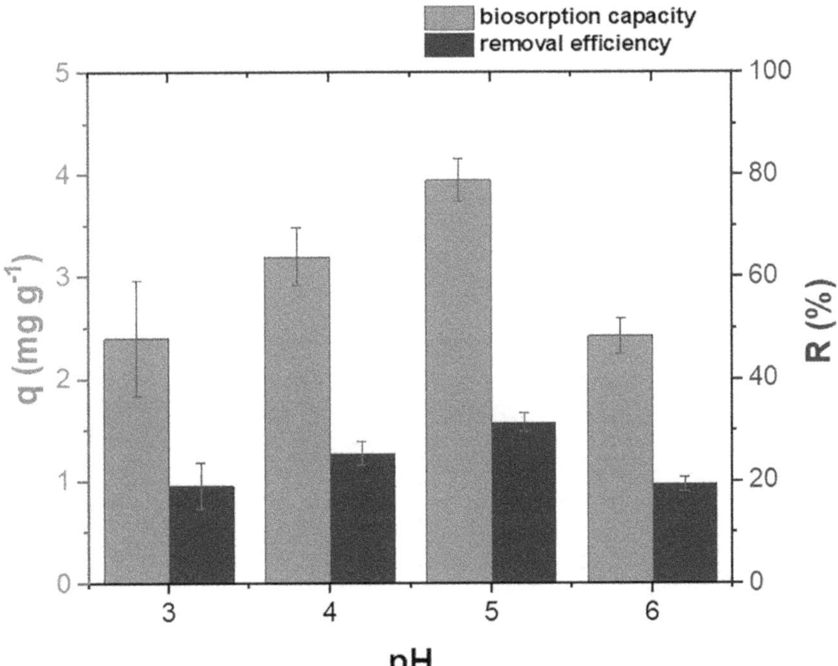

FIGURE 11.8 Effect of pH on Cu(II) biosorption by the commercial yeast *S. cerevisiae* Perlage® BB.

efficiency was 5. The pH 5 is usually a much studied range and of better performance in biosorption of potentially toxic metals [49,70,71].

11.2.4.4 Contact Time

The contact time in general is related to the time available for the biosorption process to occur. The biosorption capacity is not directly affected by the contact time of the biosorbent and adsorbate, however, it can act as a limiting factor. Therefore, increasing the contact time may allow the biosorbent to present the maximum biosorption capacity. When the biosorbent reaches its maximum biosorption capacity, its active binding sites are fully saturated, in which case increasing the contact time would have no additional effect. Therefore, it is very important to investigate the best time for the biosorption process, in this way the tests will occur until the system reaches biosorption equilibrium [46,54,72].

Contact time can be used to model the biosorption process in a series of three successive stages, which are surface biosorption, film diffusion, and intraparticle diffusion. In addition to providing a detailed insight into the kinetics of the removal rate of the biosorption process [2].

The metal-biomass interaction time can very well be regarded as equivalent to the equilibrium time. In the initial phase of the process (first minutes of contact), the biosorptive process usually presents high removal rates due to the increase in the concentration gradient between the adsorbate present in the solution and the biosorbent and a greater number of active binding sites available in the start of the process. With increasing contact time, more and more active binding groups participate in the biosorption of metal ions until biosorption equilibrium is reached. After reaching equilibrium, no significant change is noticed in the metal concentration of the solution [61].

In Figures 11.9 and 11.10, the influence of contact time on copper and gold biosorption is shown by comparing the profile of a fungus (*Geotrichum candidum*), a bacterium (*Pseudomonas aeruginosa*), and yeast (*Saccharomyces cerevisiae*), respectively. The graphs show that the kinetic profile of the three biomasses is fast (30 minutes), becoming constant over the evaluated time (30–240 minutes) for both metallic species evaluated, copper (Figure 11.9) and gold (Figure 11.10). The best performance was found in the bacterium for the copper and gold metallic species (Figures 11.9 and 11.10) [65].

11.2.4.5 Temperature

Temperature can be expressed in biosorption as a measure of the average collision or kinetic energy of molecules. The increase in temperature is the net result of the increase in intrinsic internal energy due to the higher rate of collision frequency of molecules. Thus, the higher temperature can also improve the surface-related properties, increasing the pores on the surface of the biosorbent and

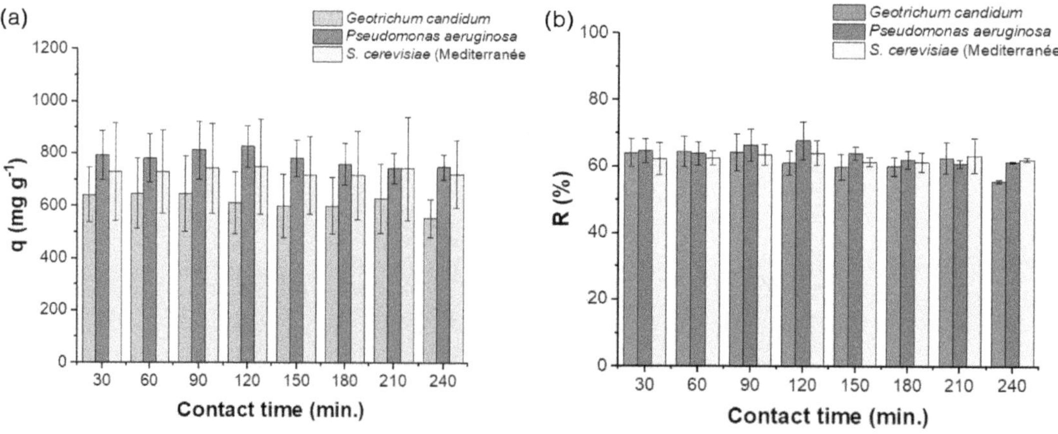

FIGURE 11.9 Effect of contact time on Cu(II) biosorption comparing the biosorption potential of a fungus, a bacterium and a yeast. (a) biosorption capacity; (b) biosorption efficiency.

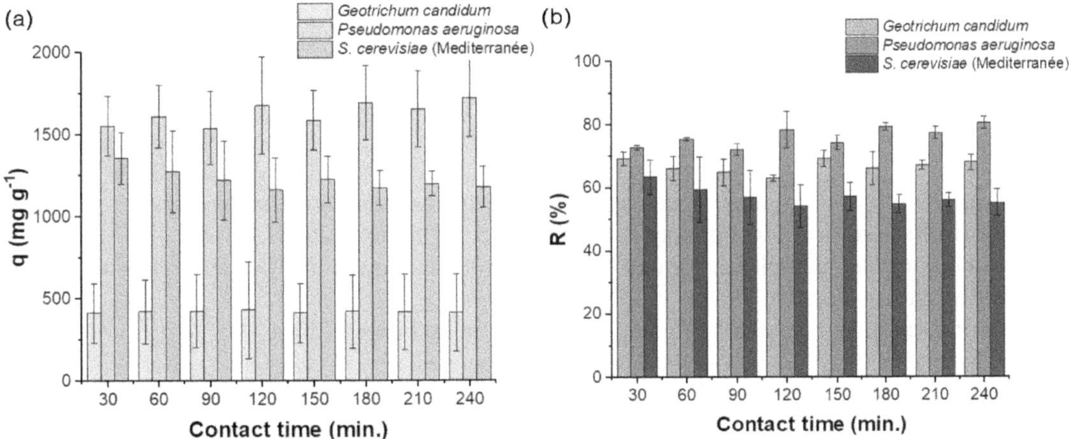

FIGURE 11.10 Effect of contact time on Au(III) biosorption comparing the biosorption potential of a fungus, a bacterium and a yeast. (a) biosorption capacity; (b) removal efficiency.

thus reducing the intraparticle resistance offered by the biosorbents. However, structural degradation of biosorbents, destruction of living biomass or metabolic action, and loss of functional groups can also occur with increasing temperature, and in this case, the desorption process can occur [2].

Temperature affects the surface activity of the biosorbent and, consequently, the biosorption capacity. The effect of temperature on the biosorption process depends on the nature of the process. The increase in temperature in an exothermic process would result in a decrease in the removal of metal ions (low efficiency), the opposite would occur for endothermic biosorption processes. Likewise, lowering the temperature for an exothermic biosorption process would increase metal ion removal, the opposite would occur for endothermic biosorption processes. In general, biosorption processes with microorganisms and lignocellulosic biomass are evaluated in the range of 25°C–30°C [46,49,70].

According to Silva et al. [66], biosorption is not necessarily an exothermic reaction like other physical adsorption reactions. However, the temperature range for biosorption is relatively narrow, normally between 10°C and 70°C, and is a function of the type of biosorbent used. Studies have shown that in the range of 5°C–35°C, the temperature has little effect on biosorption in aqueous media [66]. Other factors can also influence the biosorption process, such as:

- **Surface area:** As biosorption is a surface phenomenon, the larger the dimension and the lower the depth of the biosorbent, the more effective the biosorptive process will be [73].
- **Biosorbent property:** The physicochemical nature of the biosorbent is a determining factor since the capacity and speed of the biosorption process depend on the specific surface area, porosity, specific pore volume, pore size distribution, functional groups present, and the nature of the biomass [74].
- **Adsorbate property:** According to Domingues [74], the size of the molecule is always important when the rate of biosorption is dependent on intramolecular transport. The polarity of the adsorbate is another important factor, as a polar species will have an affinity for the solvent or polar biosorbents.

11.2.5 Biosorption Mechanism

Biosorption is a process that occurs when a substance (adsorbate), present in a liquid, remains retained on the surface of a solid (biosorbent) due to intermolecular, electrostatic, dipolar, Van der Waals, or other interactions, a set of these [75]. These interactions depend on the nature of the species involved, they can occur in isolation or simultaneously, so the stronger the interactions, the more effective the biosorption process will be [75].

Biosorption comprises several mechanisms that quantitatively and qualitatively differ according to the biomass used, its origin, and the processing method [76]. Overall, six different mechanisms can work in synergism [76]. The possible mechanisms of biosorption have already been investigated and neatly cataloged regarding the interaction between metallic species and biosorbents, classified as follows [66]:

- **Complexation:** Formation of a complex through the association of two species;
- **Coordination:** Covalent bonding of a central atom with other atoms;
- **Chelation:** The complexes formed by an organic compound are joined by at least two active sites;
- **Ionic exchange:** Formation of complexes from the exchange of ions;
- **Adsorption:** Sorption across the surface of the biomass;
- **Inorganic precipitation:** Modification in the conditions of the aqueous medium, thus generating a variation in pH, leading to metallic precipitation.

According to the aforementioned mechanisms, any one of these or a combination can occur in the biosorption process, immobilizing one or several metallic species on the surface of the biomass [66]. The ions are attracted by active sites on the surface of the biomass, where there are different functional groups responsible for joining them to the surface of the particle, such as phosphate, carboxyl, sulfide, hydroxyl, and amino [47,77].

Therefore, understanding the mechanisms by which biosorbents accumulate metallic species is important for the development of processes for the removal and recovery of metallic species in aqueous solutions. Numerous strains of microorganisms can bind to metallic species, however, there is a great difference in the responses of microbial species when exposed to metallic solutions [47,66,76].

The walls of microorganisms such as bacteria, fungi, and algae are efficient biosorbents of metallic species, however, in many cases, the initial binding can be followed by inorganic deposition of increasing amounts intracellularly. Covalent and ionic bonds may be involved in biosorption, with constituents such as proteins and polysaccharides. It is very important to know the chemical or physiological reactions during biosorption, which could enable the specification and control of process parameters to increase the speed, quantity, and specificity of metallic accumulation. That is, to model and predict its performance in removing metallic species in aqueous media [43,47,68,78,79].

11.3　AGRO-INDUSTRIAL WASTE AS BIOSORBENTS

Many researchers focus their research on the use of biosorbents from agricultural residues, fruit peels, and industrial by-products and among others (Table 11.2). The use of low-cost biosorbents becomes an alternative to the use of activated carbon, which is one of the most used adsorbents but has a high cost, which makes its implementation difficult [45].

Cellulose-based biosorbents include numerous agro-industrial residues such as rice husk, rice bran, wheat bran and husk, sawdust, Pequi bark, babassu coconut mesocarp, peanut peel, apple peel, hazelnut husk, walnut husks, coconut husks, coffee beans, cottonseed husks, banana peel, grapes, tea-leaf residues, beet pulp, corn cob, water hyacinth, soybean hulls, orange zest, stalks of cotton, sunflower stalks, and others that have been studied in the biosorption of potentially toxic metals [22,23,43,49,71].

The characterizations of the composition and components present in these biosorbents revealed their potential in the biosorption of potentially toxic metals. The cellulose-rich matrix biosorbent has various functional groups such as alcohols, carbonyl (ketone), thioether, carboxyl, amino, sulfhydryl (thiol), sulfonate, esters, and others. These functional groups can bind metal ions by substituting hydrogen ions or by donating an electron from the functional group, resulting in the complexation of metal ions in the solution. The main mechanisms of cellulosic-based biosorbents include complexation, physical adsorption, chelation, ion exchange, and chemisorption [22,43].

TABLE 11.2

Reports of the Uptake of Metallic Species by Different Biomasses

Biomasses	Ion	m (g/L)	Metal (mg/L)	q (mg/g)	R (%)	Reference
Pequi bark	Cd (II)	–	10	0.0012	99	[80]
	Pb (II)					
Macadamia mesocarp	Cu (II)	–	100	74, 70	82.4	[82]
Sewer sludge ceniza	Cu (II)	30	25	–	98	[83]
Quail egg shell	Cu (II)	0.1	100	40	60	[84]
Chemically modified peanut shell (*Arachis hypogaea*)	Cu (II)	0.8	200	14	43	[85]

Where m represents the concentration of the biosorbent, *metal*, the metal concentration studied, q the biosorption capacity and R the removal efficiency.

Biosorption using agricultural residues and plant materials is an environmentally correct process because it uses renewable biomass. Agro-industrial residues have good removal efficiency because they comprise cellulose, hemicellulose, and lignin, which are sources of functional groups such as amine, phenolic, alcohols, and esters [22], [45]. These groups can have the ability to act as active sites of metals by donating a pair of electrons to form complexes of metal ions in solution [38].

Among the various agro-industrial residues that can be used as biosorbents, the use of bark stands out in the literature [22,38,45]. Among others, residues from industrial processes, sediments from forestry, crustaceans, and ceramics industries also have good biosorption capacity for potentially toxic metals.

11.3.1 Fruit Peels

The fruit peels, which are common agricultural by-products, have been widely used as efficient biosorbents of potentially toxic metals [98]. The orange (*Citrus sp.*), for example, is one of the most abundant citrus fruits in India. Brazil is the world's largest producer of this fruit [99], and the orange residue is a by-product of the fruit industry used mainly as animal feed [100].

The residue from orange juice production comprises peel and pulp (the sieved fraction of the juice). This residue is largely composed of cellulose, pectin (galacturonic acid), hemicelluloses, lignin, chlorophyll pigments, and other low molecular weight compounds such as limonene. Pectin substances are the predominant type of polysaccharide present in the cell walls of the residue from orange juice production. In the literature, orange peels are recognized as agro-industrial residues with biosorptive properties and have been used in the biosorption process [100].

Another widely consumed fruit, that produces residues from its peel, is the banana (*Musa spp.*). The banana is one of the most consumed fruits worldwide. The banana peel represents 30%–40% of the total weight of the banana, generating thousands of tons of waste annually. The banana peel consists of a large amount of cellulose, which has functional groups capable of binding metals [101].

The use of a banana peel as a biosorbent for the biosorption of potentially toxic metals (Pb, Ni, Fe, Cu, Al, and Ba) proved to be a viable alternative, even at high concentrations of metals [91] Boniolo et al. [87] also studied this biomass, as well as Moringa seeds, which showed promising results in removing uranium from effluents.

The Pequi (*Caryocar brasiliense* Camb) is a native fruit of the Brazilian Cerrado that presents a great diversity in its physical and chemical characteristics [102]. In experiments conducted by Nascimento et al. [80], the bark of Pequi (*C. brasiliense*) *in nature* and modified with citric acid were used as biosorbents to remove potentially toxic metals. The results showed that the bark of Pequi showed a good biosorption capacity of Cd^{2+} and Pb^{2+} ions. However, Pb^{2+} showed better biosorption capacity than Cd^{2+}.

Sousa and Oliveira [81] also analyzed the peel biomass of Pequi *in nature* to remove Ni^{2+}, Cd^{2+}, Zn^{2+}, and Mn^{2+} ions. The percentage of Cd^{2+} removal was the highest, 81.4%. These data indicate that Pequi bark can be used as a biosorbent to remove potentially toxic metals.

The husk of green coconut (*Cocus nucifera* L.), Jabuticaba (*Plinia cauliflora*), and orange (*Citrus X sinensis*) were evaluated as chromium biosorbents. The results showed that the biomasses were effective in removing total chromium. Coconut husk was the most promising for the chromium biosorption process, with an average removal rate of 34.92% [92].

The characterization of cupuaçu (*Theobroma grandiflorum*) husk and açaí (*Euterpe oleracea*) seed through the parameters that influence the biosorption of the metal, such as pH in water and potassium chloride, quantification of acid and basic groups, and determination of the PZC, showed satisfactory results confirming their ability to remove metals [81].

11.3.2 Agricultural Waste

Waste from agriculture is a by-product of agricultural and forestry production and processing. The main types are straw, bark, slag, and bran. The huge amount of agricultural residues has many advantages as biosorbents in the treatment of effluents. The main ones are abundant sources, chemical stability, low cost, renewable cycle, short cycle of regeneration, respect for the environment, and green energy. In addition, its structure has high porosity and a large specific surface area, which facilitates the physical biosorption of potentially toxic metals. At the same time, it contains some active substances, such as pectin, cellulose, hemicellulose, and lignin, which can react with metal species and become active binding sites [103].

Agricultural waste mainly includes cellulose, lignin, and hemicellulose. The amount of cellulose, hemicellulose, pectin, and lignin varies depending on the type of waste. Cellulose, lignin, and hemicellulose contents, for example, in crop residues are in the range of 35%–50%, 20%–30%, and 15%–30%, respectively [103].

Cellulose, hemicellulose, and lignin are structural polysaccharides. Cellulose is a linear syndiotactic polymer of glucose, formed by β-(1→4)-glycosidic bonds. Hemicellulose is composed of a variety of uronic acid groups, which is a branched heteropolymer of D-xylose, L-arabinose, D-galactose, D-glucose, D-mannose, and D-glucuronic acid. These monomers are connected in the form of bonds (C-C) and bonds (C-O-C). Lignin is a cross-linked aromatic polymeric compound consisting of *p*-coumaryl alcohol, coniferyl alcohol, and sinapyl alcohol [103].

In Brazil, there are many by-products due to certain manufacturing and consumption processes, such as green coconut husk, sugarcane bagasse, banana peel, and others. However, the destination of these residues becomes an environmental issue because even if the part is used for certain purposes, such as composting, this residue is still not fully utilized, being, therefore, deposited in sanitary landfills, becoming an environmental liability or reducing the useful life of the landfill [104].

In this way, the use of agricultural residues as biosorbents in biosorption processes of potentially toxic metals, in addition to being a low-cost material, also helps to reduce the inappropriate disposal of this material, minimizing environmental damage and generating socio-economic value for them [22,49].

Experimental data from a study carried out by Freitas [86] suggest that the use of coffee grounds (*Coffea sp.*) as a hexavalent chromium biosorbent is an inexpensive, quick, and simple alternative to remove metals from aqueous waste. This biomass removal potential was also demonstrated by a study on uranium removal from effluents by biosorption [87].

Another waste generated by the industry is sugarcane bagasse (*Saccharum officinarum*), which Ferreira [88] analyzed during biosorption to remove Cd^{2+}, Cr^{3+}, and Pb^{2+} ions. The biosorption of these metal ions by sugarcane bagasse showed favorable results. Similar results were obtained with the same biomass by Albertini, Carmo, and Prado Filho [89].

Another material used as a biosorbent for the biosorption of potentially toxic metals is the egg house of laying hens. The biosorption of copper, chromium, nickel, and zinc ions by the biosorbent showed satisfactory results, with the maximum biosorption capacity for copper, zinc, nickel, and chromium ions of 25.43; 6.74; 9.79, and 6.73 mg/g, respectively [90].

11.3.3 WASTE FROM INDUSTRY

Certain operational waste from industries is a source of low-cost biosorbents, these materials are generally readily available in large quantities, therefore, they are cheap and of little economic value. Thus, biosorption emerged as an alternative process due to the characteristics of reducing the price of the biosorption material, application of the system capable of detoxifying a large volume of effluent with low operating costs, possible selectivity, and recovery of the metallic species [76].

Shale is a sedimentary rock often occurring in gangue that accompanies coal and other minerals. It is extracted with them and often treated as waste rock. The shale was tested for sorption of Cr(III) and Cr(VI) ions from aqueous solutions [93].

Regarding the industrialization of crustaceans, the activity produces a large amount of highly perishable waste, as the bodies of most animals (shells, heads, and feet) are not suitable for human consumption. Waste from crustacean processing consists of several compounds that have potential entry into different industries [105].

The main compound that can be extracted from crustacean residues is chitin, which can be easily converted to chitosan by its alkaline deacetylation [106,107]. Chitosan can be a new ally of science in cleaning up rivers and effluents discharged by industry. The material was shown to be able to extract toxic metals from water, such as copper, zinc, lead, cobalt, and cadmium [106,107]. The crab shell *Ucides cordatus* proved to be a very efficient biosorbent in removing total iron dissolved from Fe^{2+} contaminated groundwater since the removal was at least 50% [94].

Ramos [95] carried out the analysis of crab shell (*Cancer pagurus*), lobster shell (*Palinurus elephas*), crayfish shell (*Nephrops norvegicus*), and lobster shell (*Homarus gammarus*) in the treatment of copper contaminated water, where he also obtained satisfactory results regarding the removal of this metallic species.

In the vast majority of biosorption processes that take place, the removal of metal ions occurs in two steps. The first step is faster because it corresponds to the occupation of active sites that have a higher affinity for the metal and are more physically accessible, while the second step is slower, where the metal ions are more inaccessible and have reduced affinity with them until saturation. This behavior is a typical biosorption process involving weak binding forces between biomass and metal ions [108].

For very high concentrations of metal ions, equilibrium can be reached quickly, but the biosorption capacity is low. At lower metal concentrations, equilibrium is very slow, but the removal of metal ions from the solution is high. It is interesting to note that biosorbents made from forest residues are of the lignocellulosic type, as their composition is composed of approximately 60% cellulose and hemicellulose and approximately 20% lignin [109].

The study of the biosorption capacity of Cd^{2+} and Pb^{2+} ions by teak wood (*Tectona grandis*), widely used in the manufacture of furniture, diverse structures, and floors, showed that the biomass showed good biosorption capacity, and the biomass modification by citric acid does not affect the biosorption of the studied metallic species. The analysis of the metallic competition of chemical species in solution showed that the optimal biosorption capacity of Pb^{2+} ions was approximately twice that of the Cd^{2+} ion [110].

A study by Santos [96] showed that the treatment of effluents containing total chromium using araucaria, eucalyptus, and pine forest residues is effective and viable. This approach guarantees compliance with environmental restrictions, demonstrating the good performance of biosorbents, which are available in large quantities, at low cost, and require less preparation for use.

The sawdust composed of eucalyptus, cedar, pine, and noble wood used in the experiments of Albertini, Carmo, and Prado Filho [89] showed satisfactory results in the biosorption of cadmium. Leaves and residues of citronella (*Cymbopogon nardus*), lemon balm (*Cymbopogon flexuosus*), palm (*Cymbopogon martinii*), and eucalyptus (*Eucalyptus spp.*) were evaluated for the biosorption capacity of lead, cadmium and nickel ions in residual aqueous solutions. Eucalyptus leaf residue is a biomass with a better biosorption capacity for Cd^{2+} and Ni^{2+} ions and a biosorption capacity for Pb^{2+} ions similar to that of citronella (crude leaves and residues) [97].

11.4 MICROORGANISMS AS BIOSORBENTS

In the early 1980s, reports emerged of the ability of some microorganisms to accumulate metallic elements [79]. Other research has pointed to the ability to bind via metabolism independent of microbial biomass, whether alive or dead, to potentially toxic metals. As a result, biosorption studies grew, and several microorganisms were investigated until the present [79,111–113].

Microorganisms are reported as potential biosorbents of potentially toxic metals. This is mainly because the cell wall composition differs between different microbial populations, therefore, the mechanism and efficiency of biosorption using microorganisms will depend on the active binding sites present in the cellular components [43].

11.4.1 MICROORGANISMS

Before the existence of microorganisms was known, all organisms were grouped into the animal kingdom or the plant kingdom. When microscopic organisms with animal and plant characteristics were discovered at the end of the 17th century, a new classification system became necessary. Still, biologists were unable to agree on criteria for classifying these new organisms until the late 1970s. In 1978, Carl Woese developed a classification system based on the cellular organization of organisms. All organisms were grouped into three domains (Figure 11.11).

In addition to showing the differences in rRNA, the three domains differ in membrane lipid structure, transfer RNA molecules, and antibiotic sensitivity [114].

1. **Bacteria** (cell walls contain a carbohydrate-protein complex called peptidoglycan).
2. **Archaea** (cell walls, if present, lack peptidoglycan).
3. **Eukarya**, which includes the following groups:
   - Protista (mycetozoa, protozoa, and algae);
   - Fungi (unicellular yeasts, multicellular molds, and mushrooms);
   - Plants (mosses, ferns, conifers, and flowering plants);
   - Animals (sponges, worms, insects, and vertebrates).

Living organisms, like microorganisms, are classified into prokaryotes and eukaryotes. Prokaryotic and eukaryote organisms are distinguished based on their cellular structures [115,116]. The main difference between prokaryotic and eukaryotic cells is that prokaryotic cells do not have a nucleus and other membrane-bound structures called organelles, whereas eukaryotic cells have a nucleus and cellular organelles [115].

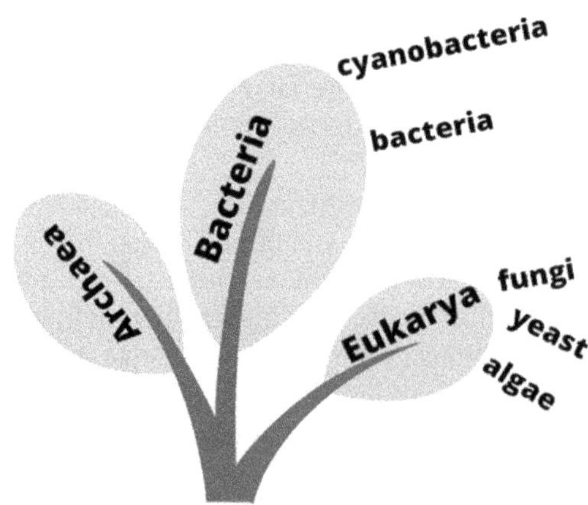

FIGURE 11.11   Three-domain system and the main microorganisms studied in biosorption.

Microorganisms comprise prokaryotes (Bacteria and Archaea Domains) and some eukaryotic organisms—Eukarya Domain [116]. Bacterial ancestors were the first living cells to appear on Earth [114]. Microorganisms are microscopic single-celled organisms that are, in general, individually too small to be visualized with the naked eye. The group includes bacteria, fungi (yeasts and molds), protozoa, and microscopic algae. It also includes viruses, acellular entities often regarded as the boundary between the living and the non-living [114]. Most microorganisms can carry out their vital processes, such as growth, energy generation, and reproduction, independently of other cells [117].

Microorganisms play central roles in human activities and all life on planet Earth (Figure 11.12). Despite being the smallest form of life, microorganisms constitute the largest mass of living matter on Earth and carry out several chemical processes necessary for other organisms [117]. Microorganisms are all around us in soil, water, and even in inhospitable places like boiling warm springs and glacial ice [117].

In the beginning, microbiology was directed at studying microorganisms that cause infectious diseases, but nowadays, it includes practical applications in science. The field of modern microbiology has expanded into many categories, such as the development of pharmaceutical ingredients, food quality control methodologies, production of by-products such as ethanol, and environmental decontamination. Microorganisms are used in the production of vitamins, amino acids, enzymes, growth supplements, and others [115].

Several microorganisms are considered foods, such as mushrooms (Fungi), as they can also be contaminants of food and food products, causing deterioration and production of toxins that can cause diseases when consumed by humans and animals, this is the case of pathogenic bacteria *Salmonella sp.* that can cause typhoid fever if it is a contaminant in foods such as poultry, eggs, milk, pork, water, shrimp and raw meats [116].

Some microorganisms such as yeasts and bacteria can transform nutrients present in food and produce important by-products for the industry, such as ethanol, carbon dioxide, lactic acid, and acetic acid, which are produced by a biological route through alcoholic and lactic fermentation, and acetic [116].

Industrial microbiology uses microorganisms that are usually cultivated on a large scale for the production of commercial products or for carrying out important chemical transformations [117]. Applied microbiology on an industrial scale began with large-scale food fermentation, which produced lactic acid from dairy products and ethanol from beer [114].

Microorganisms are also capable of acting in environmental processes. Environmental microbiology is a science that combines the application of chemical, biological, and biotechnological

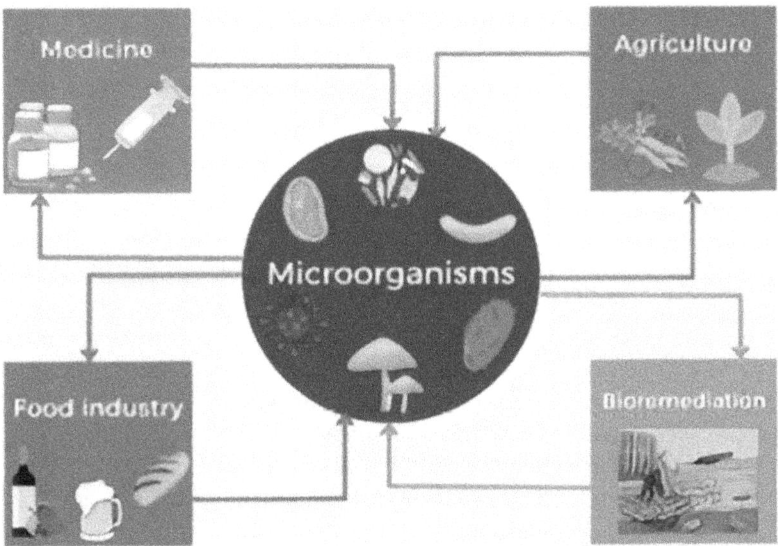

FIGURE 11.12 Importance of microorganisms in various fields of research.

concepts using microorganisms belonging to all their habitats. In addition to its impacts that can be beneficial and harmful to health and quality of life [118], [119]. Environmental microbiology is a field of study that is applied to improve the environment and benefit society [119].

The microbiological treatment of contaminated effluents has gained more and more attention, because it has high efficiency compared to conventional treatment techniques, in addition to being a low-cost and ecologically correct methodology. In this form of treatment, the exploitation of microorganisms for the removal, absorption, and degradation of pollutants are involved, thus maintaining the stabilization of the ecosystem [120].

The microbiological treatment of wastewater is one of the most promising biotechnological applications and can be done through several processes performed by microorganisms, such as bioremediation, activated sludge, bioaccumulation, and biosorption [121–123].

Biosorption is defined as the property of certain biomolecules (or types of biomass) to bind and concentrate selected metal ions or other molecules from aqueous solutions as organic compounds. Unlike a much more complex phenomenon, such as bioaccumulation, which is based on active metabolic transport, biosorption can occur through dead biomass (or by some molecules and/or their active groups) as it is a metabolically passive process and is based mainly on the "affinity" between the biosorbent and the adsorbate (species to be adsorbed) [27,40,77,124].

Biosorption can be carried out through different microorganisms. Many biosorbents have been identified as having removal capacity, including algae, fungi, bacteria, and yeasts [125]. The accumulation of potentially toxic metals is caused by numerous interactions that allow the binding of ions to functional groups on the external surface of the microorganism. This process is very efficient since, in addition to being a simple operation, there is no need for additional energy and nutrients for the microorganism [9].

11.4.1.1 Bacteria

Bacteria are relatively simple and unicellular organisms—they have a single cell. Because their genetic material is not enclosed in a special nuclear membrane, bacterial cells are called prokaryotes. The word prokaryote comes from Greek and means pre-nucleus. Prokaryotic organisms include bacteria and archaea [114]. Although bacteria and archaea look similar, their chemical composition is different. The thousands of species of bacteria are differentiated by many factors, such as morphology (shape), chemical composition, nutritional requirements, biochemical activities, and energy sources. Prokaryotes vary in size from small cells with a diameter of 0.2 μm to cells with a size of around 700 μm [117]. It is estimated that 99% of bacteria in nature exist in the form of biofilms [114].

Biofilms are a natural form of cell immobilization consisting of an agglomeration of microbial cells that results from microbial attachment to solid supports attached to a matrix of extracellular polymeric substances [27,126]. Exopolysaccharides, or *Extracellular polymeric substances* (EPS), present on the microbial cell surface can protect microorganisms against the toxic effects of metals, in addition to preventing their entry into the intracellular environment. This polymeric material consists of nucleic acids, proteins, lipids, and complex carbohydrates, which play an important role in the biosorptive process [127,128].

Bacteria make up the most abundant group of microorganisms on Earth [114], which can live in a wide range of environmental conditions due to advantages such as small size, rapid growth rate, and ease of cultivation. biosorption of potentially toxic metals [127,129].

Bacterial strains can be divided into two main classes: Gram-positive and Gram-negative. The difference is based on the Gram staining reaction and refers to the distinctions of their respective cell walls [117]. The cell wall of a Gram-positive bacterium is composed of about 90% peptidoglycan, while that of a Gram-negative bacterium is composed of only 10% peptidoglycan [117].

Peptidoglycan is a linear, peptide-bonded, multilayer polymer where the glycan units are cross-linked with meso-diaminopimelic acid, D-alanine, and D-glutamic acid. The common feature of the cell wall is cross-linking between peptide chains. Peptidoglycan also contains polymers of ribitol, glycerol, or teichoic acids linked by phosphate groups that have a high frequency of cross-bridges [43]. In Gram-negative bacteria, the cell wall comprises lipoproteins, phospholipids, glycoproteins, lipopolysaccharides, and enzymes. These components are described as acting in capturing metals from different industrial effluents [43].

The main mechanism of metal biosorption by bacterial cells is through ion exchange carried out through ligands such as oxygen, nitrogen as well as peptidoglycans, and teichoic acid present on the cell surface. In addition to being able to occur via complexation, coordination, physical adsorption, chelation, inorganic precipitation or a combination of these processes [38]. Studies have also shown that the activities of antioxidant enzymes in living bacterial cells can be increased under the stress caused by metal toxicity [130].

Bacterial biomass is considered an efficient biosorbent of potentially toxic metals because it contains numerous functional groups on its surface, such as carboxyl, imidazole, sulfhydryl, amino, phosphate, sulfate, thioether, phenol, carbonyl, amide and hydroxyl groups [27]. The negatively charged functional groups present in the peptidoglycan of the bacterial cell wall, such as the components teichoic acid and teicuronic acid from Gram-positive bacteria and peptidoglycan, such as phospholipids and lipopolysaccharides from Gram-negative bacteria, constitute the main components due to the anionic nature and affinity of bond involved between the cell wall and the metal in solution [27].

In the literature, some authors have reported that there are differences in the biosorption capacity of various bacterial biomasses according to the nature of the bacteria (Table 11.3). Gram-positive and Gram-negative bacteria have significant differences in the structure and composition of their cell walls [6].

TABLE 11.3
Reports of the Uptake of Metallic Species by Different Bacteria

Bacterium	Ion	m (g/L)	metal (mg/L)	q (mg/g)	R (%)	Reference
Ochrobacrum sp. GDOS	Cd^{2+}	–	200	34.36	–	[131]
Pseudomonas azotoformans JAW1	Cd^{2+}	2	25	–	44.67	[132]
	Cu^{2+}				63.32	
	Pb^{2+}				78.23	
Klebsiella sp. 3H1	Ag^+	0.0003	1,000	99.24	–	[133]
Ralstonia solanacearum KTSMBNL 13	Pb^{2+}	–	100	–	90	[134]
Tepidimonas fonticaldi AT-A2	Au^{3+}	–	15	1.45	71	[135]
Bacillus pumilus sp. AS1	Pb^{2+}	1	50	0.6	–	[136]
Arthrospira (spirulina) *platensis*	^{137}Cs	20	–	–	±60	[137]
	^{233}U				03	
	^{241}Am				77	
	^{237}Np				±10	
	^{239}Pu				67	
	^{90}Sr				62	
Pseudomonas sp. alive	Cd (II)	1	–	92.59	89.1	[138]
dead *Pseudomonas sp*				63.29	98.5	
Bacillus subtilis KC6	Cd (II)	–	40	–	65.33	[139]
Delftia lacustris MS3	Pb (II)	±0.37	50		59.41	[140]
Bacillus pumilus SWU7-1	Sr (II)	0.96	70	±50	94.69	[129]
Pseudomonas aeruginosa ASU 6A	Zn (II)	1	50	±45	–	[141]
Bacillus cereus AUMC B52				±40		
Lactobacillus acidophilus	Pb (II)	–	0.035	–	65.61	[130]
	Cd (II)				71.95	
Vibrio alginolyticus PBR1	Cd (II)	2	50	–	59.78	[142]
	Pb (II)				82.20	
Parapedobacter sp. ISTM3	Cr (VI)	1	20	19,03	95.10	[143]

Where m represents the concentration of the biosorbent, *metal,* the metal concentration studied, q the biosorption capacity and R the removal efficiency.

The phylum Cyanobacteria, composed of Gram-negative bacteria, cyanobacteria that are photo-autotrophic oxygen-producing prokaryotes, erroneously called blue-green algae, are widely studied as metal biosorbents [114,144].

The evaluation of the potential for biosorption and bioaccumulation of Radionuclides by the cyanobacteria *Arthrospira* (spirulina) *platensis* showed that the *A. platensis* strain CNMN–CB-02 can be efficiently used for the removal of Radionuclides from solutions, but the degree of removal depends on the type of radionuclide and the process applied for removal (biosorption or bioaccumulation). Since bioaccumulation is a metabolically dependent process. *A. platensis* CNMN–CB-02 had the lowest growth rate with the addition of ^{90}Sr to the culture medium. Whereas, the removal efficiency for ^{90}Sr was very high (90%). However, the biomass showed high absorption and biosorption capacity for ^{239}Pu, ^{90}Sr, and ^{241}Am. The studied cyanobacterium was an extremophile organism, highly tolerant to gamma rays, which makes it an excellent candidate for removal of Radionuclides present in contaminated water [137].

The dead biomass of the Gram-negative, marine, and bioluminescent bacterium *Vibrio alginolyticus* PBR1 was employed to remove two metals, cadmium (Cd) and lead (Pb), from a single aqueous system. The Box–Behnken design (BBD), and the response surface methodology (RSM), were used to optimize the biosorptive process. The three main factors influencing the process were pH, biosorbent dosage, and metal concentration. The optimal conditions for the biosorptive process were pH 5.7, a biosorbent dosage of 2.0 g/L, and a metal concentration of 50 mg L^{-1}. Removal efficiencies were 59.78% and 82.20% for Cd^{2+} and Pb^{2+}, respectively [142].

The analysis by surface methodology and RSM response of the optimization of the variables that influence the biosorption profile of Pb^{2+} and Cd^{2+} by *Lactobacillus acidophilus* showed that the best removal of Pb^{2+} was at pH 4, metal concentration of 0.035 mg/L and removal rate of 65.61% while for Cd^{2+} it was 71.95% under the same conditions [130].

Industrial, domestic, and mining effluents generally contain a mixture of potentially toxic metals. Since most contaminated areas are polluted with numerous metals. Multi-metal resistant microorganisms have been recognized as good biosorbents for biosorption [27]. Many authors isolate, select, and characterize biosorbents present in contaminated sites since this strain is already used to living with the stress caused by metallic concentrations and among other factors [27,129,139,143].

The EPS-producing strain *Parapedobacter sp.* ISTM3 was isolated from a cave, and its EPS was characterized and purified for use in Cr^{6+} biosorption studies. The evaluation of the biosorptive potential of EPS produced by *Parapedobacter sp.* in mixed solutions with the concentration of each metal (Zn^{2+}, Cu^{2+}, Pb^{2+}, Cr^{6+}, Fe^{2+}, and Cd^{2+}) of 20 mg/L showed that EPS had the highest removal efficiency and biosorption capacity for Cr^{6+} compared to other potentially toxic metals studied. Biosorption for Cr^{6+} increased under acidic conditions (pH 5.0), having a biosorption capacity of 19.03 mg/g and removal efficiency of 95.10% [143].

The Gram-negative bacterium *Delftia lacustris* MS3 was isolated from a lead, zinc mine for Pb^{2+} biosorption studies. The analyses showed that, according to the kinetic model, the system obtained a better adjustment to the pseudo-second-order model, which indicates that the removal rate control step is a function of the chemical interaction between functional groups present on the surface of the biomass and the lead ions [140].

The biosorption potential of a *Bacillus subtilis* KC6 strain isolated from a pyrite mine was investigated for Cd^{2+} removal. The tests showed that the KC6 strain has a great capacity for resistance and removal of Cd^{2+} in solution. The highest Cd^{2+} biosorption capacity was found at concentrations below 10 and 40 mg/L, being about 86.33% and 65.33%, respectively [139].

The Gram-positive bacterium *Bacillus pumilus* SWU7-1 was isolated from soil not contaminated by Sr(II) and used for Sr(II) biosorption tests. Assays with lower initial concentrations showed that live biomass can resist Sr(II) biosorption. The range of initial metallic concentrations where biosorption efficiencies above 90% were found was 54-130 mg/L. At the same time, the best biosorptive capacities were found at higher metal concentrations, around 275 mg/g at 700 mg/L of metal concentration. Indicating that the strain can be removed in high metallic concentrations.

The biosorption profile was better fitted to the Langmuir model than Freundlich, showing a maximum biosorption capacity of 299.4 mg/g [129].

Pseudomonas sp. is a Gram-negative organism, Gamma proteobacteria, belonging to the Pseudomonadaceae family, containing more than 191 species. In addition to being one of the most diverse bacterial genera, it is present in the soil, water, plants, and others. Many bacteria of the *Pseudomonas* genus can tolerate and survive in highly toxic conditions, so it is considered a good biosorbent of potentially toxic metals [145].

Several microorganisms such as *Pseudomonas sp.* are widely used for the biosorption of potentially toxic metals [132,138,141,146]. Normally, the biosorption capacities of bacteria for potentially toxic metals are in the range of 1–500 mg/g [127]. The application of microbial systems in the treatment of effluents requires little energy, as a consequence has low operating and maintenance costs [147].

The characterization of the biosorption profile of a strain isolated from soil contaminated by potentially toxic metals—*Pseudomonas sp.* 375 for cadmium removal studies showed that the maximum biosorption capacities were 92.59 and 63.29 mg/g for live and dead biomass, respectively [138]. The efficiency of Cd^{2+} removal of raw wastewater samples by *Pseudomonas sp.* 375 showed that, in a sample containing 13.34 mg/L of cadmium, only 3.34 mg/L was found, therefore the efficiency of 74.99%. Spectral analysis by FTIR showed that the functional groups involved in cadmium biosorption were $-CH_2$, $-SO_3$, $C=O$, and NH [138].

Regarding the performance of the biosorptive capacity of *Pseudomonas sp.* 375, the data showed that the live bacterial biomass has a better performance than the dead. Since in all tests, pH variation, contact time, and metallic concentration, the removal capacity was higher for live biomass. Differences in the biosorption processes of living and dead biomass may be due to intracellular accumulation, which is a complex pathway used by living biomass [138].

The Zn(II) biosorption profile by the bacteria *Pseudomonas aeruginosa* ASU 6A (Gram-negative) and *Bacillus cereus* AUMC B52 (Gram-positive) was studied as a function of several physicochemical factors, such as pH 1.0–7.0, initial concentration of metal 0.0–200 mg/L, and contact time 0–60 minutes. The optimal pH for Zn(II) removal was 6.0, while the ideal contact time was 30 minutes for both bacterial species [141].

However, Joo, Hassan, and Oh [141] found that *Pseudomonas aeruginosa* ASU 6A (Gram-negative) has a higher biosorption capacity of the potentially toxic metal Zn(II) compared to *Bacillus cereus* AUMC B52 (Gram-positive). The authors explained their finding by the fact that Gram-positive bacteria normally show low levels of surface complexation, due to the strong cross-linked peptidoglycan layer, while in Gram-negative bacteria the majority is composed of lipopolysaccharides (LPS), phospholipids, and proteins. Those are exposed on the cell surface and are responsible for the efficiency of the metal-binding capacity.

11.4.1.2 Fungi

Fungi are eukaryotes, from the Greek (true nucleus), and are organisms whose cells have a distinct nucleus containing the cellular genetic material (DNA), surrounded by a special envelope called the nuclear membrane. Organisms of the Fungi Kingdom, Eukarya Domain, can be unicellular or multicellular. Large multicellular fungi, such as mushrooms, can be similar to plants, however, they cannot carry out photosynthesis [114].

Fungi have cell walls composed mainly of a substance called chitin, which makes up 80%–90% of the cell wall. Chitin is composed of polysaccharides that contain nitrogen. Inorganic ions, polyphosphates, lipids, and proteins are also components of the fungal cell wall [43]. The unicellular forms of fungi, yeasts, are oval microorganisms that are larger than bacteria. The most common fungi are molds. Molds form visible masses, called mycelia, composed of long filaments (hyphae) that branch and intertwine [114].

Fungi are microorganisms that are widely abundant in nature. In general, filamentous fungi are known as decomposers and also parasites of animals and plants. Therefore, fungi are essential for recycling dead and decaying plants and animals [148].

Fungi are easy-to-grow microorganisms, have a high biomass yield, and have been used as metal biosorbents in aqueous solutions [38], as they can live in environments with high metallic concentration and accumulate micronutrients and toxic metals showing high removal capacity [127].

Fungi have different types of ionizable sites, in addition to several functional groups, such as amine, carboxyl, hydroxyl, phosphate, and sulfhydryl groups, that influence the biosorption capacity and specificity of fungal strains for metal removal [127]. The composition of the fungal cell wall has numerous functional groups that act as metallic active sites. Most constituents of fungal cell walls are chitin carbohydrates (3%–39%), chitosan (5%–33%), polyphosphates (2%–12%), lipids (2%–7%), and proteins (0.5%–7%). 2.5%), however, there are variations related to cell wall composition among different fungal taxonomic groups [38].

The biosorption of metals by fungal biomass is carried out through several mechanisms, such as ion exchange, complexation, chelation, adsorption, and microprecipitation [38]. Fungi and actinomycetes are microbial examples that constitute an ecological alternative for the recovery of metals. However, when compared to other microorganisms, they are considered slow biosorbents because they present a complete reduction in around 48–120 hours [149]. In the literature, several studies have evaluated the biosorption of metals via fungal biomass (Table 11.4) [150–154].

TABLE 11.4
Reports of the Uptake of Metallic Species by Different Fungi

Fungus	Ion	m (g/L)	Metal (mg/L)	q (mg/g)	R (%)	Reference
Cunninghamella elegans fungal nano–chitosan	Pb (II)	–	300	87.51	–	[20]
C. elegans fungal nano–chitosan	Cu (II)	–	300	59.62	–	
C. elegans fungal chitosan	Pb (II)	–	300	89.12	–	
C. elegans fungal chitosan	Cu (II)	–	300	57.23	–	
Penicillium blinkerium	U	20	100	3.96	79.2	[152]
Pycnoporus sanguineus	Cd (II)	40	25	–	±70	[155]
	Pb (II)	40	25	–	±80	
	Cr (III)	40	25	–	±78	
Pleurotus ostreatus	Cu (II)	2	10	–	86	[156]
Aspergillus niger PTN31	Pb (II)	50	100	20.7	±85	[157]
A. niger PTN31 (EPS)	Pb (II)	0.0525	100	291.6	±80	
Pycnoporus sanguineus immobilized in calcium alginate	Cu (II)	30	300	2.76	–	[158]
Fusarium solani	Zn (II)	1	600	5.81	–	[159]
F. solani treatment with NaOH	Zn (II)	1	600	12.5	–	
Trichoderma harzianum	Au (III)	0.4	400	1,340	±60	[160]
Pycnoporus sanguineus	Cd (II)	40	40	±5.9	±75	[110]
	Pb (II)			±3.2	±40	
	Cr (III)			±5.5	±70	
P. sanguineus modified with NaOH	Cd (II)	40	40	± 6.2	± 80	
	Pb (II)			± 3.8	± 48	
	Cr (III)			± 6.1	± 76	
P. sanguineus modified with HCl	Cd (II)	40	40	± 6.5	± 80	
	Pb (II)			± 4.1	± 49	
	Cr (III)			± 6.3	± 77	

(Continued)

TABLE 11.4 (*Continued*)
Reports of the Uptake of Metallic Species by Different Fungi

Fungus	Ion	*m* (g/L)	Metal (mg/L)	*q* (mg/g)	*R (%)*	Reference
Huenov papillotrema living biomass	Mn (II)	2	110	16.6	75.5	[50]
	Cu (II)	2	128	15.7	70.5	
Huenov papillotrema dead biomass	Mn (II)	2	110	27.1	60.3	
	Cu (II)	2	128	26.4	56.5	
Cryptococcus laurentii (AL$_{65}$)	Pb (II)	1	20	9.1	46	[161]
Pichia pastoris X33—complete cell	Cu (II)	25	100	6.2	41.1	[162]
P. pastoris—cell wall	Cu (II)	25	100	11.53	21.2	
P. pastoris—cell membrane	Cu (II)	25	100	10.97	20.7	
P. pastoris—cytoplasm	Cu (II)	25	100	8.87	18.5	
P. pastoris—treated with Proteinase–K (30 U)	Cu (II)	25	100		18.1	
P. pastoris—treated with β–mannanase (50 U)	Cu (II)	25	100		28.2	
P. pastoris—manana	Cu (II)	25	100		34	
P. pastoris—glucan	Cu (II)	25	100	–	12	
Kodamaea transpacifica	Cr (VI)	–	100	–	±38	[163]
Kodamaea transpacifica modified with (BZK)	Cr (VI)	–	100	476.19	85.8	
Saturnispora quitensis	Cr (VI)	–	100	–	±40	
Saturnispora quitensis modified with (BZK)	Cr (VI)	–	100	416.67	85.4	
Kazachstania yasuniensis	Cr (VI)	–	100	–	± 38	
Kazachstania yasuniensis modified with (BZK)	Cr (VI)	–	100	114.94	80.7	
Saccharomyces cerevisiae	Cr (VI)	–	100	–	±39	
Saccharomyces cerevisiae modified with (BZK)	Cr (VI)	–	100	120.48	75.8	
S. cerevisiae	Pb (II)	0.1	5.5	160.14	–	[164]
S. cerevisiae immobilized on chitosan magnetic nanoparticles	Cu (II)	1.5	400	140	50	[165]
S. cerevisiae in a continuous bioreactor system	Cu (II) Pb (II)	1.5	180	29.9	20	[166]
S. cerevisiae	Mn (II)	2	5.6		3.4	[167]
S. cerevisiae immobilized in sawdust	Cd (II)	10	100	1.2	–	[168]
S. cerevisiae Perlage® BB	Cu (II)	20	25		76	[63]
S. cerevisiae	Pb (II)	–	0.0528	–	91.6	[169]
	Cd (II)	–	0.0528	–	95.3	

Where *m* represents the concentration of the biosorbent, *metal,* the metal concentration studied, *q* the biosorption capacity and *R* the removal efficiency.

From mycelia of the fungus *Cunninghamella elegans* with the addition of sodium tripolyphosphate for the biosorption of potentially toxic metals Pb^{2+} and Cu^{2+}. The results showed that chitosan and nano-chitosan had high biosorption capacity, having a higher affinity for Pb^{2+} removal. Chitosan nanoparticles were more effective than fungal chitosan for the biosorption of Pb^{2+} and Cu^{2+} metals [20].

The evaluation of 57 fungal strains that were isolated from a uranium mine for uranium biosorption showed that the fungal strains (*Talaromyces amestolkiae, Verticillium leptobactrum, Phoma cf. nebulosa, Bionectria ochroleuca, Purpureocillium lilacinum, Aspergillus versicolores, Trichoderma koningiopsis nebulosa, Umbelopsis ramannian,* and others) have high strength and uranium biosorption capacity [152].

The fungus *Aspergillus niger* PTN31 that was isolated from the soil of a non-ferrous metals mine was studied in the biosorption of Pb^{2+}. The data showed that, at pH 5, the maximum biosorption capacity of the strain was 20.7 mg/g and for the EPS extracted from the fungus it was 291.6 mg/g. This difference in the capacity is because the polysaccharides and uronic acid present in the fungus EPS played the main role in the metal ion biosorption, as these components contained many functional groups, such as carboxyl and hydroxyl groups [157].

The mechanism of tolerance and biotransformation of Zn(II) ions from the fungus *Fusarium solani* KJ623702 that was previously isolated and identified was investigated. The highest biosorption potential was reported at pH 4.0 (biomass treated with 0.2 N NaOH) and pH 5.0 (untreated biomass), at a concentration of 600 mg/L of Zn(II), incubation temperature of 30°C and time contact period of 40 minutes (biomass treated with NaOH) and 6 hours (biomass without treatment). Consequently, chemical modification of the biomass reduced the contact time of the Zn(II) biosorptive process [159].

The biosorption of fungal mycelia of *Fusarium solani* without treatment and treated with NaOH showed an improvement in the biosorptive capacity with the increase of the metallic concentration of Zn(II) from 200 to 600 mg/L, the removal capacity increased from 2.79 to 7.1 mg/g in untreated biomass and 5.81 for 12.5 mg g^{-1}, for the treated one. Thus, the chemical modification of the biosorbent favored the removal process [159].

The biosorption of potentially toxic metals in a multielement system (Cd^{2+}, Pb^{2+}, and Cr^{3+}) in dry biomass of *Pycnoporus sanguineus* indicated that the fungal biomass showed a favorable behavior for the biosorption of the ions in this study and that the adequacy to the Langmuir isotherm suggested that the process occurs in a monolayer. The kinetic study showed that the metallic species under study have similar biosorption capacities among themselves, and for the contact time studied (10–120 minutes), it was observed that there was no metallic competition for the active sites of the biomass. However, the removal efficiency showed a different behavior, showing a more effective removal for Pb^{2+}, at all times worked, consequently, the fungal biomass showed more affinity for Pb^{2+} followed by Cd^{2+} and Cr^{3+}. However, in the evaluation of the influence of the biosorbent dosage, this behavior was not observed, with the best biosorption efficiency of the metal Cd^{2+}, followed by Pb^{2+} and Cr^{3+} [155].

The biomass of the fungus *P. sanguineus* was evaluated *in nature*, modified with NaOH and HCl to evaluate the biosorption potential in a competitive system of Cd^{2+}, Pb^{2+}, and Cr^{3+} ions. The results showed the fungal biomass exhibited good biosorption capacity for the metals understudy. Biomass modification with NaOH and HCl influenced the biosorption process, as well as the competition between metallic species in solution [110].

Modification with NaOH and HCl of the *P. sanguineus* fungus biomass improved the biosorption capacity of all metal ions (Cd^{2+}, Pb^{2+}, and Cr^{3+}). In particular, HCl-modified biomass performed better than NaOH-modified biomass, especially for Cr^{3+} metal. The results showed, therefore, that the chemical modification of the fungal biomass favored the biosorptive process under the conditions studied. The analysis of metallic competitiveness showed that the uptake of Pb(II) was affected, and the uptake values recorded were significantly lower than those presented by Cd(II) and Cr(III) ions [110].

The fungus *Trichoderma sp.* has been reported in the literature as a cadmium biosorbent, the biosorptive process, in this case, fits the isothermal models of Langmuir and Freundlich [127]. The fungus *Trichoderma harzianum* was used in the biosorption and consequent biosynthesis of gold nanoparticles. Studies have shown that many biosorbents can also biotransform metal ions as nanoparticles through reduction, nucleation, and subsequent stabilization. The results showed that with 0.4 g/L of fungal biomass, added to a solution of gold ions with a concentration of 400 mg/L it is possible to simultaneously carry out the processes of biosorption (metal removal) and biogenic synthesis of nanoparticles of gold, after 30 minutes of contact under agitation (200 rpm). The fungus *T. harzianum* had a high gold biosorption capacity, approximately 1,340 mg/g in 180 minutes of contact, and a removal rate above 60%, in addition to biosynthesizing gold nanoparticles with an average diameter of 30 nm [151].

Yeasts are considered low-cost biosorbents for the effective removal of potentially toxic metals, and also have many advantages, such as ease of cultivation, inexpensive growth media, feasibility for large-scale cultivation, as biomass, is safe and easy to use. In addition, they are very common residues in the industries of the food segment, which favors the reuse of this residue in the biosorption of metals [50,63].

The biomass of the psychrophilic yeast *Cryptococcus laurentii* (Al_{65}) was investigated as a biosorbent of potentially toxic metals. The highest biosorption capacity was observed at a concentration of 20 mg/L for the species Al, Cr, In, Ga, Fe, and Bi, at a concentration of 12–16 mg/g of biomass, with an optimal pH of 5 [161].

The yeast *Papiliotrema huenov* as a biosorbent of Mn^{2+} and Cu^{2+} ions presented an ideal temperature for biosorption at 30°C. The *P. huenov* biomass was excellent to remove Mn(II) and Cu(II) from aqueous solution in a short time (60 minutes) with a biosorption capacity of 27.79 and 26.6 mg/g for Mn^{2+} and Cu^{2+}, respectively. The results indicated that using yeast may be a viable option for removing Mn(II) and Cu(II) ions. The live biomass showed 75.58% and 70.5% of removal compared to 60.3% and 56.5% for dead biomass at metal concentrations of 110 mg/L Mn(II) and 128 mg/L Cu(II) [50].

The yeast *Pichia pastoris* X33 and the potentially toxic metal Cu^{2+} were used to reveal the biosorption process and mechanism of the whole cell and the main components of the microbial cell (cell wall, cell membrane, and cytoplasm) and cell wall (glucan, protein, and β-mannan). The whole-cell results showed maximum removal efficiency of 41.1% and biosorption capacity of 6.2 mg/g. The analysis of Transmission Electron Microscopy–Energy Scattered X-ray spectroscopy (TEM-EDX) confirmed the existence of Cu^{2+} on the cell surface and in the cytoplasm [162].

The maximum biosorption capacity of the cell, cell wall, cell membrane, and cytoplasm were: 16.13; 11.53; 10.97 and 8.87 mg/g, respectively. The maximum biomass removal efficiencies of *P. pastoris* treated with proteinase-K and *P. pastoris* biomass treated with β-mannanase were 18.1% and 28.2%, respectively. Maximum mannan and glucan removal efficiencies were 34% and 12%, respectively [162].

Fourier Transform Infrared Spectroscopy (FTIR) spectra showed that the amino group (N–H), a hydroxyl group (O–H), carbon-oxygen bond (C–O), –CH, C–N, and carbonyl (C=O) of a ketone or aldehyde can interact with Cu^{2+}. Thus, the work of [162] provides guidance for a better understanding of the effect of different cellular components on biosorption.

The evaluation of three species of native Ecuadorian yeasts, *Kazachstania yasuniensis*, *Kodamaea transpacifica, Saturnispora quitensis,* and yeast that is widely studied in metal biosorption, *Saccharomyces cerevisiae,* revealed that the biosorption efficiencies of the four yeasts for Cr(VI), with pretreatment with the cationic surfactant benzalkonium chloride (BZK) and without pretreatment, showed that pretreatment of the cell wall of the strains significantly improved the efficiency of sorption of Cr(VI) from all biomasses (at 124%, 140%, 111% and 101% for *K. yasuniensis, K. transpacifica, S. quitensis,* and *S. cerevisiae,* respectively). Since, after pretreatment with the cationic surfactant, an increase in the electrostatic interaction between the surface of the yeast biomass and the metal in this study can be expected [163].

Among the microorganisms that can promote the biosorption of potentially toxic metals, the yeast *S. cerevisiae* stands out for having distinct characteristics, such as its mass production through a by-product of the fermentation industry and the low cost of its culture medium, in addition to the fact that its handling is safe, which makes it quite common and useful in the food sector [63,168,170].

Despite the theoretical survey of many authors on the biosorptive potential of *S. cerevisiae* revealing that it has a medium biosorption capacity, it is considered unique to biomass in the biosorption of metals [53,63,171]. Numerous studies have demonstrated the ability of *S. cerevisiae* to remove metal from toxic and radioactive metals and to recover precious metals in an aqueous solution [170,172].

The study of the biosorptive process of the commercial yeast *S. cerevisiae* Perlage® BB showed that the biomass showed a good capacity and efficiency for the biosorption of the potentially toxic metal Cu(II) and that the investigated variables (metal concentration, pH, contact, and biosorbent dosage) influenced the biosorption process. The adsorption isotherms plotted for the Langmuir, Freundlich, and Dubinin-Radushkevich (D-R) models revealed that copper biosorption fitted the Langmuir model, which expressed a biosorption capacity of 4.73 mg/g [63].

The yeast *S. cerevisiae* biomass was used to remove Pb(II) and Cu(II) ions from aqueous solutions in continuous mode (bioreactor). Yeast biomass showed a low absorption capacity for Cu(II) ions compared to Pb(II) ions in the metal concentration range ranging from 10 to 180 mg/L. The biosorption capacity of the biomass reached a maximum of 29.9 and 72.5 mg/g for Cu(II) and Pb(II) ions, respectively, under similar experimental conditions [166].

The investigation of the Pb(II) and Cd(II) biosorption process by *S. cerevisiae* through the Taguchi screening design revealed that the three significant variables for the biosorption of lead and cadmium are the pH of the solution, the concentration of metals, and the dosage of the biosorbent. After optimizing the results by Surface and RSM the data showed that *S. cerevisiae* has a high affinity for both metals at low concentrations. The maximum removal was 91.6% and 95.3% for Pb(II) and Cd(II) ions, respectively [169].

11.4.1.3 Microalgae

Algae can be microorganisms, such as photoautotrophic eukaryotic macroscopic organisms, which are relatively simple and lack the tissues (roots, stems, and leaves) typical of plants. Identification of filamentous and unicellular algae requires microscopic examination. Most algae can be found in the oceans. The location of these organisms depends on factors such as the availability of appropriate nutrients, the wavelength of light, and the surfaces on which they grow [114,117].

Overall, microalgae and macroalgae are being evaluated in sectors such as food, cosmetics, medicines, energy, and others. In addition, algae have received much attention for effluent decontamination by biosorption. Algal biomass can be used in living or dead forms in the biosorption process. Dead biomass is more practical and easier to study because living biomass will need nutrients and favorable environmental conditions for its growth, while dead biomass will not need nutrients. Another characteristic of dead biomass is that it is not affected by the toxicity of metal ions, in addition to being able to undergo different chemical and physical pre-treatments to increase its biosorption capacity [173].

The use of microalgae is now recognized as an efficient and environmentally friendly strategy for removing contaminants from wastewater. Due to their versatility, these photosynthetic organisms can grow in a broad spectrum of wastewater, such as from agricultural, municipal, and industrial sources, while converting nutrients such as nitrogen and phosphorus into valuable products. Currently, microalgae are beginning to be exploited on a large scale for the treatment of agricultural and municipal effluents [51].

Microalgae have several applications in numerous segments that depend on the selection of the strain and suitable cultivation conditions for their growth and biomass production. Microalgae are reported in the literature as metal biosorbents. The mechanisms involved are chelation/complexation, adsorption, electrostatic interaction, surface precipitation, and ion exchange [54,174].

There are important differences in biosorption capacity between microalgae species. These differences are attributed to many factors, such as cell wall strength, polysaccharides, pectin, glycoproteins, shape, size, and functional groups interacting with pollutants. The use of microalgae biomass as a biosorbent is a good alternative because microalgae cultures require low amounts of nutrients and energy for their growth, which is reflected in a low production cost [54,64,175].

The physical and chemical characteristics of the surface of microalgae define their ability to biosorb potentially toxic metals. The presence of active sites represented by functional groups such as OH–, COOH, NH$_2$–, COO–, SH–, CO–, and HPO$_4^{2-}$ are very important in the biosorptive process. These groups are mainly found in molecules such as cellulose, alginates, glycan, and proteins that have oxygen, nitrogen, sulfur, and phosphorus. The physicochemical characteristics of the cell walls of biomass are influenced by cultivation conditions, such as light, nutrients, and stress [64]. Several studies point to the potential of microalgae in environmental decontamination (Table 11.5) [54,64].

TABLE 11.5

Reports of the Uptake of Metallic Species By Different Microalgae

Microalgae	Ion	m (g/L)	Metal (mg/L)	q (mg/g)	R (%)	Reference
Phormidium tenue	Co (II)	1	50	–	94	[54]
Chlorella vulgaris				–	87	
Chlorella vulgaris L1—light deprivation	Cd (II)		25	8.03	–	[176]
Chlorella vulgaris L2—light deprivation	Cd (II)		25	±10.6	–	
Chlorella vulgaris N1—nitrogen deprivation	Cd (II)		25	10.88	–	
Chlorella vulgaris N2—nitrogen deprivation	Cd (II)		25	11.95	–	
Desmodesmus sp. CHX1	Cu (II)	0.41	1	–	88.35	[16]
13X- zeolite spheres *C. vulgaris* -alginate (ZABs)	Cu (II)	0.4	150	85.9		[177]
blank alginate beads (BABs)	Cu (II)	0.4	150	70.2		
C. vulgaris alginate beads (CABs)	Cu (II)	0.4	150	77.3		
Spirogyra sp.	Cd (II)	4	500	22.5		[178]
	Cr (VI)			38.2		
	Cu (II)			35.6		
	Pb (II)			94.3		
Gelatinous colonies	Cu (II)	0.2	50	11.2	42.3	[179]
Cyanothece sp., Leptolyngbya spp. and	Cd (II)			16.1	63.6	
Phormidium spp	Pb (II)			80.6	31.8	
Chlamydomonas microsphaera	Cu (II)	1	50	22.2	44.5	[180]
C. microsphaera collected by PAC-flocculation	Cu (II)	1	50	35.5	70.9	
Synechocystis sp., in Fe$_2$O$_3$	Cr(VI)	0.5	80	69.8		[181]
	Pb (II)			62.6		
	Cd (II)			42.1		
	Cu (II)			38.7		
Chlorella sorokiniana residue treated with	Cu (II)		20	8.59	90	[182]
thermally expanded graphite and chitosan	Pb (II)		5	18.35	90	
Scenedesmus sp.	Cr(VI)		10	0.03	93	[183]
Planothidium lanceolatum	Cd (II)		8		±80	[184]
	Cu (II)				±85	
	Zn (II)				±90	
Acutodesmus acuminatus	Eu^{3+}		100	174.2		[185]
Botryococcus braunii dead biomass	Zn (II)		25.2	9.6		[186]
Rhizoclonium hookeri	Ni (II)	1	1000	65.8		[187]
	Pb (II)			81.7		
Chlorella pyrenoidosa air gel	U^{4+}	0.5	800	571		[188]

Where m represents the concentration of the biosorbent, *metal*, the metal concentration studied, q the biosorption capacity and R the removal efficiency.

The two microalgae *Phormidium tenue* and *Chlorella vulgaris*, isolated from the local environment (Wadi Hanifah Stream in Riyadh, Kingdom of Saudi Arabia), can be applied as biosorbents for the biosorption of Co^{2+} ions in solution. Co^{2+} biosorption by *P. tenue* was more effective than *C. vulgaris*. Under optimal Co^{2+}, removal conditions, the removal capacity of *P. tenue* was 94%, at pH 6, contact time (30 minutes), initial concentration (50 mg/L), and biosorbent dose (1 g/L), while *C. vulgaris* removed 87% under the same parameters, except pH 5.5 and contact duration (60 minutes) [54].

The study of the composition of the outer cell wall of *Chlorella vulgaris* grown in four physiological states under light and nitrogen limitation in continuous culture was used to evaluate the effect of culture conditions on the biosorption of the toxic metal Cd(II), providing a new approach based on modifying the cell wall characteristics of microalgae, rather than studying the influence of physicochemical conditions of biosorption [176].

The trials considered the study of two factors that influence the physiological response of microalgae, which were nutrient limitation and specific growth rate. The latter is equivalent to the rate of dilution in continuous culture. The nutrient limitation was achieved by light and nitrogen deprivation. These elements are fundamental for the growth of microalgae and are the most used limiting strategies, both in research and industrial operations [176].

Significant differences in terms of presence and relative amount of functional groups of the outer cell wall of the microalgae *C. vulgaris* were detected between the physiological states studied. The variation in cell wall composition had a great impact on cadmium biosorption. Nutritional limitation proved to be the factor that exerted the most significant effect on cell wall composition, while the growth rate showed little or no impact, except for amino and carbohydrate groups in nitrogen-restricted cultures. The most notable difference between the light and nitrogen-bound states was the presence of carboxyl and deacetylated amino groups in the latter [176].

The microalgae *Desmodesmus sp.* CHX1 was isolated from an oxidation pond from swine wastewater to assess its potential for Cu^{2+} biosorption. The results showed that the optimal conditions for copper removal by *Desmodesmus sp.* were pH 6, microalgae inoculation concentration of 0.41 g/L and treatment time (carbon/nitrogen ratio—C/N) of 4 days, with copper removal efficiency of 88.35% [16].

The comparative study of the biosorption potential of 13X-algal-alginate zeolite beads (ZABs), alginate blank-alginate-only beads (BABs), and *Chlorella vulgaris* alginate beads (CABs) for copper removal, according to the influence of different parameters such as contact time, pH and initial concentration of the metallic solution, reported that the maximum biosorption capacity of ZABs was 85.88 mg/g reached at 180 minutes, pH 5 and initial copper concentration of 150 mg/L while the maximum capacity obtained by BABs was 70.02 and 77.32 mg/g for CABs, respectively. Furthermore, ZABs showed greater stability than BABs and CABs in Cu^{2+}, biosorption-desorption cycles [177].

Comparison of the biosorption potential of metals cadmium (Cd), chromium (Cr), copper (Cu), and lead (Pb) by freshwater green algae (*Spirogyra sp.*) biomass as a function of contact time, initial pH, and initial concentrations of the studied metals showed that the optimal pH for biosorption of cadmium, chromium, copper, and lead was 5.5, 5.8, 5.9 and 5.0, respectively. The biosorption capacities were 22.52, 38.19, 35.59, and 94.34 mg/g for Cd, Cr, Cu, and Pb, respectively. The times required for biosorption equilibrium were 15 minutes for cadmium, 40 minutes for chromium and copper, and 50 minutes for lead. Demonstrating, therefore, significant differences in the biosorption profile as a function of the metallic species under study [178].

The gelatinous colonies are composed of the microorganisms *Cyanothece spp.* (cyanobacterium) and two filamentous algae, *Leptolyngbya spp.*, and *Phormidium spp.* were collected from rice fields and investigated to remove potentially toxic metals. The biosorption of Cu^{2+}, Cd^{2+}, and Pb^{2+} by dry biomass reached the biosorption equilibrium after 1 hour of contact time. The maximum biosorption capacity of Cu^{2+}, Cd^{2+}, and Pb^{2+} was 27.78, 28.57, and 76.92 mg/g, respectively [179].

Low-cost and sustainable use for mixed microalgae and polyaluminium chloride (PAC) waste collected during water treatment processes was evaluated for copper biosorption. The microalgae *Chlamydomonas microsphaera* in the form of dry powder (CM) and *C. microsphaera* collected by flocculation-PAC (PCM) revealed that a mixture of microalgae-PAC (PCM), which is a waste product of water treatment processes, can be successfully used as a biosorbent to remove the toxic metal copper. The biosorption efficiency depended on the pH of the solution, contact time between biosorbent and adsorbate, and concentration of the biosorbent, in addition to the presence of Fe^{3+} in the solution. The maximum biosorption capacity, according to the Langmuir model, was (79.4 mg/g) for the PCM and (57.3 mg/g) for the CM. Consequently, microalgae biomass and aluminum chloride showed better responses to the biosorption of the metal under study [180].

The production of air gel based on microalgae (*Chlorella pyrenoidosa*/chitosan, CP/CTS) was used to remove uranium from an aqueous solution. The characterization assays of the CP/CTS composite after the capture of U(VI) showed that the biosorption capacity depends largely on the ion exchange and the complexation of uranium with the functional groups ($-COOH$, $-OH$, and $-NH_2$) [188].

The investigation of the biosorption characteristics of europium (Eu) by *Acutodesmus acuminatus*, with the particular objective of recovering rare earth elements from e-waste, showed that the microalgae had a maximum adsorption capacity of 174.2 mg/g under pH 4.7 and a temperature of 40°C. The presence of intracellular europium particles that were visualized by TEM-EDX and the effect of preculture conditions suggested that the intracellular uptake of (Eu) promoted the high biosorption capacity of europium by the microalgae under study [185].

The use of metallic oxide (Fe_2O_3) as adsorptive material was used to form a composite material with microalgae, to immobilize the microalgae biomass, in addition to improving the biosorption process. The biosorption capacity for the metal ions evaluated was in the following order: Cr^{6+} (69.77 mg/g) > Pb^{2+} (62.63 mg/g) > Cd^{2+} (42.12 mg/g) > Cu^{2+} (38.68 mg/g) [181].

The treatment of residual biomass of the microalgae *Chlorella sorokiniana*, with thermally expanded graphite, and chitosan revealed that the new biosorbent produced presented a porous surface due to thermally expanded graphite and a layered structure due to chitosan. The biosorbent was used for the purification of model solutions to remove Cu^{2+} and Pb^{2+} ions under static and dynamic conditions. The maximum sorption capacity for Pb^{2+} ions was 18.35 mg/g, and for Cu^{2+} ions, it was 8.59 mg/g. The purification (removal) efficiency exceeded 90% at low concentrations of the study metals (5–20 mg/L) [182].

The Cr(VI) biosorption mediated by the microalga *Scenedesmus sp.* showed that the biomass under ideal conditions (contact time of 120 minutes, S/L ratio of 10% (w/v), the concentration of Cr(VI) of 10 mg/L, pH 1, the temperature of 30°C; particle size, $75+45\,\mu m$ and agitation of 300 rpm) showed good effectiveness in the removal being in the range 92.89% [183].

The microalgae *Rhizoclonium hookeri* was effectively applied to remove potentially toxic metals Pb(II) and Ni(II). The biosorption process was influenced by several parameters, such as initial metal concentration, contact time, pH, particle size, biosorbent dose, and temperature. The maximum biosorption capacity according to the Langmuir model was 81.7 mg/g for Pb(II) and 65.81 mg/g for Ni(II) [187].

Planothidium lanceolatum was evaluated in the biosorption of the potentially toxic metals zinc, copper, and cadmium. Assays indicated that metal biosorption by *P. lanceolatum* increases with the increasing initial concentration of metal ions in the range of 100–130 mg/L. The maximum concentrations of biosorbed metals were 104, 81, and 41 mg per $16 \times 10^8 L^{-1}$ diatom cells for Zn, Cu, and Cd, respectively. Maximum removal was obtained at pH 6 for Cu and pH 8 for Cd and Zn. The biosorption of metals was also favored with increasing temperature from 15°C to 25°C [184].

11.5 RECENT TRENDS IN METAL BIOSORPTION

11.5.1 CHEMICAL AND PHYSICAL MODIFICATIONS OF BIOMASS AND BIOCHAR PRODUCTION

New biomasses for wastewater treatment via biosorption are intensively investigated, with emphasis on agricultural and industrial residues that are evaluated in their raw or modified form, as well as production of activated carbon from fruit residues, and fruit parts such as seeds, bark in addition to parts of lignocellulosic plants. These biomasses, therefore, are abundant, cheap, and renewable biosorbents, which will contribute to the development of the local economy and environmental decontamination [49,71,189,190].

However, the selection of the ideal biosorbent to remove pollutants from an aqueous solution is one of the biggest challenges of biosorption methods. The type of raw material, the biomass activation method, the target pollutants, the interaction between the biosorbent and the adsorbate, as well as the desorption method of the adsorbed substance influence the decontamination efficiency [190].

Agro-industrial waste that is treated, both physically and chemically, can be used in the process of removing potentially toxic metals from wastewater. The chemical composition (presence of functional groups) affects the biosorption capacity, in addition to other physical characteristics such as (particle size, density, specific surface area, porosity, pore size, and distribution). The selectivity of biomasses can be improved through functionalization, which will allow the adequacy of the adsorbate affinity to the biomass surface, as well as promote its use in multi-element solutions [190].

Sawdust is a residue produced during the cutting, sewing, sharpening, and smoothing of wood. The lumber and wood processing industries are the main sources of sawdust. Sawdust is one of the residues with potential as a metal biosorbent [191–193]. Lignocellulosic biomass, such as sawdust, can be a precursor for producing various biosorbents. The main chemical composition of sawdust is generally cellulose (35%–60%), hemicellulose (15%–35%), and lignin (15%–30%) [191].

There are two different methods for preparing biosorbents using sawdust. These are the chemical and physical treatment methods. The physical treatment is divided into two stages, in the first stage the biomass is carbonized in the presence of inert gases such as nitrogen or helium, in the second stage the biomass is carbonized in the presence of air, steam or a mixture of air and steam at a temperature higher [194]. Carbonization aims to remove volatile substances from the precursor biomass at a relatively low temperature to produce carbon-rich biosorbents for activation purposes. The activation process at high temperatures ($800°C$–$1{,}200°C$) occurs with the oxidizing gas to give carbon oxides and pores in the biomass [191].

In the chemical treatment, the biomass is impregnated by an activating reagent that proceeds with the warming process in an inert condition. Activation reagents can be solutions of (KOH, $ZnCl_2$, H_3PO_4, H_2SO_4, HNO_3, $NaOH$, K_2CO_3, among others). Due to the process of degradation, dehydration, and complexation of the organic carbon present in the biomass by the activation reagents, the promotion of an increase in the number of pores in the activated biosorbents occurs [194,195]. Chemical treatment is done at a much lower temperature and without oxidizing gases compared to physical treatment. However, chemical treatment requires washing the biosorbent after activation to remove traces of the activating chemical solution. Thus, washing with hot and cold water facilitates the formation of pores in biosorbents [191].

The teak sawdust (*Tectona grandis*) modified with citric acid and *in nature* proved to be efficient in the biosorption of potentially toxic metals, Cd(II) and Pb(II), demonstrating good biosorption capacity of the metallic species in the mixed solution. However, Cd(II) showed biosorption efficiencies above 90% regardless of the variation in metallic concentration (5–50 mg/L) and biosorbent mass (2–50 g) for the citric acid-modified biomass at 240 minutes of contact [196].

The biomass of the fungus *Pycnoporus sanguineus* was evaluated *in nature*, modified with NaOH and HCl to evaluate the biosorption potential in a competitive system of Cd^{2+}, Pb^{2+}, and Cr^{3+} ions (Figure 11.13). The results revealed that the fungal biomass exhibited biosorption potential for the studied metals (Figure 11.13a–c) [110]. Modifying NaOH and HCl of the *P. sanguineus* fungus biomass improved the Cr^{3+} removal efficiency with a concentration of 30 mg/L (Figure 11.13c). Activation with HCl performed better than biomass activated with NaOH for Pb^{2+} metal (Figure 11.13b). The results showed, therefore, that the chemical modification of the fungal biomass favored the biosorptive process under the conditions studied (40 g/L of biomass, pH 5, and 24 hours of contact) for certain metals and metallic concentration ranges [110].

The evaluation of five modifications of sawdust (*Pinus caribaea* Morelet) was carried out to promote the removal of nickel ions Ni^{2+} and other trace metal ions (Co^{2+}, Cr^{3+}, Mn^{2+}) from aqueous solutions. The modifications were: (a) acidic with H_2SO_4, (b) oxidized with Fenton $_{(SF)}$—a mixture of hydrogen peroxide and iron sulfate solutions, oxidation with (c) dialdehyde (S_{CHO}), (d) dicarboxylic ($S_{COOH(2)}$) and (e) tricarboxylic ($S_{COOH(3)}$) [197].

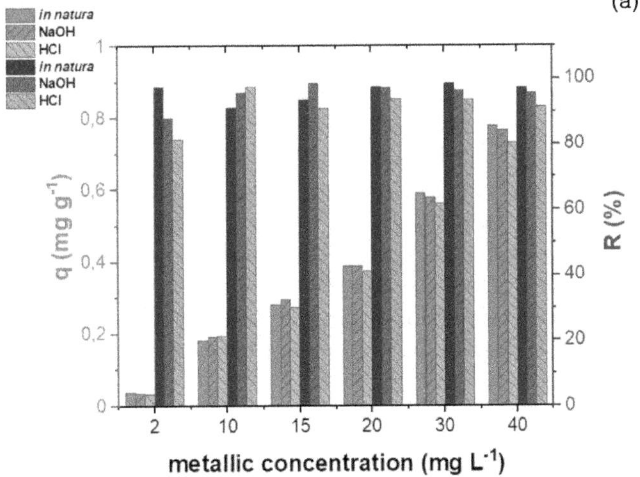

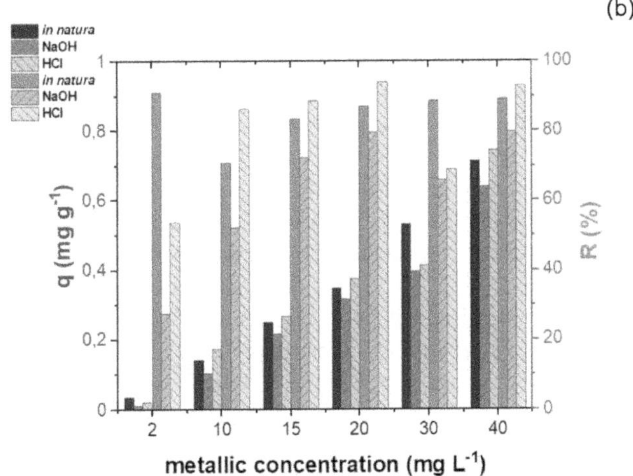

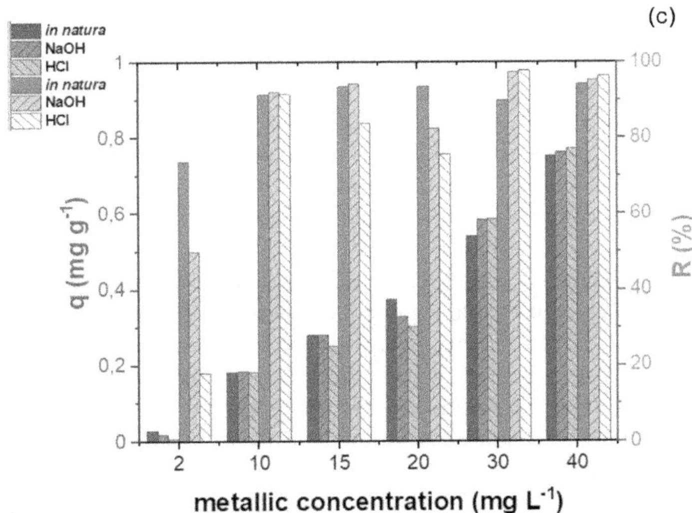

FIGURE 11.13 Biosorption capacity and efficiency of the fungus *Pycnoporus sanguineus in natura* and modified with NaOH and HCl. (a) cadmium; (b) lead and (c) chromium.

The biosorption assays showed that the sawdust removal efficiency depends on the chemical modification as well as the metal in the solution. All modifications were more effective for Ni^{2+} biosorption than natural sawdust. The acid treatment (88.14% efficiency) seems less effective than the oxidation treatments that had removal efficiency above 96%, while the unmodified biomass showed removal of 73.07%. The best removal efficiency was 98.3% for the sawdust modified with tri-carboxylic acid, evidencing that the sites of carboxylic groups linked to adjacent carbon atoms and spaced in a ring (sawdust surface) are more selective for multivalent cations [197].

However, when the metal to be analyzed is Cr^{3+}, it has a better removal rate with sawdust biomass without modification (84.1%) and only (25%) for biomass modified via Fenton oxidation [197]. The analysis of the dynamics of competitiveness for a mixed solution of (Co^{2+}, Cr^{3+}, Mn^{2+}, and Ni^{2+}) shows that natural sawdust and Fenton oxidized sawdust showed great affinity for the metals Co^{2+} and Cr^{3+} with removal efficiencies above 78% for these metallic species, while Mn^{2+} and Ni^{2+} showed removal in the range of 43%–62% [197].

Recently, several authors have also evaluated the chemical modification and production of activated carbon using agro-industrial residues [49,194,198–202]. The conversion of biomass to biochar is a promising technique because biochar is an environmentally friendly and sustainable material [203].

Biochar is a solid material that is rich in carbon and has unique physical, chemical, and biological characteristics such as high pore volume, stability, large specific surface area, as well as being enriched with surface functional groups, having a huge mineral content, and deterring high biosorption capacity. Numerous agro-industrial residues have been reported as efficient biosorbents [201,202]. Biochar production is cheap, affordable, and requires little energy. Biochar is synthesized through thermochemical processes, which include pyrolysis, hydrothermal carbonization, and gasification. The most popular thermochemical technique for producing biochar is pyrolysis, in which the breakdown of organic components occurs in an oxygen-limited atmosphere at an adjustable temperature [201,203].

Corn stalk, which has a composition of 34.4% lignin, 33.1% cellulose, and 28.9% hemicellulose, was used to manufacture activated carbon. Corn stalks were processed with two carbonization steps. The first step was carbonization at 500°C for 1.5 hours, in the second step variations of activating agents were carried out with a variety of activation temperatures, to obtain better characteristics for the final product. The activating agents were KOH and NaOH solutions. After chemical activation, the corn stalk coal was activated with N_2 gas in a heated reactor where the temperature ranged from 650°C to 750°C [194].

The characterization of activated carbon was carried out using the iodine number tests. The best characteristics of the activated carbon produced are measured by the iodine numbers to represent the surface area. The best-activated carbon was produced by chemical activation with KOH at a temperature of 750°C with an overall yield of 24.3%, and an iodine number of 602 mgI_2/g, [194].

Biochar derived from cotton stalks (CSB) was produced to biosorb As^{3+} from an aqueous solution. The results indicated that the maximum biosorption capacity of As^{3+} was 89.90 μg/g using 1 g/L of BSC, pH 6, contact time of 2 hours and As^{3+} concentration of 200 μg/L, [204].

The biosorption profile was evaluated using viable bacteria (*Bacillus cereus* RC-1) immobilized on biochar derived from rice straw (RSB), chicken manure (CMB), and sewage sludge (SSB). The best results were found for bacterial spheres immobilized on rice straw biochar (RSB), which removed 158.77 mg/g of Cd^{2+}. The assays expressed that the biosorption by cells immobilized on biochar was better than those presented by bacterial cells in suspension. Evidence that the biochar, in addition to support, also acted in the biosorptive process [205].

The performance of biochar produced from seeds of *Capsicum annuum* L. (CASB) as a Pb(II) biosorbent indicated that the biosorption equilibrium was established within 40 minutes of contact, the best dosage was 2.0 g/L of biochar. The biosorption data were well predicted by a non-linear Langmuir isotherm model with a maximum capacity of 36.43 mg/g for Pb(II) [203].

11.5.2 Biosorption and Other Biotechnological Processes

Over the years, several physical-chemical and biological methods have been developed and are used to mitigate contamination by potentially toxic metals through the treatment of effluents containing these metallic species [44,159,170,206].

The main methodologies adopted are based on physicochemical processes such as ion exchange, chemical precipitation, and reverse osmosis, however, the implementation of these technologies has a high cost as its main disadvantage [7]. Methodologies based on bioremediation have been drawing the attention of researchers due to their advantages, such as low cost and being environmentally friendly [170].

Among the bioremediation methodologies, biosorption emerges as a clean technology as it is mediated by biomass of microbial, plant, animal, or agro-industrial residues [38,207]. The growing research in biosorption is justified by the high removal rates and selectivity, low cost, minimization of chemical sludge volumes, the possibility of treating large volumes, and high efficiency in effluents with low metal concentration [9,208].

11.5.2.1 Biosorption and Bioaccumulation

In the literature, many authors investigate the biosorption and bioaccumulation of potentially toxic metals [6,159,209,210]. Biosorption is defined as a physical-chemical process of removal of organic and inorganic substances by living or dead biomass, as it is a metabolically passive process and based mainly on the "affinity" between the biosorbent and the adsorbate (species to be adsorbed). Whereas, bioaccumulation is a much more complex phenomenon that is based on active metabolic transport [42,44,77,124,210].

When the metallic species (adsorbate) present in the solution is bound to the cell surface, the process is called biosorption, but if it accumulates inside the cells (intracellular), the process is called bioaccumulation (Table 11.6). The processes occur permanently and are carried out by practically all types of biomass [44,54,211].

Biosorption is a simple physicochemical process that resembles conventional adsorption or ion exchange, the difference being that the sorbent is biological. Biosorption is a metabolically passive process. When biosorption equilibrium is reached, it can be shifted to either side: to the left in wastewater treatment or to the right if the adsorbate needs to be removed and recovered (desorption) [54,137,211].

In bioaccumulation, the first step is biosorption and subsequent steps, related to transporting metals or organic pollutants, particularly via active transport (energy required), into cells. Thus, in bioaccumulation, more active sites for removal are available, and lower concentrations of metal in the final solution can be achieved. However, for the bioaccumulation process, it is necessary to cultivate an organism in the presence of the pollutant to be removed [44,209,211,212].

TABLE 11.6

Comparison of Biosorption and Bioaccumulation Processes by Metals

Parameter	Biosorption	Bioaccumulation
Metabolism	Passive	Active
Biosorbent status	alive and dead	Alive
Nutrient requirement (cultivation)	No	Yes
metal bonding	via adsorption	Via adsorption and absorption
metal bonding area	Surface	surface and interior
Biomass metal toxicity	Non-toxic	Can be toxic and kill biomass
Biomass increase	does not occur	Biomass increase (growth) may occur

In bioaccumulation, it is possible to achieve better removal of pollutants because cells offer active sites on the surface and inside the cell. Therefore, a part of the potentially toxic metal is absorbed into the cell interior, and another part can be biosorbed on the cell surface. In addition to the fact that it is a process that occurs during cultivation, the biomass concentration eventually increases, which will provide even more active binding sites (more biomass) [44,211].

Biosorption and bioaccumulation are highly valued for removing metal cations from solutions. The history of research on biomass metal interactions dates back to the 1960s. During this period, it was discovered that, although biomass is not capable of destroying metals, it can change its properties due to the stress generated by the concentrations of metals and adapt to the environment medium [44,47,77,211].

The accumulation of potentially toxic metals causes several physiological and morphological changes in microorganisms. Metal ion toxicity can be observed at both the cellular and molecular levels. Fungi, for example, can exhibit cellular defense mechanisms that can be categorized into extracellular and intracellular environments. The extracellular inhibits the absorption and internalization of metal ions in the cell, while the intracellular reduces the toxic effect of metals through trapping by binding to biomolecules or efflux channels, thus reducing the metal concentration in the culture medium [44].

In the literature, some studies on the mechanism of metal-biomass interaction suggest that biosorption is accompanied by bioaccumulation, where the accumulation of potentially toxic metal is followed by the non-metabolic uptake of metal ions in different cellular organelles. This is possible when the biosorbent is evaluated alive and present in the culture medium [44].

Currently, many authors study both processes, so it is possible to evaluate the performance of the biosorbent according to the process of biosorption and bioaccumulation [137,174,210]. The bioaccumulation and biosorption profile of ^{137}Cs, ^{233}U, ^{239}Pu, ^{241}Am, ^{90}Sr, and ^{237}Np by the cyanobacterium *Arthrospira platensis* CNMN–CB-02 showed that the biomass growth under the influence of Radionuclides ^{241}Am, ^{239}Pu and ^{237}Np almost did not interfere with cell growth. However, ^{137}Cs and ^{233}U induced a moderate decrease in cyanobacterium growth. The lowest growth rate was observed with the addition of ^{90}Sr to the culture medium. The accumulation rate for ^{239}Pu and ^{241}Am was 98.7 and 99.4%, respectively. While, for ^{237}Np, saturation was not reached in 21 days of cultivation, the percentage of removal was 72% of ^{237}Np from the solution [137].

As the bioaccumulation process is metabolically active and involves the use of living biomass, concentrations of contaminants can be harmful to the biosorbent. Radionuclides, for example, in contact with metal ions, can damage the cell by emitting high-energy particles, which can damage DNA and destroy proteins, in addition to generating free radicals. The lower growth rate of *A. platensis* in the presence of ^{137}Cs and ^{90}Sr in the culture medium may be associated with the penetration of Radionuclides into the cell and/or the replacement of K^+ and Ca^{2+} ions. On the other hand, ^{233}U at alkaline pH (the pH of the studied medium was 9.0-9.6) and in the presence of carbonate ions in the culture medium can precipitate on the cell surface, reducing the access of nutrients to the cell [137].

The analysis of the radionuclide biosorption efficiency by *A. platensis* showed lower removal rates than those presented in the bioaccumulation assays. The removal efficiency of ^{241}Am, ^{239}Pu, and ^{90}Sr for example was 30-40% lower than that presented in the bioaccumulation results. Evidencing that, despite some elements affecting cell growth, the bioaccumulation process performed better than the biosorption process for the cyanobacterium *A. platensis* [137].

The performance of the intestinal fungus *Trichoderma brevicompactum* QYCD-6 that was isolated from *Pheretima tschiliensis* showed good potential for tolerance to potentially toxic metals. The Minimum Inhibitory Concentration (MIC) for the metals Cu(II), Cr(VI), Cd(II), and Zn(II) by *T. brevicompactum* was relatively low, ranging from 150 to 200 mg/L in the composite medium. The (MIC) represents the lowest concentration of a substance responsible for limiting growth [213].

According to the biosorption and bioaccumulation profile of *T. brevicompactum*, the authors observed that a significant amount of Zn(II) and Cu(II) was removed via biosorption, up to 68.7%

for Zn(II) and 32,7% for Cu(II). However, Pb(II) removal was mainly due to bioaccumulation (80.0%), and only (20.0%) was via biosorption. The same behavior was observed for Cd(II) and Cr(VI), which showed removal via absorption in the range of ±70% and only ±30% via biosorption. Therefore, the results indicated that the removal of metals under study was influenced by both bioaccumulation and biosorption processes simultaneously. However, the bioaccumulation profile in a mixed solution of the analyzed metals Cu(II), Cr(VI), Cd(II), Pb(II), and Zn(II) showed accumulation efficiencies below 49% in metallic concentration in the range of 30–100 mg/L [213].

The selection and isolation of indigenous heterotrophic bacteria present in tailing dam effluent (TDE) from a gold mining factory showed that the *Alishewanella agri* PMS5 strain was highly resistant to arsenic. The biosorption and bioaccumulation abilities of arsenic salts showed that the bacterium performs well in both processes, in addition to having the ability to biotransform arsenic oxyanions through oxidation and reduction (As^{3+} to As^{5+}) [214].

The bioaccumulation profile was evaluated with 1.7 µg of arsenic g^{-1} (dry weight) for 24 hours using cells from the isolated strain in medium plus As^{3+}. A comparison of the amount of adsorbate in the supernatant (24%) and the biomass (76%) showed that arsenic was probably accumulated in the cell membrane and cell debris. While the biosorption process using 10 mg/L of arsenic and 2 g/L of biomass exhibited a maximum biosorption capacity of 0.12 mg/g in 30 minutes of contact and 90% biosorption efficiency, in addition to showing biosorption equilibrium after 60 minutes [214].

The comparison of Cd(II) biosorption and bioaccumulation processes by *Scenedesmus acutus* and *Chlorella pyrenoidosa* indicated that 3% of the total cadmium removal by *C. pyrenoidosa* was by bioaccumulation and 97% by biosorption. While *S. acutus* showed 1.5% of cadmium absorption and 98.5% of removal by biosorption. Demonstrating that the two microalgae performed better in removing Cd(II) via adsorption than absorption [215].

11.5.2.2   Biosorption and Consequent Biosynthesis of Metallic Nanoparticles

Several authors investigating metallic biosorption have reported the ability of biosorbents to act in the biosynthesis of metallic nanoparticles because the mechanism involved is also related to intracellular or extracellular functional groups in the biomass [151,216–218].

Thus, the interactions between nanotechnology and biotechnology influenced the evolution of a field of research in constant development called nanobiotechnology [48,219,220]. Nanobiotechnology in line with environmental chemistry emerges as a promising technology in the biosynthesis of metallic nanoparticles through the use of biomass as bioreducing agents and biosorbents [48,219].

The research developed, as well as the synthesis and application of nanoparticles, the acronym in English *nanoparticles (NPs)*, are gaining more and more space in the industrial and technological sector. The application of nanoparticles is currently diversified, being used in different areas such as in the production of chemical catalysts, medicines, quantum dots, clothing, and cleaning products, among others [48,160,221–223].

Metal nanoparticles can be synthesized by chemical and physical processes, however, most of these processes involve toxic chemicals, in addition to being costly. Biosynthesis, also called biogenic synthesis, biological synthesis, and/or green synthesis, has drawn the attention of several researchers since its realization is given through microorganisms, plants, and plant extracts, emerging as a more environmentally adequate alternative when compared to the use of chemical reagents [48,160,223,225].

Biological methods of synthesizing nanoparticles using microorganisms, enzymes, and plant extracts have been suggested as possible alternatives because it is considered a profitable, clean, non-toxic, easy, and environmentally friendly technology when compared to physical and chemical methods [48,160,223,225,226,228]. Further study of the biochemical and molecular mechanism involved in the bioreduction of metals suggests that the ability to biosynthesize at the nanoscale varies among microorganisms [160,216,228].

A wide range of nanoparticles *(NPs)* can be manufactured using different biological groups such as bacteria, filamentous fungi, algae, yeast, and plants [48,224]. In addition, water contaminated with potentially toxic metals can also be used instead of metallic salts for its manufacture, being considered, therefore, a method for the recovery of resources and the safe use of domestic and industrial wastewater [48,224].

Each organism has its method of biochemical processing for forming *NPs*, such as oxidation, reduction of metal ions through enzymes, sugars, proteins, polyphenol groups, aldehydes, and carboxyl [48,224]. Different types of biogenically synthesized nanoparticles (*BNPs*) can be biotransformed using different metallic ions such as gold, silver, titanium, selenium, and copper, among others. However, particularly microorganisms that produce *NPs* through biological processes or biomineralization in the environment are receiving more attention [160,224]. Biological processes such as *Biologically Induced Mineralization* (BIM) and *Biologically Controlled Mineralization* (BCM) are known for the synthesis of *BNPs* [224].

The morphology and size of nanoparticles vary concerning organisms, precursors, and applied conditions. Certain aspects can affect the formation of *BNPs*, such as the type of microorganism, the concentration of the electron donor and electron acceptor mixture, temperature, contact time, and pH [160,224].

Biological methods used to synthesize nanoparticles include intracellular and extracellular methods, whose mechanisms may differ depending on the biological agent. The structures of the external surface of microorganisms play an important role in the synthesis of nanoparticles. The cell wall that is being negatively charged interacts electrostatically with the positively charged metal ions. Enzymes diffused in the cell bio-reduce metal ions into nanoparticles that diffuse through the cell wall [44,229]. Most studies on nanoparticle biosynthesis state that the presence of specific functional groups, such as hydroxyl, carboxylates, methyl, and amides I, II, and III, in addition to another variety of reducing substances generated by microbial cells, are primarily responsible for electrostatic interactions, ion exchange processes and possibly also by biosynthesis [160,224].

The mechanism of biogenic production of nanoparticles mainly incorporates bioreduction and bioprecipitation by polyphenols, peptides, amino acids, and other bioactive compounds obtained from living organisms. In addition to synthesis, these compounds act as capture and stabilization agents to prevent the agglomeration of synthesized NPs [44,230].

Three key factors are involved in synthesizing metallic nanoparticles: the reducing agent, the reaction medium, and the stabilizing agent. Generally, biosynthesized nanoparticles have spherical shapes. In biosynthesis, fungal and/or plant proteins have been reported as protective and stabilizing agents for synthesized nanoparticles [224].

In the literature, there are many reports of mycosynthesis (biogenic synthesis mediated by fungi) of metallic nanoparticles through different species such as *Fusarium sp.* [44]. The biosynthesis of metallic nanoparticles through microorganisms and plants proves to be a promising methodology and the trend is that over the years this methodology will become more used in various sectors such as nanomedicine [44,160,232–234].

Many organisms involved in biosorption are also reported in the literature as potential nanoparticle producers [44], [160,231–234]. In the biosorption process, the organisms involved can capture metal ions through several mechanisms such as ion exchange and complexation, while in biosynthesis, these agents involved can reduce the metal ion, nucleate it, and consequently agglomerate and stabilize it in nanoparticles (Figure 11.14) [218,224,227,235].

Given this and with the increasing application of nanoparticles in several segments, many research groups that were dedicated to the study of biosorption also dedicated themselves to the study of nanoparticle biosynthesis [160,232–234].

Some authors have studied the processes of biosorption and biosynthesis together. As in the studies by [234] that evaluated brown seaweed *Turbinaria conoides* in the biosorption and consequent bioreduction of gold. Biosynthesis of gold nanoparticles from biosorption mediated by the bacterium *Magnetospirillum gryphiswaldense* [233]. Through the yeast, *Pichia pastoris* was able to produce AuNPs via biosorption [232]. The fungus *Trichoderma harzianum* removed gold present in an aqueous solution and biologically synthesized AuNPs [151]. The authors proved that during the biosorptive process, biosynthesis of nanoparticles also occurred.

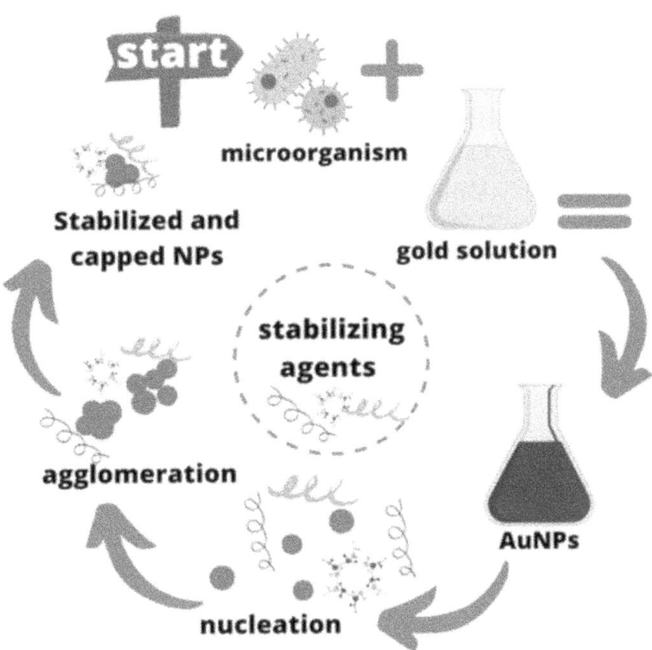

FIGURE 11.14 Schematic representation of the process of biosynthesis of NPs by microorganisms.

In the literature, several authors have studied different biomasses, whether microbial or plants and their extracts for the bioreduction of nanoparticles. It should be noted that most of the studied nanoparticles consist of silver and gold ions. While the most used biomasses represent plant extracts, this is because the process, in addition to being faster, does not require growth media to obtain the biomass, as is the case with microbial biomass. However, the biosynthesis mediated by microorganisms has also been growing because these organisms can produce nanoparticles with small diameters and are more stable [44,48,160].

11.5.2.2.1 Biosynthesis by Microorganisms

Among microorganisms, prokaryotes are drawing much attention to the biosynthesis of nanoparticles [216,236]. The biological synthesis of AuNPs from the microorganisms *Pseudomonas aeruginosa* and *Geotrichum candidum* and their respective filtrates showed that the bacterial cell-free filtrate showed 100% of the distribution by volume of particles below 50 nm, with 98% with a diameter of 7, 4 nm and 2% in the range of 42.3 nm. On the other hand, the filtrate of the fungus *G. candidum*, did not present a size distribution in the nanometric range, under the same conditions studied. What indicates the absence of aggregation of the nanoparticles biosynthesized by the bacterial filtrate is that the nanoparticles were possibly more stable by the production of rhamnolipids by the bacterial biomass [216].

Studies of gold nanoparticle biosynthesis and biosorption using the bacterium *Magnetospirillum gryphiswaldense* MSR-1 revealed in the transmission electron micrographs that the surface of the bacterium exhibited several spherical particles scattered throughout the bacterium and with irregular distribution in the range of 5 to 40 nanometers [233].

The extracellular biosynthesis of gold nanoparticles from the cell-free extract of *Staphylococcus epidermidisnm* bacteria showed, under optimized conditions, quasi-spherical and reasonably monodispersed AuNPs with sizes ranging from 20 to 25 nm [237].

The realization of bionanomining through the use of a real effluent obtained from the leaching of sulfide ore tailing was studied for the biosynthesis of copper nanoparticles through the bacterium *Pseudomonas stutzeri* as a bioreducing agent. The experiments were carried out at room temperature and pressure for 48 hours, using a copper solution prepared in the laboratory to determine the best biosynthesis conditions, which were 381 mg/L of copper concentration and 1050 mg of biomass. Using the solution obtained from the treated tailing, it was possible to obtain spherical nanoparticles with a size between 1 and 2 nm [238].

A new endophytic bacterial strain identified as *Rothia endophytica* was isolated from healthy corn roots and evaluated in the biosynthesis of silver nanoparticles. The results revealed that the biomass could biosynthesize nanoparticles ranging in size from 47 to 72 nm [239]. Screening of 30 cyanobacteria extracts showed that the cyanobacterium *Oscillatoria sp.* was able to biologically synthesize ZnO nanoparticles under the conditions of 0.02 M zinc nitrate, 10 mL of extract volume, pH 8, at 80°C for 3 hours of contact, with a size in the range of 40 nm [240].

There are reports of mycosynthesis of copper nanoparticles using the fungi *Penicillium aurantiogriseum*, *Penicillium citrinum*, and *Penicillium waksmanii* isolated from soil [241]. Fungi are reported in the literature on nanoparticle biosynthesis in recent years (Table 11.7), however, they emerge as promising because they secrete large amounts of enzymes, and have simpler growth, thus facilitating a future industrial scale [44,236].

TABLE 11.7
Reports of Microorganism-Mediated Nanoparticle Biosynthesis

Biomass	Metal	Size (nm)	Reference
Pseudomonas aeruginosa	Au	15–40	[244]
Rhodobacter capsulatus	Au	10–48	[245]
Bacillus licheniformis	Au	10–100	[246]
Pseudomonas putida MDH1	Au	12	[247]
Magnetospirillum gryphiswaldense MSR-1	Au	10–40	[233]
Pseudomonas aeruginosa MTCC 424	Au	12–150	[248]
Staphylococcus lentus	Au	5–18	[249]
	Ag	4–20	
Acinetobacter schindleri	Zn	20–100	[250]
Pseudomonas stutzeri	Cu	1–2	[238]
Rothia endophytica	Ag	47–72	[239]
Oscillatoria sp.	ZnO	40	[240]
Aspergillus niger	Ag	–	[241]
Verticilium sp.	Au	20±8	[251]
Candida albicans	Au	5	[252]
	Ag	30	
Fusarium semitectum	Ag	10–60	[253]
Cladosporium oxysporum AJP03	Au	72.32±21.80	[228]
Trichoderma asperellum	Ag	13–18	[254]
Aspergillus fumigatus	Ag	5–25	[255]
Neurospora crassa	Ag	11	[256]
	Au	32	
Alternaria sp	Ag	75.83–104.1	[242]
Endoparasitic fungus—*Cordyceps militaris* (Extract)	Ag	15	[257]
Stereum hirsutum	Cu	5–20	[243]
Rhodotorula mucilaginous	Cu	10.5	[258]
Neurospora crassa (Extract)	Ag	2–22	[259]
Cryptococcus laurentii	Ag	35	[260]
Rhodotorula glutinis	Ag	15	
Trichoderma sp. WL-Go	PbSe	10–30	[261]
Filtrate from the mycelium of the *Setosphaeria rostrata*	Ag	2–20	[262]
Fusarium oxysporum	N-Co$_3$O$_4$	25	[263]
Saccharomyces cerevisiae	Cu	10–12	[264]

In the study by [160], it was proved that it is possible to remove gold using a small amount of fungal biomass *Trichoderma harzianum* (0.4 g/L) from a solution of gold ions with a high concentration (400 mg/L) obtaining uptake of 1,340 mg of the metal per g of biomass and with a removal rate of 62% in 180 minutes. In addition to biologically producing AuNPs with an average diameter of 30 nm quickly, cheaply, and efficiently.

The formation of silver nanoparticles from *Alternaria sp.* which was isolated from the bark part of *Calophyllum apetalum* Willd showed that the fungal extract can biologically synthesize cubic silver nanoparticles with a diameter of 75.83–104.1 nm [242].

The fungus *Stereum hirsutum* was studied to evaluate its potential for use in the biosynthesis of copper/copper oxide nanoparticles. The results showed that the biosynthesized nanoparticles were mainly spherical (5–20 nm) and that the presence of amine groups attached to the nanoparticles suggests that the fungal extracellular protein is responsible for forming the nanoparticles [243].

The biological synthesis of lead selenide nanoparticles (PbSe NPs) by the fungus *Trichoderma sp.* WL-Go showed that at pH 8, with 0.5 g of fungal strain biomass, in the proportion of (1:1) mM SeO_2:$Pb(NO_3)_2$ were the ideal synthesis conditions to achieve NPs with diameters from 10 to 30 nm [261].

Silver nanoparticles produced by biological route using filtrate from the mycelium of the endophytic fungus *Setosphaeria rostrata* that was isolated from the stem of *Solanum nigrum* showed a size distribution of 2–20 nm [262].

The endophytic fungus *Fusarium oxysporum* was evaluated in the biosynthesis of N-Co_3O_4 nanoparticles. The nanoparticles synthesized by the biological route had a size of 25 nm. In addition to showing good photocatalytic activity and degrading 87% MB in 120 minutes under solar irradiation at a temperature of $32°C \pm 1°C$ [263].

Biological synthesis of copper nanoparticles by commercial yeast *Saccharomyces cerevisiae* biomass Perlage® BB showed copper nanoparticles extracellularly with more than 70% of the particles with a size between 10 and 12 nm. Using 5–25 mg of biomass added to 2 mL of a 0.001 molar copper solution for 24 hours under constant stirring (200 rpm) at a temperature of 30°C [264].

11.5.2.2.2 Biosynthesis by Other Materials

The use of different plants for the biosynthesis of metallic nanoparticles has been much researched due to their low cost and high availability [265]. The phytosynthesis method is fast, cost-effective, and with easier downstream processing than mycosynthesis. Compared to mycosynthesis, phytosynthesis is faster and the processing of biomass is easier, due to this, phytosynthesis draws a lot of attention from researchers all over the world who use plants and their extracts [266,267].

Biosynthesis of nanoparticles from plant extracts is one of the simplest techniques because plants produce functional biomolecules that actively reduce metal ions, in addition to acting as protective and stabilizing agents for nanoparticles [217,218].

The dead and dry biomass of the plant can also be used in the biosynthesis of nanoparticles, as even bioactive compounds isolated from plants such as polyphenols, alkaloids, and terpenes have demonstrated biosynthetic power. [268] and other authors in their studies with *Azadirachta indica* leaf broth suggested that the biosynthesis of gold nanoparticles is due to the presence of sugars (aldoses) and ketones present in plant extracts [48,218]. Some of these biosyntheses are shown in (Table 11.8).

The biosynthesis of gold nanoparticles was reported by [287] from fresh and aged extracts of the *Dracaena Draco* L plant. The size distribution evaluated by Dynamic Light Scattering (DLS) indicated that the fresh extract showed a dual distribution centered on the 75–300 nM, probably due to the agglomeration of the biosynthesized particles, while the aged extract showed a size distribution centered at 6–7 nm.

Zinc oxide nanoparticles (ZnO NPs) synthesized using *Butea monosperma* seed extract and exhibited nanoparticles with an average size of 24.62 nm, using 40 mL of *Butea monosperma* seed extract added to a 1 mM solution of zinc nitrate hexahydrate [288].

TABLE 11.8
Reports of Nanoparticle Biosynthesis Mediated by Other Materials

Biomasses	Metal	Size (nm)	Reference
Avena sativa	Au	25–85	[269]
Ornate turbine	Au	7–11	[270]
Mentha piperita	Au	~78	[271]
Lawsonia inermis	Ag	18	[225]
Peltophorum pterocarpum	Au	5–50	[272]
Aucus carota	Cu_2O	12	[273]
Rheum palmatum L.	CuO	10–20	[274]
Phyllanthus Embilica	Cu	15–30	[275]
Sesbania grandiflora (leaf extract)	Ag	± 20	[276]
Opuntia ficus-indica (extract)	Au	10–20	[277]
Ficus religiosa (Leaf extract)	Ag	67	[278]
Buchu Plant (Extract)	Ag	7.76–19.95	[279]
Agricultural waste	Au	2.77±6.03	[280]
mango peel extract	Au	3.67±18.01	
Kappaphycus alvarezii (Algal extract)	Fe_3O_4	14.7	[281]
Shrub—*Mussaenda erythrophylla* (Extract from the sheet)	Ag	82–88	[282]
Artocarpus altilis (leaf extract)	Ag	25–43	[283]
Delonix regia (Leaf extract)	Au	4–24	[284]
Platynereis dumerilii	Ag	4.5±16.5	[285]
Ormocarpum cochinchinense (Extract)	Ag	18–35	[286]
Dracaena Draco L Extract	Au	75–300	[287]
Butea monospermae seed extract	ZnO	24.62	[288]
Aquilegia pubiflora	Ag	19	[289]
Hylocereus polyrhizus extract and seed oil	Au	25.31	[290]

The green synthesis of silver nanoparticles with *Aquilegia pubiflora* leaf extracts biologically produced AgNPs with an average size of 19 nanometers. AgNPs have been used in various assays such as antimicrobial, antioxidant, anti-Alzheimer's, antidiabetic, antiaging, anticancer, and biocompatibility studies. The results revealed that AgNPs are promising choices for several biological applications [289].

Gold nanoparticles with a particle size of 25.31 nm were synthesized using extract of *Hylocereus polyrhizus* (red pitaya) using 50 mL of extract and 10 mL of pitaya seed oil combined with $HAuCl_4$ $3H_2O$ and kept for 4 hours contact at room temperature [290]. The process of obtaining iron oxide (Fe_2O_3) NPs using *Syzygium aromaticum* (clove) extract showed that the biosynthesized Fe_2O_3 nanoparticles had no negative (non-toxic) effect on plant and bacterial growth [291].

11.6 CONSIDERATIONS

The constant population and industrial growth and consequently the increase in the consumption of natural resources promote the production of more solid waste and effluents contaminated by potentially toxic metals. Unfortunately, planet Earth cannot mitigate the large volume of effluents discarded daily in rivers, streams, and on the ground. The need for new environmental decontamination techniques that have low cost, easy implementation, and treatment efficiency becomes essential for the perpetuation of life.

Biosorption is one of the bioremediation techniques that most emerge in the literature as promising in environmental decontamination. Several authors have studied numerous biomasses (microbial and lignocellulosic) as biosorbents of potentially toxic metals. The literature shows that there is a vast amount of biomass that can be used as biosorbents so that decontamination can be carried out even with agro-industrial residues. Thus, generating added value for a product that would previously go to landfills or be incinerated.

The future of biosorption technology is to facilitate and improve environmental decontamination, therefore, reducing the number of effluents that are discarded daily without any previous treatment in the ecosystem. In this way, we will be able to have rivers, underground water, and soils practically free of contamination from anthropogenic sources.

GLOSSARY

Bioremediation The use of microorganisms to remove or metabolize toxic substances such as metals.

Biosorption Physical-chemical process of removing chemical species from the surface of living or dead biomass, it is a metabolically passive process and is based mainly on the "affinity" between the biosorbent and the adsorbate (species to be adsorbed).

Bioaccumulation Process of removal of chemical species by living biomass through accumulation and metabolism, it is a metabolically active process.

Phytoremediation Environmental decontamination process uses plants, which absorb pollutants through their roots.

Adsorption It is the physical-chemical process in which chemical species are retained on the surface of a material, when the material is organic, it is called biosorption.

Absorption Sorption of chemical species via intracellular accumulation.

Adsorbate Species that have been adsorbed by the biosorbent, can be organic and inorganic.

Biomass Material of organic nature, it can be microbial cells, agro-industrial residues, fruit peels, and others.

Biosorbent Biomass used in the biosorption process.

Desorption Release of the adsorbate that has been adsorbed by the biosorbent.

Biochar Solid product, rich in carbon, of low density, and with a stable porous structure that is produced from different biomasses by different methods such as thermal decomposition, gasification, carbonization, and mainly via pyrolysis.

Biosynthesis of nanoparticles, biogenic synthesis or green synthesis Process of biotransformation of chemical species through the components present (functional groups, enzymes, and others) in biomasses such as microorganisms and plant extracts, the agents involved can reduce the metallic ion, nucleate it and consequently agglomerate and stabilize it in nanoparticles.

REFERENCES

[1] Janani, R.; Gurunathan, Baskar; Sivakumar, K.; Varjani, Sunita; Ngo, Huu Hao; Gnansounou, Edgard: Advancements in heavy metals removal from effluents employing nano-adsorbents: Way towards cleaner production. *Environmental Research* 203, (2022), 111815.

[2] Thirunavukkarasu, Arunachalam; Nithya, Rajarathinam; Sivashankar, Raja: Continuous fixed-bed biosorption process: A review. *Chemical Engineering Journal Advances* 8, (2021), 100188.

[3] Alluri, Hima Karnika; Ronda, Srinivasa Reddy; Settalluri, Vijaya Saradhi; Jayakumar Singh, Bondili; Suryanarayana, V.; Venkateshwar, P.: Biosorption: An eco-friendly alternative for heavy metal removal. *African Journal of Biotechnology* 6, no. 25 (2007), 2924–2931.

[4] Rehman, Kanwal; Fatima, Fiza; Waheed, Iqra; Akash, Muhammad Sajid Hamid: Prevalence of exposure of heavy metals and their impact on health consequences. *Journal of Cellular Biochemistry* 119, no. 1 (2018), 157–184.

[5] Ullah, Abid; Heng, Sun; Munis, Muhammad Farooq Hussain; Fahad, Shah; Yang, Xiyan: Phytoremediation of heavy metals assisted by plant growth promoting (PGP) bacteria: A review. *Environmental and Experimental Botany* 117, (2015), 28–40.

[6] Yaashikaa, PR; Senthil Kumar, P.; Varjani, Sunita; Saravanan, A.: Rhizoremediation of Cu(II) ions from contaminated soil using plant growth promoting bacteria: An outlook on pyrolysis conditions on plant residues for methylene orange dye biosorption. *Bioengineered* 11, no. 1 (2020), 175–187.

[7] Rai, Prabhat Kumar; Lee, Sang Soo; Zhang, Ming; Tsang, Yiu Fai; Kim, Ki Hyun: Heavy metals in food crops: Health risks, fate, mechanisms, and management. *Environment International* 125, (2019), 365–385.

[8] Kumar Awasthi, Mukesh; Ravindran, B.; Sarsaiya, Surendra; Chen, Hongyu; Wainaina, Steven; Singh, Ekta; Liu, Tao; Kumar, Sunil; Pandey, Ashok; Singh, Lal; Zhang, Zengqiang: Metagenomics for taxonomy profiling: Tools and approaches. *Bioengineered* 11, no. 1 (2020), 356–374.

[9] Razzak, Shaikh A.; Faruque, Mohammed O.; Alsheikh, Zeyad; Alsheikhmohamad, Laila; Alkuroud, Deem; Alfayez, Adah; Hossain, SMZakir; Hossain, Mohammad M.: A comprehensive review on conventional and biological-driven heavy metals removal from industrial wastewater. *Environmental Advances* 7, (2022), 100168.

[10] Duraisamy, Ramesh; Mechoro, Mihretu; Silk, Tolera; Akhtar Khan, Masood: Potential of *Mangifera indica* activated carbon for removal of chromium and iron. *Cogent Engineering* 7, no. 1 (2020), 1813237.

[11] Spain, Olivia; Plöhn, Martin; Funk, Christiane: The cell wall of green microalgae and its role in heavy metal removal. *Physiology Plantarum* 173, no. 2 (2021), 526–535.

[12] Dang, V.B.H.; Doan, H.D.; Dang-Vu, T.; Lohi, A.: Equilibrium and kinetics of biosorption of cadmium(II) and copper(II) ions by wheat straw. *Bioresource Technology* 100, no. 1 (2009), 211–219.

[13] Özer, Ayla; Özer, Dursun; Ibrahim Ekiz, H.: The equilibrium and kinetic modeling of the biosorption of copper(II) ions on *Cladophora crispata*. *Adsorption* 10, no. 4 (2005), 317–326.

[14] Mack, C.; Wilhelmi, B.; Duncan, J.R.; Burgess, J.E.: Biosorption of precious metals. *Biotechnology Advances* 25, no. 3 (2007), 264–271.

[15] Zamani Beidokhti, Majid; (Omid) Naeeni, Seyed Taghi; AbdiGhahroudi, Mohammad Sajjad: Biosorption of nickel (II) from aqueous solutions onto pistachio hull waste as a low-cost biosorbent. *Civil Engineering Journal* 5, no. 2 (2019), 447.

[16] Liu, Linhai; Lin, Xiaoai; Luo, Longzao; Yang, Jia; Luo, Jialiang; Liao, Xing; Cheng, Haixiang: Biosorption of copper ions through microalgae from piggery digestate: Optimization, kinetic, isotherm and mechanism. *Journal of Cleaner Production* 319, (2021), 128724.

[17] Suresh Kumar, K.; Dahms, Hans Uwe; Won, Eun Ji; Lee, Jae Seong; Shin, Kyung Hoon: Microalgae - A promising tool for heavy metal remediation. *Ecotoxicology and Environmental Safety* 113, (2015), 329–352.

[18] Salama, El Sayed; Roh, Hyun Seog; Dev, Subharata; Khan, Moonis Ali; Abou-Shanab, Reda AI; Chang, Soon Woong; Jeon, Byong Hun: Algae as a green technology for heavy metals removal from various wastewater. *World Journal of Microbiology and Biotechnology* 35, (2019), 1–19.

[19] Pagliaccia, Benedetta; Carretti, Emiliano; Severi, Mirko; Berti, Debora; Lubello, Claudio; Lotti, Tommaso: Heavy metal biosorption by extracellular polymeric substances (EPS) recovered from anammox granular sludge. *Journal of Hazardous Materials* 424, (2022), 126661.

[20] Alsharari, Sultan F.; Tayel, Ahmed A.; Moussa, Shaaban H.: Soil amendment with nano-fungal chitosan for heavy metals biosorption. *International Journal of Biological Macromolecules* 118, (2018), 2265–2268.

[21] Araujo, Leandro Goulart de; Borba, Tania Regina de; Ferreira, Rafael Vicente de Padua; Canevesi, Rafael Luan Sehn; Silva, Edson Antonio da; Dellamano, José Claudio; Marumo, Júlio Takehiro: Use of calcium alginate beads and *Saccharomyces cerevisiae* for biosorption of 241Am. *Journal of Environmental Radioactivity* 223–224 (2020), 106399.

[22] Othmani, Amina; Magdouli, Sara; Senthil Kumar, P.; Kapoor, Ashish; Chellam, Padmanaban Velayudhaperumal; Gökkuş, Ömür: Agricultural waste materials for adsorptive removal of phenols, chromium (VI) and cadmium (II) from wastewater: A review. *Environmental Research* 204 (2022), 111916.

[23] Rwiza, Mwemezi Johaiven; Oh, Seok Young; Kim, Kyoung Woong; Kim, Sang Don: Comparative sorption isotherms and removal studies for Pb(II) by physical and thermochemical modification of low-cost agro-wastes from Tanzania. *Chemosphere* 195, (2018), 135–145.

[24] Okoro, Hussein K.; Pandey, Sadanand; Ogunkunle, Clement O.; Ngila, Catherine J.; Zvinowanda, C.; Jimoh, Ismaila; Lawal, Isiaka A.; Orosun, Muyiwa M.; Adeniyi, Adewale George: Nanomaterial-based biosorbents: Adsorbent for efficient removal of selected organic pollutants from industrial wastewater. *Emerging Contaminants* 8, (2022), 46–58.

[25] Dean, John R.: *Bioavailability Bioaccessability and Mobility of Environmental Contaminants*. 1. Aufl. New Jersey: John Wiley & Sons (2007).

[26] Sharma, Pooja; Kumar, Sunil: Bioremediation of heavy metals from industrial effluents by endophytes and their metabolic activity: Recent advances. *Bioresource Technology* 339, (2021), 125589.

[27] Priyadarshanee, Monika; Das, Surajit: Biosorption and removal of toxic heavy metals by metal tolerating bacteria for bioremediation of metal contamination: A comprehensive review. *Journal of Environmental Chemical Engineering* 9, no. 1 (2021), 104686.

[28] Debnath, Abhijit; Singh, Prabhat Kumar; Chandra Sharma, Yogesh: Metallic contamination of global river sediments and latest developments for their remediation. *Journal of Environmental Management* 298, (2021), 113378.

[29] Schmitt, Harrison J.; Calloway, Eric E.; Sullivan, Daniel; Clausen, Whitney H.; Tucker, Pamela G.; Rayman, Jamie; Gerhardstein, Ben: Chronic environmental contamination: A systematic review of psychological health consequences. *Science of the Total Environment* 772, (2021), 145025.

[30] Cui, Yongbo; Bai, Li; Li, Chunhui; He, Zijian; Liu, Xinru: Assessment of heavy metal contamination levels and health risks in environmental media in the northeast region. *Sustainable Cities and Society* 80, (2022), 103796.

[31] Skalny, Anatoly V.; Aschner, Michael; Bobrovnitsky, Igor P.; Chen, Pan; Tsatsakis, Aristidis; Paoliello, Monica M.B.; Buha Djordevic, Aleksandra; Tinkov, Alexey A.: Environmental and health hazards of military metal pollution. *Environmental Research* 201, (2021), 111568.

[32] Singh, Simranjeet; Kumar, Vijay; Datta, Shivika; Dhanjal, Daljeet Singh; Sharma, Kankan; Samuel, Jastin; Singh, Joginder: Current advancement and future prospect of biosorbents for bioremediation. *Science of the Total Environment* 709, (2020), 135895.

[33] de Andrade, Vanda Lopes; Cota, Magdalena; Serrazina, Daniela; Mateus, Maria Luisa; Aschner, Michael; dos Santos, Ana Paula Marreilha: Metal environmental contamination within different human exposure context-specific and non-specific biomarkers. *Toxicology Letters* 324, (2020), 46–53.

[34] Hutzinger, Otto: The handbook of environmental chemistry. *Environmental Sciences Europe* 22, no. 4 (2010), 511–512.

[35] Musilova, Janette; Arvay, Julius; Vollmannova, Alena; Toth, Tomas; Tomas, Jan: Environmental contamination by heavy metals in region with previous mining activity. *Bulletin of Environmental Contamination and Toxicology* 97, no. 4 (2016), 569–575.

[36] Otaviano, Jonas Juliermerson Silva; Nascimento, Jéssica Mesquita do; Sousa, José Roberto Pereira de; Oliveira, Jorge Diniz de: Simultaneous adsorption and competitiveness of cadmium and copper in natural and anthropized soils in the city of Cajapió, Baixada Maranhense. *Ibero-American Journal of Environmental Sciences* 11, no. 5 (2020), 510–518.

[37] Yadav, Krishna Kumar; Gupta, Neha; Kumar, Vinit; Singh, Jitendra Kumar: Bioremediation of heavy metals from contaminated sites using potential species: A review. *Indian Journal of Environmental Protection* 37, no. 371 (2017), 65–84.

[38] Abdel-Ghani, Nour T.; El-Chaghaby, Ghadir A: Biosorption for metal ions removal from aqueous solutions: A review of recent studies. *International Journal of Latest Research in Science and Technology* 3, no. 1 (2014), 24–42.

[39] Nascimento, Jessica M.; de Oliveira, Jorge Diniz; Rizzo, Andrea C.L.; Milk, Selma G.F.: Biosorption Cu (II) by the yeast Saccharomyces cerevisiae. *Biotechnology Reports* 21, (2019), e00315.

[40] Qin, Huaqing; Hu, Tianjue; Zhai, Yunbo; Lu, Ningqin; Aliyeva, Jamila: The improved methods of heavy metals removal by biosorbents: A review. *Environmental Pollution* 258, (2020), 113777.

[41] Vijayaraghavan, K.; Balasubramanian, R.: Is biosorption suitable for decontamination of metal-bearing wastewaters? A critical review on the state-of-the-art of biosorption processes and future directions. *Journal of Environmental Management* 160, (2015), 283–296.

[42] Qin, Huaqing; Hu, Tianjue; Zhai, Yunbo; Lu, Ningqin; Aliyeva, Jamila: The improved methods of heavy metals removal by biosorbents: A review. *Environmental Pollution* 258, (2020), 113777.

[43] Singh, Simranjeet; Kumar, Vijay; Singh Dhanjal, Daljeet; Datta, Shivika; Singh, Satyender; Singh, Joginder: Biosorbents for heavy metal removal from industrial effluents. In *Bioremediation for Environmental Sustainability*, (2021), pp. 219–233.

[44] Priyadarshini, Eepsita; Priyadarshini, Sushree Sangita; Cousins, Brian G.; Pradhan, Nilotpala: Metal-fungus interaction: Review on cellular processes underlying heavy metal detoxification and synthesis of metal nanoparticles. *Chemosphere* 274, (2021), 129976.

[45] Carolin, C. Femina; Kumar, P. Senthil; Saravanan, A.; Joshiba, G. Janet; Naushad, Mu: Efficient techniques for the removal of toxic heavy metals from aquatic environment: A review. *Journal of Environmental Chemical Engineering* 5, no. 3 (2017), 2782–2799.

[46] Ali Redha, Ali: Removal of heavy metals from aqueous media by biosorption. *Arab Journal of Basic and Applied Sciences* 27, no. 1 (2020), 183–193.

[47] Naja, G.M.; Murphy, V.; Volesky, B.: Biosorption, metals. In *Encyclopedia of Industrial Biotechnology: Bioprocess, Bioseparation, and Cell Technology* (2009).

[48] Majhi, Krishnendu; Let, Moitri; Kabiraj, Ashutosh; Sarkar, Shrabana; Halder, Urmi; Dutta, Bhramar; Biswas, Raju; Bandopadhyay, Rajib: Metal recovery using nanobiotechnology. In *Nanobiotechnology*, (2021), pp. 283–301.

[49] Nascimento, J.M.; Oliveira, J.D.; Leite, S.G.F.: Chemical characterization of biomass flour of the babassu coconut mesocarp (*Orbignya speciosa*) during biosorption process of copper ions. *Environmental Technology & Innovation* 16 (2019), 100440.

[50] Van, P.N.; Thi Hong Truong, Hai; Pham, Tuan Anh; Le Cong, Tuan; Le, Tien; Thi Nguyen, Kim Cuc: Removal of manganese and copper from aqueous solution by yeast papiliotrema huenov. *Mycobiology* 49, no. 5 (2021), 507–520.

[51] Lutzu, Giovanni Antonio; Ciurli, Adriana; Chiellini, Carolina; DiCaprio, Fabrizio; Concas, Alessandro; Dunford, Nurhan Turgut: Latest developments in wastewater treatment and biopolymer production by microalgae. *Journal of Environmental Chemical Engineering* 9, no. 1 (2021), 104926.

[52] Beni, Ali Aghababai; Esmaeili, Akbar: Biosorption, an efficient method for removing heavy metals from industrial effluents: A review. *Environmental Technology and Innovation* 17, (2020), 100503.

[53] Huang, Haojie; Jia, Qingyun; Jing, Weixin; Dahms, Hans Uwe; Wang, Lan: Screening strains for microbial biosorption technology of cadmium. *Chemosphere* 251, (2020), 126428.

[54] Abdel-Raouf, Neveen; Sholkamy, Essam Nageh; Bukhari, Nagat; Al-Enazi, Nouf Mohammed; Alsamhary, Khawla Ibrahim; Al-Khiat, Soad Humead A.; Ibraheem, Ibraheem Borie M.: Bioremoval capacity of Co^{+2} using *Phormidium tenue* and *Chlorella vulgaris* as biosorbents. *Environmental Research* 204, (2022), 111630.

[55] Huamán, Gap; Mosque, L.M.S.; Torem, M.L.: Biosorption of Heavy Metals Using Green Coconut (Cocos nucifera) Husk Powder *Separation Science and Technology* 41(14), (2005), 3141–3153.

[56] Voss, M.; Thomas, R.W.S.P.: Sorption of copper and manganese by rhizospheric bacteria in wheat. *Rural Science* 31, no. 6 (2001), 947–951.

[57] Birth, R.F.; Lima, A.C.A.; Vidal, C.B.; Melo, D.Q.; Raulino, G.S.C.: Adsorption: *Theoretical Aspects and Environmental Applications*. Fortaleza: UFC (2014).

[58] Abbas, S. H., I. M. Ismail, T. M. Mostafa and A. H. Sulaymon. Biosorption of heavy metals: a review. *Journal of Chemical Science and Technology* 3, no. 4, (2014), 74–102.

[59] Espinoza-Quiñones, F. R., A. Módenes, L. Thomé, S. Palácio, D. Trigueros, A. Oliveira and N. Szymanski: Study of the bioaccumulation kinetic of lead by living aquatic macrophyte Salvinia auriculata. *Chemical Engineering Journal* 150, no. 2-3 (2009), 316–322.

[60] Uddin, Mohammad Kashif: A review on the adsorption of heavy metals by clay minerals, with special focus on the past decade. *Chemical Engineering Journal* 308, (2017), 438–462.

[61] Das, Nilanjana; Das, Devlina: Recovery of rare earth metals through biosorption: An overview. *Journal of Rare Earths* 31, no. 10 (2013), 933–943.

[62] Elgarahy, A.M.; Elwakeel, K.Z.; Mohammad, S.H.; Elshoubaky, G.A.: A critical review of biosorption of dyes, heavy metals and metalloids from wastewater as an efficient and green process. *Cleaner Engineering and Technology* 4, (2021), 100209.

[63] do Nascimento, Jessica M.; de Oliveira, Jorge Diniz; Rizzo, Andrea C.L.; Milk, Selma G.F.: Biosorption Cu (II) by the yeast *Saccharomyces cerevisiae*. *Biotechnology Reports* 21, (2019), e00315.

[64] Nateras-Ramirez, O.; Martínez-Macias, MR; Sánchez-Machado, DI; López-Cervantes, J.; Aguilar-Ruiz, R.J.: An overview of microalgae for Cd^{2+} and Pb^{2+} biosorption from wastewater. *Bioresource Technology Reports* 17 (2022), 100932.

[65] do Nascimento, J. M., N. D. Cruz, G. R. de Oliveira, W. S. Sá, J. D. de Oliveira, P. R. S. Ribeiro and S. G. Leite: Evaluation of the kinetics of gold biosorption processes and consequent biogenic synthesis of AuNPs mediated by the fungus Trichoderma harzianum. *Environmental Technology & Innovation* 21 (2021), 101238.

[66] Silva, J.L.B.C.; Lime Small, O.T.B.; Santana Rocha, L.K.; Araújo, E.C.O.; Marciel, Tar Barros, A.J.: Biosorption of heavy metals: A review. *Revista Saúde & Ciência Online* 3, no. 3 (2014), 137–149.

[67] Sandau, E.; Sandau, P.; Pulz, O.; Zimmermann, M.: Heavy metal sorption by marine algae and algal by-products. *Acta Biotechnologica* 16, no. 2–3 (1996), 103–119.

[68] Schiewer, S., Volesky, B.: Modeling of the proton-metal ion exchange in biosorption. *Environmental Science & Technology* 2, no. 29 (1995), 3049-3058.

[69] Pugazhendhi, Arivalagan; Ranganathan, Kuppusamy; Kaliannan, Thamaraiselvi: Biosorptive removal of copper(II) by *Bacillus cereus* isolated from contaminated soil of electroplating industry in India. *Water, Air, and Soil Pollution* 229, (2018), 1–9.

[70] Nascimento, Jessica M.; de Oliveira, Jorge Diniz; Rizzo, Andrea C.L.; Milk, Selma G.F.: Biosorption Cu (II) by the yeast *Saccharomyces cerevisiae*. *Biotechnology Reports* 21 (2019), e00315.

[71] do Nascimento, J.M.; Oliveira, J.D.: Characterizing the pequi bark (*Caryocar brasiliense* camb.) and teak sawdust (*Tectona grandis*) biomasses in natura, or modified with citric acid, according to boehm's methodology. *Virtual Journal of Chemistry* 9, no. 3 (2017), 1087–1097.

[72] Do Nascimento, Welenilton José; Landers, Richard; Gurgel Carlos da Silva, Meuris; Vieira, Melissa Gurgel Adeodato: Equilibrium and desorption studies of the competitive binary biosorption of silver(I) and copper(II) ions on brown algae waste. *Journal of Environmental Chemical Engineering* 9 (2020), 104840.

[73] Sekar, M.; Sakthi, V.; Rengaraj, S.: Kinetics and equilibrium adsorption study of lead(II) onto activated carbon prepared from coconut shell. *Journal of Colloid and Interface Science* 279, no. 2 (2004), 307–313.

[74] Domingues, V. F., G. Priolo, A. C. Alves, M. F. Cabral and C. Delerue-Matos: Adsorption behavior of α-cypermethrin on cork and activated carbon. *Journal of Environmental Science and Health Part B* 42, no. 6 (2007), 649–654.

[75] Bediako, John Kwame; Wei, Wei; Kim, Sok; Yun, Yeoung Sang: Removal of heavy metals from aqueous phases using chemically modified waste Lyocell fiber. *Journal of Hazardous Materials* 299, (2015), 550–561.

[76] Barros, D.C.; Carvalho, G.; Ribeiro, M.A.: Biosorption process to remove heavy metals from agro-industrial residues: A review. *Revista Biotecnologia & Ciência* 6 (2017), 1–15.

[77] Volesky, Bohumil: Biosorption and me. *Water Research* 41, no. 18 (2007), 4017–4029.

[78] Kuyucak, N., Volesky, B.: Accumulation of cobalt by marine algae. *Biotechnology and bioengineering* 7, no. 33 (1989), 809–814.

[79] Vijayaraghavan, K; Yun, Yeoung-sang: Bacterial biosorbents and biosorption. *Biotechnology Advances* 26, no. 3 (2008), 266–291.

[80] Nascimento, Jéssica Mesquita do; Silva, Bruno Sampaio da; Chaves, Márcia Dias; Oliveira, Jorge Diniz de: Biosorption of Cd(II) and Pb(II) ions using pequi bark (*Caryocar brasiliense* Camb) biomass modified with citric acid. *Journal of Environmental Sciences* 8, no. 1 (2014), 57–69.

[81] Santos de Sousa, Helayne; Diniz de Oliveira, Jorge: Characterization of cupuaçu bark (*Theobroma grandiflorum*) and ascara fruit (*Euterpe Oleracea*) biomasses for potentially toxic metals removal in chemistry laboratory effluents. *Journal of Interdisciplinary Debates* 2, no. 03 (2021).

[82] Boas, Naiza V.; Casarin, Juliana; Caetano, Josiane; Gonçalves Junior, Affonso C.; Tarley, César R.T.; Dragunski, Douglas C.: Biosorption of copper using the mesocarp and endocarp of natural and chemically treated macadamia nuts Biosorption of copper using the mesocarp and endocarp of natural and chemically treated macadamia. Revista Brasileira de Engenharia Agrícola e Ambiental 16 (2012), 1359–1366.

[83] Militaru, B. A., R. Pode, L. Lupa, W. Schmidt, A. Tekle-Röttering and N. Kazamer: Using sewage sludge ash as an efficient adsorbent for Pb (II) and Cu (II) in single and binary systems. *Molecules* 25, no. 11 (2020), 2559.

[84] Fola, Akinhanmi Temilade; Idowu, Adeogun Abideen; Adetutu, Adegbuyi: Removal of Cu^{2+} from aqueous solution by adsorption onto quail eggshell kinetic and isothermal studies. *Journal of Environment and Biotechnology Research* 5, no. 1 (2016), 1–9.

[85] Gupta, Nitish; Sen, Ravindra: Kinetic and equilibrium modeling of Cu (II) adsorption from aqueous solution by chemically modified Groundnut husk (*Arachis hypogaea*). *Journal of Environmental Chemical Engineering* 5, no. 5 (2017), 4274–4281.

[86] Yifira, M. T., M. B. Doda and C. K. Kanido: Coffee Husk and Coffee Ground as an Adsorbent for the Removal of Lead, Copper and Chromium from Aqueous Solution. *Pakistan Journal of Analytical & Environmental Chemistry* 24, no. 1 (2023), 106–117.

[87] Boniolo, Milena Rodrigues; Yabuki, Lauren Nozomi Marques; Mortari, Daniela Andresa; Menegário, Amauri Antonio; Garcia, Marcelo Loureiro: Brazilian biomasses applied to the removal of uranium from acid mine drainage by biosorption processes. *Holos Environment* 17, no. 1 (2017), 149.

[88] Oliveira, J. A., F. A. Cunha and L. A. Ruotolo: Synthesis of zeolite from sugarcane bagasse fly ash and its application as a low-cost adsorbent to remove heavy metals. *Journal of Cleaner Production* 229 (2018), 956–963.

[89] Albertini, Silvana; Do Carmo, Leandro Francisco; Do Prado Filho, Luiz Gonzaga: Use of sawdust and pulp from cane as adsorbents of cadmium. *Food Science and Technology* 27, no. 1 (2007), 113–118.

[90] Moreira, Débora Astoni; Souza, José Antonio Rodrigues de; Silva, Ellen Lemes; Gonçalves, Janine Mesquita; Rezende, Diego César Veloso; Oliveira, Wilson Marques; Ribeiro, Wesley Anderson Siqueira; Rezende, João Gabriel Felismino: Biosorption of heavy metals by the eggshell of laying hens. *Ibero-American Journal of Environmental Sciences* 9, no. 7 (2018), 289–295.

[91] Santana, J.S.; Santos, B.R.; Resende, B.O.: Use of banana peel as a biosorbent for the adsorption of heavy metals, enabling its use in wastewater from the galvanic industry. *INOVAE* 8 (2020), 143–157.

[92] Giri, R., N. Kumari, M. Behera, A. Sharma, S. Kumar, N. Kumar and R. Singh: Adsorption of hexavalent chromium from aqueous solution using pomegranate peel as low-cost biosorbent. *Environmental Sustainability* 4, no. 2 (2021), 401–417.

[93] Jabfonska, B: Removing of Cr (III) and Cr (VI) compounds from aqueous solutions by shale waste rocks. *Desalination and Water Treatment* 186 (2020), 234-246.

[94] Porpino, K. K. P., M. d. C. S. Barreto, K. B. Cambuim, J. R. d. Carvalho Filho, I. A. S. Toscano and M. d. A. Lima: Fe (II) adsorption on Ucides cordatus crab shells. *Química Nova* 34 (2011), 928–932.

[95] Dahiya, S., R. Tripathi and A. Hegde: Biosorption of lead and copper from aqueous solutions by pretreated crab and arca shell biomass. *Bioresource Technology* 99, no. 1 (2008), 179–187.

[96] Santos, F., L. Alban, C. Frankenberg and M. Pires: Characterization and use of biosorbents prepared from forestry waste and their washed extracts to reduce/remove chromium. *International Journal of Environmental Science and Technology* 13 (2016), 327–338.

[97] Feisther, V. A., J. Scherer Filho, F. V. Hackbarth, D. A. Mayer, A. A. U. de Souza and S. M. G. U. de Souza: Raw leaves and leaf residues from the extraction of essential oils as biosorbents for metal removal. *Journal of Environmental Chemical Engineering* 7, no. 3 (2019), 103047.

[98] Phuengphai, Pongthipun; Singjanusong, Thapanee; Kheangkhun, Napaporn; Wattanakornsiri, Amnuay: Removal of copper(II) from aqueous solution using chemically modified peels as efficient low-cost biosorbents. *Water Science and Engineering* 14, no. 4 (2021), 286–294.

[99] Feitoza, Fernando Samuel; Gasparotto, Angelita Moutin Segoria: Study on the national production of concentrated orange juice. *Interface Technological Magazine* 17, no. 1 (2020), 625–634.

[100] Dey, S.; Basha, S.R.; Babu, G.V.; Nagendra, T.: Characteristic and biosorption capacities of orange peels biosorbents for removal of ammonia and nitrate from contaminated water. *Cleaner Materials* 1, (2021), 100001.

[101] Fabre, Elaine; Lopes, Claudia B.; Vale, Carlos; Pereira, Eduarda; Silva, Carlos M.: Evaluation of banana peels as an effective biosorbent for mercury removal under low environmental concentrations. *Science of the Total Environment* 709, (2020), 135883.

[102] Alves, Aline Medeiros; Fernades, Daniela Canuto; Sousa, Amanda Goulart de Oliveira; Naves, Ronaldo Veloso; Naves, Maria Margareth Veloso: Physical and nutritional characteristics of pequi fruits from Tocantins, Goiás and Minas Gerais. *Brazilian Journal of Food Technology* 17, no. 3 (2014), 198–203.

[103] Mo, Jiahao; Yang, Qi; Zhang, Na; Zhang, Wenxiang; Zheng, Yi; Zhang, Zhien: A review on agro-industrial waste (AIW) derived adsorbents for water and wastewater treatment. *Journal of Environmental Management* 227, (2018), 395–405.

[104] Morales, E., R. Domingos and D. Angelis: Improvement of protein bioavailability by solid-state fermentation of babassu mesocarp flour and cassava leaves. *Waste and Biomass Valorization* 9 (2018), 581–590.

[105] Leffler, Merrill: Treasure from Trash, is there a profit in crab waste? *Maryland Sea Grant Publication* (1997), p. 8.

[106] Goycoolea, F.M.; Argüelles-Monal, W.; Peniche, C.; Higuera-Ciapara, I.: Chitin and chitosan. *Developments in Food Science* 41, (2000), 265–308.

[107] Kurita, Keisuke: Controlled functionalization of the polysaccharide chitin. *Progress in Polymer Science* 26, no. 9 (2001), 1921–1971.

[108] Vilar, V. J., C. M. Botelho and R. A. Boaventura: Chromium and zinc uptake by algae Gelidium and agar extraction algal waste: Kinetics and equilibrium. *Journal of Hazardous Materials* 149, no. 3 (2007), 643–649.

[109] Morais, Sérgio Antônio Lemos de; Nascimento, Evandro Afonso do; Melo, Dárley Carrijo de: Analysis of wood from *Pinus oocarpa* part I: Study of macromolecular constituents and volatile extractives. *Tree Magazine* 29, no. 3 (2005), 461–470.

[110] dos Santos Jonas Juliermerson, S. and D. de Oliveira Jorge: Competitive biosorption of Cd (II), Pb (II) and Cr (III) using fungal biomass Pycnoporus sanguineus. *Journal of Environment and Biotechnology Research* 6 (2017), 123–127.

[111] Roane, T. M., I. L. Pepper and T. J. Gentry: *Microorganisms and Metal Pollutants. Environmental Microbiology,* Elsevier, 415–439 (2015).

[112] Vendruscolo, Francielo; da Rocha Ferreira, Glalber Luiz; Antoniosi Filho, Nelson Roberto: Biosorption of hexavalent chromium by microorganisms. *International Biodeterioration and Biodegradation* 119, (2017), 87–95.

[113] Sharma, Rohit; Talukdar, Daizee; Bhardwaj, Shefali; Jaglan, Sundeep; Kumar, Rajeev; Kumar, Raman; Akhtar, M. Shaheer; Beniwal, Vikas; Umar, Ahmad: Bioremediation potential of novel fungal species isolated from wastewater for the removal of lead from liquid medium. *Environmental Technology and Innovation* 18, (2020), 100757.

[114] Tortora, Gerard J.; Funke, Berdell R.; Case, Christine L. *Micro Biology.* 12 Aufl. Porto Alegre: Artmed (2017).

[115] Schlegel, H. G. and C. Zaborosch (1993). *General Microbiology,* Cambridge University Press.

[116] Martinović, T., U. Andjelković, M. Š. Gajdošik, D. Rešetar and D. Josić: Foodborne pathogens and their toxins. *Journal of Proteomics* 147 (2016), 226–235.

[117] Madigan, M.T. et al.: *Brock's Microbiology.* 12. Aufl. Burlington, MA: Jones & Bartlett (2010).

[118] Melo, I.S.; Melo, I.S.; Azevedo, J.L. (Hrsg.): *Environmental Microbiology.* 2. Aufl. Jaguariúna: Embrapa (2008).

[119] Pepper, I., C. P. Gerba, T. Gentry and R. M. Maier: *Environmental Microbiology,* Academic press (2011).

[120] Yadu, A., B. P. Sahariah and J. Anandkumar *Microbiological Wastewater Treatment. The Future of Effluent Treatment Plants.* Elsevier (2021), 165–182.

[121] Daims, Holger; Taylor, Michael W.; Wagner, Michael: Wastewater treatment: A model system for microbial ecology. *Trends in Biotechnology* 24, no. 11 (2006), 483–489.

[122] Bitton, G. *Wastewater Microbiology.* John Wiley & Sons (2011).

[123] Chai, Wai Siong; Cheun, Jie Ying; Kumar, P. Senthil; Mubashir, Muhammad; Majeed, Zahid; Banat, Fawzi; Ho, Shih Hsin; Show, Pau Loke: A review on conventional and novel materials towards heavy metal adsorption in wastewater treatment application. *Journal of Cleaner Production* 296, (2021), 126589.

[124] Agarwal, Ankita; Upadhyay, Utkarsh; Sreedhar, I.; Singh, Satyapaul A.; Patel, Chetan M.: A review on valorization of biomass in heavy metal removal from wastewater. *Journal of Water Process Engineering* 38, (2020), 101602.

[125] Vijayaraghavan, K.; Reddy, D. Harikishore Kumar; Yun, Yeoung Sang: Improving the quality of runoff from green roofs through synergistic biosorption and phytoremediation techniques: A review. *Sustainable Cities and Society* 46, (2019), 101381.

[126] Todhanakasem, T.; Ketbumrung, K.: Using potential lactic acid bacteria biofilms and their compounds to control biofilms of foodborne pathogens. *Biotechnology Reports* 26, (2020), e00477.

[127] Yin, Kun; Wang, Qiaoning; Lv, Min; Chen, Lingxin: Microorganism remediation strategies towards heavy metals. *Chemical Engineering Journal* 360, (2018), 1553–1563.

[128] Gupta, Pratima; Diwan, Batul: Bacterial exopolysaccharide mediated heavy metal removal: A Review on biosynthesis, mechanism and remediation strategies. *Biotechnology Reports* 13, (2017), 58–71.

[129] Dai, Qunwei; Zhang, Ting; Zhao, Yulian; Li, Qiongfang; Dong, Faqin; Jiang, Chunqi: Potentiality of living Bacillus pumilus SWU7-1 in biosorption of strontium radionuclide. *Chemosphere* 260, (2020), 127559.

[130] Afraz, Vahideh; Younesi, Habibollah; Bolandi, Marzieh; Hadiani, Mohammad Rasoul: Optimization of lead and cadmium biosorption by *Lactobacillus acidophilus* using response surface methodology. *Biocatalysis and Agricultural Biotechnology* 29, (2020), 101828.

[131] Khadivinia, Elnaz; Sharafi, Hakimeh; Hadi, Faranak; Zahiri, Hossein Shahbani; Modiri, Sima; Tohidi, Azadeh; Mousavi, Amir; Salmanian, Ali Hatef; Noghabi, Kambiz Akbari: Cadmium biosorption by a glyphosate-degrading bacterium, a novel biosorbent isolated from pesticide-contaminated agricultural soils. *Journal of Industrial and Engineering Chemistry* 20, no. 6 (2014), 4304–4310.

[132] Choińska-Pulit, Anna; Sobolczyk-Bednarek, Justyna; Łaba, Wojciech: Optimization of copper, lead and cadmium biosorption onto newly isolated bacterium using a Box-Behnken design. *Ecotoxicology and Environmental Safety* 149 (2018), 275–283.

[133] Muñoz, Antonio Jesús; Espinola, Francisco; Ruiz, Encarnación: Biosorption of Ag(I) from aqueous solutions by *Klebsiella* sp. 3S1. *Journal of Hazardous Materials* 329 (2017), 166–177.

[134] Pugazhendhi, Arivalagan; Boovaragamoorthy, Gowri Manogari; Ranganathan, Kuppusamy; Naushad, Mu; Kaliannan, Thamaraiselvi: New insight into effective biosorption of lead from aqueous solution using *Ralstonia solanacearum*: Characterization and mechanism studies. *Journal of Cleaner Production* 174, (2018), 1234–1239.

[135] Han, Yin Lung; Wu, Jen Hao; Cheng, Chieh Lun; Nagarajan, Dillirani; Lee, Ching Ray; Li, Yi Heng; Lo, Yung Chung; Chang, Jo Shu: Recovery of gold from industrial wastewater by extracellular proteins obtained from a thermophilic bacterium Tepidimonas fonticaldi AT-A2. *Bioresource Technology* 239, (2017), 160–170

[136] Sayyadi, Shayan; Ahmady-Asbchin, Salman; Kamali, Kasra; Tavakoli, Nadia: Thermodynamic, equilibrium and kinetic studies on biosorption of Pb^{+2} from aqueous solution by *Bacillus pumilus* sp. AS1 isolated from soil at abandoned lead mine. *Journal of the Taiwan Institute of Chemical Engineers* 80, (2017), 701–708.

[137] Zinicovscaia, Inga; Safonov, Alexey; Zelenina, Daria; Ershova, Yana; Boldyrev, Kirill: Evaluation of biosorption and bioaccumulation capacity of cyanobacterium *Arthrospira* (*spirulina*) platensis for radionuclides. *Algal Research* 51, (2020), 102075.

[138] Xu, Shaozu; Xing, Yonghui; Liu, Song; Hao, Xiuli; Chen, Wenli; Huang, Qiaoyun: Characterization of Cd^{2+} biosorption by Pseudomonas sp. strain 375, a novel biosorbent isolated from soil polluted with heavy metals in Southern China. *Chemosphere* 240, (2020), 124893.

[139] Xie, Yanluo; He, Nan; Wei, Mingyang; Wen, Tingyao; Wang, Xitong; Liu, Huakang; Zhong, Shiqiang; Xu, Heng: Cadmium biosorption and mechanism investigation using a novel *Bacillus subtilis* KC6 isolated from pyrite mine. *Journal of Cleaner Production* 312, (2021), 127749.

[140] Samimi, Mohsen; Shahriari-Moghadam, Mohsen: Isolation and identification of *Delftia lacustris* Strain-*MS3* as a novel and efficient adsorbent for lead biosorption: Kinetics and thermodynamic studies, optimization of operating variables. *Biochemical Engineering Journal* 173, (2021), 108091.

[141] Joo, Jin-ho; Hassan, Sedky HA; Oh, Sang-Eun: Comparative study of biosorption of Zn^{2+} by *Pseudomonas aeruginosa* and *Bacillus cereus*. *International Biodeterioration & Biodegradation* 64, no. 8 (2010), 734–741.

[142] Parmar, Paritosh; Shukla, Arpit; Goswami, Dweipayan; Patel, Baldev; Saraf, Meenu: Optimization of cadmium and lead biosorption onto marine *Vibrio alginolyticus* PBR1 employing a Box-Behnken design. *Chemical Engineering Journal Advances* 4, (2020), 100043.

[143] Tyagi, Bhawna; Gupta, Bulbul; Thakur, Indu Shekhar: Biosorption of Cr (VI) from aqueous solution by extracellular polymeric substances (EPS) produced by Parapedobacter sp. ISTM3 strain isolated from Mawsmai cave, Meghalaya, India. *Environmental Research* 191, (2020), 110064.

[144] Fawzy, Mustafa A.; Hifney, Awatief F.; Adam, Mahmoud S.; Al-Badaani, Arwa A.: Biosorption of cobalt and its effect on growth and metabolites of *Synechocystis pevalekii* and *Scenedesmus bernardii*: Isothermal analysis. *Environmental Technology and Innovation* 19, (2020), 100953.

[145] Singh, Sandeep Kumar; Keshri, Jitendra; Singh, Prem Pratap; Gupta, Akanksha; Singh, Amit Kishore: Tolerance of heavy metal toxicity using PGPR strains of *Pseudomonas species*. In Singh, A.K.; Kumar, A.; Singh, P.K. (Hrsg.): *PGPR Amelioration in Sustainable Agriculture*. 1. Aufl. Amsterdam: Elsevier Inc. (2018), pp. 239–252.

[146] Palanivel, Thenmozhi Murugaian; Sivakumar, Nallusamy; Al-Ansari, Aliya; Victor, Reginald: Bioremediation of copper by active cells of *Pseudomonas stutzeri* LA3 isolated from an abandoned copper mine soil. *Journal of Environmental Management* 253, (2020), 109706.

[147] Madakka, Mekapogu; Jayaraju, Nadimikeri; Rajesh, Nambi; Subhosh Chandra, Muni Ramanna Gari: *Development in the Treatment of Municipal and Industrial Wastewater by Microorganism*. Amsterdam: Elsevier Inc. (2018).

[148] Sugiharto, Sugiharto: A review of filamentous fungi in broiler production. *Annals of Agricultural Sciences* 64, no. 1 (2019), 1–8.

[149] Nakatani, Masaki; Saitoh, Norizoh; Nomura, Toshiyuki; Konishi, Yasuhiro; Fujimori, Ryotaro; Yoshihara, Daijiro: Microbial recovery of gold from neutral and acidic solutions by the baker's yeast Saccharomyces cerevisiae. *Hydrometallurgy* 181, (2018), 29–34.

[150] Yin, Kun; Wang, Qiaoning; Lv, Min; Chen, Lingxin: Microorganism remediation strategies towards heavy metals. *Chemical Engineering Journal* 360, (2019), 1553–1563.

[151] Nascimento, Jéssica Mesquita; Cruz, Nildo Duarte; de Oliveira, Gabriel Rodrigues; Sá, Waldeemeson Silva; de Oliveira, Jorge Diniz; Ribeiro, Paulo Roberto S.; Leite, Selma G.F.: Evaluation of the kinetics of gold biosorption processes and consequent biogenic synthesis of AuNPs mediated by the fungus *Trichoderma harzianum*. *Environmental Technology and Innovation* 21, (2021), 101238.

[152] Coelho, Ednei; Reis, Tatiana Alves; Cotrim, Marycel; Mullan, Thomas K.; Corrêa, Benedito: Resistant fungi isolated from contaminated uranium mine in Brazil shows a high capacity to uptake uranium from water. *Chemosphere* 248 (2020), 126068.

[153] Shoaib, A.; Aslan, N.; Aslam, N.: *Trichoderma harzianum*: Adsorption, desorption, isotherm and FTIR studies. *Journal of Animal and Plant Sciences* 23, no. 5 (2013), 1460–1465.

[154] Ignatova, Lyudmila; Kistaubayeva, Aida; Brazhnikova, Yelena; Omirbekova, Anel; Mukasheva, Togzhan; Savitskaya, Irina; Karpenyuk, Tatyana; Goncharova, Alla; Egamberdieva, Dilfuza; Sokolov, Alexander: Characterization of cadmium-tolerant endophytic fungi isolated from soybean (*Glycine max*) and barley (*Hordeum vulgare*). *Heliyon* 7, no. 11 (2021), e08240.

[155] Nascimento, Jessica M.; Oliveira, J.D.: Biosorption of potentially toxic metals (Cd^{2+}, Pb^{2+} and Cr^{3+}) in dry biomass of *Pycnoporus sanguineus*. *Ecletica Quimica* 39, no. 1 (2014), 151–163.

[156] Buratto, Ana Paula; Costa, Raquel Dalla; Ferreira, Edilson da Silva: Application of fungal biomass of Pleurotus ostreatus in the process of biosorption of copper ions (II). *Engenharia Sanitaria e Ambiental* 17 (2012), 413–420.

[157] Dang, Chenyuan; Yang, Zhenxing; Liu, Wen; Du, Penghui; Cui, Feng; He, Kai: Role of extracellular polymeric substances in biosorption of Pb^{2+} by a high metal ion tolerant fungal strain *Aspergillus niger* PTN31. *Journal of Environmental Chemical Engineering* 6, no. 2 (2018), 2733–2742.

[158] Yahaya, Yus Azila; Mat Don, Mashitah; Bhatia, Subhash: Biosorption of copper (II) onto immobilized cells of *Pycnoporus sanguineus* from aqueous solution: Equilibrium and kinetic studies. *Journal of Hazardous Materials* 161, no. 1 (2009), 189–195.

[159] El Sayed, Manal T.; El-Sayed, Ashraf S.A.: Bioremediation and tolerance of zinc ions using *Fusarium solani*. *Heliyon* 6, no. 9 (2020), e05048.

[160] do Nascimento, Jéssica Mesquita; Cruz, Nildo Duarte; de Oliveira, Gabriel Rodrigues; Sá, Waldeemeson Silva; de Oliveira, Jorge Diniz; Ribeiro, Paulo Roberto S.; Leite, Selma G.F.: Evaluation of the kinetics of gold biosorption processes and consequent biogenic synthesis of AuNPs mediated by the fungus *Trichoderma harzianum*. *Environmental Technology and Innovation* 21, (2021), 101238.

[161] Rusinova-Videva, S.; Nachkova, S.; Adamov, A.; Dimitrova-Dyulgerova, I.: Antarctic yeast *Cryptococcus laurentii* (AL65): Biomass and exopolysaccharide production and biosorption of metals. *Journal of Chemical Technology and Biotechnology* 95, no. 5 (2019), 1372–1379.

[162] Chen, X.; Tian, Z.; Cheng, H.; Xu, G.; Zhou, H.: Adsorption process and mechanism of heavy metalions by different components of cells, using yeast (*Pichia pastoris*) and Cu^{2+} as biosorption models. *RSC Advances* 11 (2021), 17080.

[163] Campaña-Perez, Juan Fernando; Barahona, Patricia Portero; Martín-ramos, Pablo; Javier, Enrique; Belly, Carvajal; Campaña-pérez, Juan Fernando: Ecuadorian yeast species as microbial particles for Cr (VI) biosorption. *Environmental Science and Pollution Research* 26, (2019), 28162–28172.

[164] Ferreira, Joelma Morais; Luiz, Flavio; Leonor, Odelsia; Alsina, Sanchez; Conrado, Libya De Sousa; Bezerra, Eliane; Costa, Wolia: Study of balance and kinetics of Pb^{2+} biosorption by *Saccharomyces cerevisiae*. *Química Nova* 30, no. 5 (2007), 1188–1193.

[165] Peng, Qingqing; Liu, Yunguo; Zeng, Guangming; Xu, Weihua; Yang, Chunping; Zhang, Jingjin: Biosorption of copper(II) by immobilizing *Saccharomyces cerevisiae* on the surface of chitosan-coated magnetic nanoparticles from aqueous solution. *Journal of Hazardous Materials* 177, no. 1–3 (2010) 676–682.

[166] Amirnia, Shahram; Ray, Madhumita B.; Margaritis, Argyrios: Heavy metals removal from aqueous solutions using *Saccharomyces cerevisiae* in a novel continuous bioreactor-biosorption system. *Chemical Engineering Journal* 264, (2015), 863–872.

[167] Fadel, M.; Hassanein, Naziha M.; Elshafei, Maha M.; Mostafa, Amr H.; Ahmed, Marwy A.; Khater, Hend M.: Biosorption of manganese from groundwater by biomass of *Saccharomyces cerevisiae*. *HBRC Journal* 13, no. 1 (2017), 106–113.

[168] Anagnostopoulos, Vasileios A; Vlachou, Athina; Symeopoulos, Basil D: Immobilization of Saccharomyces cerevisiae on low-cost lignocellulosic substrate for the removal of Cd^{2+} from aquatic systems. Journal of Environment and Biotechnology Research 1, no. 1 (2015), 23–29.

[169] Hadiani, Mohammad Rasoul; Darani, Kianoush Khosravi; Rahimifard, Nahid; Younesi, Habibollah: Biosorption of low concentration levels of Lead (II) and Cadmium (II) from aqueous solution by Saccharomyces cerevisiae: Response surface methodology. *Biocatalysis and Agricultural Biotechnology* 15 (2018), 25–34.

[170] Massoud, Ramona; Hadiani, Mohammad Rasoul; Hamzehlou, Pegah; Khosravi-Darani, Kianoush: Bioremediation of heavy metals in food industry: Application of *Saccharomyces cerevisiae*. *Electronic Journal of Biotechnology* 37, (2019), 56–60.

[171] Saitoh, Norizoh; Fujimori, Ryotaro; Yoshimura, Tomohide; Tanaka, Hiroshi; Kondoh, Asaki; Nomura, Toshiyuki; Konishi, Yasuhiro: Microbial recovery of palladium by baker's yeast through bioreductive deposition and biosorption. *Hydrometallurgy* 196, (2020), 105413.

[172] Wang, Jianlong; Chen, Can: Biosorption of heavy metals by *Saccharomyces cerevisiae*: A review. *Biotechnology Advances* 24, no. 5 (2006), 427–451.

[173] Anastopoulos, Ioannis; Kyzas, George Z.: Progress in batch biosorption of heavy metals onto algae. *Journal of Molecular Liquids* 209, (2015), 77–86.

[174] Mustafa, Shazia; Bhatti, Haq Nawaz; Maqbool, Munazza; Iqbal, Munawar: Microalgae biosorption, bioaccumulation and biodegradation efficiency for the remediation of wastewater and carbon dioxide mitigation: Prospects, challenges and opportunities. *Journal of Water Process Engineering* 41, (2021), 102009.

[175] Melnikova, Anna; Namsaraev, Zorigto; Komova, Anastasia; Meuser, Isabel; Roeb, Marion; Ackermann, Barbara; Klose, Holger; Kuchendorf, Christina M.: AlgalTextile - a new biohybrid material for wastewater treatment. *Biotechnology Reports* 33, (2022), e00698.

[176] Soto-Ramirez, Robinson; Lobos, Maria Gabriela; Cordoba, Olivia; Poirrier, Paola; Chamy, Rolando: Effect of growth conditions on cell wall composition and cadmium adsorption in *Chlorella vulgaris*: A new approach to biosorption research. *Journal of Hazardous Materials* 411 (2021), 125059.

[177] Emami Moghaddam, Seyed Amirebrahim; Harun, Razif; Mokhtar, Mohd Noriznan; Zakaria, Rabitah: Kinetic and equilibrium modeling for the biosorption of metal ion by Zeolite 13X-Algal-Alginate Beads (ZABs). *Journal of Water Process Engineering* 33, (2020), 101057.

[178] Matei, G.M.; Kiptoo, J.K.; Oyaro, N.; Onditi, A.O.: Biosorption of selected heavy metals by green algae, spirogyra species and its potential as a pollution biomonitor. *Chemistry and Materials Research* 7, no. 7 (2015), 42–53.

[179] Tran, H.T.; Vu, N.D.; Masahito, M.; Maiko, O.; Tatsuo, K.; Kaori, O.; Shinya, Y.: Heavy metal biosorption from aqueous solutions by algae inhabiting rice paddies in Vietnam. *Journal of Environmental Chemical Engineering* 4, no. 2 (2016), 2529–2535.

[180] Jiang, Xincheng; Zhou, Xudong; Li, Caiyun; Wan, Zhenjia; Yao, Lunguang; Gao, Pengcheng: Adsorption of copper by flocculated *Chlamydomonas microsphaera* microalgae and polyaluminium chloride in heavy metal-contaminated water. *Journal of Applied Phycology* 31, (2019), 1143–1151.

[181] Shen, Li; Wang, Junjun; Li, Zhanfei; Fan, Ling; Chen, Ran; Wu, Xueling; Li, Jiaokun; Zeng, Weimin: A high-efficiency Fe$_2$O$_3$@Microalgae composite for heavy metal removal from aqueous solution. *Journal of Water Process Engineering* 33, (2020), 101026.

[182] Politaeva, N.A.; Tatarintseva, E.A.: Using adsorption material based on the residual biomass of *Chlorella sorokiniana* microalgae for wastewater purification to remove heavy metal ions. *Chemical and Petroleum Engineering* 55, no. 11 (2020), 907–912.

[183] Pradhan, Debabrata; Sukla, Lala Behari; Mishra, Bibhuti Bhusan; Devi, Niharbala: Biosorption for removal of hexavalent chromium using microalgae *Scenedesmus* sp. *Journal of Cleaner Production* 209, no. 1 (2019), 617–629.

[184] Sbihi, Karim; Cherifi, Ouafa; Bertrand, Martine; El Gharmali, Abdelhay: Biosorption of metals (Cd, Cu and Zn) by the freshwater diatom *Planothidium lanceolatum*: A laboratory study. *Diatom Research* 29, no. 1 (2014), 55–63.

[185] Furuhashi, Yasuhiro; Honda, Ryo; Noguchi, Mana; Hara-yamamura, Hiroe; Kobayashi, Shoko; Higashimine, Koichi; Hasegawa, Hiroshi: Optimum conditions of pH, temperature and preculture for biosorption of europium by microalgae *Acutodesmus acuminatus*. *Biochemical Engineering Journal* 143, no. 15 (2019), 58–64.

[186] Areco, M.M.; Haug, E.; Curutchet, G.: Studies on bioremediation of Zn and acid waters using *Botryococcus braunii*. *Journal of Environmental Chemical Engineering* 6, no. 4 (2018), 3849–3859.

[187] Suganya, S; Saravanan, A; Senthil Kumar, P.; Yashwanthraj, M.; Sundar Rajan, P.; Kayalvizhi, K: Sequestration of Pb(II) and Ni(II) ions from aqueous solution using microalgae *Rhizoclonium hookeri*: Adsorption thermodynamics, kinetics, and equilibrium studies. *Journal of Water Reuse and Desalination* 7, no. 2 (2017), 214–227.

[188] Jiang, Xiaoyu; Wang, Hongqiang; Hu, Eming; Lei, Zhiwu; Fan, Beilei; Wang, Qingliang: Efficient adsorption of uranium from aqueous solutions by microalgae based airgel. *Microporous and Mesoporous Materials* 305, (2020), 110383.

[189] Gupta, Vandana; Jaybhaye, Sandesh; Chandra, Naresh: Cauliflower leaves, an agro waste: Characterization and its application for the biosorption of copper, chromium, lead and zinc from aqueous solutions. *International Journal of Scientific Research in Science, Engineering and Technology* 5, no. 4 (2018), 169–174.

[190] Karić, Nataša; Maia, Alexandra S.; Teodorović, Ana; Atanasova, Nataša; Langergraber, Guenter; Crini, Gregorio; Ribeiro, Ana R.L.; Đolić, Maja: Bio-waste valorisation: Agricultural wastes as biosorbents for removal of (in)organic pollutants in wastewater treatment. *Chemical Engineering Journal Advances* 9 (2021), 100239.

[191] Adegoke, KA; Ololade, Oreoluwa; Okon-akan, Omolabake Abiodun; Rhoda, Oyeladun; Biodun, Abdullahi; Aderemi, Oluwaseyi; Olagoke, Halimat; Wendy, Nobanathi; Solomon Bell, Olugbenga: Sawdust-biomass based materials for sequestration of organic and inorganic pollutants and potential for engineering applications. *Current Research in Green and Sustainable Chemistry* 5, (2022), 100274.

[192] do Nascimento, J.M.; dos Santos, J.J.S.; de Oliveira, J.D.: Use of wood sawdust Teak (*Tectona grandis*) modified with citric acid in the biosorption of Cd (II) and Pb (II). *Ambiência* 12, no. 4 (2016), 955–968.

[193] Nascimento, J.M.; Santos, J.J.S.; Nascimento, R.M.; Oliveira, J.D.: Comparison of the biossorption potential of highly toxic metals Cd(II) and Pb(II) in natura biomasses from the peel of pequi (*Caryocar brasiliense* Camb), sawdust Teca wood (*Tectona grandis*) and wooden ear (*Pycnoporus sanguineus*). *Journal of Industrial Chemistry* 748 (2015), 25.

[194] Yuliusman; Farouq, Fadel Al; Sipangkar, Samson Patar; Fatkhurrahman, Mufiid; Putri, Salma Amaliani: Preparation and characterization of activated carbon from corn stalks by chemical activation with KOH and NaOH. In *AIP Conference Proceedings,* Vol. 2255 (2020).

[195] Yorgun, Sait; Yildiz, Derya: Preparation and characterization of activated carbons from Paulownia wood by chemical activation with H_3PO_4. *Journal of the Taiwan Institute of Chemical Engineers* 53, (2015), 122–131.

[196] Nascimento, J.M.; Santos, J.J.S.; Oliveira, J.D.: Use of wood sawdust Teak (Tectona grandis) modified with citric acid in the biosorption of Cd (II) and Pb (II). *Ambiência (UNICENTRO)* 12, no. 4 (2016), 955–968.

[197] Sinyeue, Cynthia; Garioud, Theophile; Lemestre, Monika; Meyer, Michael; Brégier, Frédérique; Chaleix, Vincent; Sol, Vincent; Lebouvier, Nicolas: Biosorption of nickel ions Ni^{2+} by natural and modified *Pinus caribaea* Morelet sawdust. *Heliyon* 8 (2022), e08842.

[198] Shah, Aamir Ishaq; Din Dar, Mehraj U.; Bhat, Rouf Ahmad; Singh, JP; Singh, Kuldip; Bhat, Shakeel Ahmad: Prospects and challenges of wastewater treatment technologies to combat contaminants of emerging concerns. *Ecological Engineering* 152, (2020), 105882.

[199] Cao, Yaoyao; Xiao, Weihua; Shen, Guanghui; Ji, Guanya; Zhang, Yang; Gao, Chongfeng; Han, Lujia: Carbonization and ball milling on the enhancement of Pb(II) adsorption by wheat straw: Competitive effects of ion exchange and precipitation. *Bioresource Technology* 273, (2019), 70–76.

[200] Fontana, Klaiani B.; Chaves, Eduardo S.; Sanchez, Jefferson DS; Watanabe, Erica RLR; Pietrobelli, Juliana MTA; Lenzi, Giane G.: Biossorption of Pb(II) by urucum shells (*Bixa orellana*) in aqueous solutions: Kinetic, equilibrium and thermodynamic study. *Química Nova* 39, no. 9 (2016) 1078–1084.

[201] Li, Yunchao; Xing, Bo; Ding, Yan; Han, Xinhong; Wang, Shurong: A critical review of the production and advanced utilization of biochar via selective pyrolysis of lignocellulosic biomass. *Bioresource Technology* 312, (2020), 123614.

[202] Wang, Duo; Jiang, Peikun; Zhang, Haibo; Yuan, Wenqiao: Biochar production and applications in agro and forestry systems: A review. *Science of the Total Environment* 723, (2020), 137775.

[203] Hammo, Mahmoud M.; Akar, Tamer; Sayin, Fatih; Celik, Sema; Akar, Sibel Tunali: Efficacy of green waste-derived biochar for lead removal from aqueous systems: Characterization, equilibrium, kinetic and application. *Journal of Environmental Management* 289, (2021), 112490.

[204] Ahmad, Iftikhar; Farwa, Umme; Khan, Zia Ul Haq; Imran, Muhammad; Khalid, Muhammad Shafique; Zhu, Bo; Rasool, Atta; Shah, Ghulam Mustafa; et al.: Biosorption and risk assessment of arsenic contaminated water through cotton stalk biochar. *Surfaces and Interfaces* 29, (2022), 101806.

[205] Huang, Fei; Li, Kai; Wu, Ren Ren; Yan, Yu Jian; Xiao, Rong Bo: Insight into the Cd^{2+} biosorption by viable *Bacillus cereus* RC-1 immobilized on different biochars: Roles of bacterial cell and biochar matrix. *Journal of Cleaner Production* 272, (2020), 122743.

[206] Ren, Binqiao; Jin, Yu; Zhao, Luyang; Cui, Chongwei; Song, Xiaoxiao: Enhanced Cr(VI) adsorption using chemically modified dormant *Aspergillus niger* spores: Process and mechanisms. *Journal of Environmental Chemical Engineering* 10, no. 1 (2022), 106955.

[207] dos Santos Jonas Juliermerson, S.; de Oliveira Jorge, D.: Competitive biosorption of Cd(II), Pb(II) and Cr(III) using fungal biomass *Pycnoporus sanguineus. Journal of Environment and Biotechnology Research* 6, no. 1 (2017), 123–127.

[208] Baltazar, Marcela dos Passos Galluzzi; Gracious, Louise Hase; Avanzi, Ingrid Regina; Karolski, Bruno; Tenório, Jorge Alberto Soares; do Nascimento, Claudio Augusto Oller; Perpetuo, Elen Aquino: Copper biosorption by *Rhodococcus erythropolis* isolated from the Sossego Mine–PA–Brazil. *Journal of Materials Research and Technology* 8, no. 1 (2019), 475–483.

[209] Zhao, Changsong; Li, Xiaolong; Ding, Congcong; Liao, Jiali; Du, Liang; Yang, Jijun; Yang, Yuanyou; Zhang, Dong; u. The.: Characterization of uranium bioaccumulation on a fungal isolate *Geotrichum* sp. dwc-1 as investigated by FTIR, TEM and XPS. *Journal of Radioanalytical and Nuclear Chemistry* 310, (2016), 165–175.

[210] Velez, Jessica M.Bedoya; Martínez, José Gregorio; Ospina, Juliana Tobon; Agudelo, Susana Ochoa: Bioremediation potential of *Pseudomonas genus* isolates from residual water, capable of tolerating lead through mechanisms of exopolysaccharide production and biosorption. *Biotechnology Reports* 32 (2021), e00685.

[211] Chojnacka, Katarzyna: Biosorption and bioaccumulation – the prospects for practical applications. *Environment International* 36, no. 3 (2010), 299–307.

[212] Talukdar, Daizee; Jasrotia, Teenu; Sharma, Rohit; Jaglan, Sundeep; Kumar, Rajeev; Vats, Rajeev; Kumar, Raman; Mahnashi, Mater H.; Umar, Ahmad: Evaluation of novel indigenous fungal consortium for enhanced bioremediation of heavy metals from contaminated sites. *Environmental Technology and Innovation* 20, (2020), 101050.

[213] Zhang, Ding; Yin, Caiping; Abbas, Naeem; Mao, Zhenchuan; Zhang, Yinglao: Multiple heavy metal tolerance and removal by an earthworm gut fungus *Trichoderma brevicompactum* QYCD-6. *Scientific Reports* 10, no. 1, (2020), 6940.

[214] Parsania, Somayeh; Mohammadi, Parisa; Soudi, Mohammad Reza: Biotransformation and removal of arsenic oxyanions by *Alishewanella agri* PMS5 in biofilm and planktonic states. *Chemosphere* 284, (2021), 131336.

[215] Chandrashekharaiah, PS; Sanyal, Debanjan; Dasgupta, Santanu; Banik, Avishek: Cadmium biosorption and biomass production by two freshwater microalgae *Scenedesmus acutus* and *Chlorella pyrenoidosa*: An integrated approach. *Chemosphere* 269, (2021), 128755.

[216] Nascimento, Jéssica Mesquita Do; de Oliveira, Jorge Diniz; Leite, Selma Gomes Ferreira: Comparison and preliminary evaluation of biogenic synthesis of gold nanoparticles mediated by *Pseudomonas aeruginosa* and *Geotrichum candidum*. *Journal of Bionanoscience* 12, no. 6 (2018), 803–808.

[217] Rubilar, Olga; Rai, Mahendra; Tortella, Gonzalo; Diez, Maria Cristina; Seabra, Amedea B.; Durán, Nelson: Biogenic nanoparticles: Copper, copper oxides, copper sulphides, complex copper nanostructures and their applications. *Biotechnology Letters* 35, no. 9 (2013), 1365–1375.

[218] Rai, Mahendra: Green nanobiotechnology: Biosynthesis of metallic nanoparticles and their applications as nanoantimicrobials. *Science and Culture* 65, no. 3 (2013), 44–48.

[219] Capek, I.: *Nanocomposite Structures and Dispersions: Science and Nanotechnology – Fundamental Principles and Colloidal Particles*. 1. Aufl. Amsterdam: Elsevier Science (2006).

[220] Kaur, K; Thombre, Rebecca: Nanobiotechnology: Methods, applications, and future prospects. In: Sougata Ghosh and Thomas J. Webster. Nanobiotechnology: Microbes and Plant Assisted Synthesis of Nanoparticles, Mechanisms and Applications. Amsterdam, Elsevier, (2021), pp. 1–20.

[221] Tripathi, R.M.; Chung, Sang J.: Biogenic nanomaterials: Synthesis, characterization, growth mechanism, and biomedical applications. *Journal of Microbiological Methods* 157, (2019), 65–80.

[222] Lalau, Cristina Moreira: *Evaluation of the Toxicity of Copper Oxide Nanoparticles through Macrophytes of the Species Landoltia punctata*, Federal University of Santa Catarina, Florianópolis, (2014).

[223] Iqbal, Javed; Ahsan, Banzeer; Ahmad, Riaz; Shahbaz, Amir; Anber, Syeda; Kanwal, Sobia; Munir, Akhtar; Rabbani, Atiya; Mahmood, Tariq: Biogenic synthesis of green and cost effective iron nanoparticles and evaluation of their potential biomedical properties. *Journal of Molecular Structure* 1199, (2020), 126979.

[224] Ali, Imran; Peng, Changsheng; Khan, Zahid M.; Naz, Iffat; Sultan, Muhammad; Ali, Mohsin; Abbasi, Irfan A.; Islam, Tariqul; Ye, Tong: Overview of microbes based fabricated biogenic nanoparticles for water and wastewater treatment. *Journal of Environmental Management* 230, (2019), 128–150.

[225] Ajitha, B.; Reddy, Y. Ashok Kumar; Reddy, P. Sreedhara; Suneetha, Y.; Jeon, Hwan Jin; Ahn, Chi Won: Instant biosynthesis of silver nanoparticles using *Lawsonia inermis* leaf extract: Innate catalytic, antimicrobial and antioxidant activities. *Journal of Molecular Liquids* 219, (2016), 474–481.

[226] Chandran, C.I.K.; Indira, G.: Green synthesis, characterization and antimicrobial activity of silver nanoparticles using *Morinda pubscens* J.E. Smith root extract. *Journal of Scientific and Innovative Research* 5, no. 3 (2016), 83–86.

[227] Vijayaraghavan, K.; Ashokkumar, T.: Plant-mediated biosynthesis of metallic nanoparticles: A review of literature, factors affecting synthesis, characterization techniques and applications. *Journal of Environmental Chemical Engineering* 5, no. 5 (2017), 4866–4883.

[228] Bhargava, Arpit; Jain, Navin; Khan, Mohd Azeem; Pareek, Vikram; Dilip, R. Venkataramana; Panwar, Jitendra: Utilizing metal tolerance potential of soil fungus for efficient synthesis of gold nanoparticles with superior catalytic activity for degradation of rhodamine B. *Journal of Environmental Management* 183, (2016), 22–32.

[229] Durán, Nelson; Marcato, Priscyla D.; Durán, Marcela; Yadav, Alka; Gade, Aniket; Rai, Mahendra: Mechanistic aspects in the biogenic synthesis of extracellular metal nanoparticles by peptides, bacteria, fungi, and plants. *Applied Microbiology and Biotechnology* 90, no. 5 (2011), 1609–1624.

[230] Gautam, Pavan Kumar; Singh, Anirudh; Misra, Krishna; Sahoo, Amaresh Kumar; Samanta, Sintu Kumar: Synthesis and applications of biogenic nanomaterials in drinking and wastewater treatment. *Journal of Environmental Management* 231, (2019), 734–748.

[231] Menon, Soumya; Rajeshkumar, S.; Venkat Kumar, S.: Resource-efficient technologies a review on biogenic synthesis of gold nanoparticles, characterization, and its applications. *Resource-Efficient Technologies* 3, no. 4 (2017), 516–527.

[232] Lin, Liqin; Wu, Weiwei; Huang, Jiale; Sun, Daohua; Waithera, Monica: Catalytic gold nanoparticles immobilized on yeast: From biosorption to bioreduction. *Chemical Engineering Journal* 225, (2013), 857–864.

[233] Cai, Fang; Li, Jing; Sun, Jinsheng; Ji, Yulan: Biosynthesis of gold nanoparticles by biosorption using *Magnetospirillum gryphiswaldense* MSR-1. *Chemical Engineering Journal* 175, no. 1 (2011), 70–75.

[234] Vijayaraghavan, K; Mahadevan, A; Sathishkumar, M; Pavagadhi, S; Balasubramanian, R: Biosynthesis of Au (0) from Au (III) via biosorption and bioreduction using brown marine algae *Turbinaria conoides*. *Chemical Engineering Journal* 167, no. 1 (2011), 223–227.

[235] Zikalala, Nkosingipile; Matshetshe, Kabo; Parani, Sundararajan; Oluwafemi, Oluwatobi S: Nano-structures & nano-objects biosynthesis protocols for colloidal metal oxide nanoparticles. *Nano-Structures & Nano-Objects* 16, (2018), 288–299.

[236] Das, Sujoy K.; Marsili, Enrico: A green chemical approach for the synthesis of gold nanoparticles: Characterization and mechanistic aspect. *Reviews in Environmental Science and Biotechnology* 9, no. 3 (2010), 199–204.

[237] Srinath, B.S.; Rai, V.R.: Rapid biosynthesis of gold nanoparticles by *Staphylococcus epidermidis*: Its characterization and catalytic activity. *Materials Letters* 146, (2015), 23–25.

[238] Wong-Pinto, Liey Si; Market, Ana; Chong, Guillermo; Salazar, Pablo; Ordóñez, Javier I.: Biosynthesis of copper nanoparticles from copper tailings ore – An approach to the 'Bionanomining'. *Journal of Cleaner Production* 315, (2021), 128107.

[239] Elbahnasawy, Mostafa A.; Shehabeldine, Amr M.; Khattab, Abdelrahman M.; Amin, Basma H.; Hashem, Amr H.: Green biosynthesis of silver nanoparticles using novel endophytic *Rothia endophytica*: Characterization and anticandidal activity. *Journal of Drug Delivery Science and Technology* 62, (2021), 102401.

[240] Asif, Nida; Fatima, Samreen; Aziz, Md Nafe; Shehzadi; Zaki, Almaz; Fatma, Tasneem: Biofabrication and characterization of cyanobacterium derived ZnO NPs for their bioactivity comparison with commercial chemically synthesized nanoparticles. *Bioorganic Chemistry* 113, (2021), 104999.

[241] Gade, A.K.; Bonde, P.; Ingle, A.P.; Marcato, P.D.; Durán, N.; Rai, M.K.: Exploitation of *Aspergillus niger* for synthesis of silver nanoparticles. *Journal of Biobased Materials and Bioenergy* 2, no. 3 (2008), 243–247.

[242] Chandrappa, C.P.; Govindappa, M.; Chandrasekar, N.; Sarkar, S.; Ooha, S.; Channabasava, R.; Ramesh, C.P.: Mycosynthesis of silver nanoparticles by alternaria sp isolated from bark part of *Calophyllum apetalum*. *Indo American Journal of Pharmaceutical Research* 6, (2016), 6485–6489.

[243] Cuevas, R.; Durán, N.; Diez, M.C.; Tortella, G.R.; Rubilar, O.: Extracellular biosynthesis of copper and copper oxide nanoparticles by *Stereum hirsutum*, a native white-rot fungus from Chilean forests. *Journal of Nanomaterials* 2015, (2015), 789089.

[244] Husseiny, M.I.; El-Aziz, M. Abd; Badr, Y.; Mahmoud, M.A.: Biosynthesis of gold nanoparticles using *Pseudomonas aeruginosa*. *Spectrochimica Acta - Part A: Molecular and Biomolecular Spectroscopy* 67, no. 3–4 (2007), 1003–1006.

[245] Feng, Youzhi; Yu, Yongchang; Wang, Yiming: Biosorption and bioreduction of trivalent aurum by photosynthetic bacteria *Rhodobacter capsulatus*. *Current Microbiology*, 55 (2007), 402–408.

[246] Kalishwaralal, Kalimuthu; Deepak, Venkataraman; Ram, Sureshbabu; Pandian, Kumar; Gurunathan, Sangiliyandi: Biological synthesis of gold nanocubes from *Bacillus licheniformis*. *Bioresource Technology* 100, no. 21 (2009), 5356–5358.

[247] Alam, Nurul; Sarkar, Manas; Chowdhury, Trinath; Ghosh, Dipak; Chattopadhyay, Brajadulal: Characterization of a novel MDH1 bacterium from a virgin hot spring applicable for gold nanoparticle (GNPs) synthesis. *Advances in Microbiology* 6 (2016), 724–732.

[248] Tomar, Rajesh Singh; Banerjee, Sharmistha; Kaushik, Shuchi; Singh, Sneha; Vidyarthi; A.S.: Microbial synthesis of gold nanoparticles by biosurfactant producing Pseudomonas aeruginosa. *International Journal of Advancement in Life Sciences Research* 8, no. 4 (2015), 520–527.

[249] Apte, Mugdha; Chaudhari, Prerana; Vaidya, Amogh; Kumar, Ameeta Ravi; Zinjarde, Smita: Aspects Application of nanoparticles derived from marine *Staphylococcus lentus* in sensing dichlorvos and mercury ions. *Colloids and Surfaces A: Physicochemical and Engineering Aspects* 501, (2016), 1–8.

[250] Busi, Siddhardha; Rajkumari, Jobina; Pattnaik, Subhaswaraj; Parasuraman, P.; Hnamte, Sairengpuii: Extracellular synthesis of zinc oxide nanoparticles using acinetobacter schindleri SIZ7 and its antimicrobial property against foodborne pathogens. *The Journal of Microbiology, Biotechnology and Food Sciences* 5, no. 5 (2016), 407–411.

[251] Mukherjee, P.; Ahmad, A; Mandal, D.; Senapati, S.; Sainka, SR; Khan, MI; Ramani, R.; Parischa, R.; et al.: Bioreduction of AuCl$_4$- ions by the fungus *Verticillum* species and surface trapping of gold nanoparticles formed. *Angewandte Chemie International Edition* 40, no. 19 (2001), 3585–3588.

[252] Ahmad, Tokeer; Wani, Irshad A.; Manzoor, Nikhat; Ahmed, Jahangeer; Asiri, Abdullah M.: Biosynthesis, structural characterization and antimicrobial activity of gold and silver nanoparticles. *Colloids and Surfaces B: Biointerfaces* 107, (2013), 227–234.

[253] Basavaraja, S.; Balaji, S.D.; Lagashetty, Arunkumar; Rajasab, A.H.; Venkataraman, A.: Extracellular biosynthesis of silver nanoparticles using the fungus *Fusarium semitectum*. *Materials Research Bulletin* 43, no. 5 (2008), 1164–1170.

[254] Mukherjee, P.K.; Nautiyal, C.S.; Mukhopadhyay, A.N.: Molecular mechanisms of biocontrol by *Trichoderma* spp. *Soil Biology* 15 (2008), 243–244.

[255] Bhainsa, Kuber C.; Souza, S.F.D.: Extracellular biosynthesis of silver nanoparticles using the fungus *Aspergillus fumigatus*. *Colloids and Surfaces B: Biointerfaces* 47 (2006), 160–164.

[256] Castro-Longoria, E.; Vilchis-Nestor, Alfredo R.; Avalos-Borja, M.: Biosynthesis of silver, gold and bimetallic nanoparticles using the filamentous fungus *Neurospora crassa*. *Colloids and Surfaces B: Biointerfaces* 83, no. 1 (2011), 42–48.

[257] Wang, Lei; Liu, Chong-Chong; Wang, Yi-yan; Xu, Hui; Su, Hongyan: Antibacterial activities of the novel silver nanoparticles biosynthesized using *Cordyceps militaris* extract. *Current Applied Physics* 16, no. 9 (2016), 969–973.

[258] Salvadori, Marcia R.; Ando, Rômulo A.; Oller Do Nascimento, Cláudio A.; Corrêa, Benedito: Intracellular biosynthesis and removal of copper nanoparticles by dead biomass of yeast isolated from the wastewater of a mine in the Brazilian Amazon. *PLoS One* 9, no. 1 (2014), e87968.

[259] Quester, K.; Avalos-borja, M.; Castro-longoria, E.: Controllable biosynthesis of small silver nanoparticles using fungal extract. *Journal of Biomaterials and Nanobiotechnology* 7 (2016), 118–125.

[260] Fernandez, Jorge G.; Almeida, César A.; Fernández-Baldo, Martín A.; Felici, Emiliano; Raba, Julio; Sanz, María I.: Development of nitrocellulose membrane filters impregnated with different biosynthesized silver nanoparticles applied to water purification. *Talanta* 146, no. 1 (2016), 237–243.

[261] Diko, Catherine Sekyerebea; Qu, Yuanyuan; Henglin, Zhang; Li, Zheng; Ahmed Nahyoon, Noor; Fan, Shuling: Biosynthesis and characterization of lead selenide semiconductor nanoparticles (PbSe NPs) and its antioxidant and photocatalytic activity. *Arabian Journal of Chemistry* 13, no. 11 (2020), 8411–8423.

[262] Akther, Tahira; Khan, Mohd Shahanbaj; Hemalatha, S.: Biosynthesis of silver nanoparticles via fungal cell filtrate and their anti-quorum sensing against *Pseudomonas aeruginosa*. *Journal of Environmental Chemical Engineering* 8, no. 6 (2020), 104365.

[263] Islam, S.K. Najrul; Naqvi, Syed Mohd Adnan; Parveen, Sadia; Ahmad, Absar: Endophytic fungus-assisted biosynthesis, characterization and solar photocatalytic activity evaluation of nitrogen-doped Co$_3$O$_4$ nanoparticles. *Applied Nanoscience* 11, no. 5 (2021), 1651–1659.

[264] Nascimento, Jessica M; Oliveira, Jorge Diniz De; Rizzo, Andrea C.L.; Leite, Selma G.F.: Biogenic production of copper nanoparticles by *Saccharomyces cerevisiae*. *Journal of Bionanoscience* 12, no. 5 (2018), 689–693.

[265] Khodashenas, Bahareh; Ghorbani, Hamid Reza: Synthesis of copper nanoparticles: An overview of the various methods. *Korean Journal of Chemical Engineering* 31, no. 7 (2014), 1105–1109.

[266] Mathew, Silvy; Victorio, Cristiane P.; Sidhi, Jasmine; BH, Baby Thanzeela: Biosynthesis of silver nanoparticle using flowers of *Calotropis gigantea* (L.) W.T. Aiton and activity against pathogenic bacteria. *Arabian Journal of Chemistry* 13, no. 12 (2020), 9139–9144.

[267] Lengke, Maggy F.; Sanpawanitchakit, Charoen; Southam, Gordon: Biosynthesis of gold nanoparticles: A review. In Mahendra Rai, Nelson Duran. Metal Nanoparticles in Microbiology. Berlin: Springer Berlin, Heidelberg; (2011), pp. 37–74.

[268] Shankar, S. Shiv; Rai, Akhilesh; Ahmad, Absar; Sastry, Murali: Rapid synthesis of Au, Ag, and bimetallic Au core – Ag shell nanoparticles using neem (*Azadirachta indica*) leaf broth. *Journal of Colloid and Interface Science* 275, no. 2 (2004), 496–502.

[269] Armendariz, Veronica; Herrera, Isaac; Peralta-Videa, Jose R.; Jose-Yacaman, Miguel: Size controlled gold nanoparticle formation by *Avena sativa* biomass: Use of plants in nanobiotechnology. *Journal of Nanoparticle Research* 6 (2004), 377–382.

[270] Ashokkumar, T.; Vijayaraghavan, K.: Brown seaweed-mediated biosynthesis of gold nanoparticles. *Journal of Environment & Biotechnology Research* 2, no. 1 (2016), 45–50.

[271] Ahmad, Nabeel; Bhatnagar, Sharad; Saxena, Ritika; Iqbal, Danish; Ghosh, AK; Dutta, Rajiv: Biosynthesis and characterization of gold nanoparticles: Kinetics, in vitro and in vivo study. *Materials Science and Engineering C* 78, (2017), 553–564.

[272] Balamurugan, Madheswaran; Kaushik, Saravanan; Saravanan, Shanmugam: Green synthesis of gold nanoparticles by using *Peltophorum Pterocarpum* flower extracts. *Nano Biomedicine & Engineering* 8, no. 4 (2016), 213–218.

[273] Bhardwaj, A.K.; Shukla, A.; Singh, S.C.; Uttam, K.N.; Nath, G.; Gopal, R.: Green synthesis of Cu_2O hollow microspheres. *Advanced Materials Proceedings* 2, no. 2 (2017), 132–138.

[274] Bordbar, Maryam; Sharifi-Zarchi, Zeinab; Khodadadi, Bahar: Green synthesis of copper oxide nanoparticles/clinoptilolite using *Rheum palmatum* L. root extract: High catalytic activity for reduction of 4-nitro phenol, rhodamine B, and methylene blue. *Journal of Sol-Gel Science and Technology* 81, no. 3 (2017), 724–733.

[275] Caroling, G.; Vinodhini, E.; Mercy Ranjitham, A.; Shanthi, P.: Biosynthesis of copper nanoparticles using aqueous *Phyllanthus embilica* (gooseberry) extract-characterization and study of antimicrobial effects. *International Journal of Nanomaterials and Chemistry* 1, no. 2 (2015), 53–63.

[276] Ajitha, B; Ashok Kumar, Y.; Rajesh, K.M.; Sreedhara Reddy, P.: *Sesbania grandiflora* leaf extract assisted green synthesis of silver nanoparticles: Antimicrobial activity. *Materials Today: Proceedings* 3, no. 6 (2016), 1977–1984.

[277] Alvarez, Ramon A.B.; Cortez-Valadez, M.; Bueno, L. Oscar Neira; Britto Hurtado, R.; Rocha-Rocha, O.; Delgado-Beleño, Y.; Martinez-Nuñez, C.E.; Serrano-Corrales, Luis Ivan; Arizpe-Chávez, H.; Flores-Acosta, M.: Vibrational properties of gold nanoparticles obtained by green synthesis. *Physica E: Low-Dimensional Systems and Nanostructures* 84, (2016), 191–195.

[278] Chauhan, P.S.; Shrivastava, V.; Tomar, R.S.: Phytomediated synthesis of silver nanoparticles and evaluation of its antibacterial activity against *Bacillus subtilis* and *Staphylococcus aureus*. *International Journal of Pharma and Bio Sciences* 7, (2016), 184–195.

[279] Chiguvare, Herbert; Oyedeji, Opeoluwa O; Matewu, Reuben; Aremu, Olukayode; Oyemitan, Idris A.; Oyedeji, Adebola O.; Nkeh-chungag, Benedicta N.; Songca, Sandile P.; Mohan, Sneha; Oluwafemi, Oluwatobi S.: Synthesis of silver nanoparticles using buchu plant extracts and their analgesic properties. *Molecules* 21, no. 6 (2016), 774.

[280] Yang, Ning; Weihong, Li; Hao, Lin: Biosynthesis of Au nanoparticles using agricultural waste mango peel extract and its in vitro cytotoxic effect on two normal cells. *Materials Letters* 134, (2014), 67–70.

[281] Yew, Yen Pin; Shameli, Kamyar; Miyake, Mikio; Kuwano, Noriyuki; Bahiyah, Nurul; Ahmad, Bt: Green synthesis of magnetite (Fe_3O_4) nanoparticles using seaweed (*Kappaphycus alvarezii*) extract. *Nanoscale Research Letters* 11, no. 276 (2016), 1–7.

[282] Varadavenkatesan, T.; Selvaraj, R.; Vinayagam, R.: Phyto-synthesis of silver nanoparticles from *Mussaenda erythrophylla* leaf extract and their application in catalytic degradation of methyl orange dye. *Journal of Molecular Liquids* 221, (2016), 1063–1070.

[283] Ravichandran, Veerasamy; Vasanthi, Sethu; Shalini, Sivadasan; Adnan, Syed; Shah, Ali: Green synthesis of silver nanoparticles using *Atrocarpus altilis* leaf extract and the study of their antimicrobial and antioxidant activity. *Materials Letters* 180, no. 1 (2016), 264–267.

[284] Dauthal, Preeti; Mukhopadhyay, Mausumi: Phyto-synthesis and structural characterization of catalytically active gold nanoparticles biosynthesized using Delonix regia leaf extract. *3 Biotech* 6, no. 2 (2016), 118.

[285] Garcia-alonso, Javier; Rodriguez-sanchez, Neus; Misra, Superb K.; Valsami-jones, Eugenia; Croteau, Marie-noële; Luoma, Samuel N.; Rainbow, Philip S.: Toxicity and accumulation of silver nanoparticles during development of the marine polychaete Platynereis dumerilii. *Science of the Total Environment* 476–477, (2014), 688–695.

[286] Govindarajan, Marimuthu; Benelli, Giovanni: One-pot fabrication of silver nanocrystals using *Ormocarpum cochinchinense*: Biophysical characterization of a potent mosquitocidal and toxicity on non-target mosquito predators. *Journal of Asia-Pacific Entomology* 19, no. 2 (2016), 377–385.

[287] Luna, Manuel; Zarzuela, Rafael; Mosquera, María J.; Gil, ML Almoraima; Cubillana-Aguilera, Laura M.; Delgado-Jaen, Juan J.; Palacios-Santander, José M.; Garcia-Moreno, Valme; Carmona-Jimenez, Yolanda: Biosynthesis of uniform ultra-small gold nanoparticles by aged *Dracaena Draco* L extracts. *Colloids and Surfaces A: Physicochemical and Engineering Aspects* 581, (2019), 123744.

[288] Ali, Syed Ghazanfar; Ansari, Mohammad Azam; Jamal, Qazi Mohammad Sajid; Almatroudi, Ahmad; Alzohairy, Mohammad A.; Alomary, Mohammad N.; Rehman, Suriya; Mahadevamurthy, Murali; et al.: Butea monosperma seed extract mediated biosynthesis of ZnO NPs and their antibacterial, antibiofilm and anti-quorum sensing potentialities. *Arabian Journal of Chemistry* 14, no. 4 (2021), 103044.

[289] Jan, Hasnain; Zaman, Gouhar; Usman, Hazrat; Ansir, Rotaba; Drouet, Samantha; Gigliolo-Guivarc'h, Nathalie; Hano, Christophe; Abbasi, Bilal Haider: Biogenically proficient synthesis and characterization of silver nanoparticles (Ag-NPs) employing aqueous extract of *Aquilegia pubiflora* along with their *in vitro* antimicrobial, anti-cancer and other biological applications. *Journal of Materials Research and Technology* 15, (2021), 950–968.

[290] Al-Radadi, Najlaa S.: Biogenic proficient synthesis of (Au-NPs) via aqueous extract of Red Dragon Pulp and seed oil: Characterization, antioxidant, cytotoxic properties, anti-diabetic anti-inflammatory, anti-Alzheimer and their anti-proliferative potential against cancer cell lines. *Saudi Journal of Biological Sciences* 29, (2022) 2836–2855.

[291] Jain, Ayushi; Wadhawan, Shweta; Mehta, S.K.: Biogenic synthesis of non-toxic iron oxide NPs via *Syzygium aromaticum* for the removal of methylene blue. *Environmental Nanotechnology, Monitoring & Management* 16, (2021), 100464.

12 Advances in Removal of Heavy Metals from Contaminated Soil

Mu-Hao Sung Wang, Nazih K. Shammas, and Lawrence K. Wang

ACRONYM AND NOMENCLATURE

$(CH_3)_2AsO_2H$:	Dimethylarsenic acid
$As(CH_3)_3$:	Trimethylarsine
As:	Arsenic
AsH_3:	Arsine
BDAT:	Best Demonstrated Available Technology
BEK:	Batch electrokinetic remediation process.
$Ca_3(AsO_4)_2$:	Calcium arsenate
$CaCO_3$:	Calcium carbonate
Cd:	Cadmium
CERCLA:	Comprehensive Environmental Response, Compensation, and Liability Act
CO_2:	Carbon dioxide
Cr:	Chromium
DOD:	US Department of Defense
DOE:	US Department of Energy
EBONEX:	Electrically conductive ceramic material
$H_2AsO_3CH_3$:	Methylarsenic acid
$HAs(CH_3)_2$:	Dimethylarsine
Hg:	Mercury
$K_2Cr_2O_7$:	Potassium dichromate
K_2CrO_4:	Potassium chromate
$Mn_3(AsO_4)_3$:	Manganese arsenate,
Pb:	Lead
RCRA:	Resource Conservation and Recovery Act
S/S:	Solidification/stabilization
SITE:	Superfund Innovative Technology Evaluation (SITE) program
U.S. EPA:	US Environmental Protection Agency
VOC:	Volatile organic chemicals

12.1 INTRODUCTION

12.1.1 Soil Contamination and Its Decontamination

Metals account for much of the contamination found at hazardous waste sites. They are present in the soil and groundwater at approximately 65% of the Superfund or CERCLA (Comprehensive Environmental Response, Compensation, and Liability Act) [1] sites for which the U.S. Environmental Protection Agency (U.S. EPA) has signed records of decisions [2]. The metals most frequently identified are lead, arsenic, chromium, cadmium, nickel, and zinc. Other metals often identified as contaminants include copper and mercury. In addition to the Superfund program, metals make up a significant portion of the contamination requiring remediation under the Resource Conservation and Recovery Act (RCRA) [3] and contamination present at federal facilities, notably those that are the responsibility of the Department of Defense (DOD) and the Department of Energy (DOE).

This chapter provides remedial project managers, engineers, on-scene coordinators, contractors, and other state or private remediation managers and their technical support personnel with information to facilitate the selection of appropriate remedial alternatives for soil contaminated with arsenic (As), cadmium (Cd), chromium (Cr), mercury (Hg), and lead (Pb) [4–6].

Common compounds, transport, and fate are discussed for each of the five elements. A general description of metal-contaminated Superfund soils is provided. The technologies covered are containment (immobilization), solidification/stabilization (S/S), vitrification, soil washing, soil flushing, pyrometallurgy, electrokinetics, and phytoremediation. The use of treatment trains and remediation costs are also addressed.

It is assumed that users of this chapter will, as necessary, familiarize themselves with (1) applicable or relevant and appropriate regulations pertinent to the site of interest, (2) applicable health and safety regulations and practices relevant to the metals and compounds discussed, and (3) relevant sampling, analysis, and data interpretation methods. Information on Pb battery (Pb, As), wood preserving (As, Cr), pesticide (Pb, As, Hg), and mining sites has been addressed in U.S. EPA Superfund documents [7–12]. The greatest emphasis is on remediation of inorganic forms of the metals of interest. Organometallic compounds, organic-metal mixtures, and multimetal mixtures are briefly addressed.

12.1.2 SUMMARY

This publication reviews the technologies and managerial methods for removing heavy metals from contaminated soil. The following subjects are covered in detail: (a) overview of metals, metal compounds and their physical characteristics and mineral origins; (b) description of Superfund soils contaminated with metals; (c) soil cleanup goals and technologies for remediation; (d) containment process description and its caps, vertical barriers, horizontal barriers, site requirements, applicability performance, Best Demonstrated Available Technology (BDAT) status, and Superfund Innovative Technology Evaluation (SITE) program demonstration projects; (e) various solidification/stabilization (S/S) processes (such as, ex situ cement-based S/S, in situ cement-based S/S, and polymer microencapsulation S/S), and their respective site requirements, applicability, performance and BDAT status, SITE program demonstration projects, and costs; (f) various vitrification processes (such as ex situ vitrification and in situ vitrification), and their respective site requirements, applicability, performance and BDAT status, and SITE program demonstration projects; (g) soil washing process and its site requirements, applicability, performance, BDAT status, SITE demonstrations, and emerging technologies program projects; (h) soil flushing process and its site requirements, applicability, performance, BDAT status, and SITE demonstration and emerging technologies program projects; (i) pyrometallurgy process and its site requirements, applicability, performance, BDAT status and SITE demonstration and emerging technologies program projects; (j) electrokinetics process and its site requirements, applicability and demonstration projects (by Electrokinetics, Inc., Geokinetics International, Inc., Isotron Corporation, Battelle Memorial Institute, consortium process, performance and cost, LSU-Electrokinetics, Inc., Geokinetics International, Inc. and Battelle Memorial Institute), consortium process, and electrokinetic remediation summary; (k) various phytoremediation processes (phytoextraction, phytostabilization, and rhizofiltration) and their respective future development, applicability, performance, costs, future directions, phytoremediation technology summary, site conditions and waste characteristics; and (l) use of treatment trains, and cost ranges of remedial technologies.

12.2 OVERVIEW OF METALS AND THEIR COMPOUNDS

This section provides a brief, qualitative overview of the physical characteristics and mineral origins of the five metals and the factors affecting their mobility. More comprehensive and quantitative reviews of the behavior of these five metals in soil can be found in readily available U.S. EPA Superfund documents [4,13,14].

12.2.1 OVERVIEW OF PHYSICAL CHARACTERISTICS AND MINERAL ORIGINS

Arsenic is a semimetallic element or metalloid that has several allotropic forms. The most stable allotrope is a silver-gray, brittle, crystalline solid that tarnishes in air. As compounds, mainly As_2O_3, can be recovered as a by-product of processing complex ores mined mainly for copper, lead, zinc, gold, and silver As occurs in a wide variety of mineral forms, including arsenopyrite, $FeAsS_4$, which is the main commercial ore of As worldwide.

Cadmium is a bluish-white, soft, ductile metal. Pure Cd compounds are rarely found in nature, although occurrences of greenockite (CdS) and otavite ($CdCO_3$) are known. The main sources of Cd are sulfide ores of lead, zinc, and copper. Cd is recovered as a by-product when these ores are processed.

Chromium is a lustrous, silver-gray metal. It is one of the less common elements in the earth's crust and occurs only in compounds. The chief commercial source of Cr is the mineral chromite $FeCr_2O_4$. Cr is mined as a primary product and is not recovered as a by-product of any other mining operation. There are no chromite-ore reserves nor any primary production of chromite in the United States.

Mercury is a silvery, liquid metal. The primary source of Hg is cinnabar (HgS), a sulfide ore. In a few cases, Hg occurs as the principal ore product; it is more commonly obtained as the by-product of processing complex ores that contain mixed sulfides, oxides, and chloride minerals (these are usually associated with the base and precious metals, particularly gold). Native or metallic Hg is found in very small quantities in some ore sites. The current demand for Hg is met by secondary production (i.e., recycling and recovery).

Lead is a bluish-white, silvery, or gray metal that is highly lustrous when freshly cut but tarnishes when exposed to air. It is very soft and malleable, has a high density (11.35 g/cm^3) and low melting point (327.4°C), and can be cast, rolled, and extruded. The most important Pb ore is galena, PbS. Recovery of Pb from the ore typically involves grinding, flotation, roasting, and smelting. Less common forms of the mineral are cerussite, $PbCO_3$, anglesite, $PbSO_4$, and crocoite, $PbCrO_4$.

12.2.2 OVERVIEW OF BEHAVIOR OF AS, CD, CR, PB, AND HG

Since metals cannot be destroyed, remediation of metal-contaminated soil consists primarily of manipulating (i.e., exploiting, increasing, decreasing, or maintaining) the mobility of metal contaminant(s) to produce a treated soil with an acceptable total or leachable metal content. Metal mobility depends on numerous factors. Metal mobility in soil-waste systems is determined by the following [13]:

1. Type and quantity of soil surfaces present.
2. Concentration of metal of interest.
3. Concentration and type of competing ions and complexing ligands, both organic and inorganic.
4. pH.
5. Redox status.

"Generalization can only serve as rough guides of the expected behavior of metals in such systems. The use of literature or laboratory data that do not mimic the specific site soil and waste system will not be adequate to describe or predict the behavior of the metal. Data must be site-specific. Long-term effects must also be considered. As organic constituents of the waste matrix degrade, or as pH or redox conditions change, either through natural processes of weathering or human manipulation, the potential mobility of the metal will change as soil conditions change" [13].

Cd, Cr(III), and Pb are present in cationic forms under natural environmental conditions [13]. These cationic metals generally are not mobile in the environment and tend to remain relatively close to the point of initial deposition. The capacity of soil to adsorb cationic metals increases with increasing pH, cation exchange capacity, and organic carbon content. Under the neutral to basic conditions typical of most soils, cationic metals are strongly adsorbed on the clay fraction of soils and can be adsorbed by hydrous oxides of iron, aluminum, or manganese present in soil minerals. Cationic metals will precipitate as hydroxides, carbonates, or phosphates. In acidic, sandy soils, the cationic metals are more mobile. Under conditions that are atypical of natural soils (e.g., pH < 5 or > 9 elevated concentrations of oxidizers or reducers; high concentrations of soluble organic or inorganic complexing or colloidal substances) but may be encountered as a result of waste disposal or remedial processes, the mobility of these metals may be substantially increased. Also, competitive adsorption between various metals has been observed in experiments involving various solids with oxide surfaces (g $FeOOH$, a -SiO_2, and g -Al_2O_3). In several experiments, Cd adsorption was decreased by adding Pb or Cu for all three of these solids. The addition of zinc resulted in the greatest decrease in Cd adsorption. Competition for surface sites occurred when only a few percent of all surface sites were occupied [15].

As, Cr(VI), and Hg behaviors differ considerably from Cd, Cr(III), and Pb. As and Cr(VI) typically exist in anionic forms under environmental conditions. Hg, although it is a cationic metal, has unusual properties (e.g., liquid at room temperature and easily transforms among several possible valence states).

In most As-contaminated sites, As appears as As_2O_3 or as anionic As species leached from As_2O_3, oxidized to As(V), and then sorbed onto iron-bearing minerals in the soil. As may also be present in organometallic forms, such as methylarsenic acid, $H_2AsO_3CH_3$, and dimethylarsenic acid, $(CH_3)_2AsO_2H$, which are active ingredients in many pesticides, as well as the volatile compounds arsine (AsH_3) and its methyl derivatives [i.e., dimethylarsine $HAs(CH_3)_2$ and trimethylarsine, $As(CH_3)_3$]. These As forms illustrate the various oxidation states that As commonly exhibits (—III, 0, III, and V) and the resulting complexity of its chemistry in the environment.

As(V) is less mobile and less toxic than As(III). As(V) exhibits anionic behavior in the presence of water, and hence its aqueous solubility increases with increasing pH, and it does not complex or precipitate with other anions. As(V) can form low-solubility metal arsenates. Calcium arsenate, $Ca_3(AsO_4)_2$, is the most stable metal arsenate in well-oxidized and alkaline environments, but it is unstable in acidic environments. Even under initially oxidizing and alkaline conditions, absorption of CO_2 from the air will result in the formation of $CaCO_3$ and the release of arsenate. In sodic soils, sufficient sodium is available, such that the mobile compound Na_3AsO_4 can form. The slightly less stable manganese arsenate, $Mn_3(AsO_4)_2$, forms in both acidic and alkaline environments, while iron arsenate is stable under acidic soil conditions. In aerobic environments, $HAsO_4$ predominates at pH < 2 and is replaced by $H_2AsO_4^-$, $HAsO_4^{2-}$, and AsO_4^{3-} as pH increases to about 2, 7, and 11.5, respectively. Under mildly reducing conditions, H_3AsO_3 is a predominant species at low pH but is replaced by $H_2AsO_3^-$, $HAsO_3^{2-}$, and AsO_3^{3-} as pH increases. Under still more reducing conditions and in the presence of sulfide, As_2S_3 can form. As_2S_3 is a low-solubility, stable solid. AsS_2 and AsS_2^- are thermodynamically unstable with respect to As_2S_3 [16]. Under extreme reducing conditions, elemental As and volatile arsine, AsH_3, can form. Just as competition between cationic metals affects mobility in soil, competition between anionic species (chromate, arsenate, phosphate, sulfate, and so on) affects anionic fixation processes and may increase mobility.

The most common valence states of Cr in the earth's surface and near-surface environment are +3 (trivalent or Cr(III)) and +6 (hexavalent or Cr(VI)). Trivalent Cr (discussed above) is the most thermodynamically stable form under common environmental conditions. Except in leather tanning, industrial applications of Cr generally use the Cr(VI) form. Due to kinetic limitations, Cr(VI) does not always readily reduce to Cr(III) and can remain present over an extended period of time.

Cr(VI) is present as the chromate, CrO_4^{2-}, or dichromate, $Cr_2O_7^{2-}$, anion, depending on pH and concentration. Cr(VI) anions are less likely to be adsorbed to solid surfaces than Cr(III). Most solids in soils carry negative charges that inhibit Cr(VI) adsorption. Although clays have a high capacity to adsorb cationic metals, they interact little with Cr(VI) because of the similar charges carried by the anion and clay in the common pH range of soil and groundwater. The only common soil solid that adsorbs Cr(VI) is iron oxyhydroxide. Generally, a major portion of Cr(VI) and other anions adsorbed in soils can be attributed to the presence of iron oxyhydroxide. The quantity of Cr(VI) adsorbed onto the iron solids increases with decreasing pH.

At metal-contaminated sites, Hg can be present in mercuric form (Hg^{2+}), mercurous form (Hg_2^{2+}), elemental form (Hg), or alkylated form (e.g., methyl and ethyl Hg). Hg_2^{2+} and Hg^{2+} are more stable under oxidizing conditions. Under mildly reducing conditions, both organically bound Hg and inorganic Hg compounds can be converted to elemental Hg, which can then be readily converted to methyl or ethyl Hg by biotic and abiotic processes. Methyl and ethyl Hg are mobile and toxic forms.

Hg is moderately mobile, regardless of the soil. Both the mercurous and mercuric cations are adsorbed by clay minerals, oxides, and organic matter. Adsorption of cationic forms of Hg increases with increasing pH. Mercurous and mercuric Hg are also immobilized by forming various precipitates. Mercurous Hg precipitates with chloride, phosphate, carbonate, and hydroxide. At concentrations of Hg commonly found in soil, only the phosphate precipitate is stable. In alkaline soils, mercuric Hg precipitates with carbonate and hydroxide to form a stable (but not exceptionally insoluble) solid phase. At lower pH and high chloride concentration, soluble $HgCl_2$ is formed. Mercuric Hg also forms complexes with soluble organic matter, chlorides, and hydroxides that may contribute to its mobility [13]. In strong reducing conditions, HgS, a very low-solubility compound, is formed.

12.3 DESCRIPTION OF SUPERFUND SOILS CONTAMINATED WITH METALS

Soils can become contaminated with metals from direct contact with industrial plant waste discharges, fugitive emissions, or leachate from waste piles, landfills, or sludge deposits. The specific type of metal contaminant expected at a particular Superfund site would obviously be directly related to the type of operation that had occurred there. Table 12.1 lists the types of operations that are directly associated with each of the five metal contaminants [5].

Wastes at CERCLA sites are frequently heterogeneous on a macro- and microscale. The contaminant concentration and the physical and chemical forms of the contaminant and matrix are usually complex and variable. Of these, waste disposal sites collect the widest variety of waste types; therefore, concentration profiles vary by orders of magnitude through a pit or pile. Limited volumes of high-concentration "hot spots" may develop due to variations in historical waste disposal patterns or local transport mechanisms. Similar radical variations frequently occur on the particle-size scale as well. The waste often consists of a physical mixture of very different solids, for example, paint chips in spent abrasive.

Industrial processes may result in a variety of solid metal-bearing waste materials, including slags, fumes, mold sand, fly ash, abrasive wastes, spent catalysts, spent activated carbon, and refractory bricks [17]. These process solids may be found above ground as waste piles or below ground in landfills. Solid-phase wastes can be dispersed by well-intended but poorly controlled reuse projects. Waste piles can be exposed to natural disasters or accidents, causing further dispersion.

TABLE 12.1

Principal Sources of As, Cd, Cr, Hg, and Pb Contaminated Soils

Technology Class	Specific Technology
Containment	Caps
	Vertical barriers
	Horizontal barriers
	Cement-based
S/S	Polymer microencapsulation
	Vitrification
	Soil washing
Separation/concentration	Soil flushing
	Pyrometallurgy
	Electrokinetics
	Phytoremediation

Contaminant	Principal Sources
As	Wood preserving
	As-waste disposal
	Pesticide production and application
	Mining
Cd	Plating
	Ni-Cd battery manufacturing
	Cd-waste disposal
Cr	Plating
	Textile manufacturing
	Leather tanning
	Pigment manufacturing
	Wood preserving
	Cr-waste disposal
Hg	Chloralkali manufacturing
	Weapons production
	Copper and zinc smelting
	Gas line manometer spills
	Paint application
	Hg-waste disposal
Pb	Ferrous/nonferrous smelting
	Pb-acid battery breaking
	Ammunition production
	Leaded paint waste
	Pb-waste disposal
	Secondary metals production
	Waste oil recycling
	Firing ranges
	Ink manufacturing
	Mining
	Pb-acid battery manufacturing
	Leaded glass production
	Tetraethyl Pb production
	Chemical manufacturing

Source: U.S. EPA. *Technology Alternatives for the Remediation of Soils Contaminated with AS, Cd, Cr, Hg, and Pb.* EPA/540/S-97/500, U.S. Environmental Protection Agency, Cincinnati, OH, August 1997 [5].

TABLE 12.2

Cleanup Goals (Actual and Potential) for Total and Leachable Metals

Description	As	Cd	Cr (Total)	Hg	Pb
	Total Metals	**Goals (mg/kg)**			
Background (mean)	5	0.06	100	0.03	10
Background (range)	1–50	0.01–0.70	1–1000	0.01–0.30	2–200
Superfund site goals from TRD	5–65	3–20	6.7–375	1–21	200–500
Theoretical minimum total metals to ensure TCLP Leachate<threshold (i.e., TCLP ¥ 20)	100	20	100	4	100
California total threshold limit concentration	500	100	500	20	1,000
	Leachable Metals (µg/L)				
TCLP threshold for RCRA waste	5,000	1,000	5,000	200	5,000
Extraction procedure toxicity test	5,000	1,000	5,000	200	5,000
Synthetic Precipitate Leachate	—	—	—	—	—
Multiple extraction procedure	—	—	—	—	—
California soluble threshold leachate concentration	5,000	1,000	5,000	200	5,000
Maximum contaminant level[a]	50	5	100	2	15
Superfund site goals from TRD	50	—	50	0.05–2	50

Source: U.S. EPA. *Technology Alternatives for the Remediation of Soils Contaminated with AS, Cd, Cr, Hg, and Pb.* EPA/540/S-97/500, U.S. Environmental Protection Agency, Cincinnati, OH, August 1997 [5].

[a] Maximum contaminant level=The maximum permissible level of contaminant in water delivered to any user of a public system.

— No specified level and no example cases identified.

12.4 SOIL CLEANUP GOALS AND TECHNOLOGIES FOR REMEDIATION

Table 12.2 provides an overview of cleanup goals (actual and potential) for both total and leachable metals. Based on inspection of the total metals cleanup goals, one can see that they vary considerably both within the same metal and between metals.

Similar variation is observed in the actual or potential leachate goals. The observed variation in cleanup goals has at least two implications for technology alternative evaluation and selection. Firstly, the importance of identifying the target metal(s), contaminant state (leachable versus total metal), the specific type of test and conditions, and the numerical cleanup goals early in the remedy evaluation process is apparent. Depending on which cleanup goal is selected, the required removal or leachate reduction efficiency of the overall remediation can vary by several orders of magnitude [5,18]. Secondly, the degree of variation in goals both within and between metals, plus the many factors that affect mobility of the metals, suggest that generalizations about effectiveness of a technology for meeting total or leachable treatment goals should be viewed with some caution.

Technologies that are potentially applicable to the remediation of soils contaminated with the five metals or their inorganic compounds are listed below [2,5]:

The BDAT status refers to the determination under the RCRA of the BDAT for various industry-generated hazardous wastes that contain the metals of interest. Whether the characteristics of a Superfund metal-contaminated soil (or fractions derived from it) are similar enough to the RCRA waste to justify serious evaluation of the BDAT for a specific Superfund soil must be made on a site-specific basis. Other limitations relevant to BDATs include the following: (a) the regulatory basis for BDAT standards focus BDATs on proven, commercially available technologies at the time of the BDAT determination; (b) a BDAT may be identified, but that does not necessarily preclude the use of other technologies; and (c) a technology identified as BDAT may not necessarily be the current technology of choice in the RCRA hazardous waste treatment industry.

The U.S. EPA's Superfund Innovative Technology Evaluation (SITE) program evaluates many emerging and demonstrated technologies to promote the development and use of innovative technologies to clean up Superfund sites across the country. The major focus of SITE is the Demonstration Program, which is designed to provide engineering and cost data for selected technologies.

Cost is not discussed in each technology narrative; however, a summary table is provided at the end of the technology discussion section that illustrates technology cost ranges and treatment train options.

12.5 CONTAINMENT

Containment technologies for application at Superfund sites include landfill covers (caps), vertical barriers, and horizontal barriers [4]. For metal remediation, containment is considered an established technology except for *in situ* installation of horizontal barriers.

12.5.1 PROCESS DESCRIPTION

Containment ranges from a surface cap that limits infiltration of uncontaminated surface water to subsurface vertical or horizontal barriers that restrict lateral or vertical migration of contaminated groundwater. The material provided here is primarily from U.S. EPA [5,9].

12.5.1.1 Caps

Capping systems reduce surface water infiltration, control gas and odor emissions, improve aesthetics, and provide a stable surface over the waste. Caps can range from a simple native soil cover to a full RCRA subtitle C composite cover.

Cap construction costs depend on the number of components in the final cap system (i.e., costs increase by adding barrier and drainage components). Additionally, cost escalates as a function of topographic relief. Side slopes steeper than 3 horizontal to 1 vertical can cause stability and equipment problems that dramatically increase the unit cost [4,19].

12.5.1.2 Vertical Barriers

Vertical barriers minimize the movement of contaminated groundwater off-site or limit the flow of uncontaminated groundwater on-site. Common vertical barriers include slurry walls in excavated trenches; grout curtains formed by injecting grout into soil borings; vertically injected, cement-bentonite grout-filled borings or holes formed by withdrawing beams driven into the ground; and sheet-pile walls formed of driven steel.

Certain compounds can affect cement-bentonite barriers. The impermeability of bentonite may significantly decrease when it is exposed to high concentrations of creosote, water-soluble salts (copper, Cr, and As), or fire-retardant salts (borates, phosphates, and ammonia). The specific gravity of salt solutions must be greater than 1.2 to impact bentonite [20,21]. Generally, soil-bentonite blends resist chemical attack best if they contain only 1% bentonite and 30%–40% natural soil fines. Treatability tests should evaluate the chemical stability of the barrier if adverse conditions are suspected.

Carbon steel used in pile walls quickly corrodes in dilute acids, slowly corrodes in brines or salt water, and remains mostly unaffected by organic chemicals or water. Salts and fire retardants can reduce the service life of a steel sheet pile; corrosion-resistant coatings can extend their anticipated life. Major steel suppliers will provide site-specific recommendations for cathodic protection of piling.

The construction costs for vertical barriers are influenced by the soil profile of the barrier material used and the method of placing it. The most economical shallow vertical barriers are soil-bentonite trenches excavated with conventional backhoes; the most economical deep vertical barriers consist of a cement-bentonite wall placed by a vibrating beam.

12.5.1.3 Horizontal Barriers

In situ horizontal barriers can underlie a sector of contaminated materials on-site without removing the hazardous waste or soil. Established technologies use grouting techniques to reduce the permeability of underlying soil layers. Studies performed by the U.S. Army Corps of Engineers [22] indicate that conventional grout technology cannot produce an impermeable horizontal barrier because it cannot ensure uniform lateral growth of the grout. These same studies found greater success with jet grouting techniques in soils that contain fines sufficient to prevent the collapse of the wash hole and that present no large stones or boulders that could deflect the cutting jet.

Since few *in situ* horizontal barriers have been constructed, accurate costs have not been established. Work performed by the Corps of Engineers for U.S. EPA has shown that it is very difficult to form effective horizontal barriers. The most efficient barrier installation used a jet wash to create a cavity in sandy soils into which cement-bentonite grouting was injected. The costs relate to the number of borings required. Each boring takes at least one day to drill.

12.5.2 Site Requirements

In general, the site must be suitable for a variety of heavy construction equipment, including bull-dozers, graders, backhoes, multi-shaft drill rigs, various rollers, vibratory compactors, forklifts, and seaming devices [23,24]. When capping systems are being utilized, on-site storage areas are necessary for the materials to be used in the cover. If site soils are adequate for use in the cover, a borrow area needs to be identified and the soil tested and characterized. If site soils are not suitable, it may be necessary to truck in other low-permeability soils [23]. In addition, an adequate supply of water may also be needed to achieve optimum soil density.

The construction of vertical containment barriers, such as slurry walls, requires knowledge of the site, the local soil and hydrogeologic conditions, and the presence of underground utilities [25]. Preparation of the slurry requires batch mixers, hydration ponds, pumps, hoses, and an adequate supply of water. Therefore, on-site water storage tanks and electricity are necessary. In addition, areas adjacent to the trench need to be available for the storage of trench spoils (which could potentially be contaminated) and the mixing of backfill. If excavated soils are not acceptable for use as backfill, suitable backfill must be trucked onto the site [25].

12.5.3 Applicability

Containment is most likely to be applicable to the following [5]:

1. Wastes that are low hazard (e.g., low toxicity or low concentration) or immobile.
2. Wastes that have been treated to produce low-hazard or low-mobility wastes for on-site disposal.
3. Wastes whose mobility must be reduced as a temporary measure to mitigate risk until a permanent remedy can be tested and implemented.

Situations where containment would not be applicable include the following:

1. Wastes for which there is a more permanent and protective remedy that is cost-effective.
2. Where effective placement of horizontal barriers below existing contamination is difficult.
3. Where drinking water sources will be adversely affected if containment fails, and if there is inadequate confidence in the ability to predict, detect, or control harmful releases due to containment failure.

Important advantages of containment are [5] the following:

1. Surface caps and vertical barriers are relatively simple and rapid to implement at low cost and can be more economical than excavation and removal of waste.
2. Caps and vertical barriers can be applied to large areas or volumes of waste.
3. Engineering control (containment) is achieved, and may be a final action if metals are well immobilized and potential receptors are distant.
4. A variety of barrier materials are available commercially.
5. In some cases, it may be possible to create a land surface that can support vegetation and/ or be applicable for other purposes.

Disadvantages of containment include [5] the following:

1. Design life is uncertain.
2. Contamination remains on-site, available to migrate should containment fail.
3. Long-term inspection, maintenance, and monitoring is required.
4. The site must be amenable to effective monitoring.
5. Placement of horizontal barriers below existing waste is difficult to implement successfully.

12.5.4 PERFORMANCE AND BDAT STATUS

Containment is widely accepted as a means of controlling the spread of contamination and preventing the future migration of waste constituents. Table 12.3 shows a list of selected sites where containment has been selected for remediating metal-contaminated solids.

The performance of capping systems, once installed, may be difficult to evaluate [23]. Monitoring well systems or infiltration monitoring systems can provide some information, but it is often not possible to determine whether the water or leachate originated as surface water or groundwater.

With regard to slurry walls and other vertical containment barriers, performance may be affected by a number of variables, including geographic region, topography, and material availability. A thorough characterization of the site and a compatibility study are highly recommended [25].

Containment technologies are not considered "treatment technologies" and hence no BDATs involving containment have been established.

TABLE 12.3

Containment Applications at Selected Superfund Sites with Metal Contamination

Site Name/State	Specific Technology	Key Metal Contaminants	Associated Technology
Ninth Avenue Dump, IN	Containment-slurry wall	Pb	Slurry wall/capping
Industrial Waste Control, AK	Containment-slurry wall	As, Cd, Cr, Pb	Capping/French drain
E.H. Shilling Landfill, OH	Containment-slurry wall	As	Capping/clay berm
Chemtronic, NC	Capping	Cr, Pb	Capping
Ordnance Works Disposal, WV	Capping	As, Pb	Capping
Industriplex, MA	Capping	As, Pb, Cr	Capping

Source: U.S. EPA. *Technology Alternatives for the Remediation of Soils Contaminated with AS, Cd, Cr, Hg, and Pb.* EPA/540/S-97/500, U.S. Environmental Protection Agency, Cincinnati, OH, August 1997 [5].

12.5.5 SITE PROGRAM DEMONSTRATION PROJECTS

Ongoing SITE demonstrations applicable to soils contaminated with the metals of interest include

- Morrison Knudsen Corporation (high clay grouting technology) and
- RKK, Ltd. (frozen soil barriers).

12.6 SOLIDIFICATION/STABILIZATION TECHNOLOGIES

The term "solidification/stabilization (S/S)" refers to a general category of processes that are used to treat a wide variety of wastes, including solids and liquids. Solidification and stabilization are each distinct technologies, as described below [26]:

Solidification: refers to processes that encapsulate a waste to form a solid material and to restrict contaminant migration by decreasing the surface area exposed to leaching and/ or by coating the waste with low-permeability materials. Solidification can be accomplished by a chemical reaction between a waste and binding (solidifying) reagents or by mechanical processes. Solidification of fine waste particles is referred to as microencapsulation, whereas solidification of a large block or container of waste is referred to as macroencapsulation.

Stabilization: refers to processes that involve chemical reactions that reduce the leachability of a waste. Stabilization chemically immobilizes hazardous materials (such as heavy metals) or reduces their solubility through a chemical reaction. The physical nature of the waste may or may not be changed by this process.

S/S aims to accomplish one or more of the following objectives [4]:

1. Improve the physical characteristics of the waste by producing a solid from liquid or semi-liquid wastes.
2. Reduce the contaminant solubility by formation of sorbed species or insoluble precipitates (e.g., hydroxides, carbonates, silicates, phosphates, sulfates, or sulfides).
3. Decrease the exposed surface area across which mass transfer loss of contaminants may occur by the formation of a crystalline, glassy, or polymeric framework that surrounds the waste particles.
4. Limit the contact between transport fluids and contaminants by reducing the material's permeability.

S/S technology is usually applied by mixing contaminated soils or treatment residuals with a physical binding agent to form a crystalline, glassy, or polymeric framework surrounding the waste particles. In addition to the microencapsulation, some chemical fixation mechanisms may improve the waste's leach resistance. Other forms of S/S treatment rely on macroencapsulation, where the waste is unaltered but macroscopic particles are encased in a relatively impermeable coating [27], or on specific chemical fixation, where the contaminant is converted to a solid compound resistant to leaching. S/S treatment can be accomplished primarily through the use of either inorganic binders (e.g., cement, fly ash, and/or blast furnace slag) or organic binders such as bitumen [4]. Additives may be used, for example, to convert the metal to a less mobile form or to counteract adverse effects of the contaminated soil on the S/S mixture (e.g., accelerated or retarded setting times and low physical strength). The form of the final product from S/S treatment can range from a crumbly, soil-like mixture to a monolithic block. S/S is more commonly carried out as an *ex situ* process, but the *in situ* option is available. The full range of inorganic binders, organic binders, and additives is too broad. The emphasis in this chapter is on *ex situ*, cement-based S/S, which is widely used; *in situ*, cement-based S/S, which has been applied to metals at full-scale; and polymer microencapsulation, which appears applicable to certain wastes that are difficult to treat via cement-based S/S.

Additional information and references on S/S of metals can be found in U.S. EPA documents [4,27–29]. Innovative S/S technologies (e.g., sorption and surfactant processes, bituminization, emulsified asphalt, modified sulfur cement, polyethylene extrusion, soluble silicate, slag, lime, and soluble phosphates) are addressed in U.S. EPA reports [26,30–34].

12.6.1 PROCESS DESCRIPTION

12.6.1.1 *Ex Situ*, Cement-Based S/S

Ex situ, cement-based S/S is performed on contaminated soil that has been excavated and classified to reject oversize. Cement-based S/S involves mixing contaminated materials with an appropriate ratio of cement or similar binder/stabilizer, and possibly water and other additives. A system is also necessary for delivering the treated wastes to molds, surface trenches, or subsurface injection. Off-gas treatment (if volatiles or dust are present) may be necessary. The fundamental materials used to perform this technology are Portland-type cements and pozzolanic materials. Portland cements are typically composed of calcium silicates, aluminates, aluminoferrites, and sulfates. Pozzolans are very small spheroidal particles that are formed in combustion of coal (fly ash) in lime and cement kilns, for example. Pozzolans of high silica content are found to have cement-like properties when mixed with water. Cement-based S/S treatment may involve using only Portland cement, only pozzolanic materials, or blends of both. The composition of the cement and pozzolan, together with the amount of water, aggregate, and other additives, determines the set time, cure time, pour characteristics, and material properties (e.g., pore size and compressive strength) of the resulting treated waste. The composition of cements and pozzolans, including those commonly used in S/S applications, is classified according to American Society for Testing and Materials (ASTM) standards. S/S treatment usually results in an increase (>50% in some cases) in the treated waste volume. *Ex situ* treatment provides high throughput (100–200 m^3/d/mixer).

Cement-based S/S reduces the mobility of inorganic compounds by formation of insoluble hydroxides, carbonates, or silicates; substitution of the metal into a mineral structure; sorption; physical encapsulation; and perhaps other mechanisms. Cement-based S/S involves a complex series of reactions, and there are many potential interferences (e.g., coating of particles by organics, excessive acceleration or retardation of set times by various soluble metal and inorganic compounds; excessive heat of hydration; pH conditions that solubilize anionic species of metal compounds, and so on) that can prevent the attainment of S/S treatment objectives for physical strength and leachability. While there are many potential interferences, Portland cement is widely used and studied, and a knowledgeable vendor may be able to identify, and confirm via treatability studies, approaches to counteract adverse effects by the use of appropriate additives or other changes in formulation.

12.6.1.2 *In Situ*, Cement-Based S/S

In situ, cement-based S/S has only two steps: (a) mixing and (b) off-gas treatment. The processing rate for *in situ* S/S is typically much lower than that for *ex situ* processing. *In situ* S/S is demonstrated to depths of 10 m and may be able to extend to 50 m. The most significant challenge in applying S/S *in situ* for contaminated soils is achieving complete and uniform mixing of the binder with the contaminated matrix [35]. Three basic approaches are used for *in situ* mixing of the binder with the matrix [5]:

1. Vertical auger mixing.
2. In-place mixing of binder reagents with waste by conventional earthmoving equipment, such as draglines, backhoes, or clamshell buckets.
3. Injection grouting, which involves forcing a binder containing dissolved or suspended treatment agents into the subsurface, allowing it to permeate the soil. Grout injection can be applied to contaminated formations lying well below the ground surface. The injected grout cures in-place to produce an *in situ* treated mass.

12.6.1.3 Polymer Microencapsulation S/S

Polymer microencapsulation S/S can include application of thermoplastic or thermosetting resins. Thermoplastic materials are the most commonly used organic-based S/S treatment materials. Potential candidate resins for thermoplastic encapsulation include bitumen, polyethylene and other polyolefins, paraffins, waxes, and sulfur cement. Of these candidate thermoplastic resins, bitumen (asphalt) is the least expensive and by far the most commonly used [36]. The process of thermoplastic encapsulation involves heating and mixing the waste material and the resin at elevated temperature, typically 130°C–230°C in an extrusion machine. Any water or volatile organics in the waste boil off during extrusion and are collected for treatment or disposal. Because the final product is a stiff, yet plastic resin, the treated material is typically discharged from the extruder into a drum or other container.

S/S process quality control requires information on the range of contaminant concentrations; potential interferences in waste batches awaiting treatment; and treated product properties such as compressive strength, permeability, leachability, and in some instances toxicity [28].

12.6.2 Site Requirements

The site must be prepared for the construction, operation, maintenance, decontamination, and decommissioning of the equipment. The size of the area required for the process equipment depends on several factors, including the type of S/S process involved, the required treatment capacity of the system, and site characteristics, especially soil topography and load-bearing capacity. A small mobile *ex situ* unit occupies space for two, standard flatbed trailers. An *in situ* system requires a larger area to accommodate a drilling rig as well as a larger area for auger decontamination.

12.6.3 Applicability

This section addresses expected applicability based on the chemistry of the metal and the S/S binders. The soil-contaminant-binder equilibrium and kinetics are complicated, and many factors influence metal mobility; hence there may be exceptions to the generalizations presented below.

12.6.3.1 Cement-Based S/S

For cement-based S/S, if a single metal is the predominant contaminant in soil, then Cd and Pb are the most amenable to cement-based S/S. The predominant mechanism for immobilization of metals in Portland and similar cements is precipitation of hydroxides, carbonates, and silicates. Both Pb and Cd tend to form insoluble precipitates in the pH ranges found in cured cement. They may resolubilize, however, if the pH is not carefully controlled. For example, Pb in aqueous solutions tends to resolubilize as $Pb(OH)^{3-}$ around pH 10 and above. Hg, though it is a cationic metal like Pb and cadmium, does not form low-solubility precipitates in cement; hence it is difficult to stabilize reliably by cement-based processes, and this difficulty would be expected to be greater with increasing Hg concentration and with organomercury compounds. As, due to its formation of anionic species, also does not form insoluble precipitates in the high pH cement environment, and cement-based solidification is generally not expected to be successful. Cr(VI) is difficult to stabilize in cement due to the formation of anions that are soluble at high pH. However, Cr(VI) can be reduced to Cr(III), which does form insoluble hydroxides. Although Hg and As (III and V) are particularly difficult candidates for cement-based S/S, this should not necessarily eliminate S/S (even cement-based) from consideration since (a) as with Cr(VI), it may be possible to devise a multistep process that will produce an acceptable product for cement-based S/S; (b) a non-cement-based S/S process (e.g., lime and sulfide for Hg; oxidation to As(V); and coprecipitation with iron) may be applicable; or (c) the leachable concentration of the contaminant may be sufficiently low that a highly efficient S/S process may not be required to meet treatment goals.

The above discussion on applicability also applies to *in situ* cement-based S/S. If *in situ* treatment introduces chemical agents into the ground, this chemical addition may cause a pollution problem in itself, and may be subject to additional requirements under the Land Disposal Restrictions.

12.6.3.2 Polymer Microencapsulation

Polymer microencapsulation has been mainly used to treat low-level radioactive wastes. However, organic binders have been tested or applied to wastes containing chemical contaminants such as As, metals, inorganic salts, polychlorinated biphenyls, and dioxins [36]. Polymer microencapsulation is particularly well suited for treating water-soluble salts such as chlorides or sulfates that are generally difficult to immobilize in a cement-based system [37]. Characteristics of the organic binder and extrusion system impose compatibility requirements on the waste material. The elevated operating temperatures place a limit on the quantity of water and volatile organic chemicals (VOCs) in the waste feed. Low-volatility organics will be retained in the bitumen but may act as solvents causing the treated product to be too fluid. Bitumen is a potential fuel source, so the waste should not contain oxidizers such as nitrates, chlorates, or perchlorates. Oxidants present the potential for rapid oxidation, causing immediate safety concerns, as well as slow oxidation, which results in waste form degradation.

Cement-based S/S of multiple metal wastes is particularly difficult if a set of treatment and disposal conditions that simultaneously produce low-mobility species for all the metals of concern cannot be found. For example, the relatively high pH conditions that favor Pb immobilization would tend to increase the mobility of As. On the other hand, the various metal species in a multiple metal waste may interact (e.g., the formation of low-solubility compounds by the combination of Pb and arsenate) to produce a low-mobility compound.

Organic contaminants are often present with inorganic contaminants at metal-contaminated sites. S/S treatment of organic-contaminated waste with cement-based binders is more complex than treatment of inorganics alone. This is particularly true with VOCs, where the mixing process and heat generated by cement hydration reactions can increase vapor losses [38–41]. However, S/S can be applied to wastes that contain lower levels of organics, particularly when inorganics are present and/or the organics are semivolatile or nonvolatile. Also, recent studies indicate that the addition of silicates or modified clays to the binder system may improve S/S performance with organics [27].

12.6.4 Performance and BDAT Status

Information about the use of S/S at Superfund remedial sites in 2000 indicates that S/S has been used at 167 sites since FY 1982 [26]. Figure 12.1 shows the number of projects by status for the following stages: predesign/design, design completed/being installed, operational, and completed. Data are shown for the *in situ* and *ex situ* S/S projects. In addition, information about all source control technologies is provided. With respect to S/S projects, the majority of both *in situ* and *ex situ* projects (62%) are completed, followed by projects in the predesign/design stage (21%). Overall, completed S/S projects represent 30% of all completed Superfund projects in which treatment technologies have been used for source control.

Figure 12.2 shows the types of binder materials used for S/S projects at Superfund remedial sites, including inorganic binders, organic binders, and combination organic and inorganic binders. Many of the binders used include one or more proprietary additives. Examples of inorganic binders include cement, fly ash, lime, soluble silicates, and sulfur-based binders, whereas organic binders include asphalt, epoxide, polyesters, and polyethylene. More than 90% of the S/S projects used inorganic binders. In general, inorganic binders are less expensive and easier to use than organic binders. Organic binders are generally used to solidify radioactive wastes or specific hazardous organic compounds.

Figure 12.3 shows the types of contaminant groups and combination of contaminant groups treated by S/S at Superfund remedial sites. S/S was used to treat metals only in 56% of the projects, and used to treat metals alone or in combination with organics or radioactive metals at approximately 90% of the sites. S/S was used to treat organics only at 6% of the sites [26]. Figure 12.4 provides a further breakdown of the metals treated by S/S at Superfund remedial sites. The top five metals treated by S/S are Pb, Cr, As, Cd, and Cu.

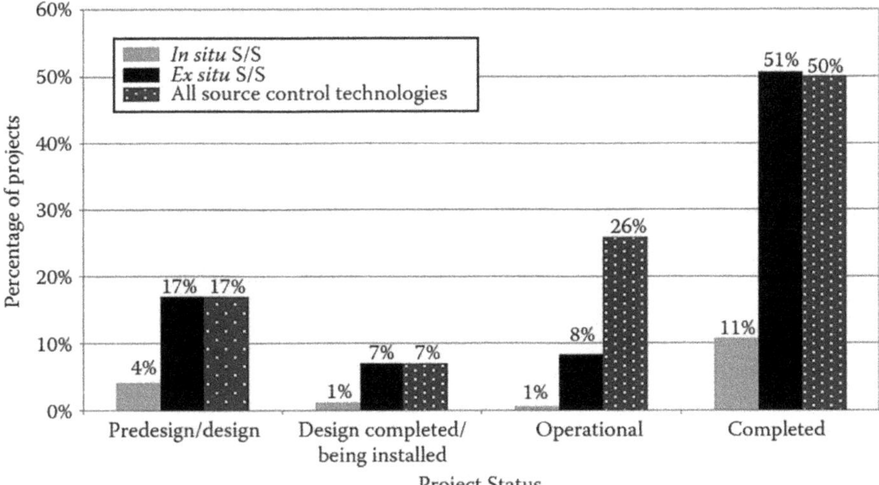

FIGURE 12.1 Percentage of Superfund remedial projects by status. (Adapted from U.S. EPA. *Solidification/ Stabilization Use at Superfund Sites*. EPA-542-R-00-010, U.S. Environmental Protection Agency, Washington, DC, September 2000 [26].)

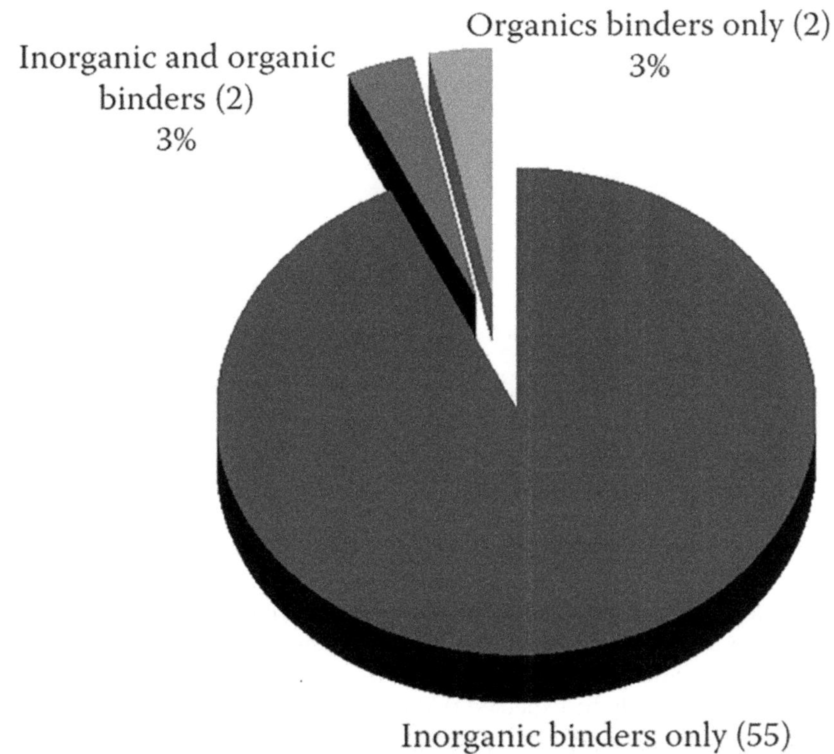

FIGURE 12.2 Binder materials used for S/S projects. (Adapted from U.S. EPA. *Solidification/Stabilization Use at Superfund Sites*. EPA-542-R-00-010, U.S. Environmental Protection Agency, Washington, DC, September 2000 [26].)

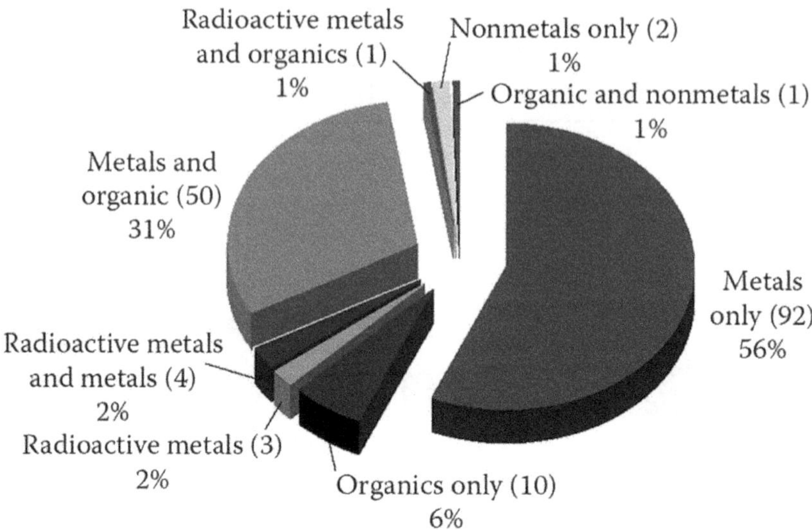

FIGURE 12.3 Contaminant types treated by S/S. (Adapted from U.S. EPA. *Solidification/Stabilization Use at Superfund Sites*. EPA-542-R-00-010, U.S. Environmental Protection Agency, Washington, DC, September 2000 [26].)

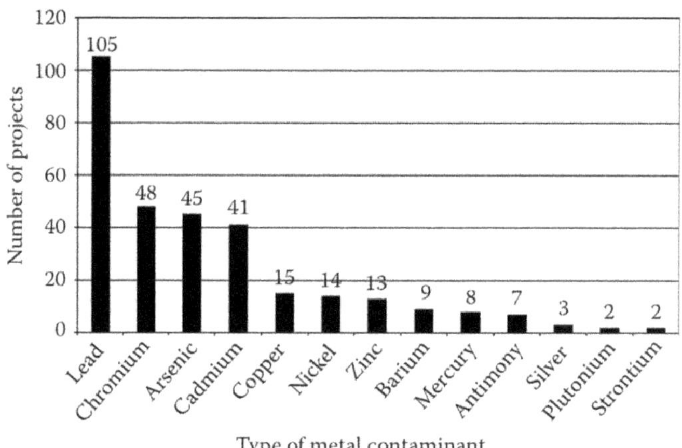

FIGURE 12.4 Number of S/S projects treating specific metals. (Adapted from U.S. EPA. *Solidification/ Stabilization Use at Superfund Sites*. EPA-542-R-00-010, U.S. Environmental Protection Agency, Washington, DC, September 2000 [26].)

S/S with cement-based and pozzolan binders is a commercially available, established technology [5]. Table 12.4 shows a selected list of sites where S/S has been selected for remediating metal-contaminated solids. Note that S/S has been used to treat all five metals (Cr, Pb, As, Hg, and Cd). Although it would not generally be expected that cement-based S/S would be applied to As- and Hg-contaminated soils, it was beyond the scope of the project to examine in detail the characterization data, S/S formulations, and performance data upon which the selections were based; hence the selection/implementation data are presented without further comment.

Applications of polymer microencapsulation have been limited to special cases where specific performance features are required for the waste matrix, and contaminants allow reuse of the treated waste as a construction material [42].

TABLE 12.4
S/S Applications at Superfund Sites with Metal Contamination

Site Name/State	Specific Technology	Key Metal Contaminants	Associated Technology
DeRewal Chemical, NJ	Solidification	Cr, Cd, Pb	GW pump and treatment
Marathon Battery Co., NY	Chemical fixation	Cd, Ni	Dredging, off-site disposal
Nascolite, Millville, NJ	Stabilization of wetland soils	Pb	On-site disposal of stabilized soils; excavation and off-site disposal of wetland soils
Roebling Steel, NJ	S/S	As, Cr, Pb	Capping
Waldick Aerospace, NJ	S/S	Cd, Cr	Off-site disposal
Aladdin Plating, PA	Stabilization	Cr	Off-site disposal
Palmerton Zinc, PA	Stabilization, fly ash, lime, potash	Cd, Pb	—
Tonolli Corp., PA	S/S	As, Pb	*In situ* chemical limestone barrier
Whitmoyer Laboratories, PA	Oxidation/fixation	As	GW pump and treatment, capping, grading, and revegetation
Bypass 601, NC	S/S	Cr, Pb	Capping, regrading, revegetation, GW pump and treatment
Flowood, MS	S/S	Pb	Capping
Independent Nail, SC	S/S	Cd, Cr	Capping
Pepper's Steel and Alloys, FL	S/S	As, Pb	On-site disposal
Gurley Pit, AR	*In situ* S/S	Pb	—
Pesses Chemical, TX	Stabilization	Cd	Concrete capping
E.I. Dupont de Nemours, IA	S/S	Cd, Cr, Pb	Capping, regrading, and revegetation
Shaw Avenue Dump, IA	S/S	As, Cd	Capping, groundwater monitoring
Frontier Hard Chrome, WA	Stabilization	Cr	—
Gould Site, OR	S/S	Pb	Capping, regrading, and revegetation

Source: U.S. EPA. *Technology Alternatives for the Remediation of Soils Contaminated with AS, Cd, Cr, Hg, and Pb.* EPA/540/S-97/500, U.S. Environmental Protection Agency, Cincinnati, OH, August 1997 [5].

S/S is a BDAT for the following waste types [5]:

- Cd non-wastewaters other than Cd-containing batteries,
- Cr non-wastewaters following reduction to Cr(III),
- Pb non-wastewaters,
- wastes containing low concentrations (<260 mg/kg) of elemental Hg-sulfide precipitation, and
- plating wastes and steelmaking wastes.

Although vitrification, not S/S, was selected as BDAT for RCRA As-containing non-wastewaters, U.S. EPA does not preclude the use of S/S for the treatment of As (particularly inorganic As) wastes but recommends that its use be determined on a case-by-case basis. A variety of stabilization techniques, including cement, silicate, pozzolan, and ferric coprecipitation, were evaluated as candidate. BDATs for As. Due to concerns about long-term stability and waste volume increase, particularly with ferric coprecipitation, stabilization was not accepted as BDAT.

12.6.5 SITE PROGRAM DEMONSTRATION PROJECTS

Completed SITE demonstrations applicable to soils contaminated with the metals of interest include [5]

- Advanced Remediation Mixing, Inc. (*ex situ* S/S),
- Funderburk and Associates (*ex situ* S/S),
- Geo-Con, Inc. (*in situ* S/S),
- Soliditech, Inc. (*ex situ* S/S),
- STC Omega, Inc. (*ex situ* S/S),
- WASTECH Inc. (*ex situ* S/S),
- Separation and Recovery Systems, Inc. (*ex situ* S/S), and
- Wheelabrator Technologies Inc. (*ex situ* S/S).

12.6.6 COST OF S/S

Information about the cost of using S/S to treat wastes at Superfund remedial sites was reported by U.S. EPA for 29 completed projects in 2000 [26]. Total costs in terms of 2007 USD [43] for S/S projects ranged from USD 86,000 to USD 18,000,000 including the cost of excavation, treatment, and disposal (if *ex situ*). The cost ranged from USD $12/m^3$ to approximately USD $1,800/m^3$. The average cost for these projects was USD $396/m^3$, including two projects with relatively high costs (approximately USD $1800/m^3$). Excluding those two projects, the average cost per cubic meter was USD 291 [26].

12.7 VITRIFICATION

Vitrification applies high-temperature treatment aimed primarily at reducing the mobility of metals by incorporation into a chemically durable, leach-resistant, vitreous mass. Vitrification can be carried out on excavated soils as well as *in situ*.

12.7.1 PROCESS DESCRIPTION

During the vitrification process, organic wastes are pyrolyzed (*in situ*) or oxidized (*ex situ*) by the melt front, whereas inorganics, including metals, are incorporated into the vitreous mass. Off-gases released during the melting process, containing volatile components and products of combustion and pyrolysis, must be collected and treated [4,44,45]. Vitrification converts contaminated soils to a stable glass and crystalline monolith [45]. With the addition of low-cost materials such as sand, clay, and/or native soil, the process can be adjusted to produce products with specific characteristics, such as chemical durability. Waste vitrification may be able to transform the waste into useful, recyclable products such as clean fill, aggregate, or higher valued materials such as erosion-control blocks, paving blocks, and road dividers.

12.7.1.1 *Ex Situ* Vitrification

Ex situ vitrification (ESV) technologies apply heat to a melter through a variety of sources such as combustion of fossil fuels (coal, natural gas, and oil) or input of electric energy by direct joule heat, arcs, plasma torches, and microwaves. Combustion or oxidation of the organic portion of the waste can contribute significant energy to the melting process, thus reducing energy costs. The particle size of the waste may need to be controlled for some of the melting technologies. For wastes containing refractory compounds that melt above the unit's nominal processing temperature, such as quartz or alumina, size reduction may be required to achieve acceptable throughputs and a homogeneous melt. For high-temperature processes using arcing or plasma technologies, size reduction is not a major factor. For the intense melters using concurrent gas-phase melting or mechanical agitation, size reduction is needed for feeding the system and for achieving a homogeneous melt.

12.7.1.2 *In Situ* Vitrification

In situ vitrification (ISV) technology is based on electric melter technology, and the principle of operation is joule heating, which occurs when an electrical current is passed through a region that behaves as a resistive heating element. Electrical current is passed through the soil by means of an array of electrodes inserted vertically into the surface of the contaminated soil zone. Because dry soil is not conductive, a starter path of flaked graphite and glass frit is placed in a small trench between the electrodes to act as the initial flow path for electricity. Resistance heating in the starter path transfers heat to the soil that then begins to melt. Once molten, the soil becomes conductive. The melt grows outward and downward as power is gradually increased to the full constant operating power level. A single melt can treat a region of up to 1,000 T. The maximum treatment depth has been demonstrated to be about 6 m. Large contaminated areas are treated in multiple settings that fuse the blocks together to form one large monolith [4]. Further information on ISV can be found in the following references [46–49].

12.7.2 Site Requirements

The site must be prepared for the mobilization, operation, maintenance, and demobilization of the equipment. Site activities such as clearing vegetation, removing overburden, and acquiring backfill material are often necessary for ESV as well as ISV. *Ex situ* processes will require areas for the storage of excavated, treated, and possibly pretreated materials. The components of one ISV system are contained in three transportable trailers: an off-gas and process control trailer, a support trailer, and an electrical trailer. The trailers are mounted on wheels sufficient for transportation to and over a compacted ground surface [50].

The field-scale ISV system evaluated in the SITE program required three-phase electrical power at either 12,500 or 13,800 V, which is usually taken from a utility distribution system [51]. Alternatively, the power may be generated on-site by means of a diesel generator. Typical applications require 800–1,000 kWh/T [46].

12.7.3 Applicability

Setting cost and implementability aside, vitrification should be most applicable where nonvolatile metal contaminants have glass solubilities exceeding the level of contamination in the soil. Cr-contaminated soil should pose the least difficulties for vitrification, since it has low volatility, and glass solubility between 1% and 3%. Vitrification may or may not be applicable for Pb, As, and Cd, depending on the level of difficulty encountered in retaining the metals in the melt, and controlling and treating any volatile emissions that may occur. Hg clearly poses problems for vitrification due to high volatility and low glass solubility (<0.1%), but may be allowable at very low concentrations.

Chlorides present in the waste in excess of about 0.5% by weight (wt) typically will not be incorporated into and discharged with the glass but will fume off and enter the off-gas treatment system. If chlorides are excessively concentrated, salts of alkali, alkaline earths, and heavy metals will accumulate in solid residues collected by off-gas treatment. Separation of the chloride salts from the other residuals may be required before or during the return of residuals to the melter. When excess chlorides are present, there is also a possibility that dioxins and furans may form and enter the off-gas treatment system. Waste matrix composition affects the durability of the treated waste. Sufficient glass-forming materials, SiO_2 (>30 wt%), and combined alkali, $Na + K$ (>1.4 wt%), are required for vitrification of wastes. If these conditions are not met, frit and/or flux additives are typically needed. Vitrification is also potentially applicable to soils contaminated with mixed metals and metal-organic wastes.

Specific situations where ESV would not be applicable or would face additional implementation problems include the following [5]:

1. Wastes containing >25% moisture content cause excessive fuel consumption.
2. Wastes where size reduction and classification are difficult or expensive.
3. Volatile metals, particularly Cd and Hg, will vaporize and must be captured and treated separately.
4. Arsenic-containing wastes may require pretreatment to produce less volatile forms.
5. Metal concentrations in soil that exceed their solubility in glass.
6. Sites where commercial capacity is not adequate or transportation cost to a fixed facility is unacceptable.

Specific situations, in addition to those cited above, where ISV would not be applicable or would face additional implementation problems include the following [5]:

1. Metal-contaminated soil where a less costly and adequately protective remedy exists.
2. Projects that cannot be undertaken because of limited commercial availability.
3. Contaminated soil <2 m and >6 m below the ground surface.
4. Presence of an aquifer with high hydraulic conductivity (e.g., soil permeability $>1 \times 10^{-5}$ cm/s) limits economic feasibility due to the excessive energy required.
5. Contaminated soil mixed with buried metal that can result in a conductive path causing short circuiting of electrodes.
6. Contaminated soil mixed with loosely packed rubbish or buried coal can start underground fires and overwhelm off-gas collection and the treatment system.
7. Volatile heavy metals near the surface can be entrained in combustion product gases and not retained in the melt.
8. Sites where the surface slope is >5% may cause the melt to flow.
9. *In situ* voids >150 m³ interrupt conduction and heat transfer.
10. Underground structures and utilities <6 m from the melt zone must be protected from heat or avoided.

Where it can be successfully applied, the advantages of vitrification include the following [5]:

1. Vitrified product is an inert, impermeable solid that should reduce leaching for long periods of time.
2. The volume of vitrified product will typically be smaller than initial waste volume.
3. Vitrified product may be usable.
4. A wide range of inorganic and organic wastes can be treated.
5. There is both an *ex situ* and an *in situ* option available.

A particular advantage of *ex situ* treatment is its better control of processing parameters. Also, fuel costs may be reduced for ESV by the use of combustible waste materials. This fuel cost-saving option is not directly applicable for ISV, since combustibles would increase the design and operating requirements for gas capture and treatment.

12.7.4 Performance and BDAT Status

ISV has been implemented at metal-contaminated Superfund sites and was evaluated under the SITE Program [52]. Some improvements are needed for melt containment and air emission control systems. ISV has been operated at a large scale 10 times, including two demonstrations on radioactively contaminated sites at the DOE's Hanford Nuclear Reservation [44,53]. Pilot-scale tests have been conducted at Oak Ridge National Laboratory, Idaho National Engineering Laboratory,

TABLE 12.5
ISV Applications at Superfund Sites with Metal Contamination

Site Name/State Key Metal Contaminants

| Parsons Chemical, MI | Hg (low) |
| Rocky Mountain Arsenal, CO | As, Hg |

Source: U.S. EPA. *Technology Alternatives for the Remediation of Soils Contaminated with AS, Cd, Cr, Hg, and Pb.* EPA/540/S-97/500, U.S. Environmental Protection Agency, Cincinnati, OH, August 1997 [5].

and Arnold Engineering Development Center. More than 150 tests and demonstrations at various scales have been performed on a broad range of waste types in soils and sludges. The technology has been selected as a preferred remedy at 10 private, Superfund, and DoD sites [54]. Table 12.5 provides a summary of ISV technology selection/application at metal-contaminated Superfund sites. A number of ESV systems are under development. The technical resource document identified one full-scale *ex situ* melter that was reported to be operating on RCRA organics and inorganics.

Vitrification is a BDAT for the As-containing wastes.

12.7.5 SITE PROGRAM DEMONSTRATION PROJECTS

Completed SITE demonstrations applicable to soils contaminated with the metals of interest include [5]

- Babcock & Wilcox Co. (cyclone furnace—ESV),
- Retech, Inc. (plasma arc—ESV),
- Geosafe Corporation (ISV), and
- Vortec Corporation (*ex situ* oxidation and vitrification process).

12.8 SOIL WASHING

Soil washing is an *ex situ* remediation technology that uses a combination of physical separation and aqueous-based separation unit operations to reduce contaminant concentrations to site-specific remedial goals [55]. Although soil washing is sometimes used as a stand-alone treatment technology, more often it is combined with other technologies to complete site remediation. Soil washing technologies have successfully remediated sites contaminated with organic, inorganic, and radioactive contaminants [55]. The technology does not detoxify or significantly alter the contaminant but transfers the contaminant from the soil into the washing fluid or mechanically concentrates the contaminants into a much smaller soil mass [56] for subsequent treatment (see Figure 12.5).

Further information on soil washing can be found in U.S. EPA innovative technology reports and programs [57,58].

12.8.1 PROCESS DESCRIPTION

Soil washing systems are quite flexible in terms of the number, type, and order of processes involved. Soil washing is performed on excavated soil and may involve some or all of the following, depending on the contaminant-soil matrix characteristics, cleanup goals, and the specific process employed [5,56]:

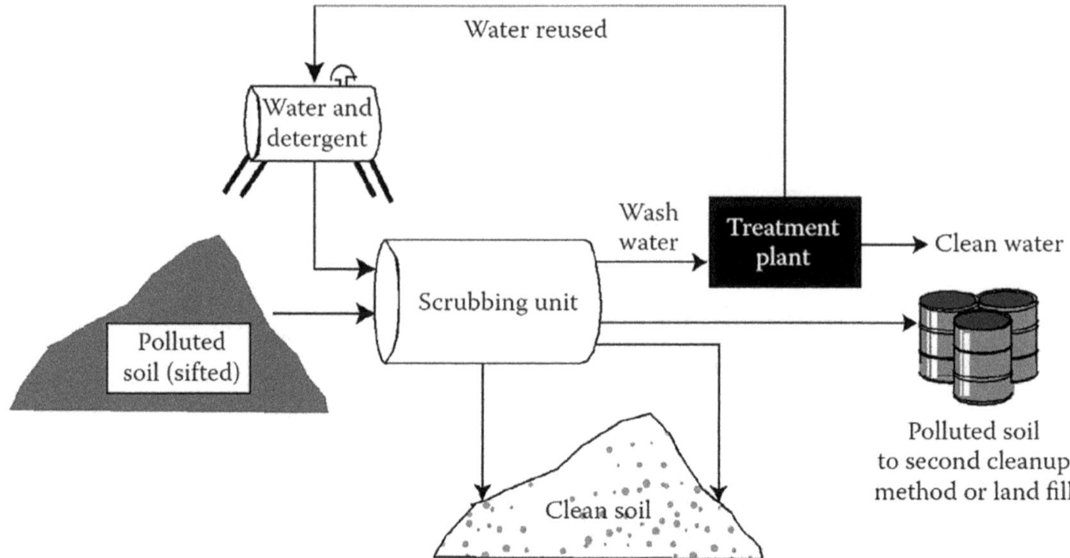

FIGURE 12.5 Soil washing operation. (Adapted from U.S. EPA. *A Citizen's Guide to Soil Washing.* EPA 542-F-01-008, U.S. Environmental Protection Agency, Washington, DC, May 2001 [56].)

1. Mechanical screening to remove various oversized materials.
2. Crushing to reduce applicable oversize to suitable dimensions for treatment.
3. Physical processes (e.g., soaking, spraying, tumbling, and attrition scrubbing) to liberate weakly bound agglomerates (e.g., silts and clays bound to sand and gravel) followed by size classification to generate coarse-grained and fine-grained soil fraction(s) for further treatment.
4. Treatment of the coarse-grained soil fraction(s).
5. Treatment of the fine-grained fraction(s).
6. Management of the generated residuals.

Treatment of the coarse-grained soil fraction typically involves additional application of physical separation techniques and possibly aqueous-based leaching techniques. Physical separation techniques (e.g., sorting, screening, elutriation, hydrocyclones, spiral concentrators, flotation) exploit physical differences (e.g., size, density, shape, color, wettability) between contaminated particles and soil particles to produce a clean (or nearly clean) coarse fraction and one or more metal-concentrated streams. Many of the physical separation processes listed above involve the use of water as a transport medium, and if the metal contaminant has significant water solubility, then some of the coarse-grained soil cleaning will occur as a result of transfer to the aqueous phase. If the combination of physical separation and unaided transfer to the aqueous phase cannot produce the desired reduction in the soil's metal content, which is frequently the case for metal contaminants, then solubility enhancement is an option for meeting cleanup goals for the coarse fraction. Solubility enhancement can be accomplished in several ways [5,59,60]:

1. Converting the contaminant into a more soluble form (e.g., oxidation/reduction and conversion to soluble metal salts).
2. Using an aqueous-based leaching solution (e.g., acidic, alkaline, oxidizing, or reducing) in which the contaminant has enhanced solubility.

3. Incorporating a specific leaching process into the system to promote increased solubilization via increased mixing, elevated temperatures, higher solution/soil ratios, efficient solution/soil separation, multiple stage treatment, and so on.
4. A combination of the above.

After the leaching process is completed on the coarse-grained fraction, it will be necessary to separate the leaching solution and the coarse-grained fraction by settling. A soil rinsing step may be necessary to reduce the residual leachate in the soil to an acceptable level. It may also be necessary to readjust soil parameters such as pH or redox potential before replacement of the soil on the site. The metal-bearing leaching agent must also be treated further to remove the metal contaminant and permit reuse in the process or discharge, and this topic is discussed below under management of residuals.

The treatment of fine-grained soils is similar in concept to the treatment of coarse-grained soils, but the production rate would be expected to be lower and hence more costly than for the coarse-grained soil fraction. The reduced production rate arises from factors including (a) the tendency of clays to agglomerate, thus requiring time, energy, and high water/clay ratios to produce leachable slurry; and (b) slow-settling velocities that require additional time and/or capital equipment to produce acceptable soil/water separation for multibatch or countercurrent treatment, or at the end of treatment. A site-specific determination needs to be made on whether the fines should be treated to produce clean fines or whether they should be handled as aresidual waste stream.

Management of generated residuals is an important aspect of soil washing. The effectiveness, implementability, and cost of treating each residual stream are important to the overall success of soil washing for the site. Perhaps the most important of the residual streams is the metal-loaded leachant that is generated, particularly if the leaching process recycles the leaching solution. Furthermore, it is often critical to the economic feasibility of the project that the leaching solution be recycled. For these closed or semiclosed loop leaching processes, successful treatment of the metal-loaded leachant is imperative to the successful cleaning of the soil. The leachant must (a) have adequate solubility for the metal so that the metal reduction goals can be met without using excessive volumes of leaching solution; and (b) be readily, economically, and repeatedly adjustable (e.g., pH adjustment) to a form in which the metal contaminant has very low solubility so that the recycled aqueous phase retains a favorable concentration gradient compared to the contaminated soil. Also, efficient soil-water separation is important prior to recovering metal from the metal-loaded leachant in order to minimize contamination of the metal concentrate. Recycling the leachant reduces logistical requirements and costs associated with makeup water, storage, permitting, compliance analyses, and leaching agents. It also reduces external coordination requirements and eliminates the dependence of the remediation on the ability to meet Publicly Owned Treatment Works (POTW) discharge requirements.

Other residual streams that may be generated and require proper handling include the following [5]:

1. Untreatable and uncrushable oversize.
2. Recyclable metal-bearing particulates, concentrates, or sludges from physical separation or leachate treatment.
3. Nonrecyclable metal-bearing particulates, concentrates, soils, sludges, or organic debris that fail toxicity characteristic leaching procedure (TCLP) thresholds for RCRA hazardous waste.
4. Soils or sludges that are neither RCRA hazardous wastes nor sufficiently clean to permit return to the site.
5. Metal-loaded leachant from systems where leachant is not recycled.
6. Rinsate from treated soil.

12.8.2 SITE REQUIREMENTS

The area required for a unit at a site will depend on the vendor system selected, the amount of soil storage space, and/or the number of tanks or ponds needed for washwater preparation and wastewater storage and treatment. Typical utilities required are water, electricity, steam, and compressed air; the quantity of each is vendor- and site-specific. It may be desirable to control the moisture content of contaminated soil for consistent handling and treatment by covering the excavation, storage, and treatment areas. Climatic conditions such as annual or seasonal precipitation cause surface runoff and water infiltration; therefore, runoff control measures may be required. Since soil washing is an aqueous-based process, cold weather impacts include freezing as well as potential effects on leaching rates.

12.8.3 APPLICABILITY

Soil washing is potentially applicable to soils contaminated with all five metals of interest. Conditions that particularly favor soil washing include the following [5]:

1. A single principal contaminant metal that occurs in dense, insoluble particles that report to a specific, small mass fraction of the soil.
2. A single contaminant metal and species that is very water or aqueous leachant soluble and has a low soil/water partition coefficient.
3. Soil containing a high proportion (e.g., >80%) of soil particles >2 mm, which is desirable for efficient contaminant-soil and soil-water separation.

Conditions that clearly do not favor soil washing include the following [5]:

1. Soils with a high (i.e., >40%) silt and clay fraction.
2. Soils that vary widely and frequently in significant characteristics such as soil type, contaminant type, and concentration, and where blending for homogeneity is not feasible.
3. Complex mixtures (e.g., multicomponent, solid mixtures where access of leaching solutions to contaminant is restricted; mixed anionic and cationic metals where the pH values of solubility maximums are not close).
4. High clay content, cation exchange capacity, or humic acid content, which would tend to interfere with contaminant desorption.
5. The presence of substances that interfere with the leaching solution (e.g., carbonaceous soils would neutralize extracting acids; similarly, high humic acid content will interfere with an alkaline extraction).
6. Metal contaminants in a very low solubility, stable form (e.g., PbS) may require long contact times and excessive amounts of reagent to solubilize.

12.8.4 PERFORMANCE AND BDAT STATUS

Soil washing has been used at waste sites in Europe, especially in Germany, the Netherlands, and Belgium [61]. Table 12.6 lists selected Superfund sites where soil washing has been selected and/or implemented.

Acid leaching, which is a form of soil washing, is the BDAT for Hg.

12.8.5 SITE DEMONSTRATIONS AND EMERGING TECHNOLOGIES PROGRAM PROJECTS

SITE demonstrations applicable to soils contaminated with the metals of interest include [5]

TABLE 12.6
Soil Washing Applications at Selected Superfund Sites with Metal Contamination

Site Name/State	Specific Technology	Key Metal Contaminants	Associated Technology
Ewan Property, NJ	Water washing	As, Cr, Cu, Pb	Pretreatment by solvent extraction to remove organics
GE Wiring Devices, PR	Water with KI solution additive	Hg	Treated residues disposed on-site and covered with clean soil
King of Prussia, NJ	Water with washing agent additives	Ag, Cr, Cu	Sludges to be land disposed
Zanesville Well Field, OH	Soil washing	Hg, Pb	SVE to remove organics
Twin Cities Army Ammunition Plant, MN	Soil washing	Cd, Cr, Cu, Hg, Pb	Soil leaching
Sacramento Army Depot Sacramento, CA	Soil washing	Cr, Pb	Off-site disposal of wash liquid

Source: U.S. EPA. *Technology Alternatives for the Remediation of Soils Contaminated with AS, Cd, Cr, Hg, and Pb.* EPA/540/S-97/500, U.S. Environmental Protection Agency, Cincinnati, OH, August 1997 [5].

- Bergmann USA (physical separation/leaching) BioGenesis™ (physical separation/ leaching),
- Biotrol, Inc. (physical separation),
- Brice Environmental Services Corp. (physical separation),
- COGNIS, Inc. (leaching), and
- Toronto Harbor Commission (physical separation/leaching).

Four SITE Emerging Technologies Program projects have been completed that are applicable to soils contaminated with the metals of interest.

12.9   SOIL FLUSHING

Soil flushing is the *in situ* extraction of contaminants from the soil via an appropriate washing solution. Water or an aqueous solution is injected into or sprayed onto the area of contamination, and the contaminated elutriate is collected and pumped to the surface for removal, recirculation, or on-site treatment and reinjection. The technology is applicable to both organic and inorganic contaminants, and metals in particular [4]. For the purpose of metals remediation, soil flushing has been operated at full scale, but for a small number of sites.

12.9.1   Process Description

Soil flushing uses water, a solution of chemicals in water, or an organic extractant to recover contaminants from the *in situ* material. The contaminants are mobilized by solubilization, formation of emulsions, or a chemical reaction with the flushing solutions. After passing through the contamination zone, the contaminant-bearing fluid is collected by strategically placed wells or trenches and brought to the surface for disposal, recirculation, or on-site treatment and reinjection. During elutriation, the flushing solution mobilizes the sorbed contaminants by dissolution or emulsification.

One key to efficient operation of a soil flushing system is the ability to reuse the flushing solution, which is recovered along with groundwater. Various water treatment techniques can be applied to remove the recovered metals and render the extraction fluid suitable for reuse. Recovered flushing fluids may need treatment to meet appropriate discharge standards prior to their release to a POTW or receiving waters. The separation of surfactants from recovered flushing fluid, for reuse in the process, is a major factor in the cost of soil flushing. Treatment of the flushing fluid results in process sludges and residual solids, such as spent carbon and spent ion exchange resin, which must be appropriately treated before disposal. Air emissions of volatile contaminants from recovered flushing fluids should be collected and treated, as appropriate, to meet applicable regulatory standards. Residual flushing additives in the soil may be a concern and should be evaluated on a site-specific basis [62]. Subsurface containment barriers can be used in conjunction with soil flushing technology to help control the flow of flushing fluids.

Further information on soil flushing can be found in references [57,62–64].

12.9.2 SITE REQUIREMENTS

Stationary or mobile soil flushing systems are located on-site. The exact area required will depend on the vendor system selected and the number of tanks or ponds needed for washwater preparation and wastewater treatment. Certain permits may be required for operation, depending on the system being utilized. Slurry walls or other containment structures may be needed along with hydraulic controls to ensure capture of contaminants and flushing additives. Impermeable membranes may be necessary to limit infiltration of precipitation, which could cause dilution of the flushing solution and loss of hydraulic control. Cold weather freezing must also be considered for shallow infiltration galleries and aboveground sprayers [65].

12.9.3 APPLICABILITY

Soil flushing may be easy or difficult to apply, depending on the ability to wet the soil with the flushing solution and to install collection wells or subsurface drains to recover all the applied liquids. The achievable level of treatment varies and depends on the contact of the flushing solution with the contaminants and the appropriateness of the solution for contaminants, and the hydraulic conductivity of the soil. Soil flushing is most applicable to contaminants that are relatively soluble in the extracting fluid, and that will not tend to sorb onto soil as the metal-laden flushing fluid proceeds through the soil to the extraction point. Based on the earlier discussion of metal behavior, some potentially promising scenarios for soil flushing would include Cr(VI), As (III or V) in permeable soil with low iron oxide, low clay, and high pH; Cd in permeable soil with low clay, low cation exchange capacity, and moderately acidic pH; and Pb in acid sands. A single target metal would be preferable to multiple metals, due to the added complexity of selecting a flushing fluid that would be reasonably efficient for all contaminants. Also, the flushing fluid must be compatible with not only the contaminant, but also the soil. Soils that counteract the acidity or alkalinity of the flushing solution will decrease its effectiveness. If precipitants occur due to interaction between the soil and the flushing fluid, then this could obstruct the soil pore structure and inhibit the flow to and through sectors of the contaminated soil. It may take long periods of time for soil flushing to achieve cleanup standards.

A key advantage of soil flushing is that contaminant is removed from the soil. Recovery and reuse of the metal from the extraction fluid may be possible in some cases, although the value of the recovered metal would not be expected to fully offset the costs of recovery. The equipment used for the technology is relatively easy to construct and operate. It does not involve excavation, treatment, and disposal of the soil, which avoids the expense and hazards associated with these activities.

12.9.4 PERFORMANCE AND BDAT STATUS

Table 12.7 lists the Superfund sites where soil flushing has been selected and/or implemented. Soil flushing has a more established history for removal of organics but has been used for Cr removal

TABLE 12.7
Soil Flushing Applications at Selected Superfund Sites with Metal Contamination

Site Name/State	Specific Technology	Key Metal Contaminants	Associated Technology
Lipari Landfill, NJ	Soil flushing of soil and wastes contained by slurry wall and cap	Cr, Hg, Pb	Slurry wall and cap
United Chrome Products, OR	Excavation from Cr impacted wetlands	Cr	Electrokinetic Pilot test, Considering *in situ* reduction

Source: U.S. EPA. *Technology Alternatives for the Remediation of Soils Contaminated with As, Cd, Cr, Hg, and Pb.* EPA/540/S-97/500, U.S. Environmental Protection Agency, Cincinnati, OH, August 1997 [5].

(e.g., United Chrome Products Superfund Site, near Corvallis, Oregon). *In situ* technologies, such as soil flushing, are not considered RCRA BDAT for any of the five metals [5].

Soil flushing techniques for mobilizing contaminants can be classified as conventional and unconventional. Conventional applications employ only water as the flushing solution. Unconventional applications that are currently being researched include the enhancement of flushing water with additives, such as acids, bases, and chelating agents, to aid in the desorption/dissolution of target contaminants from the soil matrix to which they are bound.

Researchers are also investigating the effects of numerous soil factors on heavy metal sorption and migration in the subsurface. Such factors include pH, soil type, soil horizon, particle size, permeability, specific metal type and concentration, and type and concentrations of organic and inorganic compounds in solutions. Generally, as soil pH decreases, cationic metal solubility and mobility increase. In most cases, metal mobility and sorption are likely to be controlled by the organic fraction in topsoils and the clay content in subsoils.

12.9.5   Site Demonstration and Emerging Technologies Program Projects

There are no *in situ* soil flushing projects reported to be completed either as SITE demonstrations or as Emerging Technologies Program Projects [65].

12.10   PYROMETALLURGY

Pyrometallurgy is used here as a broad term encompassing elevated temperature techniques for the extraction and processing of metals for use or disposal. High-temperature processing increases the rate of reaction and often makes the reaction equilibrium more favorable, lowering the required reactor volume per unit output [4]. Some processes that clearly involve both metal extraction and recovery include roasting, retorting, or smelting. While these processes typically produce a metal-bearing waste slag, metal is also recovered for reuse. A second class of pyrometallurgical technologies included here is a combination of high-temperature extraction and immobilization. These processes use thermal means to cause volatile metals to separate from the soil and report to the fly ash, but the metal in the fly ash is then immobilized, instead of recovered, and no metal is recovered for reuse. A third class of technologies are those that are primarily incinerators for mixed organic-inorganic wastes but that have the capability of processing wastes containing the metals of interest by either capturing volatile metals in the exhaust gases or immobilizing the nonvolatile metals in the bottom ash or slag. Since some of these systems may have applicability to some cases where metal contamination is the primary concern, a few technologies of this type are noted in the SITE program. Vitrification is addressed in a previous section. It is not considered pyrometallurgical treatment, since there is typically neither a metal extraction nor a metal recovery component in the process.

12.10.1 Process Description

Pyrometallurgical processing is usually preceded by physical treatment [5] to produce a uniform feed material and upgrade the metal content.

Solids treatment in a high-temperature furnace requires efficient heat transfer between the gas and solid phases while minimizing particulate in the off-gas. The particle-size range that meets these objectives is limited and is specific to the design of the process. The presence of large clumps or debris slows heat transfer; hence, pretreatment to either remove or pulverize oversized material is normally required. Fine particles are also undesirable because they become entrained in the gas flow, increasing the volume of dust to be removed from the flue gas. The feed material is sometimes pelletized to give a uniform size. In many cases, a reducing agent and flux may be mixed in prior to pelletization to ensure good contact between the treatment agents and the contaminated material and to improve the gas flow in the reactor [4].

Due to its relatively low boiling point (357°C) and ready conversion at elevated temperatures to its metallic form, Hg is commonly recovered through roasting and retorting at much lower temperatures than the other metals. Pyrometallurgical processing to convert compounds of the other four metals to elemental metal requires a reducing agent, fluxing agents to facilitate melting and slag off impurities, and a heat source. The fluid mass is often called a melt, but the operating temperature, although quite high, is often below the melting points of the refractory compounds being processed. The fluid forms as a lower-melting-point material due to the presence of a fluxing agent such as calcium. Depending on processing temperatures, volatile metals such as Cd and Pb may fume off and be recovered from the off-gas as oxides. Nonvolatile metals, such as Cr or nickel, are tapped from the furnace as molten metal. Impurities are scavenged by the formation of slag [4]. The effluents and solid products generated by pyrometallurgical technologies typically include solid, liquid, and gaseous residuals. Solid products include debris, oversized rejects, dust, ash, and the treated medium. Dust collected from particulate control devices may be combined with the treated medium or, depending on analyses for carryover contamination, recycled through the treatment unit.

12.10.2 Site Requirements

Few pyrometallurgical systems are available in mobile or transportable configurations. Since this is typically an off-site technology, the distance of the site from the processing facility has an important influence on transportation costs. Off-site treatment must comply with U.S. EPA's off-site treatment policies and procedures. The off-site facility's environmental compliance status must be acceptable, and the waste must be of a type allowable under their operating permits. In order for pyrometallurgical processing to be technically feasible, it must be possible to generate a concentrate from the contaminated soil that will be acceptable to the processor. The processing rate of the off-site facility must be adequate to treat the contaminated material in a reasonable amount of time. Storage requirements and responsibilities must be determined. The need for air discharge and other permits must be determined on a site-specific basis.

12.10.3 Applicability

With the possible exception of Hg, or a highly contaminated soil, pyrometallurgical processing where metal recovery is the goal would not be applied directly to the contaminated soil, but rather to a concentrate generated via soil washing. Pyrometallurgical processing in conventional rotary kilns, rotary furnaces, or arc furnaces is most likely to apply to large volumes of material containing metal concentrations (particularly Pb, Cd, or Cr) higher than 5%–20%. Unless a very concentrated feed stream can be generated (e.g., approximately 60% for Pb), there will be a charge, in addition to transportation, for processing the concentrate. Lower metal concentrations can be acceptable if the metal is particularly easy to reduce and vaporize (e.g., Hg) or is particularly valuable (e.g., gold or platinum). Arsenic is the weakest candidate for pyrometallurgical recovery since there is almost no recycling of arsenic in the United States. Arsenic is also the least valuable of the metals. The price ranges for the five metals [4] are reported here in terms of 2022 USD [43,118]:

	2022 USD/T
As	404–809 (as As trioxide)
Cd	9,865
Cr	12,980
Pb	1,160–1,280
Hg	8,760–14,823

12.10.4 Performance and BDAT Status

The U.S. EPA technical document [4] contains a list of approximately 35 facilities/addresses/contacts that may accept concentrates of the five metals of interest for pyrometallurgical processing. Sixteen of the 35 facilities are Pb recycling operations, seven facilities recover Hg, and the remainder address a range of RCRA wastes that contain the metals of interest. Due to the large volume of electric arc furnace emission control waste, extensive processing capability has been developed to recover Cd, Pb, and Zn from solid waste matrices. The available process technologies include the following [5]:

- A Waelz kiln process (Horsehead Resource Development Company, Inc.).
- A Waelz kiln and calcination process (Horsehead Resource Development Company, Inc.).
- A flame reactor process (Horsehead Resource Development Company, Inc.).
- An inclined rotary kiln (Zia Technology).

Plasma arc furnaces are successfully treating waste at two steel plants. These are site-dedicated units that do not accept outside material for processing.

Pyrometallurgical recovery is a BDAT for the following waste types [5]:

- Cd-containing batteries,
- Pb non-wastewaters in the noncalcium sulfate subcategory,
- Hg wastes prior to retorting,
- Pb-acid batteries,
- Zn non-wastewaters, and
- Hg from wastewater treatment sludge.

12.10.5 Site Demonstration and Emerging Technologies Program Projects

SITE demonstrations applicable to soils contaminated with the metals of interest include [5]

- RUST Remedial Services, Inc. (X-Trax thermal desorption) and
- Horsehead Resource Development Company, Inc. (flame reactor).

12.11 ELECTROKINETICS

Electrokinetic remediation relies on applying low-intensity direct current between electrodes placed in the soil. Contaminants are mobilized in the form of charged species, particles, or ions [2]. Attempts to leach metals from soils by electro-osmosis date back to the 1930s. In the past, research focused on removing unwanted salts from agricultural soils. Electrokinetics has been used for dewatering soils and sludges since the first recorded use in the field in 1939 [66]. Electrokinetic extraction has been used in the former Soviet Union since the early 1970s to concentrate metals and explore minerals in deep soils. By 1979, research had shown that the content of soluble ions increased substantially in electro-osmotic consolidation of polluted dredgings, while metals were not found in the effluent [67]. By the mid-1980s, numerous researchers had realized independently that the electrokinetic separation of metals from soils was a potential solution to contamination [68].

Several organizations are developing technologies for the enhanced removal of metals by transporting contaminants to the electrodes, where they are removed and subsequently treated aboveground. A variation of the technique involves treatment without removal by transporting contaminants through specially designed treatment zones created between electrodes. Electrokinetics can also be used to slow or prevent the migration of contaminants by configuring cathodes and anodes in a manner that causes contaminants to flow toward the center of a contaminated area of soil. Performance data illustrate the potential for achieving removals greater than 90% for some metals [2].

The range of potential metals is broad. The commercial applications in Europe treated copper, lead, zinc, arsenic, cadmium, chromium, and nickel. There is also potential applicability for radionuclides and some types of organic compounds. The electrode spacing and duration of remediation are site-specific. The process requires adequate soil moisture in the vadose zone; hence, the addition of a conducting pore fluid may be required (particularly due to a tendency for soil drying near the anode).

Specially designed pore fluids are also added to enhance the migration of target contaminants. The pore fluids are added at either the anode or the cathode, depending on the desired effects.

Table 12.8 presents an overview of two variations of electrokinetic remediation technology. Geokinetics International, Inc.; Battelle Memorial Institute; Electrokinetics, Inc.; and Isotron Corporation are all developing variations of technologies categorized under Approach #1, Enhanced Removal. The consortium of Monsanto, E.I. du Pont de Nemours and Company (DuPont), General Electric (GE), DOE, and the U.S. EPA Office of Research and Development is developing the Lasagna Process, which is categorized under Approach #2, Treatment without Removal [2].

12.11.1 Process Description

Electrokinetic remediation also referred to as electrokinetic soil processing, electromigration, electrochemical decontamination, or electroreclamation, can be used to extract radionuclides, metals, and some types of organic wastes from saturated or unsaturated soils, slurries, and sediments [69]. This *in situ* soil processing technology is primarily a separation and removal technique for extracting contaminants from soils.

The principle of electrokinetic remediation relies upon the application of a low-intensity direct current through the soil between two or more electrodes. Most soils contain water in the pores between the soil particles and have an inherent electrical conductivity that results from salts present in the soil [70]. The current mobilizes charged species, particles, and ions in the soil by the following processes [71]:

1. Electromigration (transport of charged chemical species under an electric gradient).
2. Electro-osmosis (transport of pore fluid under an electric gradient).
3. Electrophoresis (movement of charged particles under an electric gradient).
4. Electrolysis (chemical reactions associated with the electric field).

Figure 12.6 presents a schematic diagram of a typical conceptual electrokinetic remediation application.

Electrokinetics can be efficient in extracting contaminants from fine-grained, high-permeability soils. A number of factors determine the direction and extent of the migration of the contaminant.

TABLE 12.8
Overview of Electrokinetic Remediation Technology

General characteristics
- Depth of soil that is amenable to treatment depends on electrode placement
- Best used in homogeneous soils with high moisture content and high permeability

Approach #1	Approach #2
Enhanced removal	Treatment without removal
Description:	**Description:**
Electrokinetic transport of contaminants toward the polarized electrodes to concentrate the contaminants for subsequent removal and *ex situ* treatment	Electro-osmotic transport of contaminants through treatment zones placed between the electrodes. The polarity of the electrodes is reversed periodically, which reverses the direction of the contaminants back and forth through treatment zones. The frequency with which electrode polarity is reversed is determined by the rate of transport of contaminants through the soil
Status:	**Status:**
Demonstration projects using full-scale equipment are reported in Europe. Bench- and pilot-scale laboratory studies are reported in the United States and at least two full-scale field studies are ongoing in the United States	Demonstrations are ongoing
Applicability:	**Applicability:**
Pilot scale: lead, arsenic, nickel, mercury, copper, zinc Lab scale: lead, cadmium, chromium, mercury, zinc, iron, magnesium, uranium, thorium, radium	Technology developed for organic species and metals
Comments:	**Comments:**
Field studies are under evaluation by U.S. EPA, DOE, DOD, and EPRI	This technology is being developed for deep clay formations
The technique primarily would require addition of water to maintain the electric current and facilitate migration; however, there is ongoing work in the application of the technology in partially saturated soils	

Source: U.S. EPA. *Recent Developments for In Situ Treatment of Metal Contaminated Soils.* Contract # 68-W5-0055, U.S. Environmental Protection Agency, Washington, DC, March 1997 [2].

Such factors include the type and concentration of the contaminant, the type and structure of the soil, and the interfacial chemistry of the system [72]. Water or some other suitable salt solution may be added to the system to enhance the mobility of the contaminant and increase the effectiveness of the technology. (For example, buffer solutions may change or stabilize pore fluid pH.) Contaminants arriving at the electrodes may be removed by any of several methods, including electroplating at the electrode, precipitation or coprecipitation at the electrode, pumping of water near the electrode, or complexing with ion exchange resins [72].

Electrochemistry associated with this process involves an acid front generated at the anode if water is the primary pore fluid present. The variation of pH at the electrodes results from the electrolysis of the water. The solution becomes acidic at the anode because hydrogen ions are produced and oxygen gas is released, and the solution becomes basic at the cathode, where hydroxyl ions are generated, and hydrogen gas is released [73]. At the anode, the pH could drop to below 2, and it could increase at the cathode to above 12, depending on the total current applied. The acid front

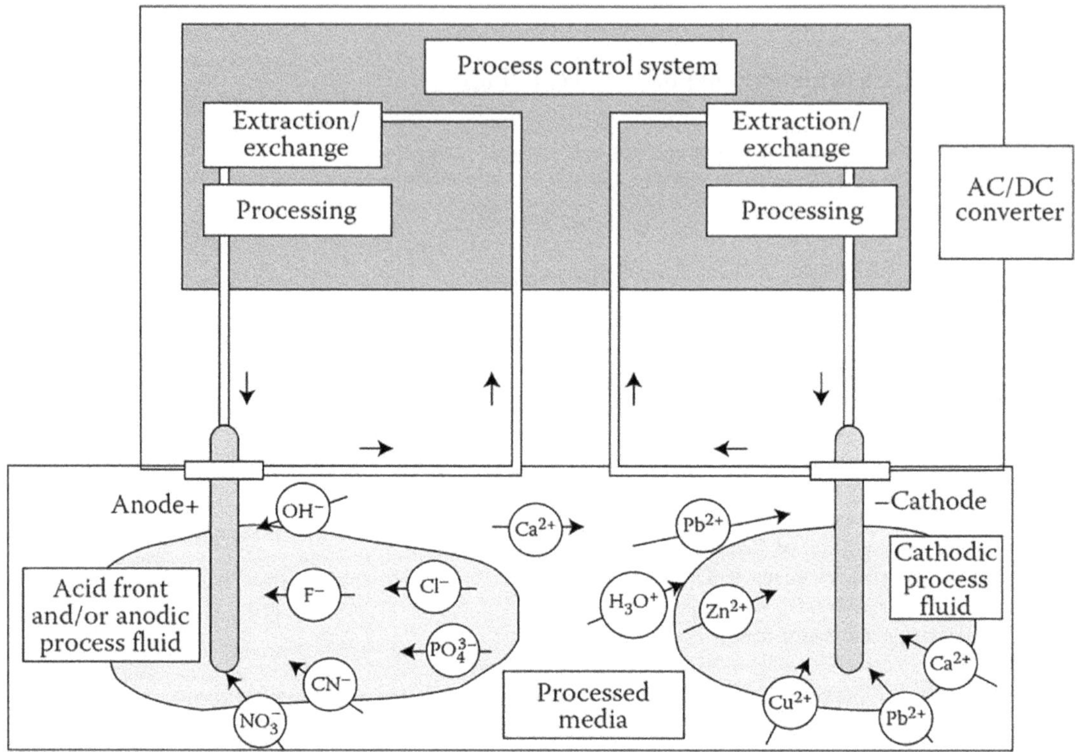

FIGURE 12.6 Diagram of one electrode configuration used in field implementation of electrokinetics. (Adapted from U.S. EPA. *Recent Developments for In Situ Treatment of Metal Contaminated Soils.* Contract # 68-W5-0055, U.S. Environmental Protection Agency, Washington, DC, March 1997 [2])

eventually migrates from the anode to the cathode. Movement of the acid front by migration and advection results in the desorption of contaminants from the soil [69]. The process leads to temporary acidification of the treated soil, and there are no established procedures for determining the length of time needed to reestablish equilibrium. Studies have indicated that metallic electrodes may dissolve as a result of electrolysis and introduce corrosion products into the soil mass. However, if inert electrodes, such as carbon, graphite, or platinum, are used, no residue will be introduced in the treated soil mass as a result of the process [2].

12.11.2 Site Requirements

Before electrokinetic remediation is undertaken at a site, a number of different field and laboratory screening tests must be conducted to determine whether the particular site is amenable to the treatment technique.

1. **Field conductivity surveys:** The natural geologic spatial variability should be delineated because buried metallic or insulating material can induce variability in the electrical conductivity of the soil and, therefore, the voltage gradient. In addition, it is important to assess whether there are deposits that exhibit very high electrical conductivity, at which the technique may be inefficient.
2. **Chemical analysis of water:** The pore water should be analyzed for major dissolved anions and cations, as well as for the predicted concentration of the contaminant(s). In addition, electrical conductivity and pH of the pore water should be measured.

3. **Chemical analysis of soil**: The buffering capacity and geochemistry of the soil should be determined at each site.
4. **pH effects**: The pH values of the pore water and the soil should be determined because they have a great effect on the valence, solubility, and sorption of contaminant ions.
5. **Bench-scale test**: The dominant mechanism of transport, removal rates, and amounts of contamination left behind can be examined for different removal scenarios by conducting bench-scale tests. Because many of these physical and chemical reactions are interrelated, it may be necessary to conduct bench-scale tests to predict the performance of electrokinetics remediation at the field scale [68,69].

12.11.3 Applicability and Demonstration Projects

Various methods, developed by combining electrokinetics with other techniques, are being applied for remediation. This section describes different types of electrokinetic remediation methods for their use at contaminated sites. The methods discussed were developed by Electrokinetics, Inc.; Geokinetics International, Inc.; Isotron Corporation; Battelle Memorial Institute; a consortium effort; and P&P Geotechnik GmbH [2].

12.11.3.1 Electrokinetics, Inc.

Electrokinetics, Inc. operates under a licensing agreement with Louisiana State University (LSU). The technology is patented by and assigned to LSU [74], and a complementing process patent is assigned to Electrokinetics, Inc. [75]. As depicted in Figure 12.5, groundwater and/or a processing fluid (supplied externally through the boreholes that contain the electrodes) serves as the conductive medium. The additives in the processing fluid, the products of electrolysis reactions at the electrodes, and the dissolved chemical entities in the contaminated soil are transported across the contaminated soil by conduction under electric fields. This transport, when coupled with sorption, precipitation/dissolution, and volatilization/complexation, provides the fundamental mechanism that can affect the electrokinetic remediation process. Electrokinetics, Inc. accomplishes extraction and removal by electrodeposition, evaporation/condensation, precipitation, or ion exchange, either at the electrodes or in a treatment unit built into the system that pumps the processing fluid to and from the contaminated soil. Pilot-scale testing was carried out with a support from the U.S. EPA that also developed a design and analysis package for the process [76].

12.11.3.2 Geokinetics International, Inc.

Geokinetics International, Inc. (GII) obtained a patent for an electroreclamation process. The key claims in the patent are the use of electrode wells for both anodes and cathodes and the management of the pH and electrolyte levels in the electrolyte streams of the anode and the cathode. The patent also includes claims for using additives to dissolve different types of contaminants [77]. Fluor Daniel is licensed to operate GII's metal removal process in the United States.

GII has developed and patented electrically conductive ceramic material (EBONEX®) that has an extremely high resistance to corrosion. It has a lifetime of at least 45 years in soil and is self-cleaning. GII has also developed a batch electrokinetic remediation (BEK®) process. The process, which incorporates electrokinetic technology, normally requires 24–48 hours for complete remediation of the substrate. BEK® is a mobile unit that remediates *ex situ* soils on-site. GII has also developed a solution treatment technology (EIX®) that removes contamination from anode and cathode solutions up to a thousand times faster than can be achieved through conventional means [2].

12.11.3.3 Isotron Corporation

Isotron Corporation participated in a pilot-scale demonstration of electrokinetic extraction supported by DOE's Office of Technology Development. The demonstration took place at the Oak Ridge K-25 facility in Tennessee. Completed laboratory tests showed that the Isotron process could affect the movement and capture of uranium present in soil from the Oak Ridge site [78].

Isotron Corporation was also involved with Westinghouse Savannah River Company in a demonstration of electrokinetic remediation. The demonstration, supported by DOE's Office of Technology Development, took place at the old TNX basin at the Savannah River site in South Carolina. Isotron used the Electrosorb® process with a patented cylinder to control buffering conditions *in situ*. An ion exchange polymer matrix called Isolock® was used to trap metal ions. The process was tested for removing lead and chromium [78].

12.11.3.4 Battelle Memorial Institute

Another method that uses electrokinetic technology is electroacoustical soil decontamination. This technology combines electrokinetics with sonic vibration. Through the application of mechanical vibratory energy in the form of sonic or ultrasonic energy, the properties of a liquid contaminant in soil can be altered in a way that increases the level of removal of the contaminant. Battelle Memorial Institute of Columbus, OH, USA, developed the *in situ* treatment process that uses both electrical and acoustical forces to remove floating contaminants, and possibly metals, from subsurface zones of contamination. The process was selected for the U.S. EPA's SITE program [79].

12.11.3.5 Consortium Process

Monsanto Company has coined the name Lasagna to identify its products and services that are based on the integrated *in situ* remediation process developed by a consortium. The proposed technology combines electro-osmosis with treatment zones that are installed directly in the contaminated soils to form an integrated *in situ* remedial process, as Figure 12.7 shows. The consortium consists of Monsanto, DuPont, and GE, with participation by the U.S. EPA Office of Research and Development and DOE.

The *in situ* decontamination process occurs as follows [2]:

1. Creates highly permeable zones in close proximity sectioned through the contaminated soil region and turns them into sorption-degradation zones by introducing appropriate materials (sorbents, catalytic agents, microbes, oxidants, buffers, and others).
2. Uses electro-osmosis as a liquid pump to flush contaminants from the soil into the treatment zones of degradation.
3. Reverses liquid flow, if desired, by switching electrical polarity, a mode that increases the efficiency with which contaminants are removed from the soil; allows repeated passes through the treatment zones for complete sorption.

Initial field tests of the consortium process were conducted at DOE's gaseous diffusion plant in Paducah, Kentucky. The experiment tested the combination of electro-osmosis and *in situ* sorption in treatment zones. Technology development for the degradation processes and their integration into the overall treatment scheme were carried out at bench and pilot scales, followed by field experiments of the full-scale process [80].

12.11.4 Performance and Cost

Work sponsored by the U.S. EPA, DOE, the National Science Foundation, and private industry, when coupled with the efforts of researchers from academic and public institutions, has demonstrated the feasibility of moving electrokinetics remediation to pilot-scale testing and demonstration stages [69].

This section describes testing and cost summary results reported by LSU, Electrokinetics, Inc., GII, Battelle Memorial Institute, and the consortium [2].

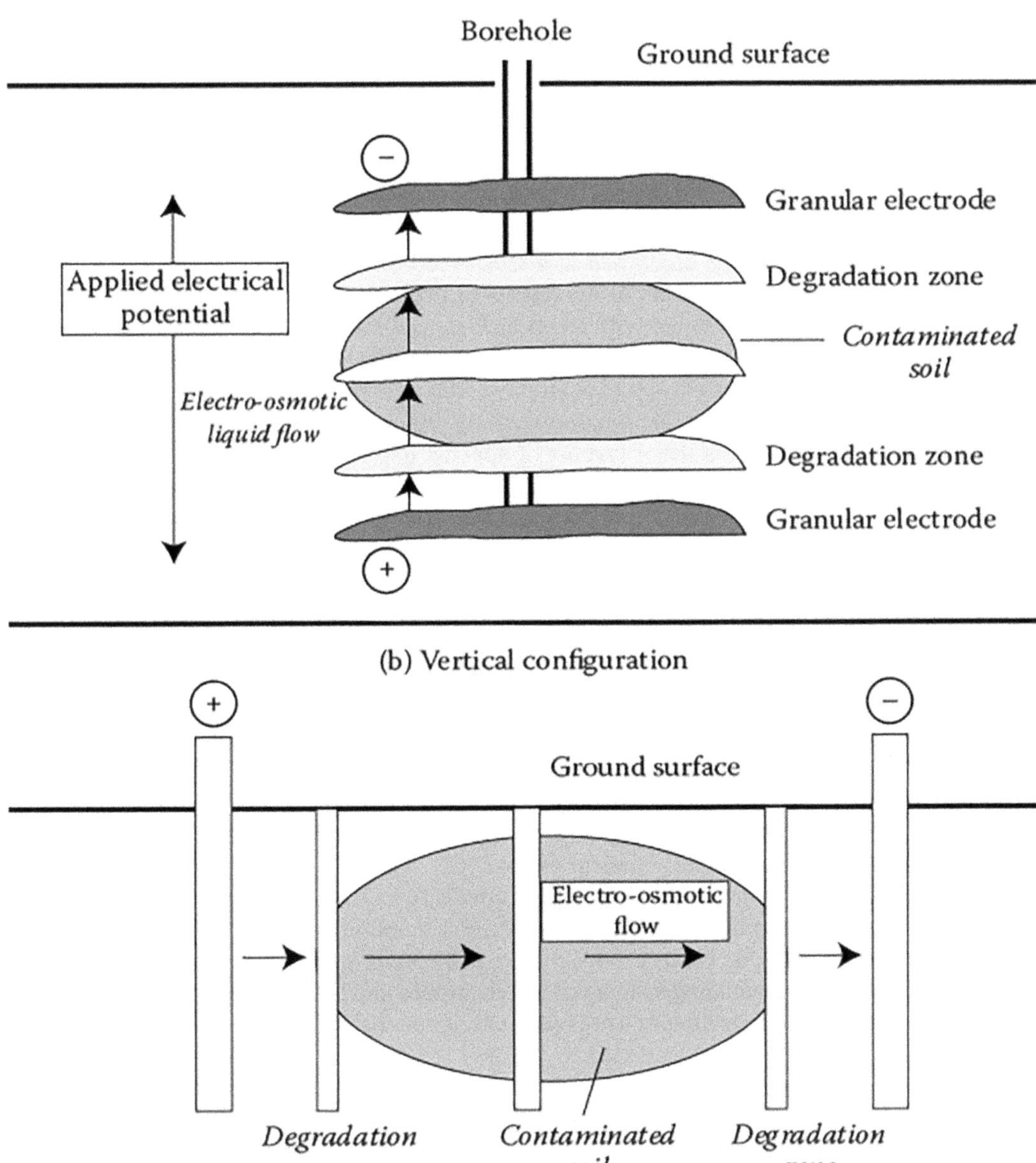

FIGURE 12.7   Schematic diagram of the Lasagna™ process. (Adapted from U.S. EPA. *Recent Developments for In Situ Treatment of Metal Contaminated Soils*. Contract # 68-W5-0055, U.S. Environmental Protection Agency, Washington, DC, March 1997 [2].)

12.11.4.1   LSU-Electrokinetics, Inc.

The LSU-Electrokinetics, Inc. group has conducted bench-scale testing on radionuclides and organic compounds. Test results have been reported for lead, cadmium, chromium, mercury, zinc, iron, and magnesium. The radionuclides tested include uranium, thorium, and radium.

In collaboration with the U.S. EPA, the LSU-Electrokinetics, Inc. group has completed pilot-scale studies of electrokinetic soil processing in the laboratory. Electrokinetics, Inc. carried out a site-specific pilot-scale study of the Electro-Klean™ electrical separation process. Pilot field studies have also been reported in the Netherlands on soils contaminated with lead, arsenic, nickel, mercury, copper, and zinc.

A pilot-scale laboratory study investigating the removal of 2,000 mg/kg of lead loaded onto kaolinite was completed. Removal efficiencies of 90%–95% were obtained. The electrodes were placed one inch apart in a 2 ton kaolinite specimen for 4 months, at a total energy cost of about 2007 USD 22/T [79].

With the support of DOD, Electrokinetics, Inc. carried out a comprehensive demonstration study of lead extraction from a creek bed at a U.S. Army firing range in Louisiana. U.S. EPA took part in independent assessments of the results of that demonstration study under the SITE program. The soils are contaminated with levels as high as 4,500 mg/kg of lead; pilot-scale studies have demonstrated that concentrations of lead decreased to less than 300 mg/kg in 30 weeks of processing. The TCLP values dropped from more than 300 mg/L to less than 40 mg/L within the same period. At the site of the demonstration study, Electrokinetics, Inc. used the CADEX™ electrode system that promotes the transport of species into the cathode compartment, where they are precipitated and/or electrodeposited directly. Electrokinetics, Inc. used a special electrode material that is cost-effective and does not corrode. Under the supervision and support of the Electric Power Research Institute (EPRI) and power companies in the southern USA a treatability and a pilot-scale field testing study of soils in sites contaminated with arsenic has been performed, in a collaborative effort between Southern Company Services Engineers and Electrokinetics, Inc. [2].

The processing cost of a system designed and installed by Electrokinetics, Inc. consists of energy cost, conditioning cost, and fixed costs associated with the installation of the system. Power consumption is related directly to the conductivity of the soil across the electrodes. The electrical conductivity of soils can span orders of magnitude, from 30 mho/cm to more than 3,000 pmho/cm, with higher values being in saturated, high-plasticity clays. A mean conductivity value is 500 pmho/cm. The voltage gradient is held to approximately 1 V/cm in an attempt to prevent adverse effects of temperature increases and for other practical reasons [69]. It may be cost-prohibitive to attempt to remediate high-plasticity soils with high electrical conductivities. However, for most deposits having conductivities of 500 pmho/cm, the daily energy consumption will be approximately 12 kWh/m³/d or about USD 1.20/m³/d (USD 0.10/kWh) and USD 36/m³/month. The processing time will depend on several factors, including spacing of the electrodes and type of conditioning scheme that will be used. If an electrode spacing of 4 m is selected, it may be necessary to process the site over several months.

Pilot-scale studies using "real-world" soils indicate that energy expenditures in the extraction of metals from soils may be 500 kWh/m³ or more at electrode spacings of 1.0–1.5 m [76]. The vendor estimates that the direct cost of about USD 50/m³ (USD 0.10/kWh) suggested for this energy expenditure, together with the cost of enhancement, could result in direct costs of USD 100/m³. If no other efficient *in situ* technology is available to remediate fine-grained and heterogeneous subsurface deposits contaminated with metals, this technique will remain potentially competitive.

12.11.4.2 Geokinetics International, Inc.

GII has successfully demonstrated *in situ* electrochemical remediation of metal-contaminated soils at several European sites. Geokinetics, a sister company of GII, has also been involved in the electrokinetics arena in Europe. Table 12.9 summarizes the physical characteristics of five sites, including size, contaminant(s) present, and overall performance of the technology at each site. GII that estimates its typical costs for "turn key" remediation projects, which are in the range of 2007 USD 160–260/m³ [2].

TABLE 12.9
Performance of Electrochemical Soil Remediation Applied at Five Field Sites in Europe

Site Description	Soil Volume (m³)	Soil Type	Contaminant	Initial Concentration (mg/kg)	Final Concentration (mg/kg)
Former paint factory	230	Peat/clay soil	Cu	1,220	<200
			Pb	>3,780	<280
Operational galvanizing plant	40	Clay soil	Zn	>1,400	600
Former timber plant	190	Heavy clay soil	As	>250	<30
Temporary landfill	5,440	Argillaceous sand	Cd	>180	<40
Military air base	1,900	Clay	Cd	660	47
			Cr	7,300	755
			Cu	770	98
			Ni	860	80
			Pb	730	108
			Zn	2,600	289

Source: U.S. EPA. *Recent Developments for In Situ Treatment of Metal Contaminated Soils.* Contract # 68-W5-0055, U.S. Environmental Protection Agency, Washington, DC, March 1997 [2].

12.11.4.3 Battelle Memorial Institute

The technology demonstration through the SITE program was completed [79]. The results indicate that the electroacoustical technology is technically feasible for removing inorganic species from clay soils [81].

12.11.4.4 Consortium Process

The Phase I field test of the Lasagna™ process has been completed. Scale-up from laboratory units was successfully achieved with respect to electrical parameters and electro-osmotic flow. Soil samples taken throughout the test site before and after the test indicate a 98% removal of trichloroethylene (TCE) from a tight clay soil (i.e., hydraulic conductivity less than 1×10^{-7} cm/s). TCE soil levels were reduced from the 100 to the 500 mg/kg range to an average concentration of 1 mg/kg [82]. Various treatment processes are being investigated in the laboratory to address other contaminants, including heavy metals [82].

12.11.5 SUMMARY OF ELECTROKINETIC REMEDIATION

Electrokinetic remediation may be applied to both saturated and partially saturated soils. One problem to overcome when applying electrokinetic remediation to the vadose zone is the drying of soil near the anode. When an electric current is applied to soil, water will flow by electro-osmosis in the soil pores, usually toward the cathode. The movement of the water will deplete soil moisture adjacent to the anode, and moisture will collect near the cathode. However, processing fluids may be circulated at the electrodes. The fluids can serve both as a conducting medium and as a means to extract or exchange the species and introduce other species. Another use of processing fluids is to control, depolarize, or modify either or both electrode reactions. The advance of the process fluid (acid or the conditioning fluid) across the electrodes assists in the desorption of species and dissolution of carbonates and hydroxides. Electro-osmotic advection and ionic migration lead to the transport and subsequent removal of contaminants. The contaminated fluid is then recovered at the cathode.

The spacing of the electrode will depend on the type and level of contamination and the selected current voltage regime. When higher voltage gradients are generated, the efficiency of the process might decrease because of increases in temperature. A spacing that will generate a potential gradient in the order of 1 V/cm is preferred. The spacing of electrodes will generally be as much as 3 m. The duration of the remediation will be site-specific. The remediation process should be continued until the desired removal is achieved. However, it should be recognized that in cases in which the duration of treatment is reduced by increasing the electrical potential gradient, the efficiency of the process will decrease [83,84].

The advantage of the technology is its potential for cost-effective use for both *in situ* and *ex situ* applications. The fact that the technique requires the presence of a conducting pore fluid in a soil mass may have site-specific implications. Also, heterogeneities or anomalies found at sites, such as submerged foundations, rubble, large quantities of iron or iron oxides, large rocks or gravel, or submerged cover material (such as seashells), are expected to reduce removal efficiencies [69].

12.12 PHYTOREMEDIATION

This technology is in the stage of commercialization for the treatment of soils contaminated with metals, and in the future may provide a low-cost option under specific circumstances. At the current stage of development, this process is best suited for sites with widely dispersed contamination at low concentrations where only treatment of soils at the surface (in other words, within the depth of the root zone) is required [2].

Phytoremediation uses plants to remove, contain, or render harmless environmental contaminants. This definition applies to all biological, chemical, and physical processes that are influenced by plants and that aid in the cleanup of contaminated substances [85]. Plants can be used in site remediation, both to mineralize and immobilize toxic organic compounds at the root zone and to accumulate and concentrate metals and other inorganic compounds from the soil into aboveground shoots [86]. Although phytoremediation is a relatively new concept in the waste management community, techniques, skills, and theories developed through the application of well-established agro-economic technologies are easily transferable. The development of plants for restoring sites contaminated with metals will require the multidisciplinary research efforts of agronomists, toxicologists, biochemists, microbiologists, pest management specialists, engineers, and other specialists [85,86]. Table 12.10 presents an overview of phytoremediation technology.

Two basic approaches for metal remediation include phytoextraction and phytostabilization. Phytoextraction relies on the uptake of contaminants from the soil and their translocation into aboveground plant tissue, which is harvested and treated. Although hyperaccumulating trees, shrubs, herbs, grasses, and crops have potential, crops seem to be most promising because of their greater biomass production. Nickel and zinc appear to be the most easily absorbed, although tests with copper and cadmium are encouraging [2]. Significant uptake of lead, a commonly occurring contaminant, has not been demonstrated on a large scale. However, some researchers are experimenting with soil amendments that would facilitate the uptake of lead by plants.

12.12.1 PROCESS DESCRIPTION

Metals that are considered essential for at least some forms of life include vanadium (V), chromium (Cr), manganese (Mn), iron (Fe), cobalt (Co), nickel (Ni), copper (Cu), zinc (Zn), and molybdenum (Mo) [86]. Because many metals are toxic in concentrations above minute levels, an organism must regulate the cellular concentrations of such metals. Consequently, organisms have evolved transport systems to regulate the uptake and distribution of metals. Plants have remarkable metabolic and absorption capabilities, as well as transport systems that can take up ions selectively from the soil. Plants have evolved a great diversity of genetic adaptations to handle potentially toxic levels

TABLE 12.10

Overview of Phytoremediation Technology

General characteristics

- Best used at sites with low to moderate disperse metals content and with soil media that will support plant growth
- Applications limited to depth of the root zone
- Longer times required for remediation compared with other technologies
- Different species have been identified to treat different metals

Approach #1	Approach #2
Phytoextraction (harvest)	Phytostabilization (root-fixing)
Description:	**Description:**
Uptake of contaminants from soil into aboveground plant tissue, which is periodically harvested and treated	Production of chemical compounds by the plant to immobilize contaminants at the interface of roots and soil. Additional stabilization can occur by raising the pH level in the soil.
Status:	**Status:**
Field testing for effectiveness on radioactive metals is ongoing in the vicinity of the damaged nuclear reactor in Chernobyl, Ukraine Field testing also is being conducted in Trenton, NJ and Butte, MT and by the Idaho National Engineering Laboratory (INEL) in Fernald, OH	Research is ongoing
Applicability:	**Applicability:**
Potentially applicable for many metals. Nickel and zinc appear to be most easily absorbed. Preliminary results for absorption of copper and cadmium are encouraging	Potentially applicable for many metals, especially lead, chromium, and mercury
Comments:	**Comments:**
Cost affected by volume of biomass produced that may require treatment before disposal. Cost affected by concentration and depth of contamination and number of harvests required	Long-term maintenance is required

Source: U.S. EPA. *Recent Developments for In Situ Treatment of Metal Contaminated Soils.* Contract # 68-W5-0055, U.S. Environmental Protection Agency, Washington, DC, March 1997 [2].

of metals and other pollutants that occur in the environment. In plants, the uptake of metals occurs primarily through the root system, in which the majority of mechanisms to prevent metal toxicity are found [87]. The root system provides an enormous surface area that absorbs and accumulates the water and nutrients essential for growth. In many ways, living plants can be compared to solar-powered pumps that can extract and concentrate certain elements from the environment [88].

Plant roots cause changes at the soil-root interface as they release inorganic and organic compounds (root exudates) in the area of the soil immediately surrounding the roots (the rhizosphere) [89]. Root exudates affect the number and activity of microorganisms, the aggregation and stability of soil particles around the root, and the availability of elements. They can increase (mobilize) or decrease (immobilize) directly or indirectly the availability of elements in the rhizosphere. Mobilization and immobilization of elements in the rhizosphere can be caused by the following [90,91]:

1. Changes in soil pH.
2. Release of complex substances, such as metal-chelating molecules.
3. Changes in oxidation-reduction potential.
4. Increase in microbial activity.

Phytoremediation technologies can be developed for different applications in environmental cleanup and are classified into three types:

1. Phytoextraction
2. Phytostabilization
3. Rhizofiltration.

12.12.1.1 Phytoextraction

Phytoextraction technologies use hyperaccumulating plants to transport metals from the soil and concentrate them in the roots and aboveground shoots that can be harvested [85,86,89]. A plant containing more than 0.1% of Ni, Co, Cu, Cr, or 1% Zn and Mn in its leaves on a dry weight basis is called a hyperaccumulator, regardless of the concentration of metals in the soil [86,92,93].

Almost all metal-hyperaccumulating species known today were discovered on metal-rich soils, either natural or artificial, often growing in communities with metal excluders [86,94]. Actually, almost all metal-hyperaccumulating plants are endemic to such soils, suggesting that hyperaccumulation is an important ecophysiological adaptation to metal stress and one of the manifestations of resistance to metals. The majority of hyperaccumulating species discovered so far are restricted to a few specific geographical locations [86,92]. For example, Ni hyperaccumulators are found in New Caledonia, the Philippines, Brazil, and Cuba. Ni and Zn hyperaccumulators are found in southern and central Europe and Asia Minor.

Dried or composted plant residues or plant ashes that are highly enriched with metals can be isolated as hazardous waste or recycled as metal ore [95]. The goal of phytoextraction is to recycle as "bio-ores" metals reclaimed from plant ash in the feed stream of smelting processes. Even if the plant ashes do not have enough concentration of metal to be useful in smelting processes, phytoextraction remains beneficial because it reduces the amount of hazardous waste to be landfilled [2] by as much as 95%. Several research efforts in using trees, grasses, and crop plants are being pursued to develop phytoremediation as a cleanup technology. The following paragraphs briefly discuss these three phytoextraction techniques.

The use of trees can result in the extraction of significant amounts of metal because of their high biomass production. However, using trees in phytoremediation requires long-term treatment and may create additional environmental concerns about falling leaves. When leaves containing metals fall or blow away, recirculation of metals to the contaminated site and migration to off-site by wind transport or through leaching can occur [2].

Some grasses accumulate surprisingly high levels of metals in their shoots without exhibiting toxic effects. However, their low biomass production results in a relatively low yield of metals. Genetic breeding of hyperaccumulating plants that produce relatively large amounts of biomass could make the extraction process highly effective [96].

It is known that many crop plants can accumulate metals in their roots and aboveground shoots, potentially threatening the food chain. For example, in May 1980, regulations proposed under RCRA for hazardous waste included limits on the amounts of cadmium and other metals that can be applied to crops. Recently, however, the potential use of crop plants for environmental remediation has been under investigation. Using crop plants to extract metals from the soil seems practical because of their high biomass production and relatively fast rate of growth. Other benefits of using crop plants are that they are easy to cultivate and exhibit genetic stability [97].

12.12.1.2 Phytostabilization

Phytostabilization uses plants to limit the mobility and bioavailability of metals in soils. Ideally, phytostabilizing plants should be able to tolerate high levels of metals and immobilize them in the soil by sorption, precipitation, complexation, or the reduction of metal valences. Phytostabilizing plants should also exhibit low levels of accumulation of metals in shoots to eliminate the possibility that residues in harvested shoots might become hazardous wastes [88]. In addition to stabilizing metals present in the soil, phytostabilizing plants can also stabilize the soil matrix to minimize erosion and migration of sediment. Dr Gary Pierzynski of Kansas State University is studying phytostabilization in poplar trees, which were selected for the study because they can be deep-planted and may be able to form roots below the zone of maximum contamination [2].

Since most sites contaminated with metals lack established vegetation, metal-tolerant plants are used to revegetate such sites to prevent erosion and leaching [98]. However, that approach is a containment rather than a remediation technology. Some researchers consider phytostabilization as an interim measure to be applied until phytoextraction becomes fully developed. However, other researchers are developing phytostabilization as a standard protocol of metal remediation technology, especially for sites at which the removal of metals does not seem to be economically feasible. After field applications conducted by a group in Liverpool, England, three varieties of grasses were made commercially available for phytostabilization [88]:

- *Agrostis tenuis, cv Parys* for copper wastes,
- *Agrosas tenuis, cv Coginan* for acid lead and zinc wastes, and
- *Festuca rubra, cv Merlin* for calcareous lead and zinc wastes.

12.12.1.3 Rhizofiltration

One type of rhizofiltration uses plant roots to absorb, concentrate, and precipitate metals from wastewater [88], which may include leachate from the soil. Rhizofiltration uses terrestrial plants instead of aquatic plants because terrestrial plants develop much longer, fibrous root systems covered with root hairs that have extremely large surface areas. This variation of phytoremediation uses plants that remove metals by sorption, which does not involve biological processes. Using plants to translocate metals to shoots is a slower process than phytoextraction [98].

Another type of rhizofiltration, which is more fully developed, involves the construction of wetlands or reed beds for the treatment of contaminated wastewater or leachate. The technology is cost-effective for the treatment of large volumes of wastewater that have low concentrations of metals [98]. Since rhizofiltration focuses on treating contaminated water, it is not discussed further in this chapter.

Table 12.11 presents the advantages and disadvantages of each of the types of phytoremediation currently being researched that are categorized as either phytoextraction on phytostabilization [88].

TABLE 12.11

Types of Phytoremediation Technology: Advantages and Disadvantages

Type of Phytoremediation	Advantages	Disadvantages
Phytoextraction by trees	High biomass production	Potential for off-site migration and leaf transportation of metals to surface Metals are concentrated in plant biomass and must be disposed of eventually
Phytoextraction by grasses	High accumulation	Low biomass production and slow growth rate Metals are concentrated in plant biomass and must be disposed of eventually
Phytoextraction by crops	High biomass and increased growth rate	Potential threat to the food chain through ingestion by herbivores Metals are concentrated in plant biomass and must be disposed of eventually
Phytostabilization	No disposal of contaminated biomass required	Remaining liability issues, including maintenance for indefinite period of time (containment rather than removal)
Rhizofiltration	Readily absorbs metals	Applicable for treatment of water only Metals are concentrated in plant biomass and must be disposed of eventually

Source: U.S. EPA. *Recent Developments for In Situ Treatment of Metal Contaminated Soils.* Contract # 68-W5-0055, U.S. Environmental Protection Agency, Washington, DC, March 1997 [2].

12.12.1.4 Future Development

Faster uptake of metals and higher yields of metals in harvested plants may become possible through the application of genetic engineering and/or selective breeding techniques. Recent laboratory-scale testing has revealed that a genetically altered species of mustard weed can uptake mercuric ions from the soil and convert them to metallic mercury, which is transpired through the leaves [2]. Improvements in phytoremediation may be attained through research and a better understanding of the principles governing the processes by which plants affect the geochemistry of their soils. In addition, future testing of plants and microflora may lead to the identification of plants that have metal accumulation qualities that are far superior to those currently known.

12.12.2 APPLICABILITY

Plants have been used to treat wastewater for more than 300 years, and plant-based remediation methods for slurries of dredged material and soils contaminated with metals have been proposed since the mid-1970s [85,99]. Reports of successful remediation of soils contaminated with metals are rare, but the suggestion of such an application is more than two decades old, and progress is being made at a number of pilot test sites [94]. Successful phytoremediation must meet cleanup standards in order to be approved by regulatory agencies.

No full-scale applications of phytoremediation have been reported. One vendor, Phytotech, Inc., is developing phytostabilization for soil remediation applications. Phytotech has also patented strategies for phytoextraction and is conducting several field tests in Trenton, New Jersey and Chernobyl, Ukraine [97]. Also, as previously mentioned, a group in Liverpool, England, has made three grasses commercially available for the stabilization of lead, copper, and zinc wastes [88].

12.12.3 PERFORMANCE AND COST

A variety of new research approaches and tools are expanding an understanding of the molecular and cellular processes that can be employed through phytoremediation [100].

12.12.3.1 Performance

Potential for phytoremediation (phytoextraction) can be assessed by comparing the concentration of contaminants and volume of soil to be treated with the particular plant's seasonal productivity of biomass and ability to accumulate contaminants. Table 12.6 lists selected examples of plants identified as metal hyperaccumulators and their native countries [92,101]. If plants are to be effective remediation systems, 1 ton of plant biomass, costing from several hundred to a few thousand dollars to produce, must be able to treat large volumes of contaminated soil. For metals that are removed from the soil and accumulated in aboveground biomass, the total amount of biomass per hectare required for soil cleanup is determined by dividing the total weight of metal per hectare to be remediated by the accumulation factor, which is the ratio of the accumulated weight of the metal to the weight of the biomass containing the metal. The total biomass per hectare (T/ha) can then be divided by the productivity of the plant (T/ha/yr) to determine the number of years required to achieve cleanup standards—a major determinant of the overall cost and feasibility of phytoremediation [100].

As discussed earlier, the amount of biomass is one of the factors that determine the practicality of phytoremediation. Under the best climatic conditions, with irrigation, fertilization, and other factors, the total biomass productivity can approach 100 T/ha/yr. One unresolved issue is the trade-off between accumulation of toxic elements and productivity [102]. In practice, a maximum harvest biomass yield of 10–20 T/ha/yr is likely, particularly for plants that accumulate metals.

These values for the productivity of biomass and the metal content of soil would limit the annual capacity for removal of metals to approximately 10–400 kg/ha/yr, depending on pollutants, species of plant, climate, and other factors. For a target soil depth of 30 cm (4,000 T/ha), this capacity amounts to an annual reduction of 2.5–100 mg/kg of soil contaminants. This rate of removal of contamination is often acceptable, allowing total remediation of a site over a period of a few years to several decades [100].

12.12.3.2 Cost

The practical objective of phytoremediation is to achieve major reductions in the cost of cleanup of hazardous sites. Salt and others [88] note the cost-effectiveness of phytoremediation with an example: Using phytoremediation to clean up one acre of sandy loam soil to a depth of 50 cm typically will cost $60,000–$100,000, compared with a cost of at least $400,000 for excavation and disposal storage without treatment [88]. One objective of field tests is to use commercially available agricultural equipment and supplies for phytoremediation to reduce costs. Therefore, in addition to their remediation qualities, the agronomic characteristics of plants must be evaluated.

The processing and ultimate disposal of the biomass generated is likely to be a major percentage of overall costs, particularly when highly toxic metals and radionuclides are present at a site. Analysis of the costs of phytoremediation must include the entire cycle of the process, from the growing and harvesting of the plants to the final processing and disposal of the biomass. It is difficult to predict the costs of phytoremediation compared with overall cleanup costs at a site. Phytoremediation may also be used as a follow-up technique after areas with high concentrations of pollutants have been mitigated or in conjunction with other remediation technologies, making cost analysis more difficult.

12.12.3.3 Future Directions

Because metal hyperaccumulators generally produce small quantities of biomass, they are unsuited agronomically for phytoremediation. Nevertheless, such plants are a valuable store of genetic and physiologic material and data [85]. To provide effective cleanup of contaminated soils, it is essential to find, breed, or engineer plants that absorb, translocate, and tolerate levels of metals in the range of 0.1%–1.0%. It is also necessary to develop a methodology for selecting plants native to the area.

Three grasses are commercially available for the stabilization of lead, copper, and zinc wastes [88]. An integrated approach that involves basic and applied research, along with consideration of safety, legal, and policy issues, will be necessary to establish phytoremediation as a practicable cleanup technology [85].

According to a DOE report, three broad areas of research and development can be identified for the *in situ* treatment of soil contaminated with metals [100]:

1. **Mechanisms of uptake, transport, and accumulation***:* Research is needed to develop a better understanding of the use of physiological, biochemical, and genetic processes in plants. Research on the uptake and transport mechanisms is providing improved knowledge about the adaptability of those systems and how they might be used in phytoremediation.
2. **Genetic evaluation of hyperaccumulators***:* Research is being conducted to collect plants growing in soils with high levels of metals and screen them for specific traits useful in phytoremediation. Plants that tolerate and colonize environments polluted with metals are a valuable resource, both as candidates for use in phytoremediation and as sources of genes for classical plant breeding and molecular genetic engineering.
3. **Field evaluation and validation***:* Research is being conducted to employ early and frequent field testing to accelerate the implementation of phytoremediation technologies and to provide data to research programs. Standardization of field-test protocols and subsequent application of test results to real problems are also needed.

Research in these areas is expected to grow because many of the current engineering technologies for cleaning the surface soil of metals are costly and physically disruptive. Phytoremediation, when fully developed, could result in significant cost savings and in the restoration of numerous sites by a relatively noninvasive, solar-driven, *in situ* method that, in some forms, can be aesthetically pleasing [85].

12.12.4 Summary of Phytoremediation Technology

Phytoremediation is in the early stage of development and is being field tested at various sites in the United States and overseas for its effectiveness in capturing or stabilizing metals, including radioactive wastes. Limited cost and performance data are currently available. Phytoremediation has the potential to develop into a practicable remediation option at sites where contaminants are near the surface, are relatively nonleachable, and pose little imminent threat to human health or the environment [85]. The efficiency of phytoremediation depends on the characteristics of the soil and the contaminants; these factors are summarized in the sections that follow.

12.12.4.1 Site Conditions

The effectiveness of phytoremediation is generally restricted to surface soils within the rooting zone. The most important limitation to phytoremediation is rooting depth, which can be 20, 50, or even 100 cm, depending on the plant and soil type. Therefore, one of the favorable site conditions for phytoremediation is contamination with metals located at the surface [100].

The type of soil, as well as the rooting structure of the plant relative to the location of contaminants, can have a strong influence on the uptake of any metal substance by the plant. Amendment of soils to change soil pH, nutrient compositions, or microbial activities must be selected in treatability studies to govern the efficiency of phytoremediation. Certain generalizations can be made about such cases; however, much work is needed in this area [85]. Since the amount of biomass that can be produced is one of the limiting factors affecting phytoremediation, optimal climatic conditions, with irrigation and fertilization of the site, should be considered for increased productivity of the best plants for the site [100].

12.12.4.2 Waste Characteristics

Sites that have low to moderate contamination with metals might be suitable for growing hyperaccumulating plants, although the most heavily contaminated soils do not allow plant growth without the addition of soil amendments. Unfortunately, one of the most difficult metal cations for plants to translocate is lead, which is present at numerous sites in need of remediation. Although a significant uptake of lead has not yet been demonstrated, one researcher is experimenting with soil amendments that make lead more available for uptake [88].

Capabilities to accumulate lead and other metals are dependent on the chemistry of the soil in which the plants are growing. Most metals, and lead in particular, occur in numerous forms in the soil, not all of which are equally available for uptake by plants [85,103]. Maximum removal of lead requires a balance between the nutritional requirements of plants for biomass production and the bioavailability of lead for uptake by plants. Maximizing the availability of lead requires low pH and low levels of available phosphate and sulfate. However, limiting the fertility of the soil in such a manner directly affects the health and vigor of plants [85].

12.13 USE OF TREATMENT TRAINS

Several of the metal remediation technologies discussed are often enhanced through the use of treatment trains. Treatment trains use two or more remedial options applied sequentially to contaminated soil and often increase the effectiveness while decreasing the cost of remediation. Processes involved in treatment trains include soil pretreatment, physical separation designed to decrease the amount of soil requiring treatment, additional treatment of process residuals or off-gases, and a variety of other physical and chemical techniques, which can greatly improve the performance of the remediation technology. Table 12.12 provides examples of treatment trains used to enhance each of the proven and commercialized metal remediation technologies [5].

TABLE 12.12
Typical Treatment Trains

	Containment	S/S	Vitrification	Soil Washing	Pyrometallurgical	Soil Flushing
Pretreatment						
Excavation	•	E, P	I, E	•	•	
Debris removal		E, P	E	•	•	
Oversize reduction		E, P	E	•	•	
Adjust pH	•	I, E, P				
Reduction [e.g., Cr(VI) to Cr(III)]	•	I, E				
Oxidation [e.g., As(III) to As (V)]	•	I, E				
Treatment to remove or destroy organics		I				
Physical separation of rich and lean fractions		I, E, P	E	•	•	
Dewatering and drying for wet sludge	•	P	E		•	
Conversion of metals to less volatile forms [e.g., As_2O_3 to $Ca_3(AsO_4)_2$]			E			
Addition of high-temperature reductants					•	
Pelletizing					•	
Flushing fluid delivery and extraction system						•
Containment barriers	•	I, E, P	I	•		•
Posttreatment/Residuals Management						
Disposal of treated solid residuals (preferably below the frost line and above the water table)		I, E, P	E		•	
Containment barriers		I, E, P	I, E			•
Off-gas treatment		I, E, P	I, E		•	
Reuse for on-site paving		P				
Metal recovery from extraction fluid by aqueous processing (ion exchange, electrowinning, etc.)				•		
Pyrometallurgical recovery of metal from sludge				•		
Processing and reuse of leaching solution				•	•	
S/S treatment of leached residual of solid process residuals(preferably below the frostline and above the water table)				•		
Containment S/S Vitrification Soil Pyrometallurgical Soil Washing Flushing Disposal						
Disposal of liquid process residuals				•		•
S/S treatment of slag or fly ash					•	
Reuse of slag/vitreous product as construction material			E	•		
Reuse of metal or metal compound					•	
Further processing of metal or metal compound					•	
Flushing liquid/groundwater treatment/ disposal						•

Source: U.S. EPA. *Technology Alternatives for the Remediation of Soils Contaminated with AS, Cd, Cr, Hg, and Pb.* EPA/540/S-97/500, U.S. Environmental Protection Agency, Cincinnati, OH, August 1997 [5].

Note: Technology has been divided into the following categories: I = *in situ* process; E = *ex situ* process; P = polymer microencapsulation *ex situ*.

12.14 COST RANGES OF REMEDIAL TECHNOLOGIES

Estimated cost ranges for the basic operation of the technology are presented in Table 12.13. The reader is cautioned that the cost estimates generally do not include pretreatment, site preparation, regulatory compliance costs, costs for additional treatment of process residuals (e.g., stabilization of incinerator ash or disposal of metals concentrated by solvent extraction), or profit [5,104]. Since the actual cost of employing a remedial technology at a specific site may be significantly different from these estimates, data are best used for order-of-magnitude cost evaluations.

12.15 CASE STUDIES

12.15.1 APPLICATION OF PHYTOREMEDIATION

The removal of heavy metals from contaminated soil can be accomplished using various physical, synthetic, and natural remediation techniques (both *in situ* and *ex situ*). The most controllable (affordable and eco-friendly) method among these is phytoremediation. The removal of heavy metal (HM) defilements can be accomplished by Priya et al. [105] using phytoremediation techniques, including phytoextraction, phytovolatilization, phytostabilization, and phytofiltration. The bioavailability of heavy metals in soil and the biomass of plants are the two main factors affecting how effectively phytoremediation works. The focus in phytoremediation and phytomining is on new metal hyperaccumulators with high efficiency. Subsequently, this study comprehensively examines different frameworks and biotechnological techniques available for eliminating heavy metals according to environmental guidelines, underscoring the difficulties and limitations of phytoremediation and its potential application in the clean up of other harmful pollutants. Additionally, the

TABLE 12.13
Estimated Cost Ranges of Metals Remediation Technologies

Type of Remediation	Cost Range 2022 USD/T
Containment[a]	17–162
S/S	108–512
Vitrification	700–1536
Soil washing	108–431
Soil flushing[b]	108–790
Pyrometallurgical	445–984
Electrokinetics[b]	81–216
Phytoremediation[c]	40–67

Source: U.S. EPA. *Recent Developments for In Situ Treatment of Metal Contaminated Soils.* Contract # 68-W5-0055, U.S. Environmental Protection Agency, Washington, DC, March 1997; U.S. EPA. *Technology Alternatives for the Remediation of Soils Contaminated with As, Cd, Cr, Hg, and Pb.* EPA/540/S-97/500, U.S. Environmental Protection Agency, Cincinnati, OH, August 1997 [2,5].

[a] Includes landfill caps and slurry walls. A slurry wall depth of 6 m is assumed.

[b] Costs reported in USD/m^3, assumed soil specific gravity of 1.6.

[c] Costs reported per acre for a soil depth of 0.50 m.

authors [105] share in-depth experience of safe removing the plants used in phytoremediation after heavy metals removal. Additional research concerning heavy metals removal by phytoremediation can be found in the literature [106–111]. The soil remediation cost and time requirement using phytoremediation (phytoextraction) technology are estimated [112] to be $15–40 per cubic meter (1997 cost data) and 18–60 months, respectively [113,114].

12.15.1.1 Application of Phytostabilization

In a process called phytostabilization, a contaminated site is simply revegetated, so the plants are used to reduce wind and water erosion that spread materials containing heavy metals. In one case study, grass or tree buffers could reduce sediment loss from the heavy metal-contaminated chat piles at a contaminated site in Galena, Kansas, USA, anywhere from 18% to 25% [115]. If all of the ground could be revegetated, sediment loss could be cut by approximately 70%. However, it would be necessary to find plants that could tolerate high levels of heavy metals.

12.15.1.2 Application of Phytoextraction

In a phytoextraction process, plant species are used to take up heavy metals and concentrate them in their tissue. The plants can be harvested, and the contaminated plant material can be disposed of safely. Sometimes, soil amendments are added to the soil to increase the ability of the plants to take up the heavy metals. One type of plant used for this purpose is called Indian mustard. This plant has been used to extract lead from soil and reduce lead contamination at various contaminated sites. Other plants used for phytoextraction include alfalfa, cabbage, tall fescue, juniper, and poplar trees [113].

12.15.1.3 Applications of Rhizofiltration

In a rhizofiltration process, heavy metals are removed directly from water by plant roots [111]. The plants are grown directly in water or water-rich materials such as sand, using aquatic species or hydroponic methods. In a case study, sunflowers on floating rafts have removed radioactive metals from water in ponds at Chernobyl, Ukraine, and other plants removed metals from mine drainage flowing through diversion troughs [111].

12.15.2 Application of Excavation

Excavation and physical removal of the heavy metals from contaminated soil is still in use at many locations, including residential areas contaminated with lead in southwestern Missouri, USA [113]. Advantages of excavation include the complete removal of the contaminants and the relatively rapid cleanup of a contaminated site [114]. Disadvantages include the fact that the contaminants are simply moved to a different place, where they must be monitored; the risk of spreading contaminated soil and dust particles during removal and transport of contaminated soil; and the relatively high cost. Excavation can be the most expensive option when large amounts of soil must be removed or disposed of as hazardous or toxic waste is required [112]. The soil remediation cost and time requirement using excavation technology are estimated [112] to be $100–$400 per cubic meter (1997 cost data) and 6–9 months, respectively.

12.15.3 Application of Stabilization

Application of stabilization for removal of heavy metals from contaminated soil on-site reduces or eliminates their ability to adversely affect human health and the environment, and has many advantages over excavation. One way of stabilizing heavy metals consists of adding chemicals to the soil that cause the formation of minerals that contain the heavy metals in a form that is not easily absorbed by plants, animals, or people [113]. This method is called in situ (in place) fixation or stabilization by which the heavy metal in soil combines with the added chemical to create a less toxic compound, so the stabilization process does not disrupt the environment or generate hazardous wastes. The heavy metal remains in the soil but in a form that is much less harmful. One case history of in situ fixation of heavy metals involves adding phosphate fertilizer as a soil amendment to soil with high amounts of heavy metal lead. Chemical reactions between the phosphate and the lead cause a mineral called lead pyromorphite. Lead pyromorphite and similar minerals called heavy metal phosphates are extremely insoluble. Accordingly, the new minerals cannot dissolve easily in water [116]. This has two beneficial effects. The minerals (and the heavy metals) cannot be easily spread by water to pollute streams, lakes, or other groundwater. Also, the heavy metal phosphates are less likely to enter the food chain by being absorbed into plants or animals that may eat soil particles. The cost of treating the soil by in situ fixation may be about half the cost of excavation and disposal of heavy metal-contaminated soil. This method is relatively rapid and takes about the same amount of time as excavation. The soil remediation cost and time requirement using stabilization technology are estimated [112] to be \$90–200 per cubic meter (1997 cost data) and 6–9 months, respectively. Liu et al. provide an overview of the process [117].

GLOSSARY

Arsenic: It is a semimetallic element or metalloid that has several allotropic forms. The most stable allotrope is a silver-gray, brittle, crystalline solid that tarnishes in air. As compounds, mainly As_2O_3, can be recovered as a by-product of processing complex ores mined mainly for copper, lead, zinc, gold, and silver As occurs in a wide variety of mineral forms, including arsenopyrite, $FeAsS_4$, which is the main commercial ore of As worldwide.

Best Demonstrated Available Technology (BDAT): BDAT status refers to the determination under the RCRA of the BDAT for various industry-generated hazardous wastes that contain the metals of interest. Whether the characteristics of a Superfund metal-contaminated soil (or fractions derived from it) are similar enough to the RCRA waste to justify serious evaluation of the BDAT for a specific Superfund soil must be made on a site-specific basis. Other limitations relevant to BDATs include the following: (a) the regulatory basis for BDAT standards focus BDATs on proven, commercially available technologies at the time of the BDAT determination; (b) a BDAT may be identified, but that does not necessarily preclude the use of other technologies; and (c) a technology identified as BDAT may not necessarily be the current technology of choice in the RCRA hazardous waste treatment industry.

Cadmium: It is a bluish-white, soft, ductile metal. Pure Cd compounds are rarely found in nature, although occurrences of greenockite (CdS) and otavite ($CdCO_3$) are known. The main sources of Cd are sulfide ores of lead, zinc, and copper. Cd is recovered as a by-product when these ores are processed.

Capping system: It is one of the containment structures which reduces surface water infiltration, controls gas and odor emissions, improve aesthetics, and provides a stable surface over the waste. Caps can range from a simple native soil cover to a full RCRA subtitle C composite cover.

Cementation: It is a cement-based solidification/stabilization (S/S) technology that reduces the mobility of inorganic compounds by the formation of insoluble hydroxides, carbonates, or silicates; substitution of the metal into a mineral structure; sorption; physical encapsulation; and perhaps other mechanisms. Cement-based S/S involves a complex series of reactions, and there are many potential interferences (e.g., coating of particles by organics, excessive acceleration or retardation of set times by various soluble metal and inorganic compounds; excessive heat of hydration; pH conditions that solubilize anionic species of metal compounds, and so on) that can prevent the attainment of S/S treatment objectives for physical strength and leachability, while there are many potential interferences, Portland cement is widely used and studied, and a knowledgeable vendor may be able to identify, and confirm via treatability studies, approaches to counteract adverse effects by the use of appropriate additives or other changes in formulation.

Chromium: It is a silver-gray metal and one of the less common elements in the earth's crust. It occurs only in compounds. the chief commercial source of chromium is the mineral chromite. Chromium is mined as a primary product and is not recovered as a by-product of any other mining operation. there are no chromite-ore reserves, nor any primary production of chromite in the United States.

Containment: It is a physical structure ranging from a surface cap that limits infiltration of uncontaminated surface water to subsurface vertical or horizontal barriers that restrict lateral or vertical migration of contaminated groundwater.

Electrokinetic remediation: (a) It is a process relying on the application of low-intensity direct current between electrodes placed in the soil. Contaminants are mobilized in the form of charged species, particles, or ions. Electrokinetics has been used for dewatering of soils and sludges since the first recorded use in the field. Electrokinetic extraction has been used to concentrate metals and to explore for minerals in deep soils. electrokinetic remediation also referred to as electrokinetic soil processing, electromigration, electrochemical decontamination, or electroreclamation, can be used to extract radionuclides, metals, and some types of organic wastes from saturated or unsaturated soils, slurries, and sediments. This *in situ* soil processing technology is primarily a separation and removal technique for extracting contaminants from soils; (b) The principle of electrokinetic remediation relies upon the application of a low-intensity direct current through the soil between two or more electrodes. Most soils contain water in the pores between the soil particles and have an inherent electrical conductivity that results from salts present in the soil. The current mobilizes charged species, particles, and ions in the soil by the following processes: (a) Electromigration (transport of charged chemical species under an electric gradient), (b) Electro-osmosis (transport of pore fluid under an electric gradient); (c) Electrophoresis (movement of charged particles under an electric gradient); and (d) Electrolysis (chemical reactions associated with the electric field).

Excavation: It is the physical removal of heavy metals from contaminated soil, which is still in use at many locations, including residential areas contaminated with lead in southwestern Missouri, USA. Advantages of excavation include the complete removal of the contaminants and the relatively rapid cleanup of a contaminated site. Disadvantages include the fact that the contaminants are simply moved to a different place, where they must be monitored; the risk of spreading contaminated soil and dust particles during removal and transport of contaminated soil; and the relatively high cost. Excavation can be the most expensive option when large amounts of soil must be removed or disposal as hazardous or toxic waste is required. The soil remediation cost and time requirement using excavation technology are estimated to be $100–$400 per cubic meter (1997 cost data) and 6–9 months, respectively.

Ex situ cement-based solidification/stabilization (S/S) technology: It is performed on contaminated soil that has been excavated and classified to reject oversize. Cement-based S/S involves mixing contaminated materials with an appropriate ratio of cement or similar binder/stabilizer and possibly water and other additives. A system is also necessary for delivering the treated wastes to molds, surface trenches, or subsurface injection. Off-gas treatment (if volatiles or dust are present) may be necessary. The fundamental materials used for this technology are Portland-type cement and pozzolanic materials. Portland cements are typically composed of calcium silicates, aluminates, aluminoferrites, and sulfates. Pozzolans are very small spheroidal particles that are formed in the combustion of coal (fly ash) in lime and cement kilns, for example. Pozzolans of high silica content are found to have cement-like properties when mixed with water. Cement-based S/S treatment may involve using only Portland cement, only pozzolanic materials, or blends of both. The composition of the cement and pozzolan, together with the amount of water, aggregate, and other additives, determines the set time, cure time, pouring characteristics, and material properties (e.g., pore size and compressive strength) of the resulting treated waste.

The composition of cements and pozzolans, including those commonly used in S/S applications, is classified according to American Society for Testing and Materials (ASTM) standards. S/S treatment usually results in an increase (>50% in some cases) in the treated waste volume. *Ex situ* treatment provides high throughput (100–200 m³/d/mixer).

Ex situ vitrification (ESV) technologies: The technologies apply heat to a melter through a variety of sources such as combustion of fossil fuels (coal, natural gas, and oil) or input of electric energy by direct joule heat, arcs, plasma torches, and microwaves. Combustion or oxidation of the organic portion of the waste can contribute significant energy to the melting process, thus reducing energy costs. The particle size of the waste may need to be controlled for some of the melting technologies. For wastes containing refractory compounds that melt above the unit's nominal processing temperature, such as quartz or alumina, size reduction may be required to achieve acceptable throughputs and a homogeneous melt. For high-temperature processes using arcing or plasma technologies, size reduction is not a major factor. For intense melters using concurrent gas-phase melting or mechanical agitation, size reduction is needed for feeding the system and for achieving a homogeneous melt.

Horizontal barrier, in situ: (a) It is one of the containment structures which can underlie a sector of contaminated materials on-site without removing the hazardous waste or soil; (b) It is an established technologies use grouting techniques to reduce the permeability of underlying soil layers.

In situ cement-based solidification/stabilization (S/S) technology: In situ, cement-based S/S has only two steps: (1) mixing and (2) off-gas treatment. The processing rate for *in situ* S/S is typically much lower than that for *ex situ* processing. *In situ* S/S is demonstrated to depths of 10 m and may be able to extend to 50 m. The most significant challenge in applying S/S *in situ* for contaminated soils is achieving complete and uniform mixing of the binder with the contaminated matrix. Three basic approaches are used for *in situ* mixing of the binder with the matrix: (a) Vertical auger mixing; (b) In-place mixing of binder reagents with waste by conventional earthmoving equipment, such as draglines, backhoes, or clamshell buckets; and (c) Injection grouting, which involves forcing a binder containing dissolved or suspended treatment agents into the subsurface, allowing it to permeate the soil. Grout injection can be applied to contaminated formations lying well below the ground surface. The injected grout cures in-place to produce an *in situ* treated mass.

In situ vitrification (ISV) technology: The technology is based on electric melter technology, and the principle of operation is joule heating, which occurs when an electrical current is passed through a region that behaves as a resistive heating element. Electrical current is passed through the soil using an array of electrodes inserted vertically into the surface of the contaminated soil zone. Because dry soil is not conductive, a starter path of flaked graphite and glass frit is placed in a small trench between the electrodes to act as the initial flow path for electricity. Resistance heating in the starter path transfers heat to the soil that then begins to melt. Once molten, the soil becomes conductive. The melt grows outward and downward as power is gradually increased to the full constant operating power level. A single melt can treat a region of up to 1,000 T. The maximum treatment depth has been demonstrated to be about 6 m. Large contaminated areas are treated in multiple settings that fuse the blocks together to form one large monolith. Further information on ISV can be found in the following references.

Lead: It is a bluish-white, silvery, or gray metal that is highly lustrous when freshly cut but tarnishes when exposed to air. It is very soft and malleable, has a high density (11.35 g/cm³) and low melting point (327.4°C), and can be cast, rolled, and extruded. The most important Pb ore is galena, PbS. Recovery of Pb from the ore typically involves grinding, flotation, roasting, and smelting. Less common forms of the mineral are cerussite, $PbCO_3$, anglesite, $PbSO_4$, and crocoite, $PbCrO_4$.

Macroencapsulation solidification: Solidification of a large block or container of waste is referred to as macroencapsulation.

Mercury: It is a silvery, liquid metal. The primary source of Hg is cinnabar (HgS), a sulfide ore. In a few cases, Hg occurs as the principal ore product; it is more commonly obtained as the by-product of processing complex ores that contain mixed sulfides, oxides, and chloride minerals (these are usually associated with the base and precious metals, particularly gold). Native or metallic Hg is found in very small quantities in some ore sites. The current demand for Hg is met by secondary production (i.e., recycling and recovery).

Microencapsulation solidification: Solidification of fine waste particles is referred to as microencapsulation

Phytoextraction: (a) It is one of the phytoremediation technologies relying on the uptake of contaminants from the soil and their translocation into aboveground plant tissue, which is harvested and treated. Although hyperaccumulating trees, shrubs, herbs, grasses, and crops have potential, crops seem to be most promising because of their greater biomass production. Nickel and zinc appear to be the most easily absorbed, although tests with copper and cadmium are encouraging. Significant uptake of lead, a commonly occurring contaminant, has not been demonstrated on a large scale. However, some researchers are experimenting with soil amendments that would facilitate the uptake of lead by plants. (b) Phytoextraction technologies use hyperaccumulating plants to transport metals from the soil and concentrate them in the roots and aboveground shoots that can be harvested. A plant containing more than 0.1% of Ni, Co, Cu, Cr, or 1% Zn and Mn in its leaves on a dry weight basis is called a hyperaccumulator, regardless of the concentration of metals in the soil; (c) Ni hyperaccumulators are found in New Caledonia, the Philippines, Brazil, and Cuba. Ni and Zn hyperaccumulators are found in southern and central Europe and Asia Minor; (d) Dried or composted plant residues or plant ashes that are highly enriched with metals can be isolated as hazardous waste or recycled as metal ore. The goal of phytoextraction is to recycle as "bio-ores" metals reclaimed from plant ash in the feed stream of smelting processes. Even if the plant ashes do not have enough concentration of metal to be useful in smelting processes, phytoextraction remains beneficial because it reduces the amount of hazardous waste to be landfilled by as much as 95%. Several research efforts in using trees, grasses, and crop plants are being pursued to develop phytoremediation as a cleanup technology; (e) In a phytoextraction process, plant species are used to take up heavy metals and concentrate them in their tissue. The plants can be harvested, and the contaminated plant material can be disposed of safely. Sometimes, soil amendments are added to the soil to increase the ability of the plants to take up the heavy metals.

Phytoremediation: (a) This technology is for treatment of soils contaminated with metals, and may provide a low-cost option under specific circumstances. This process is best suited for sites with widely dispersed contamination at low concentrations where only treatment of soils at the surface (in other words, within the depth of the root zone) is required; (b) Phytoremediation involves the use of plants to remove, contain, or render harmless environmental contaminants. This definition applies to all biological, chemical, and physical processes that are influenced by plants and that aid in the cleanup of contaminated substances. Plants can be used in site remediation, both to mineralize and immobilize toxic organic compounds at the root zone and to accumulate and concentrate metals and other inorganic compounds from soil into aboveground shoots; (c) Two basic approaches for metal remediation include phytoextraction and phytostabilization; and (d) Phytoremediation has the advantage of relatively low cost and wide public acceptance. It can be less than a quarter of the cost of excavation or in situ fixation. Phytoremediation has the disadvantage of taking longer to accomplish than other treatments.

Phytostabilization: (a) It is one of the phytoremediation technologies using plants to limit the mobility and bioavailability of metals in soils. Ideally, phytostabilizing plants should be able to tolerate high levels of metals and immobilize them in the soil by sorption, precipitation, complexation, or the reduction of metal valences. Phytostabilizing plants should also exhibit low levels of accumulation of metals in shoots to eliminate the possibility that residues in harvested shoots might become hazardous wastes. In addition to stabilizing metals present in the soil, phytostabilizing plants can also stabilize the soil matrix to minimize erosion and migration of sediment. (b) In a phytostabilization process, a contaminated site is simply revegetated, so the plants are used to reduce wind and water erosion that spread materials containing heavy metals.

Polymer microencapsulation solidification/stabilization (S/S) technology: This type of S/S can include the application of thermoplastic or thermosetting resins. Thermoplastic materials are the most commonly used organic-based S/S treatment materials. Potential candidate resins for thermoplastic encapsulation include bitumen, polyethylene, and other polyolefins, paraffins, waxes, and sulfur cement. Of these candidate thermoplastic resins, bitumen (asphalt) is the least expensive and by far the most commonly used. The process of thermoplastic encapsulation involves heating and mixing the waste material and the resin at elevated temperatures, typically 130°C–230°C in an extrusion machine. Any water or volatile organics in the waste boil off during extrusion and are collected for treatment or disposal. Because the final product is a stiff yet plastic resin, the treated material is typically discharged from the extruder into a drum or other container.

Pyrometallurgy: (a) It is a broad term encompassing elevated temperature techniques for the extraction and processing of metals for use or disposal. High-temperature processing increases the rate of reaction and often makes the reaction equilibrium more favorable, lowering the required reactor volume per unit output. Some processes that clearly involve both metal extraction and recovery include roasting, retorting, or smelting. While these processes typically produce a metal-bearing waste slag, metal is also recovered for reuse. A second class of pyrometallurgical technologies included here is a combination of high-temperature extraction and immobilization. These processes use thermal means to cause volatile metals to separate from the soil and report to the fly ash, but the metal in the fly ash is then immobilized instead of recovered, and no metal is recovered for reuse. A third class of technologies are those that are primarily incinerators for mixed organic-inorganic wastes but that have the capability of processing wastes containing the metals of interest by either capturing volatile metals in the exhaust gases or immobilizing the nonvolatile metals in the bottom ash or slag. Since some of these systems may have applicability to some cases where metal contamination is the primary concern, a few technologies of this type are noted that are in the SITE program. Vitrification is not considered pyrometallurgical treatment since there is typically neither a metal extraction nor a metal recovery component in the process; (b) Pyrometallurgical processing is usually preceded by physical treatment to produce a uniform feed material and upgrade the metal content. Solids treatment in a high-temperature furnace requires efficient heat transfer between the gas and solid phases while minimizing particulate in the off-gas. The particle-size range that meets these objectives is limited and is specific to the design of the process. The presence of large clumps or debris slows heat transfer; hence, pretreatment to either remove or pulverize oversized material is normally required. Fine particles are also undesirable because they become entrained in the gas flow, increasing the volume of dust to be removed from the flue gas. The feed material is sometimes pelletized to give a uniform size. In many cases, a reducing agent and flux may be mixed in prior to pelletization to ensure good contact between the treatment agents and the contaminated material and to improve the gas flow in the reactor.

Rhizofiltration: In a rhizofiltration process, heavy metals are removed directly from water by plant roots. The plants are grown directly in water or water-rich materials such as sand, using aquatic species or hydroponic methods. uses plant roots to absorb, concentrate, and precipitate metals from wastewater, which may include leachate from soil. Rhizofiltration uses terrestrial plants instead of aquatic plants because terrestrial plants develop much longer, fibrous root systems covered with root hairs that have extremely large surface areas. This variation of phytoremediation uses plants that remove metals by sorption, which does not involve biological processes. Using plants to translocate metals to shoots is a slower process than phytoextraction. Rhizofiltration involves the construction of wetlands or reed beds for the treatment of contaminated wastewater or leachate. The technology is cost-effective for the treatment of large volumes of wastewater that have low concentrations of metals.

Soil flushing: (a) It is the *in situ* extraction of contaminants from the soil via an appropriate washing solution. Water or an aqueous solution is injected into or sprayed onto the area of contamination, and the contaminated elutriate is collected and pumped to the surface for removal, recirculation, or on-site treatment and reinjection. The technology is applicable to both organic and inorganic contaminants, and metals in particular. For the purpose of heavy metals remediation, soil flushing has been operated at full scale, but for a small number of sites; (b) Soil flushing uses water, a solution of chemicals in water, or an organic extractant to recover contaminants from the *in situ* material. The contaminants are mobilized by solubilization, formation of emulsions, or a chemical reaction with the flushing solutions. After passing through the contamination zone, the contaminant-bearing fluid is collected by strategically placed wells or trenches and brought to the surface for disposal, recirculation, or on-site treatment and reinjection. During elutriation, the flushing solution mobilizes the sorbed contaminants by dissolution or emulsification; (c) One key to the efficient operation of a soil flushing system is the ability to reuse the flushing solution, which is recovered along with groundwater. Various water treatment techniques can be applied to remove the recovered metals and render the extraction fluid suitable for reuse. Recovered flushing fluids may need treatment to meet appropriate discharge standards prior to their release to a POTW or receiving waters. The separation of surfactants from recovered flushing fluid, for reuse in the process, is a major factor in the cost of soil flushing. Treatment of the flushing fluid results in process sludges and residual solids, such as spent carbon and spent ion exchange resin, which must be appropriately treated before disposal. Air emissions of volatile contaminants from recovered flushing fluids should be collected and treated, as appropriate, to meet applicable regulatory standards. Residual flushing additives in the soil may be a concern and should be evaluated on a site-specific basis. Subsurface containment barriers can be used in conjunction with soil flushing technology to help control the flow of flushing fluids.

Soil washing: It is an *ex situ* oil remediation technology that uses a combination of physical separation and aqueous-based separation unit operations to reduce contaminant concentrations to site-specific remedial goals. Although soil washing is sometimes used as a stand-alone treatment technology, more often it is combined with other technologies to complete site remediation. Soil washing technologies have successfully remediated sites contaminated with organic, inorganic, and radioactive contaminants. The technology does not detoxify or significantly alter the contaminant but transfers the contaminant from the soil into the washing fluid or mechanically concentrates the contaminants into a much smaller soil mass for subsequent treatment. Soil washing systems are quite flexible regarding the number, type, and order of processes involved. Soil washing is performed on excavated soil and may involve some or all of the following, depending on the contaminant-soil matrix characteristics, cleanup goals, and the specific process employed : (a) Mechanical screening to remove various oversized materials; (b) Crushing to reduce applicable oversize to suitable dimensions for treatment; (c) Physical processes (e.g., soaking, spraying, tumbling, and attrition scrubbing) to liberate weakly bound agglomerates (e.g., silts and clays bound to sand and gravel) followed by size classification to generate coarse-grained and fine-grained soil fraction(s) for further treatment; (d) Treatment of the coarse-grained soil fraction(s); (e) Treatment of the fine-grained fraction(s); and (f) Management of the generated residuals.

Solidification/Stabilization (S/S) Functions: S/S technology is usually applied by mixing contaminated soils or treatment residuals with a physical binding agent to form a crystalline, glassy, or polymeric framework surrounding the waste particles. In addition to the microencapsulation, some chemical fixation mechanisms may improve the waste's leach resistance. Other forms of S/S treatment rely on macroencapsulation, where the waste is unaltered but macroscopic particles are encased in a relatively impermeable coating, or on specific chemical fixation, where the contaminant is converted to a solid compound resistant to leaching. S/S treatment can be accomplished primarily through the use of either inorganic binders (e.g., cement, fly ash, and/or blast furnace slag) or organic binders such as bitumen. Additives may be used, for example, to convert the metal to a less mobile form or to counteract adverse S/S aims to accomplish one or more of the following objectives: (a) Improve the physical characteristics of the waste by producing a solid from liquid or semiliquid wastes; (b) Reduce the contaminant solubility by formation of sorbed species or insoluble precipitates (e.g., hydroxides, carbonates, silicates, phosphates, sulfates, or sulfides); (c) Decrease the exposed surface area across which mass transfer loss of contaminants may occur by the formation of a crystalline, glassy, or polymeric framework that surrounds the waste particle; and (d) Limit the contact between transport fluids and contaminants by reducing the material's permeability.

Solidification: It is a process that encapsulates a waste to form a solid material and restricts contaminant migration by decreasing the surface area exposed to leaching and/or by coating the waste with low-permeability materials. Solidification can be accomplished by a chemical reaction between a waste and binding (solidifying) reagents or by mechanical processes. The solidification of fine waste particles is called microencapsulation, whereas solidification of a large block or container of waste is referred to as macroencapsulation.

Stabilization: (a) It refers to processes that involve chemical reactions that reduce the leachability of a waste. Stabilization chemically immobilizes hazardous materials (such as heavy metals) or reduces their solubility through a chemical reaction. The physical nature of the waste may or may not be changed by this process; (b) Application of stabilization for removal of heavy metals from contaminated soil on-site reduces or eliminates their ability to adversely affect human health and the environment and has many advantages over excavation. One way of stabilizing heavy metals consists of adding chemicals to the soil that cause the formation of minerals that contain the heavy metals in a form that is not easily absorbed by plants, animals, or people. This method is called in situ (in place) fixation or stabilization by which the heavy metal in soil combines with the added chemical to create a less toxic compound, so the stabilization process does not disrupt the environment or generate hazardous wastes. The heavy metal remains in the soil but in a form that is much less harmful.

Superfund Innovative Technology Evaluation (SITE) program: It is a USEPA program that evaluates many emerging and demonstrated technologies to promote the development and use of innovative technologies to clean up Superfund sites across the country. The major focus of SITE is the Demonstration Program, which is designed to provide engineering and cost data for selected technologies.

Vertical barrier: It is one of the containment structures that minimizes the movement of contaminated groundwater off-site or limits the flow of uncontaminated groundwater on-site. Common vertical barriers include slurry walls in excavated trenches; grout curtains formed by injecting grout into soil borings; vertically injected, cement-bentonite grout-filled borings or holes formed by withdrawing beams driven into the ground; and sheet-pile walls formed of driven steel.

Vitrification: (a) The process applies high-temperature treatment aimed primarily at reducing the mobility of metals by incorporation into a chemically durable, leach-resistant, vitreous mass. Vitrification can be carried out on excavated soils as well as *in situ*; (b) During the vitrification process, organic wastes are pyrolyzed (*in situ*) or oxidized (*ex situ*) by the melt front, whereas inorganics, including metals, are incorporated into the vitreous mass. Off-gases released during the melting process, containing volatile components and products of combustion and pyrolysis, must be collected and treated. Vitrification converts contaminated soils to a stable glass and crystalline monolith. With the addition of low-cost materials such as sand, clay, and/or native soil, the process can be adjusted to produce products with specific characteristics, such as chemical durability. Waste vitrification may be able to transform the waste into useful, recyclable products such as clean fill, aggregate, or higher-valued materials such as erosion-control blocks, paving blocks, and road dividers.

REFERENCES

1. Federal Register, *Comprehensive Environmental Response, Compensation, and Liability Act (CERCLA or Superfund)*. 42 U.S.C. s/s 9601 et seq. (1980), United States Government, Public Laws. Available at: www.access.gpo.gov/uscode/title42/chapter103_.html, January 2004.
2. U.S. EPA. *Recent Developments for In Situ Treatment of Metal Contaminated Soils*. Contract # 68-W5-0055, U.S. Environmental Protection Agency, Washington, DC, March 1997.
3. Federal Register, *Resource Conservation and Recovery Act (RCRA)*. 42 U.S. Code s/s 6901et seq. (1976), U.S. Government, Public Laws. Available at: www.access.gpo.gov/uscode/title42/chapter82_.html, January 2004.
4. U.S. EPA. *Contaminants and Remedial Options at Selected Metal-Contaminated Sites*. EPA/540/R-95/512, U.S. Environmental Protection Agency, Washington, DC, July 1995.
5. U.S. EPA. *Technology Alternatives for the Remediation of Soils Contaminated with AS, Cd, Cr, Hg, and Pb*. EPA/540/S-97/500, U.S. Environmental Protection Agency, Cincinnati, OH, August 1997.
6. U.S. EPA. *In Situ Technologies for the Remediation of Soils Contaminated with Metals—Status Report*, U.S. Environmental Protection Agency, Cincinnati, OH, July 1996.
7. U.S. EPA. *Selection of Control Technologies for Remediation of Lead Battery Recycling Sites*. EPA/540/2-91/014, U.S. Environmental Protection Agency, Cincinnati, OH, 1991.
8. U.S. EPA. *Engineering Bulletin: Selection of Control Technologies for Remediation of Lead Battery Recycling Site*. EPA/540/S-92/011, U.S. Environmental Protection Agency, Cincinnati, OH, 1992.
9. U.S. EPA. *Contaminants and Remedial Options at Wood Preserving Sites*. EPA 600/R-92/182, U.S. Environmental Protection Agency, Washington, DC, 1992.
10. U.S. EPA. *Presumptive Remedies for Soils, Sediments, and Sludges at Wood Treater Sites*. EPA/540/R-95/128, U.S. Environmental Protection Agency, Washington, DC, 1995.
11. U.S. EPA. *Contaminants and Remedial Options at Pesticide Sites*. EPA/600/R-94/202, U.S. Environmental Protection Agency, Washington, DC, 1994.
12. U.S. EPA. *Separation/Concentration Technology Alternatives for the Remediation of Pesticide-Contaminated Soil*. EPA/540/S-97/503, U.S. Environmental Protection Agency, Washington, DC, 1997.
13. McLean, J.E. and Bledsoe, B.E. *Behavior of Metals in Soils*. EPA/540/S-92/018, U.S. Environmental Protection Agency, Washington, DC, 1992.
14. Palmer, C.D. and Puls, R.W. *Natural Attenuation of Hexavalent Chromium in Ground Water and Soils*. EPA/540/S-94/505, U.S. Environmental Protection Agency, Washington, DC, 1994.
15. Benjamin, M.M. and Leckie, J.D. Adsorption of metals at oxide interfaces: Effects of the concentrations of adsorbate and competing metals. In: Baker R. A. (Ed.), *Contaminant sand Sediments*, Chapter 16. *Volume 2: Analysis, Chemistry, Biology*. Ann Arbor Science Publishers, Inc., Ann Arbor, MI, 1980.
16. Wagemann, R. Some theoretical aspects of stability and solubility of inorganic As in the freshwater environment. *Water Res.*, 12, 139–145, 1978.
17. Zimmerman, L. and Coles, C. Cement industry solutions to waste management—the utilization of processed waste by-products for cement manufacturing. In: *Proceedings of the 1st International Conference for Cement Industry Solutions to Waste Management*, Calgary, Alberta, Canada, pp. 533–545, 1992.
18. Earth Platform. *Contaminated Soil Remediation*. Available at: https://www.earthplatform.com/contami-nated/soil/remediation, 2007.
19. Sharma, H.D. and Reddy, K.R. *Geoenvironmental Engineering: Site Remediation, Waste Containment, and Emerging Waste Management Technologies*, Wiley, Hoboken, NJ, 2004.
20. Weston, R.F. *Installation Restoration General Environmental Technology Development Guidelines for In-Place Closure of Dry Lagoons*. U.S. Army Toxic and Hazardous Materials, May 1985.

21. U.S. EPA. *Slurry Trench Construction for Pollution Migration Control.* EPA/540/2-84/001, U.S. Environmental Protection Agency, Washington, DC, February 1984.
22. U.S. EPA. *Grouting Techniques in Bottom Sealing of Hazardous Waste Sites.* EPA/600/2-86/020, U.S. Environmental Protection Agency, Washington, DC, 1986.
23. U.S. EPA. *Engineering Bulletin: Landfill Covers.* EPA/540/S-93/500, U.S. Environmental Protection Agency, Cincinnati, OH, February 1993.
24. FRTR. *Physical Barriers.* Remediation Technologies Screening Matrix and Reference Guide. Available at: https://www.frtr.gov/matrix2/section4/4-53.html, 2007.
25. U.S. EPA. *Engineering Bulletin: Slurry Walls.* EPA/540/S92/008, U.S. Environmental Protection Agency, Cincinnati, OH, October 1992.
26. U.S. EPA. *Solidification/Stabilization Use at Superfund Sites.* EPA-542-R-00-010, U.S. Environmental Protection Agency, Washington, DC, September 2000.
27. U.S. EPA. *Technical Resource Document: Solidification/Stabilization and Its Application to Waste Materials.* EPA/530/R-93/012, U.S. Environmental Protection Agency, Washington, DC, June 1993.
28. U.S. EPA. *Engineering Bulletin: Solidification/Stabilization of Organics and Inorganics.* EPA/540/S-92/015, U.S. Environmental Protection Agency, Cincinnati, OH, 1992.
29. Conner, J.R. *Chemical Fixation and Solidification of Hazardous Wastes.* VanNostrand Reinhold, New York, NY, 1990.
30. Anderson, W.C. (Ed.), *Innovative Site Remediation Technology: Solidification/Stabilization,* Volume 4, Water Environment Federation, Alexandria, VA, 1994.
31. WASTECH, American Academy of Environmental Engineers (EPA printed under license no. EPA/542-B-94-001). June 1994.
32. U.S. EPA. *A Citizen's Guide to Solidification/Stabilization.* EPA 542-F-01-024, U.S. Environmental Protection Agency, Washington, DC, December 2001.
33. U.S. ACE. *Solidification/Stabilization of Contaminated Material, Unified Facility Guide Specification.* UFGS-02160a, U.S. Army Corps of Engineers, October 2000.
34. ANL. *Fact Sheet—Solidification/Stabilization.* Drilling Waste Management Information System, Argonne National Laboratory. Available at: https://web.ead.anl.gov/dwm/techdesc/solid/index.cfm, 2007.
35. U.S. EPA. *Handbook on In Situ Treatment of Hazardous Waste-Contaminated Soils.* EPA/540/2-90/002, U.S. Environmental Protection Agency, Cincinnati, OH, 1990.
36. Arniella, E.F. and Blythe, L.J. Solidifying traps hazardous waste. *Chem. Eng.,* 97(2), 92–102, 1990.
37. Kalb, P.D., Burns, H.H., and Meyer, M. Thermo-plastic encapsulation treatability study for a mixed waste incinerator off-gas scrubbing solution. In: Gilliam, T.M. (Ed.), *Third International Symposium on Stabilization/Solidification of Hazardous, Radioactive, and Mixed Wastes,* ASTM STP 1240. American Society for Testing and Materials, Philadelphia, PA, 1993.
38. Ponder, T.G. and Schmitt, D. Field assessment of air emission from hazardous waste stabilization operation. In: *Proceedings of the 17th Annual Hazardous Waste Research Symposium,* EPA/600/9-91/002, Cincinnati, OH, 1991.
39. Shukla, S.S., Shukla, A.S., and Lee, K.C. Solidification/stabilization study for the disposal of pentachlorophenol. *J. Hazard. Mater.,* 30, 317–331, 1992.
40. U.S. EPA. *Evaluation of Solidification/Stabilization as a Best Demonstrated Available Technology for Contaminated Soils.* EPA/600/2-89/013, U.S. Environmental Protection Agency, Cincinnati, OH, 1989.
41. Weitzman, L. and Hamel, L.E. Volatile emissions from stabilized waste. In: *Proceedings of the 15th Annual Research Symposium,* EPA/600/9-90/006, U.S. Environmental Protection Agency, Cincinnati, OH, 1990.
42. Means, J.L., Nehring, K.W., and Heath, J.C. Abrasive blast material utilization in asphalt roadbed material. In: *Third International Symposium on Stabilization/Solidification of Hazardous, Radioactive, and Mixed Wastes.* ASTM STP 1240, American Society for Testing and Materials, Philadelphia, PA, 1993.
43. U.S. ACE. Yearly average cost index for utilities. In: *Civil Works Construction Cost Index System Manual,* 110-2-1304, U.S. Army Corps of Engineers, Washington, DC, p. 44. PDF file is available on the Internet at: https://www.nww.usace.army.mil/cost, 2007.
44. Buelt, J.L., Timmerman, C.L., Oma, K.H., FitzPatrick, V.F., and Carter J.G. *In Situ Vitrification of Transuranic Waste: An Updated Systems Evaluation and Applications Assessment.* PNL-4800, Pacific Northwest Laboratory, Richland, WA, 1987.
45. U.S. EPA. *Vitrification Technologies for Treatment of Hazardous and Radioactive Waste.* EPA/625/R-92/002, U.S. Environmental Protection Agency, Cincinnati, OH, May 1992.

46. U.S. EPA. *Engineering Bulletin: In Situ Vitrification Treatment.* EPA/540/S-94/504, U.S. Environmental Protection Agency, Cincinnati, OH, October 1994.
47. U.S. EPA. *Engineering Bulletin: In Situ Vitrification Treatment.* EPA/540/S-94/504, U.S. Environmental Protection Agency, Washington, DC, May 2002 (Revised).
48. FRTR. Solidification/Stabilization—In Situ Soil Remediation Technology. Remediation Technologies Screening Matrix and Reference Guide. Available at: https://www.frtr.gov/matrix2/section4/4-8.html, 2007.
49. U.S. EPA. *Geosafe Corporation In Situ Vitrification Innovative Technology Evaluation Report.* EPA/540/ R-94/520, U.S. Environmental Protection Agency, Washington, DC, March 1995.
50. FitzPatrick, V.F., Timmerman, C.L., and Buelt, J.L. In situ vitrification: An innovative thermal treatment technology. In: *Proceedings of the 2nd International Conference on New Frontiers for Hazardous Waste Management,* 305–322, 1987. EPA/600/9-87/018F. U.S. Environmental Protection Agency, EPA printed under license No. EPA/542-B-94-001, June 1994.
51. Timmerman, C.L. *In Situ Vitrification of PCB Contaminated Soils.* EPRI CS-4839, Electric Power Research Institute, Palo Alto, CA, 1986.
52. U.S. EPA. *The Superfund Innovative Technology Evaluation Program: Technology Profiles,* 4th Edition, EPA/540/5-91/008, U.S. Environmental Protection Agency, Washington, DC, 1991.
53. Luey, J., Koegler, S.S., Kuhn, W.L., Lowery, P.S., and Winkelman, R.G. *In Situ Vitrification of a Mixed-Waste Contaminated Soil Site.* The 116-B-6A Crib at Hanford, PNL-8281, Pacific Northwest Laboratory, Richland, WA, 1992.
54. Hansen, J.E. and FitzPatrick, V.F. *In Situ Vitrification Applications,* Geosafe Corporation, Richland, WA, 1991.
55. U.S. EPA. *Engineering Bulletin: Soil Washing Treatment.* EPA/540/2-90/017, U.S. Environmental Protection Agency, Cincinnati, OH, 1996.
56. U.S. EPA. *A Citizen's Guide to Soil Washing.* EPA 542-F-01-008, U.S. Environmental Protection Agency, Washington, DC, May 2001.
57. William, C.A. (Ed.). *Innovative Site Remediation Technology: Soil Washing/Flushing,* Vol. 3. American Academy of Environmental Engineers (published by EPA under EPA 542-B-93-012), November 1993.
58. U.S. EPA. *Technology Focus—Soil Washing.* Technology Innovation Program, U.S. Environmental Protection Agency, Washington, DC. Available at: https://clu-in.org/techfocus/default.focus/sec/Soil_Washing/cat/Overview, 2007.
59. Ehsan, S., Prasher, S.O., and Marshall, W.D. A washing procedure to mobilize mixed contaminants from soil. II. Heavy metals. *J. Environ. Qual.,* 35, 2084–2091, 2006.
60. Fischer, K. and Bipp, H.P. Removal of heavy metals from soil components and soils by natural chelating agents. Part II. Soil extraction by sugar acids. *Water, Air, and Soil Pollut.,* 38(1–4), 271–288, 2002.
61. U.S. EPA. *Citizen's Guide to Soil Washing.* EPA/542/F-92/003, U.S. Environmental Protection Agency, Washington, DC, March 1992.
62. U.S. EPA. *Engineering Bulletin: In Situ Soil Flushing.* EPA/540/2-91/021, U.S. Environmental Protection Agency, Cincinnati, OH, October 1991.
63. FRTR. *Soil Flushing—In Situ Soil Remediation Technology.* Remediation Technologies Screening Matrix and Reference Guide. Available at: https://www.frtr.gov/matrix2/section4/4-6.html, 2007.
64. CPEO. *Soil Flushing.* Center for Public Environmental Oversight (CPEO). San Francisco, CA https://www.cpeo.org/techtree/ttdescript/soilflus.htm, 2007.
65. U.S. EPA. *Superfund Innovative Technology Evaluation Program: Technology Profiles,* 7th Edition, EPA/540/R-94/526, U.S. Environmental Protection Agency, Washington, DC, November 1994.
66. Pamukcu, S. and Wittle, J.K. Electrokinetic removal of selected metals from soil. *Environ. Progr.* II, 3, 241–250, 1992.
67. Acar, Y.B. Electrokinetic cleanups. *Civil Eng.,* 62, 58–60, 1992.
68. Mattson, E.D. and Lindgren, E.R. Electrokinetics: An innovative technology for in situ remediation of metals. In: *Proceedings, National Groundwater Association, Outdoor Acnon Conference.* Minneapolis, MN, May 1994.
69. Acar, Y.B. and Gale, R.J. Electrokinetic remediation: Basics and technology status. *J. Hazard. Mater.,* 40, 117–137, 1995.
70. Will, F. Removing toxic substances from the soil using electrochemistry. *Chem. and Ind.,* 15, 376–379, 1995.
71. Rodsand, T. and Acar, Y.B. Electrokinetic extraction of lead from spiked Norwegian marine clay. *Geoenvironment* 2000, 2, 1518–1534, 1995.

72. Lindgren, E.R., Kozak, M.W., and Mattson, E.D. Electrokinetic remediation of contaminated soils: An update. *Waste Management 92*. Tuscon, Arizona, p. 1309, 1992.
73. Jacobs, R.A. and Sengun, M.Z. Model of experiences on soil remediation by electric fields. *J. Environ. Sci. Health*, 29A, 9, 1994.
74. Acar, Y.B. and Gale, R.J. *Electrochemical Decontamination of Soils and Slurries*. U.S. Patent No.: 5,137,608. Commissioner of Patents and Trademarks, Washington, DC. August 15, 1992.
75. Marks, R., Acar, Y.B., and Gale, R.J. *In Situ Bioelectrokinetic Remediation of Contaminated Soils Containing Hazardous Mixed Wastes*. U.S. Patent No. 5,458,747. Commissioner of Patents and Trademarks, Washington, DC, October 17, 1995.
76. Acar, Y.B. and Alshawabkeh, A.N. Electrokinetic remediation: I. pilot-scale tests with lead spiked kaolinite, II. theoretical model. *J. Geotechn. Eng.*, 122(3), 173–196, March 1996.
77. W. Pool. *Process for the Electroreclamation of Soil Material*. Patent No. 5,433,829, U.S. Patent Office. July 18, 1995.
78. U.S. EPA. *In Situ Remediation Technology Status Report: Electrokinetics*. EPA 542-K-94-007, U.S. Environmental Protection Agency, Washington, DC, 1995.
79. Editor. Innovative in situ cleanup processes. *The Hazardous Waste Consultant*, Volume 17, September/October, 1992.
80. DOE. *Development of an Integrated In-Situ Remediation Technology*. Technology Development Data Sheet, DE-AR21-94MC31185. U.S. Department of Energy, 1995.
81. U.S. EPA. *Superfund Innovative Technology Evaluation Program Technology Profiles*, 7th Edition, EPA 540-R-94-526, U.S. Environmental Protection Agency, Washington, DC, 1994.
82. U.S. EPA. *Lasagna™ Public-Private Partnership*. EPA 542-F-96-010A, U.S. Environmental Protection Agency, Washington, DC, 1996.
83. Szpyrkowicz, L., Radaelli, M., Bertini, S., Daniele, S., and Casarin, F. Simultaneous removal of metals and organic compounds from a heavily polluted soil. *Electrochim. Acta*, 52(10), 3386–3392, 2007.
84. CPEO. *Electrokinetics*, Center for Public Environmental Oversight (CPEO), San Francisco, CA. Available at: https://www.cpeo.org/techtree/ttdescript/elctro.htm, 2007.
85. Cunningham, S.D. and Berti, W.R. Remediation of contaminated soils with green plants: An overview. *In Vitro Cell. Dev. Biol.* (Tissue Culture Association), 29, 207–212, 1993.
86. Raskin, I. Bioconcentration of metals by plants. *Environ. Biotechnol.*, 5, 285–290, 1994.
87. Goldsbrough, P. Phytochelatins and metallothioneins: Complementary mechanisms for metal tolerance. In: *Fourteenth Annual Symposium 1995 in Current Topics in Plant Biochemistry, Physiology and Molecular Biology*, University of Missouri, Columbia, MO, USA, 1995.
88. Salt, D.E., et al. Phytoremediation: A novel strategy for the removal of toxic metals from the environment using plants. *Biotechnology*, 13, 468–474, 1995.
89. Kumar, P.B.A. Phytoextraction: The use of plants to remove metals from soils. *Environmen. Sci. Technol.*, 29, 1232–1238, 1995.
90. Durham, S. *Using Plants to Clean Up Soil*, U.S. Department of Agriculture (USDA), https://www.ars.usda.gov/is/pr/2007/070123.htm. January 23, 2007.
91. Morel, I.L. Root exudates and metal mobilization. In: *Fourteenth Annual Symposium 1995 in Current Topics in Plant Biochemistry, Physiology and Molecular Biology*, University of Missouri, Columbia, MO, USA, 1995.
92. Baker, A.J.M. and Brooks, R.R. Terrestrial higher plants which hyperaccumulate metallic elements—a review of their distribution, ecology, and phytochemistry. *Biorecovery*, 1, 81–126, 1989.
93. Hyperaccumulation in the Genus *Alyssum*. In: *Fourteenth Annual Symposium 1995 in Current Topics in Plant Biochemistry, Physiology and Molecular Biology*, University of Missouri, Columbia, MO, USA, 1995.
94. Baker, A.J.M. Metal hyperaccumulation by plants: Our present knowledge of ecophysiological phenomenon. In: *Fourteenth Annual Symposium 1995 in Current Topics in Plant Biochemistry, Physiology and Molecular Biology*, University of Missouri, Columbia, MO, USA. 1995.
95. Chaney, R.L., Malik, M., Li, Y.M., Brown, S.L., Angle, J.S., and Baker, A.J.M. Phytoremediation of soil metals. *Current Opin. Biotechnol.*, 8, 279–284, 1997.
96. King Communications Group, Inc. Promise of heavy metal harvest lures venture funds. *The Bioremediation Report*, 4, 1, Washington, DC, January, 1995.
97. Greger, M. and Landberg, M.T. Improving removal of metals from soil by *Salix*. In: *Proceedings of the 7th International Conference on the Biogeochemistry of Trace Elements*, Uppsala, Sweden, June 15–19, 2003.

98. Ensley, B.D. Will plants have a role in bioremediation? In: *Fourteenth Annual Symposium 1995 in Current Topics in Plant Biochemistry, Physiology and Molecular Biology,* University of Missouri, Columbia, MO, USA, 1995.

99. Cunningham, S.D. and Lee, C.R. Phytoremediation: Plant-based remediation of contaminated soil and sediments. In: *Proceedings of a Symposium of the Soil Science,* Society of America, Chicago, IL, November, 1994.

100. DOE. *Summary Report of a Workshop on Phytoremediation Research Needs.* U.S. Department of Energy, Santa Rosa, CA, July 24–26, 1994.

101. Baker, A.J.M., Brooks, R.R., and Reeves, R.D. Growing for gold...and copper...and zinc. *New Scientist,* 1603, 44–48, 1989.

102. Parry, I. Plants absorb metals. *Pollut. Eng.,* Volume 27, pp. 40–41, February, 1995.

103. USDA. Acidifying Soil Helps Plant Remove Cadmium, Zinc Metals. Agricultural Research Service. *Science Daily.* https://www.sciencedaily.com/releases/2005/06/050619192657.htm, 2007.

104. Hyman, M. and Dupont, R.R. *Groundwater and Soil Remediation: Process Design and Cost Estimating of Proven Technologies,* ASCE Publications, Reston, VA, p. 534, 2001.

105. Priya, A.K., Muruganandam, M., Sameh, S., Ali, S.S., and Michael Kornaro, M. Clean-Up of heavy metals from contaminated soil by phytoremediation: A multidisciplinary and eco-friendly approach, *Toxics* 11(5), 422, 2023. doi: 10.3390/toxics11050422.

106. Devi, P. and Kumar, P. Concept and application of phytoremediation in the fight of heavy metal toxicity. *J. Pharm. Sci. Res.,* 12, 795–804, 2020.

107. Li, Q., Xing, Y., Fu, X., Ji, L., Li, T., Wang, J., Chen, G., Qi, Z., and Zhang, Q. Biochemical mechanisms of rhizospheric bacillus subtilis-facilitated phytoextraction by alfalfa under cadmium stress–microbial diversity and metabolomics analyses. *Ecotoxicol. Environ. Saf.,* 212, 112016, 2021. doi: 10.1016/j.ecoenv.2021.112016.

108. Wang, J., Hu, X., Shi, T., He, L., Hu, W., and Wu, G. Assessing toxic metal chromium in the soil in coal mining areas via proximal sensing: prerequisites for land rehabilitation and sustainable development. *Geoderma* 405, 115399, 2022. doi: 10.1016/j.geoderma.2021.115399.

109. Yang, L., Wang, J., Yang, Y., Li, S., Wang, T., Oleksak, P., Chrienova, Z., Wu, Q., Nepovimova, E., Zhang, X., et al. Phytoremediation of heavy metal pollution: Hotspots and future prospects. *Ecotoxicol. Environ. Saf.,* 234, 113403, 2022. doi: 10.1016/j.ecoenv.2022.11340.

110. Tatu, G.L.A., Vladut, N.V., Voicea, I., Vanghele, N.A., and Pruteanu, M.A. Removal of heavy metals from a contaminated soil using phytoremediation. *MATEC Web Conf.,* 305, 00061, 2020. doi: 10.1051/matecconf/202030500061

111. US EPA. *Introduction to Phytoremediation.* EPA/600/R-99/107, US Environmental Protection Agency, USA, February 2000.

112. Schnoor, J. Phytoremediation: Groundwater Remediation Technologies Analysis Center Technology Evaluation Report TE-98-01, 37, US Environmental Protection Agency, USA, 1997. https://cfpub.epa.gov/ncer_abstracts/index.cfm/fuseaction/display.files/fileID/14295

113. Lambert, M., Leven, B.A., and Green, R.M. *New Methods of Cleaning Up Heavy Metal in Soils and Water,* Kansas State University, Manhattan, KS, USA, 2001.

114. Wood, P. Remediation methods for contaminated sites: In: *Contaminated Land and Its Reclamation,* Royal Society of Chemistry, Cambridge, UK, pp. 47–71, 1997.

115. Green, R., Erickson, L., Govin- daraju, R., and Kalita, P. Modeling the Effects of Vegetation on Heavy Metals Containment: 12; *Conference on Hazardous Waste Research,* Kansas City, Missouri, USA, pp. 476–487, 1997.

116. Lambert, M., Pierzynski, G., Erickson, L., and Schnoor, J. Remediation of lead-, zinc-, and cadmium-contaminated soils. In Hester R. and Harrison R. (Eds.), *Contaminated Land and Its Reclamation,* the Royal Society of Chemistry, Cambridge, UK, pp. 91–102, 1997.

117. Liu, L., Wei Li, W., and Guo, M. (2018). Remediation techniques for heavy metal-contaminated soils: Principles and applicability. *Sci. Total Environ.,* 633, 206–219, 2018.

118. Wang, M.H.S., Shammas, N.K. and Wang, L.K. Advances in the treatment and management of metal finishing industry wastes. In: *Control of heavy metals in the Environment* Volume 3, Wang, L.K., Wang, M.H.S., Hung, Y.T., and Chen, J.P. (editors). CRC Press, Florida, USA. pp. 346–398, 2022.

13 Advances in the Remediation of Metal Finishing Brownfield Sites

Mu-Hao Sung Wang, Lawrence K. Wang, and Nazih K. Shammas

ACRONYM

AOP:	Advanced oxidation process
ASTM:	American Society of Testing Materials
ATSDR:	Agency for Toxic Substances and Disease Registry
CERCLA:	The Comprehensive Environmental Response, Compensation, and Liability Act
CSM:	Conceptual Site Model
CWA:	Clean Water Act
DQO:	Data Quality Objectives
EPH:	Extractable petroleum hydrocarbons
ERNS	Emergency Response Notification System
FID:	Flame ionization
HM:	Heavy metals
MCL:	Maximum contaminant levels (MCLs)
NPDES:	National Pollutant Discharge Elimination System
NPL:	National Priorities List
NTIS:	National Technical Information Service
PAH:	Polycyclic aromatic hydrocarbon
PCB:	Polychlorinated biphenyls
PID:	Photo ionization detector
PLM:	Polarized light microscopy
POTW:	Publicly Owned Treatment Works
RCRA:	Resource Conservation and Recovery Act
SBLRBRA:	Small Business Liability Relief and Brownfields Revitalization Act
SDWA:	Safe Drinking Water Act
SVOC:	Semi-volatile organic compounds
TCLP:	Toxicity Characteristic Leachate Procedure
TPH:	Total petroleum hydrocarbons
TSD:	Treatment, storage, and disposal
TTO:	Total toxic organics
U.S. EPA:	U.S. Environmental Protection Agency
USACE:	U.S. Army Corps of Engineers
USGS:	U.S. Geological Service
UST:	Underground storage tanks
UV:	Ultraviolet
VCP:	Voluntary Correction Program

DOI: 10.1201/9781003541615-13

VOC:	Volatile organic compounds
VPH:	Volatile petroleum hydrocarbons.
XRF:	X-ray fluorescence

13.1 INTRODUCTION

13.1.1 Background

The Comprehensive Environmental Response, Compensation, and Liability Act (CERCLA) or Superfund [1] defines brownfields sites as "real property, the expansion, redevelopment, or reuse of which may be complicated by the presence or potential presence of a hazardous substance, pollutant, or contaminant." According to the U.S. Environmental Protection Agency (U.S. EPA), brownfields sites are abandoned, idled, or under-used industrial and commercial facilities where expansion or redevelopment is complicated by real or perceived environmental contamination [2]. Concerns about liability, cost, and potential health risks associated with brownfields sites often prompt businesses to migrate to "greenfields" outside the city. Left behind are communities burdened with environmental contamination, declining property values, and increased unemployment.

U.S. EPA's Brownfields Economic Redevelopment Initiative was established to enable states, site planners, and other community stakeholders to work together in a timely manner to prevent, assess, safely cleanup, and sustainably reuse brownfields sites [3]. With the enactment of the Small Business Liability Relief and Brownfields Revitalization Act in 2002, U.S. EPA assistance was expanded to provide greater support for brownfields cleanup and reuse. Many states and local jurisdictions also help businesses and communities to adapt environmental cleanup programs to the special needs of brownfields sites.

Preparing brownfields sites for productive reuse requires integrating many elements—financial issues, community involvement, liability considerations, environmental assessment and cleanup, regulatory requirements, and more—as well as coordination among many groups of stakeholders [4]. The assessment and cleanup of a site must be carried out in a way that integrates all these factors into the overall redevelopment process. In addition, the cleanup strategy will vary from site to site. At some sites, cleanup will be completed before the properties are transferred to new owners. At other sites, cleanup may take place simultaneously with construction and redevelopment activities.

Regardless of when and how cleanups are accomplished, the challenge to any brownfields program is to clean up sites in accordance with redevelopment goals. Such goals may include cost-effectiveness, timeliness, avoidance of adverse effects to site structures and neighboring communities, and redevelopment of land in a way that benefits communities and local economies. Regulators and site managers are increasingly recognizing the value of implementing a more dynamic approach to streamline assessment and cleanup activities at brownfields sites. This approach, referred to as the Triad, is flexible and recognizes site-specific decisions and data needs [4].

The Triad approach focuses on the management of decision uncertainty by incorporating (a) systematic project planning, (b) dynamic work planning strategies, and (c) the use of real-time measurement technologies, including innovative technologies, to accelerate and improve the cleanup process. The Triad approach can reduce costs, improve decision certainty, expedite site closeout, and positively affect regulatory and community acceptance. This approach is well aligned with brownfields site priorities, which are affected by the economics of redevelopment, community involvement, and liability considerations.

Numerous technology options are available to assist those involved in brownfields cleanup. U.S. EPA's Office of Superfund Remediation and Technology Innovation (OSRTI) encourages the use of smarter solutions for characterizing and cleaning up contaminated sites by advocating more effective, less costly technological approaches. Using innovative technologies to characterize and clean up brownfields sites provides opportunities for stakeholders to reduce cleanup costs and accelerate cleanup schedules. Often, innovative approaches are also more acceptable to communities.

TABLE 13.1
Typical Metals and Metalloids at Brownfields Sites

Metals and Metalloids

Aluminum	Calcium	Mercury	
Antimony	Chromium	Molybdenum	
Arsenic	Cobalt	Nickel	Tin
Barium	Copper Iron	Potassium	Titanium
Beryllium	Lead	Selenium	Vanadium
Bismuth	Magnesium	Silver Sodium	Zinc
Boron	Manganese	Thallium	Zirconium
Cadmium			

Source: U.S. EPA. *Road Map to Understanding Innovative Technology Options for Brownfields Investigation and Cleanup*, 4th edition, EPA 542-B-05-001, U.S. Environmental Protection Agency, Washington, DC, September 2005 [4].

The cornerstone of U.S. EPA's Brownfields Initiative is the Pilot Program. Under this program, U.S. EPA is funding more than 200 brownfields assessment pilot projects in states, cities, towns, counties, and tribes across the country [2]. The pilots, each funded at up to USD 200,000 over 2 years, are bringing together community groups, investors, lenders, developers, and other affected parties to address the issues associated with assessing and cleaning up contaminated brownfields sites and returning them to appropriate, productive use. U.S. EPA's regional brownfields coordinators can provide communities with technical assistance such as targeted brownfields assessments. In addition to the hundreds of brownfields sites being addressed by these pilots, over 40 states have established brownfields or voluntary cleanup programs (VCPs) to encourage municipalities and private sector organizations to assess, clean up, and redevelop brownfields sites.

13.1.2 Metals and Metalloids

Metals are one of the three groups of elements distinguished by their ionization and bonding properties, along with metalloids and nonmetals. Metals have certain characteristic physical properties: they are usually shiny, have a high density, are ductile and malleable, usually have a high melting point, are usually hard, and conduct electricity and heat well. Metalloids have properties that are intermediate between those of metals and nonmetals. There is no unique way of distinguishing a metalloid from a true metal, but the most common way is that metalloids are usually semiconductors rather than conductors [4].

Locations where metals and metalloids may be found include artillery and small arms impact areas, battery disposal areas, burn pits, chemical disposal areas, contaminated marine sediments, disposal wells and leach fields, electroplating and metal finishing shops, firefighting training areas, landfills, and burial pits, leaking storage tanks, radioactive and mixed waste disposal areas, oxidation ponds and lagoons, paint stripping and spray booth areas, sandblasting areas, surface impoundments, and vehicle maintenance areas. Typical metals and metalloids encountered at many sites are listed in Table 13.1.

13.1.3 Purpose

U.S. EPA has developed a set of technical guides to assist communities, states, municipalities, and the private sector to more effectively address brownfields sites. Each guide in this series contains information on a different type of brownfields site (classified according to former industrial use). In addition, a supplementary guide contains information on cost-estimating tools and resources for brownfields sites [4–6].

The overview of the technical process involved in assessing and cleaning up brownfields sites can assist planners in making decisions at various stages of the project. An understanding of land use and industrial processes conducted in the past at a site can help the planner conceptualize the site and identify likely areas of contamination that may require cleanup. Numerous resources are suggested to facilitate characterization of the site and consideration of cleanup technologies [2–6].

Specifically, the objective of this chapter is to provide decision-makers with

1. An understanding of everyday industrial processes at metal finishing facilities and the relationship between such processes and potential releases of contaminants to the environment.
2. Information on the types of contaminants likely to be present at a metal finishing site.
3. A discussion of site assessment (also known as site characterization), screening and cleanup levels, and cleanup technologies that can be used to assess and clean up the types of contaminants likely to be present at metal finishing sites.
4. A conceptual framework for identifying potential contaminants at the site, pathways by which contaminants may migrate off-site and environmental and human health concerns.
5. Information on developing an appropriate cleanup plan for metal finishing sites where contamination levels must be reduced to allow a site's reuse.
6. A discussion of pertinent issues and factors should be considered when developing a site assessment and cleanup plan and selecting appropriate technologies for brownfields, given time and budget constraints.

13.2 INDUSTRIAL PROCESSES AND CONTAMINANTS AT METAL FINISHING SITES

Understanding the industrial processes used during a metal finishing facility's active life and the types of contaminants that may be present provides important information to guide planners in the assessment, cleanup, and restoration of the site to an acceptable condition for sale or reuse. This section provides a general overview of the processes, chemicals, and contaminants used or found at metal finishing sites. Specific metal finishing brownfields sites may have had a different combination of these processes, chemicals, and contaminants. Therefore, this information can be used only to develop a framework of likely past activities. Planners should obtain facility-specific information on industrial processes at their site whenever possible. Site-specific information is also important to obtain because the site may have been used for other industrial purposes at other times in the past.

This section describes waste-generating surface preparation operations; metal finishing operations and the types of waste streams and specific contaminants associated with each process; auxiliary areas at metal finishing sites that may produce contaminants and nonprocess-related contamination problems associated with metal finishing sites. Figure 13.1 presents typical metal finishing processes and land areas, along with the types of waste streams associated with each area [7]. Table 13.2 lists the specific contaminants associated with each waste stream [2].

13.2.1 Surface Preparation Operations

Metal finishing processes are typically housed within one structure. The surface of metal products generally requires preparation (i.e., cleaning) prior to applying a finish. An initial set of degreasing tanks ([A] in Figure 13.1) are used to remove oils, grease, and other foreign matter from the surface of the metal so that a coating can be applied. Metal finishing facilities may use solvents or emulsion solutions (i.e., solvents dispersed in an aqueous medium with the aid of an emulsifying agent) in the degreasing tanks to clean and prepare the surfaces of metal parts. Wastewaters generated from cleaning operations are primarily rinse waters, which are usually combined with other metal finishing wastewaters and treated on-site by conventional chemical precipitation. These wastewaters may

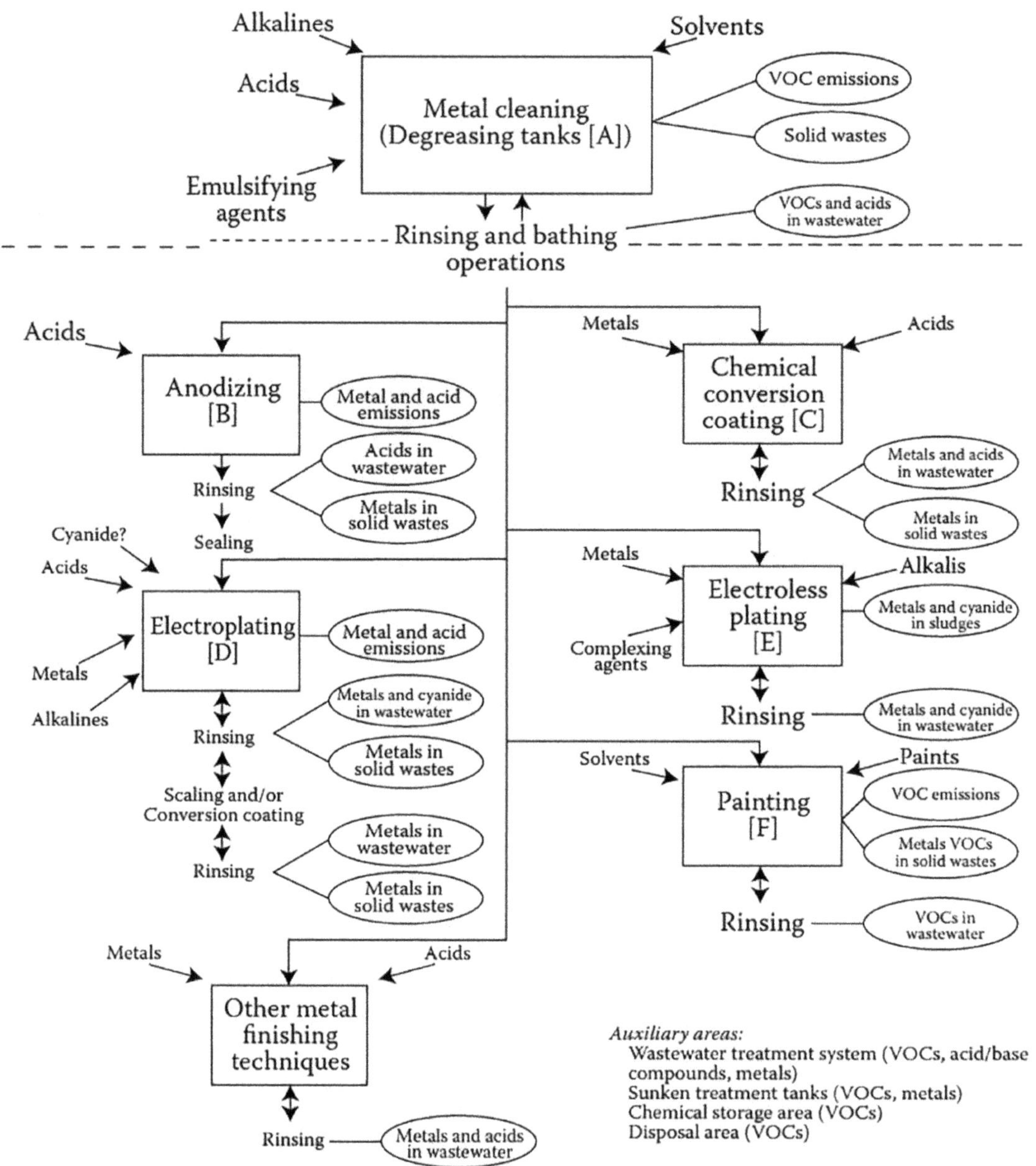

FIGURE 13.1 Typical metal finishing facility. (Adapted from U.S. EPA. *Brownfields and Land Revitalization—Tools and Technical Information.* U.S. Environmental Protection Agency, Washington, DC, 2007. [6].)

contain solvents, as listed in Table 13.2. Solid wastes such as wastewater treatment sludges, still bottoms, and cleaning tank residues may also be generated.

13.2.2 Metal Finishing Operations

Metal finishing operations are typically performed in a series of tanks (baths) followed by rinsing cycles. Acid or alkaline baths "pickle" the surface of the steel to improve the adherence of the coating. After the pickling baths, the metal products are moved to plating tanks, where the final coat is applied. Wastes generated during finishing operations derive from the solvents and cleansers

TABLE 13.2

Common Contaminants at Metal Finishing Sites

Contaminant Group	Contaminant Name
VOCs	Acetone, benzene, isopropyl alcohol, 2-dichlorobenzene, 4-trimethylbenzene, dichloromethane, ethyl benzene, freon 113, methanol, methyl isobutyl ketone, methyl ethyl ketone, phenol, tetrachloroethylene, toluene, trichloroethylene, xylene (mixed isomers)
Metals/inorganics	Aluminum, antimony, arsenic, asbestos (friable), barium, cadmium, chromium, cobalt, copper, lead, cyanide, manganese, mercury, nickel, silver, zinc
Acids	Hydrochloric acid, nitric acid, phosphoric acid, sulfuric acid

Source: U.S. EPA. *Technical Approaches to Characterizing and Cleaning Up Metal Finishing Sites under the Brownfields Initiative.* EPA/625/R-98/006, U.S. Environmental Protection Agency, Cincinnati, OH, March 1999 [2].

Note: VOCs, Volatile organic compounds.

applied to the surface and the metal-ion-bearing aqueous solutions used in acid/alkaline rinsing and bathing operations. Common metal finishing operations include anodizing, chemical conversion coating, electroplating, electroless plating, and painting. Common waste streams include metals and acids in the wastewater; metals in sludges and solid waste; and solvents from painting operations, as listed in Table 13.2. If these wastes were managed or disposed of on-site, it is possible that pollutants were released into the environment. Even at facilities where wastes were not stored on-site releases may have occurred during the handling and use of chemicals. Metal finishing operations are described below [2].

13.2.2.1 Anodizing Operations

Anodizing is an electrolytic process that uses acids from the combined electrolytic solution/acid bath tank to convert the metal surface into an insoluble oxide coating ([B] in Figure 13.1). After anodizing, metal parts are typically rinsed and then sealed. Anodizing operations produce contaminated wastewaters and solid wastes.

13.2.2.2 Chemical Conversion Coating

Chemical conversion coating ([C] in Figure 13.1) includes the following processes:

Chromating: Chromate conversion coatings are produced on various metals by chemical or electrochemical treatment. Acid solutions react with the metal surface to form a layer of a complex mixture of the constituent compounds, including chromium and the base metal.

Phosphating: Phosphate conversion coating involves the immersion of steel-, iron-, or zinc-plated steel into a dilute solution of phosphate salts, phosphoric acid, and other reagents to condition the surfaces for further processing.

Metal coloring: Metal coloring involves chemically converting the metal surface into an oxide or similar metallic compound to produce a decorative finish.

Passivating: Passivating is the process of forming a protective film on metals by immersing them in an acid solution (usually nitric acid or nitric acid with sodium dichromate).

Pollutants associated with chemical conversion processes enter the wastestream through rinsing and batch dumping of process baths. Wastewaters containing chromium are usually pretreated; this process generates a sludge that is sent off-site for metals reclamation and/or disposal.

13.2.2.3   Electroplating

Electroplating produces a surface coating of one metal upon another by electrodeposition ([D] in Figure 13.1). In electroplating, metal ions (in acid, alkaline, or neutral solutions) are reduced on the cathodic surfaces of the workpieces being plated. Electroplating operations produce contaminated wastewaters and solid wastes. Contaminated wastewaters result from work piece rinsing and process cleanup waters. Rinse waters from electroplating are usually combined with other metal finishing wastewaters and treated on-site by conventional chemical precipitation, which results in wastewater treatment sludges. Other wastes generated from electroplating include spent process solutions and quench baths that may be discarded periodically when the concentrations of contaminants inhibit their proper functions.

13.2.2.4   Electroless and Immersion Plating

Electroless plating involves chemically depositing a metal coating onto a plastic object by immersing the object in a plating solution ([E] in Figure 13.1). Immersion plating produces a thin metal deposit, commonly zinc or silver, by chemical displacement. Both produce contaminated wastewater and solid wastes. Facilities generally treat spent plating solutions and rinse waters chemically to precipitate the toxic metals; however, some plating solutions can be difficult to treat because of the presence of chelates. Most waste sludges resulting from electroless and immersion plating contain significant concentrations of toxic metals.

13.2.2.5   Painting

Painting is the application of predominantly organic coatings for protective and/or decorative purposes ([F] in Figure 13.1). Paint is applied in various forms, including dry powder, solvent-diluted formulations, and waterborne formulations, most commonly via spray painting and electrodeposition. Painting operations may result in solvent-containing waste and the direct release of solvents, paint sludge wastes, and paint-bearing wastewaters. Paint cleanup operations may also contribute to the release of chlorinated solvents. Discharge from water curtain booths generates the most wastewater. On-site wastewater treatment processes generate a sludge that is taken off-site for disposal. Other sources of wastes include emission control devices (e.g., paint booth collection systems, ventilation filters) and discarded paints. Sandblasting may be performed to remove paint and to clean metal surfaces for painting or resurfacing; this practice may be of particular concern if the paint being removed contains lead.

13.2.2.6   Other Metal Finishing Techniques

Polishing, hot dip coating, and etching are other processes used to finish metal. Wastewaters are often generated during these processes. For example, after polishing operations, area cleaning and washdown can produce metal-bearing wastewaters. Hot dip coating techniques, such as galvanizing, use water for rinses following precleaning and for quenching after coating. Hot dip coatings also generate a solid waste, oxide dross, that is periodically skimmed off the heated tank. Etching solutions are composed of strong acids or bases, which may result in etching solution wastes that contain metals and acids.

13.2.3   AUXILIARY ACTIVITY AREAS AND POTENTIAL CONTAMINANTS

13.2.3.1   Wastewater Treatment

Many of the operations involved in metal finishing produce wastewaters, which usually are combined and treated on-site, often by conventional chemical precipitation. Even though the facility would have been required to meet state wastewater discharge standards before releasing wastes, spills of process wastewater may have occurred in the area. At abandoned sites, any remaining wastewater left in tanks or floor drains could contain solvents, metals, and acids, such as those listed in Table 13.2. In addition, it is possible that wastewater sludges, which can contain metals, were left at the site in baths or tanks.

13.2.3.2 Sunken Wastewater Treatment Tank

Some metal finishing facilities have wastewater treatment tanks sunk into the concrete slab to rest on the underlying soils. This is done by design to aid facility operators in accessing the tanks. If these tanks develop leaks, the lost material, which may contain volatile organic compounds (VOCs) and metals, may be released directly to the soils beneath the building.

13.2.3.3 Chemical Storage Area

At most metal finishing sites, an area for storing chemicals used in the various operations was designated. Bulk containers stored in these areas may have leaked or spilled, resulting in discharges to floor drains or cracks in the floor. VOCs such as those listed in Table 13.2 may be found in such areas. Acids and alkaline reagents may also be found in this area.

13.2.3.4 Disposal Area

Materials, both liquid and solid, from process baths may have been disposed of at a designated area at the site. Such areas may be identified by stained soils or a lack of vegetation. These areas may contain VOCs, such as those listed in Table 13.2.

13.2.3.5 Other Considerations

Not all releases are related to the industrial processes described above. Some releases result from the associated services required to maintain the industrial processes. For example, electroplating facilities are large consumers of electricity, which requires a number of transformers. At older facilities, these transformers may have been disposed of in unmarked areas of the facility, which makes it difficult to know where leaks of polychlorinated biphenyl (PCB)-laden oils used as coolants may have occurred. Similarly, large machinery used to move metal pieces requires periodic maintenance. In the past, chemicals used for maintenance operations, such as solvents, oils, and grease, may have been flushed down the drain and sumps after use. Stormwater runoff from paved areas such as parking lots may contain petroleum hydrocarbons and oils, which can contaminate areas located downgradient. When conducting initial site evaluations, planners should expand their investigations to include these types of activities.

In addition, metal finishing facilities may have been located in older buildings that contain lead paint and asbestos insulation and tiling. Any structure built before 1970 should be assessed for the presence of these materials. They can cause significant problems during the demolition or renovation of the structures for reuse. Special handling and disposal requirements under state and federal laws can significantly increase the cost of construction.

13.3 SITE ASSESSMENT

The site investigation phase focuses on confirming whether contamination exists at a site, locating any contamination, and characterizing the nature and extent of that contamination [8]. It is essential that an appropriately detailed study of the site be performed to identify the cause, nature, and extent of contamination and the possible threats to the environment or any people living or working nearby. For brownfields sites, the results of such a study can be used in determining goals for cleanup, quantifying risks, determining acceptable and unacceptable risks, and developing effective cleanup plans that minimize delays or costs in the redevelopment and reuse of property. To ensure that sufficient information is obtained to support future decisions, the proposed cleanup measures and the proposed end use of the site should be considered when identifying data needs during the site investigation [4].

The elements of a site assessment are designed to help planners build a conceptual framework of the facility, which will aid site characterization efforts [9]. The conceptual framework should identify [2]:

1. Potential contaminants that remain in and around the facility
2. Pathways along which contaminants may move
3. Potential risks to the environment and human health that exist along the migration pathways.

This section highlights the key role that state environmental agencies usually play in brownfields projects. The types of information that planners should attempt to collect to characterize the site in a Phase I site assessment (i.e., the facility's history) are discussed. Information is presented about where to find and how to use this information to determine whether or not contamination is likely. Additionally, this section provides information to assist planners in conducting a Phase II site assessment, including sampling the site and determining the magnitude of contamination. Other considerations in assessing iron and steel sites are also discussed, and general sampling costs are included. The linking of the decision to be taken to the collected data and technologies is illustrated in Figure 13.2.

13.3.1 The Central Role of The State Agencies

A brownfields redevelopment project involves partnerships among site planners (whether private or public sector), state and local officials, and the local community. State environmental agencies often are key decision-makers and a primary source of information for Brownfields projects. Brownfields sites are generally cleaned up under state programs, particularly state voluntary cleanup or Brownfields programs; thus, planners will need to work closely with state program managers to determine their particular state's requirements for brownfields development. Planners may also need to meet additional federal requirements. Key state functions include [2]:

1. Overseeing brownfields site assessment and cleanup processes, including the management of VCPs
2. Providing guidance on contaminant screening levels
3. Serving as a source of site information, as well as legal and technical guidance.

13.3.1.1 State VCPs

State VCPs are designed to streamline brownfields redevelopment, reduce transaction costs, and provide state liability protection for past contamination. Planners should be aware of the fact that state cleanup requirements vary significantly and should contact the state brownfield manager; brownfields managers from state agencies will be able to identify their state requirements for planners and will clarify how their state requirements relate to federal requirements.

13.3.1.2 Levels of Contaminant Screening and Cleanup

Identifying the level of site contamination and determining the risk, if any, associated with that contamination level is a crucial step in determining whether cleanup is needed. Some state environmental agencies, as well as federal and regional U.S. EPA offices, have developed screening levels for certain contaminants, which are incorporated into some brownfields programs. Screening levels represent breakpoints in risk-based concentrations of chemicals in soil, air, or water. If contaminant concentrations are below the screening level, no action is required; above the level, further investigation is needed.

In addition to screening levels, U.S. EPA regional offices and some states have developed cleanup standards; if contaminant concentrations are above cleanup standards, cleanup must be pursued. The section on "Performing a Phase II Site Assessment" in this chapter provides more information on screening levels, and the section on "Site Cleanup" provides more information on cleanup standards.

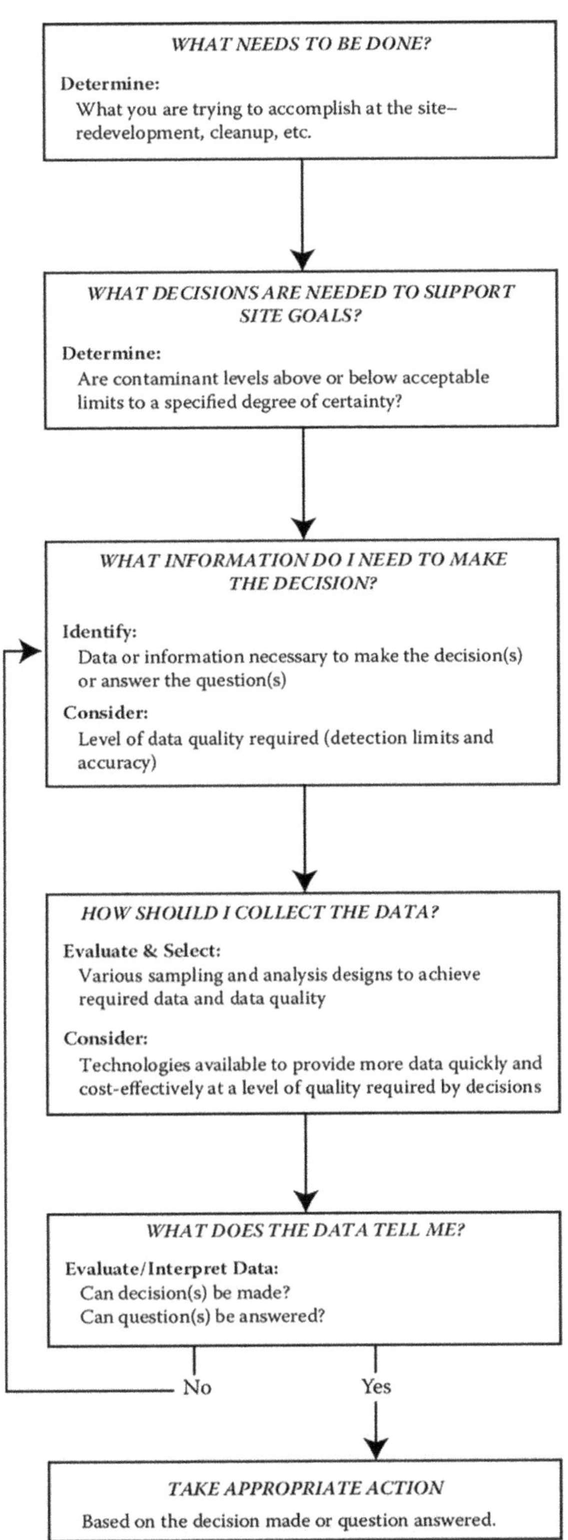

FIGURE 13.2 Linking the decision, data, and technology. (Adapted from U.S. EPA. *Road Map to Understanding Innovative Technology Options for Brownfields Investigation and Cleanup*, 4th edition, EPA 542-B-05-001, U.S. Environmental Protection Agency, Washington, DC, September 2005 [4].)

13.3.2 Performing a Phase I Site Assessment: Obtaining Facility Background Information from Existing Data

Planners should compile a history of the iron and steel manufacturing facility to identify likely site contaminants and their probable locations. Financial institutions typically require a Phase I site assessment prior to lending money to potential property buyers to protect the institution's role as a mortgage holder [10]. In addition, parties involved in the transfer, foreclosure, leasing, or marketing of properties recommend some form of site evaluation. The site history should include:

1. A review of readily available records (e.g., former site use, building plans, and records of any prior contamination events).
2. A site visit to observe the areas used for various industrial processes and the condition of the property.
3. Interviews with knowledgeable people (e.g., site owners, operators, occupants; neighbors; local government officials).
4. A report that includes an assessment of the likelihood that contaminants are present at the site.

The Phase I site assessment should be conducted by an environmental professional and may take 3–4 weeks to complete. Site evaluations are required in part as a response to concerns over environmental liabilities associated with property ownership. A property owner needs to perform "due diligence," that is fully inquire into the previous ownership and uses of a property to demonstrate that all reasonable efforts to find site contamination have been made. Because brownfields sites often contain low levels of contamination and pose low risks, due diligence through a Phase I site assessment will help to answer key questions about the levels of contamination. Several federal and state programs exist to minimize owner liability at brownfields sites and facilitate cleanup and redevelopment; planners should contact the state environmental or regional U.S. EPA office for further formation.

Information on how to review records, conduct site visits and interviews, and develop a report during a Phase I site assessment is provided below.

13.3.2.1 Facility Records

Facility records are often the best source of information on former site activities. If past owners are not initially known, a local records office should have deed books that contain ownership history. Generally, records pertaining specifically to the site in question are adequate for review purposes. In some cases, however, records of adjacent properties may also need to be reviewed to assess the possibility of contaminants migrating from or to the site based on geologic or hydrogeologic conditions. If the brownfields property resides in a low-lying area, in close proximity to other industrial facilities or formerly industrialized sites, or downgradient from current or former industrialized sites, an investigation of adjacent properties is warranted.

13.3.2.2 Other Sources of Recorded Information

Planners may need to use other sources in addition to facility records to develop a complete history. American Society for Testing and Materials (ASTM) Standard 1527 identifies standard sources such as historical aerial photographs, fire insurance maps, property tax files, recorded land title records, topographic maps, local street directories, building department records, zoning/land use records, and newspaper archives [10].

Some metal finishing site managers may have worked with state environmental regulators; these offices may be key sources of information. Federal (e.g., U.S. EPA) records may also be useful. The types of information provided by regulators may include facility maps that identify activities and disposal areas, lists of stored pollutants, and the types and levels of pollutants released. State offices and other sources where planners can search for site-specific information are presented below:

1. The state offices responsible for industrial waste management and hazardous waste should have a record of any emergency removal actions at the site (e.g., the removal of leaking drums that posed an "imminent threat" to local residents) and Resource Conservation and Recovery Act (RCRA) [11] permits issued at the site; notices of violations issued; and any environmental investigations.
2. The state office responsible for discharges of wastewater to water bodies under the National Pollutant Discharge Elimination System (NPDES) [12] program will have a record of any permits issued for discharges into surface water at or near the site. The local publicly owned treatment works (POTWs) will have records for permits issued for indirect discharges into sewers (e.g., floor drain discharges to a sanitary sewer).
3. The state office responsible for underground storage tanks (USTs) may also have records of tanks located at the site, as well as records of any past releases.
4. The state office responsible for air emissions may be able to provide information on air pollutants associated with particular types of on-site contamination.
5. U.S. EPA's Comprehensive Environmental Response, Compensation, and Liability Information System (CERCLIS) [13] of potentially contaminated sites should have a record of any previously reported contamination at or near the site.
6. U.S. EPA regional offices can provide records of sites that have hazardous substances. Information is available from the Federal National Priorities List (NPL) and lists of treatment, storage, and disposal (TSD) facilities subject to corrective action under RCRA. RCRA non-TSD facilities, RCRA generators, and Emergency Response Notification System (ERNS) information on contaminated or potentially contaminated sites can help to determine whether neighboring facilities are recorded as having released hazardous substances into the immediate environment.
7. State and local records may indicate any permit violations or significant contaminant releases from or near the site.
8. Residents and former employees may be able to provide useful information on waste management practices, but these reports should be substantiated.
9. Local fire departments may have responded to emergency events at the facility. Fire departments or city halls may have fire insurance maps or other historical maps or data that indicate the location of hazardous waste storage areas at the site.
10. Local waste haulers may have records of the facility's disposal of hazardous or other waste materials.
11. Utility records.
12. Local building permits.

13.3.2.3 Identifying Migration Pathways and Potentially Exposed Populations

Off-site migration of contaminants may pose a risk to human health and the environment; planners should gather as much readily available information on the physical characteristics of the site as possible. Migration pathways, that is, soil, groundwater, and air, will depend on-site-specific characteristics such as geology and the physical characteristics of the individual contaminants (e.g., mobility). Information on the physical characteristics of the general area can play an important role in identifying potential migration pathways and focusing environmental sampling activities, if needed. Planners should collect three types of information to obtain a better understanding of migration pathways, including topographic, soil and subsurface, and groundwater data, as described below [14–17].

13.3.3 Gathering Topographic Information

In this preliminary investigation, topographic information will be helpful in determining whether the site may be subject to contamination by adjoining properties or maybe the source of contamination of other properties. Topographic information will help planners identify low-lying areas of the facility where rain and snowmelt (and any contaminants in them) may collect and contribute both water and contaminants to the underlying aquifer or surface runoff to nearby areas. The U.S. Geological Survey (USGS) of the Department of the Interior has topographic maps for nearly every part of the country.

13.3.4 Gathering Soil and Subsurface Information

Planners should know about the types of soils at the site from the ground surface extending down to the water table because soil characteristics play a large role in how contaminants move in the environment. For example, clay soils limit the downward movement of pollutants into underlying groundwater but facilitate surface runoff. Sandy soils, on the other hand, can promote rapid infiltration into the water table while inhibiting surface runoff. Soil information can be obtained through a number of sources [2]:

1. Local planning agencies should have soil maps to support land use planning activities. These maps provide a general description of the soil types present within a county (or sometimes a smaller administrative unit, such as a township).
2. The Natural Resource Conservation Service and Co-operative Extension Service offices of the U.S. Department of Agriculture (USDA) are also likely to have soil maps.
3. Well-water companies are likely to be familiar with local subsurface conditions, and local water districts and state water divisions may have well-logging information.
4. Local health departments may be familiar with subsurface conditions because of their interest in septic drain fields.
5. Local construction contractors are likely to be familiar with subsurface conditions from their work with foundations.

Soil characteristics can vary widely within a relatively small area, and it is common to find that the top layer of soil in urban areas is composed of fill materials, not native soils. While local soil maps and other general soil information can be used for screening purposes, such as in a Phase I assessment, site-specific information will be needed in the event that cleanup is necessary.

13.3.5 Gathering Groundwater Information

Planners should obtain general groundwater information about the site area, including:

1. State classifications of underlying aquifers
2. Depth to the groundwater tables
3. Groundwater flow direction and rate.

This information can be obtained by contacting state environmental agencies or from several local sources, including water authorities, well drilling companies, health departments, and Agricultural Extension and Natural Resource Conservation Service offices.

13.3.5.1 Identifying Potential Environmental and Human Health Concerns

Identifying possible environmental and human health risks early in the process can influence decisions regarding the viability of a site for cleanup and the choice of cleanup methods used. A visual inspection of the area will usually suffice to identify on-site or nearby wetlands and water bodies that may be particularly sensitive to releases of contaminants during characterization or cleanup activities. Planners should also review available information (e.g., from state and local environmental agencies) to ascertain the proximity of residential dwellings, nearby industrial/commercial activities, and wetlands/water bodies and to identify people, animals, or plants that might receive migrating contamination; any particularly sensitive populations in the area (e.g., children; endangered species); and whether any major contamination events have occurred previously in the area (e.g., drinking water problems; groundwater contamination).

For environmental information, planners can contact the U.S. Army Corps of Engineers (ACE), state environmental agencies, local planning and conservation authorities, the USGS, and the USDA Natural Resource Conservation Service. State and local agencies and organizations can usually provide information on local fauna and the habitats of any sensitive and/or endangered species.

For human health information, planners can contact:

1. **State and local health assessment organizations**: Organizations such as health departments should have data on the quality of local well water used as a drinking water source, as well as any human health risk studies that have been conducted. In addition, these groups may have other relevant information, such as how certain types of contaminants (e.g., volatile organics, such as benzene and phenols) might pose a health risk (e.g., dermal exposure to volatile organics during site characterization); information on exposures to particular contaminants and potential associated health risks can also be found in health profile documents developed by the Agency for Toxic Substances and Disease Registry (ATSDR). In addition, ATSDR may have conducted a health consultation or health assessment in the area if an environmental contamination event that may have posed a health risk occurred in the past; such an event and assessment should have been identified in the Phase I records review of prior contamination incidents at the site if any occurred.

2. **Local water and health departments**: During the site visit (described below), when visually inspecting the area around the facility, planners should identify any residential dwelling or commercial activities near the facility and evaluate whether people there may come into contact with contamination along one of the migration pathways. Where groundwater contamination may pose a problem, planners should identify any nearby waterways or aquifers that may be impacted by groundwater discharge of contaminated water, including any drinking water wells that may be downgradient to the site, such as a municipal well field. Local water departments will have a count of well connections to the public water supply. Planners should also pay particular attention to information on private wells in the area downgradient of the facility since, depending on their location, they may be vulnerable to contaminants migrating off-site even when the public municipal drinking water supply is not vulnerable. Local health departments often have information on the locations of private wells.

In addition to groundwater sources and migration pathways, surface water sources and pathways should be evaluated since groundwater and surface waters can interface at some (or several) point(s) in the region. Contaminants in groundwater can eventually migrate to surface waters, and contaminants in surface waters can migrate to groundwater.

13.3.5.2 Community Involvement

It is important that brownfields decision-makers encourage acceptance of redevelopment plans and cleanup alternatives by involving members of the community early in the decision-making process through community meetings, newsletters, or other outreach activities. For an individual site, the community should be informed about how the use of a proposed technology might affect redevelopment plans or the adjacent neighborhood [4]. For example, the planting of trees for the use of phytoremediation may create aesthetic or visual improvements; on the other hand, the use of phytoremediation may bring about issues related to site security or long-term maintenance that could affect access to the site.

Community-based organizations represent a wide range of issues, from environmental concerns to housing issues to economic development. These groups can often be helpful in educating planners and others in the community about local brownfields sites, which can contribute to successful brownfields site assessment and cleanup activities. In addition, most state VCPs require that local communities be adequately informed about brownfields cleanup activities. Planners can contact the local Chamber of Commerce, local philanthropic organizations, local service organizations, and neighborhood committees for community input. State and local environmental groups may be able to supply relevant information and identify other appropriate community organizations. Local community involvement in brownfields projects is a key component in the success of such projects [2].

U.S. EPA can assist members of the brownfields community by directing its members to appropriate resources and providing opportunities to network and participate in the sharing of information. A number of Internet sites, databases, newsletters, and reports provide opportunities for brownfields stakeholders to network with other stakeholders to identify information about cleanup and technology options. U.S. EPA's Brownfields and Land Revitalization Technology Support Center is a valuable resource for brownfields decision-makers.

13.3.5.3 Conducting a Site Visit

In addition to collecting and reviewing available records, planners need to conduct a site visit to visually and physically observe the uses and conditions of the property, including both outdoor areas and the interior of any structure or property. Current and past uses involving the use, treatment, storage, disposal, or generation of hazardous substances or petroleum products should be noted. Current or past uses of abutting properties that can be observed readily while conducting the site visit also should be noted. In addition, readily observable geologic, hydrologic, and topographic conditions should be identified, including any possibility of hazardous substances migrating on- or off-site. Roads, water supplies, and wastewater systems should be identified, as well as any storage tanks, whether above or below ground. If any hazardous substances or petroleum products are found, their type, quantity, and storage conditions should be noted. Any odors, pools of liquids, drums or other containers, and equipment likely to contain PCBs should be noted. Additionally, indoors, heating, and cooling systems should be noted, as well as any stains, corrosion, drains, or sumps. Outdoors, any pits, ponds, lagoons, stained soil or pavement, stressed vegetation, solid waste, wastewater, and wells should be noted [10].

13.3.5.4 Conducting Interviews

In addition to reviewing available records and visiting the site, conducting interviews with the site owner and/or site manager, site occupants, and local officials is highly recommended to obtain information about the prior and/or current uses and conditions of the property, and to inquire about any useful documents that exist regarding the property. Such documents include environmental audit reports, environmental permits, registrations for storage tanks, material safety data sheets, community right-to-know plans, safety plans, government agency notices or correspondence, hazardous-waste-generator reports or notices, geotechnical studies, or any proceedings involving the property [10]. Interviews with at least one staff person from the following local government agencies are recommended: the fire department, health agency, and the agency with authority for hazardous waste disposal or other environmental matters. Interviews can be conducted in person, by telephone, or in writing.

ASTM standard 1528 [18] provides a questionnaire that may be appropriate for use in interviews for certain sites. ASTM suggests that this questionnaire be posed to the current property owner, any major occupant of the property (or at least 10% of the occupants of the property if no major occupant exists), or "any occupant likely to be using, treating, generating, storing, or disposing of hazardous substances or petroleum products on or from the property." A user's guide accompanies the ASTM questionnaire to assist the investigator in conducting interviews, as well as researching records and making site visits.

13.3.5.5 Developing a Report

Toward the end of the Phase I assessment, planners should develop a report that includes all of the important information obtained during record reviews, the site visit, and interviews. Documentation, such as references and important exhibits, should be included, as well as the credentials of the environmental professional who conducted the Phase I environmental site assessment. The report should include all information regarding the presence or likely presence of hazardous substances or petroleum products on the property and any conditions that indicate an existing, past, or potential release of such substances into property structures or into the ground, groundwater, or surface water of the property [10]. The report should include the environmental professional's opinion of the impact of the presence or likely presence of any contaminants, and a findings and conclusion section that either indicates that the Phase I environmental site assessment revealed no evidence of contaminants in connection with the property, or discusses what evidence of contamination was found.

Additional sections of the report might include a recommendations section (e.g., for a Phase II site assessment, if appropriate) and sections on the presence or absence of asbestos, lead paint, lead in drinking water, radon, and wetlands. Some states or financial institutions may require information on these substances.

If the Phase I site assessment adequately informs state and local officials, planners, community representatives, and other stakeholders that no contamination exists at the site or that contamination is so minimal that it does not pose a health or environmental risk, then those involved may decide that adequate site assessment has been accomplished and the process of redevelopment may proceed. In some cases where evidence of contamination exists, stakeholders may decide that enough information is available from the Phase I site assessment to characterize the site and determine an appropriate approach for site cleanup of the contamination. In other cases, stakeholders may decide that additional site assessment is warranted, and a Phase II site assessment would be conducted.

13.3.6 THE TRIAD APPROACH: STREAMLINING SITE INVESTIGATIONS AND CLEANUP DECISIONS

The modernization of the collection, analysis, interpretation, and management of data to support decisions about hazardous waste sites rests on the U.S. EPA's three-pronged or "Triad" approach [19–21]. The introduction of new technologies in a dynamic framework allows project managers to meet clearly defined objectives. Such an approach incorporates the elements described below [4,19–21].

Systematic planning is a common-sense approach to assuring that the level of detail in project planning matches the intended use of the data being collected. Once cleanup goals have been defined, systematic planning is undertaken to chart a course for the project that is resource-effective, as well as technically sound and defensible to reach these project-critical goals. A team of multidisciplinary, experienced technical staff works to translate the project's goals into realistic technical objectives. The Conceptual Site Model (CSM) is the planning tool that organizes the information that already is known about the site; the CSM helps the team identify the additional information that must be obtained. The systematic planning process ties project goals to individual activities necessary to reach these goals by identifying data gaps in the CSM. The team then uses the CSM to direct the gathering of needed information, allowing the CSM to evolve and mature as work progresses at the site.

A *dynamic working strategy* approach relies on real-time data to reach decision points. The logic for decision-making is identified, and responsibilities, authority, and lines of communication are established. Dynamic work strategy implementation relies on and is driven by critical project decisions needed to reach closure. It uses a decision tree and real-time uncertainty management practices to reach critical decision points in as few mobilizations as possible. Success of a dynamic approach depends on the presence of experienced staff in the field empowered to make decisions based on the decision logic and their capability to deal with new data and any unexpected issues, as they arise. Field staff maintains close communication with regulators or others overseeing the project during the implementation of dynamic work plans.

The use of on-site analytical tools, rapid sampling platforms, and on-site interpretation and management of data makes dynamic work strategies possible. Such *real-time measurement tools* are among the key streamlined site investigation tools because they provide the data that are used for on-site decision-making. The tools are a broad category of analytical methods and equipment that can be applied at the sample collection site. They include methods that can be used outdoors with hand-held, portable equipment, as well as more rigorous methods that require the controlled environments of a mobile laboratory (transportable). During the planning process, the team identifies the type, rigor, and quantity of data needed to answer the questions raised by the CSM. Those decisions then guide the design sampling modifications and the selection of analytical tools.

The Triad approach enables project managers to minimize uncertainty while expediting site cleanup and reducing project costs. For example, U.S. EPA collaborated with the Town of Greenwich, Connecticut, to implement the Triad approach to characterize a former power plant site scheduled for redevelopment as a waterfront park. The Triad approach yielded an estimated cost savings of 50%–60% when compared with a traditional approach involving two mobilizations and comprehensive analytical methods at a fixed laboratory. The City of Trenton, New Jersey, began implementing the Triad approach in 2001 as part of its program to redevelop a large number of abandoned industrial sites. Overall, the Triad approach eliminated costs associated with follow-on investigation activities while accelerating the redevelopment schedule and reducing decision uncertainty. Additional details about these and other examples are available in the U.S. EPA's Technology News and Trends newsletter [22].

13.3.7 Performing a Phase II Site Assessment: Sampling the Site

A Phase II site assessment [23] typically involves taking soil, water, and air samples to identify the types, quantity, and extent of contamination in these various environmental media. The types of data used in a Phase II site assessment can vary from existing site data (if adequate) to limited sampling of the site to more extensive contaminant-specific or site-specific sampling data. Planners should use knowledge of past facility operations whenever possible to focus the site evaluation on those process areas where pollutants were stored, handled, used, or disposed of. These will be the areas where potential contamination will be most readily identified. Generally, to minimize costs, a Phase II site assessment will begin with limited sampling (assuming readily available data do not exist that adequately characterize the type and extent of contamination on the site) and will proceed to more comprehensive sampling if needed (e.g., if the initial sampling could not identify the geographical limits of contamination).

This section explains the importance of setting Data Quality Objectives (DQOs) and provides brief guidance for doing so; describes screening levels to which sampling results can be compared; and provides an overview of environmental sampling and data analysis, including sampling methods and ways to increase data certainty.

13.3.7.1 Setting DQOs

U.S. EPA has developed a guidance document that describes key principles and best practices for brownfields site assessment quality assurance and quality control based on program experience [24–26].

U.S. EPA has adopted the DQOs Process [26] as a framework for making decisions. The DQO Process is a common-sense, systematic planning tool based on the scientific method. Using a systematic planning approach, such as the DQO Process, ensures that the data collected to support defensible site decision-making will be of sufficient quality and quantity, as well as be generated through the most cost-effective means possible. DQOs, themselves, are statements that unambiguously communicate the following:

1. The study objective
2. The most appropriate type of data to collect
3. The most appropriate conditions under which to collect the data
4. The amount of uncertainty that will be tolerated when making decisions.

It is important to understand the concept of uncertainty and its relationship to site decision-making [27–29]. Regulatory agencies, and the public they represent, want to be as confident as possible about the safety of reusing brownfields sites. Public acceptance of site decisions may depend on the site manager's being able to scientifically document the adequacy of site decisions. During negotiations with stakeholders, effective communication about the trade-offs between project costs and confidence in the site decision can help set the stage for a project's successful completion. When the limits on uncertainty (e.g., only a 5%, 10%, or 20% chance of a particular decision error is permitted) are clearly defined in the project, subsequent activities can be planned so that data collection efforts will be able to support those confidence goals in a resource-effective manner. On the one hand, a manager would like to reduce the chance of making a decision error as much as possible, but on the other hand, reducing the chance of making that decision error requires collecting more data, which is, in itself, a costly process.

Striking a balance between these two competing goals—more scientific certainty versus less cost—requires careful thought and planning, as well as the application of professional expertise [27–29].

The following steps are involved in systematic planning:

1. **Agree on intended land reuse**: All parties should agree early in the process on the intended reuse of the property because the type of use may strongly influence the choice of assessment and cleanup approaches. For example, if the area is to be a park, removal of all contamination will most likely be needed. If the land is used for a shopping center, with most of the land covered by buildings and parking lots, it may be appropriate to reduce, rather than totally remove, contaminants to specified levels (e.g., state cleanup levels; see Section 13.4 later in this chapter).

2. **Clarify the objective of the site assessment**: What is the overall decision(s) that must be made for the site? Parties should agree on the purpose of the assessment. Is the objective to confirm that no contamination is present? Or is the goal to identify the type, level, and distribution of contamination above the levels, which are specified, based on the intended land use. These are two fundamentally different goals that suggest different strategies. The costs associated with each approach will also vary.

   As noted above, parties should also agree on the total amount of uncertainty allowable in the overall decision(s). Conducting a risk assessment involves identifying the levels of uncertainty associated with characterization and cleanup decisions. A risk assessment involves identifying potential contaminants and analyzing the pathways through which people, other species of concern, or the environment can become exposed to those contaminants. Such an assessment can help identify the risks associated with varying levels of acceptable uncertainty in the site decision and can provide decision-makers with greater confidence about their choice of land use decisions and the objective of the site assessment. If cleanup is required, a risk assessment can also help determine how clean the site needs to be, based on expected reuse (e.g., residential or industrial), to safeguard people from exposure to contaminants.

3. **Define the appropriate type(s) of data that will be needed to make an informed decision at the desired confidence level**: Parties should agree on the type of data to be collected by defining a preliminary list of suspected analytes, media, and analyte-specific action levels (screening levels). Define how the data will be used to make site decisions. For example, data values for a particular analyte may or may not be averaged across the

site for the purposes of reaching a decision to proceed with work. Are there maximum values that a contaminant(s) cannot exceed? If found, will concentrations of contaminants above a certain action level (hotspots) be characterized and treated separately? These discussions should also address the types of analyses to be performed at different stages of the project. Planners and regulators can reach an agreement to focus initial characterization efforts in those areas where the preliminary information indicates that potential sources of contamination may be located. It may be appropriate to analyze for a broad class of contaminants by less expensive screening methods in the early stages of the project in order to limit the number of samples needing analysis by higher quality, more expensive methods later. Different types of data may be used at different stages of the project to support interim decisions that efficiently direct the course of the project as it moves forward.

4. **Determine the most appropriate conditions under which to collect the data**: Parties should agree on the timing of sampling activities since weather conditions can influence how representative the samples are of actual conditions.

5. **Identify appropriate contingency plans/actions**: Certain aspects of the project may not develop as planned. Early recognition of this possibility can be a useful part of the DQO Process. For example, planners, regulators, and other stakeholders can acknowledge that screening-level sampling may lead to the discovery of other contaminants on the site than were originally anticipated. During the DQO Process, stakeholders may specify appropriate contingency actions to be taken in the event that contamination is found. Identifying contingency actions early in the project can help ensure that the project will proceed even in light of new developments. The use of a dynamic workplan combined with the use of rapid turnaround field analytical methods can enable the project to move forward with a minimum of time delay and wasted effort.

6. **Develop a sampling and analysis plan that can meet the goals and permissible uncertainties described in the proceeding steps**: The overall uncertainty in a site decision is a function of several factors: the number of samples across the site (the density of sample coverage), the heterogeneity of analytes from sample to sample (spatial variability of contaminant concentrations), and the accuracy of the analytical method(s). Studies have demonstrated that analytical variability tends to contribute much less to the uncertainty of site decisions than does sample variability due to matrix heterogeneity. Therefore, spending money to increase the sample density across the site will usually (for most contaminants) make a larger contribution to confidence in the site decision, and thus be more cost-effective, than will spending money to achieve the highest data quality possible, but a lower sampling density.

   Examples of important consideration for developing a sampling and analysis plan include:

   - Determine the sampling location placement that can provide an estimate of the matrix heterogeneity and thus address the desired certainty. Is locating hotspots of a certain size important? Can composite sampling be used to increase coverage of the site (and decrease overall uncertainty due to sample heterogeneity) while lowering analytical costs?
   - Evaluate the available pool of analytical technologies/methods (both field methods and laboratory methods, which might be implemented in either a fixed or mobile laboratory) for those methods that can address the desired action levels (the analytical methods quantification limit should be well below the action level). Account for possible or expected matrix interferences when considering appropriate methods. Can field analytical methods produce data that will meet all of the desired goals when sampling uncertainty is also taken into account? Evaluate whether a combination of screening and definitive methods may produce a more cost-effective means to generate data. Can economy of scale be used? For example, the expense of a mobile laboratory is

      seldom cost-effective for a single small site but might be cost-effective if several sites can be characterized sequentially by a single mobile laboratory.

- When the sampling procedures, sample preparation, and analytical methods have been selected, design a quality control protocol for each procedure and method that ensures that the data generated will be of known, defensible quality.

7. Through a number of iterations, refine the sampling and analysis plan to one that can most cost-effectively address the decision-making needs of the site planner.

8. **Review agreements often**: As more information becomes available, some decisions that were based on earlier, limited information should be reviewed to see if they are still valid. If they are not, the parties can again use the DQO framework to revise and refine site assessment and cleanup goals and activities.

The data needed to support decision-making for brownfields sites generally are not complicated and are less extensive than those required for more heavily contaminated, higher-risk sites (e.g., Superfund sites). But data uncertainty may still be a concern at brownfields sites because knowledge of past activities at a site may be less than comprehensive, resulting in limited site characterization. Establishing DQOs can help address the issue of data uncertainty in such cases. Examples of DQOs include verifying the presence of soil contaminants and assessing whether contaminant concentrations exceed screening levels.

13.3.7.2 Screening Levels

In the initial stages of a Phase II site assessment, an appropriate set of screening levels for contaminants in soil, water, and/or air should be established. Screening levels are risk-based benchmarks that represent concentrations of chemicals in environmental media that do not pose an unacceptable risk. Sample analyses of soils, water, and air at the facility can be compared with these benchmarks. If on-site contaminant levels exceed the screening levels, further investigation will be needed to determine if and to what extent cleanup is appropriate.

Some states have developed generic screening levels (e.g., for industrial and residential use).

These levels may not account for site-specific factors that affect the concentration or migration of contaminants. Alternatively, screening levels can be developed using site-specific factors. While site-specific screening levels can more effectively incorporate elements unique to the site, developing site-specific standards is a time- and resource-intensive process. Planners should contact their state environmental offices and/or U.S. EPA regional offices for assistance in using screening levels and in developing site-specific screening levels.

Risk-based screening levels are based on calculations/models that determine the likelihood that exposure of a particular organism or plant to a particular level of a contaminant would result in a certain adverse effect. Risk-based screening levels have been developed for tap water, ambient air, fish, and soil. Some states or U.S. EPA regions also use regional background levels (or ranges) of contaminants in soil and maximum contaminant levels (MCLs) in water established under the Safe Drinking Water Act (SDWA) [30] as screening levels for some chemicals. In addition, some states and/or U.S. EPA regional offices have developed equations for converting soil screening levels to comparative levels for the analysis of air and groundwater.

When a contaminant concentration exceeds a screening level, further site assessment (such as sampling the site at strategic locations and/or performing more detailed analysis) is needed to determine that: (a) the concentration of the contaminant is relatively low and/or the extent of contamination is small and does not warrant cleanup for that particular chemical, or (b) the concentration or extent of contamination is high, and that site cleanup is needed.

Using state cleanup standards for an initial brownfields assessment may be beneficial if no industrial screening levels are available or if the site may be used for residential purposes. U.S. EPA's soil screening guidance is a tool developed by the U.S. EPA to help standardize and accelerate the evaluation and cleanup of contaminated soils at sites on the NPL where future residential land use is anticipated. This guidance may be useful at corrective action or VCP sites where site conditions are similar. However, the use of this guidance for sites where residential land use assumptions do not apply could result in overly conservative screening levels.

13.3.7.3 Environmental Sampling and Data Analysis

Environmental sampling and data analysis are integral parts of a Phase II site assessment process. Many different technologies are available to perform these activities, as discussed below.

13.3.7.4 Levels of Sampling and Analysis

There are two levels of sampling and analysis: screening and contaminant-specific. Planners are likely to use both at different stages of the site assessment.

Screening: Screening sampling and analysis use relatively low-cost technologies to take a limited number of samples at the most likely points of contamination and analyze them for a limited number of parameters. Screening analyses often test only for broad classes of contaminants, such as total petroleum hydrocarbons (TPHs), rather than for specific contaminants, such as benzene or toluene. Screening is used to narrow the range of areas of potential contamination and reduce the number of samples requiring further, more costly analysis. Screening is generally performed on-site, with a small percentage of samples (e.g., generally 10%) submitted to a state-approved laboratory for a full organic and inorganic screening analysis to validate or clarify the results obtained.

Some geophysical methods are used in site assessments because they are noninvasive (i.e., do not disturb environmental media as sampling does). Geophysical methods are commonly used to detect underground objects that might exist at a site, such as UST, dry wells, and drums. The two most common and cost-effective technologies used in geophysical surveys are ground-penetrating radar and electromagnetics [31].

Contaminant specific: For a more in-depth understanding of contamination at a site (e.g., when screening data are not detailed enough), it may be necessary to analyze samples for specific contaminants. With contaminant-specific sampling and analysis, the number of parameters analyzed is much greater than that for screening-level sampling, and analysis includes more accurate, higher-cost field and laboratory methods. Such analyses may take several weeks.

Computerization, microfabrication, and biotechnology have permitted the recent development of an analytical equipment that can be generated in the field, on-site in a mobile laboratory and off-site in a laboratory. The same kind of equipment might be used in two or more locations.

13.3.8 Increasing the Certainty of Sampling Results

One approach to reducing the level of uncertainty associated with site data is to implement a statistical sampling plan. Statistical sampling plans use statistical principles to determine the number of samples needed to accurately represent the contamination present. With the statistical sampling method, samples are usually analyzed with highly accurate laboratory or field technologies, which increase costs and take additional time. Using this approach, planners can negotiate with regulators and determine in advance specific measures of allowable uncertainty (e.g., an 80% level of confidence with a 25% allowable error).

Another approach to increasing the certainty of sampling results is to use lower-cost technologies with higher detection limits to collect a greater number of samples. This approach would provide a more comprehensive picture of contamination at the site but with less detail regarding the specific contamination. Such an approach would not be recommended to identify the extent of contamination by a specific contaminant, such as benzene, but it may be an excellent approach for defining the extent of contamination by total organic compounds with a strong degree of certainty. Planners will find that there is a trade-off between scope and detail. Performing a limited number of detailed analyses provides good detail but less certainty about overall contamination, while performing a larger number of general analyses provides less detail but improves the understanding and certainty of the scope of contamination.

13.3.9   SITE ASSESSMENT TECHNOLOGIES

This section discusses the differences between using field and laboratory technologies and provides an overview of applicable site assessment technologies [32,33]. In recent years, several innovative technologies that have been field-tested and applied to hazardous waste problems have emerged. In many cases, innovative technologies may cost less than conventional techniques and can successfully provide the needed data. Operating conditions may affect the cost and effectiveness of individual technologies.

13.3.9.1   Field versus Laboratory Analysis

The principal advantages of performing field sampling and field analysis are that results are immediately available and more samples can be taken during the same sampling event; also, sampling locations can be adjusted immediately to clarify the first round of sampling results if warranted. This approach may reduce costs associated with conducting additional sampling events after receipt of laboratory analysis. Field assessment methods have improved significantly over recent years; however, while many field technologies may be comparable to laboratory technologies, some field technologies may not detect contamination at levels as low as laboratory methods, and may not be contaminant specific. To validate the field results or to gain more information on specific contaminants, a small percentage of the samples can be sent for laboratory analysis. The choice of sampling and analytical procedures should be based on DQOs established earlier in the process, which determine the quality (e.g., precision, level of detection) of the data needed to adequately evaluate site conditions and identify appropriate cleanup technologies.

13.3.9.2   Sample Collection and Analysis Technologies

Tables 13.3 and 13.4 list sample collection technologies for oil in subsurface and groundwater, which may be appropriate for metal finishing brownfields sites. Technology selection depends on the medium being sampled and the type of analysis required based on DQOs. Soil samples are generally collected using spoons, scoops, and shovels. The selection of a subsurface sample collection technology depends on the subsurface conditions (e.g., consolidated materials, bedrock), the required sampling depth and level of analysis, and the extent of sampling anticipated. For example, if subsequent sampling efforts are likely, then installing semipermanent well casings with a well drilling rig may be appropriate. If limited sampling is expected, direct push methods, such as cone

penetrometers may be more cost-effective. The types of contaminants will also play a key role in the selection of sampling methods, devices, containers, and preservation techniques.

Table 13.5 lists analytical technologies that may be appropriate for assessing metal finishing sites, the types of contamination they can measure, applicable environmental media, and the relative cost of each. The final two columns of the table contain the applicability (e.g., field and/or laboratory) of analytical methods and the technology's ability to generate quantitative versus qualitative results. Less expensive technologies that have rapid turnaround times and produce only qualitative results generally should be sufficient for many brownfields sites.

13.3.10   ADDITIONAL CONSIDERATIONS FOR ASSESSING METAL FINISHING SITES

When assessing a metal finishing brownfields site, planners should focus on the most likely areas of contamination. Although the specific locations vary from site to site, this section provides some general guidelines.

13.3.10.1   Where to Sample

Most metal finishing facilities perform all operations indoors. Consequently, most site assessment activities should focus on contamination inside and underneath the facility. Outdoor assessment activities should evaluate points where drain pipes may have carried contaminated wastewater or spilled materials.

TABLE 13.3

Soil and Subsurface Sampling Tools

Technique/Instrumentation	Media		Relative Cost per Sample	Sample Quality
	Soil	Groundwater		
Drilling Methods				
Cable	X	X	Mid-range expensive	Soil properties will most likely be altered
Casing advancement	X	X	Most expensive	Soil properties will likely be altered
Direct air rotary with rotary hammer	X	X	Mid-range expensive	Soil properties will most likely be altered
Direct mud rotary	X	X	Mid-range expensive	Soil properties may be altered
Directional drilling	X	X	Most expensive	Soil properties may be altered
Hollow-stem auger	X	X	Mid-range expensive	Soil properties may be altered
Jetting methods	X	X	Least expensive	Soil properties may be altered
Rotary diamond drilling	X	X	Most expensive	Soil properties may be altered
Rotating core	X		Mid-range expensive	Soil properties may be altered
Solid flight and bucket augers	X	X	Mid-range expensive	Soil properties will likely be altered
Sonic drilling	X	X	Most expensive	Soil properties will most likely not be altered
Split and solid barrel	X		Least expensive	Soil properties may be altered
Thin-wall open tube	X		Mid-range expensive	Soil properties will most likely not be altered
Thin-wall piston/specialized thin wall	X		Mid-range expensive	Soil properties will most likely not be altered
Direct Push Methods				
Cone penetrometer	X	X	Mid-range expensive	Soil properties may be altered
Driven wells		X	Mid-range expensive	Soil properties may be altered
Hand-Held Methods				
Augers	X	X	Least expensive	Soil properties may be altered
Rotating core	X		Mid-range expensive	Soil properties may be altered
Scoop, spoons, and shovels	X		Least expensive	Soil properties may be altered
Split and solid barrel	X		Least expensive	Soil properties may be altered
Thln-wall open tube	X		Mid-range expensive	Soil properties will most likely not be altered
Thin-wall piston/specialized thin wall	X		Mid-range expensive	Soil properties will most likely not be altered
Tubes	X		Least expensive	Soil properties will most likely not be altered

Source: U.S. EPA. *Technical Approaches to Characterizing and Cleaning Up Metal Finishing Sites under the Brownfields Initiative.* EPA/625/R-98/006, U.S. Environmental Protection Agency, Cincinnati, OH, March 1999 [2].

TABLE 13.4
Groundwater Sampling Tools

Technique/Instrumentation	Contaminants	Relative Cost per Sample	Sample Quality
Portable Grab Samplers			
Bailers	Metals, VOCs	Least expensive	Liquid properties may be altered
Pneumatic depth-specific samplers	Metals, VOCs	Mid-range expensive	Liquid properties will most likely not be altered
Portable *In Situ* Groundwater Samplers/Sensors			
Cone penetrometer samplers	Metals, VOCs	Least expensive	Liquid properties will most likely not be altered
Direct drive samplers	Metals, VOCs	Least expensive	Liquid properties will most likely not be altered
Hydropunch	Metals, VOCs	Mid-range expensive	Liquid properties will most likely not be altered
Fixed *Situ* Samplers			
Multilevel capsule samplers	Metals, VOCs	Mid-range expensive	Liquid properties will most likely not be altered
Multiple-port casings	Metals, VOCs	Least expensive	Liquid properties will most likely not be altered
Passive multilayer samplers	VOCs	Least expensive	Liquid properties will most likely not be altered

Source: U.S. EPA. *Technical Approaches to Characterizing and Cleaning Up Metal Finishing Sites under the Brownfields Initiative.* EPA/625/R-98/006, U.S. Environmental Protection Agency, Cincinnati, OH, March 1999 [2].

Note: VOCs, Volatile organic compounds.

The typical metal finishing facility is comprised of one or more large, warehouse-type buildings that contain the bath tanks, chemical storage areas, and wastewater treatment system. The floors are likely to be a continuous concrete slab containing several drains leading to a central storm drain or sewer access. In older facilities, the feed lines from the bath to wastewater tanks are underneath the floor slab. In newer facilities, the bath tanks and/or the wastewater tanks will likely be partially submerged in the floor slab and positioned directly on the ground.

A visual inspection of the site should identify the most likely points of potential contaminant releases. These include the areas surrounding:

1. Floor drains in chemical storage and process bath areas
2. Sludges left in process bath and wastewater treatment tanks
3. Pipes underneath the floor slab
4. Tanks set through the floor slab
5. Cracks in the floor or stains in low spots on the floor.

Solvents can be highly mobile on release, and can seep into and through the concrete flooring, which is porous. The inspection of the facility floor should look not only for cracks through which solvents could migrate but also for stained areas where spilled solvents may have pooled. Wipe samples should be taken along the walls of the facility, as solvent vapors may have penetrated wall materials.

Since metal finishing operations are typically conducted inside the facility, outside points of potential release are likely to be limited to:

TABLE 13.5

Sample Analysis Technologies

Technique/ Instrumentation	Analytes	Soil	Media Groundwater	Gas	Relative Cost per Analysis	Application	Produces Quantitative Data
Laser-induced breakdown spectrometry	Metals	X			Least expensive	Usually used in field	Additional effort required
Titrimetry kits	Metals	X	X		Least expensive	Usually used in laboratory	Additional effort required
Particle-induced X-ray emissions	Metals	X	X		Mid-range expensive	Usually used in laboratory	Additional effort required
Atomic adsorption spectrometry	Metals	X*	X	X	Most expensive	Usually used in laboratory	Yes
Inductively coupled plasma-atomic emission spectroscopy	Metals	X	X	X	Most expensive	Usually used in laboratory	Yes
Field bioassessment	Metals	X	X		Most expensive	Usually used in field	No
X-ray fluorescence	Metals	X	X	X	Least expensive	Laboratory and field	Yes (limited)
Chemical calorimetric kits	VOCs	X	X		Least expensive	Can be used in field, usually used in laboratory	Additional effort required
Flame ionization detector (hand-held)	VOCs	X	X	X	Least expensive	Immediate, can be used in field	No
Explosimeter	VOCs	X	X*	X	Least expensive	Immediate, can be used in field	No
Photo ionization detector (hand-held)	VOCs	X	X	X	Least expensive	Immediate, can be used in field	No
Catalytic surface oxidation	VOCs	X*	X	X	Least expensive	Usually used in laboratory	No
Near IR reflectance/ trans spectroscopy	VOCs	X			Mid-range expensive	Usually used in laboratory	Additional effort required
Ion mobility spectrometer	VOCs	X*	X	X	Mid-range expensive	Usually used in laboratory	Yes
Raman spectroscopy/ SERS	VOCs	X	X	X*	Mid-range expensive	Usually used in laboratory	Additional effort required
Infrared spectroscopy	VOCs	X	X	X	Mid-range expensive	Usually used in laboratory	Additional effort required

(Continued)

TABLE 13.5 (*Continued*)
Sample Analysis Technologies

Technique/ Instrumentation	Analytes	Soil	Media Groundwater	Gas	Relative Cost per Analysis	Application	Produces Quantitative Data
Scattering/ absorption lidar	VOCs	X*	X	X	Mid-range expensive	Usually used in laboratory	Additional effort required
FTIR spectroscopy	VOCs	X	X	X	Mid-range expensive	Laboratory and field	Additional effort required
Synchronous luminescence/ fluorescence	VOCs	X	X		Mid-range expensive	Usually used in laboratory, can be used in field	Additional effort required
Gas chromatography (GC) (can be used with numerous detectors)	VOCs	X*	X	X	Mid-range expensive	Usually used in laboratory, can be used in field	Yes
UV-visible spectrophotometry	VOCs	X	X	X	Mid-range expensive	Usually used in laboratory	Additional effort required
UV fluorescence	VOCs	X	X	X	Mid-range expensive	Usually used in laboratory	Additional effort required
Ion trap	VOCs	X	X*	X	Most expensive	Laboratory and field	Yes
Other chemical reaction-based test papers	VOCs, metals	X	X		Least expensive	Usually used in field	Yes
Immunoassay and calorimetric kits	VOCs, metals	X	X		Least expensive	Usually used in laboratory, can be used in field	Additional effort required

Source: U.S. EPA. *Technical Approaches to Characterizing and Cleaning Up Metal Finishing Sites under the Brownfields Initiative.* EPA/625/R-98/006, U.S. Environmental Protection Agency, Cincinnati, OH, March 1999 [2].

Notes: X* indicates that there must be extraction of the sample to gas or liquid phase; VOCs, volatile organic compounds.

1. Points of discharge from effluent pipes
2. Waterways, canals, and ditches at points of pipe discharge
3. Areas where process bath materials may have been dumped.

While discharge points may be visually obvious, areas of dumping may be less apparent. Often, these areas are marked by stained soils and a lack of vegetation. Low-lying areas should also be investigated, as they make natural dumping areas, and contaminants may drain to these points.

13.3.10.2 How Many Samples to Collect

Samples should be taken in and around the areas of potential release mentioned above [34]. Planners should expect that two to three samples will be required in each area, depending on DQOs. A cost-effective approach is to perform screening analyses using field methods on all samples and then submit one sample to a laboratory for analysis using an accepted U.S. EPA method. Although the screening analyses can be conducted for broad contaminant groups, such as total organics, a contaminant-specific analysis should be conducted as a full screen for organic and inorganic contaminants and to validate the screening analyses. Contaminant-specific analyses may be conducted either in the field using appropriate technologies and protocols or in a laboratory.

13.3.10.3 What Types of Analysis to Perform

The selection of analytical procedures will be based on the DQOs established. Generally, the following analyses may be appropriate at metal finishing sites:

1. Residuals taken from drain sumps in storage areas should be screened for total organics and acids. Screening analyses for these contaminants can be performed inexpensively using a photo ionization detector (PID) or flame ionization detector (FID) for total organics.
2. Residuals taken from drains in the process and wastewater treatment areas should be screened for a similar range of organic contaminants, but additional analyses should be performed to screen for the presence of inorganic contaminants, such as the metals used in the metal finishing process. Immunoassays are an inexpensive field technology that can be used to perform screening analyses for organic contaminants and mercury. X-ray fluorescence (XRF) is another innovative technology that can be used to perform either field or laboratory analyses.
3. Soil gas should be collected at points underneath the floor slab, particularly near any tanks that are set through the floor slab, to detect the presence of solvents and other organic contaminants. These samples can be analyzed using the PID/FID technology described above. Corings of the floor slab may need to be taken and sent to a laboratory to determine whether contaminants have penetrated floor slabs.
4. Wipe samples taken from walls should be analyzed for organic compounds. These analyses can be performed using the same technologies that are used to analyze residual samples.
5. Soils and sediments at points of pipe discharge should be screened for both organic and inorganic contaminants using the PID/FID technology. XRF can be used for field or laboratory analyses.
6. Water samples collected in swales, canals, and ditches should be screened for organics. Inorganic contamination can sometimes be detected in water samples, but conditions do not always allow it.

In addition, as discussed earlier, many older structures contain lead paint and asbestos insulation and tiling. Numerous kits are readily available to test for lead paint. Experienced professionals may be able to visually identify asbestos insulation, but specialized equipment may be needed to confirm the presence of asbestos in other areas. Core or wipe samples can be analyzed for asbestos using polarized light microscopy (PLM). Local and state laws regarding lead and asbestos should be consulted to determine how they may affect the selection of DQOs, sampling, and analysis.

13.3.11 General Sampling Costs

Site assessment costs vary widely, depending on the nature and extent of the contamination and the size of the sampling area. The sample collection costs discussed below are based on an assumed labor rate of USD 54/h plus USD 12 per sample for shipping and handling. All costs have been updated to 2022 USD using the USACE Yearly Average Cost Index for Utilities [35]. The 2022 USACE cost Index can also be found in the literature [80].

13.3.11.1 Soil Collection Costs

Surface soil samples can be collected with tools as simple as a stainless-steel spoon, shovel, or hand auger. Samples can be collected using hand tools in soft soil for as low as USD 16 per sample (updated 2022 cost assuming that a field technician can collect 10 samples/h). When soils are hard or deeper samples are required, a hammer-driven split spoon sampler or a direct push rig is needed. Using a drill rig equipped with a split spoon sampler or a direct push rig typically costs more than $943/d for rig operation [36], with the cost per sample exceeding USD 47 (assuming that a field technician can collect 2 samples/h). Labor costs generally increase when heavy machinery is needed.

13.3.11.2 Groundwater Sampling Costs

Groundwater samples can be extracted through conventional drilling of a permanent monitoring well or using the direct push methods listed in Table 13.3. The conventional, hollow stem auger-drilled monitoring well is more widely accepted but generally takes more time than direct push methods. Typical quality assurance protocols for the conventional monitoring well require the well to be drilled, developed, and allowed to achieve equilibrium for 24–48 hours. After the development period, a groundwater sample is extracted. With the direct push sampling method, a probe is either hydraulically pressed or vibrated into the ground, and groundwater percolates into a sampling container attached to the probe. The direct push method costs are contingent upon the hardness of the subsurface, depth to the water table, and permeability of the aquifer. Costs for both conventional and direct push techniques are generally more than USD 63 per sample (assuming that a field technician can collect 1 sample/h); well installation costs must be added to that number.

13.3.11.3 Surface Water and Sediment Sampling Costs

Surface water and sediment sampling costs depend on the location and depth of the required samples. Obtaining surface water and sediment samples can cost as little as USD 47 per sample (updated 2022 cost assuming that a field technician cam collect 2 samples/h). Sampling sediment in deep water or sampling a deep level of surface water, however, requires the use of larger equipment, which drives up the cost. Also, if surface water presents a hazard during sampling and protective measures are required, costs will increase greatly.

13.3.11.4 Sample Analysis Costs

The updated 2022 Costs for analyzing samples in any medium can range from as little as USD 43 per sample for a relatively simple test (e.g., an immunoassay test for metals) to greater than USD 633 per sample for a more extensive analysis (e.g., for semivolatiles) and up to USD 1887 per sample for dioxins [32]. Major factors that affect the cost of sample analysis include the type of analytical technology used, the level of expertise needed to interpret the results, and the number of samples to be analyzed. Planners should make sure that laboratories that have been certified by state programs are used.

For information on costs for brownfields cleanup, the reader is referred to U.S. EPA document [37], guide [38], and remediation cost compendium [39].

13.4 SITE CLEANUP

The purpose of this section is to guide planners in the selection of appropriate cleanup technologies. The principal factors that will influence the selection of a cleanup technology include [2]:

1. Types of contamination present;
2. Cleanup and reuse goals;
3. Length of time required to reach cleanup goals;
4. Post-treatment care needed;
5. Budget.

The selection of appropriate cleanup technologies often involves a trade-off between time and cost. The U.S. EPA document on cost-estimating tools and resources [37] provides information on cost factors and developing cost estimates. In general, the more intensive the cleanup approach, the more quickly the contamination will be mitigated and the more costly the effort. In the case of brownfields cleanup, this can be a major point of concern, considering the planner's desire to return the facility to the point of reuse as quickly as possible. Thus, the planner may wish to explore a number of options and weigh carefully the costs and benefits of each. One effective method of comparison is the cleanup plan, as discussed below. Planners should involve stakeholders in the community in the development of the cleanup plan.

The intended future use of a brownfields site will drive the level of cleanup needed to make the site safe for redevelopment and reuse. Brownfields sites are by definition not Superfund NPL sites; that is, brownfields sites usually have lower levels of contamination present and therefore generally require less extensive cleanup efforts than Superfund NPL sites. Nevertheless, all potential pathways of exposure, based on the intended reuse of the site, must be addressed in the site assessment and cleanup; if no pathways of exposure exist, less cleanup (or possibly none) may be required.

Some regional U.S. EPA and state offices have developed cleanup standards for different chemicals, which may serve as guidelines or legal requirements for cleanups. It is important to understand that screening levels are different from cleanup levels. Screening levels indicate whether further site investigation is warranted for a particular contaminant. Cleanup levels indicate whether cleanup action is needed and how extensive it needs to be. Planners should check with their state environmental office for guidance and/or requirements for cleanup standards.

This section contains information on developing a cleanup plan, various alternatives for addressing contamination at the site (i.e., institutional controls and containment and cleanup technologies), using different technologies for cleaning up metal finishing sites, and postconstruction issues that planners need to consider when considering alternatives.

13.4.1 DEVELOPING A CLEANUP

If the results of the site evaluation indicate the presence of contamination above acceptable levels, planners will need to have a cleanup plan developed by a professional environmental engineer that describes the approach that will be used to contain and possibly clean up the contamination present at the site. In developing this plan, planners and their engineers should consider a range of possible options, with the intent of identifying the most cost-effective approaches for cleaning up the site, given time and cost concerns. The cleanup plan can include the following elements [2,4,40,41]:

1. A clear delineation of environmental concerns at the site. Areas should be discussed separately if the cleanup approach for an area is different from that for other areas of the site. Clear documentation of existing conditions at the site and a summarized assessment of the nature and scope of contamination should be included.

2. A recommended cleanup approach for each environmental concern that takes into account expected land reuse plans and the adequacy of the technology selected.
3. A cost estimate that reflects both expected capital and operating/maintenance costs.
4. Postconstruction maintenance requirements for the recommended approach.
5. A discussion of the assumptions made to support the recommended cleanup approach, as well as the limitations of the approach.

Planners can use the framework developed during the initial site evaluation and the controls and technologies described below to compare the effectiveness of the least costly approaches for meeting the required cleanup goals established in the DQOs. These goals should be established at levels that are consistent with the expected reuse plans. A final cleanup plan may include a combination of actions, such as institutional controls, containment technologies, and cleanup technologies, as discussed below.

13.4.1.1 Institutional Controls

Institutional controls may play an important role in returning a metal finishing brownfields site to a marketable condition. Institutional controls are mechanisms that control the current and future use of, and access to, a site. They are established, in the case of brownfields, to protect people from possible contamination. Institutional controls can range from a security fence prohibiting access to a certain portion of the site to deed restrictions imposed on the future use of the site. If the overall cleanup approach does not include the complete cleanup of the facility (i.e., the complete removal or destruction of on-site contamination), a deed restriction will likely be required that clearly states that hazardous waste is being left in place within the site boundaries. Many state brownfields programs include institutional controls.

13.4.1.2 Containment Technologies

Containment technologies, in many instances, will be the likely cleanup approach for landfilled waste and wastewater lagoons (after contaminated wastewaters have been removed) at metal finishing facilities. The purpose of containment is to reduce the potential for off-site migration of contaminants and possible subsequent exposure. Containment technologies include engineered barriers such as caps [42] for contaminated soils, slurry walls [43], and hydraulic containment. Often, soils contaminated with metals can be solidified [44,45] by mixing them with cement-like materials, and the resulting stabilized material can be stored on-site in a landfill. Like institutional controls, containment technologies do not remove or destroy contamination but mitigate potential risks by limiting access to it.

If contamination is found underneath the floor slab at metal finishing facilities, leaving the contaminated materials in place and repairing any damage to the floor slab may be justified. The likelihood that such an approach will be acceptable to regulators will depend on whether potential risk can be mitigated and managed effectively over the long term. In determining whether containment is feasible, planners should consider [2,4]:

1. **Depth to groundwater**: Planners should be prepared to prove to regulators that groundwater levels will not rise, due to seasonal conditions, and come into contact with contaminated soils.
2. **Soil types**: If contaminants are left in place, the native soils should not be highly porous, as are sandy or gravelly soils, which enable contaminants to migrate easily. Clay and fine silty soils provide a much better barrier.
3. **Surface water control**: Planners should be prepared to prove to regulators that rainwater and snowmelt cannot infiltrate under the floor slab and flush the contaminants downward.
4. **Volatilization of organic contaminants**: Regulators are likely to require that air monitors be placed inside the building to monitor the level of organics that may be escaping upward through the floor and drains.

13.4.1.3 Types of Cleanup Technologies

Cleanup may be required to remove or destroy on-site contamination if regulators are unwilling to accept the level of contamination present or if the types of contamination are not conducive to the use of institutional controls or containment technologies. Cleanup technologies fall broadly into two categories—*ex situ* and *in situ*—as described below.

1. *Ex situ:* An *ex situ* technology treats contaminated materials after they have been removed and transported to another location. After treatment, if the remaining materials, or residuals, meet cleanup goals, they can be returned to the site. If the residuals do not yet meet cleanup goals, they can be subjected to further treatment, contained on-site, or moved to another location for storage or further treatment. A cost-effective approach to cleaning up a metal finishing brownfields site may be the partial treatment of contaminated soils or groundwater, followed by containment, storage, or further treatment off-site [2]. For example, it is common practice for operating metal finishing facilities to treat wastewaters to an intermediate level and then send the treated water to the local POTW.

2. *In situ:* The use of *in situ* technologies has increased dramatically in recent years. *In situ* technologies treat contamination in place and are often innovative technologies. Examples of *in situ* technologies include bioremediation [46], soil flushing [47], oxygen-releasing compounds [48], air sparging [49], and treatment walls [50]. In some cases, *in situ* technologies are feasible, cost-effective choices for the types of contamination that are likely at metal finishing sites. Planners, however, do need to be aware that clean up with *in situ* technologies is likely to take longer than with *ex situ* technologies.

Maintenance requirements associated with *in situ* technologies depend on the technology used and vary widely in both effort and cost. For example, containment technologies such as caps and liners will require regular maintenance, such as maintaining the vegetative cover and performing periodic inspections to ensure the long-term integrity of the cover system. Groundwater treatment systems will require varying levels of postcleanup care. If an *ex situ* system is in use at the site, it will require regular operations support and periodic maintenance to ensure that the system is operating as designed.

13.4.2 Keys to Technology Selection and Acceptance

Innovative technologies and technology approaches offer many advantages in the cleanup of brownfields sites [51–56]. Stakeholders in such sites, however, must first accept the technology. Brownfields decision-makers should consider the following elements to increase the likelihood that the technology will be accepted, thereby facilitating the cleanup of the site [4]:

1. **Focus on the decisions that support site goals**: The Triad approach systematic planning is an important element of all cleanup activities. Clear and specific planning to meet explicit decision objectives is essential in managing the process of cleaning up contaminated sites: site assessment, site investigation, site monitoring, and remedy selection. With good planning, brownfields decision-makers can establish the cleanup goals for the site, identify the decisions necessary to achieve those goals and develop and implement a strategy for addressing the decision needs. Technology decisions are made in the context of the requirements for such decisions. All cleanup activities are driven by the project goals. An explicit statement of the decisions to be made and the way in which the planned approach supports the decisions should be included in the work plan.

2. **Build consensus**: Investing time, before the site work begins, in developing decisions that are acceptable to all decision-makers will foster more efficient site activities and make successful cleanup more likely. Conversely, allowing work to begin at a site before a common understanding and acceptance of the decisions have been established increases the

likelihood that the cleanup process will be inefficient, resulting in delays and inefficient use of time and money. Further, decision-makers must understand that there is uncertainty in all scientific and technical decisions. Clearly defining and accepting uncertainty thresholds before making decisions about the site remedy will build consensus. Decisions also should be made in the context of applicable regulatory requirements, political considerations, budget available for the project, and time constraints.

3. **Understand the technology**: A thorough knowledge of a technology's capabilities and limitations is necessary to secure its acceptance. All technologies are subject to limitations in performance. Planning for the strengths and weaknesses of a technology maximizes understanding of its benefits and its acceptance. "Technology approvers," typically regulators, community groups, and financial service providers are likely to be more receptive of a new approach if the proposer provides a clear explanation of the rationale for its use and demonstrates confidence in its applicability to specific site conditions and needs. This latter point underscores the importance of carefully selecting an experienced, multidimensional team of professionals who have the expertise necessary to plan, present, and implement the chosen approach.

4. **Allow flexibility**: Streamlining site activities, whether site assessment, site investigation, removal, treatment, or monitoring, requires a flexible approach. Site-specific conditions, including various physical conditions, contamination issues, stakeholder needs, uses of the site, and supporting decisions, require that all decision-makers understand the need for flexibility. Although presumptive remedies, standard methods, applications at other sites, and program guidance can serve as the basis for designing a site-specific cleanup plan and can help decision-makers avoid "starting from scratch" at each site, decision-makers should be wary of depending too heavily on "boilerplate language" and prescriptive methodologies, as well as standard operating procedures and "accepted" methods. While such tools provide excellent starting points, they lack the flexibility to meet site-specific goals. To ensure an efficient and effective cleanup, the actual technology approach, whether established or innovative, must focus on decisions specific to the site.

5. Narrow the list of potential technologies that are most appropriate for addressing the contamination identified at the site and that are compatible with the specific conditions of the site and the proposed reuse of the property:
   - Network with other brownfields stakeholders and environmental professionals to learn about their experiences and to tap their expertise.
   - Determine whether sufficient data are available to support the identification and evaluation of cleanup alternatives.
   - Evaluate the options against a number of factors, including toxicity levels, exposure pathways, associated risks, future land use, and economic considerations.
   - Analyze the applicability of a particular technology to the contamination identified at a site.
   - Determine the effects of various technology alternatives on redevelopment objectives.

6. Continue to work with appropriate regulatory agencies to ensure that regulatory requirements are addressed properly:
   - Consult with the appropriate federal, state, local, and tribal regulatory agencies to include them in the decision-making process as early as possible
   - Contact the U.S. EPA regional brownfields coordinator to identify and determine the availability of U.S. EPA support programs.

7. Integrate cleanup alternatives with reuse alternatives to identify potential constraints on reuse and time schedules and to assess cost and risk factors.

8. To provide a measure of certainty and stability to the project, investigate environmental insurance policies, such as protection against cost overruns, undiscovered contamination, and third-party litigation, and integrate their cost into the project financial package.

9. Select an acceptable remedy that not only achieves cleanup goals and addresses the risk of contamination but also best meets the objectives for redevelopment and reuse of the property and is compatible with the needs of the community.

10. Communicate information about the proposed cleanup option to brownfields stakeholders, including the affected community.

13.4.3   Summary of Technologies for Treating Metals/Metalloids at Brownfield Sites

Chemical treatment: Chemical treatment, also known as chemical reduction/oxidation (redox) [48], typically involves redox reactions that chemically convert hazardous contaminants into compounds that are nonhazardous, less toxic, more stable, less mobile, or inert. Redox reactions involve the transfer of electrons from one compound to another. Specifically, one reactant is oxidized (loses electrons), and one reactant is reduced (gains electrons). The oxidizing agents used for the treatment of hazardous contaminants in soil include ozone, hydrogen peroxide, hypochlorites, potassium permanganate, Fenton's reagent (hydrogen peroxide and iron), chlorine, and chlorine dioxide. This method may be applied *in situ* or *ex situ* to soils, sludges, sediments, and other solids and may also be applied to groundwater *in situ* or *ex situ* chemical treatment using pump and treat technology. Chemical treatment may also include the use of ultraviolet (UV) light in a process known as UV oxidation.

Electrokinetics: Electrokinetics is based on the theory that a low-density current will mobilize contaminants in the form of charged species [57]. A current passed between electrodes is intended to cause aqueous media, ions, and particulates to move through soil, waste, and water. Contaminants arriving at the electrodes can be removed by means of electroplating or electrodeposition, precipitation or coprecipitation, adsorption, complexing with ion exchange resins, or pumping of water (or other fluid) near the electrodes.

Flushing: For flushing, a solution of water, surfactants, or cosolvents is applied to soil or injected into the subsurface to treat contaminated soil or groundwater [47]. When soil is being treated, injection is often designed to raise the water table into the contaminated soil zone. Injected water and treatment agents are recovered together with flushed contaminants.

Permeable reactive barriers: Permeable reactive barriers, also known as passive treatment walls, are installed across the flow path of a contaminated groundwater plume, allowing the water portion of the plume to flow through the wall [50]. These barriers allow passage of water while prohibiting movement of contaminants by means of treatment agents within the wall such as zero-valent metals (usually zero-valent iron), chelators, sorbents, compost, and microbes. The contaminants are either degraded or retained in a concentrated form by the barrier material, which may need to be replaced periodically.

Physical separation: Physical separation processes use physical properties to separate contaminated and uncontaminated media or to separate different types of media [58–60]. For example, different-sized sieves and screens can be used to separate contaminated soil from relatively uncontaminated debris. Another application of physical separation is the dewatering of sediments or sludge.

Phytoremediation: Phytoremediation is a process in which plants are used to remove, transfer, stabilize, or destroy contaminants in soil, sediment, or groundwater. The mechanisms of phytoremediation include enhanced rhizosphere biodegradation (which takes place in soil or groundwater immediately around plant roots), phytoextraction (also known as phytoaccumulation, the uptake of contaminants by plant roots and the translocation and accumulation of contaminants into plant shoots and leaves), phytodegradation (metabolism of contaminants within plant tissues), and phytostabilization (production of chemical compounds by plants to immobilize contaminants at the interface of roots and soil). The term phytoremediation applies to all biological, chemical, and physical processes that are influenced by plants (including the rhizosphere) and that aid in the cleanup of contaminated substances [61–64]. Phytoremediation may be applied *in situ* or *ex situ* to soils, sludges, sediments, other solids, or groundwater.

Environment Canada [64] studied the effectiveness of phytoremediation in Quebec's climate using herbaceous plants (Indian mustard and fescue) and shrubs (willow) to absorb heavy metals (lead, copper, and zinc). They reported that metal concentration levels in the leaves reached 1,500–2,300 mg/kg resulted in total extraction of between 2 and 13 kg of metal per ha, per growth period.

Pump and treat: Pump and treat involves extraction of groundwater from an aquifer and treatment of the water above the ground. The extraction step is usually conducted by pumping groundwater from a well or trench [65]. The treatment step can involve a variety of technologies, such as adsorption, air stripping, bioremediation, chemical treatment, filtration, ion exchange, metal precipitation, and membrane filtration [58–60].

Soil washing: For soil washing, contaminants sorbed onto fine soil particles are separated from bulk soil in a water-based system based on particle size [66]. The wash water may be augmented with a basic leaching agent, surfactant, or chelating agent or by adjustment of pH to help remove contaminants. Soils and wash water are mixed *ex situ* in a tank or other treatment unit. The wash water and various soil fractions are usually separated by means of gravity settling [58].

Solidification/stabilization: Solidification/stabilization (S/S) reduces the mobility of hazardous substances and contaminants in the environment through both physical and chemical means [44,45]. The S/S process physically binds or encloses contaminants within a stabilized mass. S/S can be performed both *ex situ* and *in situ*. *Ex situ* S/S requires excavation of the material to be treated, and the treated material must be disposed of. *In situ* S/S involves use of auger or caisson systems and injector head systems to add binders to contaminated soil or waste without excavation, and the treated material is left in place [67,68].

Solvent extraction: Solvent extraction involves the use of an organic solvent as an extractant to separate contaminants from soil. The organic solvent is mixed with contaminated soil in an extraction unit. The extracted solution is then passed through a separator, where the contaminants and extractant are separated from the soil [69].

Vitrification: Vitrification involves the use of an electric current to melt contaminated soil at elevated temperatures (1,600°C–2,000°C or 2,900°F–3,650°F). Upon cooling, the vitrification product is a chemically stable, leach-resistant glass and crystalline material similar to obsidian or basalt rock. The high-temperature component of the process destroys or removes organic materials. Radionuclides and heavy metals are retained within the vitrified product. Vitrification may be conducted *in situ* or *ex situ* [70].

13.4.4 Cleanup Technologies Options for Metal Finishing Sites

Table 13.6 presents the technologies that may be appropriate for use at metal finishing sites. In addition to more conventional technologies, a number of innovative technology options are listed. Many possible cleanup approaches use institutional controls and one or a combination of the technologies described in Table 13.6. Whatever cleanup approach is ultimately chosen, planners should explore a number of cost-effective options.

Cleanup at metal finishing facilities will most likely entail removing a complex mix of contaminants, primarily organic solvents and metals. The cleanup will usually require more than one technology, or treatment train, because single technologies tend not to address both metal and organic contaminants. S/S can address metal contamination by limiting mobility (solubility) and thereby limiting risk. Approaches at metal finishing sites depend on local conditions. At larger metal finishing sites, one approach may be to excavate and stabilize the contaminated material with either on-site or off-site disposal or treatment of the material. Access to contaminated soils may be limited at smaller sites requiring excavation and off-site treatment or disposal. The stabilized material can be placed on-site or sent to an U.S. EPA-approved landfill.

TABLE 13.6

Cleanup Technologies for Metal Finishing Brownfields Sites Sample Analysis Technologies

Applicable Technology	Description	Examples of Applicable Land/ Process Areas	Contaminants Treated by This Technology	Limitations
		Containment Technologies		
Sheet piling	Steel or iron sheets are driven into the ground to form a subsurface barrier Low-cost containment method Used primarily for shallow aquifers	Metal cleaning, rinsing and bathing operations, chemical storage, wastewater treatment	Not contaminant specific	Not effective in the absence of a continuous aquitard Can leak at the intersection of the sheets and the aquitard or through pile wall joints
Grout curtain	Grout curtains are injected into subsurface soils and bedrock Forms an impermeable barrier in the subsurface	Metal cleaning, rinsing and bathing operations, chemical storage, wastewater treatment	Not contaminant specific	Difficult to ensure a complete curtain without gaps through which the plume can escape; however, new techniques have improved continuity of curtain
Slurry walls	Consist of a vertically excavated slurry-filled trench The slurry hydraulically shores the trench to prevent collapse and forms a filtercake to reduce groundwater flow Often used where the waste mass is too large for treatment and where soluble and mobile constituents pose an imminent threat to a source of drinking water Often constructed of a soil, bentonite, and water mixture	Metal cleaning, rinsing and bathing operations, chemical storage, waste water treatment	Not contaminant specific	Contains contaminants only within a specified area Soil-bentonite backfills are not able to withstand attack by strong acids, bases, salt solutions, and some organic chemicals Potential for the slurry walls to degrade or deteriorate over time
Capping	Used to cover buried waste materials to prevent migration Made of a relatively impermeable material that will minimize rainwater infiltration Waste materials can be left in place Requires periodic inspections and routine monitoring Contaminant migration must be monitored periodically	Anodizing, solid wastes from anodizing, electroplating, electroplating wastewaters and solid wastes, finishing wastewaters, chemical conversion coating wastewaters and solid wastes, electroless plating, electroless plating wastewaters, solid wastes from painting, wastewater treatment system, sunken treatment tank	Metals	Costs associated with routine sampling and analysis may be high Long-term maintenance may be required to ensure impermeability May have to be replaced after 20-30 years of operation May not be effective if ground water table is high

(Continued)

TABLE 13.6 (*Continued*)

Cleanup Technologies for Metal Finishing Brownfields Sites Sample Analysis Technologies

Applicable Technology	Description	Examples of Applicable Land/ Process Areas	Contaminants Treated by This Technology	Limitations
		***Ex Situ* Technologies**		
Excavation/ off-site	Removes contaminated material to an disposal EPA-approved landfill	Wastes from painting, wastewater treatment system, sunken treatment tanks, chemical storage, disposal	Not contaminant specific	Generation of fugitive emissions may be a problem during operations The distance from the contaminated site to the nearest disposal facility will affect cost Depth and composition of the media requiring excavation must be considered Transportation of the soil through populated areas may affect community acceptability Disposal options for certain waste (e.g., mixed waste or transuranic waste) may be limited. There is currently only one licensed disposal facility for radioactive and mixed waste in the United States
Chemical oxidation/ reduction	Reduction/oxidation (redox) reactions chemically convert hazardous contaminants to nonhazardous or less toxic compounds that are more stable, less mobile, or inert Redox reactions involve the transfer of electrons from one compound to another The oxidizing agents commonly used are ozone, hydrogen peroxide, hypochlorite, chlorine, and chlorine dioxide	Wastes from anodizing, electroplating, finishing, chemical conversion coating, electroless plating, painting, rinsing operations, wastewater treatment system, sunken treatment tank	Metals Cyanide	Not cost-effective for high contaminant concentrations because of the large amounts of oxidizing agent required Oil and grease in the media should be minimized to optimize process efficiency
UV oxidation	Destruction process that oxidizes constituents in wastewater by the addition of strong oxidizers and irradiation with UV light Practically any organic contaminant that is reactive with the hydroxyl radical can potentially be treated	Wastes from metal cleaning, painting, rinsing operations, wastewater treatment system, sunken treatment tank, chemical storage area, disposal area	VOCs	The aqueous stream being treated must provide for good transmission of UV light (high turbidity causes interference) Metal ions in the wastewater may limit effectiveness VOCs may volatilize before oxidation can occur precautions

(Continued)

TABLE 13.6 (*Continued*)

Cleanup Technologies for Metal Finishing Brownfields Sites Sample Analysis Technologies

Applicable Technology	Description	Examples of Applicable Land/ Process Areas	Contaminants Treated by This Technology	Limitations
	The oxidation reactions are achieved through the synergistic action of UV light in combination with ozone or hydrogen peroxide Can be configured in batch or continuous flow models, depending on the throughput rate under consideration			Contamination source is not removed The presence of multiple metal species may lead to removal difficulties
Precipitation	Involves the conversion of soluble heavy metal salts to insoluble salts that will precipitate Precipitates can be removed from the treated water by physical methods such as clarification or filtration Often used as a pretreatment for other treatment technologies where the presence of metals would interfere with the treatment processes Primary method for treating metal-laden industrial wastewater	Wastes from anodizing, electroplating, Metals finishing, chemical conversion coating, electroless plating, painting, rinsing operations, wastewater treatment system, sunken treatment tank	Metals	Discharge standard may necessitate further treatment of effluent Metal hydroxide sludges must pass TCLP criteria prior to land disposal Treated water will often require pH adjustment
Liquid-phase carbon adsorption	Groundwater is pumped through a series of vessels containing activated carbon, to which dissolved contaminants adsorb Effective for polishing water discharges from other remedial technologies to attain regulatory compliance Can be quickly installed High contaminant-removal efficiencies	Wastes from metal cleaning, painting, rinsing operations, wastewater treatment system, sunken treatment tank, chemical storage area, disposal area	VOCs	The presence of multiple contaminants can affect process performance Metals can foul the system Costs are high if used as the primary treatment on waste streams with high contaminant concentration levels Type and pore size of the carbon and operating temperature will impact process performance Transport and disposal of spent carbon can be expensive Water-soluble compounds and small molecules are not adsorbed well

(Continued)

TABLE 13.6 (*Continued*)

Cleanup Technologies for Metal Finishing Brownfields Sites Sample Analysis Technologies

Applicable Technology	Description	Examples of Applicable Land/ Process Areas	Contaminants Treated by This Technology	Limitations
Air stripping	Contaminants are partitioned from groundwater by greatly increasing the surface area of the contaminated water exposed to air Aeration methods include packed towers, diffused aeration, tray aeration, and spray aeration Can be operated continuously or in a batch mode, where the air stripper is intermittently fed from a collection tank Off-gas may require treatment Costs may be higher than competing technologies because of energy needs Handling and storage of oxidizers require special safety The batch mode ensures consistent air stripper performance and greater efficiency than continuously operated units because mixing in the storage tank eliminates any inconsistencies in feed water composition	Wastes from metal cleaning, painting, VOCs rinsing operations, wastewater treatment system, sunken treatment tank, chemical storage area, disposal area	VOCs	Potential for inorganic (iron greater than 5 ppm, hardness greater than 800 ppm) or biological fouling of the equipment, requiring pretreatment of groundwater or periodic column cleaning Consideration should be given to the Henry's law constant of the VOCs in the water stream and the type and amount of packing used in the tower Compounds with low volatility at ambient temperature may require preheating of the groundwater Off-gases may require treatment based on mass emission rate and state and federal air pollution laws

(*Continued*)

TABLE 13.6 (*Continued*)
Cleanup Technologies for Metal Finishing Brownfields Sites Sample Analysis Technologies

Applicable Technology	Description	Examples of Applicable Land/Process Areas	Contaminants Treated by This Technology	Limitations
		***In Situ* Technologies**		
Natural attenuation	Natural subsurface processes such as dilution, volatilization, biodegradation, adsorption, and chemical reactions with subsurface media can reduce contaminant concentrations to acceptable levels Consideration of this option requires modeling and evaluation of contaminant degradation rates and pathways Sampling and analyses must be conducted throughout the process to confirm that degradation is proceeding at sufficient rates to meet cleanup objectives	Metal cleaning, metal cleaning wastewaters, painting, painting wastewaters and solid wastes, wastewater treatment system, sunken treatment tank, chemical storage area, snd disposal area	VOCs	Intermediate degradation products may be more mobile and more toxic than original contaminants Contaminants may migrate before they degrade The site may have to be fenced and may not be available for reuse until hazard levels are reduced Source areas may require removal for natural attenuation to be effective Modeling contaminant degradation rates, and sampling and analysis to confirm modeled predictions extremely expensive
Soil vapor extraction	A vacuum is applied to the soil to induce controlled air flow and remove contaminants from the unsaturated (vadose) zone of the soil The gas leaving the soil may be treated to recover or destroy the contaminants The continuous air flow promotes *in situ* biodegradation of low-volatility organic compounds that may be present	Metal cleaning, metal cleaning wastewaters, painting, painting wastewaters and solid wastes, wastewater treatment system, sunken treatment tank, chemical storage area, disposal area conversion coating wastewaters and solid wastes, electroless plating, electroless plating wastewaters, solid wastes from painting, wastewater treatment system, sunken treatment tank	VOCs	Tight or extremely moist content (>50%) has a reduced permeability to air, requiring higher vacuums Large screened intervals are required in extraction wells for soil with highly variable permeabilities Air emissions may require treatment to eliminate possible harm to the public or environment Off-gas treatment residual liquids and spent activated carbon may require treatment or disposal Not effective in the saturated zone and be released into the atmosphere

(Continued)

TABLE 13.6 (*Continued*)

Cleanup Technologies for Metal Finishing Brownfields Sites Sample Analysis Technologies

Applicable Technology	Description	Examples of Applicable Land/ Process Areas	Contaminants Treated by This Technology	Limitations
Soil flushing	Extraction of contaminants from the soil with water or other aqueous solutions Accomplished by passing the extraction fluid through in-place soils using injection or infiltration processes Extraction fluids must be recovered with extraction wells from the underlying aquifer and recycled when possible	Metal cleaning, metal cleaning wastewaters, painting, painting wastewaters and solid wastes, wastewater treatment system, sunken treatment tank, chemical storage area, disposal area	VOCs	Increased pressure in the vadose zone can build up vapors in basements, which are generally low-pressure areas
Air sparging	*In situ* technology in which air is injected under pressure below the water table to increase groundwater oxygen concentrations and enhance the rate of biological degradation of contaminants by naturally occurring microbes Increases the mixing in the saturated zone, which increases the contact between groundwater and soil Air bubbles traverse horizontally and vertically through the soil column, creating an underground stripper that volatilizes contaminants Air bubbles travel to a soil vapor extraction system Air sparging is effective for facilitating extraction of deep contamination, contamination in low-permeability soils, and contamination in the saturated zone	Appropriately selected location for the wall Low-permeability soils are difficult to treat Surfactants can adhere to soil and reduce effective soil porosity Reactions of flushing fluids with soil can reduce contaminant mobility Potential of washing the contaminant beyond the capture zone and the introduction of surfactants to the subsurface		The system requires control of pH levels When pH levels within the passive treatment wall rise, it reduces the reaction rate and can inhibit the effectiveness of the wall Depth and width of the plume For large-scale plumes, installation cost may be high Cost of treatment medium (iron) Biological activity may reduce the permeability of the wall

(*Continued*)

TABLE 13.6 (*Continued*)

Cleanup Technologies for Metal Finishing Brownfields Sites Sample Analysis Technologies

Applicable Technology	Description	Examples of Applicable Land/ Process Areas	Contaminants Treated by This Technology	Limitations
Passive treatment walls	A permeable reaction wall is installed inground, across the flow path of a contaminant plume, allowing the water portion of the plume to passively move through the wall Allows the passage of water while prohibiting the movement of contaminants by employing agents such as iron, chelators (ligands selected for their specificity for a given metal), sorbents, microbes, and others Anodizing, solid wastes from metals anodizing, electroplating, electroplating wastewaters and solid wastes, finishing wastewaters, chemical contaminants are typically completely degraded by the treatment wall	Depth of contaminants and specific site geology must be considered Air flow through the saturated zone may not be uniform A permeability differential such as a clay layer above the air injection zone can reduce the effectiveness Vapors may rise through the vadose zone	VOCs Metals	Walls may lose their reactive capacity, requiring replacement of the reactive medium
Biodegradation	Indigenous or introduced microorganisms degrade organic contaminants found in soil and groundwater Used successfully to remediate soils, sludges, and groundwater Especially effective for remediating low-level residual contamination in conjunction with source removal	Metal cleaning, metal cleaning wastewaters, painting, painting wastewaters and solid wastes, wastewater treatment system, sunken treatment tank, chemical storage area, disposal area	VOCs	Cleanup goals may not be attained if the soil matrix prevents sufficient mixing Circulation of water-based solutions through the soil mayincrease contaminant mobility and necessitate treatment of underlying groundwater Injection wells may clog and prevent adequate flow rates Preferential flow paths may result in nonuniform distribution of injected fluids

(Continued)

TABLE 13.6 (*Continued*)

Cleanup Technologies for Metal Finishing Brownfields Sites Sample Analysis Technologies

Applicable Technology	Description	Examples of Applicable Land/ Process Areas	Contaminants Treated by This Technology	Limitations
				Should not be used for clay, highly layered, or heterogeneous subsurface environments
				High concentrations of heavy metals, highly chlorinated organics, long-chain hydrocarbons, or inorganic salts are likely to be toxic to microorganisms
				Low temperatures slow bioremediation Chlorinated solvents may not degrade fully under certain subsurface conditions

Source: U.S. EPA. *Technical Approaches to Characterizing and Cleaning Up Metal Finishing Sites under the Brownfields Initiative.* EPA/625/R-98/006, U.S. Environmental Protection Agency, Cincinnati, OH, March 1999 [2].

13.4.5 Postconstruction Care

Many of the cleanup technologies that leave contamination on-site, either in containment systems or because of the long periods required to reach cleanup goals, will require long-term maintenance and possibly operation. If waste is left on-site, regulators will likely require long-term monitoring of applicable media (i.e., soil, water, and/or air) to ensure that the cleanup approach selected is continuing to function as planned (e.g., residual contamination, if any, remains at acceptable levels and is not migrating). If long-term monitoring is required (e.g., by the state), periodic sampling, analysis, and reporting requirements will also be involved. Planners should be aware of these requirements and provide for them in cleanup budgets. Postconstruction sampling, analysis, and reporting costs in their cleanup budgets can be a significant problem as these costs can be substantial.

13.5 CASE HISTORIES AND EXAMPLES

13.5.1 Brownfield Contamination Sites Identification

Brownfields are real properties, the expansion, redevelopment, or reuse of which may be complicated by the presence or potential presence of a hazardous substance, pollutant, or contaminant. Obvious brownfields include metal finishing factories, research laboratories, abandoned mills, gasoline and service stations, manufacturing companies, dry cleaners, etc. Less obvious brownfields include: commercial malls, strip malls, hair and nail salons, home improvement buildings, paint store, print store, doctor office, dentist office, veterinary clinic, etc. Soil, water, and air contamination occur from a variety of sources and activities. Contaminants are any physical, chemical, biological, or radiological substances that have an adverse effect on air, water, or soil. The most common contaminants found in Rhode Island include the following: (a) heavy metals (HM); (b) volatile organic compounds (VOCs); (c) semi-volatile organic compounds (SVOCs), such as polycyclic aromatic hydrocarbons (PAHs); (d) polychlorinated biphenyls (PCBs); (e) petroleum hydrocarbons, such as total petroleum hydrocarbons (TPH), extractable petroleum hydrocarbons (EPH), volatile petroleum hydrocarbons (VPH). The Rhode Island State Department of Environment Management (RISDEM) provides the following link to fact sheets containing a variety of information on brownfields, including where they may be found and frequently reported contaminants [71]: https://www. epa.gov/brownfields/understanding-brownfields

13.5.2 Brownfields Economic Redevelopment Initiative

In the USA, the U.S. Environmental Protection Agency (U.S. EPA) has taken a Brownfields Economic Redevelopment Initiative (BERI) to empower states, communities, and other stakeholders in economic redevelopment to work together to assess, safely clean up, and sustainably reuse brownfields. USEPA has funded a certain number of Brownfields Pilot Projects every year, at up to $200,000 each, to support creative 2-year explorations and demonstrations of brownfield solutions. [72,73].

Since most of the brownfield remediation projects in the USA are managed by State environmental agencies, The facility owners should contact the local State government for initial planning and getting advice. Many states have also enacted programs that offer companies technical, financial, or legal incentives to expedite mandated cleanups and undertake voluntary investigation and cleanup at contaminated sites. Most of these programs are targeted at brownfields, but several apply to sites not on a Superfund list or subject to a regulatory' agency order. These voluntary programs are unique to each state. The advantages of a voluntary program may include: (a) limited or no oversight by the regulatory agency; (b) an ability to withdraw from the process without penalty; (c) flexibility in implementation schedules; (d) availability of grants or low-interest loans; (e) tax advantages; and (f) protection, via covenants-not-to-sue, from future liability for past contamination that is either undiscovered or newly regulated. Some states offer limited liability protection to responsible parties. Most states only offer such protection to lenders, municipalities, or nonresponsible parties. The availability and provisions of covenants-not-to-sue and mechanisms for cleanup approval vary widely and necessitate expert legal counsel. Several northeastern have delegated some level of oversight authority to privately licensed professional engineers; others have set time frames within which cleanups must be deemed approved.

Most business transactions involving real property occur without government intervention. When an owner of a small electroplating job shop or mid-sized manufacturing firm needs capital to finance business expansion or environmental cleanup, the preferred choice of financier generally is the local bank although the environmental liabilities associated with potentially contaminated properties make conventional lending difficult at best.

13.5.3 Successful Brownfield Site Redevelopment

Successful brownfield site redevelopment falls into these five categories: (a) green space: land for cultivation and agriculture, community parks, nature parks, wildlife habitat, trails, sports fields, and facilities, and open space for recreational activities; (b) residential: multi-family homes (e.g., apartments and condominiums, single-family residences and other residential purposes); (c) commercial: municipal buildings, offices, retail stores, restaurants, and other businesses; (d) industrial: manufacturing facilities, warehouses, distribution facilities, research centers, and development parks; and (e) mixed-use: multiple reuses such a residential building with shops and ground floor offices [82].

13.5.4 Practical Cleaning-Up Technologies for Brownfield Remediation

The amount and degree of clean up required will vary depending on the chemicals and pollutants identified at the brownfield site and how the site will be repurposed. Section 13.4.3 presents a summary of technologies for treating metals/metalloids at brownfield sites. These special technologies for remediation of metal finishing brownfield sites include (a) Chemical or physicochemical treatment; (b) Electrokinetics; (c) Flushing; (d) Permeable reactive barriers; (e) Physical separation; (f) Phytoremediatio; (g) Pump and treat; (h) Soil washing; (i) Solidification/stabilization; (j) Solvent extraction; and (k) Vitrification. Table 13.6 presents additional technologies for site remediation of metal finishing brownfields.

Some common technologies for cleaning up any contaminated brownfield sites are: (a) excavation: it is a mechanical process that digs up contaminated soil on the surface or subsurface to be transported for treatment or for disposal in a landfill; (b) UST removal: it is mechanical process that digs up soil that is contaminated by gasoline or other fuels to expose underground gasoline storage tanks and piping systems; (c) capping: it is mechanical process that adds a hard layer of clean soil or sometimes geotextile at the top of the soil to serve as a barrier between the surface and the contaminants, so the polluted soil can be contained or isolated; (d) on-site or "in situ"chemical treatment: it is a chemical process that injects chemicals in the soil to make the contaminants less hazardous; (e) bioremediation: it is a biological treatment process that enriches soil with nutrients, oxygen, or compounds that release oxygen in order to stimulate microbial growth, in turn, to remove mainly organic pollutants by carbonaceous oxidation, nitrification, and denitrification; and (f) lead and asbestos abatement: it is a technology by which the lead and asbestos-contaminated sites are inspected, and the hazardous lead and asbestos are removed by licensed contractors with specialized equipment [64–82]

13.5.5 Cleanup Standards for Site Remediation:

The following are the cleanup standards applied to general site remediation, including brownfield site remediation [83]. Additional technical information on brownfield remediation can be found in other US Environmental Protection Agency reports [84–91].

13.5.5.1   Cleanup Standards, Cleanup to Precontamination (i.e., Background) Levels.

This criterion requires removing or treating the contaminant until the area contaminated by the release has the same characteristics as an area unaffected by a release. The definition of background levels is usually subject to much debate and is typically agreed to on a case-by-case basis.

13.5.5.2   Cleanup Standards, Levels Based on Health-Based Criteria

This approach requires cleanup to levels that protect human health and the environment. Sufficient cleanup is conducted to ensure that exposure to any remaining contaminants is at levels deemed "safe" or "acceptable" by regulatory agencies.

13.5.5.3   Cleanup Standards, Levels Specified by Existing Standards and Guidelines

This approach sets cleanup levels based on standards that have been established specifically for cleanups or for other purposes, such as Safe Drinking Water Standards, CERCLA Soil. Screening Levels, or RCRA Corrective Action Rules.

13.5.5.4   Cleanup Standards, Limits of Analytical Detection or Lower

This criterion requires cleanup to a concentration that is at or below the concentration that can be detected with current analytical technology.

13.5.5.5   Cleanup Standards, PCB Spill Cleanup Requirements

Under the Toxic Substances Control Act (TSCA), which banned the manufacture of PCBs (polychlorinated biphenyls), any release of PCBs in concentrations greater than 50 ppm (parts per million) warrants corrective action. The 1987 rule categorizes releases by weight and type of release area.

13.5.5.6   Cleanup Standards, Precedent Levels

This criterion requires cleanup to levels achievable by or specified in remedial action plans for other similar sites.

13.5.5.7   Cleanup Standards, RCRA Corrective Action Levels

As proposed in 1990, this rule establishes processes to develop site-specific levels for contaminants in soil and groundwater. Remedial action is required when contamination exceeds these levels. MCLs serve as the action levels for substances with MCLs. For other hazardous substances, the action level is based on increased cancer risks (with one standard for known and probable carcinogens and a second standard for possible carcinogens). The resulting standards are generally as stringent as the site-specific levels often agreed to for a Super and cleanup.

13.5.5.8   Cleanup Standards, Superfund Soil Screening Levels (SSLs)

In 1993, USEPA proposed specific soil concentrations for 30 contaminants. These concentrations are used as a screening tool to determine which sites will likely require further remedial investigation. Contamination levels below those listed in the SSLs usually do not require cleanup. However, some states have established more stringent screening and/or cleanup levels. In April 1996, USEPA expanded the original list of 30 contaminants to 110 contaminants. The current SSLs are available through the National Technical Information Service (NTIS) (Document Numbers: PB96-963505, PB96-963502)

13.5.5.9   Cleanup Standards, Technology-Based Levels

This criterion requires cleanup to the extent that it is technologically possible using proven remediation technologies. USEPA tracks and reviews remediation and treatment technologies to establish or revise specific performance standards.

13.6 CONCLUSION

Brownfields redevelopment contributes to the revitalization of communities across the U.S. Reuse of these abandoned, contaminated sites spurs economic growth, builds community pride, protects public health, and helps maintain our nation's "greenfields," often at a relatively low cost. This chapter provides brownfields planners with the technical methods that can be used to achieve successful site assessment and cleanup, which are the two key components in the brownfields redevelopment process.

While the general guidance provided in this chapter will be applicable to many brownfields projects, it is important to recognize the heterogeneous nature of brownfields work. That is, no two brownfields sites will be identical, and planners will need to base site assessment and cleanup activities on the conditions at their particular site. Some of the conditions that may vary by site include the type of contaminants present, the geographic location and extent of contamination, the availability of site records, hydrogeological conditions, and state and local regulatory requirements. Based on these factors, as well as financial resources and desired timeframes, planners will find different assessment and cleanup approaches appropriate.

Consultation with state and local environmental officials and community leaders, as well as careful planning early in the project, will assist planners in developing the most appropriate site assessment and cleanup approaches. Planners should also determine early on if they are likely to require the assistance of environmental engineers. A site assessment strategy should be agreeable to all stakeholders and should address the following:

1. The type and extent of contamination, if any, present at the site
2. The types of data needed to adequately assess the site
3. Appropriate sampling and analytical methods for characterizing contamination
4. An acceptable level of data uncertainty.

When used appropriately, the site assessment methods described in this chapter will help to ensure that a good strategy is developed and implemented effectively.

Once the site has been assessed and stakeholders agree that cleanup is needed, planners will need to consider cleanup options. Many different types of cleanup technologies are available. The guidance provided in this chapter on selecting appropriate methods directs planners to base cleanup initiatives on-site- and project-specific conditions. The type and extent of cleanup will depend in large part on the type and level of contamination present, reuse goals, and the budget available. Certain cleanup technologies are used on-site, while others require off-site treatment. Also, in certain circumstances, containment of contamination on-site and the use of institutional controls may be important components of the cleanup effort. Finally, planners will need to include budgetary provisions and plans for postcleanup and postconstruction care if it is required at the brownfields site. By developing a technically sound site assessment and cleanup approach that is based on site-specific conditions and addresses the concerns of all project stakeholders, planners can achieve brownfields redevelopment and reuse goals effectively and safely.

GLOSSARY OF REMEDIATION OF METAL FINISHING BROWNFIELD SITES

Abrasive jet machining: It is a mechanical process for cutting hard, brittle materials. It is similar to sand blasting but uses much finer abrasives carried at high velocities (150–910 m/s [500–3,000 ft/s]) by a liquid or gas stream. Uses include frosting glass, removing metal oxides, deburring, and drilling and cutting thin sections of metal.

Acidification of cyanide waste streams: It is a chemical process in which cyanide is acidified in a sealed reactor that is vented to the atmosphere through an air emission control system. Cyanide is converted to gaseous hydrogen cyanide, treated, vented, and dispersed.

Advanced oxidation process (AOP) using ozone and UV: One interesting variation of ozonation technology is augmentation with UV radiation. This is a technology that has been applied to wastes in the coke byproduct manufacturing industry. A significant development has been made that has resulted in significantly less ozone consumption through the use of UV radiation. UV absorption has the following effects: (a) ozone and cyanide are raised to higher energy status; (b) free radicals are formed; (c) more rapid reaction; and (d) less ozone is required.

Alkaline chlorination (metal finishing industry): This technology is generally applicable to wastes containing less than 1% cyanide, generally present as free cyanide. It is conducted in two stages: the first stage is operated at a pH greater than 10 and the second stage is operated at a pH range of 7.5–8. Alkaline chlorination is performed using sodium hypochlorite and chlorine.

Anodizing: It is an electrolytic oxidation process that converts the surface of the metal to an insoluble oxide.

Assembly: It is the fitting together of previously manufactured parts or components into a complete machine, unit of a machine, or structure.

Barrel finishing or tumbling: It is a controlled method of processing parts to remove burrs, scale, flash, and oxides, as well as to improve surface finish.

Bioremediation: It is a biological treatment process that enriches soil with nutrients, oxygen, or compounds that release oxygen in order to stimulate microbial growth, in turn, to remove mainly organic pollutants by carbonaceous oxidation, nitrification, and denitrification.

Brazing: It is the process of joining metals by flowing a thin, capillary thickness layer of nonferrous filler metal into the space between them. Bonding results from the intimate contact produced by the dissolution of a small amount of base metal in the molten filler metal without the fusion of the base metal. The term "brazing" is used where the temperature exceeds 425°C (800°F).

Brownfield: (a) The Comprehensive Environmental Response, Compensation, and Liability Act (CERCLA) or Superfund defines brownfields sites as "real property, the expansion, redevelopment, or reuse of which may be complicated by the presence or potential presence of a hazardous substance, pollutant, or contaminant." According to the U.S. Environmental Protection Agency (U.S. EPA), brownfields sites are abandoned, idled, or under-used industrial and commercial facilities where expansion or redevelopment is complicated by real or perceived environmental contamination. Concerns about liability, cost, and potential health risks associated with brownfields sites often prompt businesses to migrate to "greenfields" outside the city. Left behind are communities burdened with environmental contamination, declining property values, and increased unemployment; (b) Brownfields are real properties, the expansion, redevelopment, or reuse of which may be complicated by the presence or potential presence of a hazardous substance, pollutant, or contaminant. Obvious brownfields include metal finishing factories, research laboratories, abandoned mills, gasoline and service stations, manufacturing companies, dry cleaners, etc. Less obvious brownfields include: commercial malls, strip malls, hair and nail salons, home improvement buildings, paint store, print store, doctor office, dentist office, veterinary clinic, etc. Soil, water, and air contamination occur from a variety of sources and activities; (c) Contaminants are any physical, chemical, biological, or radiological substances that have an adverse effect on air, water, or soil. The most common contaminants found in Rhode Island include the following: (c1) heavy metals (HM); (c2) VOCs; (c3) semi-volatile organic compounds (SVOCs), such as polycyclic aromatic hydrocarbons (PAHs); (c4) polychlorinated biphenyls (PCBs); (c5) petroleum hydrocarbons, such as total petroleum hydrocarbons (TPH), extractable petroleum hydrocarbons (EPH), volatile petroleum hydrocarbons (VPH); (d) Although the expansion, redevelopment, or reuse of brownfield site may be complicated by the presence or potential presence of a hazardous substance, pollutant, or contaminant, cleaning up and reinvesting in these brownfield properties increases local tax bases, facilitates job growth, utilizes existing infrastructure, takes development pressures off of undeveloped, open land, and both improves and protects the environment.

Burnishing: It is a process of finish sizing or smooth finishing a workpiece (previously machined or ground) by displacement, rather than removal, of minute surface irregularities. It is accomplished with a smooth point or line contact and fixed or rotating tools.

Calibration: It is the application of thermal, electrical, or mechanical energy to set or establish reference points for a component or complete assembly.

Capping: It is mechanical process that adds a hard layer of clean soil or sometimes geotextile at the top of the soil to serve as a barrier between the surface and the contaminants, so the polluted soil can be contained or isolated.

Carbon adsorption: it is a physicochemical process: (a) method: a packed-bed throwaway system to remove organic pollutants from the oily waste stream; (b) system component: contactor system and a pump station designed for a contact time of 30 min and a hydraulic loading of 162 L/min/m^2 (4 gpm/ft^2).

Certification: It is a statement of professional opinion based on knowledge and belief.

Chemical conversion coatings: They are the coatings that are applied to previously deposited metal or basis material for increased corrosion protection, lubricity, preparation of the surface for additional coatings, or formulation of a special surface appearance. This operation includes chromating, phosphating, metal coloring, and passivating.

Chemical or physicochemical treatment: Chemical or physicochemical treatment, also known as chemical reduction/oxidation (redox), typically involves redox reactions that chemically convert hazardous contaminants into compounds that are nonhazardous, less toxic, more stable, less mobile, or inert. Redox reactions involve the transfer of electrons from one compound to another. Specifically, one reactant is oxidized (loses electrons), and one reactant is reduced (gains electrons). The oxidizing agents used for the treatment of hazardous contaminants in soil include ozone, hydrogen peroxide, hypochlorites, potassium permanganate, Fenton's reagent (hydrogen peroxide and iron), chlorine, and chlorine dioxide. This method may be applied *in situ* or *ex situ* to soils, sludges, sediments, and other solids and may also be applied to groundwater *in situ* or *ex situ* chemical treatment using pump and treat technology. Chemical treatment may also include the use of ultraviolet (UV) light in a process known as UV oxidation.

Chemical precipitation and clarification (sedimentation or DAF): The most important technology in metals treatment is chemical precipitation and clarification. It is accomplished through the addition of a chemical reagent to form metal precipitants, which are then removed as solids in a clarification. The options available to a facility as precipitation reagents are lime Ca(OH)$_2$, caustic NaOH, carbonate CaCO$_3$ and Na$_2$CO$_3$, sulfide NaHS and FeS, and sodium borohydride NaBH$_4$.

Chromium reduction: It is a chemical process: (a) method: chemical reduction of hexavalent chromium by sulfur dioxide under acid conditions for the continuous operating system and by sodium bisulfite under acid conditions for the batch operating system. The reduced trivalent form of chromium is subsequently removed by precipitation as the hydroxide; (b) system component: reaction tanks, a reagent storage and feed system, mixers, sensors, and controls for continuous chromium reduction. A single aboveground concrete tank with a retention time of 45 minutes is provided. For batch operation, dual aboveground concrete tanks with 4 hours retention time are provided.

Chromium-containing waste treatment: (a) Wastes containing trivalent chromium can be treated using chemical precipitation and clarification (sedimentation or dissolved air flotation); (b) There are three treatment methods applicable to wastes containing hexavalent chromium: (b1) sulfur dioxide reduction; (b2) sodium metabisulfite reduction; and (b3) ferrous sulfate reduction. Hexavalent chromium reduction through the use of sulfur dioxide and sodium metabisulfite has found the widest application in the metal finishing industry. It is not truly a treatment step but a conversion process in which the hexavalent chromium is converted to trivalent chromium. All three reduction process converts Cr (VI) to Cr (III) which in turn Cr (III) can be removed by chemical precipitation and clarification.

Clean Water Act: The Clean Water Act requires USEPA to set permissible levels of pollutants in waters used for public water supply, recreation, fish and wildlife protection, and/or agricultural and industrial use. USEPA established pretreatment standards and effluent guidelines for 65 toxic constituents. All but a few of the standards are based on the best available technology.

Cleaning: It is an operation that involves the removal of oil, grease, and dirt from the surface of the basis material using water with or without a detergent or other dispersing materials.

Cleanup Standards, Cleanup to Precontamination (i.e., Background) Levels. This criterion requires removing or treating the contaminant until the area contaminated by the release has the same characteristics as an area unaffected by a release. The definition of background levels is usually subject to much debate and is typically agreed to on a case-by-case basis.

Cleanup Standards, Levels Based on Health-Based Criteria. This approach requires cleanup to levels that protect human health and the environment. Sufficient cleanup is conducted to ensure that exposure to any remaining contaminants is at levels deemed "safe" or "acceptable" by regulatory agencies.

Cleanup Standards, Levels Specified by Existing Standards and Guidelines. This approach sets cleanup levels based on standards that have been established specifically for cleanups or for other purposes, such as Safe Drinking Water Standards, CERCLA Soil. Screening Levels, or RCRA Corrective Action Rules.

Cleanup Standards, Limits of Analytical Detection or Lower. This criterion requires cleanup to a concentration that is at or below the concentration that can be detected with current analytical technology.

Cleanup Standards, PCB Spill Cleanup Requirements. Under the Toxic Substances Control Act (TSCA), which banned the manufacture of PCBs (polychlorinated biphenyls), any release of PCBs in concentrations greater than 50 ppm (parts per million) warrants corrective action. The 1987 rule categorizes releases by weight and type of release area.

Cleanup Standards, Precedent Levels. This criterion requires cleanup to levels achievable by or specified in remedial action plans for other similar sites.

Cleanup Standards, RCRA Corrective Action Levels. As proposed in 1990, this rule establishes processes to develop site-specific levels for contaminants in soil and groundwater. Remedial action is required when contamination exceeds these levels. MCLs serve as the action levels for substances with MCLs. For other hazardous substances, the action level is based on increased cancer risks (with one standard for known and probable carcinogens and a second standard for possible carcinogens). The resulting standards are generally as stringent as the site-specific levels often agreed to for a Super and cleanup.

Cleanup Standards, Superfund Soil Screening Levels (SSLs). In 1993, USEPA proposed specific soil concentrations for 30 contaminants. These concentrations are used as a screening tool to determine which sites will likely require further remedial investigation. Contamination levels below those listed in the SSLs usually do not require cleanup. However, some states have established more stringent screening and/or cleanup levels. In April 1996, USEPA expanded the original list of 30 contaminants to 110 contaminants. The current SSLs are available through the National Technical Information Service (NTIS) (Document Numbers: PB96-963505, PB96-963502).

Cleanup Standards, Technology-Based Levels. This criterion requires cleanup to the extent that it is technologically possible using proven remediation technologies. USEPA tracks and reviews remediation and treatment technologies to establish or revise specific performance standards.

Complexation: Complexation is a phenomenon that involves a coordinate bond between a central atom (the metal) and a ligand (the anions). In a coordinate bond, the electron pair is shared by the metal and the ligand. A complex containing one coordinate bond is referred to as a monodentate complex. Multiple coordinate bonds are characteristic of polydentate complexes. Polydentate complexes are also referred to as chelates. An example of a monodentate forming ligand is ammonia. Examples of chelates are oxylates (bidentates) and EDTA (hexadentates).

Complexed metals: (a) The complexing agents react with soluble metals to form complexed metals. In metal finishing industry complexed metals may occur in a number of unit operations but come primarily from electroless and immersion plating. The most commonly used metals in these operations are copper, nickel, and tin. Wastewaters containing complexing agents must be segregated and treated independently of other wastes in order to prevent further complexing of free metals in the other streams. (b) The metals in these waste streams are tied up or complexed by particular complexing agents whose function is to prevent metals from coming out of the solution. This counteracts the technique employed by most conventional solids removal methods. Therefore, segregated treatment of these wastes is necessary. The treatment method well suited to treating complex metal wastes is high pH precipitation.

Container: It is any portable device in which a material is stored, transported, disposed of, or otherwise handled.

Contingency plan: It is a document setting out an organized, planned, and coordinated course of action to be followed in case of a fire, explosion, or release of hazardous waste or hazardous waste constituents that could threaten human health or the environment.

Cyanide oxidation: It is a chemical process: (a) method: cyanide is destroyed by reaction with sodium hypochlorite under alkaline conditions; (b) system component: reaction tanks, a reagent storage and feed system, mixers, sensors, and controls. Two identical reaction tanks sized as aboveground cylindrical tanks with a retention time of 4 hours are provided. chemical storage consists of covered concrete tanks to store 60 days supply of sodium hypochlorite and 90 days supply of sodium hydroxide.

Cyanide reduction with hydrogen peroxide: It is a chemical process that is effective in reducing cyanide. The reduction of cyanide (CN^-) with peroxide ($H_2O_{2)}$ occurs in the first step to cyanate (CNO^-) and in the second steps to CO_2 and ammonia:

Cyanide: It is hazardous inorganic chemical containing the chemical group of CN-, which has been used extensively in the surface finishing industry for many years; The use of cyanide in plating and stripping solutions stems from its ability to weakly complex many metals typically used in plating. Metal deposits produced from cyanide plating solutions are finer-grained than those plated from an acidic solution. In addition, cyanide-based plating solutions tend to be more tolerant of impurities than other solutions, offering preferred finishes over a wide range of conditions; Cyanide is generally destroyed by oxidation. Chlorine, in either elemental or hypochlorate form, is the primary oxidation agent used in industrial waste treatment to destroy cyanide. Alternative treatment techniques for the destruction of cyanide include oxidation by ozone, ozone with ultraviolet (UV) radiation (oxyphotolysis), hydrogen peroxide, and electrolytic oxidation:

Cyanide-containing wastes treatment: There are eight methods applicable to the treatment of cyanide wastes for metal finishing: (a) alkaline chlorination; (b) electrolytic decomposition; (c) ozonation; (d) UV/ozonation;(e) hydrogen peroxide; (f) thermal oxidation; (g) acidification and acid hydrolysis; and (h) ferrous sulfate precipitation.

Designated facility: It is the permitted hazardous waste treatment, storage, or disposal facility where a shipment of waste is sent.

Discarded material: It is any substance that is abandoned, recycled, or is (in EPA's judgment) inherently waste-like.

Discharge: It is any spilling, leaking, pumping, pouring, emitting, emptying, or dumping of hazardous waste into or on any land or water.

Disposal facility: It is a permitted facility for the disposal of hazardous waste. See **Disposal.**

Disposal: The application of hazardous waste into or on any land or water so that it may enter the environment (through either the air, water, or land).

Electrical discharge machining: It is a process that can remove metal with good dimensional control from any metal. It cannot be used for machining glass, ceramics, or other nonconducting materials. Electrical discharge machining is also known as spark machining or electronic erosion. The operation was developed primarily for machining carbides, hard nonferrous alloys, and other hard-to-machine materials.

Electrochemical machining: It is a process based on the same principles used in electroplating, except that the workpiece is the anode, and the tool is the cathode. Electrolyte is pumped between the electrodes and a potential applied, resulting in rapid removal of metal.

Electrokinetics: Electrokinetics is based on the theory that a low-density current will mobilize contaminants in the form of charged species. A current passed between electrodes is intended to cause aqueous media, ions, and particulates to move through soil, waste, and water. Contaminants arriving at the electrodes can be removed by means of electroplating or electrodeposition, precipitation or coprecipitation, adsorption, complexing with ion exchange resins, or pumping of water (or other fluid) near the electrodes.

Electroless plating: It is a chemical reduction process that depends on the catalytic reduction of a metallic ion in an aqueous solution containing a reducing agent and the subsequent deposition of metal without the use of external electric energy.

Electrolytic decomposition technology: It is a process that can be applied to cyanide-containing wastes. The basis of this technology is the electrolytic decomposition of the cyanide compounds at an elevated temperature ($200°F$) to yield nitrogen, CO_2, ammonia, and amines.

Electron beam machining: It is a thermoelectric process in which heat is generated by high-velocity electrons impinging the workpiece, converting the beam into thermal energy. At the point where the energy of the electrons is focused, the beam has sufficient thermal energy to vaporize the material locally. The process is generally carried out in a vacuum. The process results in X-ray emission, which requires that the work area be shielded to absorb radiation. At present, the process is used for drilling holes as small as 0.05 mm (0.002 in.) in any known material, cutting slots, shaping small parts, and machining sapphire jewel bearings.

Electropainting: It is the process of coating a workpiece by either making it anodic or cathodic in a bath that is generally an aqueous emulsion of the coating material. The electrodeposition bath contains stabilized resin, dispersed pigment, surfactants, and sometimes organic solvents in water.

Electroplating: It is a chemical engineering process that produces a thin coating of one metal on another by electrodeposition.

Electrostatic painting: It is the application of electrostatically charged paint particles to an oppositely charged workpiece followed by thermal fusing of the paint particles to form a cohesive paint film. Both waterborne and solvent-borne coatings can be sprayed electrostatically.

Emulsion breaking and oil separation: It is a chemical process: (a) method: emulsion broken by mixing oily waste with alum and a chemical emulsion breaker, followed by gravity oil separation in a tank; (b) system component: Small mixing tank, two chemical feed tanks, a mixer, and a large tank equipped with an oil skimmer and a sludge pump. The mixing tank has a retention time of 15 minutes, and the oil skimming tank has a retention time of 2.5 hours.

EPA ID number: The ID number that US EPA assigns to a hazardous-waste-generator or a treatment, storage, or disposal facility.

Etching and chemical milling: The are chemical processes that are used to produce specific design configurations and tolerances on parts by controlled dissolution with chemical reagents or etchants.

Excavation: It is a mechanical process that digs up contaminated soil on the surface or subsurface to be transported for treatment or disposal in a landfill.

Flame spraying: It is the process of applying a metallic coating to a workpiece using finely powdered fragments of wire and suitable fluxes, which are projected together through a cone of flame onto the workpiece.

Flushing: For flushing, a solution of water, surfactants, or cosolvents is applied to soil or injected into the subsurface to treat contaminated soil or groundwater. When soil is being treated, injection is often designed to raise the water table into the contaminated soil zone. Injected water and treatment agents are recovered together with flushed contaminants.

Grinding: It is a unit process of removing stock from a workpiece by the use of a tool consisting of abrasive grains held by a rigid or semirigid binder. The processes included in this unit operation are sanding (or cleaning to remove rough edges or excess material), surface finishing, and separating (as in cutoff or slicing operations).

Hazardous-waste-generator: It is any entity whose act or process produces hazardous waste.

Hazardous waste number: The number or numbers assigned by to a solid waste on the basis of its hazardous characteristics or listing as a hazardous substance. Any waste to which you assign a hazardous waste number is hazardous waste.

Hazardous waste: It is any solid waste that has at least one Hazardous Waste Number assigned to it. In general, Hazardous Waste is solid waste that is harmful to human health or the environment.

Heat treating: It is the thermal modification of the physical properties of a workpiece through the application of controlled heating and cooling cycles. Such operations as tempering, carburizing, cyaniding, nitriding, annealing, normalizing, austenizing, quenching, austempering, siliconizing, martempering, and malleabilizing are included in this definition.

Hexavalent chrome alternatives: Trivalent chromium plating solutions can be used for decorative plating to replace hexavalent chromium. The use of trivalent chromium eliminates the reduction step in the wastewater treatment process and eliminates the problems associated with hexavalent chromium bath misting. In addition, trivalent chromium baths operate with a lower viscosity and lower concentration than hexavalent baths, thus decreasing bath drag-out. Chrome conversion coatings for paint pretreatments can sometimes be replaced by nonchrome conversion coatings or no-rinse chrome chemistries.

Hot dip coating: It is the process of coating a metallic workpiece with another metal by immersion in a molten bath to provide a protective film. Galvanizing (hot dip zinc) is the most common hot dip coating.

Impact deformation: It is the physical process of applying an impact force to a workpiece such that the workpiece is permanently deformed or shaped. Impact deformation operations include shot peening, forging, high-energy forming, heading, and stamping.

Incompatible waste: It is any hazardous waste that is unsuitable for mixing with another waste because the mixture might produce heat or pressure, fire or explosion, violent reaction, or flammable or toxic dusts or gases.

Laminating: It is the process of adhesive bonding of layers of metal, plastic, or wood to form a part.

Laser beam machining: It is the process of using a highly focused, monochromatic, collimated beam of light to remove material at the point of impingement on a workpiece. Laser beam machining is a thermoelectric process, and material removal is largely accomplished by evaporation, although some material is removed in the liquid state at high velocity. Since the metal removal rate is very small, this process is used for such jobs as drilling microscopic holes in carbides or diamond wire drawing dies and for removing metal in the balancing of high-speed rotating machinery.

Lead and asbestos abatement: It is a technology by which the lead and asbestos-contaminated sites are inspected, and the hazardous lead and asbestos are removed by licensed contractors with specialized equipment.

Leak detection system: It is a system capable of detecting the failure of a containment structure or the presence of the accumulation of liquid outside the containment structure.

Lime precipitation and sedimentation: It is a physicochemical process: (a) method: chemical precipitation of dissolved and complexed metals by reaction with lime and subsequent removal of the precipitated solids by gravity settling in a clarifier. Alum and polyelectrolyte are added for coagulation and flocculation; and (b) system component: the continuous treatment system includes reagent storage and feed equipment, a mix tank for reagent feed addition, sensors and controls, and clarification basin with associated sludge rakes and pumps. Lime is fed as 30% lime slurry prepared by using hydrated lime. The mix tank is sized for a retention time of 45 minutes, and the clarifier is sized for hydraulic loading of 1360 L/m^2 and a retention time of 4 hours. Batch treatment includes dual reaction-settling tanks sized for 8 hours of retention time and sludge pumps.

Machining: It is a general process of removing stock from a workpiece by forcing a cutting tool through the workpiece, removing a chip of basis material. Machining operations such as turning, milling, drilling, boring, tapping, planing, broaching, sawing and cutoff, shaving, threading, reaming, shaping, slotting, hobbing, filing, and chamfering are included in this definition.

Manifest: It is the shipping document (EPA form 8700-22) that accompanies a hazardous waste shipment from the time it leaves the generating facility to the time it reaches its designated facility.

MCLs. Under the Safe Drinking Water Act, USEPA has established maximum contaminant levels (MCLs) for public drinking water that, if exceeded, may adversely affect public health. MCLs are based on the scientific/technological ability to detect, measure, and cost-effectively remove or treat a specific existing, or potential, public water supply contaminant.

Multimedia filter: it is a mechanical process: (a) method: polishing treatment after chemical precipitation and sedimentation by filtration through a bed of particles of several distinct size ranges; (b) system component: filter beds, media, backwash mechanism, pumps, and controls. filter beds sized for hydraulic loading of 81 $L/min/m^2$ (2 gpm/ft^2).

Neutralization: It is a chemical process that neutralizes the acid wastes or the alkaline waste to pH 6–8 and eliminates the requirements of manifesting and/or further treatment.

On-site or 'in situ' chemical treatment: It is a chemical process that injects chemicals into the soil to make the contaminants less hazardous.

Ozonation for cyanide oxidation: It is a chemical process that can be used to oxidize cyanide, thereby reducing the concentration of cyanide in wastewater. Ozone, with an electrode potential of +1.24 V in alkaline solutions, is one of the most powerful oxidizing agents known. Cyanide oxidation with ozone is a two-step reaction similar to alkaline chlorination. In the first step, cyanide (CN) is oxidized by ozone (O_3) to cyanate (CNO^-), with ozone reduced to oxygen (O_2). In the second step, cyanate (CNO^-) is hydrolyzed in the presence of water and oxidized in the presence of excess ozone (O_3) to bicarbonate (HCO_3^-) and nitrogen (N_2). The reaction time for complete cyanide oxidation is rapid in a reactor system with 10–30 min retention time being typical. The second-stage reaction is much slower than the first-stage reaction. The reaction is typically carried out in the pH range 10–12, where the reaction rate is relatively constant. Temperature does not influence the reaction rate significantly.

Paint stripping: It is the process of removing an organic coating from a workpiece. The stripping of such coatings is usually performed with caustic, acid, solvent, or molten salt.

Painting: It is the process of applying an organic coating to a workpiece. This process includes the application of coatings such as paint, varnish, lacquer, shellac, and plastics by methods such as spraying, dipping, brushing, roll coating, lithographing, and wiping. Other processes included under this unit operation are printing, silk screening, and stenciling.

Permeable reactive barriers: Permeable reactive barriers, also known as passive treatment walls, are installed across the flow path of a contaminated groundwater plume, allowing the water portion of the plume to flow through the wall. These barriers allow passage of water while prohibiting movement of contaminants by means of treatment agents within the wall such as zero-valent metals (usually zero-valent iron), chelators, sorbents, compost, and microbes. The contaminants are either degraded or retained in a concentrated form by the barrier material, which may need to be replaced periodically.

Physical separation: Physical separation processes use physical properties to separate contaminated and uncontaminated media or to separate different types of media. For example, different-sized sieves and screens can be used to separate contaminated soil from relatively uncontaminated debris. Another application of physical separation is the dewatering of sediments or sludge.

Phytoremediation: Phytoremediation is a process in which plants are used to remove, transfer, stabilize, or destroy contaminants in soil, sediment, or groundwater. The mechanisms of phytoremediation include enhanced rhizosphere biodegradation (which takes place in soil or groundwater immediately around plant roots), phytoextraction (also known as phytoaccumulation, the uptake of contaminants by plant roots and the translocation and accumulation of contaminants into plant shoots and leaves), phytodegradation (metabolism of contaminants within plant tissues), and phytostabilization (production of chemical compounds by plants to immobilize contaminants at the interface of roots and soil). The term phytoremediation applies to all biological, chemical, and physical processes that are influenced by plants (including the rhizosphere) and that aid in the cleanup of contaminated substances. Phytoremediation may be applied *in situ* or *ex situ* to soils, sludges, sediments, other solids, or groundwater. Environment Canada [64] studied the effectiveness of phytoremediation in Quebec's climate using herbaceous plants (Indian mustard and fescue) and shrubs (willow) to absorb heavy metals (lead, copper, and zinc). They reported that metal concentration levels in the leaves reached 1,500–2,300 mg/kg resulted in a total extraction of between 2 and 13 kg of metal per ha per growth period.

Plasma arc machining: It is the process of material removal or shaping of a workpiece by a high-velocity jet of high-temperature ionized gas. A gas (nitrogen, argon, or hydrogen) is passed through an electric arc causing it to become ionized and raising its temperatures in excess of 16,000°C (30,000°F). The relatively narrow plasma jet melts and displaces the workpiece material in its path.

Polishing: It is an abrading operation used to remove or smooth out surface defects (scratches, pits, tool marks, etc.) that adversely affect the appearance or function of a part. The operation, usually referred to as buffing, is included in the polishing operation.

Precious metals: The major constituents are silver and gold, which are much more commonly used in metal finishing industry operations than palladium and rhodium. Because of their high cost, precious metals are of special interest to metal finishers.

Pressure deformation: It is the physical process of applying force (at a slower rate than an impact force) to permanently deform or shape a workpiece. Pressure deformation includes operations such as roiling, drawing, bending, embossing, coining, swaging, sizing, extruding, squeezing, spinning, seaming, staking, piercing, necking, reducing, forming, crimping, coiling, twisting, winding, flaring, or weaving.

Pump and treat: Pump and treat involves extraction of groundwater from an aquifer and treatment of the water above the ground. The extraction step is usually conducted by pumping groundwater from a well or trench. The treatment step can involve a variety of technologies, such as adsorption, air stripping, bioremediation, chemical treatment, filtration, ion exchange, metal precipitation, and membrane filtration.

Resource Conservation and Recovery Act (RCRA): It is the law from which federal hazardous waste management regulations derive their authority.

Salt bath descaling: It is the process of removing surface oxides or scale from a workpiece by immersion of the workpiece in a molten salt bath or a hot salt solution. The work is immersed in the molten salt (temperatures range from 400°C to 540°C [750°F–1,000°F]), quenched with water, and then dipped in acid. Oxidizing, reducing, and electrolytic baths are available, and the particular type needed depends on the oxide to be removed.

Sand blasting: It is the process of removing stock, including surface films, from a workpiece by the use of abrasive grains pneumatically impinged against the workpiece. The abrasive grains used include sand, metal shot, slag, silica, pumice, or natural materials such as walnut shells.

Shearing: It is the process of severing or cutting a workpiece by forcing a sharp edge or opposed sharp edges into the workpiece, stressing the material to the point of shear failure and separation.

Sintering: It is the process of forming a mechanical part from a powdered metal by fusing the particles together under pressure and heat. The temperature is maintained below the melting point of the basis metal.

Sludge drying beds: it is a thermal process: (a) method: sludge dewatered using gravity drainage and natural evaporation; (b) system component: beds of highly permeable gravel and sand underlain by drain pipes.

Small Business Liability Relief and Brownfields Revitalization Act (SBLRBRA): It is the amendment to the Comprehensive Environmental Response, Compensation, and Liability Act (CERCLA) that describes a brownfield as the "real property, the expansion, redevelopment or reuse of which may be complicated by the presence of a hazardous substance, pollutant, or contaminant," and assists the conversion of contaminated brownfield to useful real properties.

Soil washing: For soil washing, contaminants sorbed onto fine soil particles are separated from bulk soil in a water-based system based on particle size. The wash water may be augmented with a basic leaching agent, surfactant, or chelating agent or by adjustment of pH to help remove contaminants. Soils and wash water are mixed *ex situ* in a tank or other treatment unit. The wash water and various soil fractions are usually separated using gravity settling.

Soldering: It is the process of joining metals by flowing a thin, capillary thickness layer of nonferrous filler metal into the space between them. Bonding results from the intimate contact produced by the dissolution of a small amount of base metal in the molten filler metal without the fusion of the base metal. The term "soldering" is used where the temperature range falls below 425°C (800°F).

Solid waste: In general, any solid, liquid, semi-solid, or contained gaseous material that is discarded. See Appendix 5-1 for more detailed criteria.

Solidification/stabilization: Solidification/stabilization (S/S) reduces the mobility of hazardous substances and contaminants in the environment through both physical and chemical means. The S/S process physically binds or encloses contaminants within a stabilized mass. S/S can be performed both *ex situ* and *in situ*. *Ex situ* S/S requires excavation of the material to be treated, and the treated material must be disposed of. *In situ* S/S involves use of auger or caisson systems and injector head systems to add binders to contaminated soil or waste without excavation, and the treated material is left in place].

Solvent degreasing: It is a process for removing oils and grease from the surfaces of a workpiece by the use of organic solvents, such as aliphatic petroleum, aromatics, oxygenated hydrocarbons, halogenated hydrocarbons, and combinations of these classes of solvents. However, ultrasonic vibration is sometimes used with liquid solvent to decrease the required immersion time with complex shapes. Solvent cleaning is often used as a precleaning operation, such as prior to the alkaline cleaning that precedes plating, as a final cleaning of precision parts, or as a surface preparation for some painting operations.

Solvent extraction: Solvent extraction involves the use of an organic solvent as an extractant to separate contaminants from soil. The organic solvent is mixed with contaminated soil in an extraction unit. The extracted solution is then passed through a separator, where the contaminants and extractant are separated from the soil].

Sputtering: It is the process of covering a metallic or nonmetallic workpiece with thin films of metal. The surface to be coated is bombarded with positive ions in a gas discharge tube, which is evacuated to a low pressure.

Storage: It is the holding of hazardous waste for a temporary period, at the end of which the hazardous waste is treated, disposed of, or stored elsewhere.

Tank: It is a stationary device designed to contain an accumulation of hazardous waste. Typically constructed out of wood, concrete, steel, or plastic.

Testing: It is the application of thermal, electrical, or mechanical energy to determine the suitability or functionality of a component or complete assembly.

Thermal cutting: It is the thermal process of cutting, slotting, or piercing a workpiece using an oxy-acetylene oxygen lance or electric arc cutting tool.

Thermal infusion: It is the process of applying fused zinc, cadmium, or other metal coating to a ferrous workpiece by imbuing the surface of the workpiece with metal powder or dust in the presence of heat.

Thermal oxidation: It is a thermal process that is useful for destroying cyanide. Thermal destruction of cyanide can be accomplished through either high-temperature hydrolysis or combustion. At temperatures between 140°C and 200°C and a pH of 8, cyanide hydrolyzes quite rapidly to produce formate ($HCOO^-$) and ammonia (NH_3). Pressures up to 100 bar are required, but the process can effectively treat waste streams over a wide concentration range and is applicable to both rinsewater and concentrated solutions. In the presence of nitrates, formate, and ammonia can be destroyed in another reactor at 150°C to form carbon dioxide and water.

Total toxic organics (TTO): TTO is the summation of all quantifiable values greater than 0.01 mg/L for the toxic organics specified by the US Environmental Protection Agency (U.S. EPA).

Toxicity Characteristic Leachate Procedure (TCLP): It is a laboratory procedure by which you can evaluate whether or not a solid waste possesses the hazardous characteristic of toxicity.

Transporter: It is a person engaged in the off-site transportation of hazardous waste by air, rail, highway, or water.

Treatment: It is any method, technique, or process designed to change the nature of hazardous waste in such a way is to render it nonhazardous or less hazardous.

Ultrafiltration: it is a physicochemical process: (a) method: process used for oily waste stream after emulsion breaking-gravity oil separation; (b) system component: filter modules sized based on hydraulic loading of 1 L/min/m^2.

Ultrasonic machining: It is a mechanical process designed to remove material using abrasive grains, which are carried in a liquid between the tool and the work and which bombard the work surface at high velocity. This action gradually chips away minute particles of material in a pattern controlled by the tool shape and contour. Operations that can be performed include drilling, tapping, coining, and the making of openings in all types of dies.

Underground storage tank (UST) removal: It is a mechanical process that digs up soil that is contaminated by gasoline or other fuels to expose underground gasoline storage tanks and piping systems

Vacuum metalizing: It is the process of coating a workpiece with metal by flash heating metal vapor in a high-vacuum chamber containing the workpiece. The vapor condenses on all exposed surfaces.

Vapor plating: It is the process of decomposition of a metal or compound upon a heated surface by reduction or decomposition of a volatile compound at a temperature below the melting point of either the deposit or the basis material.

Vitrification: Vitrification involves the use of an electric current to melt contaminated soil at elevated temperatures (1,600°C–2,000°C or 2,900°F–3,650°F). Upon cooling, the vitrification product is a chemically stable, leach-resistant glass and crystalline material similar to obsidian or basalt rock. The high-temperature component of the process destroys or removes organic materials. Radionuclides and heavy metals are retained within the vitrified product. Vitrification may be conducted *in situ* or *ex situ.*

Welding: It is the process of joining two or more pieces of material by applying heat, pressure, or both, with or without filler material, to produce a localized union through fusion or recrystallization across the interface. Included in this process are gas welding, resistance welding, arc welding, cold welding, electron beam welding, and laser beam welding.

Wetting agents: Wetting agents are the chemicals that lower the surface tension of process baths. Lowering the surface tension reduces the drainage time and minimizes the edge effect (the bead of liquid adhering to the part edge); thus, there is less drag-out. Plating baths such as nickel and heavy copper cyanide also use wetting agents to maintain grain quality and provide improved coverage.

REFERENCES

1. Federal Register, *Comprehensive Environmental Response, Compensation, and Liability Act (CERCLA or Superfund)* 42 U.S.C. s/s 9601 et seq. (1980), United States Government, Public Laws, January 2004.
2. U.S. EPA. *Technical Approaches to Characterizing and Cleaning Up Metal Finishing Sites under the Brownfields Initiative.* EPA/625/R-98/006, U.S. Environmental Protection Agency, Cincinnati, OH, March 1999.
3. U.S. EPA. *Brownfields Home Page.* U.S. Environmental Agency, 2007. Available at: https://www.epa.gov/ brownfields.
4. U.S. EPA. *Road Map to Understanding Innovative Technology Options for Brownfields Investigation and Cleanup,* 4th Edition, EPA 542-B-05-001, U.S. Environmental Protection Agency, Washington, DC, September 2005.
5. U.S. EPA. *Brownfields Tool Kit,* U.S. Environmental Protection Agency, Cincinnati, OH, 2007.
6. U.S. EPA. *Brownfields and Land Revitalization—Tools and Technical Information,* U.S. Environmental Protection Agency, Washington, DC, 2007.
7. U.S. EPA. *Profile of the Fabricated Metal Products Industry.* EPA 3 10-R-95-007, U.S. Environmental Protection Agency, Washington, DC, 1995.
8. Brebbia, C.A. (Ed.), *Brownfields III: Prevention, Assessment, Rehabilitation and Development of Brownfield Sites.* WIT Transactions on Ecology and the Environment Series, vol. 94, p. 228. Wessex Institute of Technology (WIT), UK, 2006.
9. CERP. *Brownfields Identification,* The Community Environmental Resource Program (CERP), St Louis, Missouri, USA, 2007. Available at:
10. ASTM. *Standard Practice for Environmental Site Assessments: Phase I Environmental Site Assessment Process.* E 1527-00, American Society for Testing and Materials, West Conshohocken, PA, 2003.
11. Federal Register, *Resource Conservation and Recovery Act (RCRA),* 42 U.S. Code s/s 6901 et seq. (1976), U.S. Government, Public Laws, January 2004. US Environmental Protection Agency, Washington DC, USA.
12. U.S. EPA. *National Pollutant Discharge Elimination System (NPDES),* U.S. Environmental Protection Agency, Washington, DC, 2007.
13. U.S. EPA. *Comprehensive Environmental Response, Compensation, and Liability Information System (CERCLIS),* U.S. Environmental Protection Agency, Washington, DC, 2007.
14. ASTM. *Standard Guide for Process of Sustainable Brownfields Development.* E 1984-03, American Society for Testing and Materials, West Conshohocken, PA, 2003.
15. NEWMOA. Improving decision quality: Making the case for adopting next-generation site characterization practices. Northeast Waste Management Officials' Association. *Remediation,* spring, 2003.
16. U.S. EPA. *Quality Assurance Guidance for Conducting Brownfields Site Assessments.* EPA 540-R-98-038, U.S. Environmental Protection Agency, Washington, DC, 1998.

17. Pediaditi, K., Wehrmeyer, W., and Chenoweth, J. Sustainability indicators for brownfield redevelopment projects. In: *Proceedings of Sustainable Urban Environments: EPSRC Conference,* University of Birmingham, UK, February 2005.
18. ASTM. *ASTM Standard Practice for Environmental Site Assessments: Transaction Screen Process.* E 1528-00, American Society for Testing and Materials, West Conshohocken, PA, 2000.
19. U.S. EPA. *Brownfields Technology Primer: Using the Triad Approach to Streamline Brownfields Site Assessment and Cleanup.* EPA 542-B-03-002, U.S. Environmental Protection Agency, Washington, DC, 2003.
20. U.S. EPA. *Improving Sampling, Analysis, and Data Management for Site Investigation and Cleanup.* EPA 542-F-04-001a, U.S. Environmental Protection Agency, Washington, DC, 2004.
21. U.S. EPA. *The Triad Resource Center,* 2007. U.S. Environmental Protection Agency, Washington, DC.
22. U.S. EPA. *Technology News and Trends,* 2007. U.S. Environmental Protection Agency, Washington, DC.
23. ASTM. *ASTM Standard Guide for Environmental Site Assessments: Phase II Environmental Site Assessment Process.* E1903-97, American Society for Testing and Materials, West Conshohocken, PA, 2002.
24. U.S. EPA. *Quality Assurance Guidance for Conducting Brownfields Site Assessments.* EPA 540-R-98-038, U.S. Environmental Protection Agency, Washington, DC, 1998.
25. U.S. EPA. *Clarifying DQO Terminology Usage to Support Modernization of Site Cleanup Practices.* EPA 542-R-01-014, U.S. Environmental Protection Agency, Washington, DC, 2001.
26. U.S. EPA. *Data Quality Objective Process for Hazardous Waste Site Investigations.* EPA 600-R-00-007, U.S. Environmental Protection Agency, Washington, DC, 2000.
27. U.S. EPA and U.S. ACE. Managing uncertainty in *environmental decisions. Environmental Science and Technology,* American Chemical Society, Washington, DC, 2001.
28. U.S. ACE. *Engineering and Design: Requirements for the Preparation of Sampling and Analysis Plans.* EM 200-1-3, U.S. Army Corps of Engineers, Washington, DC, 2001.
29. U.S. EPA-OSRTI. In search of representativeness: Evolving the environmental data quality model *Quality Assurance, 9,* 179–190, 2002.
30. Federal Register, *Safe Drinking Water Act (SDWA),* 42 U.S.C. s/s 300f et seq. (1974), United States Government, Public Laws, January, 2002. Full text available at: https://frwebgate.access.gpo.gov/cgi-bin/getdoc.cgi?dbname=browse_usc&docid=Cite:+42USC300f.
31. U.S. EPA. *Subsurface Characterization and Monitoring Techniques: A Desk Reference Guide.* EPA/625/R-93-003a, U.S. Environmental Protection Agency, Washington, DC, 1993.
32. Robbat, A., Jr. *Dynamic Workplans and Field Analytics: The Keys to Cost Effective Site Characterization and Cleanup.* Tufts University under Cooperative Agreement with the U.S. Environmental Protection Agency, October 1997.
33. U.S. EPA. *Field Analytical and Site Characterization Technologies: Summary of Applications.* EPA 542- R-97-011, U.S. Environmental Protection Agency, Washington, DC, 1997.
34. U.S. EPA. *Electroplating,* U.S. Environmental Protection Agency, Mid-Atlantic Brownfields, 2007. U.S. Environmental Protection Agency, Washington, DC.
35. USAC (2020) *Civil Works Construction Cost Index System Manual,* 110-2-1304, U.S. Army Corps of Engineers, Washington, DC, pp. 44, (2020-Tables Revised 31 March) (2020).
36. Geo-Environmental Solutions. *Rental Rate Sheet,* Geoprobe Systems, Inc, September 15, 1998. Available at: https://www.gesolutions.com/assess.htm.
37. U.S. EPA. *Cost Estimating Tools and Resources for Addressing the Brownfields Initiatives.* EPA 625-R-99-001, U.S. Environmental Protection Agency, Washington, DC, 1999.
38. U.S. EPA. *Guide to Documenting and Managing Cost and Performance Information for Remediation Projects.* EPA 542-B-98-007, U.S. Environmental Protection Agency, Washington, DC, 1998.
39. U.S. EPA. *Remediation Technology Cost Compendium—Year 2000.* EPA 542-R-01-009, U.S. Environmental Protection Agency, Washington, DC, September 2001.
40. Al-Tabbaa, A. Impact of and response to climate change in UK brownfield remediation. Paper presented to the *Chartered Institute of Water and Environmental Management Hong Kong,* The Hong Kong Institution of Engineers, Hong Kong, May 2007.
41. Catney, P., Yount, K., Henneberry, J., and Meyer, P. Can we really compare brownfield regulation and redevelopment in the United States and European Union? *Revit & Cabernet 2nd International Conference on Managing Urban Land,* Theaterhaus Stuttgart, Germany, April 2007.
42. U.S. EPA. *Capping. EPA 542-F-01-022,* U.S. Environmental Protection Agency, Washington, DC, 2001.

43. U.S. EPA. *Evaluation of Subsurface Engineered Barriers at Waste Sites.* EPA 542-R-98-005, U.S. Environmental Protection Agency, Washington, DC, 1998.

44. U.S. EPA. *Solidification/Stabilization Use at Superfund Sites.* EPA 542-R-00-010, U.S. Environmental Protection Agency, Washington, DC, 2000.

45. U.S. EPA. *Solidification/Stabilization.* EPA 542-F-01-024, U.S. Environmental Protection Agency, Washington, DC, 2001.

46. U.S. EPA. *Bioremediation.* EPA 542-F-01-001, U.S. Environmental Protection Agency, Washington, DC, 2001.

47. U.S. EPA. *In Situ Flushing.* EPA 542-F-01-011, U.S. Environmental Protection Agency, Washington, DC, 2001.

48. U.S. EPA. *Chemical Oxidation.* EPA 542-F-01-013, U.S. Environmental Protection Agency, Washington, DC, 2001.

49. U.S. EPA. *Soil Vapor Extraction (SVE) and Air Sparging.* EPA 542-F-01-006, U.S. Environmental Protection Agency, Washington, DC, 2001.

50. U.S. EPA. *Permeable Reactive Barriers.* EPA 542-F-01-00, U.S. Environmental Protection Agency, Washington, DC, 2001.

51. U.S. EPA. *Site Remediation Technology InfoBase: A Guide to Federal Programs, Information Resources, and Publications on Contaminated Site Cleanup Technologies,* 2nd Edition. EPA 542-B-00-005, U.S. Environmental Protection Agency, Washington, DC, 2000.

52. U.S. EPA. *Innovative Remediation Technologies: Field-Scale Demonstration Projects in North America,* 2nd Edition. EPA 542-B-00-004, U.S. Environmental Protection Agency, Washington, DC, 2000.

53. U.S. EPA. *Brownfields Technology Primer: Requesting and Evaluating Proposals that Encourage Innovative Technologies for Investigation and Cleanup.* EPA 542-R-01-005, U.S. Environmental Protection Agency, Washington, DC, 2001.

54. NATO/CCMS. *North Atlantic Treaty Organization/Committee on the Challenges of Modern Society (NATO/CCMS) Pilot Study Evaluation of Demonstrated and Emerging Technologies for the Treatment of Contaminated Land and Groundwater* (Phase III). Annual report, EPA 542-R-02-010, U.S. Environmental Protection Agency, Washington, DC, 2002.

55. U.S. EPA. *Treatment Technologies for Site Cleanup: Annual Status Report,* 11th Edition. EPA 542-R-03-009, U.S. Environmental Protection Agency, Washington, DC, 2003.

56. U.S. EPA. *Innovative Remediation and Site Characterization Technologies Resources.* EPA 542-C-04-002, U.S. Environmental Protection Agency, Washington, DC, 2004.

57. GWRTAC, *Electrokinetics,* Technical Overview Report, TO-97-03, Groundwater Remediation Analysis Center, Pittsburgh, PA, 1997.

58. Wang, L.K., Hung, Y. T., and Shammas, N. K. (Eds), *Physicochemical Treatment Processes,* Humana Press, Totowa, NJ, , 2005.

59. Wang, L.K., Hung, Y. T., and Shammas, N. K. (Eds), *Advanced Physicochemical Treatment Processes,* Humana Press, Totowa, NJ, 2006.

60. Wang, L.K., Hung, Y. T., and Shammas, N. K. (Eds), *Advanced Physicochemical Treatment Technologies,* Humana Press, Totowa, NJ, 2007.

61. U.S. EPA. *Phytoremediation.* EPA 542-F-01-002, U.S. Environmental Protection Agency, Washington, DC, 2001.

62. U.S. EPA. *Brownfields Technology Primer: Selecting and Using Phytoremediation for Site Cleanup.* EPA 542-R-01-006, U.S. Environmental Protection Agency, Washington, DC, July 2001.

63. U.S. EPA. *Use of Field-Scale Phytotechnology for Chlorinated Solvents, Metals, Explosives and Propellants, and Pesticides—Status Report.* EPA 542-R-05-002, U.S. Environmental Protection Agency, Washington, DC, April 2005.

64. Environment Canada. *Phytoremediation of Soil Containing Heavy Metals and Hydrocarbons.* Environmental Protection, Quebec Region, 2007. Available at: https://www.qc.ec.gc.ca/dpe/Anglais/dpe_main_en.asp?innov_cemrs_200409b.

65. U.S. EPA. *Pump and Treat.* EPA 542-F-01-025, U.S. Environmental Protection Agency, Washington, DC, 2001.

66. U.S. EPA. *Soil Washing.* EPA 542-F-01-008, U.S. Environmental Protection Agency, Washington, DC, 2001.

67. Harbottle, M.J. and Al-Tabbaa, A. Combining stabilization/solidification with biodegradation to enhance long-term remediation performance. In: *Proceedings of the 2nd IASTED International Conference on Advanced Technology in the Environmental Field,* Lanzarote, Spain, pp. 222–227, 2006.

68. Harbottle, M.J., Al-Tabbaa, A., and Evans, C.W. The technical sustainability of *in-situ* stabilization/solidification. In: Al-Tabbaa A. and Stegemann, J. (Eds.), *Proceedings of the International Conference on Stabilization/Solidification Treatment and Remediation.* Cambridge, UK, pp. 159–170, April 2005.

69. U.S. EPA. *Solvent Extraction.* EPA 542-F-01-009, U.S. Environmental Protection Agency, Washington, DC, 2001.

70. U.S. EPA. *Vitrification.* EPA 542-F-01-017, US Environmental Protection Agency, Washington, DC, 2001.

71. RISDEM (2023). *What Is a Brownfield,* Rhode Island State Department of Environmental Management, Providence, RI.

72. USEPA (1996). *Metal Finishing Guidance Manual: A Practical Guide to Environmental Compliance and Continuous Improvement,* US Environmental Protection Agency, Washington, DC.

73. USEPA (2021). *Preliminary Effluent Guidelines Program Plan,* US Environmental Protection Agency, Washington, DC, 15 September 2021

74. Wang, LK, Wang, MHS, Hung, YT, Shammas, NK, and Chen, JP (2016). *Handbook of Advanced Industrial and Hazardous Wastes Management,* CRC Press, Boca Raton, FL, 1174. Pages.

75. Wang, LK, Hung, YT, and Shammas, NK (2010). *Handbook of Advanced Industrial and Hazardous Wastes Treatment.* CRC Press, Boca Raton, FL, 1378 pages.

76. Wang, LK, Shammas, NK and Hung, YT (2009). *Waste Treatment in the Metal Manufacturing, Forming, Coating and Finishing Industries,* CRC Press, Boca Raton, FL, 494 pages.

77. Chen, JP, Wang, LK, Wang, MHS, Hung, YT, and Shammas, NK (2017). *Remediation of Heavy Metals In the Environment,* CRC Press, Boca Raton, FL, 580 pages.

78. Wang, LK, Chen, JP, Hung, YT and Shammas, NK (2009). *Heavy Metals in the Environment,* CRC Press, Boca Raton, FL, 514 pages.

79. Wang, LK and Wang, MHS (1992). *Handbook of Industrial Waste Treatment,* Marcel Dekker, NY, 391Pages.

80. Wang, LK, Wang, MHS, and Hung, YT (2022). *Waste Treatment in the Biotechnology, Agricultural ad Food Industries,* Volume 1, Springer Nature Cham, Switzerland, 485 pages.

81. State of Connecticut (2021). *Brownfields Success Stories,* State of Connecticut DEEP, CT. https://portal.ct.gov/DEEP/Remediation--Site-Clean-Up/Brownfields/Brownfields-Success-Stories.

82. Bauto, L (2023). Repurposing Brownfield: Opportunities for More Housing, Safety Culture. https://safetyculture.com/topics/brownfields/

83. USEPA (1988) *S-EPA Title III Section 313 Release Reporting Guidance: Estimating Chemical Releases from Electroplating Operations,* US Environmental Protection Agency; EPA 560/4-88-004g, January 1988.

84. USEPA (2023) *Understanding Brownfields,* US Environmental Protection Agency, Washington, DC. https://www.epa.gov/brownfields/understanding-brownfields

85. USEPA (1999) *Brownfields Title VI Case Studies: Summary Report,* U.S. Environmental Protection Agency, Washington, DC.

86. USEPA (2021) *Brownfields and Environmental Justice Statement,* U.S. Environmental Protection Agency, Washington, DC.

87. USEPA (2021) *Overview of EPA's Brownfields Program,* U.S. Environmental Protection Agency, Washington, DC.

88. USEPA (2023) *2020 Environmental Benefits of Brownfields,* U.S. Environmental Protection Agency, Washington, DC.

89. USEPA (2021) *Building a Sustainable Future,* U.S. Environmental Protection Agency, Washington, DC.

90. USEPA (2023). *Overview of EPA's Brownfields Program,* U.S. Environmental Protection Agency, Washington, DC.

91. USEPA (2016) *The Environmental Protection Agency's Brownfields Pilot,* U.S. Environmental Protection Agency, Washington, DC.

14 Biological Treatment of Mercury-Polluted Wastewater

Amanpreet Kaur, Vijay Singh, Satnam Singh, Yung-Tse Hung, and Saurabh Gupta

NOMENCLATURES (ACRONYMS)

BOD	Biological oxygen demand
CaCl$_2$	Calcium Chloride
Cd	Cadmium
CNS	Central nervous system
COD	Chemical oxygen demand
CW	Constructed wetlands
DNA	Deoxyribonucleic acid
EC	Electrical conductivity
EDTA	Ethylenediaminetetraacetic acid
FSH	Follicle-stimulating hormone
H$_2$SO$_4$	Sulfuric acid
HCl	Hydrochloric acid
Hg	Mercury
Hg$^+$	Mercury Ion
Hg^{++}	Mercuric
Hg0	Mercury vapor
Hg^{2+}	Mercuric cation
Hg^{2++}	Mercurous
Hg$_2$Cl$_2$	Mercury chloride
HgCl$_2$	Mercuric chloride
HgO	Mercuric oxide
HNO$_3$	Nitric acid
KMnO$_4$	Potassium permanganate
LH	luteinizing hormone
NH$_2$OH.HCl	Hydroxylamine hydrochloride
Ni	Nickel
Pb	Lead
PM	Particulate matter
SBR	Sequence batch reactor
SnCl$_2$	Vegetated submerged bed
TDS	Total dissolved solids
TS	Total solids
TSS	Total suspended solids

DOI: 10.1201/9781003541615-14

14.1 INTRODUCTION

14.1.1 DESCRIPTION

Mercury is a naturally occurring element that originates from geological resources but eagerly distributes into the air, water, soil, and biomass of the environment (1). Mercury (known as quicksilver) has the atomic number 80. It acts as a global pollutant in drinking water that causes kidney complications (2,3). Besides its serious consequences, it is used in various products like batteries, thermometers, dentistry as an amalgam for filling, and in the pharmaceutical industry (4). Mercury enters the water through various natural sources and man-made activities. Natural sources include degassing of the earth's crust and volcanoes naturally (5); however, anthropogenic sources include alkali and metal processing, incineration of coal, medical and other waste, and mining of gold and Hg (6). Along with the water, mercury pollution enters the environment through the air. The evidence of anthropogenic and household sources through which airborne mercury enters the environment is given by Lin in 2007 and Kim in 2016 (8). The studies done by them suggested that sources include industrial facilities like coal-fired power stations, metal extraction, waste incineration, or chemical industries (7) and daily household activities, like cooking or smoking, or from the large variety of devices containing Hg (8).

14.1.2 APPLICATIONS

There are so many applications of mercury. It is used in thermometer (92), manometer (93), barometer (94), sphygmomanometer (95), mercury switches (96), fluorescent lamps (97), mercury relays (98), and many other devices. In addition to these uses, mercury is also a vital part of amalgam for dental restoration (99).

14.1.3 LIMITATIONS

Different forms of mercury present in the surroundings are fundamental mercury (HgO), inorganic mercurous (Hg^+) s, and mercuric (Hg^{2+}) salts and as organic compounds (e.g. methyl-, ethyl, and phenyl-mercury); every structure possesses diverse physicochemical property and toxicity profile (9;10). Methylated forms (monomethyl mercury and dimethyl mercury) are produced by natural processes accumulated in marine fish. Humans consume mercury from seafood and inhale it from mercury-polluted air (11). It is highly toxic as it causes neurotoxicological disorders. In invertebrates (Molluscs, Crustaceans, Echinoderms), it may cause peroxidation of lipids and also affect cell metabolism. In vertebrates (Fish), it may lead to exhaustion, dysfunction of coordination, loss of appetite, and death (12). Further, dangerous effects also include neuronal disorders (inability to talk, see, smell, and move), cardiovascular system, kidneys, bones, etc. (13)

The Redox state of mercury is mainly responsible for its toxicity (9), and its most toxic form is Hg^{2+} which binds to amino acid cysteine in protein and results in the inactivation of enzymes. Mercury enters the bloodstream because it can be easily inhaled, consequently affecting the body by converting oxidized form to reactive Hg^{2+}. Inhaled elemental mercury vapors are absorbed easily through mucus membranes and the lung which rapidly oxidizes to other forms, whereas methyl mercury is deposited in many tissues as well as in the gut, but does not cross the blood-brain barrier efficiently like elemental mercury. However, on entering the brain it is progressively demethylated to elemental mercury (14). Mercury salts, in contrast, tend to be insoluble, relatively stable, and poorly absorbed. The toxicity of mercury varies from instant to many months depending upon the cellular metabolism (15). Mercurous Mercury Calomel (Hg_2Cl_2) is still used in some regions of the world as a laxative. Although poorly absorbed, some are converted to mercuric mercury, which is absorbed and induces toxicity as expected with mercuric mercury.

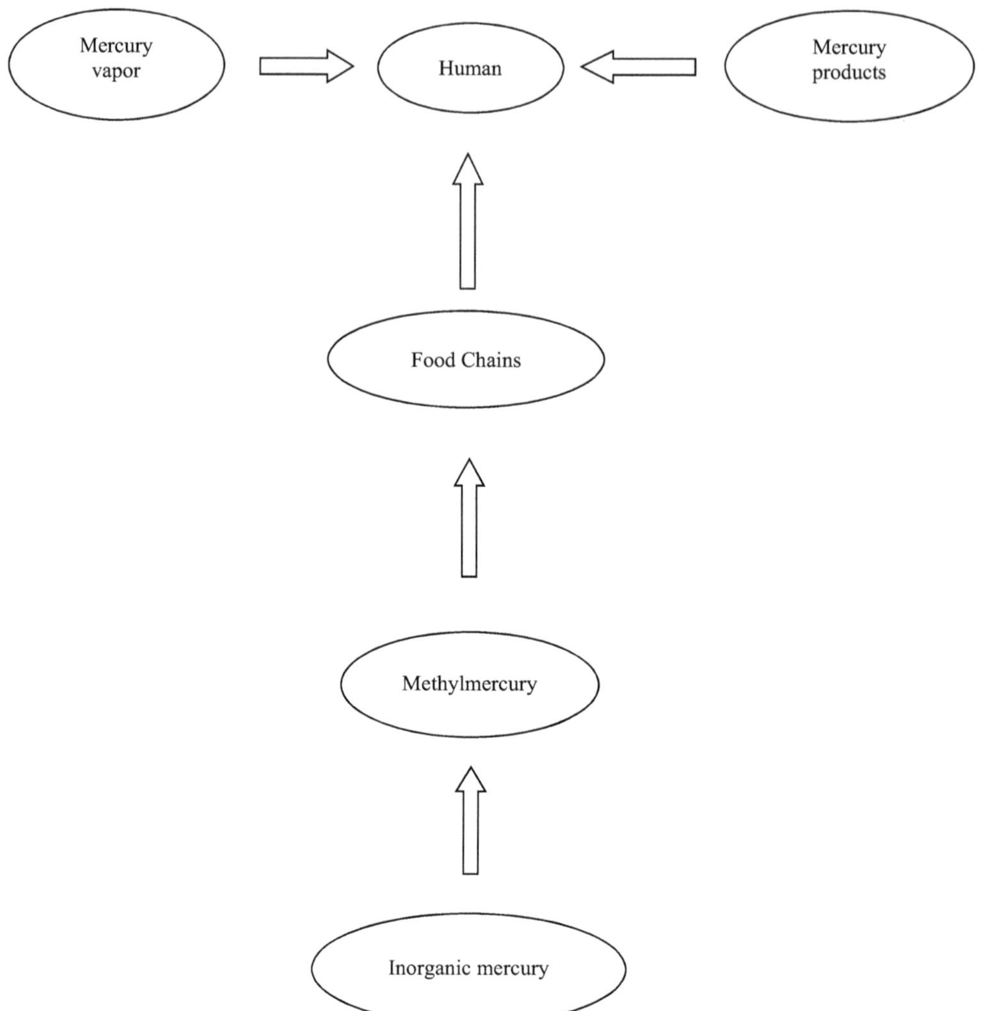

FIGURE 14.1 Routes of mercury entry in the human beings.

According to bioaccumulation prospects, mercury enters the top of the food chain via aquatic bacteria in the form of methylmercury, leading to a neurological condition, "Minamata disease" (16–18). Humans get mercury in different ways that are mentioned in Figure 14.1.

Pollution created by mercury affects every part of the atmosphere. It occurs in the atmosphere in three forms: gaseous elemental (GEM), reactive gaseous or oxidized (RGM), and particulate bound (PBM). The sum of GEM and RGM is indicated as total gaseous Hg (TGM) (19). GEM, the predominant form of TGM in the atmosphere is gaseous elemental mercury with >95% (20), which is characterized by high bioavailability, stability, and volatility; among the three atmospheric forms, and has been experimentally identified with spectroscopic methods (21). The reactive gaseous or oxidized form of mercury is troublesome because it includes a large variety of gaseous inorganic (i.e., $HgCl_2$ and HgO) and organic compounds with sufficiently high vapor pressure to exist as gas-phase (22). The third form of particulate-bound mercury contains all airborne particulate containing Hg, including both stable condensed and gaseous forms adsorbed on atmospheric particulate matter (PM). It is estimated through sampling by pulling air through a glass fiber or a quartz filter (23), which is later acid-digested and analyzed for Hg. All those particles with a diameter

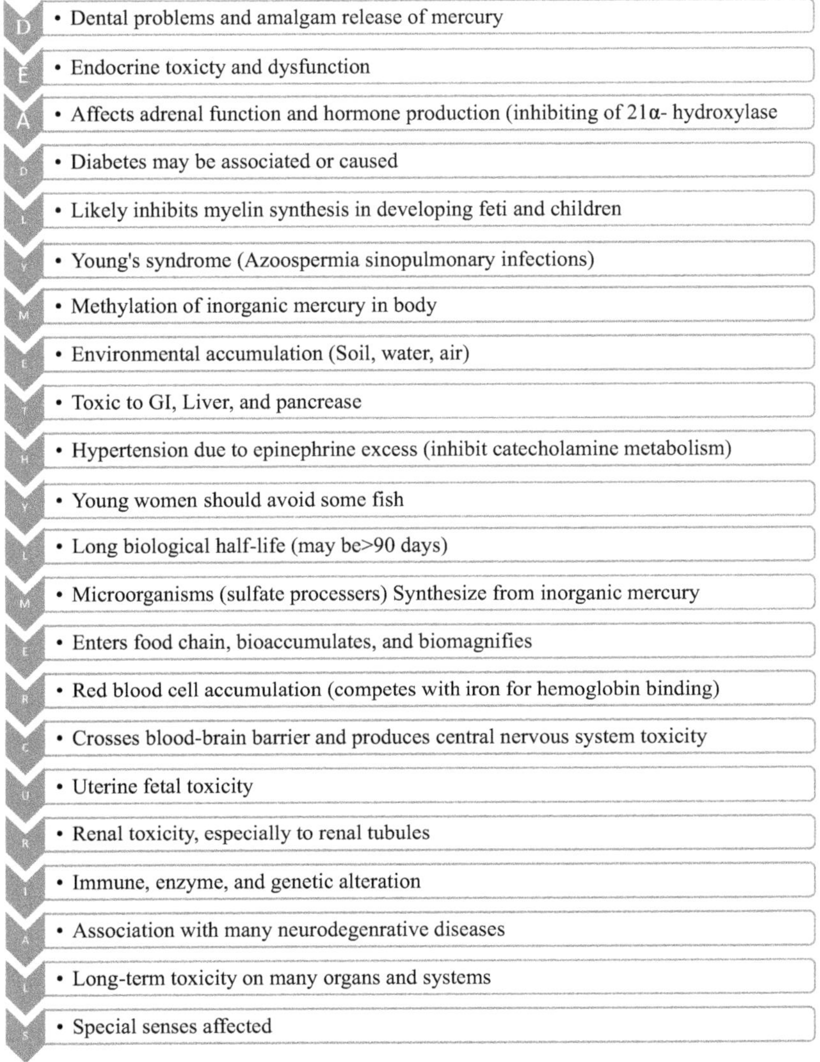

FIGURE 14.2 Elaboration of METHYLMERCURICALS letters.

<2.5 μm are often known as PBM, even though, the characterization only depends on the pore size of the filter used for its collection (24).

Methyl mercury is also known as Deadly Methylmercurials. The explanation of the said word is coated below in Figure 14.2.

14.2 SOURCES OF WATER POLLUTION BY MERCURY

14.2.1 Natural Sources

The major part of environmental mercury is introduced by natural sources (25,26).

14.2.2 Anthropogenic Sources

Heavy metals accumulate in water depending on the type of industries in the region, people's way of life, and careless disposal of wastes (27–29). Human activities in urban areas like sewage water, include household effluents, drainage water, business effluents (e.g., car washes and dental uses), atmospheric deposition, and traffic-related emissions (vehicle exhaust, brake linings, tires, asphalt wear, gasoline/oil leakage, etc.) are playing a major role in water pollution (30,31).

14.3　EFFECT OF MERCURY

14.3.1　DESCRIPTION

The human body has no significant physiological use of mercury (32) even though it has a toxic effect in low concentrations (micrograms per liter) when it is in inorganic form, but methyl mercury present in animal tissues (95% of mercury in fish is methylated) is a cause for concern (33). The reason for the accumulation of methylmercury in tissues is its soluble nature in lipids. It can damage the nervous system, and chromosomal and also cause abnormalities in the fetus even at nanograms per liter (34). The impact of small exposure to mercury is still controversial according to clinical aspects.

14.3.2　MERCURY TYPES AND ITS DRASTIC EFFECTS

Different types of mercury present in the environment include inorganic mercury (metallic mercury, mercury vapor ($Hg0$), mercurous (Hg^{2++}) or mercuric (Hg^{++})) salts, and organic mercury (compounds in which mercury is bonded to a structure containing carbon atoms which are methyl, ethyl, phenyl, or similar groups). The behavior as well as the clinical significance of the divergent form of mercury change with chemical structure (35). The toxicity effect on humans of this metal depends upon the dose and rate of exposure. Mercury vapor majorly attacks the brain. Mercurous and mercuric salts destroy the gut lining and kidney, whereas methyl mercury is widely distributed throughout the body. Exposure in large amounts causes severe pneumonitis, which in extreme cases can be fatal (14). All forms of mercury act as a poison in all forms. It alters the tertiary and quaternary structure of proteins by binding with sulfhydryl and selenohydryl groups, which affect cellular function.

14.3.3　ORGANS EFFECTED BY MERCURY

In addition to its chief target organ (brain), mercury vapor affects peripheral nerve function, renal function, immune function, endocrine and muscle function, several types of dermatitis (36), erosive bronchitis, and bronchiolitis potentially leading to respiratory failure may be accompanied by CNS symptoms such as tremor or erethism (37). At low concentrations of mercury, nonspecific symptoms produced are weakness, fatigue, anorexia, weight loss, and gastrointestinal disturbance (38) on the other hand, higher exposure levels are associated with mercurial tremor: fine muscle fasciculations punctuated every few minutes by coarse shaking, Erethism: severe behavior and personality changes, emotional excitability, loss of memory, insomnia, depression, fatigue, and in severe cases delirium and hallucination (39), Gingivitis and copious salivation (14) and persistent neurological symptoms are common (40). Mercuric salts (typically $HgCl_2$) generally damage the gastrointestinal tract and the kidneys, which results in extensive precipitation of enterocyte proteins, with abdominal pain, vomiting, and bloody diarrhea with potential necrosis of the gut mucosa. Death is also the consequence of these salts by peritonitis, septic or hypovolemic shock. However, surviving patients suffer from renal tubular necrosis with anuria. These salts rarely cause chronic poisoning. (40). Immune dysfunctions include asthma dermatitis, various types of autoimmunity (41–43), suppression of natural killer cells (44–47), and disruption of various other lymphocyte subpopulations (14). Next in the list of toxicity effects include thyroid dysfunction (which seems associated with inhibition of the 5 deiodinases, with decreased free T3 and increased reverse T3 (48), inhibition of spermatogenesis (due to accumulation in the testicles (49), Atrophy and capillary damage in thigh muscle (50).

Moreover, it is believed that mercury interferes with DNA transcription and protein synthesis (51) with the destruction of the endoplasmic reticulum and the disappearance of ribosomes (52,53). Moving further, the association of mercury with sulfhydryl groups of the mitochondrial oxidative phosphorylation complex destroys mitochondrial membranes, which may contribute to chronic fatigue syndrome (54). Mercury also acts as a causative factor in mononucleosis and is involved in leukemia, and Hodgkin's disease (55–57). In the adrenal cortex, mercury exposure is associated with lowered plasma levels of corticosterone (58), which results in adrenal hyperplasia. In the anterior pituitary, it affects the estrogen and progesterone levels, leading to ovarian dysfunction, painful or irregular menstruation, premature menopause, tipped uterus (59), and menstrual disorders including abnormal bleeding, short and long periods (60), whereas, in males, mercury can have adverse effects on spermatogenesis (61), epididymal sperm count, and testicular weight, erectile dysfunction (62). In the nervous system, the most devastating effect of mercury on nerve cells is to either die or live in a state of chronic malnutrition through interference with energy production, which is used to impair cellular detoxification processes (63). In addition to this, inorganic mercury is absorbed by the nerve terminal which can impair the synthesis of tubulin and actin in the peripheral nervous system (64). Primary sensory neuropathy is a hallmark of MeHg poisoning. The endocrine system of animals and mortals can even be affected by a low-level exposure to mercury which results in the disruption of the pituitary, thyroid, adrenal glands, and pancreas (65). The effects of mercury on the human body are described in Table 14.1.

TABLE 14.1
Mercury-Associated Disease and Affected Organ

S.No.	Effected Body Organ or System	Disease	Reference
1.	Cellular Effects	DNA damage, alteration in membrane permeability, mitochondrial dysfunction, alterations in calcium homeostasis and increased lipid peroxidation, alterations in calcium homeostasis and increased lipid peroxidation	Naganuma et al., 2005; Wang and Jia, 2005; Flora et al., 2008; Lund et al., 1993; Peraza et al., 1998
2.	Cardiovascular Effects	chest pain or angina, inhibit the cardioprotective activity of paraoxonase 1	Frustaci et al., 1999; Drescher et al., 2014
3.	Hematological	anemia including hemolytic anemia, aplastic anemia, impaired hemoglobin formation, mononucleosis, and involved in leukemia, and Hodgkin's disease	Kinjo et al., 1996; Robinson, 1952; Flanders, 1992
4.	Pulmonary Effects	Young's syndrome, bronchitis, and pulmonary fibrosis	Hendry et al., 1993; Tchounwou et al, 2003; Haddad and Stenberg, 1963
5.	Digestive And Renal Systems	inhibit the production of the digestive trypsin, chymotrypsin, and pepsin along with the function of xanthine oxidase and dipeptidyl peptidase IV, abdominal pain, indigestion, inflammatory bowel disease, ulcers, bloody diarrhea, destruction of intestinal flora, reduced resistance to pathogenic infection, e kidney damage, acute tubular necrosis, glomerulonephritis, chronic renal disease, renal cancer and nephrotic syndrome, kidney injuries including subacute-onset nephrotic syndrome, tubular dysfunction, secondary focal segmental glomerulosclerosis, syncreticatic nephrotic syndrome, nephritic syndrome, nephrotic-range proteinuria, glomerular disease, and membranous glomerulonephritis	Tchounwou et al., 2003; Vojdani et al, 2003; Summers et al., 1993; Li et al., 2010; Park and Zheng, 2012; Oliveira et al., 1987

(*Continued*)

TABLE 14.1 (*Continued*)
Mercury-Associated Disease and Affected Organ

S.No.	Effected Body Organ or System	Disease	Reference
6	Immune System	inhibiting the ability to destroy foreign substances, allergic disease, amyotrophic lateral sclerosis, arthritis, autoimmune thyroiditis, autism/attention deficit hyperactivity disorder, eczema, epilepsy, psoriasis, multiple sclerosis, rheumatoid arthritis, schizophrenia, scleroderma, and systemic lupus erythematosus	Wada et al., 2009; Gardner, 2010; Warren 1989; Singh, 2009; Schofield, 2005; Johnson and Atchison, 2009; Hybenova et al, 2010; Landrigan et al., 2010
7	Nervous System	impair cellular detoxification processes, neuronal problems through blockage of the P-450 enzymatic process, increased tissue oxidative damage, impair the synthesis of tubulin and actin, damage to the blood-brain barrier, depression, paranoia, extreme irritability, hallucinations, an inability to concentrate, memory loss, tremors of the hands, head, lips, tongue, jaw, and eyelids, weight loss, perpetually low body temperature, drowsiness, headaches, insomnia, and fatigue, Autism	Kazantzis, 2002; Frustaci et al,1999; Solt and Bornstein, 2010; Bhardwaj et al., 2009
8	Endocrine System	disruption of the pituitary, thyroid, adrenal glands, and pancreas, inhibits catecholamine degradation, which results in hypersalivation and hypertension, adrenal hyperplasia, Addison's disease, hypothyroidism, thyroid inflammation and depression, dysregulation of blood glucose levels	Wada et al., 2009; McGregor and Mason, 1991; Chen et al., 2006
9	Reproductive System	altering the circulating of levels of follicle-stimulating hormone (FSH), luteinizing hormone (LH), inhibin, estrogen, progesterone, and the androgens, Reduced fertility, epididymal sperm count, testicular weight, erectile dysfunction, menstrual disorders including abnormal bleeding, short, long, irregular cycles, and painful period	Davis et al., 2001; Schrag and Dixon, 1985; Rowland, 1994; Colquitt, 1995
10	Fetotoxicity	miscarriage, spontaneous abortions, stillbirth, low birth weights, neural tube defects, craniofacial malformations, delayed growth, inhibit fetal brain development	Yoshida, 2002; Castoldi et al., 2001; Myers and Davidson, 1998

14.4　SAMPLING AND ANALYSIS METHODS

14.4.1　DESCRIPTION

There are different methods available for the measurement of mercury pollutants, but the art. 225 ("Specific measures of protection and prevention") obligate the employer to "measure the pollutants because it may present a risk to workers' health, using standardized methods reported in an indicative list in Annex XLI". This list includes nine European Standards methods implemented by the Italian legislation, specifically, UNI EN 481, 482, 689, 838, 1076, 1231, 1232, 1540, and 12919. These standards deal with the sampling methods and operational requirements for the assessment of the pollutants in the workplace atmosphere. Analytical Methods database (66) on hazardous substances by IFA (Information System on Hazardous Substances of the German Social Accident Insurance), an online information system whose contents are continuously updated after the publication of new regulations. A list of internationally validated methods for the detection of mercury and mercury-associated compounds is presented in Table 14.2.

TABLE 14.2

Method and Associated Principal Published in Analytical Methods Database

S.No	Source Method	Principal of Method
1	ISO 17733—Determination of mercury and inorganic mercury compounds	Pumped sampling: Mercury vapor trapped using a sorbent tube containing Hydrar (with a QF prefilter for particulate mercury, if present). Diffusive sampling: Mercury vapor was sampled using a badge containing Hydrar. Dissolution with HNO_3 and HCl at room temperature. Analysis by CVAAS after reduction using $SnCl_2$
2	MDHS 16/2—Mercury and its inorganic divalent compounds in air	Pumped sampling: Mercury vapor trapped using a sorbent tube containing Hydrar (with a QF prefilter for particulate mercury, if present). Diffusive sampling: Mercury vapor was sampled using a badge containing Hydrar. Dissolution with HNO_3 and HCl at room temperature. Analysis by CVAAS after reduction using $SnCl_2$
3	MetroPol 024—Mercure sur tubes de charbon actif	Mercury vapor is trapped using a sorbent tube containing activated charcoal. Thermal desorption at 500°C under N_2 into acidic $KMnO_4$ solution; then treatment of an aliquot by addition of saturated $KMnO_4$ solution, $1+1$ HNO_3, $1+1$ H_2SO_4 15 g/L $NH_2OH.HCl$. Analysis by CVAAS after reduction using $SnCl_2$
4	MetroPol 079—Mercure sur tubes Hydrar®	Mercury vapor is trapped using a sorbent tube containing hopcalite (with a QF prefilter for particulate mercury, if present). Ultrasonic dissolution with HNO_3 and HCl at room temperature. Analysis by CVAAS after reduction using $SnCl_2$
5	BIA 8530—Quecksilber	Mercury vapor is trapped using a sorbent tube containing hopcalite (with a QF prefilter for particulate mercury, if present). Ultrasonic dissolution with HNO_3 and HCl at room temperature. Analysis by CVAAS after reduction using $SnCl_2$
6	BIA 8530—Quecksilber	Mercury vapor is trapped using a sorbent tube containing two sections of hopcalite separated by a QF filter. Dissolution with HNO_3 and HCl at 80°C (1st sorbent section and QF filter treated together; 2nd backup sorbent section treated separately). Analysis by CVAFS after reduction using $SnCl_2$
7	NIOSH 6009—Mercury	Mercury vapor is trapped using a sorbent tube containing hopcalite. Dissolution with HNO_3 and HCl at room temperature. Analysis by CVAFS after reduction using $SnCl_2$
8	OSHA ID 140—Mercury vapor in workplace atmospheres	Pumped sampling: Mercury vapor is trapped using a sorbent tube containing Hydrar or hopcalite. Diffusive sampling: Mercury vapor was sampled using a badge containing Hydrar or hopcalite. Dissolution with HNO_3 and HCl at room temperature. Analysis by CVAAS after reduction using $SnCl_2$
9	OSHA ID 145—Particulate mercury in workplace atmospheres	Particulates trapped on a cellulose ester membrane filter in a 37-mm cassette filter holder. Dissolution with HNO_3, 5M H_2SO_4, 5% $KMnO_4$, and $NH_2OH.HCl$. Analysis by CVAAS after reduction using $SnCl_2$

14.5 TREATMENTS FOR REMOVAL OF MERCURY

14.5.1 DESCRIPTION

The main important factors responsible for heavy metal removal are the metal species and concentration, the reactivity of the available biopolymers or biomass, and the composition of other wastewater components (28,67,68). The physicochemical and biological processes leading to the removal and/or sequestration of Hg are bio-remedial strategies proposed to alleviate systems polluted with Hg (69–71). Pitfalls of non-biological approaches to Hg removal (70) make a microbial-based technology for the detoxification of Hg in polluted systems a better remediation option.

Biological and physico-chemical conditions are responsible not only for the determination of heavy metal removal in biological wastewater treatment but also for the process operating conditions and design. However, despite these conditioning factors, various studied metals can be removed in biological wastewater treatment processes. Moreover, during wastewater treatment sludge is also produced (about 1% of the wastewater treated). Since metals present in the untreated wastewater stream become concentrated in sludge, disposal of heavy metal-laden sludge means an environmental hazard.

14.5.2 Physical/Chemical/Biological Methods

There are various physical and chemical methods have been used to treat mercury in wastewater.

14.5.2.1 Physical method

14.5.2.1.1 Sedimentation

Sedimentation refers to the process during which particles suspended in water tend to settle out.

14.5.2.1.2 Screening

Screening is a wastewater pre-treatment that prevents coarse solids, such as plastics, rags, and other trash, from entering a sewage system or treatment plant. Solids get trapped by inclined screens or barracks.

14.5.2.1.3 Filtration

Water filtration is the process of removing or reducing the concentration of PM.

14.5.2.1.4 Membrane Separation

This method uses the membrane's micropores to filter and makes use of the membrane's selective permeability to separate certain substances.

14.5.2.2 Chemical Methods

14.5.2.2.1 Chemical Precipitation

Chemical precipitation is a pH adjustment process mainly used to remove dissolved metals from aqueous wastes.

14.5.2.2.2 Adsorption

Adsorption is a process that involves the accumulation of a substance in molecular species in higher concentrations on the surface. Adsorption is the adhesion of atoms, ions, or molecules from a gas, liquid, or dissolved solid to a surface.

14.5.2.2.3 Coagulation

Coagulation is a somewhat simple chemical process that involves bringing insoluble materials together by manipulating the charges of particles by adding iron or aluminum salts, such as aluminum sulfate or ferric sulfate, to a wastewater stream.

14.5.2.2.4 Ion exchange

Ion exchange is a reversible chemical water treatment process for the selective removal of charged inorganic species from water using an ion-specific resin and replacing them with other similarly charged ions.

14.5.2.2.5 Neutralization

Neutralization of wastewater is required before releasing it into water bodies or reuse it in various technological processes. Wastewater containing acids or alkalis must be neutralized. Water with a pH of 6.5–8.5 is considered neutral. Neutralization is required where pH is below 6.5 and above 8.5.

14.5.2.2.6 *Solvent Extraction*

The process of isolating an organic compound from its aqueous solution by shaking it with solvent is known as solvent extraction.

14.6 BIOLOGICAL TREATMENTS

14.6.1 DESCRIPTION

From ancient times, mercury accumulated in habitats at a high level (72). So, the microbial detoxification mechanism is thought to be a very old one (73). Instead of the toxicity of Hg, some microorganisms exhibit resistance to this metal. Microorganisms including various genera of bacteria and fungi have been reported to tolerate Hg (69,74–76,91). Several reports have articulated the use of microorganisms to efficiently remove Hg from contaminated matrixes via biological reduction (74 and 75), biosorption (69,71), bioaccumulation (71), and bio sequestration (69). Bacterial strains that harbor merA genes are known to reduce Hg^{2+} to volatile $Hg0$ (74–76). Thereby, removing Hg from the local environment is achieved by using mercury-resistant microorganisms.

There are different ways through which mercury is treated from wastewater by microorganisms. These are given as follows.

14.6.2 BIOSORPTION

Biosorption can be defined as the ability of biological materials to accumulate heavy metals from wastewater through metabolically mediated or Physio-chemical pathways of uptake. In yeast cells, as the diagram shows, there are four different ways through which metal is accumulated on the membrane of the microbial cells. These are

14.6.2.1 Adsorption

Adsorption is an extracorporeal process in which molecules or metals dissolved in a liquid bind to the membrane structure.

14.6.2.2 Complexation

The process of making an atom or compound forms a complex with another.

14.6.2.3 Ion Exchange

It is a reversible Interchange of one kind of ion present in an insoluble solid with another of like charge present in a solution surrounding

14.6.2.4 Precipitation

The deposition of minerals from solution by the agency of an organism.

Yeast cells are also known to be mediocre due to metal biosorption (77). *Yarrowia* species showed resistance to toxic concentrations of several heavy metals (78,79). The study done by Oyetibu in 2014 proved the same scenario as the one above (80). He used the resting cell and growing cells of yeast (*Yarrowia* species) to remove mercury. Results of this experiment indicate that resting cells of strain Idd2 exhibited a higher maximum Hg^{2+} removal capacity (59 mg mercury per g dry cell weight (gdw–1) by adsorption than those of resting cells of strain Idd1 (32 mg gdw–1). The resting cells of strain Idd2 exhibited a higher Hg^{2+} desorption capacity using $CaCl_2$ (68%) and EDTA (48%) than strain Idd1, depicting weaker binding of Hg^{2+} onto strain Idd2, unlike strain Idd1. The actively growing yeast cells showed opposite Hg removal characteristics to those of the resting cells. Strain Idd1 adsorbed less Hg^{2+} from the culture medium supplemented with Hg^{2+} than strain Idd2. However, the growing strain Idd1 reduced and vaporized 27% of supplemented Hg^{2+} as metallic mercury ($Hg0$), while the growing strain Idd2 vaporized 15% of the supplemented Hg^{2+}. These two yeast strains are potential biotechnological tools for the eventual bioremediation of polluted aquatic systems (Figure 14.3).

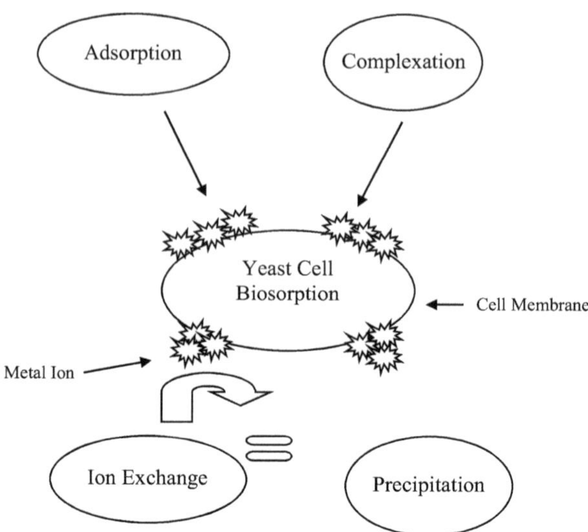

FIGURE 14.3   Four different ways of biosorption by yeast cells.

14.6.3 ENZYMATIC REDUCTION

Mercury-resistant bacteria are universal. The bacterial cytoplasmic enzyme mercuric reductase, encoded by the merA gene, reduces soluble Hg (II) to insoluble metallic Hg (0) (81,82). Mercury reduced by the bacteria can be eluted from microbial biofilm produced because it accumulates in the form of small droplets (83). Based on this property, Wagner-Döbler used seven strains of mercury mutant *Pseudomonas* to remove mercury from chemical wastewater. The strains he used were four subspecies of *Pseudomonas putida* (Spi3, Spi4, Kon12, Elb2), two subspecies of *Pseudomonas stutzeri* (Ibu¨3, Ibu¨8), and one subspecies of *Pseudomonas fulva* (Spi11), which were immobilized on carrier material inside a 700 L packed bed bioreactor. After that, mercury with neutralized chloralkali was continuously fed into the bioreactor. The outcome he observed was that a mercury retention efficiency of 97% was obtained within 10 hours of inoculation of the bioreactor (84).

14.6.4 SEQUENCING BATCH REACTOR (SBR)

Sequencing batch reactors (SBRs) or sequential batch reactors are used for the treatment of wastewater and this is the type of activated sludge process for the treatment of wastewater. Organic matter is reduced by the bubbling of oxygen measured as biochemical oxygen demand (BOD) and chemical oxygen demand (COD). The treated effluent may be suitable for discharge to surface waters or possibly for use on land.

Malakahmad and co-workers have treated synthetic petrochemical wastewater containing mercury and cadmium with a lab-scale SBR using the above method. After acclimatization of the system which lasted 60 days, 76%–90% of Hg^{2+} and 96%–98% of Cd^{2+} were removed (85).

14.6.5 ACTIVATED SLUDGE TREATMENT

Activated sludge treatment is done in a bioreactor with the help of microorganisms. This treatment is divided into two essential parts. The first is a biological reactor, and the second is a phase separator (86). Initially, in the reactor, large populations of microorganisms grow under aerobic conditions in liquor, which oxidize the dissolved or suspended organic compounds present in wastewater and convert them into carbon dioxide, water, and cellular material. The bacteria and other microorganisms form aggregates or flocs, a process known as flocculation. The mixed liquor then passes into

Influent ⟹ **Retention tank** ⟹ **SRB Phases** ⟹ **Effluent**

FIGURE 14.4 Flow diagram of SBR system.

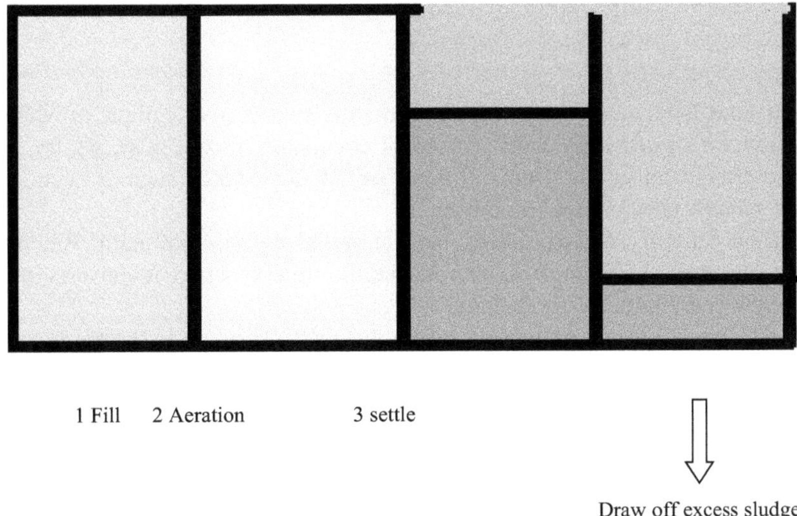

FIGURE 14.5 SBR phases.

the phase separator or settling tank. The flocs are separated by sedimentation from the liquid phase and are returned to the reactor (Figure 14.4).

The supernatant (final effluent) is then discharged to a watercourse. Any substance that is adsorbed or absorbed by the bacterial flocs is removed from the water. SBR phases include four phases in a cycle (Figure 14.5):

1. Fill
2. Aeration
3. Settle
4. Draw off excess sludge

passing through a sewage treatment plant. A combination of flocculation and settling is therefore the mechanism by which metal removal is achieved in the activated sludge process. Many industries used this method to treat their effluents composed of mercury (87–89).

14.6.6 Constructed Wetlands (by Using Natural Bioremediation)

Three integrated constructed wetlands (CW), considered efficient and cost-effective natural bioremediation systems, were used by Fazal, Zhang, and Mehmood for the removal of contaminants from wastewater. Of these three different wetlands, one was planted with water hyacinth, one with sludge (anaerobic wetland), and the third acted as the control. The wetland with water hyacinth was efficient in reducing EC (74.9%), turbidity (92.5%), COD (83.7%), TSS (91.8%), TDS (62.3%), TS (80%), nitrates (67.5%), ammonia (71.6%), phosphates (90.2%), Cd (97.5%), Ni (95.1%), Hg (99.9%) and Pb (83.4%). The anaerobic sludge was efficient in reducing EC (61.5%), turbidity (80.0%), COD (76.35%), TSS (91.9%), TDS (78.4%), TS (80%), nitrates (79%), phosphates (59.6%), ammonia (51.7%), Cd (100%), Ni (92.2%), Hg (100%) and Pb (66.75%). CW with water hyacinth was more efficient in reducing EC, turbidity, COD, ammonia, phosphates, Ni, and lead than CW with sludge, whereas sludge was more efficient in removing nitrates, Cd, and Hg as compared to water hyacinth (90).

14.7 CONCLUSION

There are many ways to treat mercury-polluted water, including physical, chemical, and biological methods. The efficient way is biological methods. There are different biological methods used by many researchers and got satisfactory removal of mercury from water.

GLOSSARY

Biological treatment Biological wastewater treatment is a modern technique in which wastewater is treated with microorganisms instead of chemicals. In this way, we try to prevent the adverse effects caused by chemical treatment of wastewater, such as chemical accumulation in water bodies or algal blooming.

Mercury pollution Mercury is a persistent, bioaccumulative, toxic pollutant. When released into the environment, it accumulates in water-laid sediments where it converts into toxic methylmercury and enters the food chain.

Metal pollution Being persistent pollutants, heavy metals accumulate in the environment and consequently contaminate the food chains. Accumulation of potentially toxic heavy metals in biota causes a potential health threat to their consumers including humans.

Harmful effects of mercury on the environment When released into the environment, it accumulates in water-laid sediments where it converts into toxic methylmercury and enters the food chain.

Bioaccumulation Bioaccumulation is a process of accumulation of chemicals in an organism that takes place if the rate of intake exceeds the rate of excretion.

Hodgkin's disease Hodgkin lymphoma (once known as Hodgkin's disease) is a group of blood cancers that affects your lymphatic system. These blood cancers start in your lymphocytes. Lymphocytes are white blood cells that are part of your lymphatic system. Every year, more people are living longer after treatment.

Cellular detoxification processes Cellular detoxification utilizes three phases and five principles to be effective. The five R's are removing the source, regenerating the cell membrane, restoring cellular energy, reducing cellular inflammation, and re-establishing methylation.

Enzymatic Reduction Enzymatic reduction can be a valid alternative because the selectivity displayed by enzymes is usually greater. In addition, the possibility of using enzymes as catalysts adds an interesting environmental advantage to the preparation.

Sequencing Batch Reactor The sequencing batch reactor (SBR) is a fill-and-draw activated sludge system for wastewater treatment. In this system, wastewater is added to a single "batch" reactor, treated to remove undesirable components, and then discharged.

REFERENCES

1. Berry, M. J., & Ralston, N. V. Mercury toxicity and the mitigating role of selenium. *EcoHealth*, 5(4), 456–459 (2008).
2. Momidi, B. K., Tekuri, V., & Trivedi, D. R. Selective detection of mercury ions using benzothiazole based colorimetric chemosensor. *Inorganic Chemistry Communications*, 74, 1–5 (2016).
3. Yu, Y., Lin, L. R., Yang, K. B., Zhong, X., Huang, R. B., & Zheng, L. S. p-Dimethylaminobenzaldehyde thiosemicarbazone: A simple novel selective and sensitive fluorescent sensor for mercury (II) in aqueous solution. *Talanta*, 69(1), 103–106 (2006).
4. Annar, S. The characteristics, toxicity and effects of heavy metals arsenic, mercury and cadmium: A review. *International Journal of Multidisciplinary Educational Research*, 11, 35–43 (2022).
5. Kollur, S. P., Shivamallu, C., Prasad, S. K., Veerapur, R., Patil, S. S., Cull, C. A., Coetzee, J.F., & Amachawadi, R. G. Recent advances on the development of chemosensors for the detection of mercury toxicity: A review. *Separations*, 8(10), 192 (2021).
6. Levin, L. Mercury in the environment. *Mercury Control: for Coal-Derived Gas Streams*, 1–12 (2014).
7. Lin, L. F., Lee, W. J., Li, H. W., Wang, M. S., & Chang-Chien, G. P. Characterization and inventory of PCDD/F emissions from coal-fired power plants and other sources in Taiwan. *Chemosphere*, 68(9), 1642–1649 (2007).

8. Kim, K. H., Kabir, E., & Jahan, S. A. A review on the distribution of Hg in the environment and its human health impacts. *Journal of Hazardous Materials, 306*, 376–385 (2016).

9. Clarkson, T. The toxicology of mercury. *Critical Reviews in Clinical Laboratory Sciences, 34*(4), 369–403 (1997).

10. Goldman, L. R., Shannon, M. W., & Committee on Environmental Health. Technical report: Mercury in the environment: Implications for pediatricians. *Pediatrics, 108*(1), 197–205 (2001).

11. Torres, F. G., & De-la-Torre, G. E. Mercury pollution in Peru: Geographic distribution, health hazards, and sustainable removal technologies. *Environmental Science and Pollution Research, 29*(36), 54045–54059 (2022).

12. Nemova, N. N., Kantserova, N. P., & Lysenko, L. A. The traits of protein metabolism in the skeletal muscle of teleost fish. *Journal of Evolutionary Biochemistry and Physiology, 57*(3), 626–645 (2021).

13. Sigel, A., & Sigel, H. (Eds.). *Metal Ions in Biological Systems, Mercury and Its Effects on ENVIRONMENT and biology,* vol. 34. Dekker, New York (1997).

14. Berlin, M., & Zalups, R. K. Mercury. In G. F. Nordberg, B. A. Fowler, M. Nordberg and L. T. Friberg (Eds.), *Handbook on the Toxicology of Metals,* 3rd edn. Elsevier, New York (2007).

15. Strasdeit, H., von Döllen, A., Saak, W., & Wilhelm, M. Intracellular degradation of diorganomercury compounds by biological thiols—insights from model reactions. *Angewandte Chemie International Edition, 39*(4), 784–786 (2000).

16. Liu, D., Qu, W., Chen, W., Zhang, W., Wang, Z., & Jiang, X. Highly sensitive, colorimetric detection of mercury (II) in aqueous media by quaternary ammonium group-capped gold nanoparticles at room temperature. *Analytical Chemistry, 82*(23), 9606–9610 (2010).

17. Xu, Y., Deng, L., Wang, H., Ouyang, X., Zheng, J., Li, J., & Yang, R. Metal-induced aggregation of mononucleotides-stabilized gold nanoparticles: An efficient approach for simple and rapid colorimetric detection of Hg (II). *Chemical Communications, 47*(21), 6039–6041 (2011).

18. Wu, J., Li, L., Zhu, D., He, P., Fang, Y., & Cheng, G. Colorimetric assay for mercury (II) based on mercury-specific deoxyribonucleic acid-functionalized gold nanoparticles. *Analytica Chimica Acta, 694*, 115–119 (2011).

19. Fu, X. W., Feng, X., Shang, L. H., Wang, S. F., & Zhang, H. Two years of measurements of atmospheric total gaseous mercury (TGM) at a remote site in Mt. Changbai area, Northeastern China. *Atmospheric Chemistry and Physics, 12*(9), 4215–4226 (2012).

20. Fitzgerald, W. F., Engstrom, D. R., Lamborg, C. H., Tseng, C. M., Balcom, P. H., & Hammerschmidt, C. R. Modern and historic atmospheric mercury fluxes in northern Alaska: Global sources and Arctic depletion. *Environmental Science & Technology, 39*(2), 557–568 (2005).

21. Edner, H., Faris, G. W., Sunesson, A., & Svanberg, S. Atmospheric atomic mercury monitoring using differential absorption lidar techniques. *Applied Optics, 28*(5), 921–930 (1989).

22. Munthe, J., Wängberg, I., Iverfeldt, Å., Lindqvist, O., Strömberg, D., Sommar, J., Gårdfeldt, K., Petersen, G., Ebinghaus, R., Prestbo, E., & Siemens, V. Distribution of atmospheric mercury species in Northern Europe: Final results from the MOE project. *Atmospheric Environment, 37*, 9–20 (2003).

23. Lynam, M. M., & Keeler, G. J. Comparison of methods for particulate phase mercury analysis: Sampling and analysis. *Analytical and Bioanalytical Chemistry, 374*(6), 1009–1014 (2002).

24. Bank, M. S. (Ed.). *Mercury in the Environment: Pattern and Process.* University of California Press, Berkeley, CA (2012).

25. Tomiyasu, T., Matsuyama, A., Eguchi, T., Fuchigami, Y., Oki, K., Horvat, M., Rajar, R. and Akagi, H. Spatial variations of mercury in sediment of Minamata Bay, Japan. *Science of the Total Environment, 368*(1), 283–290 (2006).

26. Tomiyasu, T., Matsuyama, A., Eguchi, T., Marumoto, K., Oki, K., & Akagi, H. Speciation of mercury in water at the bottom of Minamata Bay, Japan. *Marine Chemistry, 112*(1–2), 102–106 (2008).

27. Cheng, S. Heavy metal pollution in China: Origin, pattern and control. *Environmental Science and Pollution Research, 10*(3), 192–198 (2003).

28. Chipasa, K. B. Accumulation and fate of selected heavy metals in a biological wastewater treatment system. *Waste Management, 23*(2), 135–143 (2003).

29. Wang, C., Hu, X., Chen, M. L., & Wu, Y. H. Total concentrations and fractions of Cd, Cr, Pb, Cu, Ni, and Zn in sewage sludge from municipal and industrial wastewater treatment plants. *Journal of Hazardous Materials, 119*(1–3), 245–249 (2005).

30. United States. Environmental Protection Agency. Office of Water Regulations, & Standards. *Report to Congress on the Discharge of Hazardous Wastes to Publicly Owned Treatment Works (the Domestic Sewage Study).* US Environmental Protection Agency, Office of Water Regulations and Standards (1986).

566   Control of Heavy Metals in the Environment

31. Sörme, L., & Lagerkvist, R. J. S. O. T. T. E. Sources of heavy metals in urban wastewater in Stockholm. *Science of the Total Environment*, 298(1–3), 131–145 (2002).

32. Sears, M. E., Kerr, K. J., & Bray, R. I. Arsenic, cadmium, lead, and mercury in sweat: A systematic review. *Journal of Environmental and Public Health*, 2012(1), 184745 (2012).

33. McEntire, F. E., & Neufeld, R. D. Microbial methylation of mercury: A survey. Water Pollution Control, 74(4), 465–470 (1975).

34. Hoffbrand, A. V. *Postgraduate Hematology*, John Wiley & Sons, (2015).

35. Bernhoft, R. A. Mercury toxicity and treatment: A review of the literature. *Journal of Environmental and Public Health*, 2012(1), 460508 (2012).

36. Berlin, M. *Dental Materials and Health.* Statens Offentliga Utredningar, Stockholm (2003).

37. Garnier, R., Fuster, J. M., Conso, F., Dautzenberg, B., Sors, C., & Fournier, E. Acute mercury vapour poisoning (author's transl). *Toxicological European research. Recherche europeenne en toxicologie*, 3(2), 77–86 (1981).

38. Friberg, L., & Nordberg, G. F. (Eds.). *Mercury in the Environment.* CRC Press, Boca Raton, FL (1972).

39. Berglund, A., Pohl, L., Olsson, S., & Bergman, M. Determination of the rate of release of intra-oral mercury vapor from amalgam. *Journal of Dental Research*, 67(9), 1235–1242 (1988).

40. Baldi, G., Vigliani, E. C., & Zurlo, N. Mercury poisoning in hat industry. *La Medicina del Lavoro*, 44(4), 161–198 (1953).

41. De Vos, G., Abotaga, S., Liao, Z., Jerschow, E., & Rosenstreich, D. Selective effect of mercury on Th2-type cytokine production in humans. *Immunopharmacology and Immunotoxicology*, 29(3–4), 537–548 (2007).

42. Tanigawa, T., Takehashf, H., & Nakata, A. Naïve (CD4+ CD45RA+) T cell subpopulation is susceptible to various types of hazardous substances in the workplace. *International Journal of Immunopathology and Pharmacology*, 17(2_suppl), 109–114 (2004).

43. Kostial, K., Kello, D., Jugo, S., Rabar, I., & Maljković, T. Influence of age on metal metabolism and toxicity. *Environmental Health Perspectives*, 25, 81–86 (1978).

44. Ilbäck, N. G., Sundberg, J., & Oskarsson, A. Methyl mercury exposure via placenta and milk impairs natural killer (NK) cell function in newborn rats. *Toxicology Letters*, 58(2), 149–158 (1991).

45. Park, S. H., Araki, S., Nakata, A., Kim, Y. H., Park, J. A., Tanigawa, T., Yokoyama, K., & Sato, H. Effects of occupational metallic mercury vapour exposure on suppressor-inducer (CD4+ CD45RA+) T lymphocytes and CD57+ CD16+ natural killer cells. *International Archives of Occupational and Environmental Health*, 73(8), 537–542 (2000).

46. Santarelli, L., Bracci, M., & Mocchegiani, E. In vitro and in vivo effects of mercuric chloride on thymic endocrine activity, NK and NKT cell cytotoxicity, cytokine profiles (IL-2, IFN-γ, IL-6): Role of the nitric oxide-L-arginine pathway. *International Immunopharmacology*, 6(3), 376–389 (2006).

47. Vimercati, L., Santarelli, L., Pesola, G., Drago, I., Lasorsa, G., Valentino, M., Vacca, A., & Soleo, L. Monocyte-macrophage system and polymorphonuclear leukocytes in workers exposed to low levels of metallic mercury. *Science of the Total Environment*, 270(1–3), 157–163 (2001).

48. Ellingsen MD, D. G., Efskind MD, J., Haug MD, E., Thomassen MSc, Y., Martinsen MSc, I., & Gaarder MD, P. I. Effects of low mercury vapour exposure on the thyroid function in chloralkali workers. *Journal of Applied Toxicology: An International Journal*, 20(6), 483–489 (2000).

49. Rao, M. V., & Sharma, P. S. N. Protective effect of vitamin E against mercuric chloride reproductive toxicity in male mice. *Reproductive Toxicology*, 15(6), 705–712 (2001).

50. Nadorty-López, E., Torres, S. H., Finol, H., Mendez, M., & Bello, B. With occupational exposure to mercury vapours. *Histol Histopathol*, 15, 673–682 (2000).

51. Gruenwedel, D. W., & Lu, D. S. Changes in the sedimentation characteristics of DNA due to methylmercuration. *Biochemical and Biophysical Research Communications*, 40(3), 542–548 (1970).

52. Herman, S. P., Klein, R., Talley, F. A., & Krigman, M. R. An ultrastructural study of methylmercury-induced primary sensory neuropathy in the rat. *Lab Invest*, 28(1), 104–118 (1973).

53. Lund, B. O., Miller, D. M., & Woods, J. S. Studies on Hg (II)-induced H2 O2 formation and oxidative stress in vivo and in vitro in rat kidney mitochondria. *Biochemical Pharmacology*, 45(10), 2017–2024 (1993).

54. Gruenwedel, D. W., & Cruikshank, M. K. Effect of methylmercury (II) on the synthesis of deoxyribonucleic acid, ribonucleic acid and protein in HeLa S3 cells. *Biochemical Pharmacology*, 28(5), 651–655 (1979).

55. Kinjo, Y., Akiba, S., Yamguchi, N., Mizuno, S., Watanabe, S., Wakamiya, J., Futatsuka, M., & Kato, H. Cancer mortality in minamata disease patients exposed to methymercu through fish diet. *Journal of Epidemiology*, 6(3), 134–138 (1996).

56. Robinson, M. M. Dermatitis medicamentosa simulating Hodgkin's disease due to mercury compounds. *Annals of Allergy, 10*(1), 21–23 (1952).

57. Flanders, R. A. Mercury in dental amalgam-a public health concern? *Journal of Public Health Dentistry, 52*(5), 303–311 (1992).

58. Iavicoli, I., Fontana, L., & Bergamaschi, A. The effects of metals as endocrine disruptors. *Journal of Toxicology and Environmental Health, Part B, 12*(3), 206–223 (2009).

59. Chen, Y. W., Huang, C. F., Tsai, K. S., Yang, R. S., Yen, C. C., Yang, C. Y., Lin-Shiau, S.Y., & Liu, S. H. Methylmercury induces pancreatic β-cell apoptosis and dysfunction. *Chemical Research in Toxicology, 19*(8), 1080–1085 (2006).

60. Davis, B. J., Price, H. C., O'Connor, R. W., Fernando, R. R. A. S., Rowland, A. S., & Morgan, D. L. Mercury vapor and female reproductive toxicity. *Toxicological Sciences, 59*(2), 291–296 (2001).

61. Boujbiha, M. A., Hamden, K., Guermazi, F., Bouslama, A., Omezzine, A., Kammoun, A., & El Feki, A. Testicular toxicity in mercuric chloride treated rats: Association with oxidative stress. *Reproductive Toxicology, 28*(1), 81–89 (2009).

62. Schrag, S. D., & Dixon, R. L. Occupational exposures associated with male reproductive dysfunction. *Annual Review of Pharmacology and Toxicology, 25*(1), 567–592 (1985).

63. Rice, K. M., Walker Jr, E. M., Wu, M., Gillette, C., & Blough, E. R. Environmental mercury and its toxic effects. *Journal of Preventive Medicine and Public Health, 47*(2), 74 (2014).

64. Kazantzis, G. Mercury exposure and early effects: An overview. *Medicina del Lavoro, 93*(3), 139–147 (2002).

65. Minoia, C., Ronchi, A., Pigatto, P., & Guzzi, G. Effects of mercury on the endocrine system. *Critical Reviews in Toxicology, 39*(7), 627–627 (2009).

66. GESTIS. GESTIS-Database on Hazardous Substances: Information System on Hazardous Substances of the German Social Accident Insurance. Available at https://amcaw.ifa.dguv.de [accessed 2020 May 27] (2015).

67. Battistoni, P., Fava, G., & Ruello, M. L. Heavy metal shock load in activated sludge uptake and toxic effects. *Water Research, 27*(5), 821–827 (1993).

68. Aklil, A., Mouflih, M., & Sebti, S. Removal of heavy metal ions from water by using calcined phosphate as a new adsorbent. *Journal of Hazardous Materials, 112*, 183–190 (2004).

69. François, F., Lombard, C., Guigner, J. M., Soreau, P., Brian-Jaisson, F., Martino, G., Vandervennet, M., Garcia, D., Molinier, A.L., Pignol, D. & Rebuffat, S. Isolation and characterization of environmental bacteria capable of extracellular biosorption of mercury. *Applied and Environmental Microbiology, 78*(4), 1097–1106 (2012).

70. Hussein, H., Ibrahim, S. F., Kandeel, K., & Moawad, H. Biosorption of heavy metals from waste water using *Pseudomonas* sp. *Electronic Journal of Biotechnology, 7*(1), 30–37 (2004).

71. Velásquez, L., & Dussan, J. Biosorption and bioaccumulation of heavy metals on dead and living biomass of *Bacillus sphaericus*. *Journal of Hazardous Materials, 167*(1–3), 713–716 (2009).

72. Stoffers, P., Hannington, M., Wright, I., Herzig, P., & De Ronde, C. Elemental mercury at submarine hydrothermal vents in the Bay of Plenty, Taupo volcanic zone, New Zealand. *Geology, 27*(10), 931–934 (1999).

73. Osborn, A. M., Bruce, K. D., Strike, P., & Ritchie, D. A. Distribution, diversity and evolution of the bacterial mercury resistance (*mer*) operon. *FEMS Microbiology Reviews, 19*(4), 239–262 (1997).

74. Von Canstein, H., Li, Y., Timmis, K. N., Deckwer, W. D., & Wagner-Dobler, I. Removal of mercury from chloralkali electrolysis wastewater by a mercury-resistant *Pseudomonas putida* strain. *Applied and Environmental Microbiology, 65*(12), 5279–5284 (1999).

75. Von Canstein, H., Kelly, S., Li, Y., & Wagner-Döbler, I. Species diversity improves the efficiency of mercury-reducing biofilms under changing environmental conditions. *Applied and Environmental Microbiology, 68*(6), 2829–2837 (2002).

76. Huang, C. C., Narita, M., Yamagata, T., Itoh, Y., & Endo, G. Structure analysis of a class II transposon encoding the mercury resistance of the Gram-positive bacterium Bacillus megaterium MB1, a strain isolated from Minamata Bay, Japan. *Gene, 234*(2), 361–369 (1999).

77. Voleski, B. Advances in biosorption of metals: Selection of biomass types. *FEMS Microbiology Reviews, 14*, 291–302 (1994)

78. Bankar, A. V., Kumar, A. R., & Zinjarde, S. S. Removal of chromium (VI) ions from aqueous solution by adsorption onto two marine isolates of Yarrowia lipolytica. *Journal of Hazardous Materials, 170*, 487–494 (2009).

79. Strouhal, M., Kizek, R., Vacek, J., Trnkova, L., & Nemec, M. Electrochemical study of heavy metals and metallothionein in yeast Yarrowia lipolytical. *Biogeosciences, 60*, 29–36 (2003).

80. Oyetibo, G. O., Ishola, S. T., Ikeda-Ohtsubo, W., Miyauchi, K., Ilori, M. O., & Endo, G. Mercury bio-removal by *Yarrowia strains* isolated from sediments of mercury-polluted estuarine water. *Applied Microbiology and Biotechnology, 99*(8), 3651–3657 (2015).

81. Summers, A. O. Organization, expression, and evolution of genes for mercury resistance. *Annual Reviews in Microbiology, 40*(1), 607–634 (1986).

82. Ji, G., & Silver, S. Bacterial resistance mechanisms for heavy metals of environmental concern. *Journal of Industrial Microbiology, 14*(2), 61–75 (1995).

83. Brunke, M., Deckwer, W. D., Frischmuth, A., Horn, J. M., Lünsdorf, H., Rhode, M., Röhricht, M., Timmis, K.N., & Weppen, P. Microbial retention of mercury from waste streams in a laboratory column containing merA gene bacteria. *FEMS Microbiology Reviews, 11*(1–3), 145–152 (1993).

84. Wagner-Döbler, I., Von Canstein, H., Li, Y., Timmis, K. N., & Deckwer, W. D. Removal of mercury from chemical wastewater by microoganisms in technical scale. *Environmental Science & Technology, 34*(21), 4628–4634 (2000).

85. Malakahmad, A., Hasani, A., Eisakhani, M., & Isa, M. H. Sequencing Batch Reactor (SBR) for the removal of Hg2+ and Cd2+ from synthetic petrochemical factory wastewater. *Journal of Hazardous Materials, 191*(1–3), 118–125 (2011).

86. Neufeld, R. D., & Hermann, E. R. Heavy metal removal by acclimated activated sludge. *Journal (Water Pollution Control Federation), 47*(2), 310–329 (1975).

87. Leung, W. C., Wong, M. F., Chua, H., Lo, W., Yu, P. H. F., & Leung, C. K. Removal and recovery of heavy metals by bacteria isolated from activated sludge treating industrial effluents and municipal wastewater. *Water Science and Technology, 41*(12), 233–240 (2000).

88. Santos, A., & Judd, S. The fate of metals in wastewater treated by the activated sludge process and membrane bioreactors: A brief review. *Journal of Environmental Monitoring, 12*(1), 110–118 (2010).

89. Yang, Y., Wang, L., Xiang, F., Zhao, L., & Qiao, Z. Activated sludge microbial community and treatment performance of wastewater treatment plants in industrial and municipal zones. *International Journal of Environmental Research and Public Health, 17*(2), 436 (2020).

90. Fazal, S., Zhang, B., & Mehmood, Q. Biological treatment of combined industrial wastewater. *Ecological Engineering, 84*, 551–558 (2015).

91. Yavuz, H., Denizli, A., Güngüneş, H., Safarikova, M., & Safarik, I. Biosorption of mercury on magnetically modified yeast cells. *Separation and Purification Technology, 52*(2), 253–260 (2006).

92. Pecoraro, V., Petri, D., Costantino, G., Squizzato, A., Moja, L., Virgili, G., & Lucenteforte, E. The diagnostic accuracy of digital, infrared and mercury-in-glass thermometers in measuring body temperature: A systematic review and network meta-analysis. *Internal and Emergency Medicine, 16*(4), 1071–1083. (2021).

93. Alasia, F., Capelli, A., Cignolo, G., & Sardi, M. A new generation of mercury manometers at the IMGC. *Metrologia, 30*(6), 571. (1994).

94. Koyun, M., Akman, S., & Güven, A. G. Mercury intoxication resulting from school barometers in three unrelated adolescents. *European Journal of Pediatrics, 163*(3), 131–134 (2004).

95. Shahbabu, B., Dasgupta, A., Sarkar, K., & Sahoo, S. K. Which is more accurate in measuring the blood pressure? A digital or an aneroid sphygmomanometer. *Journal of Clinical and Diagnostic Research, 10*(3), LC11. (2016).

96. Simon, J., Saffer, S., & Kim, C. J. A liquid-filled microrelay with a moving mercury microdrop. *Journal of Microelectromechanical Systems, 6*(3), 208–216 (1997).

97. Hu, Y., & Cheng, H. Mercury risk from fluorescent lamps in China: Current status and future perspective. *Environment International, 44*, 141–150 (2012).

98. Simon, J., Saffer, S., & Kim, C. J. A liquid-filled microrelay with a moving mercury microdrop. *Journal of Microelectromechanical Systems, 6*(3), 208–216 (1997).

99. Spencer, A. J. Dental amalgam and mercury in dentistry. *Australian Dental Journal, 45*(4), 224–234 (2000).

15 Bioassay Technology, Terminologies, Applications, and Examples: Heavy Metals and Phenol Investigations

Lawrence K. Wang and Mu-Hao Sung Wang

ACRONYMS AND NOMENCLATURES

ADI	acceptable daily intake for a specific chemical or substance(s), kg/person/day
BAF	bioaccumulation factor, L/kg, or [(mg chemical/kg organism)/(mg chemical/L water)] in natural water environment
BCF	bioconcentration factor, L/kg, or [(mg chemical/kg organism)/(mg chemical/L water)] in laboratory water environment
CONC$_{organism}$	concentration of a chemical in the tissues of an aquatic organism, mg/kg
CONC$_{water}$	concentration of the same chemical in laboratory water, mg/L.
LIWT	Lenox Institute of Water Technology
PAC	powdered activated carbon
RATE$_{chemical}$	acceptable consumption rate of a chemical, mg/kg/day
USEPA	US Environmental Protection Agency
WEIGHT$_{person}$	a person's weight, kg/person
WRRF	water resource recovery facility

15.1 INTRODUCTION

15.1.1 SUMMARY AND MEMOIR

The technical information [1–13] of bioassay technology, terminologies, applications, and design examples mainly from the US Environmental Protection Agency (USEPA) [7–13] and the Lenox Institute of Water Technology (LIWT) [6] are presented.

The sulfide precipitation process was used for emergency control of chemical spills that contain soluble heavy metals, such as zinc, copper, cobalt, lead, cadmium, nickel, and silver. Powdered activated carbon (PAC) and granular activated carbons (GAC) were used for emergency control of chemical spills that contain soluble toxic organic compounds, such as phenol. The produced heavy metal precipitates and the phenol-adsorbed PAC and GAC need to be properly handled and disposed of. Their toxicities are investigated by a bioassay study in the laboratory for demonstration.

The analytical data for the bioassay investigations of seven metal sulfide precipitates (zinc sulfide, copper sulfide, cobalt sulfide, lead sulfide, cadmium sulfide, nickel sulfide, silver sulfide), and two phenol-adsorbed carbons (PAC and GAC) are reported.

A bioassay report on the water quality is included, as well as data on the pH, conductivity, alkalinity, dissolved oxygen, and hardness of test water. The level of free metal ions in experiments using heavy metal sulfide precipitates was determined throughout the tests and is included in this report.

There are three bioassay design examples: (a) determination of bioconcentration factor and bioaccumulation factor; (b) determination of the acceptable daily intake; and (c) determination of the criterion for a chemical in drinking water.

This publication was one of the lecture materials of the Graduate School of the LIWT, MA, USA. It is updated using the data from Cornell Aeronautical Laboratory (CAL), USEPA, and LIWT as a technology transfer report in memory of the authors' good friend and old colleague, Dr. Robert C. Ziegler, who was the Environmental Systems Division Head of Cornell Aeronautical Laboratory, Cornell University, New York, USA.

15.1.2 Bioassay Technology

15.1.2.1 Bioassay Classifications

A bioassay is a toxicity test using selected organisms (subject) to determine the acute or chronic effects of a chemical pollutant or whole effluent (stimulus). Technically speaking it is a biochemical technology to determine the potency or effect of a sample compound (stimulus) by its effect on living animals or plants (in vivo), or on living cells or tissues (in vitro) (subject). The corresponding response (e.g., death or sickness or mutation) of the subject is thereby triggered and measured. A bioassay can be either quantal bioassay or quantitative bioassay, direct bioassay, or indirect bioassay.

If the measured response is binary, the assay is a quantal bioassay because it is only a determination of whether or not an event occurs (e.g., death, sickness, and color change of the subject). If it is not a quantal bioassay, then it is a quantitative bioassay in which the measurement of the response to the stimulus is on a quantitative scale (e.g., blood sugar content and degree of color change in the cell growth medium).

In a direct bioassay, the stimulus applied to the subject is specific and directly measurable, and the response to that stimulus is recorded with the response of interest (e.g., death, sickness, appearance change, and loss of functions) of the subject; if it is not a direct assay, then it is an indirect bioassay.

15.1.2.2 Environmental Bioassay

Environmental bioassays are generally a broad-range survey of toxicity in the environment. A toxicity identification evaluation is conducted usually using a bioindicator to determine what the relevant man-made and/or natural toxicants are. These environmental bioassay procedures, called whole effluent toxicity tests, include acute toxicity tests as well as chronic test methods that involve exposing living bioindicator-type aquatic organisms to samples of wastewater for a specific length of time. For instance, the bioassay ECOTOX uses the microalgae *Euglena gracilis* as a bioindicator to test the toxicity of water samples.

15.1.2.3 Predictive Bioassay

Predictive bioassay is one type of environmental bioassays to be carried out with a simulated plant waste as close as possible in quality to the anticipated operating waste under the worst possible conditions. The predictive bioassay should contain three elements of the food chain, i.e., primary producers such as algae, invertebrates, and fish. Just because waste doesn't kill fish directly, it may ultimately prevent their survival due to its toxicity to an intermediate in the food chain. It is important to protect other elements of the aquatic community besides fish. For example, the loss of algae may impair the ability of the aquatic system to receive and transform organic wastes.

When conducting predictive bioassays, the water quality of the proposed site location should be used since there are documented examples of synergistic interactions resulting in increased toxicity. Predictive bioassays can be useful in helping identify potentially hazardous interactions and to see if preliminary waste treatment design is likely to be adequate, and if not, it can be used to estimate the degree of additional treatment required.

The detailed terminologies of bioassay technology are presented in Section 15.5 (Bioassay Technology and Terminologies) of this publication.

15.2 EXPERIMENTS USING METAL SULFIDE PRECIPITATES

15.2.1 Heavy Metal Sulfide Bioassay Study

The bioassay investigation reported in this section is an environmental bioassay with an experimental set-up similar to that illustrated in Figure 15.1.

Fifty fathead minnows were exposed to each of seven metal sulfide precipitates over a period of 30 days. On the 30th day, ten live fish were removed for chemical analysis and the test continued with the remaining fish. Each precipitate was mixed with mixed with sand and allowed to settle to the bottom of the aquarium before fish were added to the system. A 12-hour light exposure and a controlled temperature of 60°F ± 1°F in the water completed the experimental conditions.

The test water was circulated gently by a Dynaflo filter, without filtration, and was aerated in the filter itself so that the sand and precipitate mixture was not disturbed. One healthy hornwort (*Ceratophyllum*) plant was added to each aquarium system. The growth of both the plant and algae was observed.

One aquarium served as a control, excluding the metal precipitate including sand, fish, and a plant. The remaining seven aquaria had the following amounts of metal precipitate mixed with approximately 3 kg of white filter sand and settled at the bottom of 60 L of water (see Table 15.1).

The amount of metal sulfide added to an aquarium was calculated from the toxicity range of each free metal to fish. This toxicity range was estimated from at least three sources in the literature.

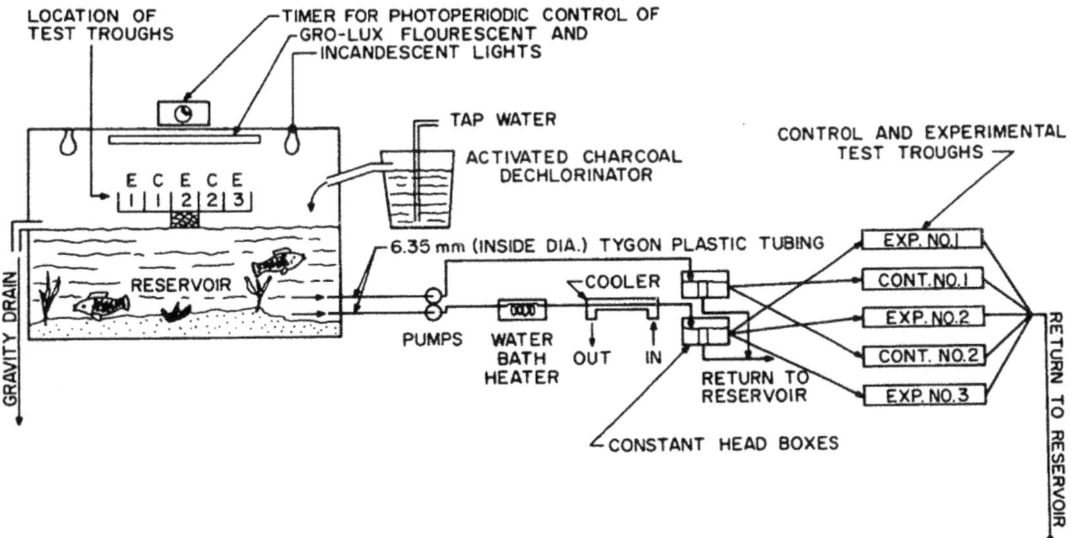

FIGURE 15.1   Experimental Design. (USEPA.)

TABLE 15.1
Bioassay Research Design

Tank No.	Metal Sulfides	The resulting Concentration of Metals if all Sulfide Dissociated
1	CuS	1.0 ppm
2	Ag_2S	1.0 ppm
3	CdS	5.0 ppm
4	CoS	15.0 ppm
5	NiS	15.0 ppm
6	ZnS	15.0 ppm
7	PbS	20.0 ppm

Note: Tank 0 is bioassay control having fish but containing no heavy metal sulfides.

15.2.2 EXPERIMENTAL RESULTS

Table 15.2 summarizes the data for each of the eight aquaria, including the concentration of free metal initially and at 1, 7, 15, 22, 30, 45, and 60 days. It should be noted that the added metal sulphides on initial O-day did not reach their equilibrium concentrations immediately.

Table 15.2 also includes the equilibrium concentration of free metal as calculated from the solubility of each metal sulfide in water. It can be seen that cobalt sulfide and nickel sulfide dissociated beyond the expected level.

After 7 days, the pH, the dissolved oxygen, hardness, and alkalinity of the control tank were determined: (a) pH $= 8.7$; (b) alkalinity $= 116$ mg/L as $CaCO_3$; (c) dissolved oxygen (DO) $= 9.85$ mg/L; Hardness $= 116$ mg/L as $CaCO_3$.

Fish were observed for behavioral patterns and survival. Interesting observations were made on the health of plants in the test waters and the growth of algae in aged tanks.

One of the first observations made in these tests was the repression of algal growth in the metal sulfide test tanks. At 5 days, growth in the control tank began. None of the metal sulfide tanks supported an algal growth until 12 days into the experiment, when a trace of algae could be seen in the lead, copper, and silver tanks. Algae began to grow in the cadmium test tank at 20 days and in the zinc tank at 30 days.

Algae was not seen in the nickel test tank and may have been beginning in the cobalt tank at 60 days when the tests were terminated. Photographs are on file to document some of these findings.

The condition of the Ceratophylum plants was also observed to vary. At 7 days into the experiment, the plant in the cadmium tank had lost its foliage and appeared to be dying. One week later, the same observation was made in the silver tank. All other plants appeared healthy at 15 days. Photographs documenting these findings are also on file.

At the termination of the experiment, plants from the control, lead, zinc, and cobalt sulfide tanks appeared healthy and green. The silver and copper plants were 20%–30% covered with foliage. Much of each green stem was bare. The plant recovered from the cadmium sulfide tank looked as it did at 7 days. Only a green stem remained; none of the foliage was recovered. The plant in the nickel sulfide test tank had disintegrated. Only a brown glob of material was recovered from the bottom of the tank.

The source of test water for all bioassay experiments now and in the past has been tap water, which was aerated for 24 hours to remove chlorine before the initiation of each test. An analysis of this water for various parameters is included in the Appendix, Chemical Analysis of Test Water Source and Washed Sand.

TABLE 15.2

Summary of Bioassay Experimental Data of Exposing Fish to Metal Sulfides

Time	Conc. of Metal in Test Water (ppm)	Equilibrium Cone, of Metal (From the Solubility of sulfide) (ppm)	Toxic Range from Literature (ppm)	Resistivity (ohm-cm)	pH	% Survival of Fish
			Control			
Initial					8.01	100%
1 day					- - -	100%
7 days					8.70	100%
15 days					- - -	100%
22 days	0.0	0.0			9.18	100%
30 days				1,360	9.91	98%
45 days					- - -	96%
60 days					9.71	96%
			Zinc			
Initial	0.60 (+5.0)	5.0	0.3–15	1,230	8.01	100%
1 day	0.30				- - -	100%
7 day	1.20				8.10	100%
15 days	1.13				- - -	100%
22 days	0.16				8.49	100%
30 days	0.17				8.60	100%
45 days	0.01				9.70	100%
60 days	0.08				9.52	100%
			Copper			
Initial	0.05 (+0.2)	0.2	0.01–3.3		8.01	100%
2 days	0.05				- - -	94%
8 days	0.09				8.50	94%
15 days	0.08				- - -	94%
21 days	0.10				9.50	92%
29 days	0.11			1,360	8.30	92%
45 days	0.10				10.08	92%
60 days	0.30				9.26	92%

(*Continued*)

TABLE 15.2 (*Continued*)

Summary of Bioassay Experimental Data of Exposing Fish to Metal Sulfides

Time	Conc. of Metal in Test Water (ppm)	Equilibrium Cone, of Metal (From the Solubility of sulfide) (ppm)	Toxic Range from Literature (ppm)	Resistivity (ohm-cm)	pH	% Survival of Fish
			Cobalt			
Initial	2.4 (+3.0)	3.0	7–16		8.01	100%
1 day	3.9				- - -	100%
7 days	5.3				8.15	98%
15 days	6.1				- - -	96%
22 days	7.0				8.20	96%
30 days	5.8			1,150	8.42	96%
45 days	5.5				7.85	91%
60 days	4.8				6.12	80%
			Lead			
Initial		0.8	0.1–25		8.01	100%
1 day	analyses not completed					98%
7 days					8.70	96%
15 days					9.77	96%
22 days				1,350	8.62	94%
30 days					9.95	94%
45 days;					9.65	94%
60 days						94%
			Cadmium			
Initial	0.70 (+1.0)	1.0	0.01–7		8.01	100%
1 day						100%
7 days					8.10	100%

(Continued)

TABLE 15.2 (*Continued*)

Summary of Bioassay Experimental Data of Exposing Fish to Metal Sulfides

Time	Conc. of Metal in Test Water (ppm)	Equilibrium Cone, of Metal (From the Solubility of sulfide) (ppm)	Toxic Range from Literature (ppm)	Resistivity (ohm-cm)	pH	% Survival of Fish
15 days						100%
22 days	0.55				8.16	100%
30 days	0.51			1,220	8.85	98%
45 days	0.52				8.99	98%
60 days	0.51				8.16	98%
	0.005 analyses not completed					
		Nickel				
Initial	2.90 (+2.4)	2.4	0.08–16		8.01	100%
2 days	3.90					94%
8 days	5.50				8.10	88%
15 days	8.00					86%
21 days	7.75				7.88	82%
29 days	6.95				7.90	72%
45 days	6.75				8.33	49%
60 days	8.15				8.07	41%
				1,150		
		Silver				
Initial	0.00 (+0.1)	0.1			8.01	100%
2 days						98%
8 days	II					90%
15 days	II				8.75	84%
21 days	II			1,500	9.80	78%
29 days	II				9.32	78%
45 days	11				9.70	75%
60 days	11				9.01	75%

15.3 EXPERIMENTS EXPOSING FISH TO POWDERED ACTIVATED CARBON WITH PHENOL ADSORBED ONTO THE SURFACE

15.3.1 PHENOL-PAC BIOASSAY STUDY

For this experiment, 92 g of phenol were adsorbed on 1,000 g of carbon (powdered activated carbon; Aqua Nuchar A), This mixture was placed into a 20-gallon aquarium and allowed to settle with approximately 3 kg of white filter sand. The water was determined to have a phenol concentration of approximately 1.96 to 2.6 ppm at the initiation of the test (when fish were added). It is known that this level of phenol is safe for fathead minnows. Previous experiments with phenol and minnows at the author's research facility have shown 10 ppm phenol to be a safe level, resulting in no abnormal behavior and 100% survival of fish. Samples of the aquarium water were taken initially at 1, 2, 3, 6, 7, 9, 15, and 30 days (Table 15.3). The results of analysis for phenol in the test water throughout the 30-day test period follow.

15.3.2 EXPERIMENTAL RESULTS

Table 15.4 summarizes the percent survival of fish with time compared to the control. Since phenol was not present in the test water throughout the experiment and since the control fish survived, it is probable that actual physical contact with the carbon produced direct effects from the concentrated phenol adsorbed on its surface. Powdered activated carbons (PAC) are easily disturbed by the swimming movements of the fish as they prefer to live at the bottom of the aquarium. This occasional contact may have caused the deaths of 32% of the test fish.

TABLE 15.3
Phenol Analysis of Test Water throughout the Test Period

Sample	ppm Phenol
Initial	1.96 to 2.6
1 day	0.3
2 days	0.0
3, 6, 7, 9, 15, 30 days	0.0

TABLE 15.4
Percent Survival of Fish with Time During Exposure To a Phenol-PAC Sediment

Time	Phenol Test	Control
Initial - 5 days	100%	100%
6–12 days	96%	100%
13–21 days	92%	100%
22 days	88%	100%
23 days	80%	100%
24 days	72%	100%
25 days	68%	100%
28 days	68%	98%
29 and 30 days	68%	98%

At the initiation of the test, the fish entered their new environment with panicky swimming motions, which caused a great deal of powdered activated carbon to be resuspended in the tank's lower four inches (10.16 cm). This contact with the PAC caused it to adhere to the gills and skin of the minnows. For the first 24 hours, they exhibited the familiar symptoms of phenol poisoning. Comparing these symptoms to those exhibited by minnows in acute toxicity tests with phenol, it was expected that the entire tank of fish would not survive 40 hours. However, the minnows adjusted to the new environment and quieted, living approximately 1 inch (2.54 cm) above the PAC layer. Most were successful in avoiding the irritating poison, except for the initial and occasional contact, which must have caused 32% to die. After 24 hours, PAC was not seen on the fish bodies, and the fish appeared to have recovered from the phenol poisoning that was seen initially. The pH of this aquarium was measured after 16 days and found to be 7.46.

15.4 EXPERIMENT EXPOSING FISH TO GAC WITH PHENOL ADSORBED ONTO CARBON SURFACE

15.4.1 PHENOL-GAC BIOASSAY STUDY

One hundred grams of phenol were adsorbed onto 1,090 g of GAC (8X 30 mesh; granular) in preparation for an experiment to test the ability of fish to live over a treatment such as this one. Initially, phenol in the test water was 0.1 to 0.3 ppm after the addition of GAC and sand to an aquarium. After 1 day (24 hours), the phenol concentration was nil <0.10 ppm. Between 15 and 30 days, the phenol increased to 0.9 ppm at 30 days. The data are summarized in Table 15.5. The physical and chemical properties of phenol under this investigation are presented in Table 15.6.

15.4.2 EXPERIMENTAL RESULTS

Performing a chi square test on the above data gives statistical evidence to support the contention that phenol did cause a significant change in the ability of fish to survive beyond the 0.01 level of significance.

Read Section 15.5 for the calculation of a bioconcentration factor (BCF).

15.4.3 PHYSICAL AND CHEMICAL PROPERTIES OF PHENOL

The physical and chemical properties of phenol can be found from Table 15.6.

TABLE 15.5
Percent Survival of Fish Exposed to Phenol-GAC Sediment

Time	Amount of Phenol in Water	% Survival in Phenol	% Survival in Control
Initial	0.3 ppm	100%	100%
7 days	Nil < 0.10 ppm	95%	97%
15 days	Nil < 0.10 ppm	78%	86%
30 days	0.05 ppm	45%	65%

TABLE 15.6

Physical and Chemical Properties of Phenol (NIOSH, 1975)

Formula:	C_6H_5OH
Molecular weight:	94.11
pK_a:	9.9
Melting point:	40–41°C
Boiling point:	181.75°C
Vapor pressure @ 25°C	0.35 mm Hg
Specific gravity: solid @ 25°C	1.071
liquid @ 25°C	1.049
Relative vapor density: (air = 1.0)	3.24
Solubility: (X = mole fraction)	Also soluble in ether,
Phenol in water: −log X =	alcohol, acetic acid,
0.375 log(66 − T) + 1.15.	glycerol, liquid sul-
Water in phenol: −log X =	fur dioxide, benzene.
−0.62 log(66 − T) + 0.99	
Color:	Colorless to light pink solid
Odor:	Sweet; threshold = 1ppm
Flashpoint: open cup	85°C
closed cup	79°C
Ignition temperature:	715°C
Light sensitivity:	Darkens on exposure to light
Saturated vapor concentration (25°C):	461 ppm

15.5 BIOASSAY DATA CALCULATION EXAMPLES

15.5.1 EXAMPLE: DETERMINATION OF BIOCONCENTRATION FACTOR AND BIOACCUMULATION FACTOR

Bioconcentration is a biochemical process by which there is a net accumulation of a chemical (such as phenol) directly from water into aquatic organisms (such as fish), resulting from simultaneous uptake and elimination (Figure 15.3).

Bioconcentration factor (BCF) is the ratio of the concentration of a chemical in the tissues of an aquatic organism to its concentration in the laboratory water if the organism and the food chain are exposed. BCF can be calculated by the following equation:

$$BCF = CONC_{organism} / CONC_{water} \tag{15.1}$$

where

BCF=bioconcentration factor, L/kg, or [(mg chemical/kg organism)/(mg chemical/L water)]
$CONC_{organism}$=concentration of a chemical in the tissues of an aquatic organism, mg/kg
$CONC_{water}$=concentration of the same chemical in laboratory water, mg/L.

Bioaccumulation factor (BAF) is the ratio of the concentration of a chemical in the tissues of an aquatic organism to its concentration in its natural water if the organism and the food chain are exposed. BAF can be calculated by the following equation:

$$BAF = CONC_{organism}/CONC_{water} \tag{15.2}$$

where

BAF=bioaccumulation factor, L/kg, or [(mg chemical/kg organism)/(mg chemical/L water)]
$CONC_{organism}$=concentration of a chemical in the tissues of an aquatic organism, mg/kg
$CONC_{water}$=concentration of the same chemical in the organism's natural water, mg/L.

The fish tissue in the bioassay study in Section 15.4.1 was analyzed and found to be 0.42 mg of phenol per kg fish tissue. The initial phenol concentration was 0.3 mg/L. What will be the bioconcentration factor (BAC) and the bioaccumulation factor (BAC)?

Solution
Since the bioassay study was conducted in the laboratory using laboratory water (not the natural water of the fish), only the bioconcentration factor (BAC) is to be calculated using Equation (15.3).

$$BCF = CONC_{organism}/CONC_{water} \tag{15.1}$$

where

BCF=bioconcentration factor, L/kg, or [(mg chemical/kg organism)/(mg chemical/L water)]
$CONC_{organism}$=concentration of a chemical in the tissues of an aquatic organism=0.42 mg/kg
$CONC_{water}$=concentration of the same chemical in laboratory water=0.3 mg/L.

$$BCF = CONC_{organism}/CONC_{water}$$

$$BCF = (0.42 \text{ mg/kg})/(0.3 \text{ mg/L}) = 1.4 \text{ L/kg}$$

15.5.2 EXAMPLE: DETERMINATION OF THE ACCEPTABLE DAILY INTAKE

Based on chronic toxicity data in animals and man, an estimated acceptable daily intake (ADI) for a specific chemical or substance(s) can be calculated by Equation (15.3).

$$ADI = (RATE_{chemical})(WEIGHT_{person}) \tag{15.3}$$

where

ADI = acceptable daily intake for a specific chemical or substance(s), kg/person/day
$RATE_{chemical}$ = acceptable consumption rate of a chemical (such as phenol), mg/kg/day (i.e., mg phenol per kg person weight per day)
$WEIGHT_{person}$ = a person's weight, kg/person

Determine the ADI for phenol if the acceptable consumption rate of phenol is 0.1 mg per kg person weight per day (0.1 mg/kg/day), and the average body weight of a person is 70 kg.

Solution
It has been established that phenol is absorbed rapidly by all routes and is subsequently rapidly distributed. If a tenfold safety factor is applied to the projected doses absorbed from inhalation at the TLV, then the projected acceptable level would be 0.2 mg/kg/day for phenol. Adoption of only 0.1 mg/kg/day is on the safety side, and acceptable.

Using Equation (15.3), the calculation will be as follows:

$$ADI = (RATE_{chemical})(WEIGHT_{person}) \qquad (15.3)$$

$$ADI = (0.1 \text{ mg/kg/day})(70 \text{ kg/person}) = 7 \text{ mg/day/person for phenol}$$

16.5.3 Example: Determination of the Criterion for a Chemical in Drinking Water

Equation (15.5) for calculating the criterion for a chemical (such as phenol) content in drinking water can be derived based on the following material balance Equation (15.4):

$$ADI = (VOL_{drinking})(CHEM) + (CHEM)(EAT_{organism})\,(BCF) \qquad (15.4)$$

$$CHEM = ADI/[VOL_{drinking} + (EAT_{organism})\,(BCF)] \qquad (15.5)$$

A similar Equation (15.6) can be developed for using BAF

$$CHEM = ADI/[VOL_{drinking} + (EAT_{organism})\,(BAF)] \qquad (15.6)$$

where

ADI = acceptable daily intake for a specific chemical or substance(s), kg/person/day
$VOL_{drinking}$ = daily volume of drinking water consumption rate per person, L
BCF = bioconcentration factor, L/kg 4 [(mg chemical/kg organism)/(mg chemical/L water)]
BAF = bioaccumulation factor, L/kg 4 [(mg chemical/kg organism)/(mg chemical/L water)]
$EAT_{organism}$ = amount of organism (such as fish) consumed, kg per person per day
CHEM = maximum permissible concentration of chemical (such as phenol in this example), mg/L

Determine the criterion for the phenol concentration of drinking water (CHEM) given the following technical information:

ADI = acceptable daily intake for a specific chemical or substance(s) = 7 mg/person/day for phenol
$VOL_{drinking}$ = daily volume of drinking water consumption rate per person = 2 L/person

BCF = bioconcentration factor = 1.4 L/kg, or 1.4 [(mg chemical/kg organism)/(mg chemical/L water)]

$EAT_{organism}$ = amount of organism (such as fish) consumed = 0.0065 kg per person per day

Solution

Using Equation (15.5)

$$CHEM = ADI/[VOL_{drinking} + (EAT_{organism}) (BCF)] \qquad (15.5)$$

$$CHEM = (7 \text{ mg/person/day})/[2 \text{ L/person} + (0.0065 \text{ kg per person per day}) (1.4 \text{ L/kg})]$$

$$CHEM = 3.48 \text{ mg/L}$$

Conclusion: The estimate from animal phenol data is 0.1 mg/kg/day. Based on chronic toxicity data in animals and a person, an estimated acceptable daily intake for phenol in a person should be 0.1 mg/kg/day or 7 mg/person/day (ADI = 7 mg/person/day), assuming a 70 kg body weight. It is concluded that assuming 100% gastrointestinal adsorption of toxic phenol, the consumption of 2 L of water daily per person ($VOL_{drinking}$ = 2 L/person) and 6.5 g of contaminated fish ($EAT_{organism}$ = 0.0065 kg/person/day) having a bioconcentration factor of 1.4 (BCF = 1.4 L/kg) would result in a maximum permissible concentration of 3.5 mg/L (CHEM = 3.5 mg/L) for the ingested water.

15.6 BIOASSAY TECHNOLOGY AND TERMINOLOGIES

The bioassay technology and terminologies are presented in this section using the technical information from the literature [1–13].

Absolute toxicity It is defined as the laboratory-determined toxic concentration of the complex effluent, while relative toxicity is defined as the toxicity of the effluent after mixing with the receiving water. Absolute toxicity is estimated with short-term Complex Effluent Toxicity Testing Program (CETTP) tests [1–13].

Acute effect It is a health effect that is short-lasting and usually reversible.

Ames assay A well-known short-term test that measures a chemical's ability to cause mutations in a specially engineered strain of the bacteria *Salmonella typhimurium*.

Ames test In an Aems test, a strain of *Salmonella* that requires histidine to grow is put on two plates with growth medium containing minimal amounts of histidine and some rat liver extract (to mimick liver metabolism). A suspected *mutagen* is added to one plate. If the plate with the suspected mutagen grows more visible colonies, it is probably mutagenic: a mutagen might cause the strain of the bacterium to regain the ability to make its own histidine (Figure 15.2).

Bioaccumulation It is a biochemical process of a chemical accumulating in a biological food chain by being passed from one organism to another as the contaminated organism is preyed upon by another organism (Figure 15.3).

Bioaccumulation factor (BAF) It is the ratio of the concentration of a chemical in the tissues of an aquatic organism to its concentration in the natural water if the organism and the food chain are exposed. BAF = (concentration of a chemical in the tissues of an aquatic organism)/(concentration of the same chemical in natural water) (Figure 15.3).

Bioassay (a) A bioassay is a toxicity test using selected organisms (subject) to determine the acute or chronic effects of a chemical pollutant or whole effluent (stimulus); (b) It is a biochemical technology to determine the potency or effect of a sample compound (stimulus) by its

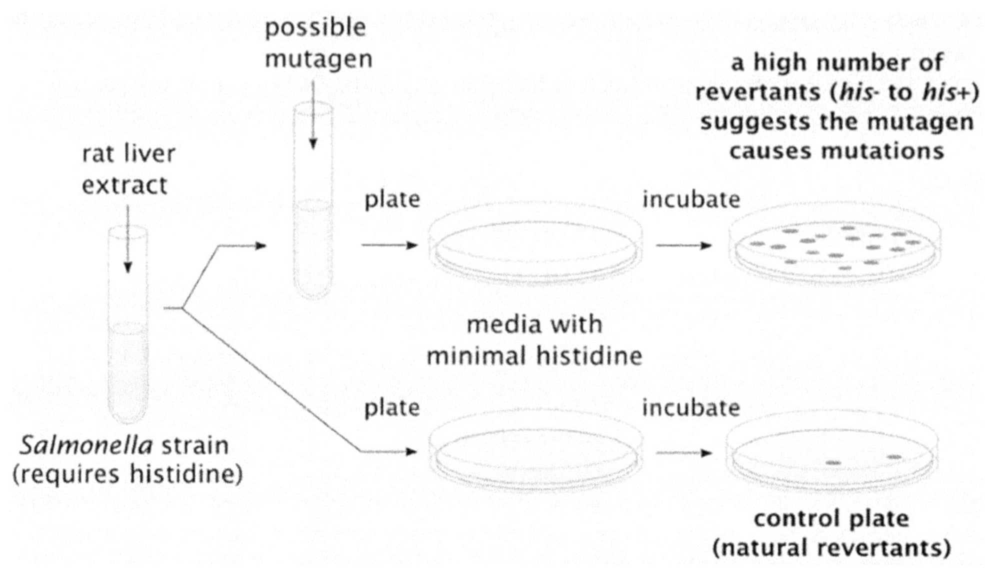

FIGURE 15.2 Ames Bioassay Test. (https://en.wikipedia.org/wiki/Bioassay.)

Ways to Determine the Bioaccumulation

$$\text{Bioaccumulation} = \frac{\text{Concentration in fish tissue}}{\text{Concentration in water}}$$

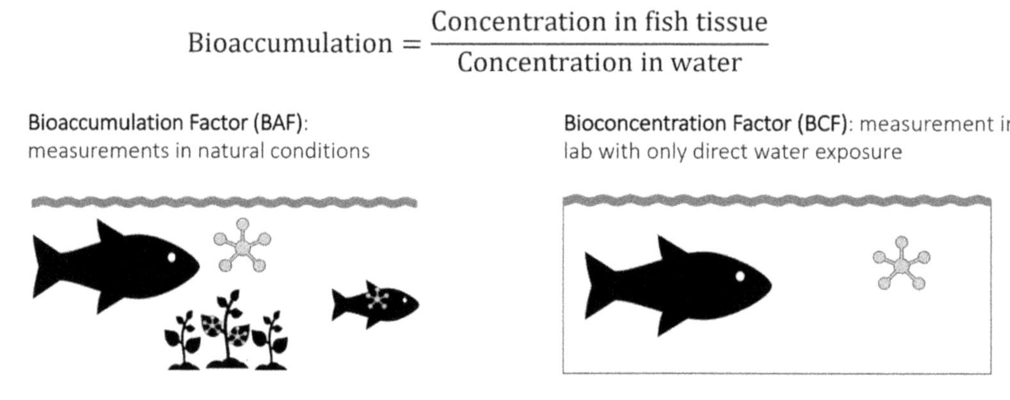

FIGURE 15.3 Bioaccumlation, Bioaccumulation Factor and Bioconcentraton Factor. (USEPA.)

effect on living animals or plants (in vivo), or on living cells or tissues (in vitro) (subject). The corresponding response (e.g., death or sickness or mutation) of the subject is thereby triggered and measured. A bioassay can be either quantal bioassay or quantitative bioassay, direct bioassay or indirect bioassay; (c) If the measured response is binary, the assay is quantal bioassay because it is only a determination of whether or not an event occurs (e.g., death, or sickness, color change, etc. of the subject); if it is not a quantal bioassay, then it is a quantitative bioassay in which the measurement of the response to the stimulus is on a quantitative scale (e.g., blood sugar content, degree of color change in cell growth medium, etc.); (d) In a direct bioassay, the stimulus applied to the subject is specific and directly measurable, and the response to that stimulus is recorded with response of interest (e.g., death, or sickness, appearance change, loss of functions, etc.) of the subject; if it is not a direct assay, then it is an indirect bioassay.

Bioavailability It is a measurement of the physiochemical access that a toxicant has to the biological processes of an organism. The less the bioavailability of a toxicant, the less its toxic effect on an organism.

Biochemical oxygen demand (BOD) It is an empirical test measuring the oxygen required for the biochemical degradation of organic material and the oxygen used to oxidize inorganic materials, such as sulfides and ferrous iron.

Bioconcentration It is a biochemical process by which there is a net accumulation of a chemical directly from water into aquatic organisms resulting from simultaneous uptake and elimination (Figure 15.3).

Bioconcentration factor (BCF) It is the ratio of the concentration of a chemical in the tissues of an aquatic organism to its concentration in the laboratory water if the organism and the food chain are exposed. BCF = (concentration of a chemical in the tissues of an aquatic organism)/(concentration of the same chemical in laboratory water) (Figure 15.3).

Bioindicator It is a species or ecological community that is used as an indicator in an environmental bioassay investigation because it is so closely associated with particular environmental conditions that its presence indicates these conditions in a specific environment.

Carcinogenic It is a situation that is able to cause cancer.

Chromosome aberrations It is a condition or result that changes the number, shape, or structure of chromosomes.

Chromosome It is a form of DNA (deoxyribonucleic acid) organization found in higher cells and organisms.

Chronic effect It is a prolonged health effect that may involve irreversible change or damage.

Chronic wasting disease (CWD) CWD is also called zombie deer disease, which is a transmissible spongiform encephalopathy (TSE) disease affecting deer. TSEs are a family of diseases thought to be caused by misfolded proteins called prions and include similar diseases such as BSE (mad cow disease) in cattle and Creutzfeldt–Jakob disease (CJD) in humans.

Complex Effluent Toxicity Testing Program (CETTP) The CETTP program in the USA includes tests that estimate "safe" or "no effect" concentrations for toxic effluents and thus predict allowable effluent concentrations that should provide for the normal propagation of fish and other aquatic life in marine receiving waters.

Complex effluent toxicity testing To develop National Pollutant Discharge Elimination System (NPDES) permits for the control of toxic effluent discharges to marine/estuarine environments, regulators require a quick, reliable, and inexpensive procedure to estimate effluent toxicity. To accomplish this requires an estimate of both the effluent's absolute and relative toxicity. Absolute toxicity is defined as the laboratory-determined toxic concentration of the complex effluent, while relative toxicity is defined as the toxicity of the effluent after mixing with the receiving water. Absolute toxicity is estimated with short-term Complex Effluent Toxicity Testing Program (CETTP) tests. Relative toxicity can be estimated in two practical ways. First, relative toxicity can be estimated by sampling the receiving water at various sites and testing the samples with the same CETTP tests used to assess absolute toxicity. Alternatively, relative toxicity can be assessed with in situ biomonitoring by determining the uptake of various toxicants and the corresponding biological effects on organisms that are transplanted into the receiving water.

Complex mixtures It is a grouping of several different chemicals or substances.

Composite sampling (a) The water quality sampling is generally conducted automatically with a collection apparatus placed adjacent to a discharge; (b) A 5-L water sample is often collected over a 24-hour period by collecting small volumes of water at regular intervals.

Direct assay (bioassay) In a direct bioassay, the stimulus (such as drugs, chemicals, pollutants, etc.) applied to the subject (such as animals, plants, humans, tissues, etc.) is specific and directly measurable, and the subject's response to that stimulus is recorded. The variable of interest is the specific stimulus and its quality required to produce a response of interest (e.g., death or sickness of the subject).

DNA (deoxyribonucleic acid) It is a large molecule that contains the genetic information responsible for cell growth, function, and reproduction.

Environmental bioassays They are generally a broad-range survey of toxicity in the environment. A toxicity identification evaluation is conducted usually using a bioindicator to determine what the relevant man-made and/or natural toxicants are. These environmental bioassay procedures, called whole effluent toxicity tests, include acute toxicity tests as well as chronic test methods involve exposing living bioindicator-type aquatic organisms to samples of wastewater for a specific length of time. For instance, the bioassay ECOTOX uses the microalgae Euglena gracilis as a bioindicator to test the toxicity of water samples.

Enzyme It is a protein that acts as a catalyst to allow a specific chemical reaction to take place in a cell.

Epidemiology It is a science of correlating exposure to a substance with the appearance of a specific disease or other effect in a human population group.

False negative It is a test result which indicates that a chemical is harmless when it is actually hazardous.

False positive It is a test result that indicates that a chemical is hazardous when it is actually harmless.

Fractionation It is a physicochemical or biochemical process of separating a complex mixture into a series of simpler mixtures (fractions).

Gene mutation It is a mutation in a single gene.

Gene It is a portion of DNA that directs the formation of a single product.

Genotoxic It means that it is able to damage genetic material.

Germinal cell It is a reproductive cell (i.e., sperm, egg).

Hazard assessment It is an evaluation process for determining if a substance is hazardous to humans.

Heritable It means that it is capable of being passed from one generation to another.

In vitro It pertains to a procedure that takes place in an artificial medium (such as, lab equipment, or glassware) without the use of live animals.

In vivo It pertains to a biological reaction or test which occurs within the body of a live animal.

Indicator system It may be a cell or organism that shows (or indicates) a specific effect.

Indirect assay (bioassay) Any bioassay that is not a direct assay is an indirect assay. In a direct assay, the stimulus applied to the subject is specific and directly measurable, and the response to that stimulus is recorded. In an indirect assay, it is different from that of a direct assay.

Indirect bioassay In an indirect bioassay, the stimulus is fixed in advance and the response is measured in the subjects. The variable of interest in the experiment is the response to a fixed stimulus of interest.

In situ biomonitoring The in situ biomonitoring program in the USA includes toxicity tests with aquatic plants and animals as well as biological and chemical tests of bioaccumulation. The mussels *M. edulis* and *M. californicus* are indigenous to and abundant in bays and estuaries around much of the coastline of the United States. Mussels feed by filtering particulates out of surrounding seawater, so they are excellent indicators of ambient water quality. In addition, mussels feed continuously throughout the year (without a dormant period), making them a particularly convenient species for physiological study. Finally, mussels are excellent for transplanting because large, healthy populations of animals can be found and distributed easily into field cages at various sampling sites.

LD_{50} value The LD_{50} value is a common measure of acute toxicity that describes the dose at which a substance is lethal to 50% of tested animals. This toxicology test is a bioassay test. Animals or cell cultures may be put under a number of levels of a suspected toxin to ascertain whether the substance causes harmful changes and at what level it does so.

Malignant It refers to the cancerous cells or tumors that may grow, proliferate, and eventually kill the organism.

Metabolic activation It is a process whereby an inactive material is changed into an active one (in the context of short-term testing, this involves the conversion of a procarcino¬gen to a carcinogen or a promutagen to a mutagen).

Metabolism It is a physicochemical or biochemical process in an organism that transforms chemicals into simpler or more complex forms.

Metabolite It is a product of metabolism.

Microbes They are microorganisms such as bacteria or yeast.

Mutagenic It means that it is able to cause mutations.

Mutation It is a stable change in the genetic material.

Oncogenic transformation It is a cancer-like change that can be brought about in isolated mammalian cells by chemical treatment.

Oncogenic It means that it is able to cause tumors.

Predictive bioassay (a) It is a bioassay to be carried out with a simulated plant waste as close as possible in quality to the anticipated operating waste under the worst possible conditions; (b) the predictive bioassay should contain three elements of the food chain, i.e., primary producers such as algae, invertebrates, and a fish. Just because a waste doesn't kill fish directly, it may ultimately prevent their survival due to its toxicity to an intermediate in the food chain. It is important to protect other elements of the aquatic community besides fish. For example, the loss of algae may impair the ability of the aquatic system to receive and transform organic wastes; (c) When conducting predictive bioassays, water quality of the proposed site location should be used since there are documented examples of synergistic interactions resulting in increased toxicity; (d) Predictive bioassays can be useful in helping identify potentially hazardous interactions, and to see if preliminary waste treatment design is likely to be adequate, and if not, can be used to estimate the degree of additional treatment required.

Procarcinogen It is a substance which is converted into a carcinogen by an organism's metabolic processes.

Promutagen It is a substance which is converted into a mutagen by an organism's metabolic processes.

Protein It is a large biological molecule essential for many cell structures and functions.

Quantal response (bioassay) The response is binary; it is a determination of whether or not an event occurs (e.g., death, sickness, and color change of the subject).

Quantal response The response is binary; it is a determination of whether or not an event occurs (e.g., death of the subject).

Quantitative response (bioassay) The measurement of the response to the stimulus is on a quantitative scale (e.g., blood sugar content, degree of color change in cell growth medium).

Quantitative response The measurement of the response to the stimulus is on a quantitative scale (e.g., blood sugar content, degree of color change in cell growth medium).

Relative toxicity It can be estimated in two practical ways. First, relative toxicity can be estimated by sampling the receiving water at various sites and testing the samples with the same Complex Effluent Toxicity Testing Program (CETTP) tests used to assess absolute toxicity. Alternatively, relative toxicity can be assessed with in situ biomonitoring by determining the uptake of various toxicants and the corresponding biological effects on organisms that are transplanted into the receiving water.

Somatic cell It is any nonreproductive cell in a multicellular organism.

Stamen hairs It is the part of the flower that produces pollen.

Stimulus Stimulus are the substances, such as drugs, chemicals, mutated meats, animal internal organs, and genetically altered milks applied to a subject (such as animals, plants, humans, tissues, and cells). The corresponding response (e.g., death, sickness, and appearance change) of the subject is thereby triggered and measured.

Subjects Subjects are living things, such as animals, plants, humans, tissues, and cells, that accept a stimulus (such as drugs, chemicals, mutated meats, animal internal organs, and genetically altered milks) in a bioassay test. The corresponding response (e.g., death, sickness, appearance change, etc.) of the subject is thereby triggered and measured.

Target cells They are isolated cells or cells within an organism that react in a specific manner to a toxic chemical or other stimulus.

Toxic It is a situation or condition that is able to produce an adverse effect.

APPENDIX: CHEMICAL ANALYSIS OF TEST WATER SOURCE AND WASHED SAND

An analysis for various water quality parameters was performed, and the results included herein

Metal	100 mL Wash Water	Tap Water
Cu	0.1 ppm	nil < 0.005 ppm
Ni	0.012 ppm	nil < 0.01 ppm
Cr	nil < 0.01 ppm	nil < 0.01 ppm
Co	0.05 ppm	nil < 0.002 ppm
Zn	0.023 ppm	nil < 0.002 ppm
Cd	0.015 ppm	0.01 ppm
Ag	nil <	nil <
Fe	nil < 0.01 ppm	nil < 0.01 ppm

A determination of metal content was made on water used to wash a sample of clean white filter sand, similar to that which had been used as a sediment in the experiments with metal sulfide precipitates.

This was done to be sure that the sand itself was not adding free metal ions to the test water. The results of this analysis can be found in the above table along with an analysis of tap water used as test water.

REFERENCES

1. Hoskins, W. M.; Craig, R. (1962). "Uses of Bioassay in Entomology". *Annual Review of Entomology.* 7 (1): 437–464. doi:10.1146/annurev.en.07.010162.002253.

2. Laska, E. M.; Meisner, M. J. (1987). "Statistical Methods and Applications of Bioassay". *Annual Review of Pharmacology and Toxicology.* 27 (1): 385–397.

3. Prinsloo, G.; Papadi, G.; Hiben, M. G.; de Haan, L.; Louisse, J.; Beekmann, K.; Vervoort, J.; Rietjens, I. M. (2017). "In Vitro Bioassays to Evaluate Beneficial and Adverse Health Effects of Botanicals: Promises and Pitfalls". *Drug Discovery Today.* 22 (8): 1187–1200.

4. Saha, G. M. (2002). *Design and Analysis for Bioassays,* Indian Statistical Institute, Kolkata, pp. 61–76.

5. Le, C. (2021). "Direct Bioassays & Estimation of Ratios". Retrieved 15 June 2021. BIOSTATISTICS FOR BIOLOGICAL ASSAYS (umn.edu)

6. Chao, A. C. and Wang, L. K. (1986). *Use of the Ames Mutagenecity Bioassay as a Water Quality Monitoring Method,* Lenox Institute of Water Technology, MA, Technical Manual No. LIR/07-86/184, 25 pages. Conference of Taiwan Water Works Association, Taipei, Taiwan, ROC, July 23, 1986.

7. USEPA (2021). *Permit Limits-Whole Effluent Toxicity (WET), National Pollutant Discharge Elimination System (NPDES)*, U.S. Environmental Protection Agency (EPA), Washington, DC, 2021-10-11.
8. USEPA (1978). *Bioassay for Toxic and Hazardous Materials*, US Environmental Protection Agency, Washington DC, EPA-430/1-78-001.
9. USEPA (1989). *Biomonitoring for Control of Toxicity in Effluent Discharges to the Marine Environment*, US Environmental Protection Agency, Washington DC, EPA/625/9-89/015.
10. USEPA (1979). *Short-Term Tests Carcinogens, Mutagens and Other Genotoxic Agents*, US Environmental Protection Agency, Cincinnati, OH, EPA/625/0-79/003.
11. USEPA (1986). *Reclamation and Redevelopment of Contaminated Land*, US Environmental Protection Agency, Washington DC, EPA/600/2-86/066.
12. USEPA (1980). *Ambient Water Quality Criteria for Phenol*, US Environmental Protection Agency, Washington DC, EPA 460/5-80-005.
13. USEPA (2023). *National Recommended Water Quality Criteria - Aquatic Life Criteria Table*, US Environmental Protection Agency, Washington DC.

16 Pollutants, Technologies, and Terminologies for Emission Control in Transportation Industry

Mu-Hao Sung Wang and Lawrence K. Wang

ACRONYM

AA	Atomic absorption
AR4	Fourth Assessment Report
BC	Black carbon
BEV	Battery-powered electric vehicles
CAA	Clean Air Act
CCl$_4$	Carbon tetrachloride
CCS	Carbon capture and sequestration
CFC	Chlorofluorocarbon
CH$_3$OCH$_3$	Dimethyl ether
CH$_4$	Methane
CNG	Compressed natural gas
CO$_2$	Carbon dioxide
DCC	Diesel catalytic converter
DMC	Dimethyl carbonate
DME	Dimethyl ether
DOC	Diesel oxidation catalyst
DPF	Diesel particulate filter
DPM	Diesel particulate matter
DPV	Diesel-powered vehicle
ETBE	Ethyl tertiary butyl ether
EV	Electric vehicles
FCC	Fluid catalytic cracking gasoline
Fe(C$_5$H$_5$)$_2$	Ferrocene
Fe(CO)$_5$	Iron pentacarbonyl
GCM	Global climate models
GHG	Greenhouse gases
GWP	Global warming potential
HBFC	Hydrobromofluorocarbon
HC	Hydrocarbon
HCFC	Hydrochlorofluorocarbon
HEV	Fuel cells hydrogen-powered hybrid electric vehicles
HFC	Hydrofluorocarbons
HFO	Hydrofluoroolefin

DOI: 10.1201/9781003541615-16

IARC	International Agency for Research on Cancer
ICAO	International Civil Aviation Organization
ICEV	Internal combustion engine vehicles
ICP	Inductively coupled plasma
IMO	International Maritime Organization
IPCC	Intergovernmental panel on climate change
LNG	Liquefied natural gas
LPG	Liquefied petroleum gas
MMT	Methylcyclopentadienyl manganese tricarbonyl
MTBE	Methyl tertiary butyl ether
N$_2$O	Nitrous oxide
NMVOCs	Non-methane volatile organic compounds
NOx	Nitrogen oxides
O$_3$	Ozone
ODP	Ozone-depletion potentials
ODS	Ozone-depleting substance
PFC	Perfluorinated carbons
PFC	Perfluorocarbons
PGE	Platinum group elements
PM	Particulate matter
TEL	Tetraethyl lead
USEPA	US Environmental Protection Agency
UCC	Upper continental crust
WMO	World Meteorological Organization

16.1 INTRODUCTION

Transportation air emissions constitute gaseous and particulate air pollution sources globally, which have become progressively more important in recent years due to increasing transportation activities. The annual transportation-based emissions of heavy metals, particulates, acid rain gases, carbon dioxide, other greenhouse gases, etc., have generated great burdens on both the regional and global environment in addition to negative impacts on human health.

This publication introduces the terminologies of air emission control in the transportation industry, covering the industries of automobile, trucking, shipping, train, aviation, and recreation (cruise).

16.2 POLLUTANTS, TECHNOLOGIES, AND TERMINOLOGIES FOR EMISSION CONTROL IN TRANSPORTATION INDUSTRY

THE POLLUTANS, TECHNOLOGIES, AND TERMINOLOGIES FOR EMISSION CONTROL IN TRANSPORTATION INDUSTRY ARE INTRODUCED IN THIS SECTION USING GLOSSARY TERMS IN ALPHABETICAL ORDER [1–8].

Aerosol (a) A small droplet or particle suspended in the atmosphere, typically containing sulfur. Aerosols are emitted naturally (e.g., in volcanic eruptions) and as the result of human activities (e.g., by burning fossil fuels). There is no connection between particulate aerosols and pressurized products, also called aerosols. (b) A product that relies on a pressurized gas to propel substances out of a container. Consumer aerosol products in the United States have not used ozone-depleting substances (ODS) since the late 1970s because of voluntary switching followed by federal regulation. The Clean Air Act and the US Environmental Protection Agency (USEPA) regulations further restricted the use of ODS for non-consumer products. All consumer products, and most other aerosol products, now use propellants that do not deplete the ozone layer, such as hydrocarbons and compressed gases. (c) Aerosols are fine solid or liquid particles, caused by people or occurring naturally, that are suspended in the atmosphere. Aerosols can cause cooling by scattering incoming radiation or by affecting cloud cover. Aerosols can also cause warming by absorbing radiation. (d) Small particles or liquid droplets in the atmosphere that can absorb or reflect sunlight depending on their composition.

Air pollution The addition into natural air of harmful or objectionable substances or extreme energy (heat, chill, noise, radioactivity) in large enough quantities to adversely affect the natural air's usefulness.

Air pollution control The limitations, prevention, reduction, or management of air pollution, so there will be no harmful or objectionable substances or extreme energy (heat, chill, noise, radioactivity) in large enough quantities to be added to natural air, in turn, not to adversely affect the natural air's usefulness.

Aircraft It is a device that is used or intended to be used for flight in the air.

Aircraft engine It means a propulsion engine that is installed on or that is manufactured for installation on an airplane for which certification under 14 CFR is sought.

Aircraft gas turbine engine It means a turboprop, turbojet, or turbofan aircraft engine.

Airplane It has the meaning given in 14 CFR 1.1, an engine-driven fixed-wing aircraft heavier than air that is supported in flight by the dynamic reaction of the air against its wings.

Alternative fuels Other than unleaded gasoline, the use of alternative fuels is one of the important ways for eliminating or controlling the emission of metals from internal combustion engines. The main alternative fuels that merit consideration from the air pollution control standpoint include liquefied petroleum gas (LPG), liquefied natural gas (LNG), compressed natural gas (CNG), hydrogen, dimethyl ether (DME), dimethyl carbonate (DMC), and biofuels such as ethanol. The significance of some of the representative alternative fuels is discussed below.

Alternative gasoline additives When lead is reduced, or removed from a gasoline pool, the octane increment formerly provided by lead must be replaced by a combination of increasing the proportion of high octane blendstocks in the pool and increasing the octane of at least some blendstocks. More specifically, to avoid using lead, the technical options for replacing the octane provided by lead include: (a) Increasing the octane of reformate by increasing reformer severity within the limits of sustainable operations. In some cases, to achieve the necessary increase in reformer severity will call for revamping and modernizing the reformer; (b) Increasing the production of high-octane blendstocks (reformate, fluid catalytic cracking (FCC) gasoline, alkylate, isomerate, or oxygenate) in the refinery. It is known that oxygenate blending adds oxygen to the fuel in oxygen-bearing compounds such as methyl tertiary butyl ether (MTBE), ethyl tertiary butyl ether (ETBE), and ethanol, and thus reduces the amount of carbon monoxide and unburned fuel in the exhaust gas; (c) Blending MTBE into the gasoline pool raised serious concerns. Consequently, MTBE use has been phased out due to issues with contamination of ground water. In some places, it is already banned. Ethanol and to a lesser extent the ethanol-derived ETBE are common replacements; and (d) Reducing the volume of light naphtha in the gasoline pool is another technical option to consider. These technical options may be applied in any combination that is technically feasible in the refinery.

Alternative vehicles They include: (a) battery-powered electric vehicles (BEV); (b) hybrid vehicles; (c) fuel cells hydrogen-powered hybrid electric vehicles (HEV).

Anthropogenic Made by people or resulting from human activities. Usually used in the context of emissions produced due to human activities.

Antiknock agents (a) Most of them are gasoline additives that are added to gasoline to increase gasoline's octane rating or act as corrosion inhibitors or lubricators, thus allowing the use of higher compression ratios for greater efficiency and power; (b) They are metal-containing additives, such as tetraethyl lead (TEL), methylcyclopentadienyl manganese tricarbonyl (MMT), ferrocene, iron pentacarbonyl, and so on.

Atmosphere The gaseous envelope surrounding the Earth. The dry atmosphere consists almost entirely of nitrogen (78.1% volume mixing ratio) and oxygen (20.9% volume mixing ratio), together with a number of trace gases, such as argon (0.93% volume mixing ratio), helium, radiatively active greenhouse gases such as carbon dioxide (0.035% volume mixing ratio), and ozone. In addition the atmosphere contains water vapor, whose amount is highly variable but typically 1% volume mixing ratio. The atmosphere also contains clouds and aerosols. The atmosphere can be divided into a number of layers according to its mixing or chemical characteristics, generally determined by temperature. The layer nearest the Earth is the troposphere, which reaches an altitude of about 8 km (about 5 miles) in the polar regions and up to 17 km (nearly 11 miles) above the equator. The stratosphere reaches to an altitude of about 50 km (31 miles) and lies above the troposphere. The mesosphere extends up to 80–90 km and is above the stratosphere, and finally, the thermosphere, or ionosphere, gradually diminishes and forms a fuzzy border with outer space. There is very little mixing of gases between layers.

Battery-powered electric vehicles (BEV) or (EV) Today, battery-powered electric vehicles (BEVs or EVs) still have a relatively limited driving range and higher initial cost relative to conventional automotives. However, the major attraction is that no air pollutants such as toxic metal vapors are directly emitted and no tailpipe is used. From an environmental perspective, battery-powered vehicles fulfill their promise of "zero emissions" along the roads and highways where they are driven. However, battery-powered vehicles have indirect impacts because of their demand for electricity, the main sources of which are coal, gas, and nuclear power, causing significant environmental impacts.

Black carbon Soot is produced from coal burning, diesel engines, cooking fires, wildfires, and other combustion sources. These particles absorb solar energy and have a warming influence on the climate.

Black carbon aerosol Black carbon (BC) is the most strongly light-absorbing component of particulate matter (PM) and is formed by the incomplete combustion of fossil fuels, biofuels, and biomass. It is emitted directly into the atmosphere in the form of fine particles (PM2.5).

Carbon capture and sequestration (CCS) Carbon capture and sequestration (CCS) is a set of technologies that can greatly reduce carbon dioxide emissions from new and existing coal- and gas-fired power plants, industrial processes, and other stationary sources of carbon dioxide. It is a three-step process that includes the capture of carbon dioxide from power plants or industrial sources; transport of the captured and compressed carbon dioxide (usually in pipelines); and underground injection and geologic sequestration, or permanent storage, of that carbon dioxide in rock formations that contain tiny openings or pores that trap and hold the carbon dioxide.

Carbon capture and storage The process of capturing carbon dioxide and injecting it into geologic formations underground for long-term storage. "Carbon capture and storage" is the same as "carbon capture and sequestration."

Carbon cycle (a) All parts (reservoirs) and fluxes of carbon. The cycle is usually thought of as four main reservoirs of carbon interconnected by pathways of exchange. The reservoirs are the atmosphere, terrestrial biosphere (usually includes freshwater systems), oceans, and sediments (including fossil fuels). The annual movements of carbon, the carbon exchanges between reservoirs, occur because of various chemical, physical, geological, and biological processes. The ocean contains the largest pool of carbon near the surface of the Earth, but most of that pool is not involved with rapid exchange with the atmosphere. (b) Circulation of carbon atoms through the Earth systems as a result of photosynthetic conversion of carbon dioxide into complex organic compounds by plants, which are consumed by other organisms, and return of the carbon to the atmosphere as carbon dioxide as a result of respiration, decay of organisms, and combustion of fossil fuels.

Carbon dioxide A naturally occurring gas, and also a by-product of burning fossil fuels and biomass, as well as land use changes and other industrial processes. It is the principal human-caused greenhouse gas that affects the Earth's radiative balance. It is the reference gas against which other greenhouse gases are measured and therefore has a global warming potential of 1.

Carbon dioxide emission and global warming The impact of carbon dioxide emission on the environment is significant. The carbon emissions contribute to the warming of the planet, leading to melting ice caps, rising sea levels, and extreme weather events. According to the USEPA, transportation accounts for about 28% of greenhouse gas emissions in the United States, and cars and trucks make up the majority of those emissions.

Carbon dioxide equivalent A metric measure used to compare the emissions from various greenhouse gases based on their global warming potential (GWP). Carbon dioxide equivalents are commonly expressed as "million metric tons of carbon dioxide equivalents ($MMTCO_2Eq$)." The carbon dioxide equivalent for a gas is derived by multiplying the tons of the gas by the associated GWP. $MMTCO_2Eq$ = (million metric tons of a gas) * (GWP of the gas)

Carbon dioxide fertilization The enhancement of the growth of plants as a result of increased atmospheric CO_2 concentration. Depending on their mechanism of photosynthesis, certain types of plants are more sensitive to changes in atmospheric CO_2 concentration.

Carbon footprint The total amount of greenhouse gases that are emitted into the atmosphere each year by a person, family, building, organization, or company. A person's carbon footprint includes greenhouse gas emissions from fuel that an individual burns directly, such as by heating a home or riding in a car. It also includes greenhouse gases that come from producing the goods or services that the individual uses, including emissions from power plants that make electricity, factories that make products, and landfills where trash gets sent.

Carbon sequestration (a) Storage of carbon through natural or technological processes in biomass or deep geological formations. (b) Terrestrial, or biologic, carbon sequestration is the process by which trees and plants absorb carbon dioxide, release the oxygen, and store the carbon. Geologic sequestration is one step in the process of carbon capture and sequestration (CCS) and involves injecting carbon dioxide deep underground, where it stays almost permanently.

Carbon tetrachloride (CCl_4) A compound consisting of one carbon atom and four chlorine atoms. Carbon tetrachloride was widely used as a raw material in many industrial uses, including producing chlorofluorocarbons and as a solvent. Solvent use ended when it was discovered to be carcinogenic. It is also used as a catalyst to deliver chlorine ions to certain processes. Its ozone-depletion potential is 1.2.

Cargo planes They are airplanes that mainly transport goods and commodities around the world in the air. Cargo planes are more fuel-efficient than passenger planes in terms of the amount of carbon dioxide emission per ton of load transported but are less fuel-efficient than cargo ships.

Cargo ships They include container ship, oil ship, etc., that carry goods and commodities around the world by sea. According to the International Maritime Organization (IMO), shipping accounts for approximately 2.5% of global greenhouse gas emissions, with cargo ships being the largest contributor. A single large container ship can emit as much pollution as 50 million cars. This is because cargo ships run on heavy fuel oil, which is a low-quality fuel that emits high levels of sulfur dioxide and nitrogen oxides. In terms of carbon dioxide emission-ton per cargo load-ton per transportation distance-mile (ton/ton-mile), cargo ships are the best (i.e., lowest) in comparison with cargo trains, cargo planes, and trucks. The impact of cargo ship emissions on the environment can be reduced in the future through the use of cleaner fuels and the implementation of energy-efficient technologies and policies. Several shipping companies have taken steps to reduce their emissions, have set a target of net-zero emissions by 2050, and have invested in alternative fuels such as liquefied natural gas (LNG), biofuels, and hydrogen. CMA CGM, the fourth-largest container shipping company, has committed to becoming carbon-neutral by 2050 and has invested in LNG-powered ships.

Catalytic converter A catalytic converter is a device used to reduce the toxicity of emissions from an internal combustion engine. Catalytic converters are used to reduce emissions of hydrocarbons, carbon monoxide, and nitrogen oxides, not to solve the lead discharge. Since lead in gasoline destroys the catalytic converters, the introduction of catalytic converters called for the use of unleaded gasoline. The catalyst itself is most often a precious metal. Platinum is the most active catalyst and is widely used. However, it is not suitable for all applications because of unwanted additional reactions and/or cost. Palladium and rhodium are two other precious metals that are used. Platinum and rhodium are used as a reduction catalyst, whereas platinum and palladium are used as an oxidization catalyst. Nickel and copper are also used, although each has its own limitations. Nickel is not legal for use in the European Union (due to reaction with carbon monoxide). While copper can be used, its use is illegal in North America due to the formation of dioxin. The catalytic converter is also a potential source of heavy metals due to the aging of catalysts by thermal effects, which could be emitted with other exhaust components]

Characteristic level It is a calculated emission level for each pollutant based on a statistical assessment of measured emissions from multiple tests.

Chlorofluorocarbon (CFC) A compound consisting of chlorine, fluorine, and carbon. Chlorofluorocarbons (CFCs) are very stable in the troposphere. They move to the stratosphere and are broken down by strong ultraviolet light, where they release chlorine atoms that then deplete the ozone layer. CFCs were commonly used as refrigerants, solvents, and foam-blowing agents. The most common CFCs were CFC-11, CFC-12, CFC-113, CFC-114, and CFC-115. CFCs have been phased out in the United States, with a few exceptions. CFCs are the gases covered under the 1987 Montreal Protocol and used for refrigeration, air conditioning, packaging, insulation, solvents, or aerosol propellants. Since they are not destroyed in the lower atmosphere, CFCs drift into the upper atmosphere where, given suitable conditions, they break down ozone. These gases are being replaced by other compounds: hydrochlorofluorocarbons, an interim replacement for CFCs that are also covered under the Montreal Protocol, and hydrofluorocarbons, which are covered under the Kyoto Protocol. All these substances are also greenhouse gases.

Class I ozone-depleting substance One of several groups of chemicals with an ozone-depletion potential of 0.2 or higher. Class I ODSs listed in the Clean Air Act include chlorofluorocarbons, halons, carbon tetrachloride, methyl chloroform hydrobromofluorocarbons, and methyl bromide.

Class II ozone-depleting substance A chemical with an ozone-depletion potential of less than 0.2. Currently, all hydrochlorofluorocarbons are Class II ODSs.

Climate (a) Climate in a narrow sense is usually defined as the "average weather," or more rigorously, as the statistical description in terms of the mean and variability of relevant quantities over a period of time ranging from months to thousands of years. The classical period is three decades, as defined by the World Meteorological Organization (WMO). These quantities are most often surface variables such as temperature, precipitation, and wind. Climate in a wider sense is the state, including a statistical description, of the climate system. (b) The average weather (usually taken over a 30-year time period) for a particular region and time period. Climate is not the same as weather, but rather, it is the average pattern of weather for a particular region. Weather describes the short-term state of the atmosphere. Climatic elements include precipitation, temperature, humidity, sunshine, wind velocity, phenomena such as fog, frost, and hail storms, and other measures of the weather.

Climate change (a) Changes in average weather conditions that persist over multiple decades or longer. Climate change encompasses both increases and decreases in temperature, as well as shifts in precipitation, changing risk of certain types of severe weather events, and changes to other features of the climate system. (b) Climate change refers to any significant change in the measures of climate lasting for an extended period of time. In other words, climate change includes major changes in temperature, precipitation, or wind patterns, among others, that occur over several decades or longer.

Compressed natural gas (CNG) CNG is a fossil fuel substitute for gasoline, diesel, or propane fuel. CNG is made by compressing natural gas, mainly composed of 90% methane (CH_4) and small amounts of ethane and other hydrocarbons, to less than 1% of its volume at standard atmospheric pressure. Natural gas has an octane value of 130, considerably higher than gasoline, with an octane value between 84 and 97, providing good engine performance characteristics. The toxic emissions with CNG, without exception, are lower than that of any other hydrocarbon fuel. This environmental benefit is because CNG is a single hydrocarbon, methane. The use of CNG in internal combustion engines permits operation with decreased NOx with little solid residues, but it still has an emission potential of volatile heavy metals such as Hg, which is indicated in the CNG quality standards. However, the measured data of mercury emission from CNG engine is not available yet.

Compression ignition It is an ignition process to operate an engine based on the diesel cycle by which fuel is injected after the air is compressed in the combustion chamber causing the fuel to ignite.

Deforestation The change of forested lands to non-forest uses. This is often cited as one of the major causes of the enhanced greenhouse effect for two reasons: (a) trees that are burned or decomposed release carbon dioxide; and (b) trees that are cut no longer remove carbon dioxide from the atmosphere.

Desertification Land degradation in arid, semi-arid, and dry sub-humid areas results from various factors, including climatic variations and human activities. Further, the UNCCD (The United Nations Convention to Combat Desertification) defines land degradation as a reduction or loss, in arid, semi-arid, and dry sub-humid areas, of the biological or economic productivity and complexity of rain-fed cropland, irrigated cropland, or range, pasture, forest, and woodlands resulting from land uses or from a process or combination of processes, including processes arising from human activities and habitation patterns, such as: (a) soil erosion caused by wind and/or water; (b) deterioration of the physical, chemical, and biological or economic properties of soil; (c) long-term loss of natural vegetation, such as conversion of forest to non-forest.

Diesel engines They are the engines that use compression ignition (based on the diesel cycle, a process by which fuel is injected after the air is compressed in the combustion chamber, causing the fuel to ignite. Diesel) to operate.

Diesel particulate filter (DPF) A DPF is a device designed to remove diesel particulate matter (DPM) or soot from the exhaust gas of a diesel engine. DPF is far more effective at reducing metal emissions associated with DPM than the diesel oxidation catalyst (DOC), which can oxidize compounds existing in the gas phase of the engine exhaust system. The latter device is ineffective at reducing the solid soot particles in DPM by any appreciable amount. DPFs are quite efficient in reducing PM emissions. For example, wallflow DPFs usually remove 85% or more of the soot, and can at times (heavily loaded condition) attain soot removal efficiencies of close to 100%. A diesel-powered vehicle (DPV) equipped with a functioning filter will emit no visible smoke from its exhaust pipe.

Diesel particulate matter (DPM) It is a black soot produced from the exhaust of a diesel engine; DPM also known as diesel soot consists of unburned carbon compounds together with those impurities of heavy metals bound within those PMs as well. The composition of heavy metals from diesel vehicle emissions is strongly affected by the vehicle's operating conditions, such as driving conditions, driving speed, and so on.

Diesel particulate matter (DPM) Diesel engines during combustion of the fuel/air mix produce a variety of particles generally classified as DPM due to incomplete combustion. The metal composition of the particles varies widely depending on the engine type, age, and the emissions specification the engine was designed to meet. Two-stroke diesel engines produce more DPM per horsepower output than do four-stroke diesel engines, as they burn the fuel/air mix less completely.

Dimethyl ether (DME) Dimethyl ether (CH_3OCH_3), also called DME, is currently attracting worldwide attention because it is a clean fuel that can be synthesized from various materials such as natural gas, coal, biomass, and so on. DME can be used as a fuel in diesel engines, gasoline engines (30% DEM/70% LPG), and gas turbines. It works particularly well in diesel engines due to its high cetane number, which is greater than 55 compared with diesel with cetane numbers 38–53. Only moderate modifications are needed to convert a diesel engine to run on DME. The simplicity of this short-carbon-chain compound leads, during combustion, to very low emissions of airborne PM, NO_x, CO, and no SO_x, meeting even the most stringent emission regulations in Europe, the United States, and Japan. Low emission of airborne particulate matter can also reduce the emission of metals bound in particles.

Ecosystem (a) All the living things in a particular area as well as components of the physical environment with which they interact, such as air, soil, water, and sunlight. (b) Any natural unit or entity including living and non-living parts that interact to produce a stable system through cyclic exchange of materials. (c) The complex of a community of organisms and the community environment functioning as an ecological unit.

Emission index It means the quantity of pollutants emitted per unit of fuel mass used.

Emission It is the release of pollutants into the air from a source.

Emissions The release of a substance (usually a gas when referring to the subject of climate change) into the atmosphere.

Energy Efficiency Design Index (EEDI) It is one of the primary regulations established by the International Maritime Organization's (IMO) for reducing emissions from cargo ships The EEDI sets minimum energy efficiency requirements for new ships and aims to reduce emissions by encouraging the use of more efficient technologies and designs.

Enhanced greenhouse effect The concept that the natural greenhouse effect has been enhanced by increased atmospheric concentrations of greenhouse gases (such as CO_2 and methane) emitted by human activities. These added greenhouse gases cause the Earth to warm.

Environment The complex of physical, chemical, and biotic factors (such as climate, soil, and living things) that act upon an organism (a living thing) or an ecological community (a collection of living things) and ultimately determine its form and survival. The circumstances, objects, and conditions that surround each of us.

Ethanol (alternative fuel) It is a biofuel manufactured from the conversion of carbon-based feedstocks such as sugar cane, sugar beet, switch grass, corn, and barley. Ethanol fuel can be combined with gasoline at different percentages or can be used in its pure form as 100%. As a matter of fact, not every vehicle can run on 100% ethanol, but most run on small percentages of ethanol blends. Ethanol has become more common because it is currently being used as an oxygenated additive, which could help achieve the reduction of soot emissions to some extent without the use of a metal-containing additive [3]. Ethanol fuel is a sustainable energy resource intended to provide a more environmentally and economically friendly alternative to fossil fuels such as diesel and gasoline. However, there are many debates surrounding the environmental friendliness of ethanol and the production viability.

Exhaust emission It means substances emitted to the atmosphere from exhaust discharge nozzles, as measured by the test procedures specified by the government.

Exhaust emission control (heavy metals) Control of exhaust emission from internal combustion engines has followed two routes: (a) fuels could be modified in terms of reduction of metal contents such as less metallic additives added, or (b) could be replaced by alternative fuels with less metal content as well without compromising the engine performance; the pollutants could be minimized from the combustion chamber by installing some particulate metal trap systems.

Ferrocene It is one kind of antiknock gasoline additive with a formula of $Fe(C_5H_5)_2$ that is an iron-based organometallic compound to be once used as antiknock agents in the fuel for gasoline engines for reducing soot formation inside engines; although it is relatively safer than TEL, it not been used widely due to its emissions from engines and its toxic nature.

Fluorinated gases Powerful synthetic greenhouse gases such as hydrofluorocarbons, perfluorocarbons, and sulfur hexafluoride that are emitted from a variety of industrial processes. Fluorinated gases are sometimes used as substitutes for stratospheric ODSs (e.g., chlorofluorocarbons, hydrochlorofluorocarbons, and halons) and are often used in coolants, foaming agents, fire extinguishers, solvents, pesticides, and aerosol propellants. These gases are emitted in small quantities compared to carbon dioxide (CO_2), methane (CH_4), or nitrous oxide (N_2O), but because they are potent greenhouse gases, they are sometimes referred to as High Global Warming Potential gases (high GWP gases).

Fluorocarbons Carbon-fluorine compounds often contain other elements such as hydrogen, chlorine, or bromine. Common fluorocarbons include chlorofluorocarbons (CFCs), hydrochlorofluorocarbons (HCFCs), hydrofluorocarbons (HFCs), and perfluorocarbons (PFCs).

Forcing Factors that affect the Earth's climate. For example, natural factors such as volcanoes and human factors such as the emission of heat-trapping gases and particles through fossil fuel combustion.

Fossil fuel (a) A general term for a fuel formed in the Earth from plant or animal remains, including coal, oil, natural gas, oil shales, and tar sands. (b) A general term for organic materials formed from decayed plants and animals that have been converted to crude oil, coal, natural gas, or heavy oils by exposure to heat and pressure in the earth's crust over hundreds of millions of years.

Friction material The friction material in vehicle brake linings consists of a wide range of compounds with, for instance, fibers of steel, glass, and plastic serving as reinforcements. Some compounds are used for their heat-conducting properties (brass chips) and good filling properties (antimony compounds). The heavy metal emissions from brake linings are Cd, Cr, Cu, Ni, Pb, Sb, and Zn. The effects of this material substitution have been shown in some studies as increased copper levels close to roads.

Gasoline additives They are the materials that are added to gasoline to increase gasoline's octane rating or act as corrosion inhibitors or lubricators, thus allowing the use of higher compression ratios for greater efficiency and power; however, some gasoline additives carry heavy environmental risks. Those metal-containing additives mainly refer to antiknock agents such as TEL, methylcyclopentadienyl manganese tricarbonyl (MMT), ferrocene, iron pentacarbonyl, and so on.

Glacier (a) A very large body of ice moving slowly down a slope or valley or spreading outward on a land surface. (b) A multi-year surplus accumulation of snowfall in excess of snowmelt on land and resulting in a mass of ice at least $0.1\,km^2$ in area, shows some evidence of movement in response to gravity. A glacier may terminate on land or in water. Glacier ice is the largest reservoir of fresh water on Earth and second only to the oceans as the largest reservoir of total water. Glaciers are found on every continent except Australia.

Global average temperature An estimate of Earths mean surface air temperature averaged over the entire planet.

Global change Changes in the global environment may alter the capacity of the Earth to sustain life. Global change encompasses climate change, but it also includes other critical drivers of environmental change that may interact with climate change, such as land use change, the alteration of the water cycle, changes in biogeochemical cycles, and biodiversity loss.

Global Climate Models (GCM) Mathematical models that simulate the physics, chemistry, and biology that influence the climate system.

Global warming (a) The recent and ongoing global average increase in temperature near the Earth's surface. (b) The observed increase in average temperature near the Earth's surface and the lowest layer of the atmosphere. In common usage, "global warming" often refers to the warming that has occurred as a result of increased emissions of greenhouse gases from human activities. Global warming is a type of climate change; it can also lead to other changes in climate conditions, such as changes in precipitation patterns.

Global warming potential (GWP) (a) A measure of the total energy that a gas absorbs over a particular period of time (usually 100 years), compared to carbon dioxide. (b) A number that refers to the amount of global warming caused by a substance. The GWP is the ratio of the warming caused by a substance to the warming caused by a similar mass of carbon dioxide (CO_2). Thus, the GWP of CO_2 is 1.0. Chlorofluorocarbon (CFC)-12 has a GWP of 8,500; CFC-11 has a GWP of 5,000; hydrochlorofluorocarbons and hydrofluorocarbons have GWPs ranging from 93 to 12,100; and water has a GWP of 0.

Good engineering judgment It involves making decisions consistent with generally accepted scientific and engineering principles and all relevant information, subject to the provisions of

Greenhouse effect (a) The effect produced as greenhouse gases allows incoming solar radiation to pass through the Earth's atmosphere but prevents most of the outgoing infrared radiation from the surface and lower atmosphere from escaping into outer space. This process occurs naturally and has kept the Earth's temperature about 60 degrees Fahrenheit warmer than it would otherwise be. Current life on Earth could not be sustained without the natural greenhouse effect. (b) Trapping and build-up of heat in the atmosphere (troposphere) near the Earth's surface. Some of the heat flowing back toward space from the Earth's surface is absorbed by water vapor, carbon dioxide, ozone, and several other gases in the atmosphere and then reradiated back toward the Earth's surface. If the atmospheric concentrations of these greenhouse gases rise, the average temperature of the lower atmosphere will gradually increase.

Greenhouse gas (GHG) (a) Any gas that absorbs infrared radiation in the atmosphere. Greenhouse gases include water vapor, carbon dioxide (CO_2), methane (CH_4), nitrous oxide (N_2O), halogenated fluorocarbons (HCFCs), ozone (O_3), perfluorinated carbons (PFCs), hydrofluorocarbons (HFCs), water vapor and sulfur hexafluoride. (b) Gases that absorb heat in the atmosphere near the Earth's surface, preventing it from escaping into space. If the atmospheric concentrations of these gases rise, the average temperature of the lower atmosphere will gradually increase, a phenomenon known as the greenhouse effect.

Habitat The place or environment where a plant or animal naturally or normally lives and grows.

Hazardous air pollutants (HAPs) Substances, usually chemicals, that cause serious health and environmental effects. Hazardous air pollutants are released by sources such as chemical plants, dry cleaners, printing plants, and motor vehicles (cars, trucks, buses, etc.).

Heavy metals (environmental) (a) Environmental heavy metals are trace elements found in low concentrations in the environment, such as air, soil, and water, which are nutrients needed for plant or animal growth in low concentrations, but can be toxic to humans, animals and plants in high concentrations. The Federal and State Regulations have established standards for controlling the following nine heavy metals (or trace elements): arsenic (As), cadmium (Cd), copper (Cu), lead (Pb), mercury (Hg), molybdenum (Mo), nickel (Ni), selenium (Se), and zinc (Zn); (b) It is a metallic element with a specific gravity greater than 5, such as cadmium (Cd), copper (Cu), lead (Pb), and zinc (Zn), etc.(c) It is a metallic element or metalloid that has an atomic density more prominent than $4000\,kg/m^3$.

Hybrid vehicle A hybrid vehicle is a modified vehicle that uses two or more distinct power sources (EV operation, internal combustion engine, etc.) to propel the vehicle. The hybrid vehicle typically achieves greater fuel economy and lower emissions than conventional internal combustion engine vehicles (ICEVs). Fewer metal emissions are primarily achieved by the following design: (a) Shutting down the gasoline or diesel engine during traffic stops or while coasting or other idle periods; (b) Using low rolling resistance tires. Hybrid cars use special tires that are more inflated than regular tires and stiffer, which reduces the drag by about half, improving fuel economy by relieving the stress of the engine; (c) Relying on both the gasoline (or diesel engine) and the electric motors for peak power needs, resulting in a smaller gasoline or diesel engine sized more for average usage rather than peak power usage. These features make a hybrid vehicle particularly efficient for city traffic with frequent stops, coasting, and idling periods. However, the overall cost of a hybrid car is still higher than a comparable gasoline-powered vehicle.

Hydrobromofluorocarbon (HBFC) An ODS consisting of hydrogen, bromine, fluorine, and carbon.

Hydrocarbon (HC) A compound consisting of carbon and hydrogen. Hydrocarbons include methane, ethane, propane, cyclopropane, butane, and cyclopentane. Although they are flammable, hydrocarbons (HCs) may offer advantages as substitutes to ODSs because they have zero ozone-depletion potential, low toxicity, and with the exception of methane, have low global warming potentials. Fossil fuels are made up of hydrocarbons.

Hydrochlorofluorocarbon (HCFC) A compound consisting of hydrogen, chlorine, fluorine, and carbon. Hydrochlorofluorocarbons (HCFCs) are one class of chemicals used to replace chlorofluorocarbons (CFCs). They contain chlorine and thus deplete stratospheric ozone, but to a much lesser extent than CFCs. HCFCs have ozone-depletion potentials (ODPs) ranging from 0.01 to 0.1. In conclusion, although ODSs, they are less potent at destroying stratospheric ozone than chlorofluorocarbons (CFCs). They have been introduced as temporary replacements for CFCs and are also greenhouse gases. Production of HCFCs is being phased out.

Hydrofluorocarbon (HFC) A compound consisting of hydrogen, fluorine, and carbon. Hydrofluorocarbons (HFCs) are a class of replacements for chlorofluorocarbons. Because they do not contain chlorine or bromine, they do not deplete the ozone layer. All HFCs have an ozone-depletion potential of 0. Some HFCs have high GWP. In conclusion, they were introduced as alternatives to ODSs in serving many industrial, commercial, and personal needs. HFCs are emitted as by-products of industrial processes and used in manufacturing. They do not significantly deplete the stratospheric ozone layer, but they are powerful greenhouse gases with global warming potentials ranging from 140 (HFC-152a) to 11,700 (HFC-23).

Hydrofluoroolefin (HFO) A compound consisting of hydrogen, fluorine, and carbon. Hydrofluoroolefins (HFOs) are alternatives to ODSs that typically have very low global warming potentials.

Hydrogen-powered hybrid EV (HEV) Electric vehicles (EVs) powered by fuel cells are another promising new concept for early 21st-century automobiles. A fuel cell can be thought of as a gas-powered battery in which a continuous flow of hydrogen and oxygen gases replaces the solid electrodes of a conventional car battery. Hydrogen is the most abundant element on the planet and the cleanest burning fuel on the basis of carbon atoms per fuel molecule. It also has the potential to produce only water when reacting with oxygen. Carbon emissions and metal pollutants from a hydrogen engine are virtually nonexistent. A hydrogen-powered hybrid EV (HEV) can reduce petroleum demands and emissions. However, hydrogen fuel cells are costly to produce and are quite fragile at present. It is still under study to produce inexpensive fuel cells that are robust enough to survive the bumps and vibrations that all automobiles experience. In addition, many designs require rare substances such as platinum as catalysts, which can again become contaminated by impurities in the hydrogen supply.

Hydrologic cycle The process of evaporation, vertical and horizontal transport of vapor, condensation, precipitation, and the flow of water from continents to oceans. It is a major factor in determining climate through its influence on surface vegetation, clouds, snow and ice, and soil moisture. The hydrologic cycle is responsible for 25% to 30% of the mid-latitudes' heat transport from the equatorial to polar regions.

Hydrosphere The component of the climate system comprises liquid surface and subterranean water, such as oceans, seas, rivers, fresh water lakes, and underground water.

Ice core A cylindrical section of ice removed from a glacier or an ice sheet to study climate patterns of the past. By performing chemical analyses on the air trapped in the ice, scientists can estimate the percentage of carbon dioxide and other trace gases in the atmosphere at a given time. Analysis of the ice itself can provide some indication of historic temperatures.

Indicator An observation or calculation that allows scientists, analysts, decision makers, and others to track environmental trends, understand key factors that influence the environment, and identify effects on ecosystems and society.

Indirect emissions Indirect emissions from a building, home, or business are those emissions of greenhouse gases that occur as a result of the generation of electricity used in that building. These emissions are called "indirect" because the actual emissions occur at the power plant that generates the electricity, not at the building using the electricity.

Industrial Revolution A period of rapid industrial growth with far-reaching social and economic consequences, beginning in England during the second half of the 18th century and spreading to Europe and later to other countries including the United States. The Industrial Revolution marks the beginning of a strong increase in the combustion of fossil fuels and related emissions of carbon dioxide.

Infrared radiation Infrared radiation consists of light whose wavelength is longer than the red color in the visible part of the spectrum, but shorter than microwave radiation. Infrared radiation can be perceived as heat. The Earth's surface, the atmosphere, and clouds all emit infrared radiation, which is also known as terrestrial or long-wave radiation. In contrast, solar radiation is mainly short-wave radiation because of the temperature of the Sun.

Intergovernmental Panel on Climate Change (IPCC) The IPCC was established jointly by the United Nations Environment Programme and the WMO in 1988. The purpose of the IPCC is to assess information in the scientific and technical literature related to all significant components of the issue of climate change. The IPCC draws upon hundreds of the world's expert scientists as authors and thousands as expert reviewers. Leading experts on climate change and environmental, social, and economic sciences from some 60 nations have helped the IPCC to prepare periodic assessments of the scientific underpinnings for understanding global climate change and its consequences. With its capacity for reporting on climate change, its consequences, and the viability of adaptation and mitigation measures, the IPCC is also considered the official advisory body to the world governments on the state of the science of the climate change issue. For example, the IPCC organized the development of internationally accepted methods for conducting national greenhouse gas emission inventories.

Iron pentacarbonyl It is one kind of antiknock gasoline additive with a formula of $Fe(CO)_5$ that is an iron-based organometallic compound to be once used as an antiknock agent in the fuel for gasoline engines for reducing soot formation inside engines; although it is relatively safer than TEL, it not been used widely due to its emissions from engines and its toxic nature.

Kyoto protocol An international treaty sets obligations and industrialized countries to lower their emissions of greenhouse gases.

Lead air emission Since lead has been removed from gasoline air emissions of lead from the transportation sector, motor vehicle exhaust is no longer the major source of lead emissions to the air, and lead air emissions have greatly declined. However, lead was emitted in large amounts from vehicles when leaded gasoline was used in the past, lead is present in many soils (especially urban soils) and can get resuspended into the air as fine lead particles. At the national level in the USA, major sources of lead in the air are ore and metals processing and piston-engine aircraft operating on leaded aviation fuel. Other sources are waste incinerators, utilities, lead-acid battery manufacturers, and lead dust on urban roads. The highest air concentrations of lead are usually found near lead smelters. There is no US National Ambient Air Quality Standard on lead. The State of California Ambient Air Quality Standard on lead is 1.5 $\mu g/m^3$

Lead toxicity Lead is a relatively soft and chemically resistant metal. Lead forms compounds with both organic and inorganic substances. As an air pollutant, lead is present in small particles. Exposure to high concentrations of lead, particularly in young children, can result in damage to the central nervous system and renal organs and may be associated with high blood pressure in adults. Human exposure to lead typically occurs via inhalation of air and ingesting lead in food, soil, water, or dust. Consequently, to get the lead out of gasoline seemed to be essential for the sake of environmental protection and human health.

Lead-based paint pollution The lead-based paint in old buildings could arise from paint flecks or chips or sanding during home renovations, in turn, causing lead pollution. In these cases, the lead-containing particles are generally too large to be inhaled. Instead, exposure is primarily through ingestion by young children or factory workers due to hand-to-mouth transfer of paint flecks or sanded paint dusts that are then swallowed. This can lead to an elevated level of lead in the human body, increasing the risk of having neurological, behavioral, and learning problems or even cancer. Care should be exercised when renovating older homes to prevent inhalation or ingestion of lead-based paint residue.

Leaded gasoline phase-out Due to the toxicity of lead, leaded gasoline was banned in most of the countries for its application as an automobile fuel. The phase-out period varies between countries. The U.S. EPA (United States Environmental Protection Agency) began working to reduce lead emissions soon after its inception, issuing the first reduction standards in 1973, which called for a gradual phase-down of lead to one-tenth of a gram per gallon by 1986. The average lead content in gasoline in 1973 was 2–3 g per gallon or about 200,000 tons of lead per year. In 1975, passenger cars and light trucks were manufactured with a more elaborate emission control system that included a catalytic converter that required lead-free. All of these efforts to remove lead from regular use resulted in an over 70% decline in blood-lead levels in Americans between 1978 and 1990. Other developed nations have followed the United States. The European countries such as Germany, France, and the United Kingdom began phase-out policy of leaded gasoline since the early 1980s, when the concentrations in leaves and human blood have steadily declined. However, as the industrial nations legislate lead's demise, the world's lead makers have been pushing to expand new markets, primarily in the developing countries. The complete phase-out (100% unleaded) worldwide still has a very long way to go.

Leaded gasoline It is a gasoline fuel containing a lead-containing additive, such as such as TEL. Lead (or TEL) in gasoline enhances engine performance since it has the property of increasing the octane rating/number in gasoline, which makes the fuel resist knocking better; lead also serves as a lubricant for the exhaust valves (valve seats). Leaded gasoline was phased out due to the lead toxicity to humans, especially to children. Many classic cars' engines have to make modifications to use lead-free gasoline due to the gradual unavailability of leaded gasoline.

Liquefied petroleum gas (LPG) LPG is a mixture of gases produced commercially from petroleum or natural gas and stored under pressure to keep it in a liquid state. LPG is composed primarily of propane with some butane, propylene, butylene, and other hydrocarbons, unlike gasoline, which is a complex mixture of hydrocarbons. LPG's average octane value is 104, which is higher than gasoline's range of 84–97, and can produce significantly better vehicle performance than the lower octane gasoline. When prepared as fuel, unlike gasoline, LPG is used as a dry gas without fuel additives, which just burns with little air pollution and little solid residues such as soot and PM with heavy metals. Even though LPG has been considered less polluting than gasoline and diesel because it contains less sulfur and emits less hydrocarbons, NO_x, PM, and CO, it has been reported that LPG has a high emission potential of volatile heavy metals such as mercury (Hg).

Major modification It is any physical change or method of operation change at an existing major stationary source that results in a significant net emissions increase, as defined in the regulations, of a regulated pollutant.

Major stationary source It is a term used to determine the applicability of Prevention of Significant Deterioration (PSD) and new source regulations. In a nonattainment area, a major source is one that has the potential to emit more than 100 tons per year. In PSD areas, the cutoff level may be either 100 or 250 tons, depending on the type of source.

Maximum Achievable Control Technology (MACT) Emissions limitations based on the best-demonstrated control technology or practices in similar sources to be applied to major sources emitting one or more of a listed toxic pollutant.

Mesosphere Earth's atmosphere is divided into five main layers: the exosphere, the thermosphere, the mesosphere, the stratosphere, and the troposphere. The stratosphere is the region of the atmosphere above the troposphere. The mesosphere is the region of the atmosphere above the stratosphere but below the thermosphere.

Metal emission sources (on-road mobile) They usually come from (a) exhaust fumes that are derived from the oxidation of fossil fuels and the aging processes of engines and catalysts. Combustion of leaded gasoline was the major source of Pb until about a decade ago. The use of lead (Pb) as an antiknocking additive in gasoline was phased out in many countries due to its toxicity, and (b) non-exhaust fumes, such as brake lining, tires, etc. As a result of the asbestos ban, producers of brake linings were forced to substitute materials during the 1980s. The heavy metal emissions from brake linings are Cd, Cr, Cu, Ni, Pb, Sb, and Zn.

Metal emissions (exhaust fumes) They are derived from the oxidation of fossil fuels and the aging processes of engines and catalysts. Combustion of leaded gasoline was the major source of Pb until about a decade ago. The use of lead (Pb) as an antiknocking additive in gasoline was phased out in many countries due to its toxicity. However, there are still some emissions of Cd, Cr, Cu, Ni, Pb, and V from fossil fuels.

Methane (CH_4) (a) Colorless, odorless, flammable hydrocarbon (CH_4) that is a product of the decomposition of organic matter and of the carbonization of coal. Methane is one of the greenhouse gas chemical compounds. (b) A hydrocarbon that is a greenhouse gas with a GWP most recently estimated at 25 times that of carbon dioxide (CO_2). Methane is produced through anaerobic (without oxygen) decomposition of waste in landfills, animal digestion, decomposition of animal wastes, production and distribution of natural gas and petroleum, coal production, and incomplete fossil fuel combustion.

Methyl tertiary butyl ether (MTBE) It is known that oxygenate blending adds oxygen to the fuel in oxygen-bearing compounds such as methyl tertiary butyl ether (MTBE), ethyl tertiary butyl ether (ETBE), and ethanol, and thus reduces the amount of carbon monoxide and unburned fuel in the exhaust gas; (c) Blending MTBE into the gasoline pool raised serious concerns. Although MTBE may be good for air quality, it has proven very bad for other parts of the environment, especially groundwater. Over the past few years, monitoring has detected MTBE in lakes, streams, and groundwater. If MTBE gets into a drinking water supply, it creates a bad smell and may pose health concerns. Because gasoline is so widely used, MTBE finds its way into almost every part of the environment. MTBE can get into water supplies from gasoline leaks, storage tanks, pipelines, and spills. It may also get into surface waters from boats and personal watercraft. MTBE evaporates into the air, but it is believed that most MTBE in air breaks down to other components. However, when MTBE gets into groundwater, it does not readily evaporate or break down. It dissolves in the groundwater and can move through an aquifer as a "plume." Consequently, MTBE use has been phased out due to issues with contamination of ground water. In some places, it is already banned.

Methylcyclopentadienyl manganese tricarbonyl (MMT) (a) It is one kind of antiknock gasoline additive; (b) MMT is an organometallic compound with the formula $(CH_3C_5H_4)Mn(CO)_3$, which was initially used as a supplement to the gasoline additive TEL to increase the fuel's octane rating and was later used in unleaded gasoline. Although banned as a gasoline additive in the United States from 1977 to 1995, MMT has been used in Canadian gasoline since 1976 and was recently introduced in Australia. (3) Combustion of the organomanganese compound MMT may be a significant source of contamination by inorganic Mn in urban areas, and it was reported that the contribution of Mn from MMT source relative to total Mn emissions was 28% (334 tons of Mn from MMT in 1999 in all the Canadian provinces versus 1,225 tons total emissions) in 1999.

Mobile sources They are moving objects that release pollution; mobile sources include cars, trucks, buses, planes, trains, motorcycles, and gasoline-powered lawn mowers.

Mobile vehicle It includes all cars and trucks.

Monitoring (monitor) Measurement of air pollution is referred to as monitoring. USEPA, state, and local agencies measure the types and amounts of pollutants in community air.

National Ambient Air Quality Standards (NAAQS) The standards established by USEPA that apply to outside air throughout the country.

National Emissions Standards for Hazardous Air Pollutants (NESHAP) These are the emissions standards set by USEPA for an air pollutant not covered by National Ambient Air Quality Standards that may cause an increase in deaths or is a serious, irreversible, or incapacitating illness.

Natural gas Underground deposits of gases consisting of 50%–90% methane (CH_4) and small amounts of heavier gaseous hydrocarbon compounds such as propane (C_3H_8) and butane (C_4H_{10}).

Natural variability Variations in the mean state and other statistics (such as standard deviations or statistics of extremes) of the climate on all time and space scales beyond that of individual weather events. Natural variations in climate over time are caused by internal processes of the climate system, such as El Niño, as well as changes in external influences, such as volcanic activity and variations in the output of the Sun.

Net emissions increase For an existing facility, It is the amount of any new emissions minus reductions in emissions caused by modifications to the facility.

New Source Performance Standards (NSPS) National USEPA air emission standards that limit the amount of pollution allowed from new sources or modified existing sources.

Nitrogen cycle The natural circulation of nitrogen among the atmosphere, plants, animals, and microorganisms that live in soil and water. Nitrogen takes on a variety of chemical forms throughout the nitrogen cycle, including nitrous oxide (N_2O) and nitrogen oxides (NOx).

Nitrogen oxides (NOx) It consist of one molecule of nitrogen and varying numbers of oxygen molecules. Nitrogen oxides are produced by the emissions of vehicle exhausts and from power stations. In the atmosphere, nitrogen oxides can contribute to the formation of photochemical ozone (smog), impair visibility, and have health consequences; they are thus considered pollutants.

Nitrogen oxides (NOX) It is a criteria air pollutant. Nitrogen oxides are produced from burning fuels, including gasoline and coal. Nitrogen oxides are smog-formers; they react with volatile organic compounds (VOC) to form smog. Nitrogen oxides are also major components of acid rain.

Nitrous oxide (N_2O) A powerful greenhouse gas with a GWP of 298 times that of carbon dioxide (CO_2). Major sources of nitrous oxide include soil cultivation practices, especially the use of commercial and organic fertilizers, fossil fuel combustion, nitric acid production, and biomass burning. The GWP is from the IPCC's Fourth Assessment Report (AR4). Natural emissions of N_2O are mainly from bacteria breaking down nitrogen in soils and the oceans. Nitrous oxide is mainly removed from the atmosphere through destruction in the stratosphere by ultraviolet radiation and associated chemical reactions, but it can also be consumed by certain types of bacteria in soils.

Noble metal catalyst Normally the manufacturing of automotive catalytic converters requires precious metals (noble metals) of palladium, platinum, or rhodium, which could have significant environmental effects such as accumulation in the ecosystem.

Nonattainment area It is a geographic area in which the level of a criteria air pollutant is higher than the level allowed by the federal standards. A single geographic area may have acceptable levels of one criterion air pollutant but unacceptable levels of one or more other criteria air pollutants; thus, an area can be both attainment and nonattainment at the same time. It has been estimated that 60% of Americans live in nonattainment areas.

Non-methane volatile organic compounds (NMVOCs) Organic compounds, other than methane, that participate in atmospheric photochemical reactions.

Nonnoble metal catalyst Since the manufacturing of automotive catalytic converters requires precious metals (noble metals) of palladium, platinum, or rhodium, which could have significant environmental effects such as the accumulation in the ecosystem, novel formulations to operate pollution-dampening catalytic converters without the need of expensive and toxic noble metals are quite necessary. Nonnoble metal transition metal catalysts can replace platinum in the oxidation-reduction reaction. In addition, a nonnoble metal, perovskite type of catalyst could be used to achieve the conversion of pollutants. Perovskites are one of the fascinating groups of catalytic materials having densely packed cubic lattices.

Non-volatile particulate matter (nvPM) It means emitted particles that exist at a gas turbine engine exhaust nozzle exit plane that do not volatilize when heated to a temperature of 350°C.

Nutrients Chemicals (such as nitrogen and phosphorus) that plants and animals need to live and grow. At high concentrations, particularly in water, nutrients can become pollutants.

Ocean acidification (a) Increased concentrations of carbon dioxide in seawater, causing a measurable increase in acidity (i.e., a reduction in ocean pH). This may lead to reduced calcification rates of calcifying organisms such as corals, mollusks, algae, and crustaceans. (b) The process by which ocean waters have become more acidic is due to the absorption of human-produced carbon dioxide, which interacts with ocean water to form carbonic acid and lower the ocean's pH. Acidity reduces the capacity of key plankton species and shelled animals to form and maintain shells.

Offset It is a method used in the 1990 Clean Air Act to give companies that own or operate large (major) sources in nonattainment areas flexibility in meeting overall pollution reduction requirements when changing production processes. If the owner or operator of the source wishes to increase the release of a criteria air pollutant, an offset (reduction of a somewhat greater amount of the same pollutant) must be obtained either at the same plant or by purchasing offsets from another company.

Operating permit It is a permit covering the air pollution requirements for existing facilities. These permits are typically more comprehensive than construction permits; air pollution sources typically have only one operating permit for their entire facility. Operating permits have an expiration date and must be renewed periodically (typically every 5 years). Operating permits are revised when a source becomes subject to new air pollution requirements or makes changes to the facility that require the permit to be updated.

Ozone Ozone is the triatomic form of oxygen (O_3). It is a gaseous atmospheric constituent. In the troposphere, it is created by photochemical reactions involving gases resulting both from natural sources and from human activities (photochemical smog). In high concentrations, tropospheric ozone can be harmful to a wide range of living organisms. Ozone is a bluish gas that is harmful to breathe. Tropospheric ozone acts as a greenhouse gas. Nearly 90% of the Earth's ozone is in the stratosphere and is referred to as the ozone layer. In the stratosphere, ozone is created by the interaction between solar ultraviolet radiation and molecular oxygen (O_2). Stratospheric ozone plays a decisive role in the stratospheric radiative balance. Since ozone absorbs a band of ultraviolet radiation called UVB that is particularly harmful to living organisms, the ozone layer prevents most UVB from reaching the ground. Depletion of stratospheric ozone, due to chemical reactions that may be enhanced by climate change, results in an increased ground-level flux of ultraviolet (UV-) B radiation.

Ozone "Hole" A large area of the stratosphere with extremely low amounts of ozone.

Ozone-depleting substance (ODS) A family of man-made compounds that includes, but are not limited to, chlorofluorocarbons (CFCs), bromofluorocarbons (halons), methyl chloroform, carbon tetrachloride, methyl bromide, and hydrochlorofluorocarbons (HCFCs). These compounds have been shown to deplete stratospheric ozone and therefore are typically referred to as ODSs.

Ozone layer depletion Chemical destruction of ozone molecules in the ozone layer. Depletion of this ozone layer by ODSs will lead to higher UVB levels (a band of ultraviolet radiation), which in turn will cause increased skin cancers and cataracts and potential damage to some marine organisms, plants, and plastics.

Particulate filter It is a filter that is designed for airborne particulate removal. The exhaust emissions from vehicle engines are the most difficult to control, within which most of the heavy metals from fuels could be bound and emitted with particulates. Consequently, the removal of the airborne particles by the particulate filter before they are exhausted into the atmospheric environment is another important control method for heavy metals originating from vehicles. The use of the particulate filter on diesel engine vehicles represents a new exhaust post-treatment technology that removes solid particles from the exhaust gases. Particulate filters have been used on on-road machines since the 1980s and in automobiles since 1996 in the USA.

Particulate matter (PM) Very small pieces of solid or liquid matter such as particles of soot, dust, fumes, mists or aerosols. The physical characteristics of particles, and how they combine with other particles, are part of the feedback mechanisms of the atmosphere.

Particulates; particulate matter (PM-10) It is a criteria air pollutant. Particulate matter includes dust, soot, and other tiny bits of solid materials that are released into and move around in the air. Particulates are produced by many sources, including the burning of diesel fuels by trucks and buses, incineration of garbage, mixing and application of fertilizers and pesticides, road construction, industrial processes such as steelmaking, mining operations, agricultural burning (field and slash burning), and operation of fireplaces and woodstoves. Particulate pollution can cause eye, nose, and throat irritation and other health problems.

Permit It is a document that resembles a license, required by the Clean Air Act for big (major) sources of air pollution, such as power plants, chemical factories and, in some cases, smaller polluters. Usually permits will be given out by states, but if USEPA has disapproved part or all of a state permit program, USEPA will give out the permits in that state. The 1990 Clean Air Act specifies requirements for permit applications, including provisions for members of the public to participate in state and USEPA reviews of permit applications. Permits will have, in one place, information on all the regulated pollutants at a source. Permits include information on which pollutants are being released, how much the source is allowed to release, and the program that will be used to meet pollutant release requirements. Permits are required both for the operation of plants (operating permits) and for the construction of new plants. The 1990 Clean Air Act introduced a nationwide permit system for air pollution control.

Photosynthesis The process by which plants take CO_2 from the air (or bicarbonate in water) to build carbohydrates, releasing O_2 in the process. There are several pathways of photosynthesis with different responses to atmospheric CO_2 concentrations.

Phytoplankton Microscopic plants that live in salt and fresh water environments.

Pollutants (air pollution) They are unwanted chemicals or other materials found in the air. Pollutants can harm health, the environment, and property. Many air pollutants occur as gases or vapors, but some are very tiny solid particles: dust, smoke, or soot.

Precession The wobble over thousands of years of the tilt of the Earth's axis with respect to the plane of the solar system.

Precipitation Rain, hail, mist, sleet, snow, or any other moisture that falls to the Earth.

Prevention of Significant Deterioration (PSD) It is a USEPA program in which state and/or federal permits are required to restrict emissions from new or modified sources in places where air quality already meets primary and secondary ambient air quality standards.

Primary standard It is a pollution limit based on health effects. Primary standards are set for *criteria for air pollutants.*

Radiation Energy transfer in the form of electromagnetic waves or particles that release energy when absorbed by an object.

Radiative forcing (a) A change in the balance between incoming solar radiation and outgoing infrared radiation. (b) A measure of the influence of a particular factor (e.g., greenhouse gas (GHG), aerosol, or land use change) on the net change in the Earth's energy balance.

Rated output (rO) It means the maximum power or thrust available for takeoff at standard day conditions as approved for the engine by the Federal Aviation Administration (FAA) of the US Department of Transportation (USDOT), including reheat contribution where applicable, but excluding any contribution due to water injection. Rated output is expressed in kilowatts for turboprop engines and kilonewtons for turbojet and turbofan engines to at least three significant figures.

Rated pressure ratio (rPR) It means the ratio between the combustor inlet pressure and the engine inlet pressure achieved by an engine operating at rated output, expressed to at least three significant figures.

Secondary standard It is a pollution limit based on environmental effects such as damage to property, plants, visibility, etc. Secondary standards are set for criteria for air pollutants.

Ship Energy Efficiency Management Plan (SEEMP) The SEEMP is established by the International Maritime Organization's (IMO); SEEMP requires ships to develop and implement plans to improve their energy efficiency and reduce emissions.

Smog It is a criteria air pollutant. A mixture of *pollutants,* principally ground-level *ozone,* produced by chemical reactions in the air involving smog-forming chemicals. A major portion of smog-formers comes from burning petroleum-based fuels such as gasoline. Other smog-formers, VOC, are found in products such as paints and solvents. Smog can harm health, damage the environment, and cause poor visibility. Major smog occurrences are often linked to heavy motor vehicle traffic, sunshine, high temperatures, and calm winds or temperature inversion (weather conditions in which warm air is trapped close to the ground instead of rising). Smog is often worse away from the source of the smog-forming chemicals since the chemical reactions that result in smog occur in the sky while the reacting chemicals are being blown away from their sources by winds.

Smoke It means the matter in exhaust emissions that obscures the transmission of light, as measured by the test procedures specified in § 1031.140.

Smoke number It means a dimensionless value quantifying smoke emissions as calculated according to ICAO Annex 16.

Soil Complex mixture of inorganic minerals (i.e., mostly clay, silt, and sand), decaying organic matter, water, air, and living organisms.

Soil carbon A major component of the terrestrial biosphere pool in the carbon cycle. The amount of carbon in the soil is a function of the historical vegetative cover and productivity, dependent on climatic variables. The amount of carbon in the soil is a function of the historical vegetative cover and productivity, which in turn is dependent in part upon climatic variables.

Solar energy It is also called solar radiation, an energy from the Sun. Also referred to as short-wave radiation. Of importance to the climate system, solar radiation includes ultraviolet radiation, visible radiation, and infrared radiation. It also includes indirect forms of energy such as wind falling or flowing water's hydropower, ocean thermal gradients, and biomass, which are produced when direct solar energy interacts with the Earth.

Solar radiation Radiation emitted by the Sun. It is also referred to as short-wave radiation. Solar radiation has a distinctive range of wavelengths (spectrum) determined by the temperature of the Sun. Of importance to the climate system, solar radiation includes ultraviolet radiation, visible radiation, and infrared radiation.

Source (a) It is any process or activity that releases a greenhouse gas, an aerosol, or a precursor of greenhouse gas into the atmosphere; (b) It is any place or object from which pollutants are released. A source can be a power plant, factory, dry cleaning business, gas station, or farm. Cars, trucks, and other motor vehicles are sources, and consumer products and machines used in industry can be sources, too.

Standard day conditions It means the following ambient conditions: Temperature $= 15°C$, specific humidity $= 0.00634\,kg\,H_2\,O/kg$ dry air, and pressure $= 101.325\,kPa$.

State implementation plan (SIP) It is a detailed description of the programs a state will use to carry out its responsibilities under the Clean Air Act (CAA). State implementation plans are collections of the regulations used by a state to reduce air pollution. The CAA requires that USEPA approve each state implementation plan. Members of the public are given opportunities to participate in the review and approval of state implementation plans.

Stationary source It is a place or object from which pollutants are released that does not move around. Stationary sources include power plants, gas stations, incinerators, houses, etc.

Stratosphere (a) The region of the atmosphere above the troposphere and between the troposphere and mesosphere. The stratosphere extends from about 8–50 km (6–31 miles) in altitude. Specifically it has a lower boundary of approximately 8 km at the poles to 15 km at the equator and an upper boundary of approximately 50 km. Depending upon latitude and season, the temperature in the lower stratosphere can increase, be isothermal, or even decrease with altitude, but the temperature in the upper stratosphere generally increases with height due to the absorption of solar radiation by ozone. So, the stratosphere gets warmer at higher altitudes. In fact, this warming is caused by ozone-absorbing ultraviolet radiation. Warm air remains in the upper stratosphere, and cool air remains lower, so there is much less vertical mixing in this region than in the troposphere. Commercial airlines fly in the lower stratosphere; (b) It is a part of the atmosphere of gases that encircle the Earth. The stratosphere is a layer of the atmosphere 9–31 miles above the Earth. Ozone in the stratosphere filters out harmful Sun rays, including a type of sunlight called ultraviolet B, which has been linked to health and environmental damage.

Subsonic It means relating to an aircraft that has not been certificated under 14 CFR to exceed Mach 1 in normal operation.

Sulfur dioxide It is a criteria air pollutant. Sulfur dioxide is a gas produced by burning coal, most notably in power plants. Some industrial processes, such as production of paper and smelting of metals, produce sulfur dioxide. Sulfur dioxide is closely related to sulfuric acid, a strong acid. Sulfur dioxide plays an important role in the production of acid rain.

Supersonic airplane It means an airplane for which the maximum operating limit speed exceeds a Mach number of 1.

System losses It means the loss of particles during transport through a sampling or measurement system component or due to instrument performance. Sampling and measurement system loss is due to various deposition mechanisms, some of which are particle-size dependent. Determining an engine's actual emission rate depends on correcting for system losses in the nvPM measurement.

Terrestrial radiation The total infrared radiation emitted by the Earth and its atmosphere in the temperature range of approximately 200–300 kelvin. Terrestrial radiation provides a major part of the potential energy changes necessary to drive the atmospheric wind system and is responsible for maintaining the surface air temperature within limits of livability.

Tetraethyl lead (TEL) (a) It is one kind of antiknock gasoline additive. (b) It is an organometallic compound with the formula $(CH_3CH_2)_4Pb$, which has been a common antiknock additive in gasoline, causing the lead. TEL usage was largely discontinued because of the toxicity of lead and its deleterious effects on catalytic converters, but is still used as an additive in aviation fuel for piston-engine-powered aircraft.

Trains Train is one type of transportation that is powered by a locomotive, and moves on railroads. There are two types of trains: passenger trains and cargo trains. Cargo trains are one of the most environmentally friendly modes of transportation because trains emit significantly less carbon dioxide per ton (of cargo load transported) per mile (of transportation distance) than trucks, and they are also more fuel-efficient.

Troposphere (a) The region of the atmosphere closest to the Earth. The troposphere extends from the surface up to about 10 km (6 miles) in altitude, although this height varies with latitude. Almost all weather takes place in the troposphere. Mt. Everest, the highest mountain on Earth, is only 8.8 km (5.5 miles) high. Temperatures decrease with altitude in the troposphere. As warm air rises, it cools, falling back to Earth. This process, known as convection, means there are huge air movements that mix the troposphere very efficiently. (b) The lowest part of the atmosphere from the surface to about 10 km in altitude in mid-latitudes (ranging from 9 km in high latitudes to 16 km in the tropics on average) where clouds and "weather" phenomena occur. In the troposphere, temperatures generally decrease with height. All weather processes take place in the troposphere. Ozone that is formed in the troposphere plays a significant role in both the greenhouse gas effect and urban smog. The troposphere contains about 95 percent of the mass of air in the Earth's atmosphere.

Turbofan engine It means a gas turbine engine designed to create its propulsion from exhaust gases and from air that bypasses the combustion process and is accelerated in a ducted space between the inner (core) engine case and the outer engine fan casing.

Turbojet engine It means a gas turbine engine that is designed to create its propulsion entirely from exhaust gases.

Turboprop engine It means a gas turbine engine that is designed to create most of its propulsion from a propeller driven by a turbine, usually through a gearbox.

Turboshaft engine It means a gas turbine engine that is designed to drive a rotor transmission system or a gas turbine engine not used for propulsion.

Vehicle brake linings They are the lining (friction material) of a vehicle brake system that consists of a wide range of compounds with, for instance, fibers of steel, glass, and plastic serving as reinforcements. Some compounds are used for their heat-conducting properties (brass chips) and good filling properties (antimony compounds). The heavy metal emissions from brake linings are Cd, Cr, Cu, Ni, Pb, Sb, and Zn.

Weather Weather is the specific atmosphere condition at a particular place and time. It is measured in terms of such things as wind, temperature, humidity, atmospheric pressure, cloudiness, and precipitation. In most places, weather can change from hour-to-hour, day-to-day, and season-to-season. Climate in a narrow sense is usually defined as the "average weather", or more rigorously, as the statistical description in terms of the mean and variability of relevant quantities over a period of time ranging from months to thousands or millions of years. The classical period is 30 years, as defined by the WMO. These quantities are most often surface variables such as temperature, precipitation, and wind. Climate in a wider sense is the state, including a statistical description, of the climate system. A simple way of remembering the difference is that climate is what you expect (e.g., cold winters), and 'weather' is what you get (e.g., a blizzard).

REFERENCES

1. Wang, MHS and Wang, LK (2021). Glossary of Natural Resources and Environmental Pollution Control. In: *Environmental and Natural Resources Engineering*, Wang, LK, Wang, MHS, Hung, YT and Shammas, NK (eds.), Springer Nature, Cham, Switzerland, pp. 421–494.
2. Wang, MHS and Wang, LK (2014). Glossary and Conversion Factors for Water Resources Engineers. In: *Modern Water Resources Engineering*, Wang, LK and Yang, CT (eds.), Humana Press, Totowa, NJ, pp. 759–851.
3. Wang, LK, (1974). *Environmental Engineering Glossary,* Calspan Corporation, Buffalo, NY, p. 439.
4. Wang, MHS and Wang, LK (2016). Glossary of Land and Energy Resources Engineering. In: *Natural Resources and Control Processes*, Wang, LK, Wang, MHS, Hung, YT and Shammas, NK (eds.), Springer Nature, Cham, Switzerland, pp. 493–621.
5. Wang, LK, Chen, JP, Hung, YT and Shammas, NK (2009). *Heavy Metals in the Environment*, CRC Press, Boca Raton, FL. pp. 475–489.
6. Wang. L.K., Balasubramanian, R., He, J., and Wang, M.H.S. Control and management of air emissions from the transportation industry. In: Control of Heavy Metals in the Environment Volume 2. Wang, L.K., Wang, M.H.S., Hung, Y.T. and Chen, J.P. (editors). CRC Press, Florida, USA. pp. 392–416, 2025.
7. Wang, M.H.S., Wang, L.K., and Shammas. Glossary of climate change, global warming and ozone layer protection. In: Handbook of Environment and Waste Management Volume 3. Hung, Y.T., Wang, L.K. and Shammas, N.K. (editors). World Scientific, Singapore. pp. 689–718, 2019.
8. Wang, M.H.S., Wang, L.K., and Shammas. Glossary of acid rain management and environmental protection. In: Handbook of Environment and Waste Management Volume 3. Hung, Y.T., Wang, L.K. and Shammas, N.K. (editors). World Scientific, Singapore. pp. 719–749, 2019.

Index

Note: **Bold** page numbers refer to tables and *italic* page numbers refer to figures.

absolute toxicity 581, 583, 585
absorption 31, 33, 35, 36, 56, 127, 301, 308, 377, 394, 396, 402, 410, 411, 417, 436, 470, 603, 606
acid cleaning of metal surfaces 296, 318
acid mine drainage (AMD) 51, 75, 120
acid pickling 67, 296–298, 301–314, 316, 318, 319
acid pickling operation 296, *310*, 318
activated carbon 41, 51, *157*, 159, 174, 178, 204, 278–282, 378, 388, 405, 408, 437
acute effect 31, 581
adsorbate 380, 381, 383–387, 394, 405, 406, 409, 411, 417
adsorbent resins 265–266
adsorption 3, 8, 11, 15, 17, 48–50, 83–85, 90, 98, 104, 123, 124, 126, 127, 130, 131, 136, 139, 149, 150, 155–161, **163**, 163–165, 170, 174, 178, 181, **181**, 185–189, 192, 200, 206, 207, 209–211, 219, 230, 231, 233, 237–243, 277–282, 348, 349, 356, 376, 378–380, 385, 387, 388, 395, 402, 405, 409
    and desorption 149, 261
aerosol 33, 35, 306, 307, 589–591, 593, 604–606
agricultural waste 288, 390
aircraft 590, 600, 606, 607
    engine 590
    gas turbine engine 590
air emission from acid pickling tanks using hydrochloric acid 306–314
airplane 16, 590, 592, 606
air pollution 14, 75, 589, 590, 600, 602–604, 606
air pollution control 68–69, 590, 604
algae, microbes, and microbially mediated processes 286–288
alternative fuels 590, 593, 595
alternative gasoline additives 590
alternative vehicles 590
aluminosilicate 126, 136, 138, 139, 260
alum sludges 257–258
AMD *see* acid mine drainage (AMD)
Ames assay 581
Ames test 581, *582*
analysis of mercury 558, **559**
anodizing operations 366, 497
anthropogenic 51, 116, 164, 165, 226, 227, 377, 417, 553, 555–556, 590
antiknock agents 590, 596, 599
applicability 103, 252, 257, 358, 434, 445, 459, 462, 484, 513, 523, 601
applications of mercury 553
arsenic (As)
    biomethylation 120, 130
    case studies in Southern Asia and Northern China 132–136, **133**, **136**, *137*
    chemistry 225–226
    competing sorption by anions 127
    contaminated countries 41
    contamination of groundwater 118–119, 138
    enhanced sorption by cations 127

geochemical processes and behavior in As-affected aquifers 124–132
mobilization
    in aquifers 134–135
    and release in shallow aquifers 132–137
occurrence and distribution 116–120
    in groundwaters 118–119, **118–119**, *120*
    in rocks, sediments, and soils 116–118, **117**
occurrence, distribution, and source in natural environment 116–120
problems 227–228
related diseases 228–230
sorption to clay minerals 126–127
sorption to Fe, Al, and Mn (hydro)oxides 126
source and biogeochemical cycle in environment 120
speciation and toxicity 41
species in natural waters 120–124
toxicology 228–230
arsenic (As) pollution 225–226
    in Bangladesh 227–228
    worldwide 227
arsenic (As) removal 230–244
    by chlorine 231
    by oxidation 230–231
    by ozone 232
    by permanganate 231–232
article output and distribution in journals 4
As *see* arsenic (As)
atmosphere 14, 34, 38, 39, 54, 120, 138, 150, 225, 227, 281, 290, 352, 408, 537, 554, 558, 589–595, 597, 599, 601, 602, 604, 606, 607

bacteria 35, 52, 56, 87, 130, 131, 139, 226, 257, 279, 282, 287, 376, 388, 393–397, 408, 411, 413, 554, 561, 562, 602
batch treatment *253*, 253–255, 357, 544
Battelle Memorial Institute 434, 462, 465, 466, 469
battery-powered electric vehicles (BEV) or (EV) 590, 591
Best Availability Technology 18, 232, 317, 358, 540
Best Available Technology Economically Achievable (BAT) 367, 368, 370
Best Conventional Pollutant Control Technology (BCT) 367, 368, 370
Best Demonstrated Available Technology (BDAT) 434, 480
Best Management Practice (BMP) 368
Best Practicable Control Technology (BPT) 358, 359
Best Practicable Control Technology Currently Available (BPT) 359, 367, 368, 370
BET adsorption isotherm 86–87
bioaccumulation **12**, 15, 83, 90, 200, 221, 287, 377, 394, 396, **409**, 409–411, 417, 554, 561, 564, 570, 581, *582*, 584
bioaccumulation factor (BAF) 570, 578–582
bioassay 569–572, 579, 581, 582, 585, 586